Peter Grindal saw active service during the 1962 Brunei rebellion and held sea commands in the ranks of lieutenant and commander before becoming Training Commander at the Royal Naval College, Dartmouth. After a period on the MoD Naval Staff, he commanded a frigate squadron, and was in charge of the task group protecting shipping in the Gulf during the Iran–Iraq War. He was subsequently Captain of a major training establishment before his final appointment, as a commodore, to the NATO post of Commander of the UK/Netherlands Amphibious Task Group. He retired from the Royal Navy in 1992 and was made a CBE in the same year.

"An awesome amount of research underpins this comprehensive narrative of a little-known part of Royal Naval history. The author's easy yet authoritative style makes the description of this brutally hard and unglamorous six-decade campaign a compelling read – not just for maritime history enthusiasts, but for any student of the slave trade and its suppression and the related diplomatic and domestic politics."

ADMIRAL OF THE FLEET THE LORD BOYCE, KG, GCB

"*Opposing the Slavers* provides a major revision of the existing literature… It will transform the study of abolition, the anti-slavery patrol and the nineteenth-century Royal Navy."

ANDREW LAMBERT, LAUGHTON PROFESSOR
OF NAVAL HISTORY, KINGS COLLEGE LONDON

"An illuminating and highly detailed account of how the Royal Navy, not least through the actions of its individual ships' captains, brought its tradition of victory to bear. And not just for country but for humanity. Through a war-fighting sailor's eye, Peter Grindal expertly charts the long anti-slavery campaign which changed the Royal Navy, the reverberations of which can still be felt in operations today."

ADMIRAL SIR GEORGE ZAMBELLAS, GCB, FIRST SEA LORD

"Peter Grindal has produced a deeply researched and detailed account of the Royal Navy's longest, proudest and least-known campaign, against the appalling slave trade. He has done so with great authority, in a way that brings it to life for us and allows us to draw, if we will, lessons for combined international action today to halt such crimes against humanity; some 35 million people worldwide are still estimated to be in effective slavery. It is an extraordinary, and most important and valuable, achievement."

VICE ADMIRAL SIR JEREMY BLACKHAM,
EDITOR, *THE NAVAL REVIEW*

"*Opposing the Slavers* is a masterful account of the Royal Navy West Africa Squadron's attempt to interdict the transatlantic slave trade during the first half of the nineteenth century. No other work comes close to its comprehensive perspective and mastery of the Admiralty records. This is the indispensable source of information for anyone seeking to understand Great Britain's naval war against the slave trade."

RAFE BLAUFARB, PROFESSOR OF HISTORY,
FLORIDA STATE UNIVERSITY

OPPOSING *the* SLAVERS

The Royal Navy's
Campaign against
the Atlantic
Slave Trade

PETER GRINDAL

New paperback edition published in 2018 by
I.B.Tauris & Co. Ltd
London • New York
www.ibtauris.com

First published in hardback in 2016 by I.B.Tauris & Co. Ltd

Copyright © 2016 Peter Grindal

The right of Peter Grindal to be identified as the authors of this work has been asserted by the author in accordance with the Copyright, Designs and Patents Act 1988.

All rights reserved. Except for brief quotations in a review, this book, or any part thereof, may not be reproduced, stored in or introduced into a retrieval system, or transmitted, in any form or by any means, electronic, mechanical, photocopying, recording or otherwise, without the prior written permission of the publisher.

Every attempt has been made to gain permission for the use of the images in this book. Any omissions will be rectified in future editions.

References to websites were correct at the time of writing.

ISBN: 978 1 78831 286 8
eISBN: 978 0 85773 938 4
ePDF: 978 0 85772 595 0

A full CIP record for this book is available from the British Library
A full CIP record is available from the Library of Congress

Library of Congress Catalog Card Number: available

Text designed and typeset by Tetragon, London

For Julie, Francesca and Alexander

This account is dedicated to the memory of the very many officers and men of the Royal Navy and Royal Marines, and their Krooman shipmates, who gave their lives, or who suffered permanently disabling wounds or disease, in the long struggle to deliver the peoples of West Africa from slavery.

Contents

List of Illustrations	xi
Maps	xiii
Acknowledgements	xxi
Foreword	xxiii
Preface	xxxi
Prologue · *The Beginning of a Great Matter*	1

PART ONE · THE TRADE

Chapter 1 · *Three Centuries of Transatlantic Slaving*	21
Chapter 2 · *The Realities of the Trade*	40
Chapter 3 · *The Slave Coasts and Seas*	62

PART TWO · THE SUPPRESSION CAMPAIGN

Chapter 4 · *Confused First Steps, 1807–11*	101
Chapter 5 · *The Tip of the Iceberg, 1811–15*	138
Chapter 6 · *Into a Legal Minefield, 1815–20*	183
Chapter 7 · *The Most Evident Falsehoods, 1820–4*	233
Chapter 8 · *A Lonely Furrow: Eastern Seas, 1824–8*	283
Chapter 9 · *Tenders and Tablecloths: Eastern Seas, 1828–31*	328
Chapter 10 · *Gallant Pinpricks: Western Seas, 1824–31*	384
Chapter 11 · *To the Cape Station: Eastern Seas, 1831–5*	411
Chapter 12 · *An Uncertain Sound: Western Seas, 1831–5*	466
Chapter 13 · *The Spanish Equipment Clause: Eastern Seas, 1835–8*	503
Chapter 14 · *Obduracy and Obfuscation: Western Seas, 1835–8*	559
Chapter 15 · *High-Handed Action: Eastern Seas, 1838–9*	615
Chapter 16 · *Forbearance Exhausted: Western Seas, 1838–9*	678

PART THREE · CONCLUSION

Summary · *Taking Stock* 727
Epilogue · *Until It Be Thoroughly Finished* 747

Guide to Abbreviations used in the Notes, Bibliography and Appendices 759
Appendix A · Suspected Slave Vessels Detained 1807–39 by Royal
  Navy Cruisers, Colonial Vessels and Letters of Marque Vessels 761
Appendix B · Treaties, Conventions and Conferences Concerning
  Atlantic Slave-Trade Suppression, 1807–39 784
Appendix C · Acts of Parliament and Orders in Council Relating to
  Suppression of the Atlantic Slave Trade, 1807–40 789
Appendix D · Commanders-in-Chief, Commodores and Senior
  Officers Appointed to Conduct Slave-Trade Suppression
  Operations, 1807–39 792
Appendix E · The Equipment Clause: Criteria for Arrest under the
  Equipment Clause in Various Bilateral Treaties 794
Appendix F · Declarations and Certificates: Forms Ordered to be Used
  by Captors on Detention of Suspected Slave Vessels 796
Appendix G · Mortality in the West Africa Squadron, 1825–39 799
Appendix H · Annual Imports of Slaves to Cuba and Brazil, 1807–39 800
Appendix I · Comparative Values of Sterling, 1807–39 802
Appendix J · Pivot-Gun Arrangement in HMS *Lynx* 803

Glossary 805
Endnotes 809
Bibliography 823
General Index 835
Index of Naval Personnel 848
Index of Naval Vessels 856
Index of Places 861

List of Illustrations

(All figures © National Maritime Museum, unless otherwise indicated)

1.	Thomas Clarkson, leading abolitionist	7
2.	Slaver schooner *L'Antonio*, a typical Baltimore clipper	184
3.	Capture of the Spanish brigantine *Dolores* by HM Brig *Ferret* on 4 March 1816	194
4.	Commodore Sir James Lucas Yeo KCB	195
5.	HMS *Inconstant*, Fifth Rate frigate	196
6.	Commodore Sir George Collier, Bt	213
7.	Capture of the Spanish schooner *Esperanza* by the boats of HM Sloop *Morgiana* on 10 December 1819	226
8.	Slave stowage plan of the French brig *Vigilante* arrested by the boats of HMS *Iphigenia* and HMS *Myrmidon* on 15 April 1822	262
9.	Commodore Francis Augustus Collier CB	319
10.	Capture of the Spanish brig *El Almirante* by HM Brig *Black Joke* on 1 February 1829	348
11.	Capture of the Spanish ship *Veloz Passagera* by HM Sloop *Primrose* on 7 September 1830	372
12.	Capture of the Spanish schooner *Voladora* by HM Schooner *Pickle* on 6 June 1829	398
13.	Capture of the Spanish brig *Midas* by HM Schooner *Monkey* on 27 June 1829	400
14.	Profile of the ten-gun brigantine *Griffon*	450
15.	Capture of the Spanish brig *Formidable* by HM Brigantine *Buzzard* on 17 December 1834	458

16.	Vice-Admiral The Rt Hon. Sir George Cockburn GCB, Commander-in-Chief on the West Indies, Halifax and Newfoundland Station	475
17.	Rear-Admiral Sir Thomas Baker, Commander-in-Chief on the South America Station	486
18.	Rear-Admiral Sir Michael Seymour, Bt, KCB, Commander-in-Chief on the South America Station	490
19.	HM Barque *Columbine* off Lisbon	533
20.	HMS *Champion*, 18-gun sloop	564
21.	HM Brig *Ringdove*, 16-gun Symondite sloop	593
22.	HM Brig *Rapid*, 10-gun "Coffin Brig"	597
23.	Capture of the Spanish schooner *Opposiçao* by HM Sloop *Pearl* on 28 April 1838	684
24.	A Royal Navy landing force destroying a slave factory © Illustrated London News/Mary Evans	746

Maps

(All maps © Mark Myers, 2015)

MAP 1.	West Africa Slave Coast, Northern Part	xiii
MAP 2.	West Africa Slave Coast, the Bights	xiv
MAP 3.	West Africa Slave Coast, Southern Part	xv
MAP 4.	The West Indies	xvi
MAP 5.	Cuba	xvii
MAP 6.	The Coast of Brazil	xviii
MAP 7.	Atlantic Ocean Currents	xix
MAP 8.	Atlantic Ocean Winds: January, February, March	xx
MAP 9.	Atlantic Ocean Winds: July, August, September	xx

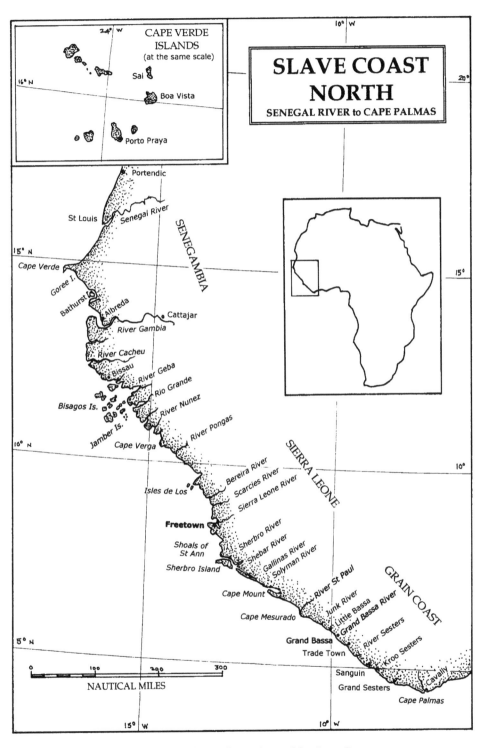

MAP 1. West Africa Slave Coast, Northern Part

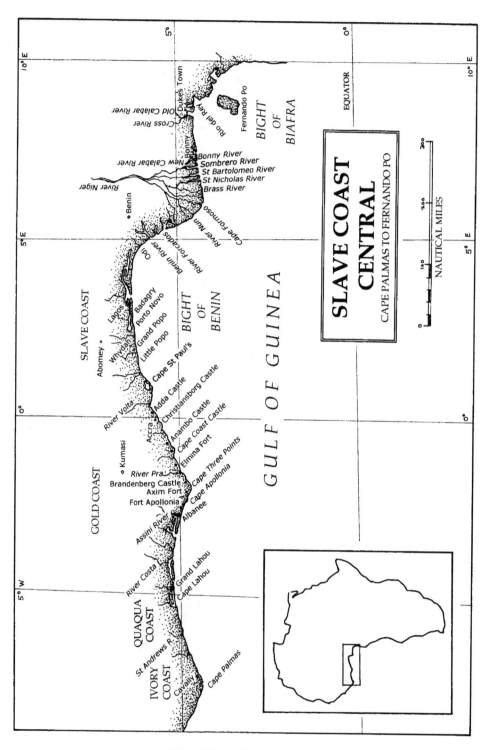

MAP 2. West Africa Slave Coast, the Bights

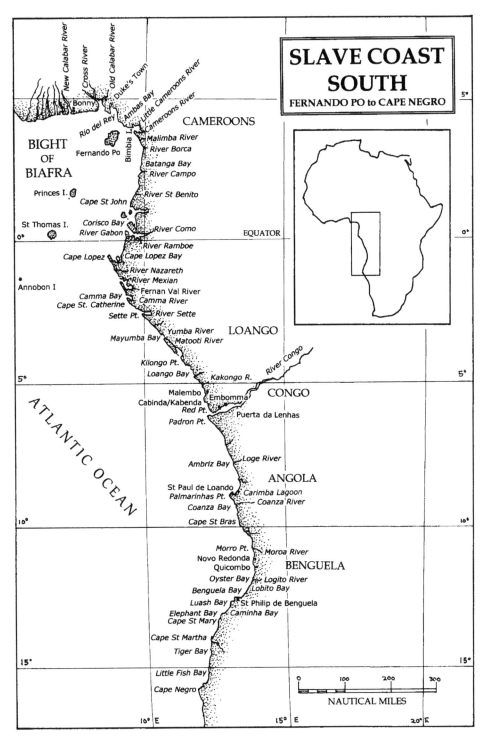

MAP 3. West Africa Slave Coast, Southern Part

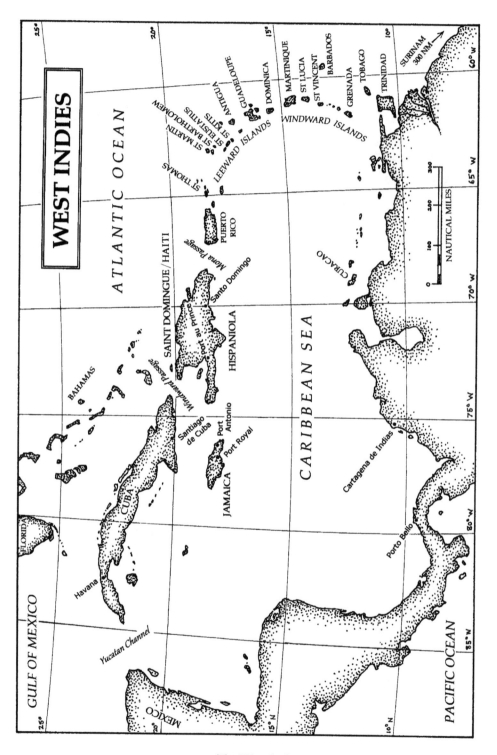

MAP 4. The West Indies

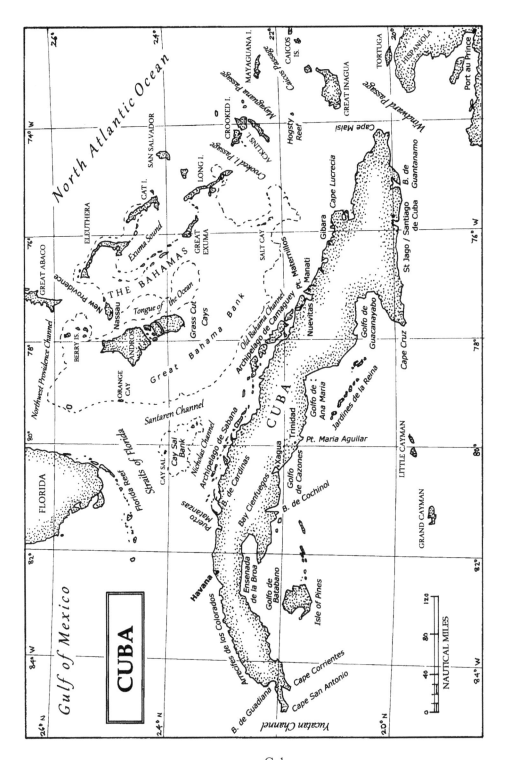

MAP 5. Cuba

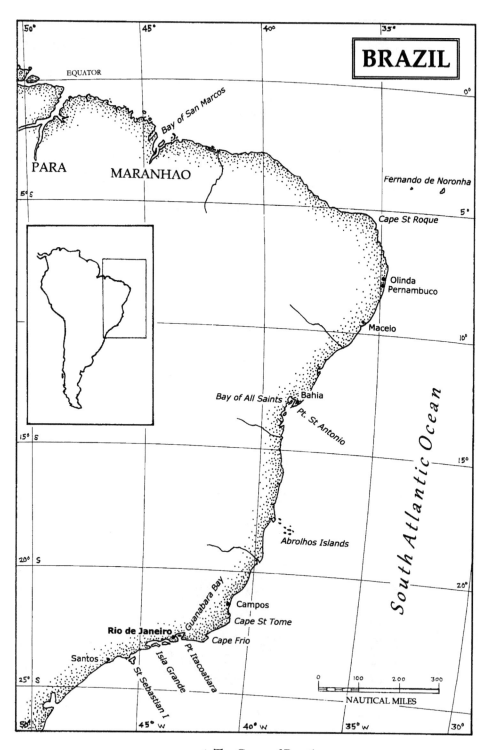

MAP 6. The Coast of Brazil

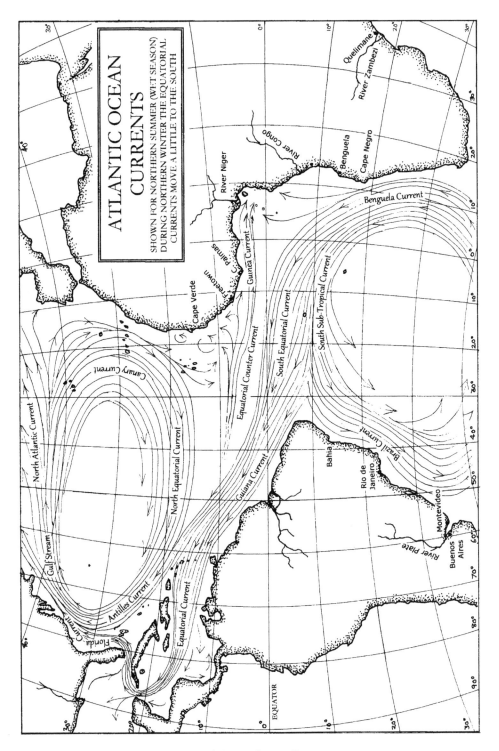

MAP 7. Atlantic Ocean Currents

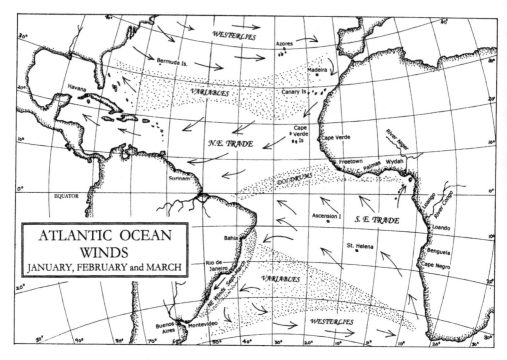

MAP 8. Atlantic Ocean Winds: January, February, March

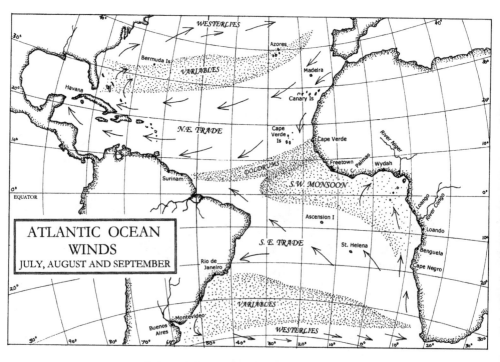

MAP 9. Atlantic Ocean Winds: July, August, September

Acknowledgements

RESEARCH AND WRITING are solitary and time-consuming tasks, and I wish, first of all, to express my gratitude to Julie, my wife, not only for her encouragement but also for her forbearance during my long periods of submersion in this project. My thanks also go to her for applying her keen lawyer's eye to proofreading, and for her valuable help with the most tedious task in the authorship of the book: compiling its long index.

The manuscript would never have reached publication had it not been for the extraordinary generosity of my brother Robert, and to him I will ever be enormously grateful.

My thanks then go to the staff of the National Archives (lately the Public Record Office) at Kew who have been unfailingly efficient and willing in their support of my research, and I am particularly grateful to Guy Grannum and William Spencer for their patient help. I also wish to thank Jenny Wraight and Ian Mackenzie at the Admiralty Library and Adrian Webb of the United Kingdom Hydrographic Office for their enthusiastic guidance through their treasures, and my thanks go also to Dr Quintin Colville and Dr John McAleese at the National Maritime Museum for their interest and advice.

My extensive research at the National Archives would not have been possible without accommodation within striking distance of Kew, and I greatly appreciate the hospitality so generously and repeatedly provided in London by my friend Peter Thompson and my stepdaughter Francesca.

Other friends have been open-handed with their help: Neil Blair has cast his kindly but critical eye over every word of the draft, Dr Brian Kirkpatrick and Dr Patricia Mathers have fielded my questions on tropical diseases, and Tank Nash, an experienced author, has encouraged me throughout with his enthusiasm and counsel. I greatly appreciate their invaluable contributions.

I offer my gratitude and deep admiration to that superlative marine artist, Mark Myers, for his splendid watercolour for the book's dust cover, a painting which he has generously allowed me to use without charge, and for his excellent maps.

David Stonestreet, Senior Editor of I.B.Tauris, has been the linchpin in bringing my efforts to fruition, and I am enormously appreciative of his morale-boosting enthusiasm, his championship of the manuscript, and his patient guiding hand along the publishing path.

The production process has been in the capable hands of David Campbell of I.B.Tauris and Alex Billington of Tetragon Publishing, to both of whom I am most grateful, and my particular thanks go to Sarah Terry, the copy-editor (hyphen, no capitals!), for her painstaking attention to detail and her patience with my linguistic foibles and nautical vocabulary.

Finally, I wish to record my especial gratitude to Professor Andrew Lambert of King's College, who gave this amateur naval historian encouragement to embark on what seemed a daunting undertaking, and who has since given me essential and unstinting guidance and support.

PETER GRINDAL
Olveston
October 2015

Foreword

TWO YEARS AFTER TRAFALGAR, the Royal Navy took on a new and far less honourable foe. On 25 March 1807 Parliament declared the trade in slaves illegal. It fell to the Royal Navy, the right arm of the British state, to translate that legislation into action, attempting to end an ancient and well established trade. While British ships soon stopped carrying human cargo, and they had been the majority at a time when the ships of France and Spain were excluded by the ongoing Napoleonic conflict, this only meant that the Trade moved into other ships, from other countries. During the Vienna Congress of 1815, and thereafter, Britain made a major diplomatic effort to secure international agreements that would extend abolition to all maritime nations. However, while many nations signed treaties abolishing the Trade, most were unwilling, or unable, to impose the law on their subjects. This was hardly surprising: there was money to be made in breaking the law. Furthermore, slavery itself did not become illegal in the British colonies until the 1830s, and much later elsewhere. Consequently a market existed for slaves, and at the right price shipping could be found to service that market. The main customers for slaves after 1807 were the burgeoning sugar plantations of Cuba and Brazil. The main suppliers were African kingdoms. The Trade persisted while there was a ready market and an adequate supply. The Royal Navy could only reduce the flow until the market disappeared.

Britain did not find much support for her moral campaign. Smaller powers demanded compensation for giving up the Trade, while powerful countries like France and the United States refused to let the British search slave ships flying their flag. They opposed abolition for economic or ideological reasons, and in so doing provided slavers with a convenient loophole. Each slaver carried multiple sets of papers, claiming a range of nationalities, to complicate the task of enforcement. The task facing the young naval officers on the anti-slavery patrol was complex. They needed to be lawyers and diplomats as well as seamen and warriors. Considerable responsibility was necessarily given to officers in their early twenties, and many would go on to have distinguished careers. This was a service for young men and small ships. However, the heirs of Nelson proved to be resourceful men, and made good use of the initiative they were given. To encourage officers and men, Britain provided a results-based payment, paying a set tariff for every slave liberated. Many of those freed by the Royal Navy were landed at Sierra Leone. Initially the patrol used sailing brigs and schooners, often captured slavers, but when steamers became available they proved particularly effective.

A permanent anti-slavery patrol operated on the West African coast between 1819 and 1869. It was a large and costly commitment of men, ships, money and lives. Tropical diseases took a terrible toll on British sailors. To make matters worse the cause of these diseases was unknown, making prevention largely a matter of guesswork. Despite such problems, the Royal Navy captured over 500 slave ships in the first thirty years of the campaign alone. The key to ending the Trade lay in changing attitudes in the countries that provided and purchased slaves. New trades and markets eventually made slavery uneconomic. While this was happening the Royal Navy stopped and searched suspicious ships, condemning any found with slaves or slaving equipment, from manacles and chains to outsize cooking pots.

This remarkable book examines the history of the patrol in the first thirty years, and provides a major revision of the existing literature. Modern studies of Atlantic slavery, Abolition and the associated political and diplomatic processes are numerous, with a significant spike in publication occurring in 2007, the bicentenary of the abolition of the British slave trade, but the mechanics of suppression have largely been ignored.[1] Existing studies of the naval campaign to suppress the Trade – those of Lloyd and Ward – are long out of date, piecemeal and unsystematic in approach.[2] They date to an era when West Africa was still dominated by European colonial empires, or their immediate aftermath, and Africans, enslaved or otherwise, had no historical voice. The British abolitionists and naval crews were necessarily good, other Europeans were mostly malign, and Americans frequently obstructive. While these texts addressed British naval activity they did so in an episodic manner, and offer nothing comparable to this thorough, systematic account. The table of captures alone is a work of the first importance.

By contrast, works focused on the wider aspects of Abolition have tended to relegate the naval campaign to the margins, often dismissing the British effort as self-serving or ineffective, without detailed analysis or any understanding of the complexities of the seagoing effort. Any attempt to address this campaign requires a knowledge of ships, seamanship, navigation and climate far beyond that to be found among land-based scholars, even those with some experience of seafaring. During his naval career Peter Grindal served in small craft, on dangerous operations close to war, under tropical conditions, and commanded ships of war. He is a sure guide to the seamanship, command and logistics that provide the bedrock of any durable system of maritime interdiction. His mastery of the archival resources has enabled him to track other young naval officers through the vicissitudes of their business, and make balanced, appropriate judgements on their actions. Without such insight and expertise most analysis of the anti-slavery patrols, both at the time and subsequently, is little better than banal.

The subtle correction of less professional assessments is one of the continuing pleasures of the text. The richly coloured and sophisticated discussions of ship-handling, sailing tactics and ship design are critical to any understanding of how the campaign was handled, the choices made by the Admiralty, and the conduct of the junior officers engaged.

At the heart of the text is a detailed analysis of the campaign, and a sympathetic treatment of the young officers and men sent out to conduct it, with too few ships, many inappropriate to the task, in a region where lethal diseases were rife, and their targets both unscrupulous and violent. If, as is noted, they occasionally overstepped their authority, resorted to drink or became boorish, the only surprise is that so few such cases emerged. The young officers who commanded the brigs and sloops involved had an opportunity to earn rewards and promotion – most tried to do so, some of them built their careers on successful commissions, and others paid the ultimate price for their determination. The Navy profited from their work, developing the last generation of sailing warships and the ship-handling skills needed to exploit them.

Yet the naval campaign was never going to stop the Atlantic Trade as long as open markets existed to buy slaves in the Americas, and with ample supplies of enslaved peoples in Africa the means would be found to bring the two together. However, as this text emphasises, the naval campaign forced the slave traders to expend ever greater sums to conduct their business, buying faster and more complex ships, larger crews, and devising ever more complex paper trails of variant ownership, not to mention the significant, but necessarily unquantifiable, economic losses inflicted by capture and condemnation.

In addition, the Royal Navy built up a significant body of local expertise in navigation, healthcare, logistics and political contacts, which would prove highly effective in the next phase of the campaign, in the 1850s and 1860s, when slave barracoons were destroyed and local rulers were actively encouraged to develop legitimate commercial alternatives.

The conduct of the campaign was hampered by the failure of the British Foreign Office to concert and apply a systematic approach to the problem of those slaving nations that failed to live up to their treaty obligations. For example, the Portuguese Trade to Brazil was illegal under the 1817 Convention, yet the Foreign Office failed to exploit this stipulation to address the Trade from African coasts south of the equator. The slow or partial transmission of information and advice to the Squadron compromised several captures, while the failure of both the Foreign Office and the Admiralty to support officers who were denied justice by biased courts called into question the seriousness of the British campaign. Both government departments were small, short of staff and under a wide range of other pressures: they simply could not handle this complex issue competently and consistently.

The nuanced conclusions of this book expand older, simpler judgements of the efficacy and impact of the Royal Navy's campaign against the Atlantic slave trade between 1807 and 1839. The cut-off date is significant: the Act of Parliament of that year which addressed the failure of Portugal to live up to its treaty obligations marked a major shift in the nature of the campaign. British patience with Portugal – a small, weak state which depended on British naval and diplomatic support – was at an end. Brazil would find the same logic applied in the mid-1840s.

The campaign against the Atlantic slave trade after 1840 would be dominated by the issue of the United States flag being used to cover slave ships of many nations. While United States resistance to any concession of a Right of Search to determine the legitimacy of the flag was understandable, given the bitter legacy of the War of 1812, it was also directly related to domestic political tensions over the question of slavery, and the very large economic stake that some sections of the country had in the continuance of the Trade, as suppliers of ships, crew and stores, and as investors in the illegal movement of slaves into Texas and the United States.[3] If the Right of Search had been conceded in 1840 the Trade could have been ended within months – as the references to opinion on the Havana waterfront make all too clear. Ultimately the Atlantic slave trade was an international collaborative enterprise, one in which many nations were complicit, including Britain. Like operators in any profitable illegal activity, those involved proved flexible, determined, and frequently resorted to violence.

By 1840 British diplomats had secured agreements with the powers involved in the Trade, agreements that appeared to provide for effective action. However, fundamental structural problems limited the impact of British action. Several nations either refused to allow British cruisers to search suspected slavers under their flag, or actively connived in the violation of international agreements signed with Britain. British cruisers could not stop American-flagged vessels, while those of Portugal and Brazil were safe below the equator. To make matters worse the standing patrol off the West African coast was small, and many of the warships used were too slow to catch the latest generation of slavers – usually swift American-built schooners. While Lord Palmerston, Foreign Secretary (1830-4, 1835-41, 1846-51) and a committed abolitionist, worked hard to improve the diplomatic context of the campaign, he recognised the limits of national power and the wider interests that were involved in any sustained effort. The states that resisted British demands did so for powerful reasons. Many believed that behind the smokescreen of moral fervour and righteous indignation the British were serving their own commercial ends. After the bitter experience of being the only significant neutral carrier for most of the French Revolutionary and Napoleonic Wars (1793-1815), the United States refused to allow the British to exercise a right of search on American-flagged vessels.[4] Freedom of the Seas had become an American mantra, one that would only be ended by the Civil War of 1861-5. Furthermore, while the United States had outlawed the importation of slaves many years before, the country remained deeply divided on the issue of slavery. The delicate balance between free and slave states before 1861 ensured no effective action was ever taken against American slave ships, or the foreign vessels that fraudulently hoisted American colours. As President Tyler declared in his 1841 Annual Message to Congress, the Stars and Stripes was being "grossly abused by the abandoned and profligate of other nations".[5] By blaming the vice on Iberian rogues, he could turn a blind eye to the builders and operators of the ships in question, most of whom were American citizens, and avoid taking any action. Suitably embarrassed by British seizures of such fraudulent "American" ships, an American anti-slavery patrol was set up in 1842, but it was never intended to be effective (to avoid upsetting sectional interests) and only became moderately active after slavers openly landed slaves in Georgia in 1858.[6] The American position proved to be the final obstacle to the closure of the Atlantic Trade because the United States could not be coerced, and pre-1861 American society was too deeply divided on the issue to be persuaded. American shipping sustained the Cuban slave market into the 1860s, with significant leakage into the United States. The intransigence of Washington limited the success of British anti-slavery, short of war.

By contrast, Brazil and Portugal, despite signing treaties with Britain outlawing the Trade, connived to continue it for economic and political reasons. Both powers feared Britain planned to seize Portugal's African settlements under the pretext of Abolition and establish alternative sources for colonial produce (sugar, coffee and cotton), thus cutting Brazil's markets. At the same time both countries were anxious for British political support, and closely tied to the British economy. Both were vulnerable to British coercion.

Further pressure for action came from the British West Indies, where the economic woes of the newly emancipated colonies had been ameliorated by a preferential import duty regime that enabled their primary export, sugar, to compete with "slave-grown" crops from Brazil and Cuba.[7] This concession to the declining political influence of the West Indian planters was attacked by doctrinaire free traders, led by the Anti-Corn Law League, whose middle-class factory-owning leaders pressed for cheap food, including sugar, as a way of lowering the wages of the working classes and increasing industrial profits. In the event neither the compensation provided for abolishing slavery nor preferential duties could help the heavily indebted West Indian planters, or the local economy. Economic power passed into the hands of merchants and bankers.[8]

The abolition of the slave trade must be seen in the wider context of British domestic and external policy. This reality reminds us that successive British governments, committed to abolition, recognised the links between abolition and wider national interests. Political and economic concerns in Africa, the West Indies and America were deeply enmeshed with European and even Asian concerns. Ending the slave trade was not an absolute priority; it had to be accomplished while maintaining peace and stability in Europe and North America and sustaining the economic interests of the state. In consequence the patrol frequently took a back seat. The diplomatic and legal complexities of conducting what were, in essence, warlike operations on the high seas against the merchant ships of many nations ensured that the abolition of the slave trade would cost the Royal Navy a great deal of blood, sweat and treasure over the 60 years that followed the British legislation. Turning a noble gesture into a concrete fact was hard work. For those 60 years the anti-slave-trade patrol on the coast of West Africa was exposed to virulent tropical diseases, especially yellow fever and malaria, entirely unaware that the vector was the humble mosquito. Various miasma theories were developed to explain why ships and men who spent the night inshore or upriver were stricken by these lethal complaints. Men died in droves: the human cost ultimately ran into thousands, prompting some liberal, humanitarian politicians to dispute the value of the effort. At the same time Quaker abolitionists questioned the morality of using force to achieve the object.

After a major political debate on the efficacy and value of the anti-slavery patrol in the late 1840s, in which the liberal and radical Members of Parliament opposed the effort, the Whig government ensured it was carried through to the end. The end came sooner than had been expected because the slave owners lost political power in Brazil in the late 1840s, allowing the Royal Navy to operate more effectively. In 1850 Brazil began actively suppressing the Trade and, by 1852, that branch of the Trade was finished. Ending demand was the key. By 1858 only Spanish Cuba and the United States held out. They would not let the British search ships flying the American flag, even if the vessels were Cuban, were acting illegally, and had slaves on board.[9]

The campaign against the Cuban Trade was compromised by the need to keep the strategically vital island out of American hands. Politicians from the American South, desperate to increase the number of slave-owning states in the Union, supported frequent, illegal attempts to invade the Spanish island. While Britain opposed the slave trade, it recognised that cutting it might lead Cubans to welcome an American invasion, while the loss of Cuba, the richest province of the Empire, might bring down the shaky Spanish monarchy, seriously damaging British interests in Europe. To make matters even more complex, free trade meant that Britain was buying much of its sugar from Cuba, where it was grown by slaves and processed by British steam-powered machinery! In these circumstances British naval power could not stop the Cuban slave trade; instead, it ensured Cuba remained Spanish. Ultimately the resolution of the Cuban and American Trades depended on larger issues, and these could only be resolved within the United States. Until then the Royal Navy and British diplomacy could only limit the Atlantic Trade, and attack the supply side of the equation.

The inability of Britain to stop American shipping carrying slaves to Cuba gave added emphasis to the work of the West Africa Squadron. Naval action alone could not stop the Trade: the key to success lay in the steady extension of colonial authority on the West African coast. Fortunately the main slave supply sites were few in number and well known. If they could be blockaded, destroyed or occupied by the British authorities, the slavers would be forced to use smaller, less well-developed bases, reducing the supply while increasing the cost and risk of the operation. The destruction of the slave depots at Lagos in 1851 was a major blow; the annexation of Lagos in 1861 was critical. British diplomatic and economic pressure ensured Portugal limited the markets south of the equator, while new trades replaced human trafficking because the Brazilian market had closed; that of Cuba soon followed.

The American Civil War of 1861–5 ended the Atlantic Trade. President Lincoln finally rendered existing American legislation effective, cutting the supply of shipping that had been carrying slaves (most ships running slaves into Cuba were American-built and -operated), and agreeing an Anglo-American Treaty giving the right of mutual search. This document, signed on 7 June 1862, marked a complete reversal of the American position. Lincoln's administration adopted British practice on naval blockades at the same time, to suppress the Southern Confederacy. In addition, Lincoln had an American slaver captain hanged, under existing American law, as a pirate, but only one. Without American ships and markets Cuba was too small a market to support the Trade. In 1869 the West Africa Squadron was abolished, because there were no slave ships to catch.[10] This fact, observed on the African coast, was easily explained: the Cuban slave market closed in the same year. The Royal Navy simply redeployed the ships and spent the next twenty years eradicating the Arab slave trade on the East African coast.[11]

Ultimately the transatlantic slave trade ended when there were no more slaves to carry, and no more slave states to buy them. If British government anti-slavery was always tempered by realpolitik, and seen in a wider context than that of the passionate men and women who campaigned at home, it was nevertheless genuine and heartfelt.

Sometimes morality has a place in diplomacy, but it is a currency too rare to be employed often, and too easily debased to be used in haste. When, at the end of a remarkable political career, Palmerston claimed that the abolition of the slave trade and the establishment of secure national defences were his proudest boasts, he was referring to two sides of the same coin. Ultimately the British stand against the slave trade was only possible because the dominating power of the Royal Navy – the only strategic instrument with a truly global reach before 1945 – made Britain a unique world power among regional players. "For Palmerston sea-power, based on an unrivalled battle fleet, was a flexible and irresistible instrument that could be used in many ways to advance his diplomatic aims."[12]

This was the longest and hardest campaign the Royal Navy ever waged. The sustained effort across two generations demonstrated just how deeply the conscience of the nation had been engaged in the mission.

Opposing the Slavers is a clear and commanding analysis, based on an impressive engagement with the archives, the environment and the navigational demands of the service. Backed by appendices and a glossary, which will ensure every landlubber knows their ropes, it will transform the study of abolition, the anti-slavery patrol and the nineteenth-century Royal Navy. Having transformed the history of the patrol it will inform future research, tying together the naval, legal, diplomatic and political campaigns for abolition.

ANDREW LAMBERT
Laughton Professor of Naval History
Kings College London

Preface

IN THE EARLY HOURS of 20 March 1850, Lord John Russell, the Prime Minister of Great Britain, rose to his feet in the House of Commons to speak in defence of a great humanitarian cause, a policy of the British Government for over forty years, which had come under attack in Parliament. "It appears to me," he said, "that if we give up this high and holy work, and proclaim ourselves no longer fitted to lead in the championship against the curse and the crime of slavery, that we have no longer a right to expect a continuance of those blessings which, by God's favour, we have so long enjoyed." The "high and holy work" to which he referred was not, in fact, the fight against slavery itself but the long struggle to halt the international traffic which underpinned the continuation of slavery in the Americas: the Atlantic slave trade from Africa.

In recent years the people of Britain have become uncomfortably conscious of the leading role that their country played in the Atlantic slave trade during the eighteenth century. Few of them, however, are equally aware of the campaign fought by successive foreign secretaries and the Royal Navy against the continuation of Atlantic slaving throughout the sixty years which followed the passing of Britain's Slave Trade Abolition Act of 1807.

In this book I seek to redress the balance by showing how Britain, in its foreign policy and in the work of its Navy, made the central, costly, mostly lonely, inevitably unpopular and always thankless stand against the great evil of slaving from West Africa during the nineteenth century.

It is not my intention either to consider the institution of slavery, on which there is an extensive library, or to examine in any depth the Atlantic slave trade itself, which has already been admirably analysed. However, some understanding of the history and nature of the slave trade, and of the environment in which it was conducted, is clearly necessary to an appreciation of the suppression work of the Navy and the diplomats. Therefore, after a Prologue which outlines the background to British abolition, Part One of the book, *The Trade*, gives a brief history of the Atlantic slave traffic, outlines its characteristics immediately before the suppression campaign, and describes the coasts and seas upon which it took place.

The main theme of the book is the Royal Navy's anti-slaving activity at sea in the Atlantic and on the coasts of West Africa and the Americas after the Abolition Act of 1807. However, these naval operations were inseparably linked with the work of the Foreign Office, and so I have given emphasis to the diplomatic struggle and its crucial relationship with events and tactics at sea. Many other factors had a significant bearing on the progress of the naval campaign: parliamentary debates, ship development, legal wrangling, disease, and changes in the economics of the slave trade (the Trade), among others. Naval operations and all of these directly relevant aspects are covered in the main body of the book, Part Two, entitled *The Suppression Campaign*.

For the narrative in Part Two, I have drawn almost entirely on primary sources, principally the reports and despatches written at the time by commanding officers, senior officers, commanders-in-chief and the British representatives at the Mixed Commissions, as well as the records of the various courts dealing with slaving cases. Memoirs of those involved I have used very sparingly, bearing in mind the distorting effects of hindsight.

Although the campaign spanned a 60-year period, surely the longest in the history of the Royal Navy, this volume describes, in depth, only the first 32 years of it. The progression of operations until the ending of the Trade in 1867 (or thereabouts) is seamless, but in 1839 the beginning of a change in emphasis and attitude can be detected. That change, however, is not the primary reason for ending this account at the end of that year. The loss by neglect, in Victorian times, of almost all of the Admiralty archives for the period 1840–55, and the disappearance of relevant Vice-Admiralty court records have rendered a similarly detailed account of the second half of the campaign impossible.

Nevertheless, in Part Three, *Conclusion*, after summarising and assessing the events of the first 32 years, I give an overview of the latter half of the campaign in order to bring the story to completion. Finally, I offer a brief evaluation of Britain's part in bringing the Atlantic slave trade to an end.

PROLOGUE

The Beginning of a Great Matter

O Lord, when thou givest to thy servants to endeavour in any great matter, grant us also to know that it is not the beginning but the continuing of the same until it be thoroughly finished that yieldeth the true glory.

SIR FRANCIS DRAKE

A BILL CALLING for the prohibition of "all manner of dealing and trading in the purchase, sale, barter or transfer of slaves [...] on, in, at, to or from any part of the coast of Africa" was introduced in January 1807 in the House of Lords by the Prime Minister, Lord Grenville. The bill declared that the Atlantic slave trade, in which the British had been leading exponents for the past century, was "contrary to the principles of justice, humanity and sound policy". For good measure, Grenville added his opinion that the Trade was not only detestable but also criminal.

Having been passed by the Lords, not without opposition from some of those who owed their seats to the slave trade, the bill went to the Commons, where there also remained an element of trenchant resistance. Nevertheless, in the early hours of 24 February, the abolitionists gained a crushing victory, and, on the announcement of the result, the entire House rose to applaud William Wilberforce. A month later King George III gave his assent to the bill, and on 1 May 1807 the African slave trade became illegal in the British Empire, with a penalty of a £100 fine for every slave found and confiscation of any vessel concerned.

The Act (47 Geo. III, c. 36) was indeed a triumph for Wilberforce, who had led the long struggle of the abolitionist cause in Parliament, stoutly supported by William Pitt as Prime Minister, who did not live to see its success. However, richly deserving though he was of applause, Wilberforce was by no means the only man worthy of praise for this immense achievement.

There had been occasional criticism of slavery by English Protestants since the beginning of serious English participation in the Trade late in the seventeenth century, notably by the Puritan Richard Baxter. However, the first figure to make a significant impact on British public consciousness of the evils of slavery was the Englishwoman Aphra Behn who, in her 1688 book *Oroonoko; or the Royal Slave*, wrote of a slave from the Gold Coast killed after leading a failed slave rebellion in the Dutch colony of Surinam. Her novel was dramatised by Thomas Southern in 1696, and his play was performed to great effect before audiences in England and France for much of the following century. Aphra Behn's powerful influence was particularly apparent in the work of the poets and authors of the early eighteenth century: James Thomson (author of "Rule, Britannia!"), Daniel Defoe, Richard Savage, William Shenstone, Alexander Pope, Sir Richard Steele and Laurence Sterne all condemned slavery on humanitarian grounds. These writers helped to create a reaction against the Trade in England, the European country with the strongest instinct for freedom as well as the leading slave-trading nation for most of the eighteenth century.

The Vatican occasionally spoke out against slave trading, but to little apparent effect, and it was among the Quakers of English America that hostility to slavery was being more frequently and forcefully voiced. In 1676, William Edmundson, an associate of the founder of the Society of Friends, wrote from Newport, Rhode Island, that slavery should be unacceptable to a Christian, and in 1688 a group of German Quakers in Philadelphia signed a petition against the concept of slavery. This was a hesitant beginning to a Quaker movement against the slave trade which took more than half a century to reach any coherence, and it was a Presbyterian judge in Boston, Massachusetts, Samuel Sewall, who wrote the first reasoned criticism of the Trade and of the institution of slavery. At last, in 1754, the annual meeting of the Society of Friends in Philadelphia published an open letter which declared that "to live in ease and plenty by the toil of those whom violence and cruelty have put in our power" was inconsistent with Christianity and common justice, and urged Quakers to make the cause of the Africans their own. Thanks largely to years of devoted work by John Woolman, the 1758 Quaker meeting in Philadelphia agreed an even stronger condemnation of slavery. Also in that year the Quaker meeting in London adopted the same theme with equal determination, and the London meeting of 1761 declared that the Society would disown Quakers who would not desist from the traffic in slaves.

Meanwhile, a quite different abolitionist influence was growing in British North America. There was concern that skilled slaves were taking work needed by white tradesmen, and a more powerful fear that an excessive number of African slaves in the colonies would lead to revolts of the kind experienced in Jamaica. This fear equalled philanthropy as the motivation of the American abolitionist movement in the late eighteenth century, and caused prohibitive duties to be imposed on slave imports in several colonies and an outright ban on the traffic in Rhode Island.

Slower to gain hold was the humanitarian concern which was later to become the mainspring of abolitionism. There were isolated statements by European writers such as the Portuguese Frei Manuel Ribeiro da Rocha who, in a book published in 1758, demanded an end to the slave trade, which he regarded as illegal, and which "ought to be condemned as a deadly crime against Christian charity and common justice", a denunciation far in advance of anything written in Britain or British America. Such statements brought no evident benefit, however, and even the powerful intellects of the French Enlightenment, all hostile to slavery, had no idea what might usefully be done in practice. Marivaux, Voltaire, Montesquieu, Diderot and Rousseau either denounced or mocked slavery, but mistakenly assumed that they had merely to launch ideas and governments would then take action. Nevertheless, anti-slavery took a place in French radical thinking, and, coincidentally, this aligned with the views of the Roman Catholic Church, which had prohibited slavery in a declaration by Pope Benedict XIV in 1741. None of this discouraged the slave traders.

In England and North America, however, popular poetry and the journalism of the periodical press was, by the middle of the eighteenth century, having an effect on cultivated opinion regarding slavery. It was the existence of a free press and the relative ease of correspondence which largely explain why the abolitionist cause gained ground and became a political issue in these two countries before any other. The *Gentleman's Magazine*, the *Weekly Miscellany* and the *London Magazine* all played an influential part in England, but the debate was not yet general, and it would be some time before any parliamentarian became involved. Indeed, as Horace Walpole wrote to a friend in 1750, "We, the temple of liberty, and bulwark of Protestant Christianity, have this fortnight been [in Parliament] pondering methods to make more effectual that horrid traffic of selling Negroes." Critical comments in Walpole's correspondence demonstrated, however, that educated Englishmen were now discussing the morality of slavery and the legal aspects of freedom.

The argument was developing in Scotland too. In *A System of Moral Philosophy* (1755), Francis Hutcheson, Professor of Philosophy at Glasgow, concluded that "all men [without exception] have strong desires of liberty and property" and that "no damage done or crime committed can change a rational creature into a piece of goods void of all right". Adam Smith, one of Hutcheson's former pupils, wrote in *The Theory of Moral Sentiments* (1759), "There is not a Negro from the coast of Africa who does not [...] possess a degree of magnanimity which the soul of his sordid master is scarce capable of conceiving." Such views were readily absorbed by the students of these two professors, and the cause was taken up by other intellectuals, including the lawyer George Wallace, who was influenced by Montesquieu, and Adam Ferguson, Professor of Philosophy at Edinburgh.

In 1765, the judge Sir William Blackstone published *Commentaries on the Laws of England* in which he stated the case against slavery and declared that the law of England "abhors and will not endure the state of slavery within this nation". Subsequent editions of this work were more equivocal, probably owing to intervention by Blackstone's benefactor Lord Mansfield, Lord Chief Justice of England, but Blackstone's *Commentaries* were, nevertheless, successful and influential.

The Quakers then began to carry the abolitionist argument beyond their own movement, chiefly through the work of Anthony Benezet of Philadelphia. He, with the dedicated John Woolman, had greatly influenced the earlier anti-slavery decisions of the Philadelphia Quakers, but he was now reaching out to the rest of North America, and to England. In his writing he quoted material from Montesquieu, Hutcheson and Wallace, used first-hand accounts of African slaving, and in his 1767 edition of *A Caution and Warning to Great Britain and the Colonies* he quoted a comment of a West Indies visitor:

> it is a matter of astonishment how a people [the English] who, as a nation, are looked upon as generous and humane [...] can live in the practice of such extreme oppression and inhumanity without seeing the inconsistency of such conduct.

In his campaign of conversion, Benezet wrote not only to fellow Quakers but also to Edmund Burke, John Wesley and the Archbishop of Canterbury, and he encouraged the Presbyterian Dr Benjamin Rush to form, in Philadelphia, the first abolition society.

At this stage the interest of English abolitionists was focused on the legal rather than the religious aspects of the issue. The law was confused concerning the status of slavery in England, and a number of cases in the late seventeenth century and early eighteenth century had produced conflicting judgments. An opinion was given in 1729 by Walpole's Attorney-General, Sir Philip Yorke, that a slave in England was not automatically free, nor did baptism "bestow freedom on him, nor make any alteration in his temporal condition in these kingdoms", and this was given credence for a period of over forty years until it was challenged in court by Granville Sharp. Sharp, a clerk in the Ordnance Office and grandson of the Archbishop of York, was an idealist who became a key figure in the English abolitionist movement. He initially came to attention by befriending three escaped slaves and resisting their repossession on the basis of the 1679 Act of Habeas Corpus. The first of Sharp's cases, that of Jonathan Strong in 1765, was not pursued by the owners. Then, in 1771, the case of Thomas Lewis came to court, but ownership was not established and Lewis was freed. At last, in 1772, the question of the legal position on slavery was addressed in the case of James Somerset, tried in the Court of King's Bench before Lord Chief Justice Mansfield. After lengthy procrastination Mansfield decided that, in the absence of positive law on the subject, slavery was so odious that nothing should be done to support it, and Somerset was released. Nevertheless this judgment established only that "the master had no right to compel the slave to go into a foreign country."

Lord Mansfield's judgment did, however, sway public opinion, and the case brought Sharp and Benezet into correspondence with each other, strengthening the transatlantic dimension of abolitionism. It was Benezet who inspired John Wesley to publish his *Thoughts upon Slavery* in 1774, the most serious attack to date on slavery and the slave trade, and, with Methodism gaining a wide appeal, a most influential one too. Dr Johnson, always an opponent of slavery, added his typically pungent comments to the controversy, and further criticism, on different lines, came from Adam Smith in *Wealth of Nations*, published in 1776. Summarising his views on slave labour, Smith wrote: "it appears […] from the experience of all ages and nations […] that the work done by freemen comes cheaper in the end than that performed by slaves." Fallacious though this assertion might have been, it was temporarily influential.

Prompting by the Quakers led to the appointment in 1775 of a House of Commons commission to take evidence on the slave trade, and this led to a debate in the House on the motion that "the slave trade is contrary to the laws of God and the rights of men". This put slavery on Britain's political agenda, but the 1776 Revolution delayed an equivalent debate in America. Slaves comprised 22 per cent of the population of the 13 colonies, and slavery and the slave trade were taken for granted by most of the American rebels. The irony of that acceptance was not lost on Dr Johnson: "How is it that we hear the loudest yelps for liberty among the drivers of Negroes?" Even Benjamin Franklin, although critical of slavery, was cautious on the issue of abolition, and the Quakers were discredited for a time because they refused to support armed resistance against the English. There was some talk of the slave trade in discussions leading to the Declaration of Independence, but it came to nothing, although slave trading with Britain, together with other commerce with the mother country, was banned by the colonies at the beginning of the rebellion. With the resumption of peace, however, the Quakers regained their standing and immediately initiated public discussion of the slavery issue.

In 1779 in Liverpool, the last public sale of a black slave in England took place, but humanitarian opinion was generally not that slavery or the Trade should be abolished. Instead, the view was that they should be made less cruel, although thinking in Scotland was more aligned with abolition. Then, in 1781, a scandalous incident concerning the Liverpool slaver *Zong*, and the court cases which followed, illustrated for the general public some of the horrors of the Trade and the state of the law concerning slaves, giving a boost to the abolitionist cause in England. *Zong* had sailed for the Caribbean with a cargo of 442 slaves from the West African island of St Thomas, and had become lost. With water running short and many slaves dead or sick, her master had ordered 133 of the survivors to be thrown overboard on the pretext that if they were to die of natural causes the loss would be to the shipowners, but if, for reasons affecting the safety of the crew, they were to be thrown alive into the sea, the underwriters would carry the loss. The insurers refused to pay, and the owners sued them, with the backing of the King's Bench. Then the underwriters petitioned the Court of Exchequer. In a second trial allowed by Lord Mansfield, the owners' counsel argued successfully that: "So far from a charge of murder lying against these people, there is not the least imputation – of cruelty, I will not say – but of impropriety." When Granville Sharp then tried without success to bring a charge of murder before the Court of Admiralty, the Solicitor-General declared that a vessel's master could drown slaves without "a surmise of impropriety".

Thanks largely to his correspondence with Benezet, Sharp now enjoyed the support of most of the English bishops, and after the *Zong* trial the abolitionist case was articulated with ever-increasing effectiveness, aided by contact between the various opponents of slavery. In 1783 a bill was introduced in the House of Commons to prevent officials of the Royal African Company from selling slaves, and that encouraged the Society of Friends to appeal for a total ban on the Trade. The following year the town of Bridgwater, the first to do so, petitioned the Commons for an end to the Trade. These attempts were unsuccessful, but were valuable achievements in that they generated debate in Parliament.

At the same time, a Scottish retired Royal Navy Surgeon, Dr James Ramsay, published two anti-slaving pamphlets and became an abolitionist supporter of importance on two counts: first, he had, unlike Benezet and Sharp, first-hand experience of the West Indies, and second, he knew the abolitionist Captain Sir Charles Middleton, Controller of the Navy and the future Lord Barham, who had once been his Commanding Officer at sea. Another pamphlet, *The Case of Our Fellow Creatures, the Oppressed Africans*, was then published by the ageing Benezet, and more than 10,000 copies were printed. A new committee of English Quakers, formed to educate the public more effectively about the slave trade, distributed the pamphlet to MPs and conducted an extensive series of lectures in the great public schools.

A new and crucial phase for the movement then began with the appearance of Thomas Clarkson on the abolitionist scene. Clarkson, a 24-year-old Cambridge graduate who had studied Benezet and the papers of a deceased slave trader, won a university prize for a Latin essay on the subject of whether it was lawful to enslave men against their will. On his way to London to have the essay published in English, he came to the realisation that "[i]f the contents of the essay were true, it was time some person should see these calamities to their end." In London the Quaker bookseller James Phillips not only published Clarkson's essay but also introduced him to the major figures of the abolitionist movement, including Ramsay, Sharp and Benezet's pupil William Dillwyn. After a long discussion with Dr Ramsay at Sir Charles Middleton's house, Clarkson decided to abandon his plans for ordination and to devote himself to the abolition cause. The campaign then started in earnest.

1. Thomas Clarkson, leading abolitionist

The Committee for Effecting the Abolition of the Slave Trade was formed in London in 1787 under Clarkson's leadership, and this marked the development into a national movement of what had until then been essentially a Quaker cause. The campaign adopted as its emblem a picture of a chained and kneeling slave bearing the legend: "Am I not a man and a brother?", an inspired piece of propaganda designed by one of its supporters, Josiah Wedgwood. Clarkson was an inspirational leader who had valuable contacts and was able to stimulate interest among a wide cross-section of educated society. He also understood what information was needed to support the campaign, and possessed the courage and diligence to gather it. However, Sharp and Ramsay disagreed with Clarkson and his Quaker colleagues on the emphasis given by the Committee to the abolition of the slave trade rather than that of slavery itself. Clarkson was convinced that if the Trade was banned slavery would collapse, and there was a concern that an attempt to abolish slavery would be seen to threaten the institution of property, a very sensitive matter which might even risk sending the British West Indies along the same road as the American colonies. Clarkson's logic may have been flawed, but his view prevailed.

In May 1787 Clarkson met William Wilberforce, Member of Parliament for Hull, and, at the urging of Sir Charles Middleton, now a rear-admiral and an MP, asked him to assume political leadership of the campaign. Wilberforce, aged 28, was well educated, eloquent, independent, charming and rich, and, like his close friend William Pitt, the Prime Minister, he was a Cambridge graduate. He had once been described as "the wittiest converser in England", and, on account of his particularly attractive voice, was known as "the nightingale of the House of Commons". He was also a Christian with an evangelical inclination, and a natural leader. Although he had already read Ramsay's pamphlets and had met John Newton, the slaving captain turned clergyman, he had neither depth of knowledge of the slave trade nor commitment to the abolitionist cause until he had talked with Pitt in February 1787. As a result of that crucial conversation, both men had become convinced that the slave trade was an evil which should be brought to an end without delay.

Although Wilberforce agreed to lead the abolitionist cause in Parliament, Clarkson remained the driving force behind the campaign. He knew that the Committee for Abolition would need to present rational arguments and comprehensive data if its crusade was ever to succeed, and he spent much of 1787 gathering information. Thanks largely to Pitt, he gained access to state documents, and he visited Bristol and Liverpool to interview slavers and slave merchants, at appreciable personal risk. He caused such a stir that a committee of the Privy Council was set up early in 1788 to investigate the slave trade, and Clarkson, helped by Pitt and Wilberforce, prepared the abolitionist case. Well-worn arguments were presented, but Clarkson also expressed a new concern about the suffering of the slave-ship crews, an aspect which he rightly thought might catch the imagination of Parliament. At the same time, the Committee for Abolition launched a major public campaign, the first in any country for a philanthropic cause, distributing publications by Benezet, Clarkson, Ramsay, Cowper, John Newton and Hannah More, founding local committees, recruiting support and conducting research.

Simultaneous with the start of Clarkson's campaign was a proposal to establish an African colony as a refuge for freed slaves living in London. An attempt had already been made to establish settlements in Nova Scotia and New Brunswick for loyalist ex-slaves who had fought for Britain in the Revolutionary War, but with little success, largely on account of the climate. A more promising suggestion, made in 1786 by the botanist Dr Henry Smeathman, who had twice visited West Africa, was that a colony might be founded in Sierra Leone, and Granville Sharp was particularly enthusiastic about the concept. The Government gave its support, and in April 1787 a convoy carrying 290 black men, 41 black women and 70 white women, including 60 prostitutes, sailed from Plymouth for Sierra Leone under the command of Commander T. Bouldon Thompson in the sloop HMS *Nautilus*. A tract of land of about ten miles by twenty between the Sherbro and Sierra Leone slave rivers had been bought from the local chief, "King Tom", and the intention was to establish a settlement regulated by "the ancient English frankpledge". Thanks to disease, drink, idleness, rain and fighting with local people, the result was disastrous. Half of the 377 settlers who had landed were dead within a year, and others deserted to join local slave dealers. Then in 1789 King Tom's successor burned the settlement of Granville Town, and the remaining colonists fled.

By the efforts of Sharpe and Alexander Falconbridge, agent of an association of benefactors and financial backers called the St George's Bay Company, 64 of the survivors were gathered together in 1791 to build a new Granville Town. In the same year, the St George's Bay Company was incorporated as the Sierra Leone Company, with directors including Wilberforce, Clarkson, Sharp and Middleton, and, at the instigation of the company, 1,190 free black people sailed from Nova Scotia in January 1792 to reinforce the settlement. These people had been involved in the unsatisfactory Nova Scotia and New Brunswick settlement project, and the Government offered them the alternative of Sierra Leone. They had been gathered together by John Clarkson, a half-pay Lieutenant of the Royal Navy and brother of Thomas Clarkson, who led the transatlantic convoy and then briefly governed Sierra Leone until ill health obliged him to return home. These new immigrants immediately cleared land for their new capital, which they named Free Town, and the settlement became a colony under a governor appointed by a committee chaired by Sharp. A marked improvement began with the arrival of Governor Zachary Macaulay in 1794, despite the pillage of Freetown (as it had become) in that same year by the French, but the thriving community he had built declined rapidly after his departure in 1799. Descent into chaos was prevented only by the arrival from Jamaica in 1800 of 550 maroons, descendants of escaped slaves. The colony became a full dependency of the Crown in 1808.

In the United States of America, the new Constitution, signed in 1787, allowed a continuation of the slave trade for another 20 years, but established a requirement for a debate on the matter in 1807. This compromise was accepted primarily to avoid a secession of the southern states, but, although there was some anti-slaving dissent in the Constitutional Convention and from other men of influence, there was no unity or commonality of objectives among American abolitionists. Nevertheless, the Constitution did not prevent individual states from taking action against slavery and the slave trade, and Pennsylvania had already led the way with a conditional and gradual ban on slavery in 1780. By the end of 1788, Rhode Island, Massachusetts, Connecticut and New York had all prohibited the Trade, New Jersey had abolished slavery, and even Virginia had decided to free illegally imported slaves. Delaware soon followed with a prohibition on African slavery. However, it was found to be extremely difficult to enforce this state legislation, and traders in the northern states continued to deal with customers in North and South Carolina and Georgia, in all of which the Trade remained legal. In 1789 the House of Representatives debated the imposition of duties on slave imports, reaching no conclusion, but in 1790 Congress prohibited foreigners from fitting-out slave ships in the USA and banned the American slave trade to foreign ports.

William Pitt, meanwhile, was bringing the slave trade abolition issue into the mainstream of British politics. He was unable to achieve a majority in favour of abolition in either his cabinet or the Tory Party, and so, in May 1788, while Wilberforce was ill, he introduced a Private Member's Bill, with a free vote. Summing up his case in the Commons, he said:

> When it was evident that this execrable traffic was as opposite to expediency as it was to the dictates of mercy, of religion, of equity, and of every principle that should actuate the breast [...] how can we hesitate a moment to abolish this commerce in human flesh which has for too long disgraced our country and which our example would no doubt contribute to abolish in every corner of the globe?

Although Pitt's bill failed, another, introduced by Sir William Dolben to restrict the number of slaves carried at sea according to ship tonnage, fared better. It was carried in the Commons, heavily amended in the Lords, passed again in the Commons by a good majority, and finally scraped through the Lords thanks to backstage intervention by Pitt. This Act had an appreciable effect on the British slave trade, increasing the number of vessels involved but reducing the overcrowding of slaves.

Unbeknown to (or unheeded by) the House of Lords, the Committee for Abolition had achieved astonishing success in arousing public sympathy for the cause. Even in the leading slaving port of Liverpool there was an abolitionist meeting, and two-thirds of Manchester's male population signed a petition for abolition of the Trade, an example followed by a hundred towns. Wesley preached a famous anti-slaving sermon in Bristol in 1788, Newton published *Thoughts upon the African Slave Trade*, and George Morland made an impact on fashionable society with his sentimental painting *The Slave Trade* at the Royal Academy.

In 1789 the House of Commons debated the Privy Council's admirable report on its investigation into the slave trade, which contained a mass of information, largely provided through Clarkson's meticulous enquiries. Supported by Pitt and Edmund Burke, Wilberforce introduced the debate in a speech later praised by Fox. They were vociferously opposed by the strong slaving influence in the House, chiefly the Members for Liverpool and London, on a now-familiar basis of prejudice, ignorance and self-interest. There was clearly a widespread fear among those sectors of the population involved in the Trade, or in the manufacture of trade goods, that legislation would follow this debate, and many petitions against abolition were immediately presented to the House of Commons. As well as those directly involved in slave trafficking, the opposition to abolition included people concerned with the sugar and cotton trades, conservative landowners, most of the admirals and some members of the Royal Family, probably including the King. These disparate elements had formed an alliance in Parliament by the time of a further debate on the Trade in early 1790, and its main representative and spokesman in the Commons, as he had been in 1789, was Bamber Gascoyne of Liverpool.

The 1790 debate came to nothing, although, despite obstruction from Gascoyne, the Commons set up a Committee of Inquiry which produced a thorough report detailing for Members of Parliament the cruelties of the Trade and of slavery in the West Indies. Early in 1790, Clarkson visited France, where he was warmly welcomed by the abolitionists among the revolutionaries. A subsequent debate in the Constituent Assembly was followed by an inadequate investigation report on the slave trade, much impassioned talk in Paris, and turmoil in the French slave-holding colonies. The National Assembly condemned slavery in principle in 1791, but took no practical measures against it, and, as the Revolution descended into chaos, interest in the issue was lost.

The murderous events in France had hardened political opposition to the British anti-slavery movement, and in April 1791 Wilberforce's introduction of his bill to abolish the Trade came at a bad time. In preparation, Clarkson had worked tirelessly to gather further information, visiting 320 ships in English ports and travelling 7,000 miles during 1790. During this debate, Parliamentary opposition changed direction from the earlier attempts to justify the slave trade and to hail its benefits, moving towards concentration on the impracticability and folly of abolition. There were fine speeches in support of Wilberforce, not only from

Pitt and Burke but also by Charles Fox. The motion was lost by 88 votes to 163, and when the news reached Bristol, Burke's constituency, the church bells rang and there was a bonfire and fireworks. Horace Walpole wrote that: "Commerce chinked its purse, and that sound is generally prevalent with the majority."

The British abolitionists were, however, much encouraged early in 1792 by the news that the Danish Government had decided to abolish the import of slaves from Africa to its West Indian islands. The ban was to come into effect at the end of 1802, and, until then, there was no limitation on imports to the Danish colonies by the slavers of any nation, although re-exportation of slaves from the Danish islands was to cease forthwith. The Danes had shipped slaves on only a small scale, but between 1792 and 1802 their slave trade flourished, largely on the illegal re-export of slaves from the Danish colonies of St Croix and St Thomas to Cuba.

Heartened by developments in Denmark, Wilberforce made another attempt to carry an abolition bill in April 1792, and this was the occasion for what has been regarded as one of history's greatest legislative debates. The decisive speech was by Henry Dundas, the Treasurer of the Navy, who exercised an extraordinary influence over Pitt. He proposed a middle way, agreeing that the Trade should eventually be abolished, but "by moderate measures, which should not invade the property of individuals, nor shock too suddenly the prejudices of our West India islands." He further proposed two other amendments: a ban on the import of elderly slaves, and an abolition of foreign slaving to British possessions. Dundas had once rescued an escaped slave from recapture, but, despite his apparent reasonableness, his sympathies now clearly rested with the West Indian planters. Fox's speech mocked Dundas, and Pitt, beginning to speak at five o'clock in the morning, concluded the debate with perhaps the finest performance of his career. The Commons voted 230 to 85 in favour of Dundas's amended motion: "that the slave trade ought to be gradually abolished". Although this fell well short of what the abolitionists wanted, it represented real progress from the previous year's result. As Fox, Charles Grey and William Windham walked home, they reflected that Pitt's speech had been "one of the most extraordinary displays of eloquence they had ever heard", and they felt that they had been present at the supreme moment of parliamentary democracy. The subsequent debate in the Lords was of lower quality, and included a comment by Lord Barrington that the slaves appeared to him so happy that he often wished himself in their situation. Nevertheless, the House supported the Dundas amendments.

It became clear that the intention of the main amendment was an indefinite delay in enacting abolition, and, although a final date of 1796 was agreed, there was no expectation of implementation. The abolitionist leaders had been outmanoeuvred, and by the time that Parliament reassembled in January 1793 the topic at the forefront of members' and the Government's minds was war with revolutionary France, not abolition of the slave trade. To make matters worse, Wilberforce lost much support because of his opposition to the war, and the campaign for abolition was equated with the French call for equality and fraternity. With the Trade continuing to expand, thanks largely to the emerging market in Cuba, there was little to hearten the abolitionists, although Bristol suffered a financial crash which ruined several of its leading slave merchants and the war at sea had led to a severe reduction in Liverpool's slaving traffic. Wilberforce persisted, however, and in 1794 he persuaded the Commons to support a bill to prohibit British merchants from selling slaves to foreign markets. The same protagonists contributed to the debate, although the new name of Robert Peel appeared in opposition, and another Member introduced the argument that, "since all Europe was in a state of confusion, it would be highly imprudent to adopt any untried expedient." Despite this success in the Commons, the bill was defeated in the Lords where the prejudiced Lord Chancellor, Edward Thurlow, remained implacably hostile.

With admirable determination and persistence, Wilberforce returned to the attack in 1795 with a motion to enforce abolition at the agreed date of 1796, but reaction to the war, the revolution in France and a slave rebellion in the French colony of Saint-Domingue had strengthened opposition. Pitt, despite the distractions of the war, spoke strongly in support of Wilberforce, as did Fox, but Dundas argued for further delay. As George Canning, newly arrived in Parliament, wrote in a letter, the motion "to the disgrace of the House, was negatived by […] 78 to 61, in defiance of plain justice and humanity." The margin of defeat was closing, however, and one firm of Glasgow slave merchants was so convinced that the bill would be carried that it engaged in speculative purchase of slaves and went bankrupt as a result, the worst financial disaster in the history of the British slave trade.

In 1796 Wilberforce secured permission to introduce yet another bill, calling, this time, for abolition of the Trade from the beginning of 1797. Pitt, Fox, Canning, Sheridan and Francis remained staunchly behind Wilberforce, but the speeches of the abolitionists were becoming repetitious, partly because Clarkson, the driving force and workhorse of the movement and their primary source of new information, had collapsed from exhaustion. The bill was lost, but by only four votes.

Public interest in the slaving issue was restimulated to an extent at this point by a lawsuit similar to the *Zong* case. A Liverpool ship, with 168 slaves embarked, was delayed on the Middle Passage by adverse weather, spending six months on passage to the West Indies with food supplies for only the expected six to nine weeks. Starvation claimed most of the 128 slaves who died, and the owners, arguing that this was attributable to the perils of the sea, claimed against the insurers. Having ascertained that the master of the vessel had not starved to death, the new Lord Chief Justice found against the shipowners.

Wilberforce suffered further defeats in 1797 and 1798, and, although new names appeared in support and in opposition, the only novel proposition made in either debate was that the governors of the West Indian colonies should be instructed to encourage the legislatures of the islands to improve plantation conditions so that the natural increase of slaves would render the Trade unnecessary. Pitt was briefly tempted by this notion, but was soon squarely behind Wilberforce again. Another attempt in 1799 saw the conversion of Dundas from an advocate of gradual abolition to an outright enemy of the cause. He argued that the issue was so important that it should be decided by the legislatures in the colonies, a view effectively ridiculed by young Canning, now in the Government. Failure on this occasion was by only 82 to 74. Another Wilberforce bill to exclude slave traders from Sierra Leone was passed in the Commons, but even such an obvious measure as this was defeated in the Lords, where opposition was led by the Duke of Clarence, who was rewarded with the freedom of the city of Liverpool.

A bill to restrict further the number of slaves per ton permitted in British ships did indeed become law in 1799, and henceforth each slave would be allowed about eight square feet of deck space instead of five or six. This modest advance seemed little recompense for over a decade of struggle by the abolitionist movement in Parliament, a period which had seen the British slave trade at its most lucrative. Between 1791 and 1800 nearly 400,000 slaves had been landed in the Americas by British vessels on about 1,340 voyages, and the British economy seemed to be increasingly dependent on slavery and slave-produced imports. The abolition campaigners were tiring, Pitt's energies were absorbed by the war, and the Navy's many anti-abolition senior officers, including Nelson, whose service in the Caribbean had given them an affinity with the planters, exerted appreciable influence.

Abolition prospects in America appeared equally unpromising. A new federal law in 1794 had prohibited US citizens from carrying slaves to other nations, and the three states in which importation of slaves had been legal after 1787 had, at various stages, banned that trade. Then, in 1800, it became illegal for any US citizen to have any share in a slave vessel en route to a foreign country. On the face of it, this was significant progress, but in fact little effort was made to enforce any of the anti-slaving legislation. A mixture of legal loopholes, corruption, nepotism and intimidation ensured that the traders of Rhode Island in particular were able to keep the slaveholding states well supplied, and between 1796 and 1807 America dominated the Cuban slave trade. It was reckoned that by 1806 the slaving fleet of the USA was almost three-quarters the size of Britain's, and it was unrestrained by any legislation on carrying capacity. When Mississippi was established as a state in 1798 it was exempt from the anti-slaving clause of the Constitution, and Napoleon's sale of Louisiana to the United States in 1803 added another slave state to the Union.

Then came two glimmers of encouragement. When, at the Peace of Amiens in 1802, Napoleon revived the French slave trade and reintroduced slavery into the French colonies, the abolitionist cause in Britain lost its revolutionary stigma. That was followed by confirmation by the Danes of their promised abolition of the Trade at the end of 1802. It was enough to hearten Wilberforce, and in 1804 he introduced his fourth abolition bill to the House of Commons. Thanks to the votes of the new Irish members, this attempt succeeded by a margin of 49 to 24, but the House of Lords remained the stumbling block, with the continuing opposition of Lord Chancellor Thurlow and others now bolstered by the arrival of Dundas transfigured as Lord Melville.

Despite this further failure, the abolitionist horizon was now looking much brighter. The propaganda of Clarkson, Sharp and James Stephen, Clarkson's successor and Pitt's brother-in-law, had apparently worked successfully on the new generation of Members of Parliament. Lord Chancellor Thurlow was dying, and Melville was impeached in 1805 for mismanagement of funds while Treasurer of the Navy. In August 1805 Pitt, anxious to reduce the risk of slave revolts in Britain's newly acquired territories in the Caribbean, used an Order in Council to prohibit slave imports to "any of the Settlements, Islands, Colonies, or Plantations which have been surrendered to His Majesty's arms during the present War". To this prohibition was added a clause to prevent the fitting-out of foreign slave ships in British ports. Wilberforce was therefore disappointed when a new version of his bill was successfully obstructed in the Commons in 1805, the first such occasion on which the preoccupied Pitt declined to speak.

In December 1806 in the United States, President Thomas Jefferson condemned the "violations of human rights which have been so long continued on the unoffending inhabitants of Africa", and he urged Congress to use the ending of the constitutional limitation in 1807 to abolish the slave trade. The following day, Senator Bradley of Vermont introduced an abolition bill. The subsequent debate concentrated on the important details of implementation rather than on issues of principle, and the bill passed the Senate on 27 January 1807, cleared the House of Representatives on 11 February, and was signed by the President on 2 March. It prohibited, from 1 January 1808, the introduction into the United States of any "Negro, mulatto, or person of colour, as a slave", and made it illegal for any US citizen to equip or finance any slave vessel to operate from any port in the United States. Penalties included heavy fines, imprisonment and forfeiture of ships and slaves. However, the value of this major legislative achievement was seriously undermined by the failure to institute any machinery for its enforcement.

Following the death of the exhausted William Pitt early in 1806, the new Government, led by Lord Grenville, held discussions with the abolitionists. Grenville had long supported the abolitionist cause, and his cabinet, including Charles Fox as Foreign Secretary, were, with one exception, like-minded men. Shortly thereafter, the Attorney-General, Sir Arthur Pigott, quietly introduced a bill which was to be a stepping stone to full abolition, but which roused only muted opposition because its significance was not initially appreciated. It proposed to prohibit slave importation by British subjects to foreign territories, both from Africa and from British dominions; use of British ships, capital, credit or insurance in any foreign slave trade; supply by British subjects of slaves to foreign ships or factories on the coast of Africa, and exportation of slaves from one British territory to another without licence. It proposed penalties of fines and forfeitures of ships and slaves for offences against these measures, and it also reiterated the prohibitions of the 1805 Order in Council. This wide-ranging and highly significant bill was passed in the Commons by 35 votes to 13, and even won a majority of 43 to 18 in the Lords, becoming law on 23 May 1806 (46 Geo. III c. 52). The abolitionists, including the now recovered Clarkson, were overjoyed.

There were further debates in mid-1806, and then Grenville and Fox moved resolutions in both Houses to abolish the slave trade with "all practicable expedition". Both Houses agreed to urge the Government to negotiate with other nations for a general abolition of the Trade, and a motion in favour of abolition was carried by 114 votes to 15 in the Commons, and by 41 to 20 in the Lords, with the help of the new Lord Chancellor, Thomas Erskine. At this stage, there was some easing of opposition from the West Indies planters and their friends on account of a sugar surplus and saturation of the slave market in Britain's older island colonies, and a measure to forbid the employment of any new vessel on the slave trade was enacted without delay.

In January 1807 Grenville felt that the time was ripe to introduce his bill for abolition, and, as a peer, he did so in the Upper House. By pure coincidence this step was taken in London in the same month in which a similar event was taking place in Washington. In the Lords, Grenville faced opposition from, among others, the Duke of Clarence and Earl St Vincent, a reflection of the Navy's persistent but poorly founded concern that abolition of the slave trade would empty the reservoir of seamen upon which the Navy relied for its manpower in wartime. Another royal duke, Gloucester, spoke in support of the bill, and it passed with a comfortable majority of 100 to 34. In the following Commons debate, fears were expressed that propagation of ideas of liberty would give rise to slave revolts and that the bill was "fraught with ruin to the colonies and to the empire", and John Fuller, Member for Sussex, remarked that, "We might as well say, 'Oh, we will not have our chimney swept, because it is a little troublesome to the boy.'" The tide had well and truly turned, however, and, with an unusually large number of Members voting, the result was victory for the Government and the abolition movement by a margin of 283 to 16.

Critics sought to represent this remarkable volte-face in British policy as an exercise in self-interest. France, Spain and the United States were suspicious that it was a measure to extend British maritime superiority, and, rather easier to understand, Britain was accused of attempting to steal a march on its trading competitors under the guise of philanthropy. More recently it has been argued that British abolition was merely a logical outcome of economic developments. It is true that in the years immediately preceding the Abolition Act, the profitability of trade with the British West Indies had declined thanks to the war's denial of European markets and to the exhaustion of the soil on the sugar islands; sugar beet was becoming available in Europe as an alternative to West Indies cane, and many investment options were emerging to compete with sugar. There was consequently a rise in the influence of new industrial magnates at the expense of the West Indies interest, and that undoubtedly led to a decisive decline in the strength of parliamentary opposition to abolition.

In 1807, however, the British slave trade remained attractively profitable to the traders and to the suppliers of trade goods, and the events of the next sixty years were to demonstrate most forcefully that the international slave trade was not yet in decline. The British West Indies market may have been near its natural demise, but the demand for slaves in Brazil and Cuba was to be voracious for many decades to come, and British merchants would have been admirably placed to exploit that opportunity as Europe emerged from war. There seems good reason to suppose that, without the endeavours of the abolitionist movement, the British slave trade would have continued, and many wished it to do so.

In the long series of parliamentary debates on the issue, no argument of potential economic or strategic advantage was ever offered by the abolitionists in support of their cause. The anti-slaving campaign in Britain was fought on purely humanitarian and altruistic grounds. Its heroes: Sharp, Clarkson, Wilberforce and Pitt, aided by their many allies, including Barham, Wesley, Fox, Burke and Sheridan, struggled long and valiantly against entrenched self-interest, and their success, gained at the height of a war of national survival, was a victory for Christian morality.

Much remained to be achieved, however. London, unlike Washington, had already appreciated that the abolition of the international slave trade would be at best incomplete unless the other slave-trading nations followed the example of the British and American denunciations. Also, it was scarcely in Britain's interest to allow its commercial rivals to continue to supply their colonies with slaves, having banned the Trade itself. Britain's diplomats therefore put their shoulders to the task of convincing other governments that they too should abolish their slave trades, in the hope that this would eventually lead to the end of slavery itself. British gold might aid persuasion, and the Royal Navy, dominant at sea and with a very long reach, was expected to enforce not only British abolition legislation but also the agreements yet to be reached with the other slaving nations.

As they jointly embarked on this crusade, the Foreign Office and the Navy can scarcely have imagined how long and tortuous it would be, or what an expenditure of resolution, patience, fortitude, treasure and lives would be necessary to bring it to a successful conclusion.

Convention Note. Throughout the ensuing narrative the use of capital letters has been kept to a minimum, but they have been applied to naval warrant ranks – Surgeon, Master, Carpenter, etc – in order to distinguish them, where necessary, from the more general, or civilian, usage of those descriptions.

PART ONE

The Trade

CHAPTER I

Three Centuries of Transatlantic Slaving

> O, fie, fie, fie!
> Thy sin's not accidental, but a trade.
>
> WILLIAM SHAKESPEARE

THE TRADE which Britain now sought to eradicate had been an accepted part of European commerce for nearly three hundred years, and its roots lay even deeper in the history of Europe and Africa.

Slavery had been common practice in many civilisations since at least 8000 BC, and black Africans, slave and free, were familiar in the Greek and Roman empires. However, it was in the fifteenth century that the seeds of the transatlantic trade in African slaves were sown by the southward expansion of Portuguese exploration, trade and influence, a process led by the half-English Prince Henry the Navigator. Becoming aware of the Moorish trans-Saharan trade in gold and slaves from, he believed, the Guinea coast, Henry determined that Portugal should benefit from these riches. His captains took small numbers of Africans to Portugal in 1442 and 1443, more as curiosities than slaves, but an expedition in 1444 raided the African coast and returned to Lagos with 235 Africans* for sale. The Atlantic traffic in slaves from West Africa had begun, but, as yet, the market was only in Europe.

On Henry's orders, the Portuguese soon began buying slaves from the black Muslim traders of the Guinea coast instead of kidnapping them, and Portuguese trade, including that in slaves, closely followed the southward progress of exploration. By 1461 the Portuguese had built a fortified trading post on an island in the Bay of Arguin, the forerunner of many castles built by Europeans on the slave coast, and in the middle decades of the century about 1,000 slaves per year were being sent to the Azores, Madeira and to Portugal itself. Before long Portuguese traders began to settle on the coast, intermarrying with native women and cementing commercial relationships.

Inevitably, there was constant conflict between Portugal and Spain for the commercial benefits of West Africa, but, by clever diplomatic manoeuvring,

---

* 'Throughout the text, "African" indicates black African. Africans of other description will be identified as such.

Prince Henry gained papal approval for Portugal's trading and slaving activity on the African coast. Motivated largely by the powerful threat of Islam, and by the wish to convert native peoples to Christianity, three successive Popes issued bulls between 1442 and 1456 which authorised the Portuguese to enslave and convert pagans, gave them sole trading rights on the coast of West Africa, and confirmed Portuguese acquisitions south of Cape Bojador as perpetual conquests. These declarations were of massive commercial significance to Portugal, and it took determined measures to enforce them against interlopers, particularly those from Spain.

After Henry's death in 1460, Portugal's commercial development of its new territories temporarily assumed a higher priority than exploration, and sugar cane was successfully introduced to Madeira, where African slaves and sugar became allied for the first time. A treaty signed by Portugal and Spain in 1480, ending their confrontation on the African coast, allowed the Spaniards to achieve a similar success with slave-produced sugar in the Canaries. Meanwhile the Portuguese pushed eastward into the Gulf of Guinea, and in 1481 built a great fortress at Elmina on the Gold Coast to defend their trading interests. The Bight of Benin and the islands of the Gulf of Guinea had been explored in 1471–2, the mighty River Congo was discovered in 1483, and in 1486 the Portuguese began to settle and grow sugar on the island of St Thomas (São Tomé).

With five years or more to run before Columbus embarked on his first voyage to the New World, Portugal had established trading posts and negotiated trading agreements for a wide range of commodities along the whole coast from Senegal to the Congo, that 3,000-mile sector of West Africa which was to provide slaves for centuries to come. Only a decade later, the Portuguese had visited the future transatlantic slaving ports of south-east Africa, and seen the already flourishing slave traffic of that area. Furthermore, many of the features which were in due course to define the Atlantic slave trade were by then in place. These included a momentous ruling by Pope Alexander VI in 1493 that the zones of influence of Portugal and Spain should be divided by a north–south line through the Atlantic Ocean, Portugal having a monopoly to the east and Spain to the west. This decision, which confirmed earlier papal bulls, led to the Treaty of Tordesillas, which enshrined the division of the world outside Europe between the two Roman Catholic powers, giving a demarcation which, at least as interpreted by Portugal, would later place eastern Brazil as well as Africa under Portuguese influence. All that was now required for expansion of Europe's traffic in African slaves was a more voracious market than could be provided by Europe itself.

The first known transatlantic cargo of slaves was carried from west to east: Taino Indians sent home to Spain from Santo Domingo by Columbus in 1493 in order to justify his exploration. He foresaw that Hispaniola could provide a regular supply of these captives, but the scheme was short-lived, partly because the Indians could not tolerate the Spanish climate. As Spain began to exploit its Caribbean possessions of Puerto Rico and Hispaniola early in the sixteenth century, and sugar cane was introduced from Madeira, the native Indians were forced to work in the fields and the gold mines. By 1509, however, the indigenous populations were in serious decline. The Indians had neither the resilience nor the robustness of the Africans, and they were dying from overwork and loss of hope. Smallpox was later to take a heavy toll. Kidnapping from neighbouring islands did not provide a solution, and King Ferdinand of Spain was concerned about the output from the gold mines of Hispaniola. Accordingly, in 1510, he authorised the export from Spain of 250 African slaves, and subsequent licences were granted for about 50 per year. The sale of the slaves was to be closely regulated, and export licences had to be bought from the Crown.

During the next decade, however, there was a complete collapse of the native Caribbean population, and pressure grew for unrestricted import of Africans, not only from those who were desperate for their labour but also from some who were concerned at the plight of the native Indians. In 1518, the youthful King Charles of Spain, later the Holy Roman Emperor Charles V, gave a licence for the importation of 4,000 African slaves to the Spanish Caribbean islands, which by then included Cuba, taking them directly from Africa if necessary. With their monopoly of commerce on the African coast, the Portuguese were the only traders who could provide such a large number of slaves, and they therefore acted as agents for the licensee.

This pattern was followed for 20 years, with merchants of several nationalities involved; the slaves were shipped from Africa, still via Europe, by the Portuguese under licences controlled by the Spanish Crown. By 1530, Spain was well established on the mainland of Central America, and in that year, for the first time, a cargo of slaves was sent, as permitted by the license of 1518, directly from Africa to the Spanish colonies. Others began to follow this example, and King João III of Portugal gave permission for shipments to be taken from the Cape Verde Islands and the island of St Thomas directly to the Spanish Americas. For the first quarter of the sixteenth century the trade in slaves to Europe, to the Atlantic islands and along the African coast remained more important than the transatlantic traffic, but the Atlantic slave trade was now truly launched.

Up to this point the Portuguese had found most of their slaves on the Guinea coast, but supply difficulties encouraged them to turn their attention to the Congo, and by 1530 they were exporting over four thousand slaves per year from this plentiful new source. Slaving from the Congo became increasingly important following Portugal's decision in 1530 to conquer the great territory of Brazil. Between 1550 and 1575 sugar production took firm hold in Brazil and the initial trickle of African slaves into the colony, to supplement the labour of native Indians, grew to a total of 10,000 in that quarter-century. Nevertheless, the main market for the transatlantic Trade remained the Spanish empire, which probably received about 25,000 in the same period, still supplied by Portuguese traders.

By the end of the sixteenth century the Christian kingdom of the Congo was effectively a Portuguese dependency, and, as a first step towards colonising the neighbouring Angola, Portugal had built the fortified coastal town of São Paulo de Luanda. Thereafter the slave trade was the economic mainstay of Angola and the Congo, and the main beneficiary was Brazil. With the indigenous slaves dying from epidemics of influenza and dysentery, and African slaves expected to succumb after ten years of labour, a continuous supply of captives was essential if Brazil was to sustain itself as the main sugar producer for Europe. Sugar was regarded at the time primarily as a medication, but it was beginning to be valued by the European rich as a sweetener even though tea, coffee and chocolate had yet to make their appearance. Brazil supplied the great Amsterdam market with sugar, began the New World sugar revolution and established the pattern of production on sugar plantations which was later to be replicated in the Caribbean.

Spain's slave trade, such as it was, had virtually collapsed by 1570, but when Portugal's royal line died out in 1580 the two nations became united. Thenceforth Spain, now dominating world trade, was able to make direct use of the experienced Portuguese merchants to supply its empire with African slaves. The merchants, however, found that they were unable to fulfil their contracts, and King Philip II of Spain decided in 1595 to sell a trading monopoly to a single merchant who, in turn, could sell licences to other traders. Through this contract, the *asiento*, over 75,000 slaves were carried from Africa to Spanish America during the first quarter of the seventeenth century, and in the same period 100,000 went to Brazil and 12,500 to St Thomas. Only a few hundred were taken to Europe. In addition there was a lively contraband trade, particularly to the Caribbean colonies.

In Africa this booming trade led to the expansion of coastal villages in the Niger Delta into city states with economies based on slave trading with Europeans. Bonny and New Calabar, for example, became strong monarchies, and Old Calabar and Brass developed into powerful commercial republics. However, the slave markets in the Bights never became as important in the Trade as those in Angola and the Congo.

The Portuguese and Spaniards regarded the Atlantic as a private lake during the sixteenth century, but they were troubled by interlopers, mostly French and English, and in 1562 John Hawkins initiated the English slave trade. With the financial backing of powerful men, and with the approval of Queen Elizabeth I despite the protests of Spain, he made three expeditions to the coast of Sierra Leone. Obtaining slaves by a combination of kidnapping, piracy and trade, he carried them to Spanish America, and, with the exception of his disastrous third voyage, returned with a handsome profit. After this brief and discreditable English foray it was the turn of the Dutch to enter the Guinea Trade, with conspicuously greater success.

By 1600 the Netherlands was building the greatest merchant fleet in the world, and, having invested heavily in Brazilian sugar production, had secured half the carrying trade between Brazil and Europe. In 1612 the Dutch Guinea Company built Fort Nassau as its headquarters on the Gold Coast only 15 miles from Portugal's great fortress of Elmina. The Netherlands already had a precarious foothold at Loango just north of the Congo, and in 1617 it purchased from the Portuguese, and fortified, the strategically valuable island of Goree at Cape Verde. The Dutch West India Company was re-established in 1621 and given a 24-year monopoly of Dutch trade to Africa and the West Indies, but it resisted involvement in the slave trade, partly on moral grounds.

In 1623 the Dutch West India Company launched a spectacular assault on Portugal's South Atlantic empire. Having temporarily taken Benguela, south of the Congo, the company's fleet successfully attacked the Brazilian sugar port of Bahia. It then recrossed the Atlantic in an unsuccessful attempt to take Luanda, Portugal's primary source of slaves for Brazil. A Portuguese–Spanish force soon recaptured Bahia, but the Dutch in Brazil then moved north and seized Pernambuco and Olinda. As the company developed new sugar plantations in this north-east corner of Brazil, it had to reconsider its stance on slave trading. Dismissing its moral qualms, it began to ship Africans into its South American possessions, to the New Netherlands in North America and, shortly afterwards, to the new Dutch Caribbean colonies of Curaçao, St Eustatius and St Thomas. In order to secure a more reliable supply, the Dutch governor-general of Brazil sent an expedition to the Guinea coast in 1637, capturing the Portuguese fortresses at Elmina and Axim. Gaining momentum after a slow start in the slave trade, the Dutch decided that if their Brazilian plantations were to be adequately provided with labour, they would have to seize control of Portugal's slave supply in Angola. Consequently, in 1641, the Dutch West India Company despatched a fleet to West Africa in a second, and successful, attempt to capture Loando, taking St Thomas and recapturing Benguela into the bargain.

Soon the Dutch were not the only Europeans making incursions into the West African slave trade of Spain and Portugal. France had already formed a number of companies to trade on the African coast from Senegal to the Bights, although not initially for slaves. By 1645, however, any scruples the French may have had about slaving had been overcome by the demands of their West Indies planters in the recently acquired islands of St Christophe, Tortuga, Guadeloupe, Martinique, St Lucia, St Vincent, the Grenadines and Grenada. By that stage too, European experiments to meet the colonial requirement for agricultural workers by introducing indentured white labourers and convicts were judged to have failed.

The English too were encroaching on the Portuguese–Spanish slaving preserve. By 1632 their colonial possessions in the New World included Bermuda, Virginia, Massachusetts, Barbados, Antigua, Nevis and Montserrat, and some of these colonies felt they needed slave labour. A monopoly of British trade with Africa had been granted by King James I in 1618 through formation of the Company of Adventurers to Guinea and Benin, but, like the French, the English were ambivalent about slaving. Nevertheless, a few slaves were beginning to arrive in Virginia in English ships by 1628, and in 1632 King Charles I licensed a new syndicate to carry slaves from Guinea. This syndicate built a fort at Cormantine on the Gold Coast, which remained the English headquarters until 1661. An English attempt to build another at Cape Coast in 1650 was thwarted by the Swedes, who themselves built the first Cape Coast Castle, which, like Elmina, was to feature so strongly in West African slaving.

The Spanish Crown, in the meantime, continued to sell the *asiento* to a succession of Portuguese merchants. Most of these were Jews supposedly converted to Catholicism, and the Inquisition conducted a campaign against these *converso*s, with the result that the Portuguese slave-trading network serving Spain in the Americas was ruined. In 1640 Portugal and Catalonia rebelled against Spain, and the Portuguese regained their independence. The Portuguese slavers' cooperation with the Spanish Crown ceased, and the *asiento* was suspended. Ruin threatened the Spanish empire in the New World, but interlopers, mainly Dutch, maintained an adequate supply of slaves. The Dutch were now the undisputed leaders in the Atlantic slave trade, and they built a string of castles on the Gold Coast.

The Portuguese, however, no longer constrained by Spanish authority, began to stage a recovery. Their first measure, adopted in 1643, was to reach eastward to Mozambique for a fresh source of supply. Then the Portuguese colonists who had stayed in Brazil under Dutch rule rebelled. Not only did they expel the Dutch from all but a small enclave around Pernambuco, but they also sent an expedition across the Atlantic in 1648 to retake Loando and St Thomas, and to drive the Dutch from the Congo.

Despite these upheavals, transatlantic slaving continued steadily. During the second quarter of the seventeenth century, about 100,000 Africans were shipped to Brazil, 50,000 to Spanish America, 22,000 to the British West Indies (a few of whom were re-exported to the North America colonies), and 2,500 to the French Caribbean islands. Angola remained the main source, and the Dutch, whose domination of the West Indies slave trade continued in the 1650s, were soon re-established in the Congo region, where they traded Angolan slaves and began to threaten the Portuguese Angolan traffic. Although they finally lost all their Brazilian possessions in 1654, the Dutch held territory in the north of South America as well as several Caribbean islands. Their island of Curaçao became an important slaving centre in the 1650s, selling illicitly to the Spaniards, English and French, and the Dutch colonies in North America were also importing slaves, mostly from Curaçao.

Many of the Dutch planters expelled from Brazil found refuge in Barbados, taking their sugar-producing skills with them, and at this point the Caribbean began to supplant Brazil as the major sugar-growing area of the Americas. By 1670 sugar had transformed the economies of the British West Indian islands, and the French joined in the conversion of much of the archipelago to sugar production. Europe's appetite for sugar appeared to be limitless, stemming partly from the growing fashion in the 1650s for drinking sweetened coffee, tea and chocolate, but mostly from the craving of poor folk for sugar as an affordable source of calories and as a means of making a dull and monotonous diet more appetising. Rum became popular in Britain, and so did jam. The Caribbean was to supply this vast market for the next 200 years, and although it was the English and French planters who produced the sugar in the early days, it was the Dutch who initially provided most of the slaves who worked the plantations.

The importance of this trade was judged by the colonising nations to be so great, and the need for regular supplies of slaves to be so crucial, that they followed the Dutch example of forming national companies which combined African and Caribbean interests. In all of these enterprises sugar was the dominant commodity, colonies were required to trade exclusively with their parent countries, and trade with its colonies was conducted entirely through the national monopoly company of the parent country. Planters and merchants disliked the constraints that these monopolies imposed, and British and Dutch captains became adept at smuggling slaves and other merchandise into (especially) the Spanish colonies.

In London a new Guinea Company was founded in 1651 to trade principally on the Gold Coast, but, following Parliament's victory in the Civil War, the company succumbed under attacks on its ships, largely carried out by a piratical monarchist fleet in the West Indies, in alliance with the Portuguese. Nevertheless, a steady trade in slaves by London ships was underway, and a new company, the Royal Adventurers into Africa, was launched after the Restoration of 1660 with the Duke of York as its president and Samuel Pepys among its shareholders. The company's principal objective was to trade in gold, with slaves as a secondary commodity, and a new coin, popularly called the "guinea", was struck from some of the gold brought from the Gold Coast. Not surprisingly, the new company clashed with the Dutch: it restored the English forts in West Africa, seized the Cape Verde Islands, took Cape Coast and several other west-coast possessions from the Dutch, and captured the North American city of New Amsterdam in New Holland, renaming it after the company's royal President. Cape Coast Castle became the permanent English headquarters in West Africa.

The company undertook to supply Africans to Barbados and the new British colony of Jamaica, and, as its trade expanded, the company planned to re-export large numbers of slaves from the British islands to the Spanish colonies. This, however, challenged the Dutch ambition for exclusive European rights to trade on the Guinea coast, and was a major cause of the Second Dutch War. Soon after the outbreak of hostilities Admiral de Ruyter recaptured most of the Guinea coast forts and built Fort Amsterdam at Cormantine, at the heart of the Gold Coast. With this loss of control the Adventurers Company failed to meet its commitments to the planters, and slave prices doubled. The demand grew for a free trade in English slaving, and the company was obliged to sell licences to private merchants to trade within the monopoly.

The Adventurers were ruined by the Second Dutch War and the company was wound up in 1672, but a replacement, the Royal African Company, was immediately formed. Its sphere of operations on the African coast stretched from Cape Blanco to the Cape of Good Hope, and, despite its intention to import gold and other valuable African produce into England, its principal trade was, from the start, slaves to the West Indies. The Royal African Company continued to use Cape Coast as its African headquarters, maintained several posts on the Gold Coast and traded slaves as far south as the Dutch reserve of Loango Bay. It had difficulties from the beginning, not least the challenge of illegal competition from interloper merchants and captains, many of them from Bristol, which was becoming the leading sugar and West Indies port in Britain. Nevertheless, it exported 74,000 slaves in its first 15 years, and by 1700 slaving accounted for 60 per cent of the company's trade. In the final quarter of the seventeenth century the British West Indies imported about 175,000 slaves from one source or another, over 100,000 more than in the previous quarter-century.

The decade before the Glorious Revolution of 1688 had seen the Royal African Company competing successfully with the Dutch, but after the loss of royal support in 1688 the company's trade increasingly fell prey to interlopers. Ten years later the Government finally gave way to the pleas of manufacturers, merchants and planters, and removed the company's monopoly. The interlopers were given legal status as "separate traders", and the new legitimacy of its erstwhile interlopers allowed Bristol properly to enter the slave trade. It was joined by many of the small English ports, and between 1698 and 1707 the separate traders carried 75,000 slaves against the Royal African Company's 18,000.

The French also founded a succession of monopoly companies for trade in West Africa, and, in a more extreme form of the English experience, these in turn collapsed. Those few French investors who were interested in the Trade preferred to back interlopers rather than the national companies, and, when all else failed, the French colonists were given tacit permission to buy their slaves from the British and Dutch Caribbean islands. Nevertheless, in the period 1675 to 1700, the French shipped about 55,000 slaves into their Caribbean possessions, and they improved their position on the African coast at the expense of the Dutch and the Portuguese.

For a quarter of a century after its loss of Portugal in 1640, Spain pursued a policy of issuing individual slaving licences as an alternative to the *asiento*, but this arrangement was fatally undermined by illegal imports, primarily by the Dutch and English, to meet the requirements of the frustrated Spanish colonists. Although the Spanish Crown must have realised that its only effective means of supplying its American empire with slaves was to conduct its own trade directly with Africa, it made no effort to persuade the Vatican to revise its fifteenth-century concession to Portugal to control trade in Africa. Instead, Spain reintroduced the *asiento* in 1663, and for the next 20 years the monopoly frequently changed hands, from Genoese to Portuguese to Italian merchants, without ever meeting the promised output. The Spanish colonists generally continued to rely more on their illicit Dutch and English suppliers. Finally, after a period in the hands of a Dutch banker and then a Portuguese company, the *asiento* was bought back by the French-born King of Spain and offered to France.

Portugal's break from Spain made little difference to its slave trade with Brazil. After expelling the Dutch, and acting through its colony of Angola, it established a powerful influence over the kingdom of Congo and brought other minor kingdoms in the region into its slaving sphere. Increasingly Angola became a commercial dependency of the Portuguese colony of Brazil, and the Brazilian gold rush at the end of the seventeenth century gave a powerful stimulus to the slave trade with Africa. Although initially most of the slaves came from Angola and Congo, with a few from Mozambique, an increasingly high proportion was imported from the Gulf of Guinea, where the Portuguese had re-established trading posts. By this stage too, the merchants of Rio de Janeiro were sending ships to Africa for slaves. In the second half of the seventeenth century about 350,000 Africans were imported to Brazil, and 150,000 in the first decade of the eighteenth century.

There was now a growing Scandinavian presence on the West African coast. Early Swedish voyages were financed by the Dutch, and in 1649 a Swedish national company was formed along the lines of the Dutch West India Company. It built several fortified posts on the Guinea coast and, thanks largely to English and Dutch preoccupation with their wars, these posts were successful for some years. The Danes began trading in Africa in 1649, and not only captured the new Swedish trading posts but also built two forts of their own. Failing to recover their possessions, the Swedes left Africa, but did not entirely lose interest in the slave trade. The Danes continued successfully to supply their small Caribbean sugar island of St Thomas from their newly acquired African trading posts, exporting about 4,000 slaves between 1675 and 1700. The Germans too made a brief foray into African trade. Following a slaving expedition under the Brandenburg flag by a Dutch captain, a number of Brandenburger posts were established on the Guinea coast by another Dutch interloper. From here the Germans sold slaves to the Portuguese, Dutch and Danes, but their attempt to find a base in the Caribbean was thwarted by the Dutch, and, tiring of the whole project, they sold out to Holland in 1720.

By 1700 transatlantic slaving was a massive international trade. The total number of slaves exported between 1650 and 1675 was probably about 350,000, and between 1675 and 1700 the number rose to over 600,000, most of them to the Caribbean. Despite this expansion of the Trade, in none of the slave-trading nations had the national chartered companies achieved financial success with their monopolies; they were generally unable to attract adequate capital, they were obliged to trade whatever the commercial circumstances, and they were too easily outflanked by interlopers. They were subjected to constant criticism by their customers, independent traders and manufacturers, and, worst of all, the overheads incurred by their responsibilities for maintaining their African posts were prohibitively high. The forts and castles were most numerous along the Gold Coast, but extended north beyond the Senegal River, eastwards into the Bights of Benin and Biafra, and south to the Congo. These fortified posts were crucial to the security of European trade, in all commodities, but in most cases the land on which they were built was merely leased from local kings and chiefs, and only the Portuguese were willing to establish colonies in Africa.

Despite the obvious lack of success of the monopoly companies they continued to survive, and the supposed prize of the Spanish *asiento* became one of the major issues of the War of the Spanish Succession, which began in 1701. The French-born Philip V of Spain gave the *asiento* to France, and Portugal immediately allied itself with Britain and the Habsburgs against the Bourbons. In Spain the *asiento* treaty with France was unpopular, and Spanish officials obstructed the French traders at every turn. There was indignation in Britain at being denied the Spanish monopoly in favour of France, in spite of being the clear leader of the Trade in the northern hemisphere. Meanwhile, despite the war, the illegal slave trade prospered. By 1710 Britain was selling 10,000 slaves per year in the Caribbean, including the Spanish colonies, ten times more than the French, and many in the Spanish empire felt that only Britain could meet the colonists' demand for slaves. Manufacturers in Britain petitioned Parliament to ensure that the slave trade would not be restricted at the end of the war, and, at its insistence, Britain was granted the Spanish *asiento* by the Treaty of Utrecht in 1713.

News of the grant was celebrated in London with a torchlight procession, and, for £7.5 million, the government sold the *asiento* monopoly to the South Sea Company, which had been formed in 1711 to trade specifically with the Spanish empire. The company was required to provide 4,800 slaves annually for 30 years, and to make sizeable payments and loans to the King of Spain for its monopoly privileges. Philip V and Queen Anne were allocated half of the company's stock between them. The arrangement was not entirely popular in England and the Bristol merchants, among others, disliked it because it seemed to favour London. It displeased the Jamaican planters too, because it threatened their very profitable illegal trade with the Spanish colonies.

The South Sea Company agreed to buy its slaves from the Royal Africa Company in Africa, and its ships traded mainly from Loango, the Gold Coast and Dahomey, with occasional voyages from Senegambia and as far afield as Mozambique and Madagascar. Exports to the Spanish colonists of Buenos Aires went directly from Africa, but slaves destined for the Spanish Caribbean empire were "refreshed" in Barbados and Jamaica before sale. Most of the eminent figures of the period were shareholders in the South Sea Company, but the company was not the great success that had been expected. It did not take advantage of the many applications from independent merchants for licences, and it was troubled by illegal trading, as all its predecessors had been. Bristol ships were the leading culprits, and men-of-war were not beyond blame. Moreover, the Dutch, French and Portuguese continued to trade illegally with the Spanish colonies whenever opportunity offered, regardless of the *asiento*. Added to these difficulties were the depredations of the pirates lurking in the West Indies. In 1718 and 1727 war with Spain interrupted the company's trading, and its property on enemy territory was seized for two years on each occasion. In 1720 there was wild speculation in the company's shares, and when the share price plummeted that autumn, powerful men and institutions were ruined. Because of a close connection between the company and the national debt, the country almost faced bankruptcy, but the company survived, and succeeded in selling about 64,000 slaves between 1715 and 1731.

The *asiento* traffic apart, the British slave trade as a whole was thriving. Nearly 150 vessels were involved in 1720, carrying well over 100,000 slaves between 1710 and 1720. About the same number were exported in the next decade, mostly to the British and Spanish colonies in the Caribbean and Central America, and a further 10,000 went to the British North American colonies. At this stage London was still the principal English slaving port, but in the 1730s it was overtaken by Bristol, which was pioneering the Trade to Virginia. The Act of Union in 1707 had allowed Scottish ports into the Trade, and Glasgow joined the West Indies and Americas traffic. Between 1730 and 1740 British ships carried about 170,000 slaves, about 60,000 of them intended for the Spanish empire and perhaps 40,000 going to the southern colonies of British North America. This total for the decade, for the first time, probably exceeded the Portuguese trade to Brazil.

In addition to its legitimate *asiento* trade, the South Sea Company sponsored contraband traffic to the Spanish colonies in manufactured goods as well as slaves, to the serious detriment of Spain's imperial economy. This led in 1739 to the War of Jenkins' Ear, which then merged into the War of the Austrian Succession, and it was 1748 before the company's contract was renewed. Spain, meanwhile, made alternative supply arrangements for its empire, although it had no alternative but to buy slaves from the British and Dutch markets in the Caribbean. The South Sea Company regained its monopoly at the end of the war, but it decided, in agreement with the Government, that the *asiento* was no longer a promising proposition and renounced the contract in 1750 on payment of £100,000 by Spain.

In France, earlier in the century, the Mississippi Company had absorbed six other trading companies to become the largest commercial concern in the world. It suffered speculation similar to that involving the South Sea Company, but it survived the subsequent collapse in its share value, and was given the monopoly of trading Guinea coast slaves. Otherwise French trade with Africa was open to all, although tax was charged on slaves imported to the French West Indies, and only five "privileged ports" were permitted to participate. Throughout the eighteenth century the French slave trade was centred on Saint-Domingue, which achieved remarkable success as a sugar island, and Nantes, which sent about 800 slave ships to Africa between 1715 and the French Revolution. By 1741 almost all French ports were allowed to engage in the slave trade, but Nantes had established an unassailable pre-eminence. In the 1720s the French traded at least 85,000 slaves, and over 100,000 in the 1730s. Nantes alone exported 55,000 between 1738 and 1745.

Bristol retained its position as the leading British slaving port for only 20 years before being displaced by Liverpool. Liverpool had entered the Trade in the 1690s, and was soon prospering from an illegal slave traffic to the Spanish empire. In 1740 it sent 33 ships to Africa, and the number continued to grow thereafter. Liverpool was better located than London for Atlantic trade, and less vulnerable than Bristol to interference during war with France. It also had the deep-water harbour which Bristol lacked, and its ships could avoid marauders in the Channel approaches by taking passage north of Ireland. Homeward-bound Liverpool ships often avoided customs duty by landing goods at the Isle of Man, and were able when outward bound to load some of their slaving trade goods at the island. It seems too that Liverpool owners treated their ships' crews less generously than was the practice in London and Bristol, and while the Bristol merchants continued conservatively to gather their slaves from the Gold Coast and Angola, the Liverpool traders adventurously sought fresh sources of supply in Sierra Leone, Gabon and the Cameroons. Added to these advantages was the immense benefit of the proximity of the manufacturing industries of Manchester, which provided in abundance an array of slaving trade goods. The cost of transport between Manchester and Liverpool was drastically reduced by the opening of the Bridgewater Canal in 1772, and the annual value of Manchester's exports rose from £14,000 in 1739 to more than £300,000 by 1779, half of which went to Africa for purchasing slaves.

The Netherlands was no longer an important Atlantic power in the 1700s, but it retained four colonies on the north coast of South America, to which it carried 190,000 slaves during the century, mostly to the new cotton plantations of

Surinam. The Dutch West India Company turned the small island of St Eustatius into a slave mart, but the company lost its monopolies in Africa and the Caribbean in the 1730s and independent traders were thenceforth free to operate on payment of a fee. Neutrality during the War of the Austrian Succession allowed Holland a period of prosperous slave trading, but a decision in the 1750s to ally itself on the Guinea coast to enemies of the increasingly powerful Ashanti nation was to give Dutch traders serious supply difficulties.

Following the British example, Denmark expanded its involvement in the Trade early in the eighteenth century, and from 1725 the Danish West India Company allowed imports by private traders. There was a lively illegal Danish slave traffic to the Spanish colonies, and the demand from the plantations of the Danish Caribbean islands greatly increased with the acquisition of St John in 1719 and its conversion from cotton to sugar production. The Caribbean island of St Thomas became a slave transit camp, and the supply to the sugar islands grew to 9,000 in 1755 and to 24,000 by 1775.

Until overtaken by Britain in the 1730s, Portugal remained the major Atlantic slaving nation. All of its slaves were shipped to Brazil: 150,000 in the 1720s, and 160,000 in the following decade. More than half of them were exported from Angola, mostly originating from far inland, and the remainder came from the Guinea coast. The slaves' destinations, via the ports of Rio de Janeiro and Bahia, were generally the gold and diamond mines, and, as the indigenous Brazilian tribes were destroyed by enslavement and disease, the emphasis on African slaves increased. The traffic was now almost entirely between Brazil and Africa without direct contact with Portugal itself, and the preferred trade goods were English manufactures and Indian products imported to Africa by the English and Dutch East India Companies. The Lisbon government was finally persuaded to abandon indirect administration from Portugal of the Brazil trade, and powerful slave merchants became established in Angola, operating a transatlantic traffic.

In the early eighteenth century British North America finally emerged as a significant participant in the slave trade. Until then the only requirement of the colonists had been for a few domestic slaves, but the planters began to appreciate the potential of indigo and rice in the Carolinas and tobacco in Virginia. It seemed to the planters that African labour was essential, and British independent traders began to take advantage of this in the first decade of the century. For a time the colonial assemblies, concerned that the rising numbers of African slaves would incur a risk of rebellion, imposed import duties, but by the end of the 1720s an appreciable traffic into Pennsylvania and Boston had begun. American merchants were soon in competition with the English traders, sending ships to Africa from

Boston, Salem, New York and harbours in Maryland and South Carolina. The customers for slaves were mostly in the southern colonies, but the shippers were generally in New England, and Rhode Island became the pre-eminent slaving colony. By the mid-1700s Newport was a major port whose captains were selling slaves throughout Britain's North American and Caribbean colonies.

The most valuable single import to Britain and France in the late eighteenth century was sugar, and by the end of that period the purchase of sugar would cost a typical poor English family as much as 6 per cent of its income. Supply depended upon the plantations of the Caribbean, and they in turn depended on regular cargoes of African slaves. Britain continued to dominate the Trade, and during the 1740s its ships carried over 80,000 slaves to Jamaica and Barbados, 60,000 to Virginia and the Carolinas, and 60,000 to other colonies. The single year 1749 saw 70 slave ships sail from Liverpool, nearly 50 from Bristol, eight from London and about 20 from minor ports such as Glasgow and Lancaster. In 1750 the traffic was opened to all British subjects, and a new holding company, the Company of Merchants Trading to Africa, was formed to maintain the forts and trading posts in West Africa.

British ships carried a further 200,000 slaves in the 1750s, and the outbreak of the Seven Years' War in 1756 brought marked advantage to British slaving. The conflict almost halted the French slave trade, and the French sugar islands and West African slave factories fell into British hands. Britain also took Havana from Spain, an event which was to have an immense influence on the slave trade. Until then, Cuba had been involved in sugar-growing and slavery in only a small way, but the nine months of British occupation brought a transformation. Slaves flooded into the island, and the Cuban planters bought vast quantities of British cloth, utensils, clothing and sugar equipment. Britain relinquished many of its war gains at the Treaty of Paris in 1763, but by then Cuba had been launched on its career path as the world's greatest sugar producer.

Meanwhile, following Britain's surrender of the *asiento* in 1750, Spain was frustrated in its attempts to make other arrangements for the supply of slaves. In 1753, therefore, the Spanish Crown at long last gave authority for Spanish companies to carry slaves directly from Africa to Cuba, setting aside the papal rulings and the terms of the 1493 Treaty of Tordesillas, which had given Portugal sole rights for African trade. The Spanish traders, however, had no experience of the slave coast, and Spain had no castles on the Coast to support and defend its trading activities. So it was not until 1758 that its first direct slaving voyage was made. However, Spain was still unwilling to shed its faith in the *asiento*, and a new company was duly formed in 1765. Attempts to buy slaves from the English and French posts on the Guinea coast failed dismally, and the company then decided to purchase all its slaves in the Caribbean. This scheme never made money, not least because the Spanish planters wanted to buy directly from the British in Jamaica. In the 1770s Cuba was the main slave buyer in the Americas, having increased its sugar production sevenfold since the British occupation of Havana in 1762–3, and three-quarters of Cuba's slave imports came from Jamaica.

In the years leading up to the American Revolution, the slave traders in British America were growing rich. Despite a high natural increase among the slave populations of the southern colonies, a characteristic which distinguished them from those of the sugar islands, the demand for labour was growing rapidly, and a new market was opened to the slavers in 1750 when slavery was declared legal in Georgia. Slaving voyages were being made from Maryland, New York, Massachusetts, Maine and Pennsylvania, but on a relatively small scale compared with Rhode Island, which sent 165 ships to Africa from Newport in the 1760s.

After the Seven Years' War, France was determined to make good its losses, and in particular to escape from reliance on Britain for the supply of slaves to the French sugar islands. It dispensed with the national monopoly, and strengthened its position on the Guinea coast, using the licence fees from independent traders to maintain its African forts. In 1767 the French colonies, for the first time, exported more sugar than did the British islands: 77,000 tons to the British 72,000, and French ships were soon to carry 100,000 slaves in a ten-year period, also for the first time. Nevertheless, there was still a lucrative contraband trade into Saint-Domingue by British slavers.

The years before French entry into the American Revolutionary War against Britain in 1778 saw success in the slave trade for all the major nations engaged. Britain, still the leader in the Trade, shipped 250,000 slaves in the 1760s, mostly from the Bight of Benin, the Niger Delta and the Loango coast, and mostly in Liverpool ships. Loango Bay had become a particularly successful source, largely because the local rulers retained their independence and traded with all comers. By 1780 its annual output was approaching 15,000 slaves and it was providing two-thirds of the French trade.

Growing conflict between Britain and its North American colonies, culminating in the Declaration of Independence and the outbreak of war in 1776, damaged Britain's slave trade. Its exports were reduced to 200,000 in the 1770s, and, with merchants unable to pay promised wages and maintain employment, there were riots in Liverpool. The Trade was, however, little affected for France and the neutral nations, and the Dutch island of St Eustatius not only made slaves readily available to all but also supplied food to the British West Indies and to the American rebels until it was captured by Admiral Rodney in 1781.

The 1780s saw not only a postwar recovery but also a further expansion of the slave traffic. Prohibitions on trade between the British West Indies and the new United States of America caused a brief economic setback in the Caribbean, but 750,000 slaves were carried across the Atlantic during the decade, 325,000 of them by Britain. Newport, Rhode Island, was ruined by British occupation during the war, but the neighbouring Rhode Island ports of Bristol and Providence, as well as Boston, Salem, Philadelphia and Charleston, soon compensated for that. French slave exports grew to 270,000 during the same period; Marseilles had entered the Trade, and Bordeaux, sending its ships as far as East Africa for slaves, was challenging Nantes as leader of the French trade. Even Spain was expanding its slave traffic, having procured from Portugal the islands of Fernando Po and Annobón in the Gulf of Guinea to provide the African bases it had lacked.

It was estimated in 1783 that four-fifths of Britain's income from overseas came from West Indies trade, and much subsidiary industry in the slaving nations depended on the slave trade. Prospects for improving sugar production were also looking most promising with the introduction to the Caribbean of high-yielding Otaheite cane from the Pacific, and consideration was being given to how the new Birmingham-built steam engines might assist in processing sugar. Coffee was now being grown in the Caribbean in considerable quantities, especially in Saint-Domingue for the French market. Talk of slave-trade abolition in Britain, the USA and France was beginning to cause some disquiet among merchants and planters, but the Trade was as yet unaffected, and there was no sign of these novel ideas in Spain or Portugal. Spain finally despaired of monopoly companies in 1789 and gradually extended a free market, with conditions, to cover all its territories by 1795. Only the Dutch slave trade was in decline, never having recovered from the Fourth Anglo-Dutch War of the early 1780s.

By this time the slave requirements of the old Spanish American mainland colonies were much reduced thanks to the growth of the indigenous populations, but there was compensation in the increased traffic to Cuba and to the territory that would become Venezuela. A deputation of Cuban planters and businessmen visited England to buy sugar-processing equipment and to investigate British methods of sugar production and slaving, and it became clear that Spain not only had high ambitions for the future of Cuba but also would take any measures necessary to acquire the slaves needed to attain its aspirations. The island's land area under cultivation for sugar had already increased from 3,000 to 500,000 acres in the 30 years since 1762.

Having aided the rebel colonists during the American Revolutionary War, France was well placed to take advantage in the Caribbean of the severance of links between Britain and the USA. Between 1779 and 1789 production of sugar in Saint-Domingue almost doubled, putting some Jamaica planters out of business and obliging them to sell their slaves to the French. This prosperity was disrupted, however, by the consequences of the French Revolution of 1789. The French National Assembly condemned the principle of slavery in 1791, and in 1794 the Convention in Paris declared the emancipation of slaves. This decision, however, had been violently pre-empted in 1792 by the rebellion of Saint-Domingue's 450,000 black slaves and 50,000 mixed-race slaves against the colony's 40,000 whites, and by 1798 Saint-Domingue was nominally independent under a black leader. This was by no means the first slave rebellion in the Caribbean, but it was the only one completely to succeed.

French ships carried slaves into Havana illegally in the early 1790s, but the outbreak of war in 1793 virtually stopped the French Trade. With the Peace of Amiens in 1802, however, Napoleon reintroduced slavery to the French empire and revived the Trade. He had reclaimed Louisiana from Spain in 1800, and sent an expedition to re-establish French authority in its Caribbean possessions. When war resumed in Europe in 1803, Napoleon's Caribbean aspirations evaporated, and he sold Louisiana to the USA. Renewed hostilities brought the French slave trade, once again, to a halt.

When Denmark declared in 1792 that it would abolish the slave trade in 1803, the Danish Trade gained a new, if temporary, lease of life. Its main traffic was re-exportation from the Caribbean islands of St Croix and St Thomas, much of it to Cuba, which received over 12,000 slaves in this fashion. Cuba's need for slaves appeared to be insatiable, and not only were there no longer any restrictions at Havana for foreign slavers, but also in the 1790s Cuban vessels began to trade directly with Africa. In defiance of US legislation, American slavers were trading to Cuba far more than they were to their own southern states, and by 1807 the United States was dominating the Cuban Trade. Inspired by this burgeoning activity, a new breed of Cuban slave merchants appeared in Havana, sending ships to Africa or other Caribbean islands for slaves in competition with the foreign traders, and with a determination to import as many slaves as they could in the shortest possible time. Conservative estimates show 150,000 slaves entering Cuba between 1790 and 1810.

Slave imports to Brazil had risen, too: 200,000 between 1801 and 1810, three-quarters of them from Angola and the remainder from the Gulf of Guinea. Portugal desperately needed the tax income from the Brazil Trade as well as its colony's sugar, tobacco and cotton. There was no consideration of abolition in either Portugal or Brazil.

In the southern states of North America the invention of the cotton gin in 1793 had revolutionised the scale of cotton production. In 1792 the USA's output had been 138,328 lb; in 1800 it had risen to 17.8 million. The demand for slaves in the Carolinas and Georgia rose correspondingly, and South Carolina allowed 40,000 slaves from Africa to enter legally between 1803 and 1807, almost all in British and American ships.

To the depression of the abolitionists, the British slave trade was more profitable than ever. In the 1790s British ships had landed nearly 400,000 slaves in the Americas from 1,340 voyages, at an average profit per voyage of about 13 per cent. At the beginning of the nineteenth century not only was Britain shipping 50,000 slaves per year, but also its economy was relying ever more heavily on slave-produced goods. Over 90 per cent of its cotton, for instance, came from Louisiana, Brazil and the rapidly developing Demerara-Suriname, where most of the plantations were British-owned.

It was now very nearly 300 years since the King of Spain had authorised the first westward traffic of African slaves to the New World. The Atlantic slave trade had developed, at an accelerating pace, into a vast multinational commerce supporting the production in North and South America and the Caribbean islands of commodities essential to the developed nations of the western world. The stage was now set for arguably the most momentous event in the Trade's history: abolition of the slave trade by Britain, the nation which had been pre-eminent in the traffic for the past 70 years. Despite the work of the abolition movement in Britain and America, the Trade had not been discouraged in the slightest, and there was widespread scepticism that it ever would be. After all, piecemeal attempts by the states and federal government of the USA to legislate against slave trading had been virtually ignored by the slavers. Admittedly, Britain had recently placed certain restrictions on slave imports to its newly acquired territories in the Caribbean, but it was inconceivable to most people that this greatest of trading nations would voluntarily withdraw from a commerce in which it was the world leader. In any case, Britain was currently embroiled with the European powers in a world war, and would scarcely be distracted by the arguments of the abolitionists.

The sceptics were wrong. In 1807, as we have seen, Britain and the USA both enacted legislation against the slave trade. Thereafter, Britain alone not only enforced its anti-slaving law, but also threw its weight into a determined and protracted campaign to achieve universal abolition of the Trade.

CHAPTER 2

The Realities of the Trade

> Man, I can assure you, is a nasty creature.
> MOLIÈRE

THE PEOPLE of most of the slave-trading nations were almost entirely ignorant of the nature of the traffic in human beings which was underpinning their national economies in the eighteenth century. Brazil and Cuba were exceptions in this respect, but even in America the northern states saw little sign of the slaves being carried by their New England ships to the markets of the south. Slaves rarely passed through the ports of northern Europe, and the great majority of Europeans had never set eyes on a black African. Admittedly there was in England by the early 1700s a very small-scale trade in Africans for domestic work, mainly in the slaving ports; it had become fashionable for wealthy women to keep black boys as toys; and small numbers of African merchants' children were sent to England to be educated. In France, where, unlike England, slavery was legal, there were larger populations of Africans in the slaving ports, especially Nantes, and there were black minorities in the Dutch slaving port of Zeeland and in Lisbon and Seville. Nevertheless, the general lack of European contact with black people led to a supposition that Africans were brutish creatures fit only for slavery, if, indeed, the matter was given any thought at all. It was a salve to the consciences of some to believe that an African was more fortunate to be working for a civilised white master in the Americas than to be subject to the cruelties of a savage black heathen king in Africa, a conviction fostered by the slave merchants. The Vatican had gone even further in this direction: the Roman Catholic Church had long considered that slaving, together with other trade with Africa, presented an opportunity for religious conversion, and it became Portuguese and Spanish policy to baptise slaves before they were landed in the New World.

These comfortable notions would probably not have survived for long in the minds of decent Europeans had the true nature of the slave trade been apparent to all. As it was, the only white men who were familiar with the Trade's characteristics were the officers and crews of the slaving vessels, and those who handled the slave cargoes at each end of their transatlantic voyage. These people were generally hardened to their work, and, as the commander of the Royal African Company's castle at Cape Coast remarked in 1700, "Your captains and mates [...] must neither have dainty fingers nor dainty noses, few men are fit for these voyages but them that are bred up to it. It's a filthy voyage as well as a laborious [one]."[1] Early in the nineteenth century, however, the men of the Royal Navy's cruisers on the West African coast were to come face to face with the realities of this "filthy voyage" and, to some extent, with the inhumanities which preceded it.

As the eighteenth century drew to a close, the pattern of the Atlantic slave trade varied little between the major slaving nations. No longer was the state the prime mover in the Trade; it had at last been recognised almost everywhere that private enterprise, with minimum restriction, would yield the best results. At the head of the slaving hierarchy were the merchants who, from their counting houses in London, Liverpool, Lisbon, Rotterdam, Providence, Nantes, Seville and other centres of trade, financed and arranged slaving voyages, as much slavers themselves as were the crews of their ships. They were remote from the brutalities of the Trade, but were rarely entirely ignorant of them.

A typical slave merchant would have wide commercial interests, and slaves might not be his dominating concern. Some traders, particularly English and American, were owners of West Indies plantations, and a few owned property on the West Africa coast, where their cargoes were assembled. A significant proportion of slave merchants were particularly well aware of the true nature of their business, having commanded slaving ships in their younger days, and there were still a few slaving masters who owned, or had a share in, the vessels they commanded. Slave traders rarely operated entirely independently; voyages were generally financed by partnerships of about six merchants usually associating for only one voyage at a time. In the smaller ports the participants might include professional men and traders such as spinners, milliners and pawnbrokers. The most common type of slaving association was between relatives, and the Trade gave rise to many family empires on both sides of the Atlantic. Aristocrats were frequently involved in the partnerships, and complete outsiders might also seek to invest in a voyage. The highest number of voyages financed by a single partnership was probably the 80 arranged by a Nantes family; most traders were involved in only one or two slaving voyages, and of 1,130 French slave merchants in the eighteenth century only 25 family partnerships invested in more than 15 voyages.[2]

The only significant exception to this general pattern was in the massive traffic of slaves from Angola to Brazil. By the late eighteenth century this was mostly organised by Luso-Africans, descendants of Portuguese adventurers who had remained in Africa to live with the natives. Unlike their European and American counterparts, they were directly involved in all stages of the Trade leading to embarkation of slave cargoes for the transatlantic voyage.

Perhaps three-quarters of all slaving voyages followed a classic triangular track, starting in Europe and carrying manufactured goods to West Africa, where these "trade goods" were exchanged for slaves, then carrying the slaves to the Caribbean or the Americas on the so-called "Middle Passage", and finally returning to Europe with West Indian or tropical American goods purchased with the proceeds of slave sales. Such a voyage would last for at least a year and cover about 12,000 miles. Variations from this format were the direct trades between Angola and Brazil, and between the USA and West Africa. Some slaving voyages ended with sale of the vessel as well as her slaves in the West Indies or, occasionally, with return of the ship to Europe in ballast. There was also a considerable direct trade between Europe and the West Indies which had no slaving component.

Slaving vessels were referred to as "slavers", as were their owners, officers and crews, as well as the dealers in slave cargoes, and those vessels trading north of the Equator were commonly known as "Guinea-men". Until well into the nineteenth century they were typical ocean-going merchantmen of no special design or construction, and through most of the 1700s they were generally ship-rigged. Many, particularly the English ships, were of 200 tons or less, although French and Dutch slaving vessels tended to be bigger, and the leading Nantes shipbuilder of the late eighteenth century believed that the ideal slaver would be between 300 and 400 tons, with ten feet of hold and four feet four inches between decks. As the size of ocean-going merchantmen increased at the end of the eighteenth century, the lower limit of tonnage at which vessels were rigged with three masts rose sharply, and by the beginning of the suppression campaign slaving vessels of less than 200 tons or so were mostly two-masted, rigged as either brigs or schooners. Slavers were usually armed against the threat of pirates, and a vessel of 200 tons might carry as many as ten guns in peacetime. Thanks largely to hull deterioration hastened by warm seas, and damage caused by tropical squalls, only rarely would a slave ship have a career of more than six voyages or survive for more than ten years.

The manning ratio by this time was generally about one man per five tons, and most of the hands would be in their twenties, although a slaver crew would usually include some boys. The master and mates would probably be in their thirties, and specialists such as the carpenter and sailmaker might be older. In Brazilian and Rhode Island ships there were sometimes free black men among the crews, but they could never become officers. Even slaves were occasionally used as crew members. The Trade had a bad reputation among seamen, and the majority of white crews of slavers probably came from the crimping houses, where they were bought from landlords while dead drunk or, deep in debt, were seduced by false promises of high wages. The unpopularity of seafaring in slavers was well earned: discipline was harsh and often brutal, the men were liable to be paid in the reduced currency of the West Indies, and the death rate from disease was appallingly high. A higher proportion of seamen than slaves died of accident, disease or misuse on an average slaving voyage: 18 per cent against 12 per cent in the Dutch trade. Thomas Clarkson, during his researches for the abolitionist movement, produced figures for 1786 which are probably accurate: of 5,000 sailors on the triangular route from Britain, 2,320 came home, 1,130 died, 80 were discharged in Africa and unaccounted for, and 1,470 were discharged or deserted in the West Indies.[3]

A slaver master had to be a man of considerable ability if his voyage was to be a commercial success. Not only did he need the obvious skills of seamanship and leadership and the mental and physical toughness necessary to any sea captain of the age, but he also had to be adept at negotiating slave purchase prices with the wily and ruthless kings and merchants of the African coast. For it was the master who conducted business at that level of slaving, not his paymasters in Europe or America. He had to maintain discipline among a crew made up usually of particularly hard men engaged in probably the most hazardous trade afloat, and he had to care for a highly perishable cargo and cope with the constant threat of a slave rebellion. Many of these men were callous to the point of brutality, tyrants to crews and slaves, but others were intelligent, courageous and, by the standards of a hard-bitten profession, civilised and kindly officers. Slaver masters recorded their opinions on the Trade more frequently than did the merchants, and, although some of them occasionally expressed doubts or even remorse, a more general view was that the traffic was a necessary evil, and that Africans were happier as slaves in the West Indies than as slaves "subject to the caprices of their native princes". The Trade was not regarded as a permanent profession by sea captains, and few of them went to Africa more than three times. Most would subsequently command ships in other trades or become shipowners, and, in the experience of the Dutch West India Company at least, on one in ten slaving voyages the master died.

The surgeon was an important member of a slaver crew, although his inclusion was not a legal requirement, and some ships, including most Americans, economised by neglecting to have one. It was the surgeon's advice to the master which was decisive in the selection of slaves to buy, and the master naturally looked to his surgeon to keep his cargo alive and healthy during the Middle Passage. The surgeon might be the only educated man on board, and it is perhaps not surprising that one of the more comprehensive and objective first-hand descriptions of the late-eighteenth-century slave trade comes from a slaver doctor, Alexander Falconbridge, in his *Account of the Slave Trade on the Coast of Africa*. As an incidental result of the large number of slaver surgeons trained at the Liverpool Royal Infirmary, the Liverpool School of Tropical Medicine was founded.

There was a good deal of variation in the trade goods carried by slavers, depending on what was available in home ports or on the outward voyage, and on the requirements and tastes of the region of Africa from which slaves were to be purchased. For example, in the slave harbour of Whydah in 1767 a slave might be bought for

> sixteen anchors of brandy (133 imperial gallons), or twenty cabess of cowries (100,000), or 200 pounds of gunpowder, or twenty-five guns, or ten long cloths, or ten blue bafts (lengths of coarse blue Indian cotton material), or ten patten chints (chintz), or forty iron bars.[4]

The goods loaded by the slavers to exchange for slave cargoes generally accounted for about two-thirds of the cost of a voyage.

Cloth, woollen and cotton, was generally the most sought-after cargo. White was preferred on the Guinea coast and blue in Angola, but a wide choice of patterns was available by the late eighteenth century. Metals came next in popularity, and by this stage iron bars were in particular demand in West Africa both for forging into tools and as a trading medium. More important as currency was the cowrie shell, almost ideal as a form of money and much used in the slave trade. Next in significance in the Trade was weaponry. African monarchies regarded muskets as essential for self-defence and could obtain them only through slaving; 300,000 of them were imported annually from Europe in the second half of the eighteenth century, many of them of inferior quality. Obviously gunpowder was equally desirable, and Britain exported over 2 million lb of it to Africa in 1790. Alcohol was, naturally, a very popular trade product, and was also regularly used to lubricate the bargaining process. Wine, beer and cider played a part, but spirits were of much greater importance. Gin was sometimes used, but brandy from Europe, and rum from New England and sometimes from Britain, were in much higher demand. The Brazilians did a good trade with a particularly powerful cane brandy. Tobacco was long used by Brazil to buy slaves, and a low-quality tobacco treated with molasses to prevent it from crumbling became so popular in Africa that European masters often bought it from Brazilian ships to add to their own trade goods. Finally, most slavers would, in addition to their main trading cargoes, carry an array of minor goods and trivia ranging from smoked cod to silk hats, and from shaving bowls to glass bracelets. Beads were probably the most important trinkets.

The slaves destined to be bought with these trade goods would, in almost every case, have begun their cruel hardships long before they set eyes on the slave ships. The means by which Africans were enslaved were almost as varied as the goods which purchased them, and practices differed from region to region, but the places of origin of the slaves were, with very few exceptions, remote from the coastal towns and trading posts where they were sold into the ships.

In the early days of slaving it was not uncommon for slaver crews to kidnap, or "panyar", Africans on the coast and carry them on board, but masters and European traders soon realised that it was greatly to their advantage to cultivate amicable relationships with the coastal kings and chiefs, and by the eighteenth century panyaring was extremely rare and merely opportunist. With that rare exception, African slaves were enslaved by Africans. In many parts of the continent the principal means of doing so was by capture in war, a practice commonplace long before the arrival of European slavers. Argument continues about the degree to which the Atlantic slave trade exacerbated intertribal feuding and raiding, or, indeed, contributed to full-scale wars between powerful kingdoms, and there will never be a complete answer. It is clear, however, that in much of the – almost constant – low-level hostilities between some tribes slaves were a by-product, sold into the Trade if their labour was not required by their captors; the alternative was to slaughter them. Large numbers of slaves were often sold for the Guinea coast trade by the Muslim tribes of the northern interior after their frequent *jihad*s against their pagan neighbours. It is equally beyond doubt that at times warfare, and particularly raiding on neighbouring tribes, was used solely as a means of procuring slaves for the Atlantic traffic during periods when the market was good and supplies from other sources were scarce. This was particularly true of those parts of Central Africa which supplied the Trade in Angola and the Congo.

A great many slaves were victims of kidnapping, a means of capture which ranged from lone travellers being taken by single ambushers, through the seizing of children from their homes by groups of marauders while parents were at work in the fields, to the wholescale pillaging of villages, sometimes even by their own overlords. Kings and chiefs also profited greatly from the sale of those convicted of crimes or misdemeanours. Adultery and repeated theft regularly incurred enslavement, and one trader observed that the existence of the Atlantic trade resulted in more and more offences being punished with slavery: "they strain for crimes very hard, in order to sell into slavery".[5] Sometimes the tributes or revenues due to a king from his villages would be paid in slaves, and parents occasionally resorted to selling their children to alleviate extreme poverty. There were also instances of rulers maintaining slave villages where their captives awaited sale, and children were born into slavery in these places. A most unusual source provided much of the supply for the large slave market at Bonny in the Niger Delta: the local oracle Chukwa levied fines of slaves from convicted individuals or even families, and people who consulted the oracle with what were considered to be stupid questions were liable to be seized and sold.

Slaves from the African interior, the vast majority of those destined for the Atlantic trade, were generally taken initially by their captors to markets or fairs relatively close to their places of origin. Here they were sold to African merchants who might either send or take them directly to the coast or, as an interim measure, shift them from market to market if it appeared profitable to do so. In some areas the function of the slave fairs was performed by the individual kings and chiefs of the interior; King Tegbesu of Dahomey, for instance, is estimated to have made the equivalent of £250,000 a year from slave sales in the 1750s, five times the income of the richest English landowner of the period.

After sale at their final interior market the slaves would be taken to the coast, either bound hand and foot and loaded into canoes, or marched in guarded caravans. These caravans, or "coffles", would consist of 100 or more slaves, often carrying water or produce such as ivory on their heads, both as a convenient means of transport and as a discouragement from escape. Slaves in coffles, particularly the men, were usually fettered and secured together with rope or chain either in pairs by the leg or in groups by the neck. Sometimes the men would be fastened together in pairs by wooden neck-yokes. The only European to accompany a slave coffle for any length of time and to write of the experience was the explorer Mungo Park, and he recorded a march of 500 miles covering an average of 20 miles per day. However, there were reports at the time of such journeys taking as long as 80 days. Although few Europeans witnessed slaving activity in the interior, it was clear from the condition of slaves arriving on the coast that they had generally been harshly treated. Most were weakened by wounds, hunger, exhaustion, exposure and disease, and it has been estimated that, of those enslaved in the interior, half were lost to either escape or death before arrival at the coast.

Preferred trading seasons differed from place to place, but a slaving voyage from Europe to the Guinea coast would ideally begin in the early autumn to make best use of fair winds on the southerly passage, to catch the dry season on the coast and to allow escape from West Africa by February or March before the monsoon returned with its rain and squalls, as well as a foul wind for the start of the Middle Passage. That would achieve delivery of the slaves to the markets in the Americas well before the autumn sugar-cane planting and the December cane harvest. It would also permit departure for home before insurance rates rose at the beginning of the hurricane season in the Caribbean. Nevertheless, numerous factors, foreseen or unforeseen, might conspire to thwart such a plan. It might, for example, be necessary to delay sailing from Africa until the yam crop was ready in July to provide food for the slaves on the Middle Passage.

A slaver might visit a number of anchorages in order to complete her slave cargo, but by the end of the eighteenth century it became increasingly the practice for a ship to take all her slaves from one source. As the surgeon Alexander Falconbridge records, the pace of trading was fairly leisurely, and a slaver would probably strike her yards and topmasts on arrival on the slave coast, and enclose her weather decks with improvised roofing and side screens in order to give protection from sun and rain and to prevent slaves from jumping overboard. Between two and four months of trading was regarded as rapid, and between four and six as normal.

Trading procedures varied on different parts of the slave coast, but it was common practice for a newly arrived master to invite the local king or headman on board, or to call on him ashore, to seek his permission to trade for slaves. Having been given presents, or "dashes", such as liquor or pieces of cloth and silk, the King would give the master leave to "break trade". Interpreters and other local functionaries would be appointed, at a price, to facilitate dealings; the master would then be free to approach the African slave merchants, and the King might well have his own slaves to sell. At some trading posts a broker would, for a commission, act for the various local traders in negotiations with slaver masters. At major slaving centres there were likely to be free African, mixed-race or European agents who would buy the slaves from the inland traders and gather their captives in sheds, or "barracoons", at their slave "factories" close to the shore. Islands were favoured for these factories, and in some places the slaves were imprisoned in hulks lying offshore to minimise opportunities for escape. In places where factories had been established, the slaver masters would deal with the factory owners instead of individual traders, and cargoes could generally be completed more quickly. The local chiefs and kings profited from these transactions not merely by the sale of their own slaves but also by charging duty on all dealings on their territories.

With the decline in commercial importance of the national monopoly companies, the company castles on the Gold Coast played a less significant part in the slave traffic, but they still offered an alternative source of supply for the slavers. The Europeans in their coastal forts, especially the Dutch at Elmina, the British at Cape Coast Castle and Anamabo, and the Danes at their main fort of Christiansborg had, over the years, established monopoly slave-trading arrangements with the neighbouring kingdoms and tribes. The relationship between Elmina and the great inland kingdom of Ashanti was particularly strong, and over a thousand Ashanti slaves per year were exported from the castle during the 1770s. Company agents, or "factors", at the forts purchased slaves from African traders and tribal leaders, and imprisoned them in the cellar dungeons of their castles to await the arrival of slave ships offshore. Trade goods from the slavers were stored in the castles until they could be used by the factors to buy more slaves. Added to this official traffic was the private, and illegal, sale of slaves by the governors and officers of the forts.

Slaver masters and African traders missed few opportunities to cheat each other. The Europeans would water brandy and rum, add false heads to gunpowder kegs and cut lengths from the middle of bales of cloth, compensating the weight with pieces of wood. A curious adulteration, in this case to please the recipients, was mentioned by William Richardson in *A Mariner of England*:

> The brandy that we brought out for trade was very good, but the darkies thought it was not hot enough and didn't bite – as they called it; therefore, out of every puncheon we pumped out a third of the brandy, put in half a bucketful of cayenne pepper, then filled it up with water and in a few days it was hot enough for Old Nick himself, and when they came to taste it, thinking that it was from another cask, they would say "Ah, he bite."[6]

The Africans for their part were skilled in concealing infirmities in the slaves they were offering for sale, even to the extent of painting any who were sick. It was also common practice to sleek the captives with palm oil to give an impression of good health. Furthermore, the traders were ruthless in overcoming any obstacle to a sale: James Arnold, an ex-naval surgeon in a slaver told a Parliamentary Committee in 1789 that he had seen a woman brought out to his ship for sale with a child in her arms. The master refused to take her on account of the child. On the following day she was brought back to the ship without the child, who had been killed by the African trader to facilitate the sale of the mother.

Whatever tricks might be used by the traders to disguise disease or defects in their slaves, it was the business of a slaver's surgeon to check the health of the merchandise before the master committed himself to a purchase. Surgeon Alexander Falconbridge explained that when the Africans were shown to the European purchasers

> they first examine them relative to their age. They then minutely inspect their persons and inquire into the state of their health; if they are afflicted with any disease or are deformed or have bad eyes or teeth; if they are lame or weak in the joints or distorted in the back or of a slender make or are narrow in the chest; in short, if they have been or are afflicted in any manner so as to render them incapable of much labour. If any of the foregoing defects are discovered in them they are rejected. But if approved of, they are generally taken on board the ship the same evening. The purchaser has liberty to return on the following morning, but not afterwards, such as upon re-examination are found exceptionable.

Captain Richard Drake, in his autobiography *Revelations of a Slave Smuggler*, recalled from his early days at the beginning of the 1800s that the master of the slaver *Coralline*, trading in the Cameroons, had, unusually, taken his boats well upriver to buy his slaves. Here he employed the services of a mixed-race overseer who doubled as "a sort of Negro doctor", able, apparently, to tell an unsound slave almost at a glance. While the master, in shirt and duck trousers with palm-leaf hat, walked up and down the line of fettered slaves, the overseer

> handled the naked blacks from head to foot, squeezing their joints and muscles, twisting their arms and legs, examining teeth, eyes and chest, and pinching breasts and groins without mercy. The slaves stood in couples, stark naked, and were made to jump, cry out, lie down and roll and hold their breath for a long time. The women and girls were used no more gently than the men by this mulatto inspector.[7]

Captain Thomas Phillips emphasised in his journal that:

> Our greatest care of all is to buy none that are pox'd, lest they should infect the rest [...] therefore our surgeon is forc'd to examine the privities of both men and women with the nicest scrutiny, which is a great slavery, but what can't be omitted.[8]

Slaves refused by a master were likely, at best, to be beaten by the trader, no matter what their illness or defect might be. Falconbridge observed that at New Calabar, in particular, it had frequently been known for such slaves even to be put to death. He mentioned instances of a trader dropping under the stern of the slaver in his canoe and beheading his rejected and worthless merchandise in full view of the master.

Once a deal had been struck, the selected slaves were generally branded with a hot iron, indicating either the company or the private trader who had made the purchase. Practices varied from trader to trader, and the brand might be on the shoulder, the breast, the upper arm or, occasionally, the thigh or back. Silver marks were favoured by some companies because they made a sharper scar. Instructions to its captains from the South Sea Company in 1725 describe the process. The brand was to be on the "left shoulder, heating the mark red hot and rubbing the part first with a little palm or other oil and taking off the mark pretty quick, and rubbing the place again with oil".[9]

The slaves were generally made to kneel for branding and, although some endured the pain with stoicism, the scene described by Captain Drake, as he continues his account of the slaving activities of the *Coralline*, is probably more typical:

> The slaves were fetched up singly, made to lie down on their faces where they were held by a big Negro while another kept the branding irons hot in a fire close by and a third applied them between the shoulders of the shrieking wretches. At first there was horrible yelling, for the poor Negroes expected to be tortured to death.[10]

According to Captain Thomas Phillips of the *Hannibal*, an interloper slaver, the branding "caused but little pain, the mark being usually well in four or five days, appearing very plain and white after".[11]

Slaves of the Portuguese were subjected to a further process before boarding the ships. The Portuguese forbade the embarkation of any slave who had not been baptised, although the candidates were unlikely to receive any comprehensible preparation for this sacrament. In an attempt to comply with canon law requiring proper instruction of adults before baptism, it had been decreed that every Portuguese slaver should carry a priest who would initiate the slaves into Christianity during the transatlantic passage; but a shortage of priests tended to stand in the way of this aspiration. Slaves bound for Brazil from Angola and the Congo were usually baptised in perfunctory mass ceremonies before sailing, but those from the Gold and Slave Coasts were often not baptised until they reached Brazil.

The slaving vessels, meanwhile, made preparations for receiving their cargoes. Once a slaver's lower deck was empty of trade goods it was transformed into the slave deck, and, in order to increase the ship's slave-carrying capacity, the carpenter would construct an additional slave deck with timber carried for the purpose. This took the form of a platform, between about five and eight feet in width, extending inward from the ship's side on the main slave deck halfway between deck and deckhead. Slaves could then be stowed on this shelf as well as on the deck below it, allowing the slaves headroom of 30 inches or less in all but the centre of the deck. If the slave deck was not already suitably subdivided, transverse partitions were built to separate men from women and, if necessary, men from boys. Falconbridge gives us an example of the relative dimensions involved: a Liverpool ship with a length between decks of 92 feet and a beam of 25 feet divided her slave deck to give the male slaves 45 feet, the women 10 feet and the boys 22 feet; the remaining 15 feet provided a storeroom.[12] The proportions reflect the general practice of carrying women as no more than a third of the cargo. The carpenter's final preparation was to remove the solid hatches giving access to the lower deck and replace them with gratings to allow ventilation to what was now the slave deck.

The next major task for the slaver crew was to purchase and load victuals for the slave cargo, to fill and stow the water casks, and to gather wood for cooking fuel. The foodstuffs provided for the slaves varied to some extent from region to region on the coast, depending upon availability and the preferences of the people of the area. Yams, rice, horse beans and maize (known as Indian corn) all featured as main ingredients, with supplements of palm oil, pepper, flour, bread or biscuit, plantains and, occasionally, salt beef and pork, potatoes, and even oranges. Cassava was a staple food in Portuguese slavers, and oats in French ships. Also dried shrimps, coconuts and lime juice are mentioned by Captain Crow in his slaving memoirs. Some of these stores would be embarked at the slaver's home port, but it was greatly preferable to obtain them on the slave coast as short a time as practicable before they were needed. Not least of the reasons for this was the constraint on stowage space when there was a full load of trade goods on board; one slaver gave the firm recommendation that a vessel which took in 500 slaves must provide above a hundred thousand yams, "which is difficult because it is difficult to stow them as they take up much room".[13] Masters had to balance the stowage requirement for stores and water against their desire to load the maximum practicable number of slaves, and not infrequently the quantities of food and water embarked were inadequate for a voyage extended by adverse weather.

Once the major preparations had been made, slave-irons would be broken out of store ready for the captives they were designed to constrain, and the ship would be ready to receive her cargo. Embarkation in the shelter of a river estuary or creek was straightforward enough, but on parts of the coast the only anchorage available to the slavers was an open roadstead opposite a beach fully exposed to the Atlantic swell. In such circumstances there was no option but to land trade goods and embark slaves by canoe through the surf. This was inevitably a hazardous undertaking, and, although the canoes were manned by highly skilled natives, stores and lives were frequently lost, and the dangers were increased by the numerous sharks cruising inshore. At times the surf ran so high that boat work became impossible for extended periods, and trading and loading would be greatly delayed. It took far too long to complete a cargo even without such delays. The slaves were generally embarked in batches as they were purchased from various traders following the spasmodic arrival of coffles from inland, and ships sometimes moved from one anchorage to another during trading. The result was that the slaves loaded early in the process might be incarcerated on board for up to five months before the ship sailed.

It was a constant concern to the slavers that the Africans might rebel or escape while they were being held in barracoons or castle cellars, during embarkation by boat, or after they had been loaded on board. The slaves, and particularly the men, were therefore closely constrained, usually with iron manacles. Neck collars, wrist shackles and ankle-irons were all employed, depending on the circumstances. While they were being moved the slaves would usually be shackled in pairs, and, once on board, the men slaves would spend much of their time secured to the deck by their ankle-irons. Great care had also to be taken to prevent anything which might be used as a tool or weapon from falling into their hands. In some ships a barricade would be built athwartships on the upper deck to provide a defensive position for the crew in the event of a slave revolt while at anchor, and cannon might be mounted behind it to cover the forward part of the deck and the hatches to the slave deck. Despite these precautions, and sometimes thanks to drunkenness on the part of the crew or undermanning on account of sickness, slaves would occasionally make a bid to escape. Often they would jump from the boats or canoes before reaching the slaver, not infrequently to drown or to be taken by sharks, and there were numerous instances of crewmen being attacked and sometimes killed as slaves broke out of the slave deck and attempted to penetrate the ship's side-netting before leaping into the sea. These revolts and escape bids were rarely successful, and slaves who survived such incidents and remained in captivity would be punished. Major rebellions were infrequent, and those that did occur were brutally suppressed and usually followed by the execution of supposed ringleaders.

Slave losses from revolts and escapes during the waiting period ashore and afloat were invariably exceeded by the numbers lost to disease and other causes. One Bristol master, for example, bought 939 slaves in 1790 and lost 203 of them "of natural causes" while still on the coast of Africa. However, the crews of the slavers probably suffered even more severely from disease than did the slaves while their ships lay off the African shore, and they could expect scarcely more sympathy. The policy of the master of the brig *Ruby* towards his fever-stricken seamen was: "No work, no victuals", and on being told that one of his men was on the point of death, he said: "Let him die and be damned."[14] The attitude of this Captain Williams was probably not untypical.

As numbers increased on a ship's slave deck, greater care had to be taken in stowing the slaves so that the maximum practicable number could be accommodated. Indeed, they were generally so closely packed by the time that the cargo was complete, as surgeon Falconbridge wrote, "as to admit of no other position than lying on their sides". Captain Crow's memoirs describe how slaves were stowed in the *Coralline*:

> They were shackled down in tiers [...] sitting between each other's legs, fore and aft [...] the blacks were strung across in gangs of six or eight, according to size, and their ankle-bolts were secured by two iron rods running amidships and padlocked in the centre.

The men slaves would be kept shackled throughout their time on board, and would spend at least some of each day, probably depending on the master's perception of the risk of rebellion, with their ankle-irons secured to the slave deck. The women, although given no more space than the men, were not shackled. Both men and women would be stripped naked before leaving the shore, and so they would usually remain, although the occasional relatively compassionate master would allow the women scraps of material for loincloths.

Unlike the other slaving nations, Portugal had attempted in the seventeenth century to legislate against overcrowding in slavers, and, although application of the regulations was poorly enforced, the Portuguese did have a reputation for greater humanity than the other European shippers. For example, Portuguese slavers made a practice of placing about 15 slaves into the care of each sailor, most of whom were themselves black Africans, and they often provided mats for the slaves to lie on. In the ships of other nations the slaves lay on the bare planks of the deck.

Generally speaking a slaver would be able to carry two slaves per ton. For example, the English ship *Brookes* of 297 tons could load 609 slaves. However, the Dolben Act of 1788 limited British vessels to five slaves per two registered tons, up to 201 tons, and one slave per ton beyond that. Although this made little difference to the majority of British slavers, the small number of larger British ships were permitted thenceforth to carry fewer slaves than their foreign competitors of equivalent size.

Slaver crews were accommodated with little more consideration for their comfort than were their cargoes; masters had cabins, and so usually did the officers, but the hands would have to sling their hammocks in any available corner, or resort to sleeping in the boats or even on the open upper deck. The officers would often sacrifice much of their cabin space by loading personal slaves to supplement their wages, sometimes smuggling them on board.

Once her stores and cargo were complete, a slaver would dismantle her improvised deckhouse, send up her topmasts, cross her yards, bend her sails and weigh her anchor for the start of the Middle Passage. Ships sailing from the Guinea coast for the Caribbean would expect to be at sea for anything between two and three months. The slowest recorded passage in the eighteenth century was nine months, and the fastest was 25 days, both by French vessels. On the shorter voyage across the South Atlantic from the Congo, or Angola to Brazil, the Portuguese averaged about 30 days.

For the first few days at least, seasickness would be added to the discomforts of the slaves, but, for some, hope of escape did not die until the coast of Africa had finally dipped below the horizon. Some risk of slave mutiny would remain throughout the passage; indeed, there was some form of insurrection on one out of eight or ten voyages; but most of these occurred while the slaves perceived some chance of escaping ashore, and the Africans seemed to accept their fate once their homeland was no longer in sight. This meant that security could then be relaxed to some extent, and, in fair weather, the slaves would be removed in batches from the slave deck for fresh air. A typical procedure for the men slaves was to secure a long chain to a ringbolt on the upper deck, pass it through the shackles of 50 or 60 slaves and then lock the end to a second ringbolt. The slaves were encouraged to exercise, and surgeon James Arnold of the brig *Ruby* wrote that it was the business of the chief mate to make the men dance, but that they could be induced to do so only by frequent use of the cat, and, even so, could merely "jump up and rattle their chains". In some ships the master was sufficiently confident to allow the men slaves, by turns, to move unrestrained around the upper deck during daylight hours, and to use latrines on platforms built outboard from the ship's side. The women were given greater freedom than the men; they were unfettered and generally permitted to remain on the upper deck as long as they pleased during daylight. They too were encouraged to dance.

The freedom of movement given to the female slaves was to the advantage of the crew, for reasons explained by Alexander Falconbridge:

> On board some ships the common sailors are allowed to have intercourse with such of the black women whose consent they can procure [...] The officers are permitted to indulge their passions among them at pleasure and sometimes are guilty of such brutal excesses as disgrace human nature.[15]

Some captains abused their power as disgracefully as any of their men in this respect, including the brutal Williams of the *Ruby*:

> It was his general practice on the receipt of a woman slave – especially a young one – to send for her to come to his cabin so that he might lie with her. Sometimes they would refuse to comply with his desires and would be severely beaten by him and sent below.[16]

Neither the threat of punishment by a disapproving master nor lack of privacy would deter some of the men. Captain John Newton, the future clergyman and hymn-writer, recorded that: "In the afternoon, while we were off the deck, William Cooney seduced a woman slave down into the room and lay with her brutelike, in view of the whole quarter deck, for which I put him in irons."[17]

Nevertheless, slave women did at times benefit from forming relationships with members of the crew, gaining favours from them which made the voyage more bearable. Apparently such attachments could become strong, as far as the slaves were concerned, at least, and Falconbridge commented that some of the black women "have been known to take the inconstancy of their paramours so much to heart as to leap overboard and drown themselves".

In fair weather the slaves would generally be fed on the upper deck, at about ten in the forenoon and five in the evening in most English ships, although the Dutch and some ships of other nations apparently cooked three meals per day. As mentioned earlier, foodstuffs varied between the ships of the slaving nations and between the regions where stores were embarked, but a typical daily ration for a slave in an English ship was 3 pounds 10 ounces of yam, 10 ounces of biscuit, 3.5 ounces of horse-beans, 2 ounces of flour, and a portion of salt beef. A plantain and an ear of maize might be added one day in five. Whatever the ingredients, the food was boiled in large coppers, initially intensifying a common fear among the Africans that they themselves were destined to be boiled and eaten by the white men. The slaves would be grouped on the deck in messes of ten, each around a food tub from which they would eat with wooden spoons. Each slave was allowed between one and three pints of drinking water per day, except in the not uncommon event of a shortage, and this might be served either by passing around a bucket and ladle during the meal or by handing each slave his share from a barrel as he passed the slave-deck hatch on his way below after the meal. In more enlightened ships the slaves might also be given lime juice as an antiscorbutic and sticks of citrus wood for chewing to clean their teeth.[18]

Slaves occasionally attempted to commit suicide by refusing to eat, and such recalcitrants were either persuaded to change their minds by flogging or the threat of being forced to swallow red-hot coals, or had their jaws levered apart by a screw-operated device called a *speculum oris*.

Mealtimes were seen as periods of high risk for slave rebellions, and Captain Phillips of the *Hannibal* describes the precautions taken by a cautious crew:

> what of our men are not employ'd in distributing their victuals to them, and settling them, stand to their arms; and some with lighted matches at the great guns that yaun (*sic*) upon them, loaded with partridge [shot], till they have done and gone down to their kennels.[19]

Nevertheless, measures as extreme as this were probably rare once a slaver was out of sight of land, unless there were indications of unrest.

Such slave mutinies that did occur at sea were crushed ruthlessly, and very few achieved any degree of success, although there were likely to be casualties among the crew. Masters were naturally reluctant to lose more of their cargoes in punishment for a rebellion than was entirely necessary, and their reprisals tended to be limited in quantity but, for the sake of deterring further attempts, not in severity. The ringleader of a revolt in a Dutch slaver immediately had his right hand cut off and shown to every slave; on the following day his left hand was cut off and similarly exhibited, and on the third day he was beheaded and his torso was hoisted to the main yard, where it was displayed for two days. After a slave uprising in the Nantes ship *Affriquain*, the crew tied up the most guilty blacks, that is those who led the revolt, by their arms and feet and, lying them on their backs, we whipped them. As well as that we put hot plasters on their wounds to make them feel their faults the more.[20]

These slaves were then left to die of their injuries.

Instances of this sort apart, brutality in slave ships was not the normal state of affairs, although there were some masters and officers who took pleasure from gratuitous cruelty to slaves and crew. It was in the interests of all involved in the Trade to deliver to the markets in the Americas as many slaves as possible, not only alive but in good health, and in most slavers every practicable measure was taken to achieve that end. Cleanliness of the slave deck, for example, was usually taken seriously. James Barbot, supercargo in the *Albion-Frigate*, emphasised that:

> We were very nice in keeping the places where the slaves lay clean and neat, appointing some of the ship's crew to do that office constantly and several of the slaves themselves to be assistants to them and thrice a week we perfumed betwixt decks with a quantity of good vinegar in pails, and red-hot iron bullets in them, to expel the bad air, after the place had been well washed and scrubbed with brooms; after which the deck was cleaned with cold vinegar.[21]

In Captain Phillips's *Hannibal* care was taken to make the Africans "scrape the decks where they lodge every morning very clean, to eschew any distemper that may engender from filth and nastiness", and Captain Crow, supercargo of probably the last legitimate English slaver, took great pains to promote "cleanliness, for I considered that on keeping the ship clean and orderly, which was always my hobby, the success of our voyage mainly depended".[22]

All too often, however, circumstances beyond the control of masters and officers would frustrate their good intentions for the welfare of the cargo, and conditions on the slave deck would degenerate into a state to appal the senses of even hardened slavers. Foul weather would immediately cause a deterioration, most seriously by preventing ventilation of the slave deck. The supply of air to the slaves was poor at the best of times and entirely inadequate for the tropics, mainly depending as it did on the gratings of the slave-deck hatches. In some vessels a little more air was provided by a few small ports in the ship's side, and occasionally a master would take the trouble to rig windsails to increase airflow through the hatches. At the onset of heavy weather, however, the windsails came down, the ports were shut tight and the hatches were battened down with tarpaulins, rapidly causing the slave deck to become intolerably hot and the atmosphere suffocating. Then came seasickness, to which, according to Falconbridge, the Africans were particularly prone. The foulness of the air and the debilitation of the slaves then exacerbated the inadequacy of sanitary arrangements below decks:

> In each of the apartments are placed three or four large buckets, of a conical form, nearly two feet in diameter at the bottom and only one foot at the top and in depth about twenty-eight inches, to which, when necessary, the Negroes have recourse. It often happens that those who are placed at a distance from the buckets, in endeavouring to get to them, tumble over their companions, in consequence of their being shackled [...] In this situation, unable to proceed and prevented from getting to the tubs, they desist from the attempt; and as the necessities of nature are not to be resisted, they ease themselves as they lie.[23]

The slave Equiano, writing of his experiences on the Middle Passage, recalled that children often fell into such tubs and were almost suffocated.

It is scarcely surprising that the loathsome conditions caused by foul weather led to disease among the slaves, but in any case sickness in epidemic proportions was liable to strike a slaver. The worst and most common illness was dysentery, or "the flux", and its ensuing dehydration, which accounted for about a third of slave deaths on board. Next in terms of destruction came smallpox, and various types of ophthalmia occasionally caused devastation among slaves and crew. Added to these were scurvy and a variety of skin diseases. By good fortune and good management a slaver might escape these scourges almost entirely, but this was not often the case. The very high losses of earlier years on the Middle Passage, such as the 24 per cent mortality suffered by the Royal Africa Company in the late 1600s, had been reduced to an average of about 9 per cent by the end of the eighteenth century, although there were still instances of horrific death rates. Surgeon Falconbridge attempts to describe the scene on a slave deck with the hatches covered and dysentery rife:

> The deck [...] was so covered with the blood and mucus which had proceeded from them in consequence of the flux, that it resembled a slaughter house. It is not in the power of the human imagination to picture a situation more dreadful or disgusting.[24]

The sick were subjected to added suffering by lying on the bare deck:

> By this means those who are emaciated frequently have their skin and even their flesh entirely rubbed off, by the motion of the ship [...] And some of them, by constantly lying in the blood and mucus that has flowed from those afflicted with the flux and which is generally so violent as to prevent their being kept clean, have their flesh much sooner rubbed off.[25]

As an example of a voyage seriously struck by disease, Surgeon James Arnold described to a Parliamentary Committee the case of the slaver *Britannia*, which had lost 230 of her cargo of 450 to smallpox. As the sickness spread the sick berth was soon full:

> Only those who were so bad as to be incapable of moving were admitted. There they lay in one mass of scab and corruption, frequently sticking to each other and to the deck till they were separated to be thrown overboard.

A slaver's crew was hardly more fortunate in the event of an epidemic, and, as the ship became increasingly short-handed, those even barely capable of work were obliged to continue. James Stanfield, in his book *The Guinea Voyage*, relates the fate of the boatswain of one of his ships:

> He grew so bad at last that the mucus, blood and whole strings of intestines came from him without intermission. In this deplorable condition, when he could hardly stand, he was forced to the wheel [...] The poor man was placed on one of the mess tubs, not being able to stand and also so that he might not dirty the deck [...] He died that night and the body was thrown overboard.[26]

There was little that slavers' surgeons could do to alleviate these grim afflictions, but most of them seem to have done their best, and many of them sacrificed their own health, and even their lives, in the process.

"Tight-packing" of the slave cargo was assumed to have been a significant cause of high mortality, but statistics suggest that there was no benefit on that score from more humane loading; an epidemic would race through the cargo of even a lightly laden slaver. The weather, and, particularly, the length of a voyage, were more weighty factors. Storms would cause injury as well as encourage sickness, and if the ship was lost the slaves would almost certainly be prevented from escape by closed hatches. Calms would not only slow the passage and reduce airflow to the slave deck, but also would increase the risk of food and water shortages.

Arrival at their destination harbours in the islands of the Caribbean or the mainland Americas brought some relief to the slaves as they were prepared for sale to planters or dealers. In the Brazilian ports the slaves would be transferred from the ships to ware-rooms on the ground floors of the merchants' houses, and here they were cleaned, shaved, fattened with good food and, if necessary, painted to give an impression of health. Despite the apparent improvement in treatment, however, deaths continued from heat and overcrowding or from diseases brought from the ships. Usually the buyers would visit these houses to subject the slaves to physical examination and make their purchases, but slaves were often sold at auction, and some merchants would hawk their chained slaves from house to house. Buyers would generally re-brand their new slaves, and paid a 5 per cent tax to the government.

In Saint-Domingue and other Caribbean islands most slave cargoes would be taken by the owners' agents to a large field with huts, where a few days of rest and fattening would make them ready for sale. They might be made to bathe in the sea and given clothes, and their bodies would be rubbed with palm oil to improve their appearance. The French traders in Saint-Domingue usually gave their slaves strong drink to invigorate them before they were presented to buyers, and examination by French planters tended to be particularly thorough. Sales might take place in the fields where the slaves had been "refreshed", or, occasionally, in the ships to which the slaves might be returned for greater security.

Sale procedures in the English colonies were less formal and bureaucratic, and there was less emphasis on hygiene and on preparing slaves for sale. A common form of sale was the "scramble". This would take place either on the upper deck of the slaver (perhaps shaded by an awning to prevent the buyers from seeing the merchandise clearly) or in a yard ashore. The slaves would be equally priced at a figure previously agreed, and, as the gates opened or at a signal, the purchasers would rush in among the slaves and grab those they wanted. This was a terrifying experience for the Africans, and it was at this point that families who had contrived to remain together would generally be split apart.[27]

Inevitably there would be numbers of sick and emaciated slaves remaining unsold by these various methods, and they would probably be offered by public auction at knock-down prices. As in Africa, various deceptions were practised to conceal the true condition of these slaves, and Surgeon Falconbridge gives the example of a Liverpool master who boasted that he had cheated buyers by ordering his surgeon to plug with oakum the anuses of slaves suffering from dysentery. Despite these tricks, there might still be those who were unsaleable, "refuse slaves", and they were likely to be left to die uncared for on the quayside or beach.[28]

The profits to be made from this trade varied with the vagaries of supply and demand, and with the chances of disease and shipwreck. Typically contrasting experiences were reflected in the views of two neighbouring slave merchants on Chesapeake Bay in the mid-1700s: for one, "The Africa trade is quite dangerous for life and health, though most profitable", but for the other, "There are more disasters in those voyages than any other whatsoever." Records show that by the late eighteenth century the profits of 100 per cent or more frequently made by the interloper slavers of the seventeenth century and early eighteenth century were generally a thing of the past. The cost of slaves in Africa and the selling prices in the Americas were drawing closer together, and by 1780 slaver masters were having to pay African traders the equivalent of £50 for each slave, ten times the price in the late 1600s. A few voyages still produced spectacular financial results, as was shown by the 150-ton Nantes ship *Jeune Aimee* which carried 264 slaves from Angola to Saint-Domingue in 1783. The cost of ship, trade goods, crew and slaves was 156,000 livres, and the cargo was sold for 366,000 livres, a profit of 135 per cent.[29] For the Trade as a whole, however, average profits had declined to about 10 per cent, and a significant proportion of voyages resulted in financial losses. There was no sign, however, of a decline in demand for slaves in the Americas, and the slave traders of Africa were as anxious as ever to sell.

In 1788 the painter George Morland exhibited at the Royal Academy a painting depicting the slave trade, and he called it *The Execrable Traffic*. By then it was at last beginning to dawn on the educated people of Europe and North America just how appropriate this title was.

CHAPTER 3

The Slave Coasts and Seas

Beware and take care of the Bight of Benin,
There's one comes out for forty goes in.
SLAVERS' DOGGEREL

THE AREA over which the Royal Navy conducted its operations against the Atlantic slave trade during the nineteenth century was dictated, as will be seen in Part Two, by factors such as the scope of current international treaties and the resources available for the campaign, but potentially it covered the entire region over which the slavers pursued their trade. The eastern boundary of this immense area was the seaboard of West Africa between the River Senegal at latitude 16° N and Cape Negro at 15° 40′ S. From the rivers, creeks and beaches of that 3,300 miles of shore, generally known as the "slave coast",* the Trade reached across the Atlantic Ocean to the harbours of the two great slave markets of the era: Brazil and Cuba, and, to a lesser extent, to other islands of the West Indies and to the southern states of the United States of America. Ultimately the Royal Navy would hunt the slavers in almost all those parts of the Atlantic and its coasts where they might be found, but British ships were never permitted to extend their work into the waters of the USA. So the western boundary of the Navy's suppression theatre was, in effect, formed by the coasts of Brazil and Cuba. Portuguese slaving for the Brazilian market took place, to some degree, from the coast of Mozambique as well as from West Africa, but the Royal Navy very rarely ventured into the Indian Ocean as part of its Atlantic campaign, and East Africa is therefore excluded from this description of the slave coasts and seas.

The slave trade in the Atlantic, and consequently the Royal Navy's efforts to eradicate it, was crucially affected by the topography and navigational characteristics of the shores of the African slave coast, and of the approaches to the slave harbours of Cuba and Brazil. Similarly they were dictated and constrained by the winds, weather, currents and tides of the coastal waters and

---

* All distances shown in this chapter and hereafter are given in nautical miles. The international nautical mile (the distance subtended by one minute of latitude) is a standard distance of 1,852 metres at all latitudes, about 2,025 yards. One nautical mile equals ten cables.

the open Atlantic Ocean. Not least, they suffered the ravages of the diseases endemic in tropical West Africa. Underlying all of these factors was, in the early years of the suppression campaign, a dearth of accurate navigational and environmental information available to mariners plying their trade on, or patrolling, the slave coast, a part of Africa which had attracted little interest from the surveyors and hydrographers of the Royal Navy before the beginning of the campaign in 1807.

For most of the period of this account, the coastal territories of West Africa were controlled by local kings and chieftains, and, apart from the refuge for liberated slaves established in Sierra Leone by the British in 1787 and the long-standing Portuguese domination of the trading settlements of Angola and Benguela, no attempt had yet been made at European colonisation. There were no formal frontiers, and tribal boundaries were mostly ill-defined. By the early 1800s only very small numbers of white people had settled on the coast, and, with negligible exceptions, they were confined to the trading posts at the mouths of the navigable rivers and in the forts of the Gold Coast. Many of these lodgements had frequently changed hands with the ebb and flow of European wars, and with the treaties which ended them, and, although some posts lay under the influence of one trading nation or another, most had an international flavour by the beginning of the nineteenth century. Difficult terrain, unfriendly tribes and disease had effectively discouraged Europeans from venturing into the interior. The movement of trade was almost entirely confined to the rivers.

The general configuration of the slave coast was well known at the start of the anti-slaving campaign, and existing charts were accurate enough for general navigation, but the inshore surveys necessary for pilotage charts had yet to be made. Vessels trading on the coast had long relied on the local knowledge of native pilots, but the Royal Navy's new operational requirements, and the abolitionists' efforts to encourage legitimate trade in West Africa as an alternative to slaving, gave a particular focus to the extraordinary energy, fortitude and skill of the British hydrographers of the nineteenth century. Two famous figures, Captain W. F. W. Owen in HMS *Leven* and Lieutenant A. J. E. Vidal in HMS *Barracouta*, surveyed the thousand miles of shore between Cape Mount and the Bights of Benin and Biafra in 1826, and Vidal returned to the coast in HMS *Aetna* in 1838. Lesser-known hydrographers, such as Commander Belcher in 1831, also in *Aetna*, contributed to the work, and the commanding officers of patrolling cruisers made more limited surveys and observations whenever their operational duties allowed. Consequently the Navy's ships on the slave coast had excellent navigational material by the middle of the nineteenth century, but that benefit had come at a price. Vidal, for example, lost 17 men during his 1838 survey, and the 1849 *Sailing Directions* record that Cape Skyring was so named to mark the place

> where the excellent and deservedly respected Commander of that name lost his life on 22 December 1833 while he was engaged in the survey of this barbarous coast […] this promising and amiable officer fell a sacrifice to his over-forbearance towards the savage tribe of Moors who surrounded him. The adjacent bight in the shore, where the boats were lying which rescued his mangled body, was of course called Murder Bay.

* * *

The territory at the northern extremity of the African slave coast was known as Senegambia after the two great rivers of the region, the Senegal and the Gambia. French trading influence was strong in the valley of the Senegal, but the river was difficult to navigate under sail because of the strength of its current. The town and fort of Saint Louis at the mouth of the river had been retained by Britain after the Seven Years' War, but in 1779 it was regained by the French.

Between 360 miles and 520 miles off Saint Louis lies the bleak volcanic archipelago of the Cape Verde Islands, colonised and cultivated by the Portuguese in the fifteenth century and a key component in their early slaving empire. Slaving through these islands persisted into the nineteenth century with slaves being shipped from the mainland for onward passage to Brazil or Cuba.

A hundred miles to the south of the River Senegal, and standing prominently between the mouths of the two principal rivers of the region, is Cape Verde, the westernmost point of the African mainland. Close to the southern shore of this headland is the small island of Goree which had long been a major slaving centre, fortified in the European fashion and using European currency for trade, but its heyday was over by the beginning of the nineteenth century. It had frequently changed hands between the major slaving nations, but Britain had recovered it from France by the beginning of the suppression campaign.

The coastline to the south of the high headland of Cape Verde is mostly of low cliffs broken by small bays with beaches of white sand. Behind the shore, the land was forested. These characteristics continue for the 100 miles to the mouth of the River Gambia, a major waterway, tidal for 150 miles and navigable to seagoing sailing vessels for 235 miles. The English had founded slave factories upstream at Cattajar, which was over 130 miles from the sea, and at Joar, and the French retained a foothold on the estuary at Albreda. Fort James, built by the British on St Andrew's Island immediately opposite Albreda, was captured in 1779 by the French, who failed to exploit it and the British resumed control. The lower reaches of the river were otherwise largely controlled by Afro-Portuguese settlers.

From the Gambia southwards, the low and forbidding shoreline is interrupted by numerous other river mouths, principally the Casamanza, the Cacheu, the Geba and the Rio Grande, all of which had been used by the Portuguese for slaving since the fifteenth century. The region remained under Portuguese influence, and Bissau, at the mouth of the Geba, provided a valuable source of livestock, rice and fruit for visiting ships. Fresh water too was available there, and the description of its watering-place in the *Sailing Directions* of 1849 is representative of those from which the Navy's cruisers, and the slavers, would replenish with fresh water:

> The watering-place is on the beach a little to the westward of the fort: it consists of pits 3 or 4 feet deep, but their produce is so slow that not more than 30 barrels can be filled in a day; and then it requires to be filtered for drinking.

The centre of the Hispano-Portuguese slave trade in these parts had been on the Cacheu, a river navigable for 100 miles but whose entrance presented hazards, common on the slave coast, of "shallow banks and irregular shoals on many of which the sea breaks furiously." Between these shoals there was a "narrow and tolerably straight but dangerous passage into the river". The *Sailing Directions* recommended

> a look-out at the mast-head for broken or discoloured water, leadsmen in both chains, and being ready to haul out if they should call less than four fathoms […] and, if practicable, to get out a couple of boats ready to send ahead with danger flags, in case of becoming entangled among the shifting sands which occur in the mouth of all rivers of this kind […] Here it will be prudent to anchor in order to verify the latitude, and to wait for the sea-breeze, for a flowing tide, and for sufficiently clear weather.

There was hardly a river of West Africa for which this advice was not equally valid.

Lying 15 miles offshore from these river mouths are the Bisagos Islands and Shoals, a chain 120 miles in length, known to the English at the time as the Shoals of the Rio Grande. The indigenous people of these islands had long raided the mainland for slaves in 70-foot canoes, and the dominant mixed-race inhabitants of the eighteenth century maintained stocks of slaves there. Owing to the disproportionate demand for male slaves in the Americas, a large majority of the islands' people were females, and most of the trading was done by women. Mariners of the early nineteenth century were warned that "no dependence can be placed on anything that is to result from the industry or humanity of the inhabitants of the Bisagos Islands". Hen Island, lying adjacent to the southern approaches to the Cacheu River, was to achieve some notoriety as a slaving post in the 1800s.

Apart from these islands, the sandbanks at river mouths, occasional offshore shoal patches, and reefs extending from headlands, the seabed off this northern part of the Senegambian coast shelves fairly gently to the beach. The rise of tide at spring tides is only six feet or so at Cape Verde and about 12 at the Rio Grande.

South-east of the Rio Grande and the Bisagos Islands the offshore shoals and banks are more numerous, beginning with the volcanic Jamber Islands, a group "replete with dangers". The seabed shelves even more gradually than further north, with soundings of 20 fathoms at up to 50 miles offshore. The coast is generally low and swampy, with sandbanks in the offing and a "frightful reef from the Cappaches River for 16 miles". Two notable slave rivers reach the sea along this stretch of shore: the Nunez and the Pongas. They were known as the "rivers of the south", and it is estimated that up to a tenth of the total West Africa exports of slaves during the eighteenth century were shipped from these two waterways. The Pongas, a river which was to feature regularly during the suppression campaign, flowed to the sea through a network of banks and low islands which altered in form and extent with every gale and rainy season. There were, however, two channels through the breakers navigable by ships of up to 15 feet of draught, although there was "a sad scarcity of marks". Tidal rise here is 15 feet at spring tides and 9 feet at neaps.

As the coast continues to the south, the low shore of mangrove swamps with its detached breaker-line becomes, after 20 miles, a series of islands intersected with channels and creeks which constantly change with south-westerly gales. At this point, extending to 15 miles from the coast, lie the volcanic Isles de Los, a group of rocks, reefs and three inhabited islands. A London merchant owned one of these islands, and traded a variety of commodities from it, including 6,000 slaves per year in the 1780s. It became a favourite port of call of American slavers at the end of the eighteenth century. From the Isles de Los the view of the mainland in clear weather gave a panorama of "fertile shelving hills, stupendous features of distant mountains, plains covered with trees", and "the beautiful little island of Matacong" to the south. In poor visibility all that could be seen was a "low mangrove coast enveloped in mist, indistinct river openings, and columns of smoke from native villages". On the Bereira, one of these minor rivers south of the Isles de Los, the daughter of a Liverpool entrepreneur and an African had established herself as a particularly successful slave trader in the late eighteenth century. Her success was based largely on the stream of refugees from an Islamic *jihad* in the interior which had also fed the slave trade on the Nunez and Pongas rivers to the north.

Senegambia merged imperceptibly hereabouts into Sierra Leone, named after the lion-like mountain on the southern side of the estuary of the Sierra Leone River. The river mouth is about seven miles wide, but is obstructed by shoals which, however, leave a mile-wide passage to the south of a middle ground. This gave access to the embryonic British refuge for freed slaves at Freetown on the southern shore inside the river entrance. There are several islands in the estuary, including the fortified Bance Island, a famous slave-trading post of the seventeenth and eighteenth centuries, which, while in the hands of a London Scottish syndicate, had been furbished with an elegant central building and a golf course. The French reduced it to ruins in 1779, but a British family continued to trade slaves from it until the early 1800s. The estuary of the Sierra Leone provided by far the best harbour on this northern part of the coast.

The shore to the south of the Sierra Leone River is formed of sandy beaches with thick forest advancing to within a few yards of the sea. Thirty miles offshore are the extensive Shoals of St Ann, and closer to the coast lie the small groups of the Banana Isles and the Plantain Isles, both ruled by British slavers in the eighteenth century. The beach is interrupted by a number of minor rivers, but the next major opening is the estuary of the Sherbro River, 15 miles across at the sea. Despite extensive sandspits, there is an easy passage into the river, and the English established themselves on its swampy banks initially to export hardwoods. By 1700, slaves were being traded extensively from the river mouth, and slaving, in the hands of several English, Irish and mixed-race traders, continued there throughout the eighteenth century. The southern bank of the river is formed by Sherbro Island, separated from the mainland by the River Shebar, 30 miles south of the Sherbro.

Twenty miles of straight, low, sandy and shoal-free coast lead to the Gallinas River, a relatively insignificant waterway, but which, by 1780, became a notorious centre of Portuguese slaving. A combination of the actions of the surf and the currents of the Gallinas and its neighbouring river to the south-east, the Solyman (or Moa), had thrown up a narrow barrier of sand, five or six miles in length, between the sea and the shore. This formed a system of lagoons and creeks between the two rivers, and for a further two miles to the north-west of the Gallinas. The creeks and rivers were largely bordered by mangrove swamps, and the surrounding country, all of it flat, was covered with scrub or forest. Behind the sand-barrier lay several small, low islands, on some of which the slavers operated, and a number of native towns stood on the banks of the two rivers. River water reached the sea through two shifting openings in the barrier, and these gave access to boats.

Forty miles to the south-east of the Gallinas is the high and rocky headland of Cape Mount, which marks the north-western limit of the Windward Coast. The thousand-foot hills of the Cape descend to a continuous 50 miles of beach backed by low and forest-covered country. Although this would change later on, at the start of the nineteenth century there was no significant slaving on this stretch of coast, even from the "fine wide" River St Paul which reaches the sea immediately north of Cape Mesurado, nor from the similarly sandy shores between Mesurado and the Junk River, another 40 miles to the south-east. Slaves were available in small numbers, however, on the 200 miles of coast between the Junk River and the major headland of Cape Palmas.

This 200 miles of shore was known as the Grain (or Pepper) Coast, one of several regions named after their principal exports in the early days of West African trade. It has numerous minor river mouths, almost all of them blocked by very shallow sandbars. From the estuary of the Grand Bassa, 60 miles south-east of Cape Mesurado, the coast becomes more rocky with some sandy bays, and reefs extend to five cables offshore, with boat passages to shoreward of them. The next significant opening, the River Sesters (or Cestos) had been an old Portuguese slave market, and, despite a difficult entrance with shifting sandspits, the Trade there was not yet dead. A further hundred miles of low, thickly wooded coast with sandy beaches and straggling rocks passed the tiny port of Sanguin, centre of the residual slave trade in this region, and reached the river and large native town of Grand Sesters. At this point there was a three-mile lagoon inside the beach with a very narrow entrance, safe for boats only in calm weather, but off the town the swell was broken by Factory Island, which allowed boats to beach.

Fifteen miles of clean sandy beach with low, forested country behind it runs along to Cape Palmas, and this was the country of the Kroomen. These fine seamen and intelligent linguists were particularly skilled in handling boats in the surf of the open beaches and in the treacherous currents of river mouths, and they made valuable local additions to the crews of slavers and of anti-slaving cruisers. Their contribution to the suppression campaign was to be immense.

Cape Palmas, where the line of the coast finishes its south-easterly run, marked the beginning of the Ivory (or Teeth) Coast. From here the shoreline heads east-north-east before curving to the east, and this may be seen as the western entrance to the Gulf of Guinea, whose 1,300 miles of shore formed almost half of the slave coast. The Cape itself is a rocky peninsula joined to the mainland by a sandy isthmus. The 1797 chart shows the isolated danger of Coley's Rock some miles offshore, and records that Captain Coley discovered the hazard in 1795, finding ten feet of water on top of the rock and noting that "it tapers down to 7 fathoms all round, as close as you can chuck a biscuit"! Sandy beaches continue to fringe the shore, and, apart from reefs and dangerous ledges in the vicinity of the Cape, there are only occasional patches of rock. The river mouths were all barred by sand, and there were long, narrow and frequently stagnant lagoons in many places between the beach-barrier and the main shore. In many cases the rivers broke through to the sea only in the rainy season, and the height of tide at springs is a mere four feet. The barrier was wide enough in places to support trees and villages. Fifteen miles eastward of Cape Palmas the river and town of Cavally was one of only two areas of the Ivory Coast from which significant numbers of slaves were exported at the end of the 1700s. The other lay a further 80 miles along this beach-bound and surf-lashed coast, on the St Andrews River, the only notable waterway for some distance, which marked the termination of the Ivory and Windward Coasts and the beginning of the 80 miles of Quaqua Coast.

Immediately to the east of St Andrews River red cliffs rise to a height of 150 feet, and, although the sandy beach continues, there are off-lying rocks. For the next 60 miles the beach was clear, but constantly pounded by surf, and the strip of water behind the beach continued for much of this distance as far as Cape Lahou, the town of Grand Lahou and the mouth of the Grand Lahou River. Several lagoons near the town were a source of slaves during the eighteenth century and into the nineteenth century, but in the 1780s the ivory trade here was considered to be more important. The Grand Lahou River, to which the entrance is very narrow and the bar dangerous, formed the western boundary of the Gold Coast.

For a further hundred or so miles the coast changes little in character, and is interrupted by only two significant rivers, the River Costa, which was connected to large lagoons on both sides and where surf on the bar prevented safe landing, and the broader Assini River, where the French had attempted unsuccessfully to establish a trading fort. A little to the west of the Assini the lagoon behind the beach came to an end at Albanee. A line of hills running to the coast then terminates at Cape Apollonia, and from Albanee the land was heavily forested, with villages studding the coast. The first of a long series of European forts, the English Fort Apollonia, stood four miles to the east of the Cape, and from here a high beach, backed by many villages among palm groves, curved towards the south, past the Snake River and the Dutch forts of Axim and Brandenburg Castle, to the prominent headland of Cape Three Points. From Apollonia the shore is clear and sandy, except where Axim Fort stood on a low rocky point, and landing was safe there.

Cape Three Points is formed by a knot of hills on rough ground, and from here the shoreline resumes an east-north-easterly direction and the aspect of the coast changes. For the remaining 200 miles of the Gold Coast the shore becomes a succession of sandy beaches interspersed by rocky headlands girded with rocks and reefs extending several cables offshore. Forty miles to the east of Cape Three Points the only river of any size on this coast, the Pra, emerged between banks fringed with mangroves over a bar with two feet of water. Several minor rivers were pent-up and stagnant behind beaches. Atlantic rollers are either broken by the reefs or expend themselves as surf on the beaches.

Along the Gold Coast there stood in the eighteenth century a string of as many as one hundred European trading posts and forts of varying sizes and importance. Most belonged to the English and the Dutch, with a handful of Danish possessions. The Portuguese had lost interest here, and the French had failed to become established. These settlements were there under sufferance from the dozen or so small monarchies which controlled the coastal territories, and the Europeans either leased, rented, had bought or had been given their small patches of ground by these kingdoms. In exchange for monopolies on local trade, the European companies involved undertook to defend the adjacent African towns from attack, but the Europeans did not generally venture far from the security of their forts or castles. The forts had initially been built for trade primarily in gold and ivory, and it was not until about 1740 that slaves for export to the Americas had overtaken these two commodities in importance on the Gold Coast. Elmina, headquarters of the Dutch West Indies Company in Africa, remained the greatest of the fortresses, trading slaves largely supplied by the King of the Ashanti from his capital, Kumasi, 120 miles inland. Ten miles further east stood the British castle at Cabo Corso (or Cape Coast), claiming the best of a poor choice of landing places on this coast. Among its extensive facilities, Cape Coast had "slaveholds" for up to 1,500 captives. A more recent British fort had been built specifically for slaving a further 20 miles to the east at Anamabo, and the main Danish castle of Christiansborg stood 80 miles beyond that at Accra. African towns had grown up under the walls of the more substantial fortifications, and in some cases areas of forest in the vicinity had been cleared for cultivation. In addition to conducting trade on behalf of their companies, the governors of the forts had, illegally, long traded slaves on their own account, and from the mid eighteenth century the Dutch West India Company had permitted this practice. The Company's officials began to use Afro-Dutch mixed-race inhabitants to find slaves for them, and by the 1790s these *tapoeijer*s had become the most effective slave traders on the Gold Coast.

All of these European posts relied entirely on communication with the sea for trade and for survival, but none had any form of harbour for seagoing vessels. The fortifications were generally built on rocky promontories, and in most instances some shelter for boats was provided by the off-lying reefs. Landing places might be in small boat harbours or on relatively sheltered beaches. The slave vessels themselves had to lie in open anchorages a mile or more offshore. The Porguee Bank, 30 miles from the coast, is shown on the 1797 chart as a "Good Anchorage & excellent Fishing" in 14 fathoms, but such a position would have served only for waiting rather than trading. The seabed off the whole Gold Coast is fairly good holding ground on sand, although the Danes found the anchorage off Christiansborg a difficult one with sharp rocks on the bottom, and they described their landing place as "a most vile harbour". For much of the year shipping off this coast is exposed to the winds of the South-West Monsoon.

Sixty miles east of Accra the Gold Coast ended, just beyond the final European castle of Adda, at the swampy estuary of the River Volta and the headland of Cape St Paul's. This is the western extremity of the Bight of Benin, and here began the Slave Coast, that particular part of the much lengthier slave coast named by earlier traders for its principal commodity. The shoreline of the Bight turns initially to the north-east and then continues in a curve of 380 miles towards south-south-east, culminating at Cape Formoso. Throughout its length the coast is low, flat and monotonous. Nothing can be seen of it from a deck beyond 12 miles offshore. The land to the west was savannah-like and fronted by a continuous sandy beach, bright with surf. Behind the beach was a string of lagoons, of varying width, which enabled almost continuous boat transport for over 200 miles. There was only one permanent opening to the sea from these lagoons, although there were several partial breaks in the beach through which the lagoons would empty during the wet season.

Most of this region was dominated by the great inland kingdom of Dahomey which, through the market at its capital, Abomey, 120 miles to the north, supplied most of the slaves to the coast. The European nations had not been allowed to establish powerful forts here as they had on the Gold Coast, and the slave traffic on the coast was in the hands of a number of nominally independent kingdoms and autonomous towns. There were upwards of 15 slaving stations along this shore, and six of these were of appreciable significance. Sixty miles east of Cape

St Paul's stood the first of them, Little Popo, merely a cluster of huts at the top of the beach. Thirty miles beyond that was Grand Popo, behind the beach ridge, alongside one of the occasional outfalls, and invisible from seaward. Fifteen miles further on was Whydah, prominent in the Trade for many years, and standing on the north shore of the lagoon a couple of miles from the beach. Another 30 miles to the east was the aptly named Porto Novo, a small but successful station built on the beach. Then, after 18 more miles of sand and surf, came Badagry, with a small beach station, and its main town alongside the lagoon half a mile inland of the landing place. Finally, a further 45 miles to the east, there was Lagos. Standing on an island alongside the only permanent waterway through the beach, Lagos was the only Slave Coast town to boast something of a harbour, although the lagoon outfall produced "against the surf a struggle of no ordinary kind upon the bar", and at half-ebb there would be a foul scum and smell three miles offshore. Vessels of five feet draught or less could attempt entry at high water on spring tides or during the dry season, but during the wet season the rollers broke "fearfully" at a mile from the entrance. Nevertheless, this was a slaving port of major significance.

There were modest European trading posts, and sometimes slave factories, in these six places, but the Trade was generally conducted directly between African merchants and the ships. There are no shoals in the Bight, and good anchorages were available off all the towns, although the surf would often interrupt communication with the shore and the hurling action of the swell at the back of the surf could at times seize a vessel three-quarters of a mile off the beach. Except occasionally at Lagos, traffic between slaving vessels and the shore was solely by native canoe, and even in skilled hands this was a hazardous business. Despite the difficulties, this Slave Coast exported perhaps as many as 2 million slaves before the Trade finally ceased, nearly half of them by the Portuguese.

At Odi, about 80 miles beyond Lagos, the long beach of sand ends, a foreshore of mud begins, and the surf breaks at a mile offshore. Here too the lagoon system merged into the network of rivers and creeks which forms the delta of the great River Niger. The shoreline now inclines to the south-south-east, and lightly forested country gave way to a dense mass of trees which grew down to the high-water margin of the sea. The 1849 *Sailing Directions* mention 13 recognisable river mouths in the 130 miles of shore leading to Cape Formoso, but only five were recorded by the early Portuguese traders as "the slave rivers". The largest of these was the Benin River (or the "Great River Formoso"), with a north-easterly tributary leading to the inland kingdom and city of Benin. Over the Bar of Benin at its mouth there was about nine feet of water, with another seven feet at high water springs, but even then the send caused by the swell allowed only vessels of eight-feet draught or less to enter without risking smashed sternposts or broken backs. The anchorage was three miles off the bar. There was no slaving network here to compare with those of the Ashanti or Dahomey, but the French established trading factories on the lower river (although with no great success) and the English, in particular, bought slaves on the Benin and on its neighbour, the Rio dos Forcados, during the eighteenth century. Their purchases were often made from the local tribes whose long war canoes dominated the lower rivers, ensuring a regular supply of captives to all comers.

A warning in the *English Pilot* of 1753–9 concerning the River Lamas, a little to the south of the Rio dos Forcados, gives some indication of the hazards involved in approaching this coast at the time, and of the European sailors' perception of the character of the inhabitants:

> This River has often been taken for the River *Forcades*, and many pilots deceiv'd thereby; running into it, 'till they become shoal, then perceiving their Error, but too late, there the Ship is lost, and the Men endeavouring to save themselves from being swallow'd by the Sea and Mud, are devoured and eaten up by the greedy *Negroes*.

Entry was indeed impracticable at many of these minor rivers, and, although there was good holding ground on mud to seaward of the river mouths, the constant groundswell along this coast made anchorages uncomfortable. Adding to the unpleasantness, vegetable matter and silt are discharged by the rivers at every ebb, discolouring the sea and producing a filthy brown scum and a sickening smell for several miles offshore.

The River Nun emerges a little to the north of the low, wooded and poorly defined Cape Formoso, at which the coastline swings slowly to the east and into the Bight of Biafra. The next 200 miles of shore, the southern edge of the vast delta of the River Niger, is formed by a maze of rivers and creeks threading through densely forested, generally low-lying and often swampy country. The first of a succession of rivers discharging Niger waters into the Bight of Biafra was the Rio Bento (or Brass River), with the slaving town of Brass on its eastern bank. In the next 40 miles there emerge, in succession, the San Nicholas, Santa Barbara, San Bartolomeo and Sombrero rivers, all of them difficult to access, with bars of mud and sand and bordered by dense mangrove. The seashore too is covered with mangroves and fringed by a sandy beach, and at the approach to two of these river mouths there is shoal water with breakers out to five miles from the coast.

A few miles further to the east there is a much broader opening of seven miles between Touché Point and Rough Corner forming the estuary of the New Calabar and Bonny rivers, renowned Bights slaving harbours. Entry was generally made with the prevailing wind through a western channel which crossed the broad shoal extending south-west from Touché Point, and over two bars into the estuary. Departure was usually by an eastern channel without a bar. The estuary was encumbered by sandbanks, and a pilot was indispensable for visiting ships. There was an anchorage in 11 fathoms off Bonny Town which stood on the eastern shore, seven miles from the sea, and the Bonny River branches off the estuary to the north-east. The New Calabar River, with an anchorage at its mouth, enters the estuary from the north-west.

Fifteen miles east of Rough Corner is another sand-choked entrance allowing passage for boats only, and with a dangerous shoal eight miles offshore, but beyond that there is no other significant opening for 50 miles. Then appears, at the eastern edge of the Delta, by far the most extensive of its estuaries, the combined Old Calabar and Cross rivers. The opening, between Tom Shots Point and East Head, is ten miles wide, but the entrance is made typically narrow and difficult by a line of knolls, no more than two and a half fathoms in depth and stretching for over 12 miles south-east from Tom Shots Point. A few miles north of the bar the estuary broadens out, but much of it dries at low water, and it is obstructed by swampy islands. The Great Qua and Little Qua rivers empty into it from the east, and then the Old Calabar River branches off to the north-east while the Cross River continues to the north. There were three trading settlements on the southern bank of the Old Calabar: Henshawe Town, Duke's Town and Old Town, with an anchorage in three fathoms off Duke's Town, the best part of thirty miles from the sea.

The Old Calabar was almost exclusively an English slaving area during the 1600s, but the French and Dutch were increasingly to be found there during the following century, and exports never fell below 70,000 slaves in a decade. Trade in the region was controlled by a powerful native commercial brotherhood, known as the Egbe, which had developed an engaging form of English as a trading language. By the eighteenth century this association was bringing slaves from fairs a great distance inland, using canoes 80 feet long and carrying 120 people.

From East Point the coast takes a 12-mile curve around the southern side of the Backasey peninsula and recedes into a shallow bay with several broad creeks opening from it. The westernmost of these is the Rio del Rey, four and a half miles wide at its mouth. From the eastern shore of the Rio del Rey estuary the coast starts its long southerly run towards the Cape of Good Hope, beginning with a 70-mile sweep of shore around the Cameroon Mountain, much of it low cliffs perforated with caves. Thereafter it recedes into Ambas Bay, with its three islands and fairly safe landing, before opening into the Little Cameroons River with Bimbia Island at its mouth. A further 14 miles of low, mangrove-covered shore leads into the estuary of the Cameroons River, five miles wide, land-to-land, but with sandbanks which narrow the entrance to a mile and a half, and a bar in the middle of the bay. Several streams flow into the estuary, among them the Cameroons River itself from the east-north-east and the Malimba River from the east-south-east. A dangerous reef, the Dogsheads, projects from the southern headland of the bay. From here to the southward a uniformly low and thickly wooded shore, generally fringed by a sandy beach with occasional detached patches of rock, extends for a further 150 miles to Cape St John. All of it is subjected at times to heavy surf. At the mouths of the minor Borca, Campo and St Benito rivers, sandbars prevented entry to all but small vessels and boats, although anchorages were generally safe here and in the open roadstead at Batanga Bay, 40 miles south of the Cameroons River. Along this coast communication with the shore was mostly by surf boat through the breakers, and the cottonwood canoes of Batanga were celebrated for the skill with which they were handled and the lightness of their construction; a six-foot canoe weighed only 15 lb, and the vessel was "as buoyant as a water-bird". Along this Cameroons coast the merchants of Liverpool pioneered a new branch of the slave trade in the late eighteenth century.

Immediately to the south of Cape St John is Corisco Bay, 30 miles across and with a wooded shore bordered by extensive shoal water. Into it emerges the St John River, and in the middle of the bay lies Corisco Island, long used by the Dutch as a trading station, but not specifically for slaves. A further 30 miles to the south is the estuary of the River Gabon, beginning to fulfil its full potential as a slave region only in the 1780s. Although its entrance is obstructed by extensive shoals, this was regarded by the 1868 *Pilot* as the finest harbour on the west coast of Africa, accessible to the largest ships and offering a "commodious anchorage" which could give "shelter for a fleet". The vegetation on the shore was magnificent, the land was fertile, and John Newton, the slaving captain, regarded its inhabitants as "the most humane and moral people I ever met with in Africa". At 25 miles from the sea the estuary divides into two branches: the River Como to the east and the River Ramboe to the south-east, both navigable for 40 miles. Tides here are strong and irregular, and, although the rise of tide at springs is only seven feet, there is frequently an ebb of four or five knots.

Lying in the Bight of Biafra are three sizeable islands, all of them volcanic in origin, with needle-shaped peaks, luxuriant vegetation and shores of black sand. The northernmost and largest is Fernando Po, 35 miles by 17 miles in extent, and only 19 miles from the Cameroons coast. It had been a Spanish possession since 1778. The 10,200-foot Clarence Peak is visible at a hundred miles to the west in clear weather and was a valuable navigational aid to ships in the Bight. The best yams in Africa were to be found there, and so was very pure water, although that was guarded by alligators. Slavers and others had long used the island to load with these supplies, and with wood for fuel. The much smaller Princes Island (or O Principe), only ten miles by four in size, lies 111 miles south-south-west of Fernando Po and 120 miles from Cape St John on the mainland. This Portuguese possession was similarly useful to passing mariners. A further 80 miles to the south-south-west is the island of St Thomas (or São Tomé), the main slave emporium of the whole region until the early seventeenth century. From there the Portuguese traders had dominated slaving on the mainland coast between the Niger Delta and the Congo. It measures 25 miles by 17 miles and it abounded in game, fish, fruit and vegetables, as well as wood and excellent water. Despite its lack of harbours, its strong currents and its baffling winds and calms, slavers had provisioned there for many generations.

A few miles to the south of both the island of St Thomas and the estuary of the Gabon River, runs the Equator.

The climate along the slave coast to the north of the Equator is primarily governed by two major wind systems: the South-West Monsoon, which is a summer distortion of the trade wind of the southern ocean, and the North-East Trade Wind. Lying between the North-East Trade and the South-West Monsoon is the Equatorial Trough (or "Doldrums"), a low-pressure belt of calms and light variable winds, varying in width from 300 miles to almost nothing. Although the Doldrums move a little north and south with the sun, they remain north of the Equator throughout the year.

The South-West Monsoon lasts from about June until the middle of October along the more northerly and westerly stretches of the coast, bringing cloud and considerable rainfall, but near the Equator the monsoon rains persist for most of the year. The southern limit of the North-East Trade Wind pushes gradually southward along the coasts of Senegambia, Sierra Leone and the Grain Coast during late autumn, passing Cape Verde in November and reaching its most southerly point, at a latitude just north of Cape Palmas, in December and January before retreating north again. Along the easterly run of coast from Cape Palmas to the Bights the North-East Trade Wind replaces the Monsoon only for very brief periods in January. This wind, generally known on the coast as the "Harmattan", brings hot, dry, dusty air from the African interior. The periods during which these two contrasting winds prevail are naturally known as the "wet season" of the Monsoon (or "the Rains"), and the "dry season" of the Harmattan. This means that the northern extremity of the slave coast experiences a dry season normally from November to April, but the season shortens as northerly latitude reduces until it exists for only part of January in the Bights. However, a phenomenon known as the "little dry season" occurs in the low latitudes for a few weeks in August, for reasons which are not entirely clear.

The South-West Monsoon winds are generally light, five to ten knots, although from the Ivory Coast eastward there are sometimes strong westerlies in the afternoons during the wet season, and its direction is usually south-west, except off the Windward Coast, where it tends to be southerly. The Harmattan averages from 10 to 15 knots, blowing from between east and north in the Gulf of Guinea but from anything between east and north-west on the coasts of Senegambia and Sierra Leone. Gales are rare in all seasons, but the whole slave coast north of the Equator is subject to "disturbance lines" consisting of lines of squalls, often known as "tornadoes" despite the absence of any whirlwind effect. These disturbance lines are usually orientated north–south and move west at about 25 knots. They are accompanied by massive banks of cloud and usually by heavy rain and often thunder, but the barometer gives little or no warning of their approach. Gusts of up to 50 knots occur in the squalls, usually from the east, but the wind moderates fairly quickly, although the rain often persists for several hours. These disturbance lines may occur at any time during the wet season or immediately before or after it, but are most frequent at the beginning and end of the season: broadly speaking, in April and May, and October and November. In July and August the Niger Delta also suffers from heavy squalls from the south-west, backing south.

The swell is rarely heavy, or even moderate, in the seas off the slave coast, but the seasonal winds generate a low swell along most of the coastal waters. North of the Equator this swell is most apparent during the South-West Monsoon, producing a heavy surf on the exposed shores from Cape Verde to the Bights.

Torrential rainfall is a major feature of much of the slave coast. Most of the region north of the Equator frequently suffers very heavy downpours during the wet season, at times up to 10 or 12 inches of rain in a 24-hour period, although the south-east-facing Gold Coast tends to be a little less wet than the rest of the coast of the Gulf of Guinea. The whole of the coast of Senegambia, Sierra Leone and the Grain Coast is affected by thunderstorms during the wet season, and much of the region between Cape Verde and Freetown approaches the world's thunder frequency record, with a monthly average of 10–15 storms. The Bights, which experience an annual average of 80–100 such storms, are thundery for most of the year. Violent squalls are liable to occur during thunderstorms.

It is hot throughout the year. Loss of sunshine in the wet season accounts for a relatively small seasonal range, and air temperatures are at their highest in April and their lowest in August. Temperature variation during the day is significant only during the dry season. There is no fog on this coast, but poor visibility is fairly frequent. In the wet season this is caused by heavy rain, and in the dry season the Harmattan often carries a thick dust haze out to sea, reducing visibility to between six and two miles, and sometimes to less than five cables. Humidity is very high in the Monsoon, generally over 90 per cent, although a drop to the mid-70s is noticeable in the early afternoon on much of the coast. In the Bights, however, any drop is barely perceptible.

In the 1753–69 *English Pilot* there is a description of the weather to be found on the Gold Coast, no doubt compiled from the observations of slavers:

> In *January*, it begins to blow hard Sea-streams along this Coast out of the S.W. but it rises to a higher Note and blows much harder in *February*, bringing with it sometimes Rain, and sometimes a Hurricane. In the end of *March*, and Beginning of *April*, great Tempests (called by the *Portuguese Ternados*) arise both at Sea and Land, and withal great Rains, mixt with Thunder, Lightning and Earthquakes, which continue to the End of *May*. And this Weather is foreseen by the clouding of the Sky in the South-East.
>
> During the rainy Season, that is in *May* and *July*, little or no Land-Winds stir; but from Seaward it blows out of the S.W. and W.S.W. causing the Waves to rowl very high.
>
> This rainy Season begins to cease in *August*, but yet the Sea hath a rowling Motion, with tumultuous billows.
>
> The Weather grows fair in *September*, and the Air clear with gentle South Winds; and from that time till *January* it continues very fair, and the hottest Days being in *December*.

Earthquakes aside, this seems to be fairly close to the mark.

The coastal waters of the slave coast north of the Equator experience tidal streams which reverse their direction four times daily, but they are usually weak and are apparent only close inshore. They flow parallel to the coast, except near the entrances to rivers, where the rising tide sets towards river mouths and the falling tide sets away from them. This effect is appreciable at ten miles or more to seaward of the entrances of the larger rivers, amounting to as much as a knot at that distance and three or four knots at the river bars. During the rainy season the volume of river flow causes a more pronounced and longer set on the falling tide than on the rising tide.

Southward of the Equator the West African coast initially runs south and then turns to the west, in a 60-mile sweep of moderately high and level ground fronted by a tree-lined beach and intersected by several creeks, into Cape Lopez Bay. This was the territory of King Bongo, a drunken and unpredictable slaving monarch. The surf is not continuous along this eastern shore of the Bight of Biafra, and landing over the beach was sometimes comparatively easy. Cape Lopez Bay is, in any case, well sheltered from a south-westerly swell by Lopez Island, at whose northern extremity the long promontory of Cape Lopez thrusts out to the north-west. This headland is the southern limit of the Bight of Biafra, and here the line of the coast turns abruptly towards south-south-east. The shores of the bay and cape are low, sandy and covered by mangroves. The island is separated from the mainland by the River Lopez, and 20 miles to the south of the Cape the first of three delta rivers, the Nazareth, the Mexian and the Fernan Val, reach the sea. Behind the shore, these waterways formed an extensive network of creeks and swamps interspersed with forest, and the parent river flowed almost parallel to the coast for nearly 30 miles. These rivers pour into the sea, a vast quantity of fresh water which is not absorbed for four or five miles, and the sea is very heavy off the entrances on the ebb. Sandbanks at the river mouths were constantly shifting, and access was possible only for boats.

The shore continues past the shallow indentation of Camma Bay and the mouth of the River Camma until, at 80 miles south of Cape Lopez, it reaches the slight projection of Cape St Catherine, one of the minor slave-trading places on this coast. Another 40 miles to the south is Sette Point, close to which the River Sette emerges, marking the northern boundary of the kingdom of Loango and on whose bank, 50 miles from the coast, stood the town of Sette. Although there were anchorages along this stretch of shore there was no safe landing for ships' boats except at the two river mouths, and there only on exceptionally smooth days. A low shore with occasional sandy beaches and trees, sometimes to the water's edge, stretches onwards a further 30 miles to Point Pedras, a bank of sand and rock extending two miles to the south-west. Beyond that, 50 miles of continuous narrow beach backed by mangroves runs to Cape Mayumba. There was no safe anchorage on this coast south of Sette Point.

At Cape Mayumba the shore gradually rises until hills fall steeply to the sea at the headland itself. This was the beginning of Mayumba Bay, ten miles across and two miles in depth, the first of the major slaving harbours belonging to the Vili Kingdom of Loango. Close to the Cape a creek offered a safe landing place, and further along the bay the River Yumba meets the sea. Towards the southern end of the bay's fine sandy beach is the narrow and rock-barred mouth of the Matooti River, which was sometimes passable by boat. A short distance upstream the river opened out into a wide basin into which other streams flowed, and boats could be carried 200 yards across the beach to the river. A number of slave factory buildings stood at the top of the beach, and the shore was dotted with settlements with names such as Prince Jack James's Town. This was considered to be the southern limit of Great Rains, and southward of the bay the vegetation was less vigorous and arid ground began to appear. A rock- and coral-studded seabed over the next 40 miles to Banda Point prevented safe anchorage.

There are several river mouths and several off-lying shoals in the 30-mile stretch to Kilongo Point and the Kilongo River on whose bank, four miles upriver and set in the midst of beautiful park-like country, stood the town of Kilongo. The next 25 miles of wooded shore bounded by sand and rock was unsafe for ships, and then even the broad stream of the Kiloo River was practicable only for boats. At that point, however, begins Loango Bay, the premier slaving harbour of the Vili state. The bay is 12 miles across and 3 miles deep, and some shelter from the swell is afforded by the reef of Indian Bar extending from Indian Point, the southern headland. The bay gave a safe anchorage, although it is shallow and the sea frequently breaks heavily within it. Behind a narrow beach at the eastern side there was a lagoon which could be entered by coastal vessels; otherwise the shoreline was wooded. Slave factories stood at the bay's southern shore, but the town of Loango itself lay five miles inland. Despite Portuguese pressure, the Vili monarchs had managed to maintain a free market in slaves for many years, although by the end of the 1700s the King's power was declining in favour of the port officials.

Thirty-five miles south of Indian Point, beyond Black Bay and the Louise Loango River, is the "fine entrance" of the Kacongo River, which had trading stations at its mouth and for 40 miles upstream and whose waters discoloured the sea for seven miles offshore. A little further south was the stream and village of Landano, where the Portuguese would later establish a large trading station, extending their factories a long way upriver, although the beach was a bad one for landing. The coast then leads into the small bay of Malembo, curving past the town of Malembo in a westerly direction to form Malembo Point, beyond which there is a bank of over a mile in length. This was one of two smaller Vili slaving posts, and the bay, difficult to identify from seaward, provided a snug anchorage for small vessels, although deep draught vessels had to lie exposed to the swell. Rollers were frequent and strong hereabouts.

Point Cascaes, five miles south of Malembo Bay, formed the northern boundary of the second slaving harbour of the Vili Kingdom, Cabinda (or Kabenda). Here cliffs and coastal hills gave way to a broad plain studded with palms. The bay is spacious but obstructed by shoals, and, although there was a fine sandy beach and landing place, the anchorage was well offshore and exposed. Close to the town of Cabinda, the bay ends with its south-westerly headland of Cabinda Point, where verdant hills reach the sea, again in tall cliffs. Thirty-five miles of dangerous coast, mostly low and fringed with woodland, lead south-east to Red Point and the estuary of the mighty River Congo.

The mouth of the river, from Red Point to Padron Point, is 25 miles wide. A narrow beach, fronted at first by the Mona Mazeo Bank, then runs for 26 miles along the north shore to Boolambemba Point. By that stage the river's width has narrowed to about three miles, and the scouring effect of the stream was so strong that no bottom could be found with a 90-fathom lead and line. The river would rise nine feet or so above its ordinary level about six weeks after the start of the rains in early November, and currents of up to eight knots were experienced in the estuary. These carried down floating islands of bamboo and debris, which sometimes endangered ships at anchor. The country bordering the estuary is low and marshy with banks covered by mangroves, and over the next 15 miles or so of the north shore a number of side rivers and creeks meet the main stream. These openings, among them the aptly named Mosquito Creek and Pirates Creek, made useful hiding places for slaving vessels. Entry to the river under sail was always from the south in order to avoid the main outgoing current which set to the north-west, partly over shoal ground. A southerly branch of the current headed south-west from the river mouth, but then curled back to the south-east towards the shore. It was difficult for a sailing vessel to cross the stream from the north, and to enter the river it was essential to have the aid of a sea breeze between mid-morning and late evening.

The Kingdom of Congo had become weak and increasingly dependent on Portugal, and there were large mixed-race populations in the towns. Slaving had declined here during the 1700s, but by the end of the century the Zombo people had become active traders on the estuary and the Sonyo people, independent of the kingdom, were trading extensively in the country to the south of the river. About 25 miles upstream of Boolambemba Point lay the famous slave port of Puerta da Lenha, the limit of sea navigation. Above it the river divides into three branches, and shallow-draught vessels could continue to Embomma, the principal town of the river, set in picturesque and hilly country, and, by the mid nineteenth century, one of the world's great slave marts.

From Padron Point, 120 miles of unusually straight, moderately high and generally safe coast runs south-south-east to Ambriz Bay. This was the northerly boundary of Portuguese territory, and Angola's northernmost harbour, soon to be developed as a new centre for trading slaves to Brazil. From the low and sandy Loge Point it is five miles to the high white cliff of Ambriz Head, on whose summit the slave factories would be built. The Loge River, barred to all but boats, flowed through the marshy ground surrounding the bay, and at night the swamps were covered by dense mist.

The capital of Angola, and its main slave-exporting harbour, lay 60 miles south of Ambriz. St Paul de Loando (or Luanda) had been the largest European town in Africa since its foundation in 1575, although with only around four hundred Portuguese inhabitants. It had supplied up to ten thousand slaves a year during the 1700s, mostly to Brazil, and, as in all of the slaving harbours further to the south, both town and slave trade were managed by a powerful class of white or mixed-race merchants. The harbour, with an entrance one and a half miles broad, lay between the mainland to the south-east and the narrow, low-lying Loando Island, six miles in length, to the north-west. Loando Reef extends a further two miles from the northern point of the island. The entrance was commanded from the mainland shore by a fort with two tiers of guns, and there was further strong fortification of the town itself, four and a half miles to the south-west. The anchorage, an unusually well-protected one for this coast, extended three miles south-west from the harbour entrance. From the southern end of the harbour was a shallow exit to the Carimba lagoon, lying parallel to the shore and separated from the sea by a low sandy spit, but shifting sand was soon to make this exit impassable. The lagoon, with a two-mile defended opening to the sea, extended most of the 25 miles to Palmarinhas Point, where the coast turns sharply again to the south-south-east.

After a short distance the shore recedes into Coanza Bay and the mouth of the Coanza River, which marked the southern border of Angola and the beginning of the Benguela coast. Although the river mouth was two miles wide it was blocked by a very dangerous bar, passable to boats only at high water, and the anchorage was nine miles offshore. The Coanza and the other rivers and harbours of this coast provided the outlets of a huge network of waterways and paths which penetrated far into the interior, bringing to the coast slaves purchased by Luso-African traders from the monarchies of the forests.

The next stretch of coast, described on the 1794 chart as rich in slaves and ivory, is mostly moderately high with a succession of bluff headlands generally ending in high cliffs. There was a beautiful little harbour, sheltered by a spit, at Cape St Bras, and slave factories were to be found here, as well as adjacent to a good anchorage to the north of Morro Point and at the village of Kongo just south of the Moroa River. Landing was generally very difficult along the whole of this coast, but there were few offshore dangers. At 130 miles south of Coanza was Novo Redonda, standing on a fairly high point of land. It was then a town third in size and importance among the Portuguese possessions on this coast, although hampered by a hazardous landing. A little to the south were the bay and town of Quicombo, where there were more Portuguese slave factories and, unless there were rollers, a safe anchorage.

High ground continues to the south, and the shore gradually curves to the south-west. The cliffs are broken by the mouth of the Logito River, whose watering place, although difficult, was regarded as offering the best fresh mountain water on the coast. There were slave factories here too, as there were close to the south at Oyster Bay, on whose sandy beach boats could generally land through the surf. The only secure harbour was at Lobito Bay, where a pocket of water eight cables wide runs back to the west-south-west for two miles behind the shore. It appears that, until the mid-1800s, this refuge was known only to slavers. Twenty miles south of that, and 100 miles from Quicombo, is Benguela Bay, seven miles broad and two and a half miles deep, on whose shore of sandy beach backed by marshy plain stood the town of St Philip de Benguela. At the end of the eighteenth century this was about to usurp Loando's position as Portugal's premier slaving harbour, supplying a quarter of the Brazil trade. Benguela, like Loando, was short of drinking water, and slaves embarked here were often dehydrated even before they were loaded into the slavers. On the western side of this shallow bay the shore rises to the headland of Punta de Chapeo, upon which stands a curiously shaped hill called St Philip's Bonnet, recognisable from 25 miles to seaward.

Immediately beyond this headland there are several snug coves, and 30 miles further south at Luash Bay a natural mole provided another sheltered harbour for small vessels, with slave factories behind it. There is high ground backing this coast, but the beach of white sand projects a long way from it and was difficult to detect at night. Another 20 miles of shore led past Caminha Bay, with its moderate shelter and safe landing, Portuguese settlement and plentiful supplies, to the anchorage in Elephant Bay, perfectly secure and beyond the influence of rollers. There was another Portuguese slave factory here, but in years to come the Navy would regard this as a fine place to refresh a ship's company after an anti-slaving cruise, despite the absence of fresh water.

From Cape St Mary at the southern limit of Elephant Bay, 45 miles of high coast continues to Cape St Martha, where the lofty cliffs end, and then to Tiger Bay, sheltered from the south-west but difficult to detect. The next harbour, after another 50 miles of low cliffs and sandy indentations, is Little Fish Bay, whose low-lying shore falls back six miles from an opening of similar breadth. Fort San Fernando defended this completely sheltered anchorage and its town, the most southerly Portuguese settlement on the coast. Hills and high cliffs, receding to a large open bay, run for a final 30 miles to Cape Negro. This high and rugged mass resembles a black face, hence its name, and it marks a change in landscape and climate. Captain Owen, passing it from the south, noted that the desert partially ended here with the first tree that he had seen for hundreds of miles, and that "heavy seas and cold boisterous weather changed to smooth waters and mild, pleasant temperature". The Cape, carrying on its summit a marble cross erected by Bartolomeu Dias in 1486, also marked the southern limit of Portuguese domination and, at long last, the end of the slave coast.

Lying in the South Atlantic are two islands which had parts to play in the slave trade or its suppression. The first is Annobón, the fourth and, at four and a half by one and a half miles, the smallest in the string of islands emerging from the Bight of Biafra. Its nearest neighbour, St Thomas, is 110 miles away, and the mainland 200 miles. Spain bought the island, together with Fernando Po, from the Portuguese in 1778, intending to use it for slave trading, but never settled it. It produced plentiful supplies of fruit for the rare visitors to its poor anchorage. The second island, Ascension, became a British possession in 1815 and was to be of great value to the suppression campaign. A rugged volcanic peak, measuring seven and a half by five and a half miles, it lies 1,600 miles or so to the west of Loando and about 800 miles south-south-west of Cape Palmas. The 2,800-foot summit of its Green Mountain could be seen from a frigate's deck at 65 miles. Condensation on this mountain provided a good quantity of water, and the hundred acres of the mountain's upper slopes under cultivation in 1800 were an oasis in a desolation of decomposing lava. The island's shores are mostly steep-to, and the usual anchorage for visiting ships was at Clarence Bay, on the north-west side, where they were generally safe from the rollers which, at times, broke with fearful violence on the western, leeward, side of the island. These rollers arrived without warning, attaining "a terrific and awful grandeur".

The climate of the slave coast south of the Equator is dominated throughout the year by the South-East Trade Wind, as are conditions on the ocean to the west of it. Over the open sea the "South-East Trade" blows steadily over all of this vast area, apart from a narrow belt immediately to the south of the Line, varying in direction between east-south-east and south-south-east. To the north its dominance is bounded by the Doldrums, and, except during the southern summer, its north-easterly component is drawn across the Equator to become the South-West Monsoon. Its strength over open water is light or moderate in the northern part of its area, but increases at times to strong over the more southerly part. Close to the African shore its direction is less steady, and, although it tends to blow parallel to the coast, it is regularly diverted by land and sea breezes. In these coastal waters the forenoons are relatively calm, but in the afternoons the sea breeze reinforces the prevailing wind and, with the addition in places of a coastal deflection effect, the strength occasionally reaches 30 knots. Gales are rare, although the seas north of the River Congo are subject to "tornadoes", similar to those encountered north of the Equator, which bring squalls of up to 50 knots and rough conditions for an hour or more. Sometimes these squalls lack the usual rain, and are naturally known as "dry tornadoes", and there is the additional hazard to a sailing vessel of the "white squall", which suddenly whips the surface of the sea into spray.

    The annual rainfall in the Doldrums and the Bights is always heavy, up to an annual average of 160 inches at the Cameroons River, but by the southern limit of the slave coast the average falls to one and a half inches. Humidity similarly reduces from the debilitating level of the Bights. The area from the Cameroons to the River Congo experiences frequent thunderstorms, many of them violent. They usually develop inland during the afternoon and reach the coast in the evening, some of them persisting over the sea until dawn. Thunder becomes less frequent with increasing southerly latitude. Rain is torrential in these tropical thunderstorms and in most squalls, and the rain and dense cloud is liable drastically to reduce visibility.

    Very poor visibility is also caused in the coastal waters of Angola by a phenomenon known as "Cacimbo". This is fog formed by a cooling of the overlying air by the cold water brought north by the offshore current, and it is intensified by the upwelling of deep and even colder water associated with the current. It is generally at its worst between dusk and dawn.

Tides along the southern hemisphere's part of the slave coast are diurnal, and their range rarely exceeds six feet at spring tides. Consequently the tidal streams along the coast are very slight. The only exceptions are in some of the river entrances, where the range may be nine feet and a tidal stream of two or three knots may occur. In the rainy season the outflow of river water sometimes prevents any ingoing stream, and the ebb can then run for eight or ten hours at a rate of four or five knots.

Virtually the entire south-western coast of Africa experiences rollers of greater or lesser power, depending on the season and on the aspect of each particular stretch of coast. Shores more exposed to the north are most affected during the northern winter, and those facing more to the south receive a heavier pounding in the southern winter. The biggest rollers usually occur in December, January, July and August, and from July to September a very heavy swell sometimes sets in along the whole extent of the southern slave coast. When the rollers are running there is a tremendous surf on the beaches, and landing, even by native surf boat, is highly dangerous. This big swell is liable to get up without warning, usually in nothing more than a light breeze, and anchoring on the open coast was particularly unwise in a sailing vessel, especially in depths of less than ten fathoms, where the swell becomes very threatening.

* * *

Few shores in the world were more hostile to ships than the West African slave coast. It was entirely bereft of ports, had few sheltered harbours, lacked even a modicum of shelter over considerable distances, was exposed over almost its entire length to ocean swell, and was liable to vicious squalls with little warning. The climate was scarcely more welcoming; all but the most southerly part of the coast was subjected to extreme heat and torrential downpours, with respite from the rain only in the Congo, Angola and, for a couple of months in the year, the Gold Coast. The most comfortable berth that could be expected was either in the stinking waters of a sweltering creek in the Niger Delta or in an offshore anchorage with the ship pitching, rolling and tugging at her cable in the unremitting Atlantic swell. This was, however, not the end of the torments awaiting the mariner in tropical West Africa. The slave coast was also a hotbed of disease.

Most European sailors, even in well-managed vessels, were accustomed to ill health; typhus, scurvy, venereal diseases, smallpox, tuberculosis and cholera were familiar visitors to ships, the last two increasingly so as the nineteenth century progressed. West Africa, however, had a specialised, and frightening, repertoire of diseases. It was known as the White Man's Grave, and with good reason. Yaws, leprosy, elephantiasis and the guinea worm were endemic on the slave coast, and slavers imported them to the Americas in their cargoes. For European slaver crews and the men of the Royal Navy cruisers, however, it was the fevers of the coast which presented the most fearful danger. Blackwater, dengue and yellow fevers and malaria all raged around the stagnant waters with which the shores and the Niger Delta abounded, but it was the last two of these which caused the greatest havoc among mariners.

Yellow fever is an acute and frequently fatal viral infection, unlike malaria, which is a parasitic infection of rather lower lethality. Both are transmitted by the bites of female mosquitoes, the malarial variety confining her attentions to the hours of darkness, while the yellow-fever variety prefers to feed by day, but none of this was recognized until the very end of the nineteenth century. It was perceived that black people had a degree of immunity to yellow fever, unlike malaria, but it is now known that this was not an inborn characteristic, but simply a protection acquired in consequence of surviving previous exposure.

The widely accepted theory of causation, known as the miasmatic (or climatorial) theory, was described by Dr T. Winterbottom in 1807: "Though the most common cause of fever in hot climates be the air which blows over marshes, yet when the fever is once introduced among a number of people, it is very apt to become infectious."

It was realised that stagnant water had a part to play in the transmission of fever, but the general belief was that the danger lay in the mists and fogs which arose from the water rather than from the insects which bred on it. Sailors were convinced that the dense mists (or "smokes") carried to seaward from the mangrove swamps of West Africa by the offshore breezes at dawn were poisonous exhalations from the land. Nevertheless it was gradually, and correctly, appreciated that daylight brought a reduction in risk, although Dr John Wilson was mistaken when he attempted to explain the reason in 1846: "Miasm, condensed and concentrated, through the absence of light and heat, rises emanating from the debris and decomposition around, but, like the morning dew, through the influence of the sun […] soon passes away."

Surgeons floundered in the dark, frequently applying therapies which, as Captain Owen complained, were not only irrational but often fatal. Freetown could be described as "a pestiferous charnel house", and a British captain slaving from the Bonny River could expect to lose at least ten of his crew to fever in every voyage. On the Gold Coast, "There is scarcely an instance of a European arriving here who is not attacked with the endemic fever, either immediately on his arrival or within four months."

* * *

Apart from the offshore effects of the great rivers, the movement of water more than a few miles from the coast is controlled by the ocean currents: currents which, in the Atlantic, exerted a marked influence on the pattern of the Middle Passage of the slave trade. Forming part of the great clockwise gyration of the surface water of the North Atlantic, the Canary Current flows at about half a knot south-westward off the north-western coast of Africa before curving west into mid-ocean and merging into the North Equatorial Current. This south-westerly movement was of particular use to slavers bound from Europe to the Gulf of Guinea during the dry season, when not only is the effect of the main current felt further to the south but also a branch turns south-east to follow the Windward Coast, albeit unreliably, as far as Cape Palmas. Having exhausted the help of the Canary Current, a slaver bound for the Gulf could seek the aid of the Equatorial Countercurrent to carry her eastward. This current is at its most apparent in the summer when, as the west-going North Equatorial Current moves to the north, it can be detected in mid-Atlantic. In the winter (or dry season) it begins much closer to the African coast, and its summer rate of three-quarters of a knot reduces to half a knot.

As it reaches the coast in the region of Cape Palmas, the Equatorial Countercurrent, merging in the dry season with the south-easterly set along the Windward Coast, becomes the Guinea Current. This sweeps steadily eastward along the Guinea Coast at up to two knots in the wet season, rising occasionally to three knots. The powerful Guinea Current is pressed against the coast by the west-going South Equatorial Current on its southern flank, and off the Ivory Coast it is constrained to a width of only 100 or 120 miles, which accounts for the speed of its set. Once past Cape Three Points, however, it widens and slows until it reaches the Bights at only half a knot. Along the Ivory and Gold Coasts it is generally half a knot slower in the dry season thanks to a southward retreat of the South Equatorial Current. The Guinea Current was, of course, a blessing to mariners heading east for the Gold Coast or the Bights, but it was cursed by the men of naval anti-slaving cruisers and their prizes beating westward into the south-westerly winds of the Monsoon towards their base at Freetown. Loaded slavers, on the other hand, outward bound for the Americas, were more likely to drop southward across the Guinea Current and into the westward set of the

South Equatorial Current to help them at a rate of up to one and a half knots into mid-Atlantic. This assistance they would be likely to find just north of the Equator in the wet season, but in the dry season, once clear of the Guinea Current, they would have to overcome a weak northerly set, particularly in the Bights, before finding the favourable westward current a few degrees south of the Line.

Currents off the slave coast to the south of the Equator are less complex than those in the northern hemisphere, and form part of the anti-clockwise circulation of surface water in the South Atlantic. A northerly drift along the entire West Africa coast south of the Line, caused by the prevailing wind, is given a westerly component by the earth's rotation, and the set which results is called the Benguela Current. As the surface water of the current moves away from the shore it is replaced by colder water brought up from below in a phenomenon known as "upwelling". Although the Benguela Current is very extensive, its speed is only between a quarter and half a knot, and it is not consistent, especially near the coast. For brief periods the flow may be in directions other than north-west, even south-east. Only a small remnant of the current remains north of the Equator to flow, close to the coast, into the Bight of Biafra.

The variability of the Benguela Current increases near the coast north of Cape Negro, and eddies are liable to form, but river outflows enhance the north-westerly flow of the current. This enhancement is particularly evident in the case of the River Congo, and the current reaches a rate of two knots here at times. The outflow of this great river also turns the sea a yellowish olive-green colour, and both the discoloration and the more powerful set may be perceptible at up to 300 miles offshore. Many slavers were to find the strength and reliability of this current an invaluable help in making an escape from the Congo under the surveillance of British cruisers.

Over a band of sea from 150 miles north of the Equator to about 1,500 miles south of it, the Benguela Current turns to the west to form a broad set heading into mid-Atlantic. Although there is no clear distinction between them, the more northerly and stronger part is known as the South Equatorial Current, and the more southerly and weaker part as the South Sub-Tropical Current. The South Equatorial Current is perhaps 700 miles wide and consistently averages between half and three-quarters of a knot. Its more feeble and less constant neighbour, extending over 900 miles or so, has a speed mostly of a quarter of a knot. Combined with the South-East Trade Wind, these currents gave a relatively easy passage across the Atlantic to slavers outward bound from the Gulf of Guinea, the Congo and Angola. On approaching the coast of South America, vessels heading for the Caribbean would then benefit from the Guiana Current, a component of the South Equatorial Current sweeping at up to two knots along the northern coast of South America and merging with the North Equatorial Current. Those bound for the slave ports of Brazil would turn south-west with the Brazil Current, a residue of the South Equatorial Current and South Sub-Tropical Current which follows the eastern coast of the continent at half a knot as far south as the River Plate.

* * *

As they rode the South-East Trade Wind and the South Equatorial Current on passage to the slave ports of Brazil, ships approaching the eastern shores of South America were obliged to do so with abnormal caution. The coast, as the 1860 *South America Pilot* puts it, "presents a peculiarity which merits particular notice". From Cape St Roque, the north-eastern extremity of the continent, a barrier reef extends along the coast to the southward for over 350 miles. This reef of coral rock is about five miles wide at the surface and lies from three cables to five miles offshore. It is sheer on its landward face, slopes to seaward and is generally submerged, although it dries to three feet or so in places. Behind this natural breakwater are smooth water and channels navigable to coasters, and occasional gaps in the reef allowed access for ocean-going vessels to the main harbours, including the slaving port of Pernambuco (or Recife) and its near neighbour Olinda, about 180 miles south of Cape St Roque. There are no dangers more than ten miles offshore except over the most southerly section of the reef, where shoals with less than ten fathoms over them extend up to 15 miles offshore. This was a hazardous lee shore for sailing vessels, especially as there were few salient features to enable them to identify their landfalls and the Brazil Current often sets onshore.

The southern hundred miles of this reef-bound shore follow a south-westerly line. Thereafter, 200 miles of coast without significant offshore dangers and consisting mostly of barren, sandy beaches, interrupted by a number of river mouths, follow the same direction. This stretch of shore terminates at Point St Antonio, at the eastern extremity of the deep Bay of All Saints. Protected by this headland and lying on the eastern side of the bay was the major slaving harbour of Bahia (or San Salvador). From the western side of the bay the shoreline resumes a southerly direction. For 150 miles the coast is low and sandy with a number of rivers which form channels between wooded islands, giving way to cliffs and then to sandy beach. Over this stretch there are no significant offshore dangers, but for the next 300 miles or more the coast is fringed by shoal water in which there are numerous hazards, including the Abrolhos Islands, which extend over 120 miles to seaward. For the next 220 miles the line of the shore inclines more towards the south-west, and the coast is still generally low-lying, but with high ground coming down to the sea in places. As far as Cape St Tome banks with depths of less than 30 fathoms extend to 45 miles offshore, but over the next 90 miles to Cape Frio the seabed shelves a little more rapidly.

At the rugged Cape Frio the coastline turns to the west and once more becomes low-lying. Most of the next 60-mile stretch is formed by a barren sandy beach backed by lagoons, but then mountains converge on the coast and cliffs of almost sheer granite form Ponta Itacoatiara at the eastern approach to Guanabara Bay. From an entrance less than a mile wide the bay opens out to a depth of 18 miles or so, and on its western side lies Brazil's then capital city and principal slaving port, Rio de Janeiro. To the west of Guanabara Bay the 130 miles of shore to the island of St Sebastian is formed by alternating cliffs and beaches, and is heavily indented with bays. The whole coast to the west of Cape Frio is generally steep-to, with no dangers lying more than a mile off it, although there are several small offshore islands. Along this south-facing coast the sea is usually heavy, and during south-easterly winds there is a set towards the land.

Although there was a degree of secondary slave traffic from Rio de Janeiro to the Rio de la Plata, the most southerly destination of the transatlantic trade was Rio de Janeiro, and neither West Africa slavers nor anti-slaving cruisers were intimately concerned with the shores of Brazil to the west and south of St Sebastian Island.

The whole of this region is under the general influence of the South-East Trade Wind. In the more northerly parts the wind over the open sea is predominantly south-easterly throughout the year, but further south the direction is more variable between south-east and north. Mean wind-strength is about 15 knots, decreasing a little between February and April, but rising to about 18 knots between June and November. The likelihood of gales is low, although coastal squalls are liable to occur in the more northerly areas in the autumn, and further to the south, usually from the south or south-west on cold fronts, most frequently between April and September. Between the Abrolhos Islands and Cape Frio squalls from the east-south-east, known as "Abrolhos", are fairly frequent between May and August. Coastal areas north of Rio de Janeiro also experience land and sea breezes. The land breeze is generally a light offshore wind which blows from midnight until shortly after dawn, but the sea breeze builds up from mid-morning to a strength of about 15 knots towards the shore by mid-afternoon, before dying at sunset.

Rainfall is heavy along the entire coast. In the northern half of the region the annual average lies between 48 and 71 inches, with the weather at its wettest between March and July and its driest between September and January. Further south there is an even wider variation of the annual average, between 39 and 79 inches. Here the distinction between wet and dry periods is less marked, although rainfall is at its greatest in autumn and spring, and the driest month is July. In the north there are thunderstorms on about five days a year, but further south there is generally little thunder as far as Cape Frio; to the south of that thunderstorm activity increases again rapidly. The mean air temperature in the vicinity of Pernambuco is at its maximum of 27 °C in February, and at its minimum of 25 °C in August, although temperatures are more variable in coastal waters than they are over the open sea. Further south the temperatures, and humidity, become progressively more comfortable.

* * *

Slaving traffic bound for the major northern hemisphere market of Cuba would cross the Atlantic on either of two ocean routes. Vessels departing from Senegambia, Sierra Leone or, depending on the time of year, the Grain Coast, would pick up the North-East Trade Wind and the North Equatorial Current to carry them with relative ease towards the "Caribbee Islands" at anything between 13°N and 20°N, avoiding straying north into the variable winds of the "Horse Latitudes". They would then have the choice of either passing north of Hispaniola or entering the Caribbean Sea through the chain of the Windward and Leeward Islands, generally using either the Guadeloupe, Martinique or St Vincent Passage. Slavers from the Guinea Coast and the Bights, bound for either the Caribbean or Brazil, would be likely to head south into the South Equatorial Current and the South-East Trade Wind, where they would join westward-bound traffic from the Congo and Angola. Ships taking this southern route for Cuba would generally hope to make a landfall on the high island and peak of Fernando de Noronha, 190 miles off the north-eastern point of Brazil, before heading north-west at a safe distance off the north-east coast of Brazil but close enough to gain the help of the Guiana Current. Passing through the Doldrums, at their narrowest off the Brazil coast, they would then run into the North-East Trade Wind to carry them into the Caribbean, probably either between Tobago and Grenada or through the St Vincent Passage.

Cuba is an island a little over 620 miles in length and about 90 miles wide at its broadest, easterly, part, lying on an axis of west-north-west in a shallow arc with its convex side to the north. Its westerly part is narrow, and on the longitude of Havana, about 160 miles from the western tip, the island is only about 27 miles wide. From Punta Maisí, the most easterly point of land, the southern coast runs generally due west for a little over 200 miles and is steep and rocky with occasional sandy beaches. Along this stretch of shore there are several rivers navigable by boats in the rainy season, and a number of small coves as well as the large havens of Guantanamo Bay and Santiago de Cuba (or St Jago de Cuba). These two are typical of Cuban harbours: pouch-shaped, extensive and well-sheltered bays. Santiago de Cuba, about 100 miles from Punta Maisí, was the second city and port of the island, serving the eastern end, and active in the slave trade during the 1800s.

At the end of this westerly run of steep-to coast lies the headland of Cabo Cruz, from which the shore turns sharply north-east into the deep Golfo de Guacanayabo, followed by the Golfo de Ana Maria. This is an irregular shoreline of 250 miles, predominantly low-lying and fronted by sand, mud and mangrove swamps. At a distance of between 20 and 40 miles offshore there is a chain of islands, islets, cays (or keys) and reefs called Los Jardines de la Reina, and the area between that and the mainland is a mass of detached shoals and reefs, through which there are passages and channels. Deep water again reaches the coast at Punta Maria Aguilar, close to the port of Trinidad and its two harbours of Porto Masio and Porto Casilda from which merchants despatched a number of slavers to Africa. Initially the coast here is high and steep-to, but further to the west it becomes low and flat as it borders Bahia Cienfuegos and Bahia Cochinos, in some places rocky and in others swampy, but much obstructed with inshore reefs and shoals and offering safe anchorage only to small vessels. A broad tongue of deep water, the Golfo de Cazones, reaches to the coast here, ending a little to the west of Bahia Cochinos. On its seaward edge there begins another wide expanse of shoal water, much of it studded with rocks and patches of sand just awash, which stretches for 180 miles to the west, covering the Golfo de Batabanó and the deep bay of Ensenada de la Broa. This bank is marked on its southern edge by one large island, the Isle of Pines (or Isla de Pinos), and a number of cays, and it is fringed to east, south and south-west by coral reefs, through which there are several narrow openings. The shore of the Golfo de Batabano is low, swampy and skirted by mangrove cays, and from its western extremity a further 25 miles of low, flat and forested ground, bordered by steep bluffs and sand-cliffs, leads past Cabo Corrientes to the westernmost tip of the island, Cabo San Antonio.

The first 200 miles of the north coast of Cuba, running roughly west-north-west from Punta Maisí, is formed mainly of small cliffs fronted by sandy beaches and interrupted by small but deep bays, sheltered from wind and sea and making safe natural harbours. For 130 miles from Punta Maternillos and the harbour of Nuevitas the coast is fringed by the cays of the Archipiélago de Camagüey, a few of them up to 12 miles in length but all low and covered with mangroves. Narrow, shallow channels lead into expanses of shallow water between the islands and the shore, with numerous small cays and extensive mudflats, and these offer a few anchorages for small vessels. This coast forms the southern shore of the Old Bahama Channel, 13 miles in width and the main north-westerly route for Havana, although rarely attempted by sailing vessels from west to east, against the prevailing wind and current. The northern limit of the channel is set by the edge of the Great Bahama Bank. The north coast of Cuba continues with a 120-mile fringe of many small cays, the Archipiélago de Sabana, whose outer edge consists of a steep-to coral barrier, the most continuous on the island's shore. Between this barrier and the Cay Sal Bank, 24 miles to the north, passes the Nicholas Channel. At the western end of the Archipiélago de Sabana is the Bahia de Cardinas, protected to the north-west by the long Peninsula de Hicacos, whose whole seaward length is fronted by a fine beach. Another ten miles of low sandy shore leads to Puerto Matanzas, a broad and well-protected harbour and the port of departure for a number of slavers.

Forty-five miles further to the west, past a shore of low cliffs and beaches, lies the great port of Havana, the most notorious slave harbour of the nineteenth century. The city, standing on a low plain on the west side of the harbour and close to sea level, was defended by long lines of fortifications. The secure and landlocked harbour is formed by a bay with three inlets, divided by two promontories projecting from its southern shore, and its narrow entrance channel, a mile in length, was commanded by forts. The coast to the west of Havana is low and covered by mangroves with occasional interruptions of beach or bare rock, and the shore is fringed by a narrow rocky shoal on which the sea breaks in rough weather. In the first 40 miles or so there are several bays, entered by narrow channels, which provide safe anchorage. Thereafter the reefs and cays of the Arrecifes de Los Colorados lie outside the shoal, allowing only one reasonably safe channel to the bays and inlets along the coast. The final 40 miles to Cabo San Antonio is formed of low bluffs and beaches bordering the shallow waters of the Bahia de Guadiana.

In addition to the pre-eminent port of Havana, there were only three or four recognised slaving harbours to serve the huge Cuban slave market, but the coast offered many safe and secluded bays and anchorages. Consequently slavers who wished to avoid anti-slaving cruisers or the eyes of the Havana authorities had a wide choice of where they might discharge their cargoes.

Extending out to 270 miles from the north coast of Cuba are the islands of the Bahamas, the exposed tips of a great submerged ridge, generally flat and with a highest elevation of 200 feet. The numerous and irregular-shaped islands, islets and rocks of the chain, many of them only a few feet above the sea surface, are mostly situated on the edges of coral and sandbanks, some of which, the 1859 *West Indies Pilot* warns, are "of the most dangerous character". The Great Bahama Bank, over 300 miles in length and extending for 60 miles to the west and south of the islands, has a number of channels through it, but much of it is studded with innumerable coral-heads and ledges, many of which nearly dry at low tide. Further to the west and lying between the Florida Strait and the Nicholas Channel is the Cay Sal Bank, about 60 by 40 miles in extent, similar in character to the Great Bahama Bank but separated from it by the Santaren Channel.

The Bahama islands and banks gave refuge at times to slavers attempting to evade or escape pursuit by British cruisers, but this was a most hazardous area for mariners without local knowledge, especially in darkness or foul weather. At least in daylight, as the 1859 *Pilot* pointed out:

> A most remarkable feature is the exceeding clearness of the sea water, which enables the bottom to be discovered at considerable depths and at some distance from aloft; the navigation of the banks is consequently conducted almost entirely by eye.

Nevertheless, there are several relatively safe channels through the Bahamas which give access to the north coast of Cuba, either in combination with, or avoiding, the Old Bahama Channel.

The predominant wind over the Caribbean Sea and islands, and most of the Bahama Islands, is the North-East Trade Wind, blowing steadily throughout the year, generally at about 15 knots and rarely above 25. This raises a sea of three or four feet at all seasons, and a low to moderate swell with a longer period than the sea waves. During the wet season, disturbances in the trade wind's flow, known as "easterly waves", move westward at about 15 knots, bringing squalls and heavy thundery downpours in their wake. Occasionally in winter a strong northerly or north-westerly flow of cold air, a "norther", develops over the area, usually heralding its approach with a heavy bank of cloud on the horizon to the north-west. A "norther" is liable to bring violent line-squalls in the northern part of the area with gusts to over 60 knots. Close to coasts the prevailing wind is likely to be modified by a sea breeze in the afternoon and a land breeze at night.

Tropical depressions are liable to affect all parts of the Caribbean area, and five or six of them intensify into tropical storms and hurricanes in the average year. Tropical storms and hurricanes may develop at any time between June and November, but they are most likely in August, September and October. They originate either in the area itself, or in the Atlantic as far east as the African coast, and, although their tracks are unpredictable, they have a tendency to curve clockwise. With their cyclonic winds of up to 100 knots and their confused and mountainous seas, they presented an extreme hazard to sailing vessels. They are also liable to cause exceptionally high tides.

The Caribbean's annual rainfall may be anything from 28 to 160 inches depending on topography, coming usually as light showers during the dry season and as torrential thundery downpours in the wet season, which generally lasts from May until December. Temperatures are usually high throughout the year, averaging between 24 °C and 29 °C at sea, although temperatures can fall to 10 °C in a "norther". The Trade Wind gives well-broken cumulus cloud for most of the year, with a slight increase in cover in the summer, but tropical storms, hurricanes and easterly waves produce overcast skies and heavy cumulonimbus cloud.

The Caribbean Sea is affected by the Equatorial Current, a continuation of the South Equatorial Current combined with the southern part of the North Equatorial Current. It sets westward at about one and a half knots before passing through the Yucatan Channel between Cuba and Mexico, and fanning out. The major part of it turns east-north-east and joins a flow coming from the Gulf of Mexico. The combined stream, the Florida Current, flows eastward about 30 miles off the northern coast of Cuba at two to three knots, and then northward between Florida and the Bahamas. Meanwhile, the continuation of the northern part of the North Equatorial Current becomes the Antilles Current to the east of the Bahamas, where it splits. One branch sets along the north shore of Cuba and through the Old Bahama Channel at about one knot, and the other passes more slowly along the north-easterly side of the Bahamas. Both branches then join the Florida Current to form the Gulf Stream, setting north-eastward across the Atlantic at about one knot. Some parts of the Cuban coast experience currents in reverse of the main offshore flow: there is a west-going countercurrent close to the north-west coast, and along the concave southern shore there is an east-going countercurrent. There is also a minor branch of the Antilles Current setting south-westward through the Windward Passage to the east of Cuba, and an indraught towards the shore is apparent on the north Cuba coast to the east of the Old Bahama Channel. Tidal streams are weak throughout the Caribbean except in narrow channels between reefs.

The final destination for most Cuba-bound slavers was Havana, and, bearing in mind the winds and currents, masters had a choice of approach. Ships which had passed through the island chain of the Lesser Antilles could run the length of the Caribbean, helped by the Equatorial Current and the Trade Wind, pass through the Yucatan Channel between Mexico and Cuba, and then hope to gain some help from the Florida Current for a probable beat of nearly 200 miles to the north-east along the north coast of the island. Alternatively they could head out of the Caribbean through the Mona Passage to the east of Hispaniola, joining traffic which had remained north of Hispaniola before coasting north-west along the northern shore of Cuba. This route would take them through the Old Bahama Channel, between the island and the Great Bahama Bank, with the likelihood of a fair wind and the aid of a knot or so of the Antilles Current. Those hoping to avoid detection by making a less obvious but much more hazardous approach might initially head north-west outside the Bahamas, with the help of the main Antilles Current, and then feel their way south-west through the Bahama Islands and the shoals of the Great Bahama Bank. Vessels heading for the southern Cuban ports might choose to pass through the Lesser Antilles islands and have an easy run down the Caribbean with wind and current. On the other hand, they could also find a fair wind and set by keeping north of Hispaniola and then heading into the Caribbean through the Windward Passage between Hispaniola and Cuba. Departure from the island for slavers outward bound for Africa was much more problematical, particularly so if their destination was the Congo or Angola. Unless they were prepared for a long transatlantic beat against the trade winds and the Equatorial Currents, they would probably head north with the powerful Florida Current. They could then pick up the Gulf Stream and westerly winds, and then the North Atlantic Current, to carry them across the ocean to join Africa-bound European traffic in the Canary Current.

* * *

This, then, was the immense expanse of sea and coast upon which the slavers plied their vile trade during the nineteenth century, and which consequently became the theatre of operations for the Royal Navy in its campaign to suppress the Atlantic slave traffic. As will be seen, the area of activity for the Navy's ships specifically tasked with suppressing the Trade was initially limited to the West African slave coast north of the Equator, but by the close of the campaign the British cruisers had hunted their quarry at all extremities of this vast area.

PART TWO

The Suppression Campaign

CHAPTER 4

Confused First Steps, 1807–11

What is the answer? [...]
In that case, what is the question?

GERTRUDE STEIN

ON 22 OCTOBER 1807 the last of many convoys of British Guinea-men slavers sailed from West Africa. The five merchant vessels, laden with slaves from Bance Island,* were, as was usual, protected by the Royal Navy against French or Spanish interference. In this case the escort was the ship-sloop *Favorite* (16),† Acting Commander Frederick Hoffman. On the seven-week passage from Sierra Leone to Barbados, Hoffman regularly visited the slave vessels, reporting that he found them orderly and clean and that the slaves were healthy.[1] The irony of his situation could hardly have escaped Hoffman: the Slave Trade Abolition Act had come into force on 1 May 1807, nearly six months before the sailing of his convoy, but, because the Guinea-men had been cleared for sailing from their British ports before the May deadline, they were considered to be slaving legally and were consequently entitled to naval protection. Had they been cleared for their voyages after the enforcement date it would have been Hoffman's duty not to protect them but to arrest them and take them before an Admiralty court. Indeed, the first successful arrest of a British slaver had taken place four months before Hoffman's convoy sailed from Africa; the English brig *Busy* was detained with a cargo of slaves in the West Indies by the frigate *Alexandria* (32), Captain The Hon. Edward King, and condemned by the Vice-Admiralty court at Tortola in early July.

It was not until 27 July 1807 that what was probably the last legal British slaver sailed for Africa from her home port of Liverpool, her clearance certificate

---

* The island's name is now spelled "Bunce", but "Bance" appears to have been in general use in the early nineteenth century.
† The figure in brackets immediately following a vessel's name indicates her establishment of guns. In rated ships (Sixth Rate and above) carronades (see glossary) were considered to be additional to the main armament of long guns and until 1816 were not included in the establishment number. In sloops and below the main armament often consisted entirely of carronades, and they were always counted in the establishment. The name which generally follows the gun establishment is that of the Commanding Officer. He was traditionally referred to by the courtesy title of "Captain" whatever his rank, but in this account his actual rank is used for clarity.

bearing the date of 27 April. She mounted 18 guns, and her owner, Henry Clarke, had obtained for her a Letter of Marque for her voyage. She was the ship *Kitty's Amelia*, commanded by Captain Thomas Forrest. She carried another experienced slaver as supercargo: Captain Hugh Crow, widely known as "Mind your Eye Crow" on account of the loss of one of his eyes in his youth, and, on the death of Forrest, Crow took command.[2] When the ship arrived at Bonny after a voyage of seven weeks she was immediately boarded by the anxious local chieftain, King Holiday, who questioned Crow on the rumour that Britain had abolished the Trade. The King explained the dire consequences to himself and his country of this astonishing development, but concluded: "We tink trade no stop, for all we Ju-Ju men tell we so, for dem say you country no can niber pass God A'mighty." Captain Crow thought the King's remarks were not lacking in sense and shrewdness.[3] Slaves had probably accounted for three-quarters of West Africa's exports in the eighteenth century, and the explorer Mungo Park had assessed in the 1790s that three-quarters of the population of Senegambia were slaves. So it was hardly surprising that the kings and chiefs of the Guinea Coast should be incredulous at the notion of slave-trade abolition, and should continue to sell to anyone wishing to buy.

There was no shortage of buyers. The Abolition Act had effectively deterred reputable British merchants and shipowners from further involvement in the Trade, but there were still individual masters willing to risk the penalties imposed by the Act: forfeiture of their ships and fines of £100 for every slave found on board.* There would also be a few British traders ashore on the African coast for some time to come. But of far greater significance than these occasional British lawbreakers were the slavers of other nations. Any hopes that might have been entertained by the abolitionists in Britain that foreign governments would follow the British example were to be dashed. In particular, Portugal and its colony of Brazil had lost none of their zest for the Trade. In Loango, the Congo and Angola the Portuguese merchants traded with even greater intensity, fearing that British naval power would bring a speedy end to the traffic, and in 1807 alone 8,000 slaves were taken from Elmina to Bahia. The slavers of Spain, France and the Netherlands, although greatly constrained by the Royal Navy's blockade of Napoleon's ports and its harassment of all enemy merchant shipping, had every intention of resuming their activities when they could. The Spanish slavers gained this freedom in 1808 with the ending of their country's hostilities against Britain, and from then on Spain and Portugal were the largest shippers of slaves. American

---

* Equivalent contemporary values of the pound for the entire period of the campaign are given at Appendix I.

traders, shipowners and masters remained enthusiastically engaged in the traffic. Even after the Act of Congress to abolish the USA's international slave trade took effect on 1 January 1808, a combination of the weakness of the government in Washington and the influence brought to bear by southern planters, and by the merchants and shipbuilders of the north-eastern states, ensured that the law was not enforced. Finally there was a smattering of slaving vessels under a variety of other flags, including, still, the occasional Dane in contravention of her country's slave-trade abolition in 1803.

In late 1807 Brazilian commerce, including its slave trade, was given a massive boost. The British Ambassador in Lisbon had persuaded the Portuguese Prince Regent, Dom João, to reject an ultimatum from Napoleon to close his ports to the British, and he had also indicated that Britain would occupy Brazil if the Portuguese royal family were to be captured by the French. Dom João decided to flee to Rio de Janeiro with his court and many Lisbon businessmen, taking passage in British ships at the end of November 1807. In consequence of this move, much of the trade which had previously gone to Lisbon now went to Rio, and Rio was transformed into a capital city. Before long Brazil became a coequal with Portugal in the United Kingdom of Portugal, Brazil and the Algarves, with its seat of government in Rio. The volume of Brazil's slave trade quickly reflected this new prosperity, increasing from about 10,000 per year to 20,000 following the Regent's arrival, and Dom João is said to have invested in it. Even before the court's departure from Lisbon, the Governor of the Brazilian slaving port of Bahia, concerned about the consequences of British abolition, had been assured that "It is far from being the royal intention to restrict this commerce in any way. On the contrary, the royal authorities wish to promote and facilitate it in the best way possible."

In Britain, law-abiding merchants, the vast majority, set about developing new forms of trade with West Africa to replace the traffic in slaves. Hides, beeswax and acacia gum all offered possibilities, in addition to the old African commodities of ivory, gold, rice, timber, pepper and peanuts, but most promising of all was palm oil. For some time this oil had been imported into Europe for making soap and candles, but it was also an essential ingredient for lubricants, and the Industrial Revolution was leading to a massively increased demand. Some merchants converted remarkably successfully from the slave trade to palm oil from the Bights, ignoring the fact that their African suppliers were generally using slaves to tend their trees. This enthusiasm in Britain for "legitimate trade" on the slave coast was to be a continuing theme throughout the suppression campaign. The newly founded African Institution, under the presidency of the abolitionist Duke of

Gloucester, was aimed at "diffusing useful knowledge and exciting industry among the inhabitants", but aspiration and achievement were still a long way apart. The House of Commons was quick to rebuke the Company of Merchants Trading to Africa for failing to convince Africans that abolition was good for them, but the Company's committee replied:

> Can the wildest theorist expect that a mere act of the British legislature should, in a moment, inspire [the] unenlightened natives of the vast continent of Africa, and persuade them, nay more, make them practically believe and feel that it is for their interest to contribute to, or even acquiesce in, the destruction of a trade not inconsistent with their prejudices, their laws, or their notions of morality and religion, and by which alone they have been hitherto accustomed to acquire wealth and purchase all the foreign luxuries and conveniences of life?[4]

This response seems to reflect an accurate appreciation of the realities of the situation.

Of course it was not only the slaving potentates of West Africa and the slaver masters of Europe and the Americas who remained to be convinced of the benefits of abolition.* The markets for slaves, especially in Brazil and Cuba, were as voracious as they had ever been. In Brazil slaves comprised half the population, providing labour for the mines and the sugar and coffee plantations, for the transportation of produce, and, on a much smaller scale, for all domestic service. The particular demand for heavy labour led to an imbalance of men to women slaves in the country of almost four to one, and the birth rate among slaves was consequently low. As a result there was a need for constant replenishment from Africa, and the slave trade accounted for a third of all Brazil's commerce.

Slavery continued throughout Spain's possessions in the Americas, but by 1807 Cuba was its only colony employing large numbers of slaves. By comparison with Brazil its slave population was still small, but, thanks to the huge market for sugar in Europe and the United States, the island's cane production was increasing rapidly to the point where it would soon export more sugar than all of the other Caribbean islands put together. Slave labour fuelled this expansion, and the demand for fresh slaves came not only from the need to man new plantations but also from the requirement, as in Brazil, for regular

---

* The term "master" denotes the officer commanding a merchantman or slaver. "Master" is a naval warrant rank and post in a man-of-war (see glossary).

replacements for those, again the male slaves, worn out by their labours in the cane fields and sugar mills.

In the USA, the massive increase in cotton production in Georgia and the Carolinas had decisively reversed a decline in the country's slave population, but the cotton fields, by contrast with the Caribbean and Brazilian sugar plantations, were worked mostly by female slaves. Between 1800 and 1810 the number of slaves in the United States rose by a third, but, again in contrast with Cuba and Brazil, the slave trade into North America remained very small in scale, a disparity which can be explained only by the natural increase permitted by the high proportion of women slaves. Finding such an unsatisfactory market in their own country, American slavers turned their attention to Cuba and, to a lesser extent, Brazil. In a period of nine months in 1807, out of a total of 37 slaving vessels entering Havana, no fewer than 35 were US ships.

With such a volume of slave traffic continuing to cross the Atlantic there was always the possibility of chance encounters between slavers and British warships, particularly in the approaches to the West Indian islands, where there was a fairly high concentration of British men-of-war. Commanding officers of the Royal Navy's cruisers on the Jamaica Station, as elsewhere, were well aware of the abolition legislation passed in London and Washington, but they had almost certainly not been adequately briefed on which slavers they might legitimately detain. Indeed, the full implications of the new laws had yet to be tested in court. While hostilities lasted, enemy vessels, and neutrals trading war contraband goods with the enemy or his colonies, could be taken as prizes of war. British slavers, under their own or false colours, were of course liable to arrest under the 1807 Abolition Act. So much was clear, but it was also supposed that the Royal Navy had a duty to enforce the anti-slaving laws of other nations, particularly the United States, or even that it was entitled to seize the slaving vessels of any nationality. There was confusion not only in the minds of naval officers but also in those of the Vice-Admiralty courts, situated in many British territories abroad, which were authorised to judge prize cases and violations of the Abolition Acts on behalf of the High Court of Admiralty in London. It was to be some years before clarification was achieved.

* * *

Any initial sense of complacency among American slavers that they were immune to British action was quickly dispelled. As early as the autumn of 1807 the Charleston schooner *Nancy* was detained off St Thomas in the West Indies by the frigate *Cerberus* (32), Captain William Selby, and HM Hired Schooner *Venus*, Lieutenant Francis Bligh. *Nancy*, 67 feet in length, was carrying 77 slaves from the French colony of Senegal, and, despite her claim that she was in distress, a prize crew took her into Tortola on a charge of trading with a French colony, for which she was condemned on 27 November. Three other loaded American slavers were taken before the Vice-Admiralty court at Tortola early in 1808: the brig *Amedee*, intercepted in December 1807 by the gun-brig *Swinger* (12), Lieutenant James Bennett; the brig *America*, detained by the frigate *Latona* (38), Captain James Wood; and the ship *Africa*, taken by the gun-brig *Haughty* (12), Lieutenant John Mitchell. *Amedee* was condemned, but the *America* was ordered to be restored, as was the *Africa*, although this last decision was reversed two years later.

Surviving court records often omit the charges on which captured vessels were adjudged, and also the grounds for condemnation or restoration, but some of these Americans may have been detained simply on suspicion of trading with French territories. As we shall see, the case of *Amedee*, at least, was different. So was that of the Charleston brig *Tartar*, overhauled and boarded off the French island of Martinique on 2 February 1808 by the elderly 44-gun frigate *Ulysses*, Captain C. J. W. Nesham. Despite the master's claim that he was heading for Martinique merely for provisions, the *Tartar* was taken into Barbados, where she was condemned as prize not only for proceeding to a colony of the enemy but also for trading in slaves contrary to the laws of her country. There was no protest from Washington, and American slavers, concluding that the Stars and Stripes would no longer afford them protection, mostly took refuge under Spanish colours. In many cases, to reinforce this disguise, a Spaniard would be appointed as master of an all-American crew. The African slave coast was soon swarming with these bogus Spanish vessels, and arrests of American-flagged slavers in the West Indies came to an abrupt halt. Another case, that of the Swedish schooner *La Fortuna*, was successfully brought before the Tortola court by Lieutenant George Spearing of the schooner *Subtle*. Nothing is recorded of the circumstances, beyond the fact that *La Fortuna* was carrying a cargo of African slaves when she was taken on 15 June 1808, but she must have been judged to be carrying war contraband.

An appeal against the condemnation of *Amedee* provided a test case on the legitimacy of British seizures of American slavers under the abolition laws, but it was not until 1810 that the matter came before the Appeal Court of the Privy Council. The claimant admitted that *Amedee* (or *Amedie*) had been heading for Cuba in contravention of US law, but argued that she had violated no belligerent rights, and that no British court had a right to take cognisance of American municipal law.* Sir William Grant, Master of the Rolls, gave his judgment on 28 July. He said, in essence, that the British legislature had pronounced the slave trade to be contrary to the principles of justice and humanity, and that a British court therefore had the right to assert that, prima facie, the Trade was illegal. This obliged foreign claimants to show that, by their own laws, the slave trade was legal. A claimant could have no right, upon a principle of universal law, to claim the restitution of human beings carried as his slaves. In this case the law of the claimant's country, the USA, allowed no right of property such as he claimed, and therefore there could be no right of restitution. The court affirmed the judgment made in Tortola.[5]

These, however, were chance arrests, and the international slave trade was not of immediate interest to Britain at this stage. Such concern as there was following the 1807 Abolition Act concentrated on residual slaving by British citizens and British vessels. There were still small numbers of British slave dealers in the trading stations of West Africa who did not trouble much to disguise their identities, but the slaving masters continuing to operate under British colours were probably outnumbered by British subjects commanding slavers under the ensigns of other nations. Given the choice, these renegades seem initially to have favoured the Stars and Stripes, no doubt because the disguise would be easier to sustain than false Spanish or Portuguese identities, but that changed with the arrest of American slavers in the West Indies. There were also several prominent firms in Liverpool and London who continued to take part in the Trade by investing in, or owning, supposed Spanish or Portuguese vessels. Other British firms still supplied trade goods to foreign slavers, and British sailors losing their billets in British slavers naturally offered their services and expertise to foreign masters, especially Cuban-based Spaniards.

The extent of this continuing British activity is difficult to fathom. By comparison with Britain's slave trade of the late eighteenth century it was certainly very limited, but it was still at a level to concern the abolitionists. Zachary Macaulay, ex-Governor of Sierra Leone and now secretary of the African Institution, informed the Admiralty in July 1809 that 35 ships suspected to be slavers had left from Liverpool since December 1807. Most of them sailed with Portuguese masters and under Portuguese colours, but carrying British trade goods, sometimes with British supercargoes. An English vessel, the *George*, had cleared for the Old Calabar River on the pretext of loading with palm oil, but it was known

---

* Names of slaving vessels tend to appear in slightly differing forms in ships' logs, commanding officers' letters and court records. The spelling used by the courts is generally given as being most probably correct.

by the Institution that she was to lie for some time offshore as a stores ship for slave-trade goods before loading slaves and sailing for Brazil under Portuguese colours. Large orders were known to have been received in London from Havana merchants for trade goods for slaving voyages under Spanish colours. Attempts had also been made at Lloyd's Coffee House to arrange insurance for Portuguese and Spanish slavers, although the underwriters were not generally disposed to break the law. A similar bid had also been made to insure a British ship under Swedish colours.

Responsibility for enforcement abroad of Britain's new anti-slaving legislation lay, in their two spheres of activity, with the Lords Commissioners of the Admiralty and the Secretary of State for War and the Colonies. These officials and their departments, weighed down with the pressing concerns of the war against Napoleon, might (not entirely unreasonably) have regarded policing action against a handful of slave traders as a marginal issue, or even a total irrelevance. That indeed appears to have been the attitude as far as the Admiralty was concerned. Had the abolitionist Admiral Lord Barham still been in post as First Lord there might have been a more positive reaction, but he had left with the change of administration in January 1806. At the Colonial Office, however, Lord Castlereagh clearly offered a more receptive ear to the pleas of William Wilberforce.

Wilberforce did not hesitate to nag (as he put it) "My dear Lord C". He was insistent that a Vice-Admiralty court should be set up in Sierra Leone "to try and adjudge cases both of Prize negroes and negroes seized for violating the foreign slaves act". As an additional lever he referred to Castlereagh's "earnest wish" to recruit released slaves to black regiments in the British Army, pointing out that if captured slavers could be condemned by a court in Sierra Leone rather than in the West Indies it would be possible to raise entirely new regiments in Africa.[6] With a splendid disregard for the realities of naval affairs, he wrote to Castlereagh on 26 October 1807 that a ship of war "must sail to Sierra Leone in a very few days", and shortly afterwards, in a letter to the Colonial Office, he urged that "a govt Commission must be sent (to Sierra Leone) immediately". Although unrealistic on some aspects of practicality, Wilberforce seems to have appreciated the nature of the enforcement task and, to some extent, to have grasped what was needed to achieve it. Clearly, he also realised that he would have to press insistently and repetitively if slaving was to receive any attention amidst competing issues, and he was evidently prepared to employ any potentially helpful tactic which came to hand.[7]

In fact an Act (47 Geo. III, c. 44) had already been passed by Parliament in August 1807 transferring to the Crown "certain possessions and rights vested in the Sierra Leone Company, and for shortening the duration of the said company, and for preventing dealing or trafficking in the buying or selling of slaves within the Colony". By this Act Sierra Leone was taken over by the Crown in January 1808 as the centre for abolition activities, and, in accordance with Wilberforce's wishes, a Vice-Admiralty court was established at Freetown (in a "house of tarpaulins") by an Order in Council of 16 March 1808, to deal with slaving vessels detained under British legislation. Slave vessels captured in the vicinity of the west coast of Africa were to be sent into Freetown, and those condemned were to be confiscated and sold. Cargoes of slaves, legitimately captured, would be condemned to the King by the court, and would be liberated in Sierra Leone. These slaves would be maintained at Government expense for a year, but thereafter they would have to fend for themselves. They would be given the alternatives of volunteering to go to the British West Indies as apprenticed labourers or as recruits to the West Indies regiments, but, in the event, the vast majority stayed in Sierra Leone.[8]

This Order in Council also dealt with another matter of significance to the Navy. The King and his ministers could rely upon a strong sense of duty among the officers of the Fleet, but the Government had long recognised that financial incentives were necessary for recruitment and were desirable for inducing optimum performance from the Navy and the Army.* The Abolition Act of 1807 extended the prize-money principle to slave-trade suppression, introducing the payment of a bounty known as "Head Money" for slaves released through the actions of His Majesty's Ships and of privateers carrying appropriate Letters of Marque. The idea of the bounty was not entirely new; it was an extension of the current practice of paying head money on the crews of captured enemy warships and privateers. The bounty payment was to be in addition to the proceeds of the sale of the condemned slaving vessels. The Order in Council of 16 March 1808 laid out the regulations for this new incentive, and in directing the sums to be paid it distinguished between two types of capture. In the case of a slave cargo taken as prize of war, head money would be £40 for a man slave, £30 for a woman and £10 for a child not above 14 years of age. For seizures of slaves at sea for forfeiture under a slave trade abolition Act, however, the payment would be half that made for a prize of war. The decision on which circumstance applied was not necessarily clear-cut, and lay with the Admiralty court adjudicating the case. Distribution of head money was to be in accordance with existing regulations for the payment

---

* The 1805 legislation governing the payment of Prize Money was entitled *An Act for the Encouragement of Seamen, and for the better and more effectual manning of His Majesty's Navy during the present war*.

of prize money. On the face of it, the sums involved could be large, but, as will be seen in due course, the officers and men on the coast of Africa would not become rich on their bounties.⁹

Further urging by Wilberforce may have been partly responsible for a letter from Lord Castlereagh at the Colonial Office to the Lords Commissioners of the Admiralty on 15 August 1808, a letter which can probably be regarded as the initiation of the Navy's anti-slaving campaign on the west coast of Africa:

> In consequence of the Act for the Abolition of the Slave Trade, the placing [of] the settlement of the district of Sierra Leone under the management of Government and the alteration which our Commerce and Relations with the Native Powers of the Coast of Africa must experience, it has been deemed expedient that a Commission should be appointed for surveying all the Coast of Africa to which the Trade of the African Company and the Sierra Leone Company have extended.
>
> I am therefore to acquaint your Lordships, by His Majesty's Commands, that with this view he has been graciously pleased to nominate Mr Ludlam the late acting Governor of Sierra Leone and Mr William Dawes to be joint Commissioners for surveying the Coast of Africa, and I am to desire that your Lordships will lay before His Majesty the name of a Captain in the Royal Navy to be associate in the Commission.
>
> I am further to signify His Majesty's Commands that you do order to be fitted out a ship of 20 Guns to be commanded by the Officer who His Majesty shall approve to be a Commissioner, to be properly fitted with instruments and stores for the purpose of survey and that you do order everything necessary to be in readiness so as that the ship may sail from hence as early in September as practicable.
>
> When your Lordships shall have acquainted me with the name of the Officer whom His Majesty shall appoint to be third joint Commissioner the necessary Commission and Instructions will be prepared without delay and communicated to your Lordships.
>
> The Commissioners according to a vote passed in the last session of Parliament will receive each £1500 a year whilst employed in the survey.¹⁰

Despite the conventional closing courtesies of humility and obedience from the Secretary of State for War and the Colonies, this was no begging letter to the Admiralty. Lord Castlereagh, preoccupied though he must have been by military developments in the Peninsular War, clearly intended that the Colonial Office would take the lead in action abroad against the slave trade, and he presented Their Lordships with the fait accompli of a Royal Command. The reason for the pretext of a survey is not clear. Perhaps Castlereagh, with his keen interest in foreign affairs, was aware of the extent of official ignorance of conditions on the African coast and perceived the need for a degree of reconnaissance before taking enforcement action. Maybe he calculated that the Admiralty would be more amenable to the concept of a hydrographic expedition than to the idea of an open-ended policing involvement. In any event, Their Lordships replied immediately that they were "desirous of recommending Captain Edward Henry Columbine as a proper person to be appointed to the Commission", and that a ship "would be provided with all possible dispatch".

Columbine, a half-pay captain of six years' seniority, was currently employed on temporary duty with the Hydrographer of the Navy in London, and the Admiralty's choice of him for this new task was presumably based not only upon his availability but also on some experience he had of surveying. Initially no time was lost. Captain Columbine received his appointment as Commissioner on the day following receipt of Lord Castlereagh's letter, and on 6 September he took command of HMS *Solebay* at Spithead, having apparently been given to understand that he would also be allocated two smaller vessels. *Solebay* (32) was an elderly 12-pounder frigate* and, although Columbine was pleased with the appearance and character of his new ship's company, he reported that the frigate's rigging was in need of survey, that some of her guns were probably "honey-combed", and that she was in need of a refit. The Admiralty ordered that the ship should be taken into dock in Portsmouth, "all-standing if necessary" to avoid delay, and, with the work rapidly done, she was back at Spithead on 6 October. Momentum seems then to have evaporated. Columbine bombarded the Admiralty with requests for special stores to preserve the health of his crew in a hot climate, for a 12-pounder carronade for his ship's boats, for some "Portsmouth werries", which he considered more suitable than ships' boats for survey work, and even (unsuccessfully, if far-sightedly) for permission to purchase a Virginia pilot schooner. His ship, meanwhile, remained at anchor, and, at last, in early April 1809, Wilberforce interceded with Castlereagh on behalf of the impatient Columbine. On 9 April 1809 *Solebay* finally received her orders to take under her protection

---

* The term "frigate" is used in this account only for ships of the Fifth Rate and the handful of 50-gun Fourth Rate ships no longer fit to lie in the line of battle. Ships of the Sixth Rate were also officially classified as frigates, but some distinction between the rates is considered appropriate in the context of this campaign. When a number of ship-sloops were later reclassified as Sixth Rates they were known as "jackass frigates". The French referred to Sixth Rates as "corvettes", and this term gradually established itself in the Royal Navy.

merchant vessels bound for Africa, and to proceed to Sierra Leone, carrying into execution instructions received from Lord Castlereagh. Captain Columbine was also directed to hoist a broad pendant* on leaving the English coast.[11]

This long delay after completion of *Solebay*'s refit is unexplained, and it frustrated Castlereagh's intention that the survey should begin after the rainy season in 1808. The frigate was certainly ready. It may have been that Their Lordships were waiting for a worthwhile number of merchantmen to form a convoy for West Africa, and they were probably having difficulty in providing two suitable consorts for *Solebay*, as apparently promised to Columbine. Whatever the true reason, the delay gave time for half a solution to the latter problem to be found in the gun-brig *Tigress*. She was the recently captured French *Pierre César*, newly refitted, established with fourteen 12-pounder carronades and 60 men and placed under the command of Lieutenant Robert Bones. On 12 April she joined Columbine at Spithead, but it was not until 5 May, two years after the passing of the Abolition Act, that *Solebay*'s convoy of 11 sail was fully assembled, had dropped down from Spithead to St Helen's, and finally sailed for Africa.

Meanwhile, the brig-sloop *Derwent* (18), Commander Frederick Parker, was already off West Africa. She had sailed from Spithead in mid-November 1807 on a routine deployment in support of the British settlements, and six weeks later she made the first recorded arrest of a slaving vessel on the Coast† when she detained the *Minerva* at Freetown on 30 December, although neither the nationality of this vessel nor the outcome of her capture are known. A few months later *Derwent* was involved in an episode in which her conduct seems not to have been entirely creditable, and which had subsequent repercussions.

On about 18 March 1808 she had taken into Freetown two American slavers, the sloop *Baltimore* and the schooner *Eliza*, and there is good evidence to indicate that, at the behest of the Governor, 140 of the captured slaves were sold to the settlers as "apprentices". The receipts, after the Governor's commission had been deducted, were almost certainly paid to Commander Parker and his Purser. The vessels and slaves had not been condemned by the newly established Vice-Admiralty court, and although the Abolition Act had authorised apprenticeship for released slaves, sale of slaves on the pretext of apprenticeship was

---

* A broad pendant is the distinguishing flag of a commodore, a triangular or swallow-tailed flag flown at the main masthead in the officer's ship. Commodore was a temporary rank granted to a captain whose appointment required additional authority and status. A commodore might command his flagship, as was the case in the West Africa Squadron, or he might be carried additional to the commanding officer in the manner of an admiral.

† "The Coast" was used at the time by those off the west coast of Africa to denote the slave coast, and it is similarly used in this narrative.

illegal. Parker may not have been aware of just how irregular his action was, but the same can hardly be claimed for Governor Ludlam, who was *ex officio* Judge of the Freetown court. *Derwent*'s logs for much of 1808 and 1809 have been lost, but it is known that she made prize the French schooner *Marie Paul* with a cargo of 60 slaves on 23 August 1808, and the *Two Cousins*, probably American, on 6 September. These two cases were correctly handled by the Freetown court, and both vessels were condemned.

A month or so later *Derwent* boarded the sloop *São Joaquim* and the schooner *São Domingo*, both loaded with slaves and claiming to be Portuguese. In a dispatch to London, the Governor reported that both had been condemned by the Freetown court (for what reason he did not say), and he described a number of the features of the case. The sloop sank just after capture, drowning a *Derwent* officer and 17 slaves, and the schooner was barely afloat on arrival at Sierra Leone. Both vessels were formidably armed for their size. The schooner had been built with two names, one English and one Portuguese, and the court was shown a paper purporting to be her register but which was dated before the schooner was built and belonged to another vessel of the same name. The master told the court that he had purposely sailed without colours or papers because he had been taken before for having papers, and he thought it very hard that he should be taken both for having papers and for having none. These convolutions were typical of those through which cruiser commanding officers and the courts would in future have to feel their way.

In late November 1808 *Derwent* was procuring supplies for the garrison at Goree, and, that done, she left the Windward Coast, where she had made all her arrests to date, for a cruise to leeward (to the east) for three months, returning to Sierra Leone in February 1809. Her foray into the Bights bore no fruit, but she had more luck on her return, detaining two slavers of unknown nationality, the *Rapid* and the *Africaan*, although the former was released by the Freetown court. *Derwent* had expected to sail for home in April 1809, but she was not pressingly in need of refit, and it made obvious sense for her to join *Solebay*'s convoy during its passage south and complete Commodore Columbine's small squadron.

It is curious that Their Lordships should have ordered Columbine to take his instructions from Lord Castlereagh rather than issue orders of their own in the normal way. The arrangement may have indicated that the Admiralty wanted nothing to do with the work of the African survey commission, and Columbine may well have felt uncomfortable about this apparent washing of Admiralty hands. His final instructions from Castlereagh in early April contained two warrants by which he was authorised and directed to act on the coast of Africa, and he sent them to the Admiralty for the information of Their Lordships. Columbine probably surmised that his half-pay status of the previous year indicated that he had not been standing high in the Board's estimation, and it is likely that he was anxious not to be forgotten as he sailed for the naval backwater of West Africa. Such concern was probably a factor in a major decision he made immediately upon his arrival on station.

Solebay and her brood reached Goree in mid-June, and Columbine was soon engaged with Major Charles Maxwell, the Commandant of the Goree garrison, in planning an expedition against the French colony of St Louis on the River Senegal. Maxwell was concerned at the damage being done to the African coasting trade by small French privateers running out of the river, and he had probably been harbouring an ambition to root out this nuisance. He would have seen his opportunity with the arrival of Columbine's squadron, and Columbine, for his part, might have been worried that the survey and his anti-slaving work would be disrupted by the privateers. The two commanders would undoubtedly, and very reasonably, have agreed that no opportunity to strike a blow against the French should be missed, and would have hoped privately that success in such a venture would do no harm to their reputations in the Admiralty and the Horse Guards. Accordingly, by the beginning of July 1809, the Commodore had gathered a flotilla consisting of *Solebay*, *Derwent*, *Tigress*, the transport *Agincourt*, the colonial schooner *George*, and an assortment of small armed vessels. In order to give the appearance of a more powerful force, he also requisitioned several unarmed merchant vessels.

On 4 July the flotilla sailed from Goree with Major Maxwell and a contingent of 166 soldiers of the African Corps embarked, and on the evening of 7 July it anchored off the Senegal bar. The following day a landing force of 160 soldiers, 50 marines and 120 seamen was put ashore through very heavy surf, but the *George* was driven ashore, and the sloop *Mary* and a schooner were wrecked. Adding to those misfortunes, several boats were upset on the beach, at the cost of the lives of Midshipman Sealy of *Derwent*, five seamen from *Solebay* and one merchant seaman. Then, that night, a boat from *Derwent* went inshore to sound the bar but capsized in the attempt, drowning Commander Parker. The French had taken up position, with a force of 400 regulars, militia and volunteers, about ten miles upriver from the bar. So Maxwell marched his marines and soldiers along the left bank of the river towards the enemy, and then waited until the *George* was refloated and boats could be brought in to give him support from the river. On 9 July the French marched out to attack him, but finding the opposition unexpectedly strong, rapidly retired to a formidable defensive position at a battery on the southern end of the island commanding passage of the river and protected by a chain boom. Behind the boom there lay seven armed vessels. The *George* was refloated on 10 July, and the following evening *Solebay* and *Derwent* took up bombarding positions a cable apart off the bar. They achieved some success against the French position in an hour of firing, but, in shifting berth in a light breeze and heavy swell at 2000, *Solebay* first ran foul of *Derwent* and then went ashore. No lives were lost and most of her stores were recovered, but the frigate became a total loss. The Commodore shifted his broad pendant to *Derwent*, taking with him Lieutenant Tetley to replace Commander Parker in command. On the morning of 12 July the landing force was re-embarked and the flotilla proceeded upriver to make a night attack on the French position, but before that could be launched it was learned that the enemy intended to surrender. The following morning the French defences and vessels had been abandoned and the boom broken, and during the day terms were agreed for the surrender of the Senegal colony to the British.[12]

As Columbine made his way home in *Derwent* to face the inevitable court martial for the loss of *Solebay*, it seemed that Castlereagh's survey had been ended before it had started and that the opportunity for concerted action against the West Coast slavers had been lost, at least for the time being. Despite its setbacks, the Senegal episode provided some salutary experience for the Navy of operating across a West Africa river bar and into an estuary, work which would become commonplace in the anti-slaving campaign. In particular, it gave stark warning of the dangers of the surf on that coast, and provided evidence of the unsuitability of vessels as unhandy and deep-draughted as frigates for inshore operations in those difficult conditions.

Probably against his own expectations, Columbine's initiation on the Coast was not to be wasted, but, after landing Columbine at Weymouth on 22 August, *Derwent*'s Africa career was over. Acquitted at his court martial on 24 August, Columbine wrote to the Admiralty a week later to offer his services once again. Unbeknown to him, however, the Secretary of State for War and the Colonies had pre-empted the verdict of the court. On 13 August Lord Castlereagh had written to the Admiralty signifying His Majesty's Command that, in consequence of the loss of *Solebay* and the return of Captain Columbine, another vessel should take Columbine to Sierra Leone and carry troops to garrison Fort Louis in the River Senegal. The Admiralty replied, perhaps between clenched teeth, that HMS *Crocodile* would do so.

Crocodile, a Sixth Rate ship of 22 guns, would not normally have warranted a post-captain of seven years' seniority, but Columbine would have been mightily relieved to find himself in another command and was unlikely to quibble about her rate. He was, however, about to be burdened with another (and certainly unwelcome) task. In September 1809 Castlereagh left the Colonial Office following his duel with George Canning, the Foreign Secretary, and was replaced by Lord Liverpool. Liverpool, also an abolitionist, seems to have been content to continue his predecessor's policy regarding West Africa, but found himself in need of a new governor for Sierra Leone.

The Governor at the time of the Crown's assumption of control over the colony in January 1808 was Thomas Ludlam, a company appointee of course, and nominated as one of the commissioners for Castlereagh's survey. He had been invited to stay temporarily in post, but was uncomfortable with his new accountability and was anxious to hand over in July 1808 to his young, conscientious and painfully intense successor, Thomas Perronet Thompson. Unfortunately for Thompson, Ludlam initially chose to remain in Freetown to await the start of the survey, and, to make matters worse, Mr Dawes, another previous governor, arrived with Thompson in HMS *Mutine* to take up his duties as a survey commissioner. The more the new Governor saw of the way in which the company's officials had conducted their business, the more convinced he became of their corruption, particularly in the matter of the slave trade. He seized particularly on the incident of Ludlam's alleged sale of the *Derwent* slaves, and not only did he report his concerns, at great length, to London, but also he made his accusations public in the colony. The atmosphere in Freetown became increasingly poisonous, and the Governor's dispatches became increasingly shrill until, in April 1809, Castlereagh felt obliged to recall Thompson to London "to discuss with H.M. Government matters concerning the administration of the colony". Thompson could not leave his post until a successor arrived, but it seems from his acknowledgement to the Colonial Office in August that he had been instructed by Castlereagh's April letter to hand over the government to Commodore Columbine. This was well before the loss of *Solebay*, and there is no evidence that either the Admiralty or Columbine knew anything of Castlereagh's extraordinary intention.[13]

On 28 December 1809, nearly nine months after Castlereagh's letter to Governor Thompson, Lord Liverpool informed the Admiralty that it was the King's Command that Captain Columbine, as one of the commissioners for examining the coast of Africa and about to proceed to Sierra Leone, should on his arrival take upon himself the temporary administration of the government. Liverpool conceded that this would probably preclude Columbine from accompanying the other commissioners, who were therefore directed to collect information without waiting for *Columbine* and to write a joint report with him on their return to Sierra Leone.¹⁴ The only apparent reaction from Their Lordships was to tell Columbine, now loaded with the incompatible tasks of the survey, slave-trade suppression, colonial governorship and charge of a Vice-Admiralty court as well as command of *Crocodile*, that, in consequence of this alteration, it was no longer necessary that he should hoist a broad pendant.

So it was as a captain that Columbine sailed in *Crocodile* from Spithead on 13 January 1810, with a convoy of four sail, in his second attempt to fulfil the bidding of the Colonial Office on the west coast of Africa. *Tigress* had remained on the Coast following the capture of Senegal, and during Columbine's absence the Squadron had been reinforced by the arrival of the cutter *Tickler* (8), Lieutenant Richard Burton. She had left Falmouth in mid-July 1809 with a transport under convoy to Lisbon, and with instructions, by then outdated, to continue to the coast of Africa and follow the orders of *Solebay*.

A more temporary addition was provided at the beginning of October with the arrival of the brig-sloop *Hawke* (18), Commander Henry Bourchier, which had brought out a big convoy bound for the African coast and beyond. Her orders were to proceed to Cape Coast Castle and remain on the Coast for two months before returning with any merchantmen requiring escort, and this pattern was to be repeated with a number of convoy escorts over the next five years. Occasionally other warships would pass through the area with merchantmen under convoy but with no instructions to contribute to the anti-slaving work. The first of these was the ship-sloop *Dauntless* (18), typically ordered simply to touch at Madeira, Goree, Cape Coast Castle and Princes Island, and then return. Nevertheless she took such opportunities as she was offered to board suspicious vessels before arriving off Cape Coast Castle on Christmas Day 1809, and there her Commanding Officer, Commander Josiah Wittman, went ashore. She lay at anchor for a fortnight before sailing for Princes Island. Two days into the passage Commander Wittman died of fever, the first of many commanding officers to fall victim to the diseases of the slave coast during the course of the campaign.

Tigress, which was to become the Squadron's stalwart of this opening period, de-stored the wreck of *Solebay* and then cruised the coast of Senegambia for some months, making a foray north to the Canaries for supplies before returning to Sierra Leone in early December. None of her boardings led to an arrest, and her only excitement was to lose both topmasts in a gale. An inshore cruise at the end of the year took her back to Goree where, on 3 February 1810, she met *Crocodile*, which had made a swift passage from England. After again carrying away her main topmast, accompanying *Crocodile* to Freetown and seeing Captain Columbine installed as Governor, she nosed back northwards, investigating the Isles de Los and sending her boats up the Rio Pongas.

Her first stroke of luck came at midday on 24 March when, from her anchorage at the mouth of the river, she sighted a strange sail. Lieutenant Bones weighed and made sail, gave chase, and at 1645 boarded his quarry, the brig *Rayo* under Spanish colours and loaded with 129 slaves from the Pongas. Probably suspicious that the vessel was a British slaver under bogus colours, Bones put a prize crew on board her and sent her to Freetown, but there the court accepted that she was Spanish and released her. Staying in the vicinity of the Rio Pongas *Tigress* made two unsuccessful boardings before, in reward for her persistence, detaining the slaver *Lucia* on 3 April. She took her prize into Freetown where *Lucia*, unable to prove her claimed Spanish nationality, was condemned, together with her cargo. This was to be the last contribution by *Tigress* for a few months; she was due for refit, and, after re-embarking her two prize crews, she sailed for home on 3 May 1810.

Meanwhile, the cutter *Tickler*, with a length on deck of only 63 feet (making her one of the smallest cruisers to operate on the slave coast), was following a similar pattern, but with less success. She had arrived at Goree in mid-August 1808, and then cruised between Senegal and Sierra Leone, sometimes in company with *Tigress*. She made several boardings without result, and while she was again at Goree in October her boats brought back a damaged American schooner which had been fired on by the overenthusiastic gunners of the garrison as she, perfectly legitimately, left the anchorage. *Tickler*'s Carpenter made amends by repairing the shot damage. Between early November 1809 and late January 1810 she visited Tenerife for stores and made a lengthy cruise in the vicinity of the Canaries, probably in the hope of catching some of the slavers which were thought to be fitting-out among the islands. She then worked her way back to Sierra Leone, where she reported herself to Columbine and was given despatches for England. On 1 March she sailed for Spithead with little to show for her efforts against the slavers.

Hawke had parted company from her convoy by the time she arrived at Goree on 4 October 1809, and was free to hunt for slavers on her way to Cape Coast Castle. Only four days later she chased a stranger and brought her to after firing several warning shots, only to discover that she had intercepted the innocent British brig *Kingmore*, bound from Sierra Leone to Goree. No doubt *Kingmore* feared she was being pursued by a French privateer, and wild goose chases of this sort were not infrequent. *Hawke*'s inshore passage into the Bight of Benin brought her no sign of a slaver, and after a week at anchor off Cape Coast Castle at the end of October she called at Accra before heading south for Princes Island. Her three-week stay there was to have sad consequences. She weighed anchor on 27 November, and a week later, off Cape St Pauls, fever appeared on board. During the next three weeks, as she patrolled the Gold Coast, she lost a midshipman and nine seamen and marines of her complement of 121. Unable to isolate her fever cases on board, as Commander Bourchier believed was necessary, the brig took her pinnace in tow as a sickbay. As her final task *Hawke* headed south to St Thomas and Annobón, the rendezvous areas for homeward-bound merchantmen, in case any should need convoy to England. There were no clients, but she fell in with the ship-sloop *Argus* (18), and the two warships kept company as they headed for home.

With the exception of *Tigress*'s brief concentration on the Rio Pongas, the anti-slaving element of this activity appears to have been well-meaning but haphazard. With no command structure on the Coast following the loss of *Solebay*, tactical coherence was perhaps too much to expect, and lack of organisation was exacerbated by uncertainty about what, other than a British-flagged slaver, constituted a legitimate arrest. The arrival of *Crocodile* at Freetown on 10 February 1810 might well have led to an improvement, at least in the deployment of scant resources, had Captain Columbine been free to exercise command as Senior Officer, but he felt obliged to move ashore and concentrate his energies on governing a fractious colony.

Crocodile lay in the river until the end of March, doing little other than help to fight a fire in Freetown which destroyed 30 houses, but on 30 March her First Lieutenant, John Filmore, took her away on a short cruise to the north. He would have been carefully briefed by his Captain, and Columbine had undoubtedly studied reports on slaving practices along this northern part of the Coast, as well as heeded the views of the previous governors with whom he was surrounded in Freetown. One of them, Ludlam, having returned briefly to England because of the delay to the survey, had taken his passage back to Freetown in *Crocodile*, and no doubt he repeated to Columbine the comments he had made to Macaulay of the African Institution in a letter of May 1808:

> I do not think that anything except active cruisers can prevent an illicit Slave Trade. The first Scheme seems to be, to keep the slaves on shore at some independent place till every thing is ready; then to proceed on the voyage and run off the Coast as quick as possible. It is very likely that in other cases the Portuguese flag will be made use of.[15]

Lieutenant Filmore evidently intended that *Crocodile* should be an "active cruiser", and he coasted northwards for an investigation of the Isles de Los, making some unrewarding boardings en route. Arriving off the islands on 2 April he manned and armed the barge, and sent her among the islands for 24 hours while *Crocodile* herself cruised in the vicinity as close inshore as she could, lying at anchor overnight. The barge returned empty-handed on the morning of 3 April, and Filmore took the ship the short distance to an anchorage off Matacong, again sending the barge inshore. The following day the boat found a schooner which she believed to be American and a sloop, both apparently abandoned, and took possession of them. Filmore put a crew into the schooner and despatched her to Freetown, while *Crocodile* made sail for the south with the sloop, the *Polly*, in tow. The Vice-Admiralty court decided to restore the *Polly* to her owners, but the schooner *Doris* was condemned for having insufficient papers to prove her alleged Spanish ownership, and, as will be seen later, Captain Columbine had a good use for her.

After her brief spell of independence, *Crocodile* lay moored in the Sierra Leone River for a fortnight before embarking Commissioners Ludlam and Dawes for an expedition to investigate the coast to the south and into the Bights. The long-delayed survey was starting. Final instructions to the commission had reached Columbine only very shortly before *Crocodile* sailed from England, and might not have appeared at all if Lord Liverpool had not been prompted by Wilberforce. These instructions do not seem to have survived, but, at Wilberforce's suggestion, they probably reflected closely a memorandum to the Colonial Office by Zachary Macaulay which still exists, and Columbine was certainly given a copy of that memorandum. The requirements were impossibly comprehensive. All forts, trading places, settlements and rivers of significance between Goree and Gaboon were to be visited, as were the islands of Princes and St Thomas. Full and accurate information was demanded on the state of Africa in general, and "minute and particular investigation" was required into the state of British settlements. A favourable impression was to be conveyed to African chiefs of the principles behind abolition of the slave trade, and there was to be encouragement of these chiefs, and of British subjects on the Coast, to engage in agriculture and trade in the natural products of Africa. So it went on. The memorandum also emphasised the African Institution's enthusiasm for British acquisition of the Portuguese settlement of Bissau in order to curtail the slave trade on the Senegambian coast, but whether this item found its way into Lord Liverpool's instructions is uncertain.[16] Wilberforce also mentioned to Lord Liverpool that the Admiralty had directed Columbine to survey the coast "in a nautical way", but remarked: "that, it is obvious, is a mere nothing"!

With these objectives in mind, *Crocodile*, now the only cruiser on the Coast, sailed from Sierra Leone on 17 April 1810. Three days later she anchored off Cape St Ann to land the commissioners, a procedure she repeated as she worked her way along the Windward Coast and into the Gulf of Guinea. Lieutenant Filmore, again in temporary command, was not to be diverted entirely from his quest for slavers, however. From the anchorage off the Shebar River on the evening of 22 April he sent the barge and cutter, manned and armed, to search the river. They returned the following day carrying the master and papers of the schooner *Esperanza*, which was found to be an American slaver under Portuguese colours, and on 24 April Filmore sent a prize crew to seize the schooner and take her to Freetown, where she was condemned. The boats returned to the ship with the *Esperanza*'s crew, three of whom had volunteered to join *Crocodile*. A few days later, off Cape Mesurado, Filmore sent the boats inshore again, this time to no avail, and on 12 May there was a frustrating encounter off Cape Three Points. In mid-morning *Crocodile* sighted two strangers and made all sail in chase. As she slowly closed in on one of them, a sloop, *Crocodile* cleared for action, but the chase contrived to pass under the frigate's stern and fired several muskets. Thereafter the sloop gained ground on her pursuer, and *Crocodile* lost sight of her at 1800. This seems to have been the first instance of armed resistance to a boarding.

There was less difficulty when the ship sighted a brig under Spanish colours to the east of Cape Three Points on 17 May. *Crocodile* fired a gun and hoisted her colours, and the brig, the *Ana*, submitted to a boarding party. With the brig in company, *Crocodile* passed Elmina the next day and anchored off Komina Fort. With the boarding party reduced to a prize crew of a petty officer, six seamen and two marines, the brig sailed for Freetown, but it would be over seven months before she came before the court. On 19 May *Crocodile* reached the roadstead off Cape Coast Castle, where she lay for a fortnight, and while the commissioners were ashore she achieved a further success. A brig wearing Portuguese colours anchored in the vicinity on the night of 21 May, and the following morning *Crocodile*'s barge was sent to board her. She was found to be the *Donna Mariana*, and documents were discovered in a passenger's trunk which showed that she had been fitted-out at Liverpool for slaving. Filmore took 14 men out of her and replaced them with a prize crew of the same size to take her to Sierra Leone. There the court condemned her as British property under neutral colours. This was the last slaver detention for a month, and the commissioners completed their

Gold Coast visits before *Crocodile* turned back to the west. As the ship rounded Cape Palmas on 22 June she fell in with a schooner which she brought-to with a gun, and the jolly boat was despatched for the schooner's master. Filmore sent a petty officer to take charge of the vessel, whose name and nationality is not shown in *Crocodile*'s log. There is no mention of her in the court records either, so she may well have been released without adjudication.

The schooner had just been taken in tow on the morning of 23 June when the final small drama of the cruise occurred. At 0600 Thomas Ludlam, Commissioner and ex-Governor, died "of a dysantry (*sic*) after 21 days sickness", and the Carpenter began work on a coffin to preserve the body until arrival at Sierra Leone. The ship was making slow progress, however, and by the beginning of July the Surgeon, not surprisingly, had reported to Lieutenant Filmore that to keep the corpse longer "would prove injurious to the health of the Ship's Company". Accordingly, in the early hours of 2 July, Ludlam's body was committed to the deep. It was a further 11 days before *Crocodile* entered the Sierra Leone River, and there she remained for the next three months. Captain Columbine had asked the Admiralty for Lieutenant Filmore to fill a vacancy in *Solebay* in September 1808, and he had obviously taken Filmore with him to *Crocodile*. As he considered the results of this cruise, the survey apart, Columbine must have been well pleased with his choice of First Lieutenant.

He had not been idle himself on the matter of the slave trade. As he reported to the Admiralty on 25 July 1810, in the ship's absence he had kept at Sierra Leone one of *Crocodile*'s boats and a crew, and he had also manned the schooner *Doris* as a tender to *Crocodile*, planning to use both vessels for surveying the local waters. However, they had proved useful for investigating suspicious visitors to the Sierra Leone River, and they had brought in two slavers, the *Mariana* on 20 April and the *Zaragozano* on 2 June. Both were subsequently condemned, the first for having insufficient papers to prove Spanish nationality and the second, Spanish property commanded by an American, for receiving supplies for the slave trade at two British ports. The *Doris* was notable as the first condemned slaver to be fitted-out and manned as a tender to a cruiser for anti-slaving operations.

Meanwhile, developments had been taking place in the broader field. In 1809 the Cuban merchants, with the support of the Captain-General of the colony, introduced an "ourselves alone" policy, owing to a decline in slave imports by foreigners. Only 1,162 new slaves had been landed in that year, and Francisco de Arango, a leading member of the island's planter oligarchy, declared that "All our hopes centre on ourselves alone, and our entire attention must be directed to that end." Cuban slave merchants set about turning this declaration into action. They sent the brigantine *San Francisco* to London to buy trade goods for exchange for slaves on the African coast, they despatched other vessels directly to the slave coast, and they encouraged direct voyages from Africa to Cuba by Spanish ships commanded by Spaniards. They also developed close contacts with North American shipbuilders and merchants in Baltimore and Philadelphia.

Of greater significance to the suppression campaign of the next five years was a Treaty of Friendship and Alliance signed at Rio de Janeiro on 19 February 1810 between His Britannic Majesty and the Prince Regent of Portugal. Among many other issues, the treaty declared that:

> The Prince Regent, being fully convinced of the injustice and impolicy of the slave trade [...] resolved to co-operate with His Britannic Majesty in the cause of humanity and justice by adopting the most efficacious means for bringing about a gradual abolition of the slave trade throughout the whole of his Dominions.[17]

Portuguese subjects would not thenceforth be permitted to carry on the slave trade on any part of the coast of Africa which did not form part of the Prince Regent's dominions and in which the trade had been abandoned by the European powers which formerly traded there. The right was reserved, however, for Portuguese subjects to continue trading in slaves within the African dominions of the Crown of Portugal, and it was to be distinctly understood that the rights of the Crown of Portugal remained to the territories of Cabinda and Malembo, and that there was to be no restraint on the commerce of the ports on the "Costa da Mina" belonging to or claimed by the Crown of Portugal. It is evident that the Portuguese had neither wish nor intention to curtail their African slave trade, and they saw this treaty as protecting the status quo. The merchants of Brazil, and probably the Prince Regent too, were undoubtedly glad to see the ambassador sent in 1810 by the King of Dahomey to the Vice-Royalty with a reassurance to his customers that he, for his part, would maintain the slave traffic. In that year Rio recorded its largest annual import of slaves: 18,677 in 42 ships.

In London the interpretation of the treaty was quite different. The only trading places regarded as Portuguese between Senegambia and the Equator were Whydah and the islands of Princes and St Thomas, and Britain considered that any Portuguese slaver found in the Gulf of Guinea trading at other than these three places might now legitimately be seized by a British cruiser. This represented a sharp change of direction by Britain, and, as the Royal Navy proceeded to carry it into action, the slaving nations watched with either puzzlement or cynicism.

The French, Spaniards and North Americans concluded that Britain was using its new posture, presenting it as philanthropic, as a means of consolidating its navy's command of the sea.

At a distance from the slave coast, and in the course of the general interdiction of wartime merchant traffic, a trickle of slaver arrests continued to be made, although in many cases the presence of slaves on board was probably not the primary reason for detention. Of the four captures made in the approaches to the Caribbean islands in 1809 and 1810, only two resulted in condemnation. *Le Joseph* was taken to Jamaica by the old Fifth Rate two-decker *Argo* (44), Captain Digby, to be adjudged in June 1809, and a month later in the same court the *Jane* was condemned, having been detained by the brig-sloop *Satellite* (16), Commander R. Evans.

There was a higher success rate during this period at another trade choke point, the Cape of Good Hope, although here too there was no particular campaign against the slave trade. Judging by their names, almost all of the vessels detained there were French and would have been condemned as prizes of war in the Cape's Vice-Admiralty court. In 1808 there had been a spate of nine slaver captures in the same area: *La Jeune Laure*, which was condemned specifically under the Abolition Act, *La Parsiphal*, *La Souffleur*, *Marschal Dandels*, *La Prairie*, *L'Esperance*, *L'Aventure*, *Ceres* and *La Gobe Houcha*, all with fairly small cargoes, the largest being 142 slaves. Most of the men-of-war on the Cape Station seem to have been involved: the ancient battleships *Raisonable* (64) and *Leopard* (50), the more modern 50-gun ship *Grampus*, the 12-pounder frigate *Nereide* (36), the Sixth Rate *Laurel* (22), the brig-sloop *Harrier* (18), the ship-sloop *Otter* (16) and the cutter *Olympia* (10). A further five successful detentions were made on that Station in 1809: *Tilsit*, *Le Trois Amis*, *La Venus*, *La Paris* and *La Mouche*, captures which again involved *Raisonable* as well as the 18-pounder frigate *Boadicea* (38) and the captured privateer brig *Caledon*. The traffic around the Cape may well have been discouraged by these successes, and only three slavers were taken in 1810: *L'Urania*, *La Charlotte* and *L'Amazone*, detained by the ship-sloop *Stork* (16), the new 18-pounder frigate *Iphigenia* (36) and the *Otter*. In all, these captures off the Cape of Good Hope achieved the release of only about 430 slaves, but they appear to have stemmed what seems to have been an appreciable French trade from East Africa to Brazil.

Back in Sierra Leone, *Crocodile* lay moored in the river while her Captain brought order to his colony. Throughout this extended period of the ship's inactivity, however, the schooner tender, manned with 45 of *Crocodile*'s men, made cruises of up to a month in the local area, and in early August she brought in the Spanish schooner *St Jago*, although this prize was later restored on appeal. The ship's barge was occasionally manned and armed for more limited forays, but without success.

Captain Columbine's governorship was not without its difficulties, and one of his irritants was the colonial schooner *George*, fitted-out for local use at Goree and Senegal by the lieutenant-governor of those two settlements, Lieutenant-Colonel Maxwell. She was 58 feet in length with six 6-pounder guns, and although she had a small crew of a master and a few seamen, she had been placed under the command of Lieutenant Moore of the Royal African Corps, and her crew was augmented with soldiers of the Corps. Columbine complained to the Admiralty in August 1810 that the *George* had been converted to a cruiser under Maxwell's orders without legal authority. She had taken five small vessels in the Gambia, and had manned them with soldiers, setting up what amounted to a private navy, and Moore had then proceeded to cruise out of Sierra Leone in direct defiance of Columbine. He had taken the *George* into the River Pongas, put her aground, and been obliged to cut a mast away. He had then sent a vessel from the Pongas to Freetown for adjudication, been advised to release her, had refused, and, when the court had restored the vessel and awarded costs against the *George*, Moore had threatened to appeal. Columbine was not to be thwarted entirely, however. While the *George* was lying off Freetown he sent a detachment of *Crocodile*'s marines on board and removed ten seamen: four *Solebay*s and five *Derwent*s who must have been acquired after the Senegal episode, plus a deserter from *Amphion*. This would have left Moore sorely short-handed and Columbine with a salve to his annoyance.

The *George* was by no means unsuccessful in her anti-slaving efforts, and Maxwell must have been well satisfied with her, particularly as her prizes were credited to him by the court. Few details are now available of these captures, but on 15 December 1810 the Freetown court adjudged seven cases brought by Maxwell. Five of them would have been the vessels, apparently Spanish, taken in the River Gambia by the *George*, as reported by Captain Columbine. One of these was certainly the *Merced*, described by Columbine as American under false Spanish colours, and she was condemned for refitting or receiving supplies for the slave trade at a British port. Another, the *Vincedor*, was condemned on the same grounds. A third was probably the *Maria Delores*, judged to be British property engaged in the slave trade and also condemned. The other two, the *Catalina* and the *Santa Barbara*, were released. Another vessel, the *Perla*, had been seized by the garrison of Goree and was condemned for receiving supplies at a British port while engaged in the slave trade. The circumstances of the detention of the eighth vessel, the *Atrevida*, are unknown, but she was released, although subsequent intelligence from the African Institution indicated that she was American under false Spanish colours. The *Perla* was the second success of Maxwell's garrisons; they had seized the *Floridana* earlier in the year on the same grounds.

The colonial schooner was not the only anti-slaving vessel operating on the Coast outside Captain Columbine's control. Head money had inevitably attracted the attention of privateers, and two vessels had obtained Letters of Marque against slavers from the High Court of Admiralty. The first of these was the *Minerva*, commanded and probably owned by Alexander Macaulay, and she made only one appearance, detaining the Swedish schooner *Penel* on 30 July 1809. The court decided to condemn the *Penel*'s slaves as purchased from a British citizen, but released the vessel and her remaining cargo. Macaulay may have been sufficiently discouraged by a poor return from this case to give up privateering, but he seems to have been active for some time in Sierra Leone as an agent for the cruisers and as a chandler. The privateer *Dart*, Captain James Wilkin, was more persistent. Between July 1810 and February 1811 she brought five slavers into Freetown, and all were condemned. The circumstances of her captures are unknown, but the grounds for condemnation were in all cases unusual. The *Cirilla* was condemned to the *Dart* "in consequence of contempt of court and contumaciousness on the part of the claimants [the slavers]". *Hermosa Rita* was initially a simple case of false colours, but she had forcibly opposed being brought to adjudication; within 24 hours of the arrest the slaver crew had retaken the vessel, and for four days she again headed for Cuba before the prize crew managed to regain control. In March 1811 the *Mariana*, the *Santo Antonio Almos* and the *Flor Deoclerim* were the first vessels to be adjudged and condemned under the Anglo-Portuguese treaty of 1810. All of these prizes, with the exception of *Cirilla*, were condemned to the King, and *Dart*, perhaps disappointed by so little profit, sailed for England at the beginning of May 1811.

Under Captain Columbine's control Sierra Leone was now in a tranquil state, and the Governor's burden had been eased by the arrival in February 1810 of Alexander Smith as Deputy to the Governor and as Chief Judge of the Vice-Admiralty court. By now, however, Columbine was a sick man, and in all of his despatches to the Colonial Office from early autumn 1810 onward he begged to be recalled in the spring of 1811. He insisted, nevertheless, on retaining command of *Crocodile*, as he had been promised, pointing out that she would be sorely in need of refit by that summer. In mid-October 1810 he took *Crocodile* for a cruise to Goree and Senegal, advised by his doctor that the only means of saving his life was to escape the fever conditions of Sierra Leone and to go to sea.

Three further slave vessels were brought to the court by *Crocodile* during this period, but there is no record of the circumstances of their capture. The first, the *Diana*, was an American vessel which had been converted to Swedish property at St Bartholomew's, retaining her American crew. She was taken on 11 September with 84 slaves, almost certainly by the frigate's tender which had sailed for a cruise a week earlier, but she was restored on appeal. The *Emprenadadora* and *Los Dos Amigos* were both American vessels fitted for slaving, with American supercargoes and crews but under Spanish colours. Both were released. It is just possible that they were taken by *Crocodile* at the beginning of her short cruise to the northward in October, but it seems more likely that they also fell victim to the tender.

Crocodile returned to Freetown at the end of November 1810, by which time the Admiralty had told the Colonial Office that it intended to relieve Columbine of his command in the early spring. Despite three bouts of fever he had not by any means lost interest in the slavers. He reported to the Admiralty from sea that he had purchased for £850 a "copper-bottomed schooner, 74 feet long" as a tender for *Crocodile*, and that he had despatched the recently returned *Tigress* to the Bight of Benin, "as I conceive that part of the coast to be far more frequented by the slave dealers than any other".[18] The gun-brig *Protector* (12), Lieutenant George Mitchener, had arrived at Freetown in early September with instructions simply to follow the orders of Captain Columbine, and he proposed to send her too into the Bights. The tender he mentions may possibly have been in addition to the *Doris*; *Crocodile*'s log refers later to "the *St Jago* tender", but the *St Jago* was probably the *Doris* herself, purchased from the court and then renamed by Columbine. In any event, *St Jago*'s naval career was brief; she was wrecked in one of the bays of Freetown harbour in the summer of 1811. While their parent ship again lay at her moorings from December 1810 to May 1811, *Crocodile*'s tenders and boats were busy on cruises to the rivers and islands to the north and in surveying the harbour.

A few days after *Crocodile*'s return to Freetown, there came an explanation of the mysterious disappearance of her prize the *Ana*, taken in May 1810 off Cape Three Points. Thomas Bourne, master's mate, who had been put in command of the Spanish brig, returned on board *Crocodile* to report that the *Ana* had been so cranky that he had been obliged to run for Princes Island and leave her there, and three of his crew of eight had died. It is not clear how the *Ana* subsequently reached Freetown, but reach it she did, to be condemned in January 1811 for buying slaves from a British subject in a British port. Three days after Bourne's return one of the tenders brought in another Spanish ship, the *Vivilia*, which she had found in the River Nunez and which was condemned for fitting-out and loading slaving trade goods in London. The final capture credited to *Crocodile* was the *Lucy* of unknown nationality, taken on 10 January 1811, almost certainly by a tender (although the ship makes no mention of it). The grounds for this condemnation were unusual: kidnapping, selling and sending away free natives of Africa who had been released by the court.

Tigress had returned to the Coast by late August 1810 after her brief respite for refit in Portsmouth. She had wasted no time in rejoining the fray, and, as she passed the Isles de Los on her passage south, she had boarded a schooner which hoisted a Spanish ensign at *Tigress*'s approach. Although *Tigress* detained her, and had the satisfaction of returning to Freetown on 31 August with a prize at her heels, the *Pez Volador* was released on appeal. On 9 September *Tigress* was away again, this time to the Bights on Captain Columbine's orders. Her boardings of Portuguese vessels off Little Popo brought no results, but off Badagry on 22 September she found and detained the slaving ship *Marquis de Romana*, flying Spanish colours although (as the African Institution later reported) English-owned and cleared from Liverpool. Taking 23 seamen out of the slaver, and putting on board 14 of his own men under the command of his Master, Lieutenant Bones set a course for Sierra Leone with the prize in company. Deprived of a quarter of her ship's company, and with the remainder preoccupied by the need to guard a sizeable bunch of prisoners, *Tigress* was in no position to continue her cruise. Owing to Doldrums calms and the Guinea Current, the voyage to Freetown took six weeks, and after a similar period in harbour, during which she saw her prize condemned, she made an uneventful month's cruise to Goree and back. On 8 February she set off again for the Bights, but despite boarding a couple of Spaniards and a Portuguese during a fortnight's cruise in the area of St Thomas, she had no success until she reached Cape Mount on her return. On 3 April she sighted a suspicious vessel at anchor under the headland, and, anchoring alongside her, she found her to be an American slaver, the *Elizabeth*. She detained her and took her into Freetown on 12 April, and, although the court records on the case are incomplete, the *Elizabeth* is believed to have been condemned. Lieutenant Robert Bones was now about to find himself with an unexpected role which would keep his brig in harbour for most of the next three months.

Protector had reached Sierra Leone only a week astern of *Tigress* on 8 September 1810, and her first task was to take a brief look around the Banana Islands and to pay respects to the recently deceased chief of the neighbouring Plantain Islands. In the absence of *Crocodile* and Captain Columbine, *Protector* remained at Freetown during most of October and November, but Lieutenant Mitchener manned and armed his cutter and sent her away for a week to the River Scarcies under the command of his Master. The Scarcies is the next opening to the north of the Sierra Leone River, about 15 miles from Freetown, but its proximity to the settlement was no discouragement to slavers. The cutter was in luck, and on 9 November she brought back a slaver brig under Spanish colours, the *Maria*. In adjudging and condemning the prize, the court found that she was commanded by a Dane and owned by a Frenchman.

On 27 November, now with her orders from Columbine, *Protector* sailed for the Bight of Benin. Off Cape St Paul's she boarded a Portuguese slaving brig, another off Whydah, five more at Porto Novo and yet three more off Badagry, but made no arrests. On 29 December she anchored eight miles off the mouth of the Benin River, and on the following day the cutter was sent into the river to reconnoitre, again with the Master. This first recorded probing of one of the rivers of the Niger Delta in search of slavers brought back news of a vessel at anchor. The cutter was manned and armed, and went back upstream, only to find that the vessel was an apparently legitimate Portuguese schooner. It had been a long, hot day for the cutter's crew, without reward.

During January 1811 *Protector* visited Princes Island and St Thomas, making one unprofitable boarding, before returning to the Gold Coast and making an inshore passage to the westward. She boarded further Portuguese vessels off St Andrews Bay and Cape Palmas, without making an arrest, before arriving at Freetown on 5 March. There could have been little doubt in Lieutenant Mitchener's mind about which nationality was responsible for most of the slaving in the Bights. After a routine passage to Goree and back, interrupted only by boarding an American vessel, Mitchener was obliged to haul *Protector* ashore on Bance Island in the Sierra Leone River to replace some of her copper. This was probably the most extensive repair work yet undertaken at Freetown, and it was fortunate that new sheathing was available in the colony.

William Wilberforce, as the leading light of the African Institution, had continued his nagging. In May 1810 he wrote from Tenerife to the Foreign Office with his particular concern of the moment: that the island had become the rendezvous of English and American vessels carrying on the slave trade to Havana under Spanish colours. It was not unusual, he wrote, for these vessels to have two or three nominal masters to suit all circumstances which might occur. He also reported that the granting of Spanish papers was a source of considerable emolument to the Governor of Tenerife, and that the importation of new Africans into Havana was encouraged by the Spanish Government as much as possible.

His conclusions were that a frigate or two should cruise constantly between Madeira, the Canaries and the coast of Africa; that a consul should be appointed to Tenerife; and that measures were needed to stop the slaving vessels in the West Indies.[19] It says much for Wilberforce's influence that his proposal did not fall on deaf ears, and on 8 August at Spithead, Captain The Hon. Frederick Irby of the 18-pounder frigate *Amelia* (38) received orders to proceed with his ship to the bay of Santa Cruz. There he was to examine all the ships belonging to Great Britain and seize those preparing for slaving. He was then to cruise out into the Atlantic for 14 days, to the north-west of the Canaries, and return to Plymouth.

Amelia arrived at Tenerife on 11 September 1810, and found a large ship under Spanish colours lying in the bay of Santa Cruz. This vessel had sailed from Liverpool and was supposedly bound for Buenos Aires, but Irby suspected that she might be heading for Africa. *Amelia* waited until 21 September, and then sailed to patrol to leeward of the islands in order to catch the ship if she should steer eastward. On the evening of the next day, however, she diverted to intercept a ship and a schooner heading for the islands. She discovered that they were the *Gallicia* and the *Palafox*, sailing in company from Goree for Tenerife, and a letter was found on board one of them which gave Irby the evidence necessary to detain them and send them into Plymouth. The story later emerged that the *Gallicia* was an English ship, the *Queen Charlotte*, which had sailed from London to Cartagena de Indias, where she had obtained Spanish papers and a Spanish master. There too she met the *Palafox*, another English vessel, which had cleared from Jamaica as the *Mohawk* and was in Cartagena for a nominal sale. They had sailed in company to Goree, where the Commandant confiscated their papers. Escaping from Goree to obtain new papers at Tenerife, they fell into *Amelia*'s clutches, and were later condemned in England on slaving charges.

In addition to these two, *Amelia* boarded one Scottish, one American and ten English vessels between 23 August and 23 September, but made no other arrests. On her way home, to cap a short but successful cruise, she took the French privateer *Le Charles* (20) after a 13-hour chase, and reached Plymouth on 16 November. At the end of the year Irby received an appointment to the brand-new frigate *Crescent*, but asked to remain in *Amelia*, "having been three years in her and having the highest opinion of her good qualities, and should be sorry to leave her for any other frigate". The Admiralty's decision to approve this request was to be of some significance for the campaign on the west coast of Africa.

In January 1811 there took place one of the few arrests in the open Atlantic Ocean. The brig-sloop *Mutine* (18), Commander Nevinson de Courcy, had sailed from Plymouth in mid-November 1810 with a convoy for Rio de Janeiro, and on 14 January she fell in with the brig *Aragansia Castellano*, under Spanish colours. The brig's master claimed that she was bound from Bristol, Rhode Island, to Tenerife, but she had clearly diverted from that track. The boarding party found leg irons and other slave-trading equipment, and the master and mate admitted that they were in fact heading to Angola for slaves. Presumably on the grounds that he believed her to be an American under false colours, de Courcy detained the brig and sent her into Plymouth, a long haul for the prize crew.

Captain Columbine, ill though he was, remained anxious to bring the hopeless task of Castlereagh's survey to a conclusion. Finding that the deceased Ludlam's notes on the Gold Coast were extremely deficient, and that Mr Dawes had confined himself to astronomy, he intended to run down to the eastward in *Crocodile* to complete the work before returning home. He felt sure that Their Lordships would not have expected him to carry out an "actual geographical survey" with the time and resources at his disposal. The rains lasted late at the end of 1810, however, and the colony, the ship's company of *Crocodile* and Columbine himself suffered from repeated fevers. Orders had at last been despatched from the Admiralty in November that as soon as Columbine was relieved by Lieutenant-Colonel Maxwell as Governor of Sierra Leone he was to "leave instructions with the Commanders of the Gun Bts under his command for their guidance and proceedings" and return in *Crocodile* to Spithead. In late April *Crocodile* loaded a valuable cargo of 360 tusks of ivory from the *Elizabeth* schooner which she was to take in convoy, and prepared to return home. There was still no sign of Maxwell, and Columbine, in desperation, handed over temporary charge of the colony to Lieutenant Bones of *Tigress*. On 10 May 1811 *Crocodile*, briefly in company with *Protector*, sailed for England, but it was too late for Columbine. The ship's log for 19 June sadly records that at 0240 Commodore Edward Henry Columbine departed this life, and that at 0845 his body was committed to the deep with the usual ceremony, Fayal bearing S88°E 511 miles.

Although he had not been one of the stars of that illustrious period of Britain's naval history, Edward Columbine leaves the impression of a sound, conscientious and caring officer who tackled an impossible combination of tasks uncomplainingly and made the most of the inadequate resources at his command. It was Wilberforce's opinion that Columbine was a "man of high spirit and strong professional feelings, though of a steady, resolute but rather anxious temper, and of businesslike habits". As the Commanding Officer of *Crocodile* and Governor of Sierra Leone, Columbine signed himself, quite properly, as "Captain", but the logs of the vessels on the Station consistently, and rather touchingly, refer to him as "the Commodore". Clearly he was held in respect, and probably some affection, by his officers.

As *Crocodile* sailed for home, with eight condemned slaving vessels and nearly 600 freed slaves to her credit, she left only two small brigs to cruise the slave coast, and no officer on the Station above the rank of lieutenant. Before his departure from Freetown, Columbine reported to the Admiralty in pessimistic terms:

> I am sorry to acquaint Their Lordships that as far as I can judge the slave trade is carried on again to an extent nearly as great as ever, under the disguise [of] the Spanish or Portuguese flag. The nature of the coast in this neighbourhood renders it dangerous and generally impossible for a ship to go near the shore, and the vigilant caution of the slave dealers induces them to land their slaves as soon as they hear of or see a vessel of war approaching. The only method therefore by which the Abolition Act can effectively be put in force is by means of small vessels.[20]

Although his points were valid, Columbine perhaps gave himself too little credit. The Trade under British colours had been brought to an end by the presence of the Navy's cruisers, and the slavers under false identities on the Senegambia and Windward Coasts had been forced into greater caution. It later became apparent that nine out of about 24 slavers to leave Cuba for the African coast during 1809 and 1810 had been intercepted. By Captain Columbine's initiative the idea of using suitable condemned slaving vessels as tenders to the cruisers had been tried successfully, albeit without Admiralty authority, and, despite his distractions, Columbine had begun to introduce some tactical coherence to the campaign.

Although the slave trade was no longer openly carried out under British colours, it was known that British subjects were still active in the slave factories of the Senegambia and Grain Coast rivers. It was also recognised in London that there was still widespread British involvement in the Trade under foreign colours and in the areas of financial investment and supply of trade goods. Furthermore, it was appreciated that the 1807 Abolition Act was inadequate to deal with much of this activity. Consequently, in May 1811, Parliament strengthened and extended its abolition legislation with a new Act (51 Geo. III c. 23), often referred to as the Felonies Act. From 1 June 1811 not only slaving but also aiding and abetting the Trade, and fitting-out a slaver, would be treated as felonies, punishable by up to 14 years' transportation or three to five years' hard labour. For slaver petty officers and below arrested for violation of the Act, offences would be treated merely as misdemeanours, carrying a lesser sentence of imprisonment not exceeding two years. Ratings in slavers who gave information against their officers would not be liable to punishment. The Act allowed colonial governors and commanders-in-chief, as well as officers of ships of war and authorised privateers, to seize slave vessels and to benefit from seizures, giving legitimacy to the kind of action already taken by Lieutenant-Colonel Maxwell at Goree and Senegal. Existing regulations concerning forfeiture of slave vessels and slaves were unchanged, and there was no further restriction on movement of slaves from one British settlement in the West Indies to another. The severity of the punishment of transportation introduced by this Act, effectively a life sentence, finally ensured that overt and active involvement in slaving by British subjects, or anyone under British jurisdiction, was a thing of the past. Involvement behind the scenes or under false identity was another matter, however, and the question of the legitimacy of British investment in a foreign slave trade remained a matter of debate.

The powerful new deterrent of the 1811 Act was not the only helpful development in May of that year. Following a classic case of false identity, measures against British slaving under Spanish or Portuguese flags, and against illegal slave vessels which had hastily relanded their cargoes, were strongly underpinned by important legal judgments.

In July 1810, the *William and Mary* had sailed from New York under American colours and commanded by her American owner, George Trenholm. At Madeira, in order to obtain Portuguese papers for her, a false sale was made to a Portuguese clerk in an English mercantile house. The vessel was renamed *Fortuna*, and another Portuguese was appointed master, but Trenholm remained as supercargo with full control of the voyage and ship. On 6 October she sailed from Madeira, and about seven miles off Funchal she happened to fall in with the frigate *Melampus* (36), Captain Edward Hawker. The boarding party's inspection showed clearly that *Fortuna* was heading to Africa for slaves, and Hawker sent her into Plymouth instead of Freetown. Consequently her case came before the great Sir William Scott, Judge of the High Court of Admiralty. In making his adjudication on 12 March 1811, Scott stated in forceful terms that by every indication, including Trenholm's admission, the sale in Madeira had been for the sole purpose of covering the real property. The *Fortuna* was American, he said, and, by the precedent of the *Amedee* case, he condemned her and her cargo.[21]

The court went on to make another important decree. The "right of seizure was conferred not merely by the circumstance of a ship having actually traded in Slaves, but by the manifestation of an intention to trade in Slaves". It then helpfully indicated some of the legitimate grounds for suspecting that intention: a disproportionate number of water casks or quantity of provisions; bulkheads for confining slaves, or the makings of them; small tubs for messing slaves; chains and fetters; and main-deck gratings (for ventilating a slave deck). No longer would a false sale provide any protection in law, and no longer would the mere absence of slaves prevent condemnation of an evident slaver, at least in British courts.[22]

The cruisers might not have become aware of these crucial decrees had it not been for the African Institution. Zachary Macaulay produced printed accounts of the *Amedee* and *Fortuna* judgments and sent them to the Admiralty for circulation, together with the Institution's comments. Initially Their Lordships took no action, but were finally persuaded in May 1811 to distribute the sheets to commanders-in-chief and the cruisers, without the Institution's comments. Commanding officers were already being given copies of the Abolition Acts, subsequent Orders in Council and the Portuguese treaty to interpret as well as they could, but this circular from the African Institution was apparently the only additional guidance they received.

These moves in Parliament and the High Court were being matched by efforts on the diplomatic front. Pressure against the Trade had not concentrated solely on Portugal and Brazil during these early post-abolition years. Immediately after the end of hostilities between Britain and Spain in 1808, the Foreign Secretary in London had urged on the Spanish Government the desirability of gradual abolition of the slave trade throughout its empire. In 1810 William Wilberforce was in touch with Marquis Wellesley, the British Minister in Spain, on the same theme, and the following year the Foreign Office assured Wilberforce that the Minister in Cadiz had impressed upon the Cortes the "Justice and Policy" of abolishing the Trade. These overtures found little favour, and in 1811 the Spanish Ambassador in London complained about the seizure of slaving vessels under Spanish colours. In a robust response Wilberforce suggested to the Foreign Office that it should ask the Ambassador for more information on these vessels. He pointed out that in all (or at least nearly all) instances there was reason to believe that the property and ships under Spanish colours were really American or British.

Despite these unpropitious signs, the new liberal constitution in Spain allowed a legislative assembly which represented the empire as well as the parent nation, and this gave the opportunity, in March and April 1811, for the first formal proposals for abolition of the slave trade and even of slavery itself. The debate which followed struck the Cuban planters with horror, and the island's Captain-General wrote to the Cortes in July requesting the Government to treat the idea of abolition of the Trade with reserve, "in order not to lose this important island".[23]

The immediate concern to the British Government, however, was to prevent British slavers from trading with Cuba under Spanish colours. In May 1811, Wellesley was told to ask Spain "to take all necessary action" to prevent British and American slaver masters from using Spanish colours and documents. The Spanish Government was to be assured that the Royal Navy would be available to help put into effect any regulations Spain might care to introduce, detaining suspicious vessels at Tenerife for instance, and having them condemned in cases in which it was not positively shown that "the whole concern" was bona fide property of Spanish subjects. Anxious not to risk losing British support in the fight against the French, the Spaniards made a conciliatory reply, saying that the authorities in Tenerife would be instructed to act as appropriate, but they showed not the slightest enthusiasm for naval assistance.

* * *

In the four years since the passing of the 1807 Abolition Act there had been an air of uncertainty about how the legislation might be enforced, and about how far Britain could legitimately extend its action against the slave trade. Despite regular and perhaps irritating badgering by Wilberforce and the African Institution, the Government failed either to define precisely the aim of any action to be taken or to put measures firmly in place. It may have been assumed that once the Act was on the statute book the Trade would die of its own accord. If that was the case it would indicate that ministers were largely ignorant of the extent and workings of the slave trade and had failed to understand just how far it reached beyond British jurisdiction. Castlereagh's demand for a survey gives some weight to that suspicion, although he may merely have been introducing a pretext for action on the slave coast. Whatever the truth of that, the Colonial Secretary gave a lead of sorts, and it was appropriate that he should have done so. The initial requirement was simply to enforce the new law against British slavers, and it was obvious that action would need to be taken at British possessions on the African slave coast. Law enforcement in the colonies was Castlereagh's business. He had to rely on his colonial governors to exercise that responsibility, but in the case of Sierra Leone, at the centre of the key area, Governor Ludlam was a company employee. So it had clearly been necessary not only to bring Sierra Leone under direct Crown control but also to reach out into the seas of West Africa where the slavers plied their trade; and that, of course, had meant active involvement by the Admiralty.

It was hardly surprising that the Admiralty was not enthusiastic. In all consideration of these early years of the suppression campaign it must be remembered that the Abolition Act came into force only 18 months after the Battle of Trafalgar. The geographical extent of the war at sea against France and its allies was greater than ever, and, with Britain's battle fleets no longer seriously challenged, the demands of the struggle fell disproportionately on the frigates, sloops and brigs of the Royal Navy, a resource that was overstretched at the best of times. It is perhaps remarkable that Their Lordships allocated any men-of-war at all to the coast of Africa, and it is arguable that they would not have done so had it not been for the need to protect trade to and from West Africa and beyond, to give support to the sources of valuable commodities on the Gold Coast, and to deter any French incursion into the area. The Admiralty should readily be forgiven for regarding slave-trade suppression as a sideshow at this stage, but its evident failure to define the anti-slaving role of the cruisers and to issue clear instructions in that regard to the commanding officers is less easy to excuse.

Thanks to this lack of commitment, the deployment of cruisers to West Africa was piecemeal and never remotely adequate, even for a strictly limited national policing operation. This inadequacy was exacerbated by the absence on the Coast of any command structure worthy of the name, and even when a Senior Officer was sent to the Station he was diverted from his proper duty through a disgraceful economy measure by the Colonial Office. The resulting unsystematic, hit-or-miss cruising, with little apparent application of intelligence information, was concentrated mostly to the north of Cape Palmas. The occasional more distant cruises to leeward were largely for the purpose of supporting Cape Coast Castle and the other British forts on the Gold Coast, or for gathering convoys at the Bights islands. The Bight of Biafra, long a fruitful trading area for English slavers, was entirely ignored. Rarely was there any attempt to probe the tortuous rivers where British slave merchants, among others, traded openly. Most captures on the Coast were made at sea, and largely by chance. Even more fortuitous were the seizures in mid-Atlantic, and arrests in the approaches to the Caribbean islands were made by cruisers primarily concerned with intercepting war contraband and French privateers.

For a slaver the risk of capture was low. Nevertheless it was enough to deter British renegades from sailing under their own colours. The obvious and apparently safe option of assuming false identities under the ensigns of neutral or allied nations was very quickly adopted by these outlaws, and it was scarcely surprising that the British authorities, the abolitionists of the African Institution and especially the Navy's cruisers should be anxious to prevent this fraud. More difficult to understand was the erroneous belief, not only by the cruisers but also by the courts, that Britain was entitled to enforce the abolition legislation of other nations, particularly the United States, without authorisation by those nations. These two factors gave rise to what would now be termed "mission creep", and the process was given further encouragement by the woefully imprecise treaty with Portugal in 1810. For a high proportion of the Navy's arrests during this initial phase of the campaign the legal justification was, at best, very tenuous. Much of the blame for that lay with the lack of clear briefing from London, and the trend was encouraged by the inexpert Vice-Admiralty court at Freetown. The *Amedee* and *Fortuna* judgments had probably muddied the water rather than cleared it.

The legal and diplomatic uncertainties were to some extent masked by the war, but they would soon have to be grasped, and if any imagined that solutions would be readily found they were quite wrong. In other ways the cruisers had not yet been faced by the worst that operations on the slave coast could bring. So far the Squadron had been remarkably little touched by the diseases of West Africa, there had been no interference from the French and, with two minor exceptions, intercepted slavers had submitted peaceably to detention. Change for the worse would not be very long delayed.

CHAPTER 5

The Tip of the Iceberg, 1811–15

It is burning a farthing candle at Dover, to shew light at Calais.
DR SAMUEL JOHNSON

THE CAMPAIGN against the slave trade had indeed begun, but the Royal Navy was by no means clear about what role it was required to play. Strengthened legislation against involvement in the Trade by British citizens had reached the statute book, but it was becoming clear that slaving under British colours was a thing of the past, and the few Britons still directly engaged in the traffic were operating either under the cover of foreign flags or ashore in Africa and clear of British jurisdiction. Enforcement of the British abolition laws had therefore rapidly become a minor task, but realisation that other nations were not about to follow Britain's example in abolishing the slave trade was presenting a challenge of a quite different order of magnitude.

No proper consideration had been given to how this challenge should be tackled, and, despite the entirely inadequate 1810 treaty with Portugal, there was no basis in law for measures by the Royal Navy against foreign slavers. Misguided notions of "natural law" and the presumption that Britain had the right and duty to enforce American abolition legislation had led British men-of-war into illegal action. The Vice-Admiralty courts, lacking in expertise, had done nothing to discourage it. A confused situation was further clouded by belligerent rights against the carriage of war contraband by neutrals, and, still embroiled in a worldwide struggle against Napoleon, the Government had weightier matters on its mind than clarifying the rules for suppressing the international slave trade. The Admiralty, in particular, its resources of smaller warships stretched to the limit, had no desire to concern itself in any way with this unwelcome additional task.

Meanwhile, there was a hiatus not only in the command of the Royal Navy's minuscule squadron on the west coast of Africa but also in the administration of the infant colony of Sierra Leone, the Squadron's base.

* * *

Lieutenant Robert Bones of *Tigress* must have been greatly relieved that he held the post of Governor of Sierra Leone for no more than seven weeks. His tenure of office (and a period of freedom from supervision for *Tigress* and *Protector*, the two remaining British cruisers on the west coast of Africa) came to an end on 29 June 1811 with the arrival of the Sixth Rate *Thais*. Taking passage in *Thais* from Senegal to Freetown was Lieutenant-Colonel Maxwell, and he assumed the governorship of Sierra Leone on 1 July. *Thais* also brought with her a Mr Thorpe, sent from England to be Chief Justice of Sierra Leone, a much-needed introduction of legal expertise to the colony.

While her Commanding Officer had been ashore in Government House, *Tigress* had effectively been confined to Freetown, but her boats had taken possession of the slaver *Capac*, brought into the river by the merchant brig *Telegraph*, Mr E. Griffiths, and condemned as Droits of Admiralty. Meanwhile *Protector*, after saying farewell to *Crocodile* on 10 May, had made a short cruise to the Plantain Islands, at least partially to pay respects to the recently deceased chieftain of the area. Then, after a brief stay in Freetown, she sailed for the Gulf of Guinea in mid-June.

Thais (20) had recently been recategorised from ship-sloop to Sixth Rate, one of a number of ships so elevated, and so she now had a junior captain in command. This was Edward Scobell, and when he sailed with a convoy from Spithead for the Coast of Africa at the end of April 1811 he may have expected to be de facto Senior Officer on the Station. When he arrived at Goree on 12 June, however, he found the frigate *Arethusa* (38) commanded by the considerably more senior Captain Holmes Coffin. *Arethusa*'s orders were couched in terms similar to those used for a number of temporary cruisers on the coast:

> On parting with her convoy of East Indies ships in Latitude 15°N [off Cape Verde], to close with the Coast of Africa about Goree, as far to the southward as the Island of St Thomas, looking into the several Bays and Creeks for the seizure of ships carrying on the traffic in slaves, and having proceeded down the Coast as far as the Island of St Thomas to return in like manner along the Coast as far as Cape Verde on this service, and then repair to Spithead.[1]

Probably unbeknown to either Scobell or Coffin, there was already another visitor in the area, the Sixth Rate *Myrtle* (18), also newly promoted from ship-sloop, which was on a long leash from the Tagus squadron. There appears to have been no ordered pattern or policy to these deployments, and no command structure had been put in place to control the operations of what now amounted to a squadron of five cruisers.

Captain Clement Sneyd of *Myrtle* had been ordered by Admiral Berkeley, commanding the force in the River Tagus, to escort a Cape-bound transport to the southern limit of the Station, to detach to St Jago in the Cape Verde Islands for despatches, and to cruise back to the Tagus. He was not to be away for more than 42 days. A cruise to the slave coast was clearly not part of Berkeley's plan, but Sneyd intended to make the most of his freedom from tedium off Lisbon. *Myrtle* arrived in the anchorage of Porta Praya at St Jago on 13 May and immediately made her presence felt by seizing the 100-ton schooner *Roebuck* of London with slave irons and equipment concealed on board. On sailing four days later with the *Roebuck* in company, she sighted a ship making for the African coast and boarded her. This was the *Gerona*, which Sneyd detained on discovering that she had been owned and fitted-out in England before being sold to Spanish merchants a few days before sailing from Havana. This run of success continued on 18 May when another recent Spanish acquisition, the American-built *Santa Rosa*, was added to the tally. Whatever good intention Captain Sneyd may have had for heading back to Portugal was now overtaken by his need to take his prizes south to the Freetown court and to recover his prize crews, and he proposed to make good use of his passage to Sierra Leone.

Judging from the reports of natives off the mouth of the River Gambia, Sneyd suspected that there were vessels slaving in the river, and so he manned and armed his boats and sent them into the river under the command of Lieutenant Scott, his First Lieutenant. They re-emerged the next day with the *Nuestra Señora de los Dolores*, an American schooner under Spanish colours, carrying 112 slaves. Still not content, Sneyd then put Lieutenant Absolon into the *Roebuck* with twelve men and sent her to search the Rio Pongas while *Myrtle* and her three other prizes made for Freetown, arriving on 4 June. On 26 June the court condemned the *Gerona* and the *Nuestra Señora de los Dolores*, but released the *Roebuck* and the *Santa Rosa* on the grounds that their intention to trade slaves had not been clearly proved. This was perhaps a disappointing result, particularly in the case of the *Roebuck*, but news of Sir William Scott's decree on right of seizure probably did not reach the Freetown court until the arrival of *Thais* three days later. Despite that, *Myrtle* had produced an excellent three weeks' work.

Arethusa was having a less satisfactory time. Having sailed from Goree in company with *Thais* and the *George*, she parted company on 23 June, and was next heard of when her launch arrived at Freetown on 1 July with the news that the ship had grounded on a sunken rock off Factory Island in the Isles de Los and needed assistance. *Thais* sent a watch of hands to help *Tigress* prepare for sea, and Lieutenant Bones, no doubt glad to be leaving Freetown astern, took his brig northwards on 3 July, followed by *Myrtle*.

That same day, *Protector* returned to harbour with a schooner prize, the *Palomo*, and on 4 July she too set off for Factory Island. Three days later the crippled *Arethusa*, making six and a half feet of water an hour, was ready for the passage to Freetown. She arrived safely two days later, accompanied not only by *Myrtle*, *Tigress* and *Protector*, but also by two prizes she had taken before her grounding: the schooner *Hawk* and the brig *Harriet*. *Arethusa* was immediately hauled up onshore for repair, while her prizes, and *Protector*'s, were condemned to the King (although the *Harriet* was restored on appeal). Both of the frigates deployed to the Coast since the start of the campaign had grounded, giving further weight to Captain Columbine's contention that the inshore hunt for slavers was work for small vessels.

Captain Sneyd's 42-day deadline was already long past, but he was probably not greatly disappointed that he was still not free to sail for the Tagus. As *Arethusa*, now seaworthy but hardly sound, began her passage home, the rest of her mission abandoned, Captain Coffin took the precaution of keeping *Myrtle* in company. On 12 August, Sneyd was allowed a final reconnaissance of the River Gambia, and Coffin loaned him *Arethusa*'s cutter to help. It was discovered, however, that the *George* had just left the river with a slave brig, so *Myrtle*, now released by *Arethusa*, made for Goree, where she rejoined her prize *Gerona*, which she evidently intended to take north with her. Then she sailed for the Tagus to face the wrath of Admiral Berkeley, probably not displeased with herself. *Arethusa* reached Plymouth on 21 September for docking and refit.

On his return to Lisbon, Captain Sneyd reported to his Admiral what he had learned of the slave trade. There were, he wrote, American vessels under Spanish colours going into the Gambia every week, and nearly a thousand slaves were ready for shipping off when he was in the river. He had been told that there were no less than a hundred vessels slaving under Spanish or Portuguese colours, mostly in the Bights and further south. At Angola there were apparently six or seven larger ships of eight to ten guns. He had found that forged papers were generally picked up at St Jago, and that the heraldic arms for sewing onto false Spanish colours were made in England. The slave trade, particularly to leeward (the Gold Coast and beyond), was considered to be more flourishing now than before the 1807 Act. The inhabitants of Goree were apparently assisting the Gambia slave merchants with supplies, and at Sierra Leone, "many of the captured vessels have been purchased for a mere trifle and resold to slavers who have immediately carried off cargoes of slaves". Sneyd's assessment was probably not wide of the mark in any aspect, and his point about the resale of forfeited slave vessels was the first mention of what was to become a long-standing scandal in Freetown.[2]

Thais had sailed from Freetown on 16 July 1811 with the one remnant of her convoy, bound for Cape Coast Castle. On 27 July she sighted a strange sail at anchor inshore off Trade Town on the Grain Coast. She hoisted a Spanish ensign, anchored alongside the stranger, boarded her and found her to be the English-built vessel *Havanna* under Spanish colours, with 100 slaves on board. The master and mate were either English or American, and the vessel had last been fitted-out at Liverpool. Captain Scobell arrested her and sent her to Sierra Leone where, as was the general practice in a successful case against a loaded slaver, she was condemned to *Thais* and the slaves were condemned to the King for release in Sierra Leone.

After a period in the Cape Coast roadstead, where she loaded ivory and left her convoyed stores ship, *Thais* examined 13 Portuguese slavers along the Slave Coast, detaining two of them. The first, off Badagry on 30 August, was the brig *Venus*, and the second the brig *Calypso*, taken off Lagos three days later. Scobell sent them both off to Freetown and *Thais* sailed south for uneventful calls at Princes Island and St Thomas. Heading back to Cape Coast on 26 September she found the *Venus* struggling against the current on her way to Sierra Leone, and gave her a tow for two days. After loading more ivory at Cape Coast Castle and gold dust as well as ivory at Accra, a sure sign that she was about to head home, she turned westward. She called at Freetown for provisions, but did not await the arrival of her prizes, and sailed for Spithead on 13 December. The *Calypso* was restored on appeal, but the *Venus*, when she finally arrived, was condemned. The slave coast had not seen the last of *Thais*.

A week after *Thais* had left Freetown for Cape Coast, *Tigress* sailed for an abnormally long haul to the southward. After a brief stop at Annobón, she reached away to the African coast and worked slowly south from the River Sette to Cabinda, standing into the bay on 26 August. Here she found a ship, two brigs and a schooner, and Lieutenant Bones decided to detain the ship and the schooner, although it is not clear what his reasons were for doing so; they were Portuguese vessels trading legitimately for slaves at a recognised Portuguese settlement. With the Master, a midshipman and a third of his men in the prizes, and 17 Portuguese in *Tigress*, Bones was in no position to continue the cruise, and he sailed the next day for Sierra Leone with the two slavers in company. He touched again at Annobón, and was back at Freetown on 14 September. It had been an unusual cruise, prodding the Portuguese slaving heartland south of the Line for the first time, but it had been a rather pointless effort; the two prizes, the *Paquette Volante* and the *Urbano* were restored on appeal, but at least the 97 slaves were released.

After only four days in Freetown, *Tigress* sailed again for a very different adventure. The colonial schooner *George* was stranded up the Rio Pongas, and Governor Maxwell was obviously anxious to extricate her. After embarking a Rio Pongas pilot, *Tigress* weighed at sunset on 18 September in company with a schooner, probably the tender *St Jago*. Crossing the Rio Pongas bar on 22 September, she felt her way upriver, using warps and kedges, and with her cutter sounding ahead. A Mr Lawrence was already embarked to assist in negotiations, and on 23 September the local chief, Mango Cutty, and 13 of his people came on board to help in the recovery. After a night at anchor, *Tigress*'s boats took her in tow, and she came to abreast King Cutty's enclosure, a quarter of a mile from the factory of a Mr Samuel, where the *George*'s crew had been imprisoned. *Tigress* then contrived to seize a boat from the factory carrying Messrs Elkins, Curtis and Irwin, presumably British slavers, and a crew of 16. These people were held on board by Lieutenant Bones as hostages until the palaver for the *George*'s return had been satisfactorily concluded, and then Elkins and some of the boat's crew were released. King Cutty and his men were landed and *Tigress* weighed her kedge, but she had to grapple for her bower anchor, a task not eased by a constant downpour of rain with thunder and lightning.

On 25 September the master of the *George*, Mr Creedy, and his 19 men came on board and were taken back to their schooner. That evening the river was struck by a heavy tornado, but the following day *Tigress* and the schooner began their passage downriver in light airs, warping, kedging and towing with the boats. King Cutty came on board to say farewell, and he was saluted with 13 guns. Two days later, still amid heavy rain, the remaining hostages were released and Mr Lawrence was landed, but there were still three more frustrating days of towing and anchoring against a sea breeze before sail could be made over the bar. It is not clear why Lieutenant Bones missed the opportunity to arrest this bunch of slave traders, but it is reasonable to suppose that that such a course of action was opposed by King Cutty as damaging to the profitable traffic in his domain. Nevertheless, *Tigress* had achieved her objective in a week of hard work, determination and fine seamanship.[3]

After leaving the Rio Pongas, *Tigress* discharged 14 Kroomen, additional hands who had, no doubt, given invaluable help in working the brig into and out of the river. Perhaps Lieutenant Bones would have been glad of them five days later when *Tigress* had to resort to the sweeps for an abortive investigation of the Rio Grande, and again during the middle watch of 20 October when the men were once more at the sweeps for three hours in chase of a strange sail, which frustratingly proved in the morning to be an innocent English brig. On 25 October *Tigress* anchored at Freetown, where she lay throughout November.

Although the Caribbean and its approaches saw a spate of slaver captures during 1811, there is no evidence that the naval dispositions in the area were in any way influenced by anti-slaving considerations. Ten vessels carrying very small numbers of slaves, probably the private trading purchases of their officers, fell foul of what seems to have been general wartime trade interdiction during the year and were taken into Tortola. Judging from the names of vessels and masters shown in court records, they were of various nationalities including Spanish, French, American and possibly Dutch. Of these ten, *Edward*, *Falcon*, *Scourge*, *Industry* and *Sally* were released or restored to their owners on appeal some years later.* Four of the five vessels successfully prosecuted, *Porpoise*, *Espiegle*, *Regent* and *Adela*, were taken by the Bermudan-built schooner *Laura* (10), Lieutenant Charles Hunter, as close to an ideal anti-slaving cruiser as was available at the time. The other, *Gibraltar*, was arrested by the brig-sloop *Amaranthe* (18), Commander George Pringle.

There were, however, six true slavers arrested in these western waters during 1811. Available details of these captures are sparse, but the first was the *Neptune*, of unknown nationality, taken into Tortola with 140 slaves in January by the ship-sloop *Wanderer* (16), Commander Gore Willock. For some reason she was restored to her owner, but there were three successful prosecutions between March and June at the Bahamas court involving a total of 448 slaves: the Spanish *El Alrevido*, and the Portuguese *Sancta Isabel* and *Joanna*, taken respectively by the brig-sloop *Colibri* (16), Commander J. Thompson, the ship-sloop *Rattler* (16), Commander A. Gordon, and the schooner *Decouverte* (12), Lieutenant R. Williams. The sloop *St Christopher* (18), Commander McCulloch, took the Spanish schooner *San José y Anemas* into Antigua with 211 slaves, and, further north, the Bermuda court condemned the Spanish-flagged brig *Empressa* for being equipped for the Trade when arrested in March 1811 by *Colibri* and the ship-sloop *Lille Belt* (14), Commander Arthur Bingham.

With relations between Britain and the United States deteriorating rapidly, the western Atlantic was about to become a more active theatre of war, and, judging by the Vice-Admiralty court records, there followed a hiatus in anti-slaving work. Several merchantmen carrying penny-packets of slaves were arrested in 1812 and all were condemned at Tortola, although probably for carrying war contraband

---

* The captors of the five vessels taken into Tortola and restored to their owners were: *Laura*, *Arachne*, *Liberty* and *Maria*.

rather than slaving. *Laura* accounted for *Anthony* and *Augustus*; *Amaranthe* for *Penelope*; the cutter *Maria* (10), Lieutenant W. H. Dickson, for *Jonah* and *Coque Maire*; and the brig-sloop *Peruvian* (18), Commander A. J. Westopp, for *Prevoyant*. However, there was only one recorded seizure of a true slaver in the Caribbean area between 1811 and late in 1813: the Spanish schooner *El Dos de Mayo*, taken by *Maria* into Antigua with 52 slaves.

Not only did the commanders-in-chief in the western Atlantic and Caribbean not make dispositions to intercept the slave traffic but also, in the view of the African Institution, some officers in the West Indies connived in the illegal trade. In a letter of May 1811 to the Colonial Office, Zachary Macaulay reported the episode of the Spanish-flagged schooner *Jove* recently taken into Barbados by the brig-sloop *Persian* (18). The *Jove* had been released although, as Macaulay asserted, there were good grounds for her condemnation, and it was rumoured that Vice-Admiral Sir Alexander Cochrane, Governor of Guadeloupe at the time, had an interest in the slaver and was anxious to see her released. Such was the support in Barbados for the illegal traffic that the Commanding Officer of *Persian* had great difficulty in obtaining an agent for his case, and the man who finally undertook the task probably betrayed his client and managed to deprive him of his right of appeal. Cochrane, a slave owner with property in the Caribbean, was suspected of involvement in contraband slaving, particularly in dealings with the cargo of the *Amedie*, condemned and freed by the Tortola court in 1808.[4]

Macaulay also alleged that when the Saintes was captured by a British squadron in 1809 a number of French slaves were seized by the ships and taken to Martinique where, with the full knowledge of the British military authorities, they were sold at public auction for the benefit of their captors. Furthermore, slaves were being smuggled by night into the islands of St Croix and St Kitts from the Swedish island of St Bartholomew, as well as into Barbados and Demerara, and Macaulay clearly suspected that the British governors were deliberately placing no obstacles in the way of this traffic.

In October 1811 the naval interregnum on the African slave coast was drawing to an end. Early that month, perhaps influenced by the success of Captain Irby's brief cruise to the Canaries in 1810, the Admiralty ordered him to hoist a broad pendant in *Amelia* and:

> To take *Kangaroo* under his orders and direct her to take under convoy Trade bound to the Coast of Africa and proceed with it to Goree, there to join him. [...] To proceed in *Amelia*, calling at Goree, and take under his command H.M.

> Ships and Vessels on the Coast of Africa, proceed along the coast as far to the southward as Benguela, and receive on board Ivory and Gold Dust. To send one of the vessels the most in need of repair with convoy from Princes Island the latter end of December, and another before commencement of the Rains in May next. During the time he may thus be employed, which may be from 9 to 12 months, to make the Isle of St Thomas his principal Rendezvous, and at the expiry thereof to proceed thence with convoy to the Downs, leaving necessary Instructions.[5]

Obviously Their Lordships' minds were still focused on protection of trade rather than anti-slaving operations, and the specification of Benguela as the southern limit of patrol raises a suspicion that the Admiralty was vague in its understanding of the Portuguese treaty.

Commodore The Hon. Frederick Paul Irby, however, was a man to act on his own judgment. He was a son of Lord Boston, educated at Eton, a veteran of The Glorious First of June and Camperdown, and had been made Post in 1802 at the age of 23. He had taken command of *Amelia* in late 1807, and had distinguished himself in February 1809 when his ship had been largely responsible for cornering three large French frigates at Sables d'Olonne. After another two years on the coast of France, she had also assisted in destroying the frigate *Amazone*. In contrast with his predecessor on the coast of Africa, Irby was an officer high in the estimation of the Admiralty.

Amelia sailed for Africa on 15 October 1811, and three weeks later the ship-sloop *Kangaroo* (16), Commander John Lloyd, left Spithead with a convoy of five sail to join the West Africa Squadron.[*] After a month at Sierra Leone with no sign of *Kangaroo*, *Amelia* sailed for the Gold Coast on 6 December, and the Commodore clearly intended to concentrate on capturing slavers, whatever his orders indicated. He worked his way along the coast as far as Cape Coast roads without incident, although he heard of a couple of suspicious vessels on the Grain Coast, but while watering at Cape Coast on 31 December he seized his first victim. This was the brig *São João* under Portuguese colours, which appeared to be trading outside Portuguese territory, and he despatched her to Freetown before heading eastward on New Year's Day 1812. On 4 January *Amelia* stopped and examined the Portuguese brig *Bon Caminho*, finding her to be bound legally

---

[*] Christopher Lloyd's *The Navy and the Slave Trade: The Suppression of the African Slave Trade in the Nineteenth Century* (Longman Green: London, 1949; Frank Cass: London, 1968) wrongly asserts that Irby had at his disposal the Sixth Rate *Ganymede* and the brig-sloop *Trinculo*. These two vessels were in fact employed far away on quite different operations.

for Whydah to load slaves, but, to Irby's annoyance, carrying canoes procured at Cape Coast undoubtedly for the purpose of taking her cargo off the beaches.

On the following day he wrote a very tart letter to Governor White of Cape Coast Castle, reporting what he had found, making clear the action he would take in the event of a repetition of such assistance to a slaving vessel, and reminding the Governor of the penalties introduced by the 1811 Felonies Act. Irby copied his letter to the Admiralty, complaining at the same time that British settlements were allowing slavers to replenish with water. The Admiralty sent it on to the Colonial Secretary, now Lord Bathurst, who in turn passed it to the Secretary of the Africa Committee, the merchant company which controlled the British trading posts on the African coast. Smarting not only from Irby's blunt message but also from some forthright instructions from the Committee, Governor White was suitably apologetic to the Commodore.[6]

After boarding three legitimate Portuguese slavers off Whydah on 5 January, *Amelia* moved on to Porto Novo two days later and found three more Portuguese brigs which Irby judged were slaving on non-Portuguese territory. Not only that, but also canoes from Cape Coast and other British settlements were discovered on board all of them. On these grounds the Commodore arrested all three, the *Destino*, the *Dezanganos* and the *Felis Americano*, and sent them in for adjudication. Irby was obviously warming to the task, and when *Protector* joined *Amelia* at Porto Novo with news of six vessels lying off Lagos he immediately headed east to investigate. He found that three of them, all in a very sickly state, had received no slaves, and had previously been boarded, presumably by *Protector*, and permitted to trade at Lagos. Warning them to take care to abide by the Portuguese treaty, he turned his attention to the other three which had nearly completed their cargoes of slaves, and he seized them all for violation of the treaty. *Amelia* took the schooner *Flor de Porto* in tow, *Protector* took charge of the brig *Prezares*, and the colonial schooner *George*, which had conveniently appeared, was told to escort the brig *Lindeza*. This sizeable flotilla then turned to the west, and each vessel made the best of her way to Freetown, *Amelia* arriving on 21 February.

This first cruise by the new Commodore was indeed fruitful; all seven of his prizes, with a total of 549 slaves, were condemned. The judgments on the *Sao Jao*, the *Lindeza* and the *Prezares* were reversed on appeal, but this was not until 1820 in the latter two cases. Furthermore, the Bight of Benin had been made keenly aware that a new and energetic Commodore had arrived on the slave coast.

Prior to meeting *Amelia* off Porto Novo, *Protector* had ranged widely. After sailing from Freetown on 14 June 1811, with the intention of making a cruise to leeward, she made an early return with the Spanish-flagged schooner *Palomo*, which she had detained off Cape Mount on 25 June. On her arrival at Freetown with her prize, she was diverted to assist the grounded *Arethusa* at the Isles de Los, but in early August sailed for another attempt at a cruise to the south. During the best part of a year on the Coast, she had suffered perhaps more than her fair share of foul weather. Such entries as: "Fresh breezes and strong gales with thunder and lightning and heavy rain" occur all too frequently in her log, and there is regular evidence of the deterioration of her canvas, cordage and equipment in the conditions of the wet season. Repair to split sails, including the courses, was a constant labour, and in May she had parted her best bower anchor cable in a gale, recovering the anchor only after two days of sweeping by the boats. As she headed south from the Banana Islands a more serious defect came to light: the head of the mainmast was found to be much decayed, and she turned again for Freetown. Then, in the early hours of 10 August, a squall carried away her foretopmast and fore gunter mast, and that began a week-long demonstration of the seamanship skill, energy and self-reliance of this little brig's crew of 50 men.

After an hour of clearing the wreckage of her fore-rigging, *Protector* made her way to anchor off Cape Sierra Leone. By the end of the following day a new foretopmast had been prepared, sent up and rigged, and the brig had come to anchor off Freetown. On 12 August the mainmast was unrigged, the yards and the maintop were lowered to the deck, the main topmast and both topgallant masts were sent down, and the mainmast was lifted out. The following day a new mainmast from ashore, measuring 52 feet in length and 17 inches in diameter, was stepped and the maintop was sent up. Over the next three days the main rigging was set up; the main topmast was rigged and sent up, as were the fore and main topgallant masts; the rigging of the upper masts was set up; the yards were crossed; and the main course, main topsail, boom mainsail and both topgallant sails were bent. Finally the main rigging was rattled-down, and on 17 August she sailed. Her feat had been accomplished without the aid of dockyard facilities, and gave a vivid illustration of what a ship's company on a remote station was able to do and expected to achieve with its own resources.[7]

On this third attempt at a distant cruise *Protector* made an inshore passage into the Gulf of Guinea for a short visit to Cape Coast Castle before retracing her steps and patrolling the Grain Coast for three weeks. During the cruise it was found that a new pair of topgallant studdingsails had been rotted by the humidity and that the gig was worm-eaten beyond repair, and when the brig next returned to Freetown in mid-October 1811 it was discovered that her bends were worm-damaged too. Once she had made repairs with new planking from ashore, *Protector* weighed on 31 October for another, more ambitious, cruise along the Guinea coast. Passing down the Grain Coast she anchored briefly off the Gallinas River, Cape Mount, the Junk River and then Trade Town, but once past Cape Palmas she made a beeline for the Niger Delta. This was a new departure for the Squadron. *Protector* herself had investigated the Benin River a year previously, but Lieutenant Mitchener now wanted to take a look at the south-facing shore of the Delta. Anchoring initially off the River Nun, he worked his way eastward until, on 10 December, he stood over the bar into the Old Calabar estuary. Finding nothing of interest, probably because he did not penetrate far enough, he then headed south, anchoring for a short time off Fernando Po before making for Princes Island where the brig watered and, as was common practice on the Coast, embarked a bullock for subsequent slaughter. By early January *Protector* was back on the Slave Coast, and off Lagos she found and boarded six Portuguese brigs from Brazil.

Lieutenant Mitchener seems to have been rather hesitant about detaining Portuguese vessels on this coast, but, as we have already seen, his Commodore, whom he found off Porto Novo on 8 January 1812, was less reticent. Following Irby's seizures off Lagos on 10 January, *Protector* sailed for Freetown with the Portuguese brig *Prezares* in tow. Mitchener decided to take a wide southerly sweep to gain advantage from the South Equatorial Current and the South-East Trade Wind, and the only relief from the tedium of the two-month passage came on 2 February off St Thomas. Sighting a strange sail, *Protector* cast off her tow, made sail, gave chase and finally boarded the Portuguese ship *Maria Primero* with 490 slaves on board.* Removing 28 prisoners and putting a midshipman and eleven hands on board his new prize, Mitchener again took the brig in tow and, with the ship under his lee, made for Freetown. This satisfactory capture, which resulted in a condemnation for violation of the Portuguese treaty, marked *Protector*'s final contribution to the campaign. She arrived at Sierra Leone on 12 March, stored with provisions, exchanged six tons of iron ballast for six tons of ivory, and on 11 April, worn out by nearly two years on the Coast, she sailed for home with two merchantmen under convoy.

It was not long before *Tigress*, the other old hand, followed *Protector* to England. She had sailed from Freetown on 6 December 1811, initially for a look at the Isles de Los, and had then spent several days nosing among the Banana Islands with the cutter sounding ahead. After coasting as far as Cape Mesurado she returned to Sierra Leone in mid-January 1812 without seeing any sign of a slaver. At Freetown Lieutenant William Carnegie joined to relieve Robert Bones in command, and Bones left the brig at Goree on 20 February, shortly to be

---

* The figure of 490 slaves is shown in *Protector*'s log, but the court records give a number of 381.

promoted to commander. After a fruitless month of patrolling the Senegambia coast *Tigress* arrived at Freetown for the last time, and sailed on 12 April for Cape Coast Castle and Accra. Here she loaded the usual cargo of ivory and gold dust, commodities of such value that they were not willingly entrusted to merchantmen, particularly in wartime. Finally she called at Princes Island before departing for Plymouth Sound on 31 May. With only a brief respite for refit in 1810, she had spent three gruelling years on the Coast, making eight arrests and releasing nearly five hundred slaves.

The newly arrived *Kangaroo*, having shed her convoy, set off to find *Amelia*. She called at Sierra Leone where she found orders awaiting her from the Commodore for a rendezvous, and sailed again in pursuit on 8 January. At Cape Coast a fortnight later she learned of *Amelia*'s departure, and sailed for St Thomas, the Squadron's principal rendezvous. Arriving on 16 February she found one of the Portuguese brigs released by Irby in the Bight of Benin. She loaded with wood and water while she waited, and then, at daybreak on the twenty-first, her boredom was relieved by the sight of a suspicious cutter in the offing. *Kangaroo* gave chase and, with some difficulty, brought-to what proved to be the *Vigilant*, formerly of Liverpool, and claiming to be heading initially for Princes Island and then bound for Bahia with her 63 slaves. Commander Lloyd, however, was convinced that she was aiming for a rendezvous with the Portuguese brig at St Thomas to transfer the slaves, and he arrested her, landing her superfluous hands at St Thomas before sending her to Freetown, where she is believed to have been condemned.

Having done his best to find the Commodore, Lloyd now felt free to pursue his own cruise, and he sailed east for the mainland coast. Drawing a blank at Cape Lopez on 26 February, he headed north for the Gabon. Here he was told that there was a slaving brig in the river, and he tried to take *Kangaroo* after her, but grounded. So he sent in the boats, and they searched for 48 hours without success. Concluding that the brig had slipped past him during the night or was hidden in one of the "impenetrable creeks of the Gabon", Lloyd gave up the hunt and turned back for the Gold Coast. He anchored initially off Cape Apollonia on 22 March, but then encountered a schooner which refused to heave to and made a run for it. The chase lasted as far as Anamabo, where Lloyd at last managed to board the schooner and found, to his fury, that she was a London vessel called *Quiz* carrying a Letter of Marque from Governor Maxwell authorising her to cruise for slavers. To make matters worse, in the course of the boarding *Kangaroo* had dropped her best bower anchor in 35 fathoms and the cable had parted. There was little Lloyd could do about the *Quiz*, perverse and obstructive though her behaviour had been, although he does appear to have heard a complaint by the Cape Coast merchants that she had held to ransom a schooner belonging to Elmina.

Kangaroo managed to recover her anchor, and then made for Cape Coast, where she completed with water and supplies before heading back for the Bight of Benin. On arriving off Little Popo on the evening of 4 April she boarded a Portuguese ketch and learned that there had been a skirmish off Whydah a few days previously between a number of Portuguese vessels and a schooner under English colours. She sailed to investigate, and anchored in Whydah roads the following morning. There she found six Portuguese brigs, including the *Urania* which had been detained by the *Quiz* off Cape Coast Castle but had been retaken by her Portuguese crew. *Urania* was lying close inshore and surrounded by boats; despite this, *Kangaroo* dropped down close to her and sent armed boats to board. On their approach the Portuguese all retreated ashore with as much plunder as they could carry, and the boarding party found only the five men of *Quiz*'s prize crew, one of them in irons.

Kangaroo seized the *Urania* as a slaver, warped her to a safe berth and then examined the other five vessels in the anchorage. She found that four of them were without logbooks or papers, claiming that they had been stolen four days previously by a schooner which had come into the roads under English colours but had then hoisted a black flag. Presuming her to be a pirate, the Portuguese had fired on her and driven her off, but Lloyd had no doubt that this had been the *Quiz*. One of the vessels without papers, the *San Miguel Triumphante*, had bought canoes for slaving at Cape Coast, and had loaded slaves at Popo. Her master refused to come offshore, and the mate and surgeon cut away a boat and escaped. So Lloyd detained her and the cargo of 136 slaves, over half of them children, and on 7 April, with the two slavers in company, *Kangaroo* weighed for Sierra Leone, where she arrived on 15 May. Although both prizes were condemned in Freetown, the judgment on the *Triumphante* was reversed on appeal in 1821.

Lloyd reported the *Quiz* affair to the indignant Commodore, and in an angry letter to the Admiralty the following month, Irby reported the aftermath of *Kangaroo*'s encounter. The schooner *Quiz* was a vessel of 115 tons with a crew of 21, mostly Americans, mounting six guns and commanded by one George Neville. Her odious conduct on the coast induced Irby to take action against her in the Vice-Admiralty court on the grounds of a suspicion of "trading with the enemy", abuse of her Letter of Marque by plundering vessels she had detained, sometimes wearing a warship's pendant, and her master occasionally wearing a uniform resembling that of a lieutenant in the Royal Navy. She was also in the habit of hoisting a skull-and-crossbones flag. The Commodore was incensed that his action failed, a result he attributed largely to the obstructiveness of the King's Proctor of the court who, he found, was also Purser of the *Quiz*. He complained bitterly that a serious injustice had been done and that "indignities had been offered to His Majesty's Service"; he intended to appeal and he sent a copy of the court proceedings to the Admiralty. However, his efforts seem to have come to nought, and he had to content himself with seizing the offending pendant and black flag.[8]

In this instance the hunter appears to have been quite as unpleasant as the hunted, and Governor Maxwell had shown poor judgment in exercising his new authority to issue Letters of Marque. He had clearly failed to take the obvious precaution of consulting the Navy, and there was at least a whiff of corruption about the affair. It seems, though, that those with a financial interest in the *Quiz*, whether or not they included the Governor, made little profit; there is no record of any prize cases in her name in the Freetown court. In general, though, Maxwell appears to have been supportive of the Navy's work, and early in his tenure he raised with the Colonial Secretary the fraught issue of the cost of maintenance of the crews of captured vessels, which was supposedly the responsibility of the captor warships. He pointed out that, if the requirement was enforced, the outlay would be ruinous to commanding officers, and he pressed for a more satisfactory solution.[9]

He also recommended the raising of another company for the Royal African Corps from the released slaves in Sierra Leone, as was later approved. Nevertheless, he seemed fond of conducting his own slaver-chasing activities behind the Navy's back, using the colonial schooner *George*, which had been sent out to Goree in 1809 merely as a communications vessel. The Spanish Government, however, had complained about the *George's* depredations against Spanish-flagged slavers in the Gambia River in 1810, and Lord Liverpool, not yet relieved by Lord Bathurst as Colonial Secretary, had invited Maxwell to explain himself. The Governor's defence in a letter of December 1811 was robust, and he further reported to Liverpool that he now found the *George* too small for serving all three stations for which he was responsible: Senegal, Goree and Sierra Leone, and so he had transferred her to Commodore Irby as a tender and replaced her with a "130 ton, copper-bottomed schooner in every way suitable for communication between the settlements". Although Maxwell does not mention a name, this must have been the colonial schooner *Princess Charlotte*, commanded initially by Thomas Cooper and then by Lieutenant W. S. Sanders of the Royal African Corps, which was to play a valuable part in the campaign for some years.

Once *Amelia*'s prizes had been condemned, the Commodore took action to alleviate the problem of slaver crews highlighted by Maxwell. Choosing the *Lindeza* as the worst of the captured vessels he loaded her with 70 Portuguese and sent them to Bahia. The Governor agreed to victual them for the voyage, and Irby took the precaution of telling *Tigress* to escort this unsavoury cargo well clear of the African coast. The Commodore sent a similar human cargo westward in April 1812. Maxwell's recruiting efforts among the freed slaves had resulted in 50 volunteers for the West Indies regiments, and the arrival of the brig-sloop *Scorpion* at Freetown from Barbados on 6 April provided the opportunity to give them passage. Command of *Scorpion* had devolved on Lieutenant Addis when Commander Gore had lost his life in trying to save a seaman who had fallen overboard, and the Commodore ordered him back to Barbados with the recruits.

While *Kangaroo* had been cruising the Bights, *Amelia* had paid a visit to the Cape Verde Islands partly to find slavers and partly to replenish with bread and flour, but Irby gave himself another task when the ship arrived at Porto Praya on 28 April. Governor Maxwell had told him the story of an English brig which had been bound from Freetown to Goree with several free Africans as passengers, all of them British subjects. The master of the brig, however, had diverted to Porto Praya, where he had sold nine of his passengers. Most of these were known to have been taken to Cayenne, but it was thought that one remained, and Irby determined to recover him. The Portuguese Governor of the islands made every difficulty, and complained that British warships had run away with his soldiers. Indeed several of the garrison had begged *Amelia* to take them on board, but although sympathetic, Irby had refused. Finally the Commodore won his case, and the free African was saved. The frigate found no slavers among the islands, but her mission was otherwise achieved and she returned to Freetown via Goree.

The early months of 1812 saw another temporary and unintended reinforcement of the Squadron in the same pattern as *Myrtle*'s visit. This was in the shape of the Sixth Rate *Sabrina* (20), another of Admiral Berkeley's Tagus Squadron. Her Commanding Officer, Captain James Tillard, would have heard of *Myrtle*'s adventures on the slave coast, probably from Clement Sneyd himself, and when *Sabrina* was sent away on a cruise on 7 December 1811 he availed himself of a similar opportunity. He headed for the Cape Verde Islands, and while *Sabrina* was lying at anchor in the English Roads at Boa Vista on 30 December she sighted a schooner working up under Portuguese colours. The officer sent by Tillard to examine her found a cargo of slaves, and judged that the schooner, the *Princessa da Beira*, was in fact American and brand new. Tillard sent her off to Sierra Leone with a prize crew, and *Sabrina* moved on to Porto Praya, anchoring there on

2 January 1812. In company with *Sabrina* was the schooner *Vesta* (10), Lieutenant G. Miall, and Tillard despatched her, meanwhile, to take a look into some of the other harbours of the Cape Verdes. As Tillard reported to his Admiral, he was now obliged to make for Freetown to appear in the Vice-Admiralty court against his prize, and to recover his prize crew, but he obviously intended to make the most of his passage.

With *Vesta* again in company, *Sabrina* entered the River Gambia on 11 January and anchored off the small town of Jillifree late on the next evening. Here she took possession of the schooner *Il Pepe*, with nine slaves on board and under Spanish colours which were probably false. Tillard found that most of the planned cargo was due to be loaded the next morning, and obviously the contract would not be fulfilled when the slave merchant found that the schooner had been captured. Anxious to lay hands on the remaining slaves, Tillard decided that it was "necessary to adopt prompt and decisive measures", indeed dubious measures. At daybreak he seized several of the slave merchant's canoes, and sent one of the occupants ashore with the message that he would take the others away as hostages if the merchant evaded the slave contract. Eventually, 64 more slaves were sent out, and Tillard returned the canoe crews before sailing with his second prize for Sierra Leone.

In a letter to the Admiralty in early February, Tillard reported that sickness among the slaver crews was likely to delay adjudication until the end of the month, and that he proposed, in the interim, to cruise between the Isles de Los and the Shoals of the Rio Grande to intercept slave traffic from the Nunez and Pongas rivers. There was another particular task for *Vesta*. Governor Maxwell had given Tillard the descriptions of four British subjects known to be acting in violation of the Abolition Act on the Rio Pongas, and *Vesta* was given until the end of February to bring them back to Freetown.* Lieutenant Miall managed to arrest two of the suspects, Messrs Hickson and Samo, the first British slavers other than ships' officers known to have been apprehended under the Abolition Acts. By this time *Sabrina*'s two prizes had been condemned, the *Princessa da Beira* for trading under false Portuguese colours and contrary to the Portuguese treaty (charges which would appear to be mutually exclusive), and the *Il Pepe* for being American and English property.

Tillard appeared to be in no haste to return to the Tagus, however, and by April patience with him had been exhausted at the Admiralty, if not at Lisbon.

---

* The four British slavers were: Wilson, an Irish deserter from the Royal Navy; Hickson, a Scottish ex-mate of a merchantman; Lightburne, a Bermudan ex-mate of a merchantman; and Samo, an English clerk.

On 6 April Their Lordships, having seen Tillard's February letter, ordered Admiral Berkeley to send *Sabrina* home as soon as she reappeared "so that the conduct of her commander may be enquired into for proceeding so far from her station". The order was repeated 12 days later in more urgent terms and requiring *Vesta*'s presence too, informing Berkeley that the investigation would be before a court martial. To add to Captain Tillard's troubles, the Admiralty sent a copy of his letter to Sir William Scott, Judge of the High Court of Admiralty, who gave his opinion that the method used in the Gambia to extract the slaves from ashore had been highly improper, and that Tillard should be denied the head money for them.

By now Commodore Irby's squadron had been reinforced by the arrival of the gun-brig *Daring* (12), Lieutenant W. R. Pascoe, sister vessel to *Protector*. She had left Spithead in mid-March with instructions to take a convoy to Madeira and the coast of Africa, and then to join Irby at St Thomas and follow his orders. After a delay at Vigo with a leaky merchantman, she made a quick start to her anti-slaving work. By the time that she spoke with the *Princess Charlotte* off Cape Mesurado on 4 June she had captured the brig *Centinella* under Spanish colours and sent her into Freetown. She was intending to proceed to leeward, not realising that *Amelia* was astern of her at Sierra Leone, but an eventful few days on the Grain Coast obliged her to change her plans.

Finding a Spanish brig trading slaves at Trade Town, Pascoe detained her and put a prize crew of four on board. This was too few to handle the vessel and control the crew, and the Spaniards were not inclined to submit meekly. On the departure of *Daring* they retook the brig and set Pascoe's men adrift, the first time such a thing had happened but by no means the last. On this occasion, by happy chance, *Daring* found her men, but this was a sharp reminder of an ever-present danger to prize crews. It also demonstrated the wisdom of commanding officers, such as Lieutenant Mitchener, who took prisoners out of detained vessels and replaced them with strong prize crews, accepting the consequent weakening of their own ships' companies. Pascoe was more successful in his next interception, finding and detaining a large Spanish ship slaver, the *San Carlos*, in the same waters, and he was taking her back to Sierra Leone when, on 11 June, he fell in with *Amelia*, a day out of Freetown. The Commodore told him to head north once the prize had been adjudged, to cruise between Cape Verde and Cape Palmas, paying particular attention to the Pongas and Nunez rivers.

Amelia was on her way to Cape Coast Castle, but delayed to make an unsuccessful search for the culprit in *Daring*'s unpleasant incident. On his arrival at Cape Coast on 28 July, Irby learned that the British fort at Winnebah, 60 miles to the east, had been attacked by natives of the surrounding town. He took on board a detachment of the Royal African Corps and landed it with his marines at Winnebah, where it was found that the natives had deserted the town. Irby burned the town and, deciding that the best course of action was to abandon the fort, he demolished that too. This diversion over, *Amelia* set out on a 1,300-mile passage to St Helena, much of it undoubtedly close-hauled, and to no apparent advantage to the campaign. Leaving the island on 8 September, she stretched away to the African coast and worked her way inshore northward from Benguela Bay to the River Congo, investigating the Portuguese slaving places as she went. Irby obviously wanted to ferret out any British slavers engaged in the traffic to Brazil, but he was also keen to learn more of the extent of the Portuguese trade. He assessed that no less than 10,000 slaves were being taken annually from St Philip de Benguela, and at Loando he found 13 vessels slaving under Portuguese colours.[10]

Irby's information at Loando came at a price: *Amelia* maintained the grounding record of Fifth Rate frigates on the Coast, striking the sand extending to the north of Loando Island, but getting off without damage by lightening the ship. Irby calculated an annual export trade of 20,000 slaves from Loando, mostly to Rio, Bahia and Pernambuco, but some, he reckoned, in English and American ships under Portuguese colours to Havana. This suspicion was reinforced when, after leaving St Paul de Loando on 5 October, *Amelia* found a ship in Ambriz Bay, the *Andorinha*, under Portuguese colours but fitted-out at Liverpool and probably English property. Having sent the *Andorinha* with her large cargo of slaves to Freetown, Irby took his ship 60 miles up the River Congo but found nothing. He heard of a slaving schooner further upriver at Embomma, but that was not accessible to the frigate and he judged it too risky to send the boats so far, so he continued to coast north to Cape Lopez. From there he headed out to St Thomas in the hope of picking up information on slavers, but drew a blank and turned for Sierra Leone, which he reached on 8 November.

Before leaving Freetown for this long cruise, the Commodore had sent *Kangaroo* to patrol off Senegambia, and when she arrived off the River Gambia on 11 July she fell in with the colonial schooner *Princess Charlotte*. In a failed attempt to take a schooner found slaving up the river in Ventain Creek, two of *Princess Charlotte*'s men had been badly wounded by the natives of the creek. Commander Lloyd believed that if he could take *Kangaroo* into the creek and so threaten the town the slaving schooner would be given up, but no square-rigged vessel had yet attempted the passage, and the natives had always defied man-of-war boats trying to make seizures. Ordering *Princess Charlotte* to take the lead, *Kangaroo* got abreast of the town without accident, but discovered that the slaver had been shifted about 15 miles further into another creek. Lloyd threatened that he would burn the town if the schooner and her slaves were not surrendered, and the vessel and 67 slaves were handed over. The schooner's papers, colours and crew were still missing, but when Lloyd demanded them they were returned, although without the master. Lloyd then found that the vessel was the *Hope*, an American making her fourth slaving voyage to Ventain. Her master had once commanded the colonial schooner *George* at Goree, so it was no surprise that he had not given himself up.

This success came at heavy cost; so far the Squadron had been remarkably free of disease, but fever caught up with *Kangaroo* as she left the River Gambia. By the time she reached Freetown only one lieutenant and 15 men of her ship's company of 121 were fit for duty. Nine men died after arrival in the Sierra Leone River; the hospital had no room for the sick, and Lloyd rented rooms ashore for the worst cases.[11]

At the end of June 1812 the strength of the Squadron was brought back to four by the arrival of *Thais* after her refit at Portsmouth. Her re-entry to the fray had begun to the south of Goree, on her return voyage, with her arrest on 24 June of the schooner *Dolphin*, from Cuba under American colours and loaded with slaves. Pausing only briefly at Freetown, she sailed for the Bights and the Portuguese territories south of the Equator. On approaching the anchorage at Loango Bay, Captain Scobell struck his fore and mizzen royal masts and hoisted a Portuguese ensign, the second time he had disguised his ship to avoid alerting slavers; a perfectly legitimate ruse, although none of the other cruisers had so far tried it.

Thais was also unusual in the frequency of her practice with the great guns and small arms. At such a distance from likely encounters with the French, other commanding officers may have felt little need to maintain a high level of gunnery training, but Scobell clearly ran a taut ship. Nevertheless, the heat, humidity and tedium of the Coast bred tensions in *Thais* as they did in all the cruisers, and Scobell was obliged, for example, to punish James Lowe, seaman, with 48 lashes for "disobedience to the Boatswain's orders, telling him he might do it himself, telling the Boatswain's Mate on Duty he might kiss his arse, also for insolence to the Master"![12]

Thais's disguise was apparently successful. In Loango Bay on 14 August she arrested the Spanish brig *Carlotta*, although the reason for the detention and subsequent condemnation are not clear. After coasting south to Cabinda, she returned to Loango at the end of the month and picked up the Portuguese brig *Flor d'America* with her cargo of 364 slaves, for violating the Portuguese treaty and trading "contrary to the passport and to the laws of Nature and Nations", on which grounds she was later condemned. Scobell had to fumigate the slave deck of this prize and transfer to her two tons of water before taking her in tow, and he disposed of the seven Portuguese crew by giving them rations and sending them in their boat to Cabinda. At Mayumba Bay a few days later he detained the Portuguese schooner *Orizonte*, and then he headed for Cape Lopez Bay, this time under Spanish colours.

There was no further success as *Thais* made her way via St Thomas and Annobón back to the Grain Coast, but off Cape Mesurado on the forenoon of 13 October she sighted a strange sail at anchor. The stranger rapidly weighed, and *Thais* gave chase. The pursuit lasted until the following morning when both vessels were becalmed, and Scobell sent away his boats with the chase about 12–15 miles distant, but nine hours later they returned empty-handed after a gruelling day at the oars in wet season heat. On arrival at Freetown a fortnight later Scobell found that *Orizonte* and her 18 slaves had been condemned, and he was much relieved to find *Kangaroo* in harbour. *Thais* had run short of rum, a serious matter, but Commander Lloyd sent across a cask of the precious spirit and Scobell could once more serve a full allowance to his ship's company.

At about this time *Daring* and *Kangaroo* shared the capture of the unladen slaver *Nueva Constitución*, but the incident is reported only in the Vice-Admiralty court records. It is likely that she was taken between the River Pongas and Sierra Leone, where the two cruisers might occasionally have been in company during the autumn. She was probably under Spanish colours, but the court judged on 20 November that she was American property and "trading in slaves with persons under the jurisdiction of Sierra Leone in violation of the abolition laws of Great Britain and America". The court's confusion over its jurisdiction was again evident, and its decision to condemn her was subsequently reversed.[*]

The strength of four cruisers on the Coast was held only briefly. When *Amelia* returned to Freetown on 8 November 1812 the Commodore found the whole Squadron assembled, and was concerned to discover the debilitated condition of *Kangaroo*. He was in no doubt that the fever attack had resulted from the River Gambia expedition, and *Kangaroo* was clearly unfit for further work on the African coast. Another consignment of recruits for the West Indies regiments was ready to go to the Caribbean, so, fitting-out his prize *Andorinha* to carry 300 troops and preparing *Kangaroo* to take another 50, Irby sent a number of *Amelia*'s seamen to boost Lloyd's depleted crew, and ordered him to Barbados and then home.

The coast of Senegambia was giving particular concern to the Commodore, and he despatched *Daring* back to the Rio Pongas and the River Nunez, commenting

---

[*] The seizure of the *Nueva Constitución* was initially credited to *Thais*, *Kangaroo* and *Daring*, but the logs of *Thais* and *Kangaroo* make no mention of the capture. It is possible that *Daring* was the sole captor, but her log has been lost. The incident remains something of a mystery.

that: "The daring manner in which slavers carry on their inhuman traffic so immediately in the vicinity of the British settlements makes it necessary that the Rivers on this part of the Coast should be most closely watched."[13]

Lieutenant Pascoe in *Daring* had become something of a specialist on the Trade to windward of Sierra Leone. He had noted that the slavers between Cape St Ann and Cape Verde were vessels of between 50 and 80 tons, generally manned by Americans but mostly under Spanish colours. They swiftly loaded their cargoes in the Pongas and Nunez, and were gone within a day or two of arrival. He had detected that the Trade from Bissau was on the increase, and that the Grain Coast was occasionally visited by Spaniards en route to leeward in order to collect any slaves who happened to have been brought down from the interior.

Thais, refreshed by her replenishment of rum, continued her patrol of the Grain Coast for another month before setting off once more for Loango Bay, Cabinda, Ambriz, Mayumba and Annobón, a cruise lasting until February 1813. Her Commanding Officer was convinced that the Portuguese would have entirely ignored the treaty with Britain were it not for the presence of British men-of-war, but that the Trade in the Bights had been reduced by occasional visits by cruisers. Nevertheless, he estimated that slave exports from the coast between Sierra Leone and Cape Negro during 1812 had numbered 40,000–45,000.[14]

Events remote from the slave coast were about to affect the suppression campaign. After years of increasing tension, generated largely by the interruption to American trade caused by the British blockade of Europe, the US Congress had declared war on Britain on 18 June 1812. The grounds for this conflict were varied, including Britain's methods of blockade and the decisions of her prize courts, American grievance at impressment of her seamen by the Royal Navy, the practice by British warships of boarding American ships to recover deserters, and Washington's ambition to conquer Canada, although the balance was finally tipped by Napoleon's diplomatic machinations. In fact, unbeknown to Washington, American demands on the matter of seaborne trade with Europe had been met by Britain just before the declaration of war, but Congress subsequently decided to continue the war on the grounds of the impressment question alone.

Of course there was no agreement on the numbers of US seamen pressed into the Royal Navy; Castlereagh had admitted to over 3,300 Americans in the Fleet in 1811, but some American historians have claimed 20,000. The issue was confused by the facts that seamen of the period were polyglot characters belonging more to the sea than to any particular country, and that documents showing American citizenship were bought and sold on quaysides. It was ironic that this unnecessary war, fought ostensibly to protect American seamen, was utterly opposed by New England, where America's seafaring population lived. As far as the anti-slaving campaign was concerned, the obvious short-term implication of hostilities was that all US-flagged merchantmen could lawfully be made prizes of war, but of far more consequence to the struggle against the Trade was the long-term poisoning of diplomatic relationships between the two nations.

The conflict was brought to an end by the Treaty of Ghent, concluded on 24 December 1814. In its Article X the two signatories expressed their desire to continue "their efforts to promote [the slave trade's] entire abolition", and agreed "that both contracting parties shall use their best endeavours to accomplish so desirable an object". There remained, however, a major obstacle to these "best endeavours", and it was expressed in forthright terms by the US Secretary of State, John Quincy Adams, to the American Ambassador in London in 1814:

> the admission of a right in officers of foreign ships of war to enter and search the vessels of the United States in time of peace under any circumstances whatever would meet with universal repugnance in the public opinion of this country.

Unfortunately, Washington's interpretation of "vessels of the United States" encompassed any vessel which might choose to hoist an American ensign, whether she had a right to do so or not. The total incompatibility of this uncompromising attitude with Britain's insistence on the necessity of boarding and searching suspected slavers was probably the greatest single obstacle to effective action against the slave traffic. Its ill-effects were to be felt on the African coast for generations, and it is beyond doubt that it delayed immeasurably the ending of the Cuban Trade.

Commodore Irby's intention was to take *Amelia* to leeward early in 1813, and then, in accordance with his orders, to escort any waiting merchant vessels home from Princes, but, while apparently still in the vicinity of Sierra Leone on 20 December, the frigate arrested the slaver *Triumpho de O'Nia*. Little is known of this incident, probably because subsequent events disrupted Irby's routine correspondence, but the vessel was certainly condemned by the Freetown court, with 17 slaves.

The Commodore's plan was about to suffer violent disruption by the French war. In late November two new French heavy frigates, the *Arethuse* and the *Rubis*, both of 40 guns, escaped from Nantes and made for Sierra Leone with the intention of disrupting British trade. On 27 January 1813 they surprised *Daring* at the Isles de Los, and Lieutenant Pascoe ran his brig ashore and burned her to avoid capture.[15] Pascoe managed to carry this news to Freetown, where he found *Amelia*, and Commodore Irby sailed on 3 February to intercept the Frenchmen. By good fortune both French ships had run aground, and *Rubis* had become a total loss, transferring her ship's company into *Serra*, a Portuguese prize. *Arethuse*, commanded by Captain Bouvet, was refloated, however, and managed to repair her damaged rudder. Irby found the French ships at the Isles de Los on 6 February, but it was not until the following day that he was sure that he had just one frigate to deal with, and that evening *Amelia* closed to engage. The two ships were fairly well matched, but *Amelia*'s ship's company had been weakened by fever. In the moonlit night action which followed, at times a vicious hand-to-hand encounter, much of it with the gun muzzles almost touching, the two frigates fought to a standstill. All of *Amelia*'s officers were either killed or wounded, and command finally devolved on Mr De Mayne, the Master. By midnight the gentle breeze had died, and the ships drifted helplessly apart, both badly shattered and with heavy casualties. *Amelia* limped home to Spithead with her wounded Commodore, and Captain Bouvet nursed his ship back to St Malo with a string of prizes.

A month after this action *Thais* returned to Freetown, the only remaining member of Irby's Squadron. Scobell quickly discovered all that had happened up to the point of *Amelia*'s hasty departure, but even in mid-March he remained in ignorance of subsequent events. *Thais* was not now entirely alone on the Coast. After a call at Cape Coast Castle on her passage back from her cruise south of the Line she had encountered the ship-sloop *Tweed* (16), Commander T. E. Symonds, sent out with a convoy, and the two ships made for Sierra Leone in company. There Scobell found orders addressed to Commodore Irby which required *Tweed* to return immediately to England, and he sent her to collect the season's convoy home from Princes or St Thomas.

Scobell's principal concern had now become the threat to trade on the African coast, not from French commerce raiders but from American privateers. He was aware that damage had already been caused there by one such marauder, and he was concerned that this success would encourage others to follow. Indeed he had heard that two or three of them had been as far south as Cape Verde, and he reported his intention to take *Thais* in that direction to counter the threat. However, he was delayed by difficulty in obtaining provisions, the first indication of such a problem on the Coast. So far the cruisers had found supplies at Freetown to be plentiful, if limited in range: fresh beef and live bullocks, salt pork, yams, peas, rice and sugar had been available there, and there were other sources of supply at Cape Coast, Princes and St Thomas. There were vegetables and fruit too at the Bights islands, but *Amelia* had been obliged to visit the Cape Verdes to find bread and flour.

Sailing finally on 17 March 1813, *Thais* first headed south, and as she passed Cape Mesurado at daybreak a fortnight later she sighted a strange sail. In a moderate breeze she gave chase, and it is clear that Captain Scobell quickly recognised that this was almost certainly not simply another slaver that he had flushed. After pursuing for two hours, *Thais* saw that the chase, six or seven miles to the north-west, had taken to the sweeps, and she put out her own sweeps on the larboard side. Scobell cracked on with every scrap of canvas he possessed: royal studdingsails, ringtail, spritsail and spritsail topsail, but to no avail. It was evident that the stranger was equally making every effort to escape, and by midday she had opened to eight or nine miles. In mid-afternoon *Thais* saw a squall approaching and shortened sail to weather it. As the wind again moderated and the rain cleared Scobell resighted his quarry, now without her two topmasts. Less experienced on the Coast than her pursuer, she had paid the price of her lack of caution under the threat of the squall and now lay helpless as *Thais* rapidly came up with her. *Thais* fired one round as she approached, and the chase's American colours came down. As Scobell had undoubtedly suspected, she was the privateer brig *Rambler* (12) with a crew of 88, and *Thais* triumphantly took her back to Freetown. By early June Scobell was able to distribute the resulting prize money to his men.[16]

With these interruptions by war disposed of, *Thais*, now the lone warship on the Coast, redirected her attention to the slave trade. At the end of May 1813 she snapped up the Spanish sloop *Juan* from Havana, later condemned as American property engaged in the slave trade, and while *Thais* was at Freetown the Letter of Marque *Kitty*, Mr John Roach, brought in two slaver brigs, the Spanish *San Jose Triumfo* and the Portuguese *Phoenix*. *Thais* took charge of these prizes, presumably because *Kitty* was unable to cope with them, and shared in the bounties on their condemnation. She sailed again for the Grain Coast on 15 June in company with *Princess Charlotte*, and with the *Juan*, which Scobell had manned as a temporary tender. On the evening of 25 May the ship's cutter, yawl and jolly boat were armed and sent into the River Mesurado for an operation which was probably the first of its kind but which would be repeated with variations many times in the years to come.

Captain Roach of the *Kitty* had brought back information on a number of Englishmen trading slaves at factories on the rivers Mesurado and St Paul. They had threatened to kill Roach if he did not give up a slaver he had taken in the vicinity of one of the factories, and Scobell intended that his boats, commanded by Lieutenant Wilkins, should ferret them out. At 0830 the following morning *Thais* heard gunfire and musketry in the direction of a factory on the River St

Paul, and Scobell moved his small squadron close inshore, ready to give support to the boats. An hour later it could be seen that the factory, belonging to a man called Bostock, was ablaze, but no news emerged. At 1500 the gig was sent into the river for information, and returned late that night with two wounded men. The factory had been destroyed by a fire which had broken out in the spirit store, and the slaves, who had been driven into the bush, were recovered; but the two wanted slavers, Robert Bostock and John McQueen, had escaped into the forest.

Over the next two days the boats brought off 233 slaves, half of whom were judged to be American-owned, and therefore prizes of war, and the other half, being English-owned, were forfeited under the Abolition Acts. *Thais* and *Princess Charlotte* shared the head money. On 1 July, to Scobell's satisfaction, Bostock and McQueen were cornered in the woods by the natives and were hauled on board *Thais* to be taken to Freetown. There, the following day, they were convicted under the Abolition Acts not by the Vice-Admiralty court but by a Court of Oyer and Terminer before the Chief Justice, Robert Purdie, and were sentenced to 14 years' transportation. *Thais* later took them home to be handed over to the Overseer of Convicts at Portsmouth, but that would not be the last that Captain Scobell heard of Bostock and McQueen.[17] This small operation had applied force directly against the slave merchants for the first time, and it had incurred a fight in which the European traders had persuaded their native supporters to take up arms on their behalf. It had shown, however, that once the slavers were beaten and economically useless they could count on few friends among the local people.

Like *Kangaroo*, poor *Thais* was to pay a heavy price for her success. She had lost two marines to gunfire and drowning in the river action, but that was to be expected. The real blow came with five deaths from fever during the following fortnight in Freetown, three more on the day she sailed, and a further eight during her final three months on the Coast, all apparently from fever except for a seaman who fell from the main yard. After a few days at sea, perhaps in a hopeless attempt to escape the fever, she returned to Freetown, where she found the ship-sloop *Albacore* (16), Commander H. T. Davies, and spent a week in company with her on passage to Cape Coast. On her arrival there in mid-August 1813 *Thais* encountered another newcomer, HMS *Favorite*, Captain John Maxwell, the same ship which had escorted the last Guinea convoy to Barbados in 1807 but now reclassified as a 26-gun Sixth Rate. After a final call at Accra, *Thais* made for Annobón where, on 12 September, she collected two merchantmen for the passage home. In a total of two years on the Coast she had arrested ten slavers and achieved the release of nearly 950 slaves.

Information which today would be called "operational intelligence" clearly played a part in the cornering of *Rambler* and the destruction of the St Paul River slave factory, and the cruisers, particularly those with more experience on the Coast, were clearly growing more adept at gathering the intelligence necessary to achieve any degree of success. The hit-or-miss cruising of the earlier years had produced small numbers of interceptions at sea of unwary or sluggish slave vessels, but as the slavers became more alert and sharpened their own efficiency the cruisers needed to target their searches more precisely, especially with such scant resources available. This was particularly so north of Cape Palmas, where the Americans were trading under Spanish colours. During their patrols the cruisers had made a practice of questioning natives at the mouths of the slave rivers, but undoubtedly found the information they gleaned to be of variable reliability. The masters of some of the legitimate merchantmen boarded by the cruisers probably provided a more trustworthy source, but their information, although helpful in building a broad picture of the trends of the Trade, would rarely have been sufficiently fresh for operational use.

That would also have been true of news reaching the Governor of Sierra Leone through the commercial network of the Coast and passed on to any cruiser which happened to be in the river. As far as the anti-slaving forces were concerned, the only evident intelligence network worthy of the description belonged to the Africa Institution, and had probably been inherited from the Abolition movement. Its tentacles appear to have reached to most corners of the Trade, but it could not produce information of immediate use to the cruisers. There was, as yet, no evidence of any attempt to form an organisation which would do so, and the degree of general hostility on the Coast to the Navy's work would probably have made such an attempt nugatory. In the way of operational intelligence the slavers were almost certainly better served than the Navy.

Once again there was no commodore on the slave coast to direct operations, and with the departure of *Thais* all experience of the task and conditions on the Coast had been lost. Beginning with *Favorite* and *Albacore*, which arrived together in mid-July 1813, there followed a relay of seven temporary cruisers to continue the campaign until the end of 1814. Usually there were two vessels on station, occasionally there were three for a few weeks, and sometimes only one for a short period. With minor variations, the orders to all except one were to a standard format:

> Take under protection Trade bound for the Coast of Africa [...] Having seen the convoy in safely, cruise between C. Verde and C. Palmas, touching at Sierra

Leone, for 8 weeks, then run down the coast and cruise for 6 weeks between C. Palmas and the Volta. Then repair to Cape Coast Castle and proceed with any Trade to...

The deployments were obviously geared to the requirements of convoy rather than countering the slave trade, and the directive to cruise between Cape Palmas and the Volta, a coast on which there was negligible slaving, demonstrated the lack of thought being given to the campaign in the Admiralty. The absence of any coordination or continuity was a recipe for ineffectiveness, but, although some of these visitors made little contribution other than providing a brief presence on the Coast, others achieved significant success.

Captain Maxwell in *Favorite* was evidently a forceful personality. Showing little patience with his Admiralty orders, he made a search of the whole coast as far as Cape Coast, and then continued into the Bight of Benin. On 31 August 1813 he took the empty Portuguese brig *Providencia* at Porto Novo on the Slave Coast, noting that the Portuguese slavers in those parts were generally "most wretched vessels and not worth taking to Sierra Leone unless they have a cargo on", and observing that these Portuguese slavers clearly held copies of the Anglo-Portuguese treaty and delayed loading their slaves from the barracoons until the very last moment before sailing.[18] The following day he was lucky to catch the partially loaded Portuguese brig *San Josef Desforca* at Badagry, and, after a nugatory search of the Bonny River, he sent his boats up the Old Calabar River where, on the seventh, they found the Portuguese schooner *St Joseph*. The 59 slaves discovered on board were stowed so close together in irons that they were in danger of suffocation, and Maxwell was obliged to take them into *Favorite*, deciding to allow the now empty slaver to make for Princes Island because she was too small for the voyage to Sierra Leone. On 16 September the boats were again successful, returning from the River Gabon with another Portuguese schooner, the *Bon Jesu*, with 55 slaves.* Arriving back at Sierra Leone, *Favorite* paused only briefly to refit her rigging and replenish with water and provisions before heading on 7 November for the Rio Pongas. Her three prizes and the slaves were condemned before the end of the year.

Although intelligence available in Freetown was generally useless for intercepting slave vessels, it was, of course, much less perishable in the case of slave

---

* The court records give a figure of 48 slaves released from *St Joseph* and 47 from *Bon Jesu*, so Maxwell must have lost 19 to disease from these two vessels on passage to Freetown. The initial number of slaves in the *San Josef Desforca* is unknown, but 48 were condemned from her. Incidentally, Maxwell transposes the names of *St Joseph* and *San Josef Desforca* in his report, but court records clarify the matter.

factories, and Governor Maxwell's mention of factories run by Englishmen on the Pongas persuaded his namesake in *Favorite* to emulate *Thais*'s recent exploit. On 10 November Captain Maxwell sent his boats, manned and armed, into the river with "a small government schooner" and a native pilot. They emerged nearly a week later without loss, having burned nine slave factories ("some of them very fine buildings") and spiked 23 guns. The slavers, however, had made a successful last stand at the principal factory at the head of the river, a strong natural position which the boats could not approach, and none of the Englishmen were arrested.

After three days at Freetown, *Favorite* sailed on 23 November in company with the newly arrived ship-sloop *Plover* (18), Commander Colin Campbell, and she then began to pay the price of her temerity in the Pongas. For three weeks she lay becalmed within sight of Sierra Leone while fever carried off Lieutenant Moorhouse, Midshipmen Maxwell and Reid, and 13 of the ship's best men, no doubt those selected for the boats, in all 13 per cent of the ship's complement. Captain Maxwell angrily wrote that the Governor of Sierra Leone should have sent some of his "excellent black troops" (seemingly immune to yellow fever if not malaria) to deal with this nest of slavers within striking distance of Freetown instead of "exposing the valuable lives of British Seamen". By contrast with *Albacore*, which departed the Station in early November having made little impression, *Favorite*'s six months on the Coast had been eventful and valuable, if latterly costly, but Captain Maxwell's hard-earned experience was wasted when, after a final visit to Cape Coast, he sailed for home on 31 December.

Governor Maxwell was apparently not inclined to use his troops for river work while he had the skills and resources of the Navy at his disposal, and when the ship-sloop *Spitfire* (16) arrived in January 1814 he sought the help of her Commanding Officer, Commander John Ellis, with a problem in the Gallinas River.* It had been reported that the Liverpool privateer *Kitty* had been pirated by an English slave trader called Crawford in collaboration with a Spanish schooner loaded with Crawford's slaves, and that the schooner's master had murdered the master of the *Kitty*, Mr Roach. Crawford and the Spaniards had then plundered the *Kitty* and scuttled her. *Spitfire* made for the Gallinas, and on the morning of 22 February sent in the boats commanded by Lieutenant Hopkins. They re-emerged that evening with Crawford's launch and all his trade goods, and Roach's boat's crew. Crawford had been warned by a Sierra Leone sloop about *Spitfire*'s approach, and he had sent to the local chiefs to collect armed Africans to oppose a passage of the river. However, the false Spanish colours hoisted by

---

* *Spitfire* was classified as a fireship of the *Tisiphone* class, but was employed as a ship-sloop.

Ellis deceived him, and he escaped by the skin of his teeth into the bush when he sighted the boats.

Lieutenant Hopkins reported that he had found another factory belonging to a man named Crundell, and Ellis sent Hopkins back with a stronger force the following day in the hope of catching him, but the boats were attacked on their arrival and driven off. This disappointing skirmish was *Spitfire*'s only encounter with the slavers, and she was distracted in April by a fruitless chase of an American privateer showing British colours. Her intention to hunt the American in the Bights was thwarted by her inability to find bread, flour, rice or any substitute at Cape Coast or elsewhere, and Commander Ellis was obliged to sail for St Thomas to look for provisions and for his homeward-bound merchantmen.

The threat to trade from American privateers remained, but that from the French was evaporating. Napoleon had abdicated on 9 April 1814, and on 30 May the long war with France was apparently brought to an end by the First Treaty of Paris. This happy news would take a long time to reach the coast of Africa, however, and a hunt for two powerful French commerce raiders was still in progress along the slave coast. The 40-gun frigates *Etoile* and *Sultane* had left France in late 1813, and had been found at the Cape Verde Islands on 23 January by the British frigates *Creole* (36), Captain G. C. McKinzie, and *Astrea* (36), Captain B. Ashley. The French beat off their pursuers in a severe running action and then made for St Malo. The British ships, however, were concerned that their opponents had headed for the Gulf of Guinea, and, when they had made repairs, they began a hunt in that direction which lasted until August. In the course of this search they fell in with, and arrested, two laden slavers: the *Gertrudis la Preciosa* on 1 June, probably under Spanish colours, and, a month or so later, the Portuguese *Bon Successo*. Both prizes were condemned at Freetown, the former for fitting out in British territory and the latter for infringing the Portuguese treaty, and their 728 slaves were freed.

By 1813 the Portuguese merchants were complaining to their government about the arrests of their ships, and a representation was in turn made by Portugal to the Court of St James. There was even a rumour that the new Brazilian frigate *Infante Don Pedro* might sail for the African coast to protect Portuguese trade, at the instigation of the Prince Regent. In May the Foreign Secretary asked the Admiralty to prevent its cruisers from molesting bona fide Portuguese slavers, although there were negligible grounds for supposing that the treaty, as it was understood in London, had been contravened. Perhaps the Foreign Office was anxious to placate a Portuguese Government about to pass, under British influence, new legislation placing loading limits on its slaver masters. That law, passed in November 1813, allowed five slaves per ton up to 201 registered tons, and then one slave per ton above that figure. In all probability the enforcement of this rule was far from rigorous, and there is no evidence that the British cruisers used the new regulation, even if they knew about it, as a pretext for inspecting Portuguese vessels.[19]

In fact the Royal Navy brought only two vessels before the Freetown court during 1814, and only one of those was Portuguese. The industrious, and experienced, colonial schooner *Princess Charlotte* was a great deal more effective, however; she arrested nine slavers during the course of the year, but only one of those was Portuguese. Her other victims were probably all under Spanish colours and were taken between Cape Verde and Cape Palmas, in most cases in the rivers, and condemned for trading with British subjects.

More Portuguese slavers were arrested on the Leeward Islands Station than on the slave coast during 1814. In June the Portuguese 12-gun xebec *Manuella*, under Spanish colours, insured in London and with 508 slaves on board, was taken into Tortola by the brig-sloop *Mosquito* (18), Commander James Tomkinson. Tomkinson reported to his Commander-in-Chief that he had found secret papers on board his prize, showing that the *Concha*, released in May at Tortola, had indeed been a slaver, belonging to the same Portuguese owner as the *Manuella*, and that another of this slaver's vessels, the *Venus*, was about a fortnight astern of the *Manuella* and heading for Havana with 450 slaves.[20] The Commander-in-Chief, Admiral Durham, had the satisfaction of reporting that this vessel, the *Venus Havannera*, a ship of 12 guns, was caught in July by the sloop *Barbadoes* (16), Acting Commander John Fleming, and condemned at Tortola. However, he had learned that the slaves from the *Manuella* and the *Venus Havannera* were in a particularly wretched state; of the 617 landed alive, 222 had subsequently died.

On the same station the Spanish-flagged brig *Carlos*, mounting ten guns, carrying 444 slaves, and "supposed to be American", was arrested in May 1814 by the frigate *Pique* (36), Captain The Hon. A. Maitland; in June Lieutenant Dickson's cutter *Maria* took the Spanish schooner *El Josepha* into Antigua with 278 slaves; and the final slaver capture of the year, the Spanish *Candelaria*, was detained by the frigate *Barossa* (36), Captain William McCulloch, and condemned at St Thomas. There appears to have been only one slaver arrest on the Station in the first half of 1815: the Spanish *Alrevido*, a joint capture by the frigate *Ister* (36), Captain John Cramer, and the brig-sloop *Columbine* (18), Commander Richard Muddle.

Almost all of these slavers had been heading for Havana from the Bonny or Old Calabar rivers, and those taken under Spanish colours were condemned, almost certainly, for being American property. Many of the considerable number of slaves released would have been sent to the plantations as free labourers or "apprentices", where their conditions would have been hardly distinguishable from those of the slaves. Others would have been apprenticed to tradesmen, employed as domestic servants or been offered the opportunity to volunteer for the West Indies regiments. Of the 172 slaves freed from the *Candelaria*, for example, 40 were taken into the army, 124 were apprenticed, and eight died.

Also in the West Indies, there was an unusual outcome and an illuminating aftermath to the interception in June 1814 of a 180-ton Portuguese slaver by *Ulysses* (44), Captain Thomas Browne. She was bound for Havana, and on sailing from St Thomas in the Gulf of Guinea she had 432 slaves on board. By the time that she was boarded off Cuba 130 of them were dead from starvation and suffocation. Browne's initial intention was to send her to Jamaica for adjudication, but the survivors were in such a state that he realised that the only humane course of action was to allow her to continue to Havana. When he was on the African coast in 1815, Browne took the opportunity to remonstrate with the Governor of St Thomas for allowing the slaver to sail so badly provisioned and overloaded. The Governor replied that he would have been better pleased if the vessel had sailed with 1,000 slaves, because masters and owners had to pay so much per head into the Royal Fund and to the Governor.

There had been much less suppression activity on the Cape Station, although the Commander-in-Chief, Rear-Admiral Stopford, voiced his concern in early 1812 about the great assistance given to the Portuguese slavers bound from Mozambique to Rio in permitting them to water at the Cape. This opportunity for replenishment allowed the slavers to crowd their ships "in a most inhuman manner" rather than carrying sufficient water for the whole voyage, but nothing seems to have been done to stop it.[21] In 1812 Stopford detained only one of these Portuguese-flagged slavers, on the grounds of American ownership, and the next recorded arrest on the Station was in 1814. This was an embarrassing incident involving the frigates *Niger* (38) and *Laurel* (36), sent out to the western extremity of the Station to search for enemy ships. On 28 September, about 400 miles north-east of Ascension Island, they encountered the Portuguese brig *Boa União*, under American colours and bound from Cabinda to Rio with slaves. Captain Peter Ranier of *Niger* hoisted an American ensign during the chase, and as daylight faded he caught the brig and put on board a prize crew of a lieutenant, two midshipmen and ten men. However, he failed to take the elementary precaution of ensuring that they had achieved control of the slaver before the frigates made sail and parted company. The result was that the slavers retook the vessel and carried the prize crew, some of them wounded, to Pernambuco in irons. There they were marched through the streets and imprisoned for three weeks before being taken to Rio and handed over to HMS *Cherub*. Five years later the owners of the slaver had the effrontery to claim indemnity for loss and damage.

Instead of supporting the actions of its ships against the complaints of the Portuguese and the Foreign Office, the Admiralty belatedly sent out instructions concerning seizure of Portuguese vessels, although the commanding officers of the cruisers by now believed they understood the constraints from their own reading of the treaty. Within a year, however, a new treaty had been concluded between Britain and Portugal, and the 1810 document had been declared void. The preamble to the new agreement, signed on 22 January 1815, declared that:

> His Britannic Majesty and His Royal Highness the Prince Regent, equally animated by a sincere desire to accelerate the moment when the blessings of peaceful industry and innocent commerce may be encouraged throughout this extensive portion of the Continent of Africa, by its being delivered from the evils of the Slave Trade, have agreed to enter into a Treaty for the said purpose.

Portugal's subsequent behaviour made it clear that it was, in fact, animated by a sincere desire to delay the ending of the slave trade for as long as possible and, in the meantime, to extract as much money as it could from the ally who had just delivered it from Bonaparte. Nevertheless, Portugal agreed that trading for slaves on the African coast north of the Equator would become unlawful for Portuguese subjects on ratification of the treaty. There would, however, be a period of grace of six months from the date of ratification for slave vessels which had cleared from the ports of Brazil before publication of ratification. Portugal undertook, furthermore, that the Portuguese flag would not be used in the slave trade other than to supply Portugal's transatlantic possessions. It also promised to adopt, in concert with Britain, "such measures as may best conduce to the effectual execution" of the agreement. Britain, for its part, promised to prevent any interference with the Portuguese Trade south of the Line, and agreed to remit further payments on a loan of £600,000 made to Portugal in 1809.

In addition, in a convention signed on 21 January 1815, there was an acceptance that the wording of the 1810 Treaty had been less than clear, and Britain agreed to "make liberal indemnification to Portuguese subjects for property detained under the doubts aforesaid". This liberality amounted to a sum of £300,000 in discharge of claims from Portuguese merchants for slave vessels detained by the Royal Navy's cruisers before June 1814, and Britain promised not to interfere in the disposal of the money. The navy of each of the two signatories was now empowered to arrest not only slavers operating under its own flag but also slaving vessels belonging to the other nation and acting in contravention of the treaty, but it was highly improbable that Portugal would take any such action. There was a further undertaking that there would, in due course, be another treaty by which "the slave trade would be universally prohibited throughout the entire dominions of Portugal". Foreign Secretary Castlereagh probably felt that he had achieved a valuable agreement, but Wilberforce told him that the new treaty was "full of hypocrisy, wickedness and cruelty".

Following these negotiations with Portugal, which have a whiff of appeasement about them, Captain Irby wrote to the Admiralty in February 1815, on behalf of himself and other commanding officers, to represent that several condemned Portuguese slavers had been restored by the Court of Appeal without allowing the captors their expenses. These decisions, he believed, resulted partly from

> the Lords of Appeal supposing that places on the slave coast of the Bight of Benin to belong to Portugal which never did, and which had merely been claimed by Portugal in order to extend the slave trade in violation of the treaty.

He went on to complain that the time allowed for the slaver owners to appeal had been extended beyond the original one year, an outrageous concession already condemned by Wilberforce, and, wrote Irby, "when it is known in Brazil that these vessels have been restored, claimants will know that they have only to appeal in order to be restored". He asked Their Lordships that the captors in these cases should be allowed their expenses, especially as "service on the Coast of Africa is particularly arduous to the impairment of the health and strength of the crews".[22] Irby's points were undoubtedly valid and fair, and the Admiralty agreed to make an application to the Court of Appeal. That august and remote body was, as far as can be ascertained, unsympathetic.

The Portuguese treaty was the most concrete outcome to emerge from a great flurry of diplomatic activity in 1814–15 affecting the anti-slaving campaign. British public opinion was expressed in petitions to the Government, signed by three-quarters of a million people (no mean figure in a mainland population of less than 13 million) urging it to persuade the restored French monarchy to abolish the slave trade, and similar appeals were sent to the Tsar of Russia. Lord Castlereagh and the Duke of Wellington, as the British delegation to the great

Congress of Vienna, were determined to press for an abolition declaration from the assembled great powers, even though Wilberforce and the abolitionists had by 1814 succeeded in irritating even Castlereagh. Wilberforce remained very active, despatching letters to the French Foreign Minister, Talleyrand, Tsar Alexander and Wellington seeking support for international abolition. There was no enthusiasm for abolition in France, but in the First Treaty of Paris the restored French Government undertook to join Britain to do everything possible to suppress the traffic in slaves, and Talleyrand persuaded King Louis XVIII to agree to abolish the slave trade within five years. The Treaty of Ghent in December 1814, ending the war between Britain and America, included an agreement to use "best endeavours" to abolish the slave trade. Finally, in February 1815, Castlereagh at Vienna secured a declaration by the Governments of Britain, France, Spain, Sweden, Austria, Prussia, Russia and Portugal that "the commerce known by the name of the African slave trade is repugnant to the principles of humanity and universal morality" and that the colonial powers had a "duty and necessity" to abolish it as soon as possible. There was no agreement on timing, however, and an extraordinary rider was included that no nation could be made to abolish the Trade "without due regard to the interests, the habits and even the prejudices" of its subjects.

The Foreign Office also achieved its first noticeable headway with Spain, in an agreement of 5 July 1814. Madrid's first undertaking, to prohibit Spanish subjects from trading slaves to anywhere other than Spanish possessions in the Americas, was hardly necessary, but, importantly, it promised to prevent the use of the Spanish flag by foreign slavers. There was some progress with Denmark too. Earlier that year, as a by-product of the peace treaty made at Kiel in January 1814, the King of Denmark engaged to cooperate with Britain "for completion of so beneficent a work [total abolition of the slave trade] and to prohibit all his subjects, in the most efficient [way], and by the most solemn Laws, from taking any share in the Slave Trade." That document also contained perhaps the first formal declaration of Britain's position regarding the international slave trade: "His Majesty the King of the United Kingdom of Great Britain and Ireland, and the British Nation, being extremely desirous of totally abolishing for ever the Slave Trade". Having thus nailed its colours to the mast, Britain would find that it had handed to the slaving nations a lever for extracting concessions from the country in all subsequent negotiations.

While these diplomatic machinations had been underway the brig-sloop *Ariel* (18), Commander Daniel Ross, had spent an uneventful and lonely four months cruising the northern part of the Coast. She sailed for home in mid-October

1814, soon to return, and was immediately replaced by the ship-sloop *Brisk* (16), Commander Henry Higman. A fortnight later *Brisk* was joined by the Sixth Rate *Porcupine* (22), Captain Booty Harvey, and just before Christmas two more ships arrived, bringing with them a new Commodore.

Like his predecessor, Captain Thomas Browne had been made Post in 1802, but he was a decade older than Irby, and his only recent action had been as Flag Captain to Rear-Admiral T. Byam Martin in *Aboukir* (74) at the siege of Riga in 1812. He was now in command of the frigate *Ulysses* (44), and he had been ordered in October 1814 to take under his command the ship-sloop *Comus* (22), Captain John Tailour, and sail for West Africa with a stores ship and a number of transports, hoisting a broad pendant as he left the Channel. On arrival he was to:

> Take *Brisk*, and *Ariel* if she is still on that station, under his orders and employ them and *Ulysses* in cruising between Cape Verde and the Gabon, visiting the British settlements. To provide for the convoys sailing from the Island of St Thomas (calling at Cape Coast Castle 15 days previous thereto) on 30 January, 30 April, 30 July and 30 October. *Ulysses* to continue on this service whilst the health of the ship's company and supplies permit, and then proceed with one of the said convoys to Spithead.[23]

With the American war still unfinished the Admiralty was still preoccupied with convoys, but the arrival of four fresh cruisers on the Coast would rejuvenate the suppression campaign, which had been at a low ebb for the past year.

Brisk arrived at Freetown on 29 October 1814 and immediately stored ship. Although it may not have indicated a permanent improvement in logistical support, she found there everything she needed, including flour and rum. After an abortive search of the Gallinas River by her cutter at the beginning of December, she had better luck on the morning of the eighth when, in a dead calm, Higman sent the boats, manned and armed, to chase a schooner seen close inshore off Grand Bassa. The boats, under the command of the First Lieutenant, John Dewar, had a hard row of 15 miles before they boarded the schooner against determined resistance and drove many of the crew into the sea. Eight hours after leaving, they returned to *Brisk* with their prize, the Spanish slaver *Union*, and anchored her alongside *Brisk*. The *Union*, which was carrying six guns, may have been a pirate, but she was condemned for fitting-out illegally for slaving and "acting contrary to the law of Nations" by firing at *Brisk*'s boats.*

---

* This prize is shown as *La Juanna* in *Brisk*'s log.

With 21 prisoners from the *Union* on board, *Brisk* cruised eastward into the Bight of Benin and on the evening of 7 January 1815, after making all sail in chase, she boarded and detained the Portuguese slaving brig *Conceição* off Porto Novo. Landing 19 of the Portuguese crew at Porto Novo, and sending the *Conceição* to Freetown, Higman headed towards the Bight of Biafra. On 13 January, about 50 miles south of Cape Formoso, he captured the Portuguese brig *General Silveira* with 238 slaves on board, and taking her in tow, he turned his ship for Sierra Leone. Twelve days later, off the Gold Coast, he found the *Conceição* making little progress against the current, and for two months *Brisk* struggled westward in Doldrums weather with one or the other, and sometimes both, of her prizes in tow. The final 15 miles took three days. Higman deserved his satisfaction in seeing both slavers condemned for breaching the Portuguese treaty.

As her deployment drew towards its end, *Brisk* made a final cruise to the Gold Coast, arriving in mid-April at Cape Coast, where she found *Ulysses*. Following the Commodore to St Thomas, she found her convoy and sailed for home on 11 May, but she had not yet finished with the slavers. On the following day she sighted, chased and seized the Portuguese ship *Dido*, bound from the Cameroons with a cargo of 327 slaves. After towing the prize westward for 11 days, Higman finally despatched her for Sierra Leone on 29 May for successful condemnation.

On his arrival at Freetown at the end of November 1814, Captain Harvey of the *Porcupine* indicated an advanced attitude to fever prevention, a matter which appears not to have been seriously considered by other commanding officers. Before allowing a watering party ashore in Sierra Leone he served each man with one gill of water mixed with bark, Peruvian or cinchona bark, known since the seventeenth century as a preventive and cure for African fevers but rarely used. Whether or not Harvey made this dosing a regular practice is not clear, but the subsequent health of his ship's company indicates that he probably did so. During the ship's first two months on the Coast, all of it spent off Senegambia, the boats were often sent inshore, and the cutter and pinnace searched the Gambia River for a week without subsequent ill-effects, and without luck.

Apart from the questionable success of driving ashore and destroying an American privateer schooner at Boa Vista in late January 1815, a month after the end of the American war, an action which cost two British lives, *Porcupine*'s work on the Coast was mostly between Senegal and the River Gambia. Other events distracted her attention from the slavers. At the beginning of February *Ulysses* had arrived at Senegal and Goree with a number of transports. These were troopships awaiting the withdrawal of the British garrisons as soon as a French force should arrive to take over the two settlements in compliance with the First

Treaty of Paris, and the Commodore left them in the charge of *Porcupine*. On 29 April, however, a breathless *Ariel* arrived from Cork with urgent despatches announcing the startling news that Napoleon had escaped from Elba and was back in Paris. Happily this was just in time to prevent the return of Goree and Senegal to the French. *Ariel* then sped on to track down the Commodore.

During her wait for the French, *Porcupine* had allowed herself to wander as far as the River Gambia. Between 21 March and 3 April her cutter and pinnace, or pinnace and launch, were almost continuously in the river, manned, armed and usually victualled for several days. In all of their forays they were under the command of the First Lieutenant, Acting Lieutenant Robert Hagan, whose name would in due course become familiar on the Coast. For all his efforts he made only one capture. Following information from the *Princess Charlotte* that two schooners were trading slaves in the river, the boats were sent in on 23 March and returned two days later with the schooner *Sophie*, probably under Spanish colours. This sole capture by *Porcupine* was later restored by the court on payment of costs incident to the detention.

The African Institution had long been concerned about the volume of Portuguese and Spanish Trade in the Old Calabar River, and in 1811 Wilberforce wrote to Yorke, First Lord of the Admiralty, and to Castlereagh on the subject, but the cruisers on the Coast had shown scant interest in the river. Commodore Browne's attention was soon drawn to the Old Calabar by complaints from merchantmen who had loaded palm oil in the river that they had suffered long delays because of the preference given to the many slavers. *Comus* was about to head for the Bights, and Browne told Captain Tailour to take a look at the problem.

On 16 March 1815 *Comus* was running down to the entrance of the river in a thick Harmattan haze when she detected two schooners to windward. Sending an armed boat under the command of the First Lieutenant, Lieutenant P. Graham, to board one of these vessels, the ship set off in chase of the other. The boat hoisted a Union Flag and fired a shot ahead of the schooner. Two more warning shots brought no response other than the hoisting of Portuguese colours, but at the next warning shot the schooner bore up and fired her two starboard guns at the boat. Graham returned the fire and signalled *Comus* for help. The schooner fired her starboard guns again, wore round to fire her two larboard guns, and then turned on her heel to unleash another starboard broadside. As she attempted to tack to bring her larboard guns to bear the wind died, and Graham was able to board her. She was the *Dos Amigos* from Pernambuco, fully equipped for slaving, and her crew told Graham that they would not be alone in resisting search. *Comus* had been obliged to abandon her chase of the other schooner which, according to the Portuguese in the *Dos Amigos*, was the American-built *Conception* under Spanish colours.

The two schooners had sent away a boat, probably one chased into the Bonny River by *Comus* on 14 March, and Tailour was anxious to find both the *Conception* and the boat. He also wanted to discover the reason for this "Spirit of Resistance" among the subjects of a supposedly friendly power. After several days of fruitless search to windward, he ran down to the Old Calabar estuary to reconnoitre. On 24 March, taking advantage of a sea breeze and an officer of the merchantman *Liverpool* as pilot, *Comus* worked her way as far into the estuary as the wind would allow before anchoring. Realising that any further delay would sacrifice the essential element of surprise, Tailour sent his boats upriver that night, again under the command of Lieutenant Graham, and they arrested five slavers lying at anchor off Duke's Town. The first one to be approached by Graham's division of boats was the *Catalina*, the schooner claimed by the *Dos Amigos* prisoners to be the *Conception*. She resisted the boarding with musket fire, and, although Graham repeatedly shouted in English and Portuguese that the boats were British, the boarding party was obliged reluctantly to use its weapons. Of the slaver crew one was killed, the master and five were seriously wounded, and one was slightly hurt. One *Comus* seaman was seriously wounded, and three slightly so. The schooner had no slaves on board, but was fully fitted for slaving.

It took ten minutes to subdue the *Catalina*, and meanwhile the second division of boats, commanded by Lieutenant Pierce, tackled the slavers on the town side of the river. They were taken more by surprise, and although the schooner *Carmen* gave some resistance, they were taken fairly easily. Then Graham was able to go to Pierce's assistance. Apart from the Spanish-flagged *Catalina*, the prizes were two further Spanish vessels and two Portuguese: *Carmen*, *Intrepide*, *Bon Sorte* and *Estrella*, all loaded with slaves, 469 of them all told. Meanwhile, *Comus* was making her way upriver, and was disconcerted to sight a fleet of large canoes heading downstream to meet her. Captain Tailour beat to quarters, but this was, in fact, Duke Ephraim, the Old Calabar headman, with all the neighbouring chiefs and their war equipment, giving an amicable welcome to the first man-of-war to be seen in the river.

Lieutenant Graham's concern now was that his return downriver with the boats should not be resisted, and on the advice of the masters of British merchantmen in the anchorage, he called for a meeting with Duke Ephraim. The Duke gave his assurance that the prizes could be removed without molestation. With the slavers provisioned and watered, *Comus* was dropping down towards the sea on 3 April when she met a brig heading upriver. This was the *Santa Anna* from Pernambuco under Portuguese colours and fitted out for slaving, so she was added to the tally. As soon as the smallest of the seven prizes, *Dos Amigos*, had been loaded with the bulk of the slaver crews, *Comus* set off with her haul for St Thomas in the hope of finding *Ulysses*. It was far from an easy passage. *Comus* often had up to three of her convoy in tow, the Surgeon was regularly sent to treat sick slaves, and Tailour took a number of slaves on board *Comus* to relieve overcrowding and feeding difficulties. Nevertheless, there was yet another success on 23 April when *Comus* cast off her tow to chase and detain the Portuguese schooner *Maria Madelena*, bound from Princes to the Gabon for slaves.

Even this was not the end of the episode. *Comus* detached the *Maria Madelena* for Freetown on 1 May and arrived at St Thomas on the sixth, finding *Ulysses* and *Brisk* at anchor. At dawn the following day she sighted the Portuguese prize schooner *Estrella* dangerously close to the rocks, and sent her boats to give assistance. They brought the schooner to anchor alongside *Comus* after discovering that when the prize crew had gone aloft to work the sails after anchoring the previous night the Portuguese had risen against them, taken a boat and escaped ashore, together with 13 slaves who had been put aboard her as the convoy's hospital ship. The troublesome *Estrella* did go ashore on the eighth, but she was refloated next day. The *Intrepida* had already been detached, but on 20 May *Comus* sailed with the remainder of her prizes. Ten days later she left them to make their own way to Sierra Leone, and arrived there herself on 3 June.

In due course the Portuguese prizes were condemned for breach of the Portuguese treaty, and the Spaniards were adjudged guilty on a variety of charges, including illegal fitting-out and firing on the *Comus* boats.[*] Tailour was understandably delighted; he commended Lieutenant Graham in the warmest terms, reported that Lieutenant Pierce and the boats' crews merited his highest praise, and, commenting that there had been much to do for the whole ship's company in getting the prizes ready for sea, he was "proud to say that not an individual has failed to give me entire satisfaction". Nevertheless, Graham had been distressed by the need to use force, and Tailour was perturbed at what he saw as unjustifiable resistance by people supposed to be allies.[24]

Commodore Browne in *Ulysses*, meanwhile, had been busy off Senegambia and the Grain Coast. After recovering one of *Solebay*'s bower anchors, presumably to replace one of her own, she left *Porcupine* at Senegal with the transports on 19 February 1815. The next day she sighted three schooners to the south-west and gave chase under all sail. The breeze was very light, and Browne had to resort to

---

[*] There is no mention of the *Maria Madelena* in the court records, and, for reasons not explained, she may not have reached Sierra Leone.

shifting his foremost guns aft to trim the frigate, and to the old trick of wetting the sails to improve their efficiency, before he managed to board the Spanish-flagged schooner *Dolores* after a seven-and-a-half-hour chase. Indeed it was a considerable achievement for a big and a far from youthful frigate to overhaul a schooner in anything other than heavy weather. The *Dolores* was bound to the Gambia for slaves, and after exchanging seven prisoners for a prize crew of a petty officer and eight hands Browne sent her into Freetown, where she was condemned as being fitted-out by British or American subjects. After a fortnight at Sierra Leone herself, *Ulysses* sailed in company with a schooner manned as a tender, probably the *Dolores*. His destination was Cape Coast, but, whether he had planned it or not, the Commodore's next objectives were to be the Grain Coast slave factories.

Browne had heard in Freetown about the slave trade in the Grand Bassa River, and as *Ulysses* was heading south-east along the Windward Coast he decided to take a look into the river. Three boats were sent away on the morning of 20 March, and they were being hoisted that afternoon after an abortive search when a schooner was sighted at the mouth of the neighbouring St John's River. There is some discrepancy between the account in the ship's log about what followed and Browne's report to the Admiralty in May, and, if the log is to be believed, the schooner, under Spanish colours, had been detained by the tender. The log also records that the Spaniard had 22 slaves and a part cargo of trade goods on board, and it seems clear that *Ulysses* despatched two boats with a petty officer and eight seamen to take charge of her. Before the prize crew could board her, the schooner, according to Browne, headed upriver, but it is not certain whether she went into the St John's River or the Grand Bassa. As the boats attempted to follow her they were fired on from the bush by a large number of natives within pistol-shot, and forced to retreat.

A native afterwards went downriver to tell the boats that the ambush had been arranged by an Englishman called Crewe, who owned a slave factory on the river. So the following morning Browne sent his First Lieutenant, Lieutenant Phillips, into the river with all five of the frigate's boats, manned, armed and with a party of marines, to destroy the factory and seize Crewe. Owing to the swell it took some time to cross the bar, and Crewe had time to send off his 200 slaves and make good his own escape, but Phillips burned the factory and ten houses belonging to it. There is no further mention of the schooner in either the frigate's log or in Browne's correspondence, and she presumably managed to land her slaves and avoid capture.

While he was dealing with Crewe's establishment, Phillips learned of another factory at Mesurado. So on 22 March a party of marines was put into the tender, and she and the barge, again under the command of the First Lieutenant, made the passage of about 30 miles to the north to burn that one too. Unfortunately the owner had been warned by Crewe, and only four slaves were released. It was not until 30 March that the tender and the barge rejoined *Ulysses* off the River Sesters and the Commodore was able to continue his passage to Cape Coast. Browne believed that these had been the only two factories remaining to windward of Apollonia, apart from those on the River Gambia and Rio Grande which were still conducting a considerable trade, but slave factories were all too easy to rebuild.

At about this time either *Ulysses* or her tender detained *Le Cultivateur* as she was heading into the Bonny River for slaves. She appeared to be French and would seem to have been a legitimate prize of war, and Browne sent her to Sierra Leone. There the prize crew found that the Vice-Admiralty court was without a judge and was not sitting, and so it took the prize to England for adjudication in the High Court of Admiralty. The effort proved nugatory because the King's Advocate gave his opinion that the vessel was indeed French property, and, the war having finished by this time, he recommended restoration. Browne withdrew his prosecution.

As she had hoped, *Ulysses* found *Brisk* at Cape Coast, and she sent the tender back to Sierra Leone. The Commodore's priority now was to make arrangements for the April and July convoys, and he headed for St Thomas. He was also anxious to find some provisions, pending the arrival of fresh supplies from England for Sierra Leone and Cape Coast. *Ulysses* and *Brisk* reached the island in company on 28 April, and there met the schooner *Catalina*, one of *Comus*'s prizes detached by Captain Tailour. The Commodore had received some intelligence about slavers in the River Gabon, 180 miles to the east, and encountering *Catalina* gave him an unexpected opportunity to investigate. The following day he sent her away with the First Lieutenant, the Lieutenant of Marines and a crew of 45 seamen and marines, and eight days later she returned with a detained Portuguese schooner. This seems to have been the *Diligente*, listed in the Freetown court records as captured by *Ulysses* and, with 29 slaves, condemned to the Crown on appeal on the unusual charge of bartering for slaves with the cargoes of vessels condemned in the court of Admiralty.*

By then *Comus* had joined, and after seeing *Brisk* depart for England with her two-ship convoy on 11 April and instructing *Comus* to return later to St Thomas to take charge of the 30 July convoy, the Commodore took *Ulysses* as far south as

---

* Commodore Browne's reports are sparse and imprecise, and it is conceivable that the schooner involved in the Great Bassa/St John's River incident of 21 March was the *Diligente*, but, if that was the case, the schooner taken in the River Gabon in early May did not reach the Freetown court.

St Helena. It appears, however, that before departing, he manned the *Diligente* as a tender, and that she continued to operate on the frigate's behalf until at least the end of July, when *Comus* found her at St Thomas with the *Nova Fragantinha*. The records of the 1819 London Slave Trade Commission indicate that the tender illegally detained two more Portuguese vessels, the *Conceição*, which was wrecked on passage to Sierra Leone, and the *Correio*, with 345 slaves, which, for some reason, also never came before the Freetown court.

At St Helena, by chance, *Ulysses* met the season's homeward-bound fleet of East Indiamen, carrying cargoes reckoned to be worth £10 million. Hearing the news of Napoleon's escape and the renewal of the war, the masters of the eleven Indiamen begged Browne to convoy them home. No doubt bearing in mind the value of the merchantmen, as well as his instructions to "proceed with one of the said convoys to Spithead", the Commodore agreed.* The convoy sailed on 15 June, only three days before the Battle of Waterloo. Unnecessary though the frigate's protection proved to be, the East India Company was, as usual, generous in its gratitude and rewarded Browne with a valuable service of plate.[25] Now *Comus* was the only man-of-war on the slave coast, and she had no immediate expectation of support.

* * *

By the time that peace was renewed in Europe, a total of 178 suspected slaving vessels are recorded as having been arrested and their cases subsequently judged by the various Vice-Admiralty courts, 118 of them in Sierra Leone. Of the grand total, 34 were restored to their owners at the time or had their condemnations reversed on appeal, and, as will be seen, there were to be yet more, long delayed, appeals. The Royal Navy's cruisers had made 145 of the arrests, colonial vessels had accounted for 22, and Letters of Marque had achieved ten.† As a result of these captures, and of releases by *Thais* from the St Paul River factory and by the garrisons of Senegal and Goree, the Freetown court freed nearly 7,200 slaves. They were handed into the care of the Government of Sierra Leone to be employed or apprenticed, or to work the land on their own account. It seems that in cases of restoration of slave vessels the slave cargoes were generally freed, but the court

---

* It has been implied by at least one writer that Commodore Browne deserted his station in escorting the East Indies fleet to England, but that accusation cannot stand in the face of the original orders to *Ulysses* from the Admiralty.

† The circumstances of the seizure of the remaining one, the schooner *Vanganza*, condemned at Freetown in 1814 with 43 slaves, have defied investigation.

records often fail to clarify that point. The results of cases arising in the western Atlantic and Caribbean are even less clear, partly because the Vice-Admiralty courts there were dealing with prizes of two wars as well as slavers, and partly because their records are now in some disarray. The island courts had tried at least 41 slave vessels by mid-1815 and had condemned 33 of them, with 2,760 or more slaves consequently freed. At the Cape of Good Hope action against the Trade from East Africa to Brazil was inconsistent and largely ineffective. Eighteen slavers, all apparently French, were condemned by the Cape court between 1808 and 1810, and 730 slaves were released, but after that there was only one recorded detention involving an unknown number of slaves. A very few unladen slavers were taken into English ports, but the records of only two such cases are available.

In eight years a total of perhaps 10,600 slaves had been freed from the Atlantic Trade by the efforts of British naval forces and army garrisons, but that number, respectable though it might seem, has to be set against the total volume of the traffic across the Atlantic. Captain Scobell of *Thais* had made a brave calculation of between 40,000 and 45,000 slaves exported during 1812, and he was remarkably close to the truth. Actual numbers will never be known for sure, and published estimates have varied wildly, but the figure calculated for that year by David Eltis, probably the most accurate available, is 39,400. His total for the first eight years of the suppression campaign is 284,000.[26] Britain's cruisers had indeed tackled only the tip of a massive iceberg.

The deployment of forces had been inconsistent and chronically inadequate for the task encountered during the past four years, and the authority given to the cruisers' commanding officers was ill-defined for dealing with the Trade as they found it. The British slave trade was still supposedly the primary target, but the Trade under British colours had ceased to exist. By 1815 only a tiny number of British slave traders been brought to book, and there is no evidence that any court action had been taken against British subjects with financial interests in the Trade. In the event the British legislation was applied well beyond justifiable limits of jurisdiction in order to condemn slaving vessels associated, however tentatively, with British interests but under foreign flags and with foreign crews.

The campaign since 1810 had focused not only on eradicating the residue of British involvement in active slaving but also on enforcing that year's Treaty of Friendship and Alliance with Portugal. However, Britain's belief that the treaty bestowed on the Royal Navy the right to arrest Portuguese slavers, and placed such vessels under the jurisdiction of British Admiralty courts, was mistaken. The anti-slaving element of the agreement had appeared to achieve a step towards abolition, but it was, in fact, almost meaningless. Portugal had merely declared a general intention to cooperate with Britain; it had carefully avoided granting Britain permission to interrupt its lucrative supply of slaves to Brazil. The resulting arrests by the Royal Navy were scarcely legitimate. The subsequent Treaty of 1815 was an improvement, but, as time would tell, it was still seriously flawed.

Even excluding the Portuguese territories south of the Equator, the West Africa Squadron's efforts extended along more than two thousand miles of a hazardous and difficult shore, but Admiralty orders had given instructions to look into "the several Bays and Creeks" as though this was the south coast of Devon. Despite such demonstrations of a lack of understanding and interest from Their Lordships, and of guidance from above other than raw treaties and Acts of Parliament, individual commanding officers had done their best in the face of an impossible task. They were still hampered by the absence of an effective command structure, but during this last four years the tactical use of the cruisers had been more soundly founded on knowledge of the Trade. In particular, the trend towards expeditions into the slaving rivers of Senegambia and the Grain Coast had been influenced by an improved, if unorganised, flow of intelligence information.

There had been worthwhile cruises to the Bights, and, although arrests south of the Line were forbidden, the Royal Navy had shown itself off the southerly Portuguese slaving ports. In this period, however, the cruisers had encountered dangers not much apparent during the earlier years. Forays to the Coast by French frigates and American privateers, although damaging, had not much affected the suppression campaign, but the growing attitude of resistance among the slavers had been more threatening. The African fevers, which had allowed the Squadron something of a honeymoon until 1812, had struck since then, with devastating effect, at several vessels which had sent boats into the rivers. A risk of another sort was posed by frustration among commanding officers; without evident exception they loathed the Trade, and clearly they felt they had a moral duty to take every action in their power against it. On more than one occasion this had led them into ill-judged behaviour. In future the opponents of Britain's moralistic position would not be slow to exploit such errors.

Almost imperceptibly the campaign had transformed from a national policing action into a quagmire, and the difficulties of the West Africa Squadron were about to multiply.

CHAPTER 6

Into a Legal Minefield, 1815–20

It is not what a lawyer tells me I may do; but what humanity, reason and justice tell me I ought to do.

EDMUND BURKE

IT MIGHT HAVE been supposed that the ending of the long war with France would have brought a higher priority to the anti-slaving campaign and a consequent expansion of the West Africa Squadron. However, as at the end of every conflict, the Government wasted no time in reducing the Fleet to a peacetime establishment. As soon as it seemed that the French threat had been removed the cull began, and in 1814 and 1815 no fewer than 168 Fifth Rates and smaller were sent to the breakers. The resulting shortage of the types of vessels needed as cruisers on the African coast was compounded by a new task. Europe continued to be obsessed by the shadow of Napoleon, and elaborate precautions were put in place to prevent his escape from St Helena. Inevitably these measures fell primarily to the Royal Navy, and, until the prisoner's death in 1821, as many as a dozen vessels were engaged in guarding him, or on passage to and fro. This task did, however, bring an unplanned benefit to the west coast cruisers: the lonely Atlantic island of Ascension was settled by the Navy in 1816 as an outpost for the St Helena patrol, and it was subsequently to become an increasingly valuable refuge for the West Africa Squadron.

The end of the war cleared the way for a further development, which immediately made a massive impact on the suppression campaign: the return of the French slavers. During the Hundred Days in 1815 Napoleon abolished the French slave trade, and the restored Louis XVIII immediately confirmed the abolition on his return to the throne, but the French Trade was far from finished. The Government in Paris was divided on the issue: the Ministry of Foreign Affairs wished to ingratiate itself with Britain, but the Ministry of Marine was beset by the shipping interest. French slavers had been prevented by the Royal Navy's trade blockade from operating to any extent during the war, and they were now anxious to re-spread their wings. When the *Affriquain*, owned by the ex-privateer captain and national hero Surcouf, sailed for Angola on an illegal slaving voyage in August 1815, no action was taken, and French shipowners and merchants took their Government's indecision as implicit approval for a resurrection of the Trade. Three years later the French Government finally declared the slave trade illegal, but slave traders had so many friends in positions of authority that this measure merely transformed a tolerated practice into a clandestine one. By then the Guinea Coast was infested with French slavers.

The re-emergence of France as a major slave-trading nation was not the only new difficulty for the West Africa Squadron and its counterparts on the Leeward Islands and Jamaica Stations. As suppression measures became more threatening, at least north of the Equator, the slavers perceived the need for speed at sea as well as in loading, and, with the ending of the Anglo-American war in 1814, an ideal type of vessel became available in considerable numbers. This was the small, swift, two-masted "Baltimore clipper", which originated for privateering and blockade-running during the American War of Independence, probably as a derivation of the Bermuda Sloop. The type was further developed, for the same roles, during the war of 1812, but by 1815 the owners of these vessels were seeking alternative employment for them. Poor cargo capacity rendered them largely useless for legitimate trade, but they were clearly ideal for more nefarious purposes, particularly piracy and smuggling. They also quickly commended themselves to American and Cuban slave traders. For slavers their lack of hold capacity was a minor consideration by comparison with their outstanding fair-weather sailing capabilities and their relative ease of concealment in the African rivers and creeks. When the boy Theodore Canot, who was to become a leading

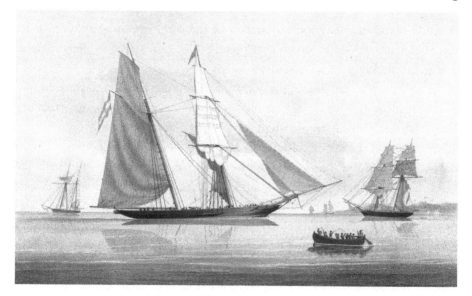

2. Slaver schooner *L'Antonio*, a typical Baltimore clipper

slave trader, first set eyes on a gathering of these Baltimore clipper slavers at Havana he was entranced: "There was something bewitching to my mind in their racehorse beauty. These dashing slavers, with their arrowy hulls and raking masts, got complete possession of my fancy."[1]

The Baltimore clipper was to become synonymous with the slave trade, and builders on the east coast of the United States and in Cuba and Brazil continued to refine the type to meet the requirements of the slavers. The conditions in which these clippers would be used could be accurately predicted: specific harbours on both sides of the Atlantic, and ocean passages in tropic or semi-tropic weather with generally light but steady winds and occasional squalls. What were required were light-weather flyers, but they also needed to stand up to hard squalls and to carry sail if under pursuit in a blow. As long as they sailed well in the conditions of the westward transatlantic passage it did not greatly matter if they were less swift on some other points of sailing. A slaver was invariably the pursued in an encounter with a cruiser, and she could choose her heading in a chase, as long as she had sea room. Almost all the Baltimore clippers were rigged as brigs or schooners, and most of the schooners carried square topsails to give better performance in open-water sailing. Masts were heavily raked, spar lengths and sail areas were increased for work in the Trade, and all manner of exotic light sails were introduced: kites, studdingsails outside studdingsails, water sails, ringtails, gunter skysails and moonrakers. Deadrise and draft were increased, ballast was carried low, waterlines became sharper, displacements were reduced and construction became lighter. Length was usually 75–100 feet between perpendiculars, and rarely exceeded 110 feet on deck, which equated to about 300 tons. Beam was generally about a quarter of length. Despite the lightness of their armament these clippers were not buoyant, and their low freeboard and scanty bulwarks made them wet in a seaway. Decks were clear and unencumbered, hatches were few and small, and it appears that, with little accommodation provided below, the officers and crew lived largely on the open decks. By merchant-ship standards the crews of these Baltimore clipper slavers had to be large to manage their cargoes and to handle their vast areas of sail; for example, one brig of 105 feet on deck carried 40 men. Rough and brutal though their officers mostly were, many of them were fine seamen and ship-handlers, and, once free to spread their canvas at sea, these beautiful vessels could usually, but not always, show a clean pair of heels to the cruisers.

It would be a great while before there was any corresponding improvement in the quality of the cruisers sent to deal with this expanding menace. The Admiralty had neither funds nor motivation for developing small vessels with the necessary characteristics, and for the next decade and more the West Africa

Squadron had to make do with heavy frigates, Sixth Rates, sloops and gun-brigs of the Napoleonic period. The heavy frigates, although having some redeeming features, were generally too ponderous and unhandy for catching the increasingly agile slavers, and with draughts of 18 feet or so they were incapable of work close inshore. The sloops and the Sixth Rates were of a more suitable size, but, with few exceptions, they lacked the speed to compete with the newer slaving vessels in open water. The only cruisers able, with care, to cross the bars protecting the major slaving rivers were the little gun-brigs. They were 80–90 feet in length,[*] more nearly the size of the ideal vessel, mounted two long 6-pounder or 12-pounder bow guns and eight or ten 18-pounder carronades, and were manned by crews of 50. Although they sailed fairly well off the wind, they were rarely a match for the Baltimore clippers, and they were not designed for lengthy cruising. Added to that, their small ships' companies were rapidly depleted to a critical level by a demand for prize crews, and, with only about five feet under the beams on their lower decks, they were particularly uncomfortable to live in, especially in the sweltering heat of the slave coast.

* * *

On the departure of Commodore Browne and *Ulysses* in June 1815, HM Ships *Porcupine*, *Ariel* and *Comus* remained briefly on the Coast. *Porcupine*, after another disappointing and tedious three months between Senegal and Freetown, sailed for Barbados in July with troops from Sierra Leone. Having delivered the news of Napoleon's escape from Elba, *Ariel* made no evident contribution to the suppression campaign, and continued on her way in early September with despatches to the Cape and Mauritius. *Comus* then found herself the only man-of-war on the Coast, and the West Africa Squadron temporarily ceased to exist.

Comus was not entirely alone during this period; the colonial schooner *Princess Charlotte* gave her spirited support and continued the struggle after *Comus* sailed for home. Few details of this schooner's characteristics survive, but she must have been the "130 ton, copper-bottomed schooner" procured by Governor Maxwell to replace the *George*, and she seems to have been armed with a mixed bag of about 14 long guns and carronades. By now she was commanded by Lieutenant W. S. Saunders of the Royal African Corps, and his particular challenge was to deal with American and Spanish slavers trading to Cuba from the northern slave

---

[*] The length given for a sailing warship is the Length On Deck (Length On The Lower Deck for large ships), unless otherwise indicated. This is roughly equivalent to the modern Length Between Perpendiculars.

rivers. They were a murderous bunch who had declared their intention to continue the traffic by force of arms, and to destroy every British vessel in their power. Their craft were fairly small, between 150 and 180 tons, but mounted long guns and carried crews of up to 80 men. If they sailed in groups the smaller cruisers had little chance of dealing with them.[2]

On the evening of 22 June 1815 *Princess Charlotte* found a brig at anchor off the Gallinas River and flying a small American jack. Closing the brig before an onshore wind, Saunders hoped to trap her between his schooner and the beach. The brig opened fire as soon as *Princess Charlotte* was within range, and a shift of wind allowed her to weigh and stand to the northward. The chase lasted for nearly two hours in constant rain and a heavy swell which rendered ineffective the fire of *Princess Charlotte*'s long guns. The brig's superior sailing qualities took her out of range, leaving the schooner with two dead, three wounded and three of her carronades unserviceable thanks to carriage failure.

This reduction in her armament, and a lack of faith in her remaining guns, put *Princess Charlotte* at a disadvantage three days later at daybreak when she encountered two schooners under Spanish colours off the River Sesters. Saunders hoisted a Spanish ensign to make his approach. By 0700 *Princess Charlotte* was within range and under her own colours, and the schooners, now one close astern of the other, opened fire. Saunders held his fire for another 20 minutes, and then opened with round shot and grape from his larboard broadside. Before long three more of his larboard carronades had failed, and he attempted to board the leeward schooner. She managed to evade, and the gun action continued in light and baffling airs. By 0810 *Princess Charlotte*'s rigging was badly cut about and two of her starboard carronades were unserviceable, but she made another attempt to board. Having again evaded successfully, the leeward schooner made signs of surrender, but as soon as she was out of the line of *Princess Charlotte*'s guns she reopened fire and crippled the colonial schooner aloft. Then she ran for it. The windward schooner, meanwhile, had been maintaining a raking fire from half pistol-shot, but on becoming the target of *Princess Charlotte*'s fire she dropped away and rejoined her consort. A few minutes later the colonial schooner's larboard 6-pounder and 9-pounder long guns failed, and, running short of ammunition, Saunders was obliged to decline further action. Having faced a total of 16 long guns, *Princess Charlotte* was perhaps fortunate to have sustained only seven wounded in this, the first pitched battle of its kind on the Coast. It was an indication of the degree of resistance likely to be offered in future by slavers under Spanish colours, and occasionally others.

Despite these two reverses, *Princess Charlotte* was by no means unsuccessful in the years 1814–16. Although the events surrounding the captures are not recorded,

Lieutenant Saunders and his schooner are given credit in the court records for the arrests of no fewer than 18 slavers, all but two of which were condemned. Her most valuable successes were the Portuguese ship *Nossa Senhora da Vitória*, with 434 slaves, and the ship *Golandrina*, caught off Cape Mount, with 144. Of her other victims one, the *Gaveo*, was Portuguese and several were probably American under Spanish colours. In all, she achieved the release of 884 slaves.

By comparison with *Princess Charlotte*'s adventures, the final few weeks spent on the station by *Comus* were relatively quiet. She lay at Sierra Leone for the latter half of June 1815, and sailed to leeward on 2 July. Off Grand Bassa Captain Tailour learned from a canoe that the two schooners which had engaged the *Princess Charlotte* off the River Sesters were still on the Coast and buying rice for slaves, but *Comus* saw no sign of them. At dawn on 15 July, however, while rounding Cape Palmas, she surprised and boarded the Portuguese brigantine *Abismo*, heading for Elmina and St Thomas for slaves. No sooner had she taken the brigantine in tow, than *Comus* sighted another suspicious sail. After a six-hour chase she arrested a schooner recorded in her log as the Spaniard *Palafox*, also bound for Elmina for slaves. Neither of these vessels had slaves on board, but their intentions were clear, and Tailour sent them both to Freetown, where they were condemned. Three days later *Comus* encountered *Ulysses*'s tender, probably the ex-slaver *Dolores*, in company with a detained Portuguese brig, the *Nova Fragatinha*, which the tender had taken off Anamabo without slaves and abandoned by her master and crew. Captain Tailour sent a prize crew to take charge of the brig, presumably because the tender was unable to provide an adequate crew herself, and it was probably on these grounds that the prize was later condemned to *Comus* instead of *Ulysses*. After a final visit to Cape Coast, *Comus* made for St Thomas to find candidates for her July convoy. She sailed on 9 August for England with four merchantmen in company, as well as the *Nova Fragantinha*, which had also made her way to St Thomas with the tender. As the convoy passed Sierra Leone on its homeward passage Tailour sent the prize brig into Freetown.*

The administration of Sierra Leone was suffering severe turbulence in mid-1815. Governor Maxwell, in poor health, had been allowed home on leave in late 1814, and in June 1815 he resigned his office. As his temporary replacement he had appointed Major Appleton, the senior military officer in the colony, but it was a very sickly season in Freetown and, suffering from fever, Appleton had to hand over to Mr Hyde, a member of the Council. Hyde held the reins until the

---

* The court records show two *Comus* captures not mentioned by *Comus* herself: the *Senor de Cano Verde*, condemned for violation of the Portuguese treaty, and the *Corrego de St Thome*, probably also Portuguese. Both were without slaves.

Lieutenant-Governor of Senegal, Lieutenant-Colonel Charles MacCarthy, arrived in late July and was subsequently appointed Governor of Sierra Leone. Lieutenant-Colonel Chisholm moved from Goree to Senegal to become Lieutenant-Governor, and a Lieutenant-Colonel Brereton arrived to take charge of Goree. The state of the Sierra Leone courts was even more chaotic. The Chief Justice, Mr Thorpe, had been accused of fraud and dismissed. Thereafter, with the few other legal men in the colony laid low by fever, there was no one even minimally qualified to act as Chief Justice and Judge of the Vice-Admiralty court. In effect, the court had not operated since January 1815. Although he appointed the Clerk of the Court as Acting Chief Justice in August, MacCarthy had to wait until March 1816 for a new Chief Justice and Judge of the Vice-Admiralty court to arrive from England. This was Dr Robert Hogan, who soon earned the Governor's high opinion, but by the end of October 1816 Hogan was dead of fever.

During his sick leave in London, Governor Maxwell heard in March 1815 that the Prince Regent had, for some reason, pardoned three British slavers: Brodie, Cook and Dunbar. The circumstances of their arrests are not known, but they had apparently been convicted at Freetown in June 1814 and sent home under sentence of 14 years' transportation. A fourth slaver, Charles Hickson, seized in 1812 by the schooner *Vesta*, had been convicted, very belatedly, at the same sessions. He, however, had remained in Sierra Leone to serve his sentence of three years' hard labour, and Maxwell suggested that he too should be considered for pardon.[3] It was not until April 1816 that approval was received to remit the remainder of Hickson's sentence, but, unbeknown of course to Maxwell, he had been released by Acting Governor Appleton when the original news of the pardons arrived. The motivation for this clemency is unexplained; the signal it conveyed to the few remaining British slave traders on the Coast would seem not to have been helpful.

By the summer of 1815 Lieutenant Saunders of the colonial schooner *Princess Charlotte* was a sick man, and in July he requested leave to return home to recover. As his replacement MacCarthy proposed Acting Lieutenant Robert Hagan, late of HMS *Porcupine*, and the Colonial Office gave its approval. Hagan, aged 22, had served for nine years as a midshipman, the last three as First Lieutenant of *Porcupine* in the Acting rank. MacCarthy had met him when taking passage in that ship, and also held Captain Harvey of *Porcupine* in high regard. With the prospect of half pay looming when the ship paid off on her return home, Hagan volunteered for the colonial schooner. Harvey had recommended him, and the Governor had agreed. MacCarthy subsequently had good reason to congratulate himself on his choice.[4]

After five months without a single man-of-war on the slave coast, a new cruiser arrived at Sierra Leone on 9 January 1816. She was the Sixth Rate *Bann* (20), Captain William Fisher. Her orders listed the tasks of trade protection and interception of pirates as well as slavers, and she was to cruise as far as the Gabon as seemed most advisable for those purposes, visiting the British forts and settlements. *Bann* had given passage to Lieutenant Hagan, and MacCarthy, anxious that the ailing Saunders should be relieved in *Princess Charlotte* without delay, had hired the small colonial sloop *Mary* to take Hagan to find his new command. It would take little time for Robert Hagan to show his mettle.

The *Mary* sailed on 14 January, and three days later fell in with an American schooner under Spanish colours leaving the Gallinas. Hagan tried unsuccessfully to take her by surprise with the boats and then, despite the *Mary*'s considerable inferiority in force, he set off in chase. Meanwhile, Captain Fisher in *Bann* had heard at Freetown that there were two slavers loading at the Gallinas and, finishing his refit in haste, he sailed early on 18 January. Later in the day *Bann* encountered *Mary* and her chase, and in a dead calm the boats of the two vessels were sent under Lieutenant Tweed of the *Bann* to examine the schooner. She opened fire on the boats with grape and musketry, and they returned it. In about 15 minutes the boats' crews carried the schooner at the cost of two wounded, and found that she was the *Rosa*, previously the wartime American privateer *Perry*. She carried six guns and a crew of 19, mostly Americans, and was bound for Havana with a cargo of 276 slaves. Fisher took her into Sierra Leone, where she was condemned jointly to *Bann* and *Mary*.

Hagan then shifted to a condemned schooner to continue his search for *Princess Charlotte*. His crew consisted mostly of Kroomen and native soldiers, with a few Americans and five English seamen, and he unhelpfully renamed his new command *Young Princess Charlotte*. Having taken the empty Spanish schooner *Guadeloupe* on 14 February, he again returned to Freetown, where he heard of another two vessels buying slaves in the Rio Pongas. Having embarked a detachment of the garrison, he sailed to find them. In Hagan's absence the Governor heard from Saunders in *Princess Charlotte* that he had intercepted two schooners off the Gallinas, but after the exchange of a few shots they had escaped. However, Saunders knew that they had landed their trade goods, and he expected them to return. Moreover, one of them was the *Compadore*, a piratical slaver which had seized two Sierra Leone vessels in the past year, and MacCarthy was anxious to stop her depredations. *Princess Charlotte* was short of provisions, and so the Governor hired one of Hagan's condemned prizes to resupply her, compounding confusion by renaming her *Queen Charlotte*. When Hagan returned with two more Spanish prizes from the Rio Pongas, the schooners *Eugenia* and *Juana* with 82 slaves, both taken on 3 March, he sailed again immediately to renew his hunt for the *Princess Charlotte*.

The three schooners met on the morning of 9 March, and Hagan transferred to the *Queen Charlotte*. A plan was made to search down the coast as far as Trade Town for the *Compadore*, and at 0200 two days later a vessel was detected standing out from under Cape Mesurado. The schooners crowded on all sail in chase, but only *Queen Charlotte*, using her sweeps, was able to stay in touch in the darkness. At daylight the other schooners were out of sight, but by 0900 *Queen Charlotte* was within range of what was seen to be a brig and not the *Compadore*. The two contestants hoisted their colours and briefly exchanged fire. *Queen Charlotte* gained some ground, but there was little wind, and Hagan manned and armed a boat to send in pursuit. The boat's carronade became dismounted and she had to return, but there was now a breeze and Hagan, trying to get alongside the chase, had closed to grapeshot range by noon. He brought his larboard broadside to bear, and within minutes the chase appeared to have been silenced, with her colours hanging half down. Thinking that she had surrendered, Hagan took the gig to board her, but finding that she was still ready to resist he returned and reopened fire. The chase returned the fire, but then the breechings of all Hagan's larboard broadside guns gave way.

By now it was 1400, and at this point Lieutenant Saunders arrived with *Princess Charlotte*'s cutter. He was joined by *Queen Charlotte*'s barge and gig, and, in a dead calm, he attempted to board while Hagan's starboard broadside gave covering fire. The barge was met by musketry and two rounds of grape at a range of 20 yards, and three-quarters of her crew were killed or wounded. The other two boats also took casualties, and Saunders was obliged to withdraw. As *Queen Charlotte* reopened fire the chase hauled down her colours and hailed for quarter. *Young Princess Charlotte*'s boat arrived just in time to take possession. The prize was found to be the French brig *Louis* bound from Martinique to Bonny for slaves, with a crew of 32 and armed with five mounted guns and a variety of swivels, musketoons, blunderbusses and small arms. The cost of her capture was high: seven dead and eleven wounded.

Hagan's justification for pursuing this capture is not entirely clear. In the prevailing calm, he may not have recognised the French colours until the range was close and shots had been exchanged.[*] He was unlikely to have fired first, but

---

[*] This would not have been as unlikely an error as now appears. French slavers would have been wearing not the Tricolore but either the white colours of the Bourbons or port flags, most of which were white and blue. The Portuguese ensign at the time was blue and white, not the modern red and green.

he was not a man to back away once he had been fired on. However, the probable explanation is that he shared what appears to have been a widely held notion that the vague declaration at the Congress of Vienna authorised action against slavers under the flags of the signatories. In any event, on 3 April the Vice-Admiralty court, by then presumably with Dr Hogan on the bench, condemned the *Louis* to *Queen Charlotte* and *Princess Charlotte*. That, however, was not the end of the story. This episode was to lead to a milestone judgment, quoted as a precedent on numerous occasions thereafter. Captain Forest of the *Louis*, after absconding from parole in Sierra Leone, appealed to the High Court of Admiralty, and in 1817 Sir William Scott reversed the decision of the Freetown court. In doing so he wrote, "I can find no authority which gives us the right of interruption to the navigation of states in amity upon the high seas." He added that no government could force the way to the liberation of Africa by trampling on the independence of other states of Europe. He also declared that "to procure an eminent good by means that are unlawful is [not] consonant with private morality".[5]

Well before the case reached the High Court, Lord Castlereagh, on advice from Doctors' Commons, had told the Admiralty that the Navy had no right to detain foreign vessels for slaving in contravention of the laws of their own nations, but Scott's decision effectively proclaimed carte blanche to the French slavers. The judgment was obviously correct, and the episode re-emphasised the worthlessness of the Vienna declaration. Castlereagh had not yet abandoned his hope of concerted multinational action against the slave trade, but this high-profile incident gave added impetus to a policy of negotiating bilateral treaties with all the slaving nations.

Ten days after delivering the *Rosa* to the Sierra Leone court, *Bann* sailed for a cruise into the Bights. At Cape Coast Castle Captain Fisher was briefly diverted by an abortive punitive expedition against the natives of Apollonia who had plundered a wrecked merchantman, but he then continued along the Slave Coast as far as Whydah. Here on 5 March he captured the slaver *Temerario*, described as a pirate by Fisher and probably under Spanish colours. Loading had only just begun when *Bann* appeared, and the boats turned back to shore, leaving only 17 slaves on board. The vessel was powerfully armed and manned with 18 guns and 80 men, and she kept up an incessant fire on *Bann* as she approached, but inflicted damage only on sails and rigging. After despatching the *Temerario* to Freetown and calling at Princes Island for water and provisions, *Bann* headed south-east on 15 March, making for the Gabon, the southern limit of her writ.

The following evening, however, she was still in sight of the island after being driven back by tornadoes, and consequently encountered the Portuguese brig

Santa Antonio Milagroso. This was a vessel of 120 tons, with nearly 600 slaves on board. In the first hundred miles of her passage from the Cameroons bound for Bahia more than 30 had died, and among the living and dying the *Bann*'s boarding party discovered a decomposing corpse. Fisher found it impossible to describe the suffering of the slaves, and he immediately shifted many of them into *Bann*. Taking the slaver in tow, he turned back for Sierra Leone, his 115-foot ship crowded with 300 people despite the depletion of prize crews. The rain was almost constant, provisions ran low and almost a third of the ship's company fell sick.[6] Notwithstanding all *Bann*'s care, another 43 slaves died during the month-long passage, but 505 were liberated on arrival. Not a man of the ship's company was lost. On sailing for Barbados and home on 25 April, Fisher decided to take a look into the Gallinas, and off Sherbro Island he detained a large armed felucca under Spanish colours. She was apparently slaving, but does not appear to have been condemned, perhaps because her captor did not return to Freetown.

Earlier in March 1816 there had been a notable success in the mid-Atlantic. On the fourth the brig-sloop *Ferret* (14), Commander J. Stirling, on her way home after escorting Bonaparte to St Helena, fell in with the Spanish-flagged brigantine *Dolores* 300 miles north of Ascension. *Ferret* was not a run-of-the-mill sloop; she had been a clipper-built schooner privateer taken from the Americans in 1812, and, although she had since been oversparred as a brig and weighed down by additional scantlings and armament, she was still fairly fleet of foot. The *Dolores* was from almost the same stable, another clipper-built ex-American privateer, and *Ferret*'s appearance seemed to give her no concern. Immediately the sloop hoisted her colours; however, the slaver turned on her heel and opened fire with her two stern long guns. The two vessels would normally have been fairly well matched for close-range firepower, but *Ferret* had landed her only two long guns for the defences of Ascension, and had been obliged to strike down to the hold four of her twelve 12-pounder carronades to make room on deck for additional water. Unable to reply to the slaver's well-aimed long-range fire, she had to keep clear and work her way round to the chase's lee quarter. Both vessels were under all sail by the wind, but *Ferret*'s fore and main topsail yards and the gaff of her boom mainsail were soon brought down. She made rapid repairs, and four hours after the start of the pursuit, despite further damage aloft, she had closed to good fighting range. When her boom mainsail was shot away, and Stirling had laid *Ferret* alongside her, the slaver surrendered. The 218-ton *Dolores* was 13 days from Bonny to Havana with 250 slaves, and in the action she had lost seven wounded to the three dead and one wounded in *Ferret*. This was a rare instance of a clipper slaver caught in open-ocean chase, and Stirling took his prize into Freetown, where she was condemned.

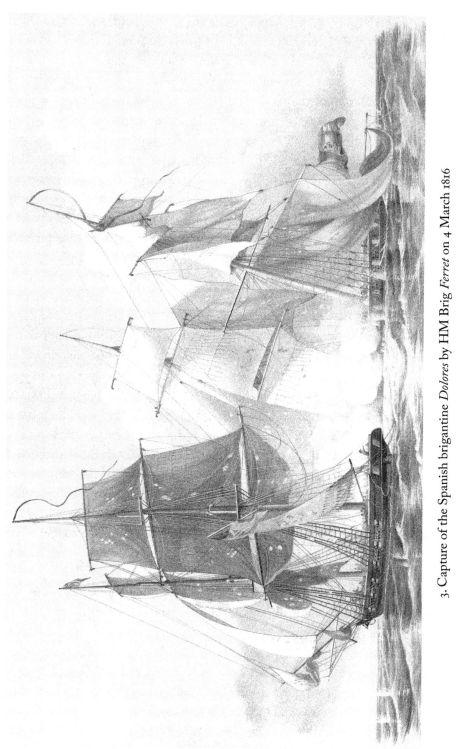

3. Capture of the Spanish brigantine *Dolores* by HM Brig *Ferret* on 4 March 1816

There was a sad outcome to another mid-ocean interception on 11 March. The brig-sloop *Zenobia* (18), Commander Nicholas Dobree, was on passage between England and the Cape Station when she encountered the American schooner *Selby* to the west of the Cape Verde Islands. The schooner immediately fled under all the canvas she could crowd on, and *Zenobia* chased her for 30 miles. Throughout the chase the American master stood on deck with a brace of pistols threatening to shoot any man who attempted to shorten sail. Nevertheless, *Zenobia* caught her, and, finding that she was using false Spanish papers and was bound from Havana to the Rio Pongas for slaves, Dobree sent her off to Plymouth with his First Lieutenant, George Chepmel, and ten men. Chepmel called at the Azores, where the British Consul-General provided him with £20 worth of stores, but the *Selby* was never heard of again. The Admiralty could find no justification for the arrest, and the £20 was later charged against Dobree's half pay.[7]

Shortly after *Bann*'s departure a new Commodore, Sir James Lucas Yeo KCB, arrived at Sierra Leone in the frigate *Inconstant* (36) on 29 April 1816. Yeo had, extraordinarily, been promoted to lieutenant at the age of 15, and in 1805, while in the frigate *Loire*, had stormed the fort at Mudros and brought out every vessel in the port. For that he was promoted to commander into one of his prizes, the sloop *Confiance*, and in her he led an Anglo-Portuguese force to conquer Cayenne in 1809. He was immediately made Post at the age of 26, and the Prince Regent of Portugal conferred on him a knighthood of the Order of St Bento d'Avis, not previously awarded to any Protestant.

Commodore Yeo's orders directed him to visit the British forts and settlements as far south as Benguela, a pointless limit in the absence of any British presence south of the Gold Coast and with no British anti-slaving remit south of the Equator, and Yeo ignored it. He was also ordered to enforce the Abolition Acts and the Portuguese treaty, but no mention was made of pirates. Finally, he was warned against interfering with merchant vessels of other nations and reminded that the Right of Search was a belligerent right only in time

4. Commodore Sir James Lucas Yeo KCB

of war. The Admiralty was aware, and was clearly concerned, that commanding officers had visited, on the High Seas, "vessels under the flags of Powers in amity with His Majesty", and had, in some cases, sent then in for condemnation upon having found them to be concerned with the slave trade. The Commodore's top priority was still to be the protection of British settlements and trading posts.[8]

Their Lordships evidently did not envisage a long deployment for *Inconstant*, and, with so few resources at his disposal, Yeo left the Windward Coast and the rivers of Senegambia to the colonial vessels and concentrated his own efforts on the Bights. After visiting the Cape Coast, Anamabo and Accra, *Inconstant* rounded Cape St Paul's on 21 May 1816 and sighted a schooner at anchor off Quitta Fort. On seeing the frigate the schooner instantly weighed and headed eastward. Two hours later *Inconstant* had closed sufficiently to open fire, and was soon alongside. The boarding party found that the schooner was the empty Spanish slaver *Carmen*, but she had no register, which threw doubt on her legitimacy, and Yeo detained her and kept her in company. Three days later the frigate's boats boarded and arrested the Portuguese schooner *Caveira* off Lagos with ten slaves on board and laden with trade goods of rum and tobacco. She was sent to Sierra Leone, where she was condemned along with her slaves.

Leaving the *Carmen* off Cape Formosa to make her way to Princes Island, Yeo headed for the island to land his prisoners from the *Caveira*, and then steered east

5. HMS *Inconstant*, Fifth Rate frigate

to take a look at the Rio St Benito and Corisco Bay before returning to Princes. After dark on 27 June the Commodore sent three boats away to bring out a slaver he had detected inshore. In the early hours the boats returned, towing the Portuguese schooner *Dos Amigos* with 247 slaves on board. The Portuguese fort fired a couple of shots in the darkness, but by 0600 *Inconstant* had the new prize in tow, and, with the *Carmen* again in company, was heading for Sierra Leone. The seizure of *Dos Amigos* was not merely unusual, it was also illegal: there had clearly been no breach of the Portuguese treaty, and it seems that the Freetown court condemned vessel and slaves largely because the defendants failed to make an appearance. The decision was ultimately reversed, and Yeo earned the Admiralty's "decided disapprobation".

On 28 July *Inconstant* made another arrest while she lay at anchor off Freetown. Sighting a suspicious sail in the offing, Yeo sent his boats to board, and they brought back the empty slaver brig *Monte de Carmo Testa* which, together with the *Carmen*, was condemned. The frigate sailed again on 31 July for another visit to the Gold Coast forts before leaving in late August for Barbados and England. She was joined at Cape Coast by the *Caveira*, which Yeo appears to have manned as a tender, and, in the absence of any note of the capture in *Inconstant*'s log, it seems likely that the final prize credited to the frigate, the *Scipiao Africano*, was taken by her tender.

For the final four months of 1816 there was again no man-of-war on the Coast, but the one remaining colonial vessel continued her sterling work throughout the year. After the *Louis* incident, Lieutenant Saunders had reported that the *Princess Charlotte* was weak, leaky and unfit for sea in the rainy season, and it appeared impracticable to repair her at Sierra Leone. Governor MacCarthy was given approval to replace her, and there was an ideal vessel available: the *Louis*. In April she became the colonial brig *Prince Regent* with Acting Lieutenant Hagan in command, and she immediately made her mark when, on about 20 April, the large Spanish-flagged schooner *La Neuve Aimable* appeared off Freetown. She had sprung a leak and been forced to bear up for Sierra Leone. Boats from *Bann* and *Prince Regent* boarded her and found 388 slaves on board before bringing her in to be condemned to Hagan and the Governor.[9]

Hagan's next victim was the piratical slaver *La Paz*, found off Cape Mesurado on the morning of 5 May. After a pursuit of seven hours and an action of 15 minutes *Prince Regent*'s men boarded in two divisions under Mr Hayman, the Master, and Mr Thompson, the First Mate. Hagan was the first on board and was stabbed in the hand by the vessel's nominal master, a Frenchman called Segure. Hagan shot him dead. The vessel had been the American *Argus* until she changed colours in

Havana; her supercargo, part-owner and effectively her master was an American, as were most of her crew of 63. *Prince Regent*'s men were outnumbered, and the 13 guns in the *La Paz* gave her a marked superiority over her captor.* Indeed her crew had been told the previous evening that they were going to seize *Prince Regent* and take her to Havana. *Prince Regent*'s casualties were one killed and two wounded, and the *La Paz* had three killed and nine wounded. The pirate had already captured the English schooner *Apollo*, which Hagan released, and was intending to attack all Sierra Leone vessels to avenge supposed wrongs to Havana merchants. The *La Paz* and her 108 slaves were condemned as English or American property, and her surviving officers may have been sent to England for trial. After this success *Prince Regent*'s striking rate dropped, but she claimed five more prizes before the year was out. The first, on 30 September, was the American brig *Triumphante*, under Spanish colours with 349 slaves. She was a powerful vessel, mounting ten guns and manned by 64 men, but she was boarded and seized in the Cameroons River by the brig's boats under the command of Mr Thompson, who had been a midshipman in *Bann*. In this assault, obstinately resisted with pikes, Thompson was wounded in the head and some months later had to resign his post. Two Portuguese vessels were taken by surprise on 7 November: the schooner *Rodeur* with 32 slaves and the brigantine *Caroline* with 18, and on the same day Hagan captured the Portuguese schooner *Santa Johanna* after disguising *Prince Regent* and hoisting foreign colours. On boarding this third vessel he found that she measured only 27 tons but had 85 people on board, of whom 72 were slaves, stowed over the water casks in a space 40 feet in length with two feet nine inches of headroom. She was, nevertheless, bound for Bahia. The final prize of the year, on 15 November, was the Portuguese brig *Ceres* with 11 slaves, and she, along with the earlier four, was condemned.

Amid some confusion, preparations were at last being made by mid-1816 for France to retake possession of Senegal and Goree as agreed in the Treaty of Paris. The first evidence on the African coast of the French expedition was the wreck of the frigate *La Meduse* near Cape Blanco on 2 July, and many of her survivors were cared for at Senegal by the British garrison. One of the surviving passengers was the Governor Designate, Colonel Julien Schmaltz.† A forerunner

---

* The armament and complement of *Prince Regent* is not known, but it is unlikely that she had substantially more than the five guns she mounted as the *Louis*, and the size of crew probably remained approximately the same at 32.

† This disaster was made famous by the painting *The Raft of the Medusa*, by Théodore Géricault. The makeshift raft on which about 150 people, mostly soldiers, escaped the wreck, was supposedly abandoned by the boats. Only ten survivors were rescued from the raft, but all 250 in the boats reached shore.

of the expedition, the corvette *L'Echo*, appeared off Senegal on 25 July, but Governor MacCarthy was starved of instructions from London, and the British garrisons did not evacuate the settlements until 25 January 1817. By mid-March MacCarthy was expressing to Lord Bathurst at the Colonial Office his concern at the revival of the slave trade in both Senegal and Goree with the return of the French. Several vessels under Spanish colours had already embarked slave cargoes there. In January 1818 he reported that all the inhabitants of Senegal and Goree were aware of the resurgence of the slave traffic, and that the French authorities were paying no more than lip service to their proclaimed abolition. Two months later there was information that Schmaltz was not only encouraging the Trade but also was putting six slaves of his own into each vessel loaded. It was also reported that one of the Senegal chiefs had pillaged two villages and enslaved 200 of his own subjects. Relations between MacCarthy and Schmaltz quickly became strained, and remained so.

Both Yeo and MacCarthy were deeply pessimistic about trends in the Trade. There seemed to be more slaving vessels at work north of the Line than there had been before Abolition. Since the peace there had been a particularly marked increase in the numbers at the Rio Pongas, Cape Mount and to leeward. Most of them were under Spanish colours, but were believed to be either American, Portuguese or French property. MacCarthy complained that at the Portuguese possessions to the north of Sierra Leone even the governors had gangs of slaves ready for market. A reliable report from the Bights indicated extensive slaving from the rivers Benin, Bonny and Calabar and the Cameroons by "foreign armed vessels of every description" sailing under Spanish colours from Havana. Furthermore, despite Britain's effort and investment over the previous decade, a renewal of slaving was reversing progress made towards legitimate trade. MacCarthy wrote that "it not only turns the minds of the natives from peaceful habits of industry, to rapine and plunder, and has actually created a famine a short distance to leeward of us, and raised the price of provisions here". He feared that British merchants, unable to compete with the slavers, would be ruined. It was Yeo's view that the shift of the traffic into Portuguese and Spanish hands had quadrupled the degree of cruelty involved, primarily because the traders feared that the Trade would be finished in a few years' time and were overloading their vessels. These slavers were making huge profits; the master of the *Caveira*, for example, had bought each of his slaves for £5.10s worth of tobacco, and expected to sell them for 400 dollars (about £100) each. It was no wonder that they intended to continue under any circumstances and at every cost.[10]

Yeo had other concerns too. The first, which MacCarthy shared, was the increasing presence of piratical slavers under Spanish colours. These were vessels heavily armed with 14 to 20 guns and with big crews of up to 80 men which, instead of trading for slaves, seized them from other slaving vessels. They were not averse to more traditional acts of piracy, and, as Yeo reported, they were "determined to fight any vessel they have a chance with". He was also dissatisfied with Sierra Leone as a base for anti-slaving operations. Apart from it having a "detestable climate" and being in a "deplorable state", it was a great distance to windward of the Bights where most of the slavers were captured, and mortality in the prizes during the passage was liable to exceed 10 per cent. He believed that the Gold Coast was a more suitable region for a settlement of emancipated slaves, although the forts were currently in a poor condition and their occupants little better than prisoners of the neighbouring chiefs. He recommended the acquisition of the Dutch fort at Axim, which had an excellent anchorage and landing place. Not surprisingly, MacCarthy disagreed. What was unarguable was his opinion that suppression could not be achieved with dull sailing ships which could not cope with the fast Spanish-flagged slavers commanded by "active and enterprising men", either French or American. The general plan of these people, Yeo explained, was to complete with water and provisions before arrival, land trade goods and supercargo when the coast was clear, and head out to sea until they reckoned that their slaves were ready. Then they would stand inshore to await a signal before returning to load and to sail within a few hours.

There was little respite for the men of *Inconstant* and *Bann*. On 19 December 1816 Yeo again left Spithead for the African coast, and this time he was soon followed by the Sixth Rate *Cherub* (26), to which Captain William Fisher and the ship's company of *Bann* had transferred. After a swift passage, *Cherub* made her first capture on 17 January 1817 off Cape Lahou on the Quaqua Coast. Her prize was the Portuguese ship *Esperanza*, loaded with 413 slaves from Whydah, and she was sent to Sierra Leone. By the time he arrived at Cape Coast on 21 January Fisher had heard numerous accounts of piracy by large armed schooners, generally showing Spanish colours but manned "by ruffians of all nations", and, when *Cherub* called at Cape Coast Castle, Governor White told him of a Spanish-flagged ship slaving in the neighbourhood which he suspected might be one of them. She was well armed and manned by 140 men, and she was in communication with the Dutch Governor Daendels at Elmina, who was supplying her. When Daendels had passed through Freetown in 1815 to take up his post he had assured Governor MacCarthy that he would immediately enforce the recent decree by the King of the Netherlands against the slave trade, but it now seemed that he might himself have a financial interest in this slaver. Two days later *Cherub* found the ship, but, although she claimed to be seeking an encounter with a British cruiser, the slaver's superior sailing qualities kept her out of the sloop's clutches.

Hearing that there were Portuguese slavers at Popo, Fisher headed east and found there on 26 January a schooner which he was assured by natives in a canoe was a slaver from Princes Island. *Cherub* gave chase, but the schooner was out of sight by the morning of the twenty-eighth. Fisher steered to cut her off from Princes, and he sighted her again two days later, but again she sailed clear. Repeating his tactics, Fisher finally found his quarry almost within gun-shot at 0300 on the thirty-first. This time *Cherub* held her own, but could not close the range. At noon it fell calm, and Fisher sent away his boats under Lieutenant James Henderson. As they approached the schooner she opened fire with grape and musketry, and at last hoisted her colours, French, but indistinguishable from Portuguese in the calm. Henderson replied with the 12-pounder in the pinnace and muskets, and, as the boats came alongside, the schooner struck her colours. Then, however, she fired two more rounds of 12-pounder grape and langridge into the pinnace and gig, and followed that with harpoons and other missiles. Of eight officers and men in the gig, seven were wounded, and in the pinnace two were killed and four wounded, one mortally. The cutter, making for the bows for a diversionary boarding, was spared. Henderson fired a round of grape through a port onto the schooner's quarterdeck, and boarded. Resistance then ceased. She was the *Louisa*, and on board were 47 villains of mixed nationality, only nine of them French. The schooner carried eight 12-pounders and one 9-pounder, and she was fitted for 300 slaves but was carrying none. Keeping the three officers and two other men in *Cherub*, Fisher allowed the remainder to escape, and sent the schooner to Sierra Leone.

That was not the end of the story, of course. Lord Castlereagh at the Foreign Office was not pleased that another French vessel had been taken. After seeking the advice of the Advocate General, he not only repeated to the Admiralty that British cruisers had no right to arrest French slavers, but also stated that it would be pointless to take criminal proceedings on the grounds of the firing on and killing of *Cherub*'s men because self-defence would be claimed. He intended to put the case before the French Government in the hope of a trial under French law. When the Admiralty's rebuke reached Fisher, he replied, with some asperity, that he had had no way of knowing that the *Louisa* was French before she opened fire, and that if she had identified herself he would not have molested her. As it was, he wrote, he would have been neglecting his duty had he failed to investigate an armed vessel which in two days of chase had not shown her colours.[11] This was the second occasion on which a French slaver had brought on an unnecessary action with a cruiser and her boats, and it is difficult to escape the conclusion that the French could not resist an opportunity to goad the Royal Navy and spill its blood, having suffered so much anguish at its hands for the past quarter-century and more.

After watering at Princes and checking the Whydah anchorage, *Cherub* returned to Cape Coast on 27 February in the hope of meeting Yeo. There was no Commodore, but Fisher was surprised to receive a petition from the Dutch Government Secretary at Elmina asking for a passage to Europe. On 3 March the man himself, a Mr Milet, arrived and pleaded for the protection of the British flag in the name of his own king. He wished to return home to inform his Government about the atrocious behaviour of Governor Daendels, and he brought with him evidence that Daendels was conducting an extensive trade with Spanish and Portuguese slavers, supplying them with goods, water, canoes and Africans. He also produced papers showing that the Spanish slavers intended to seize any British merchantman they found. Fisher willingly complied with Milet's request, and that afternoon *Cherub* sailed for Jamaica and England. As it happened, Daendels did not live to be brought to justice; he died at Elmina "of an apoplectic fit" in May 1818.

Inconstant, meanwhile, had arrived at Sierra Leone on 13 January 1817, and, after making good the defects of an unidentified tender, perhaps the *Caveira*, and despatching her to Barbados, had been making a largely uneventful cruise as far as the Gabon. At Cape Coast Yeo was told that a ship of 26 guns had appeared off the castle a few days earlier, and her master had asked the whereabouts of *Inconstant*, claiming that he had come "to blow her out of the water". She had shown no colours, but she was full of men and there was little doubt that she was a Spanish-flagged slaver. The frigate's boats searched the Gabon without success, and then she made for St Thomas, where she met the prize schooner *Carmen*, which, it seems, she had manned as another tender. On 16 March she hove to off Annobón, and Yeo sent the barge and the cutter inshore. The frigate's log gives no indication that the Commodore was aware of anything amiss, but the boats returned with the surviving crew of a prize belonging to *Cherub* which had been wrecked on the island: a midshipman, five seamen and marines, four prisoners and three Kroomen. The prize must have been the *Louisa*, taken six weeks earlier. One of the seamen died in the boats, and a Krooman joined him that evening as *Inconstant* weighed for Ascension. On the passage south and then back to the Gold Coast, two more seamen and a marine were added to the bill for the *Louisa* incident. On 1 May, with the rains about to begin, the Commodore sailed for the West Indies and home.

Despite the low level of naval activity on the slave coast at the time, the Treasury seems to have become concerned at the cost of head money payments on emancipated slaves, and the bounty entitlement was reduced by an Order in Council of 11 July 1817. The 1807 Abolition Act had set figures of £20 for a man slave, £15 for a woman and £5 for a child captured under the Act, and double those amounts for slaves condemned as prizes of war. These payments were now replaced by one of £10 for every slave. Over the years the total sums paid in prize money and bounties had fluctuated considerably: for example, £17,090 in 1810, a peak of £36,620 in 1812, and a trough of £12,269 in 1816. Much of this money did not reach the captors, being siphoned off by agents and other intermediaries, and that which did percolate through was usually very late in doing so, sometimes after the recipient ship had paid off, and not infrequently after the deaths of some of those entitled to it. This was particularly the case in the event of an appeal. Proceeds from the sale by the courts of condemned slaving vessels were likely to be paid to the captor cruisers more quickly than in prize of war cases, but the expenses of the court proceedings were deducted before payment. A typical payment was made to *Princess Charlotte* after the condemnation of the *Gaveo*: the gross proceeds were £307.17.5, expenses in Sierra Leone amounted to £112.3.9, and the schooner received the balance of £195.13.8 to be distributed in accordance with the Prize Regulations. Occasionally expenses exceeded proceeds, as in the case of the *Esperanza*, for which gross proceeds were £335.3.0 and expenses £379.15.0. *Cherub* was charged for the deficiency.

The year 1817 brought significant success to the patient diplomacy of the Foreign Office. It began with the signing in London on 28 July of an Additional Convention to the Portuguese Treaty of January 1815, and this agreement, with a term of 28 years, was to provide a format for others to be negotiated by Britain with most of the slaving nations in the next few years. The convention prohibited slaving by British ships or British subjects under any flag, by Portuguese ships in any of the African harbours or roads prohibited by the 1815 treaty, by vessels under the British or Portuguese flag on the account of any foreign subject, and by Portuguese vessels bound for any port not in Portuguese dominions. The Portuguese slave trade would continue to be permitted from Portuguese possessions south of the Equator between 8°S and 18°S, and also from territories between 5°12′S and 8°S over which the King of Portugal had retained his rights, namely Malembo and

Cabinda. Portugal undertook to promulgate within two months of ratification of the convention the punishments for illicit slaving, and to renew the prohibition on importation of slaves into Brazil under any flag other than Portugal's. Legitimate slavers would be given passports for their voyages.

Vessels suspected of carrying slaves illicitly obtained could be detained and brought to trial by two Mixed Commissions, one in Brazil and the other in Africa. These Commissions would each consist of a Commissary Judge and a Commissioner of Arbitration from each of the two nations, and a Registrar appointed by the sovereign of the country of residence of the Commission. They were required to judge, without Appeal, on legality of capture and, in the event of the liberation of a captured vessel, on the indemnification she was to receive. Sentences were to be given as summarily as possible. Ships legally detained were to be condemned as lawful prize and sold by public sale with their gear and cargo, other than slaves, for the benefit of the two governments. Slaves condemned by a Mixed Commission were to be delivered to the government on whose territory the Commission was established, and each slave was to be given a certificate of emancipation. The two governments bound themselves to guarantee the liberty of freed slaves. Compensation for losses incurred by detained vessels not condemned was to include the value of slaves released, and an allowance of 5 per cent interest per annum was made on compensation awards. The convention also laid down the detailed regulation for the conduct of the Mixed Commission courts.

It was decided that the two Anglo-Portuguese Mixed Commissions would sit at Freetown and Rio de Janeiro, but a further, temporary, Commission would be established in London. This measure stemmed from Britain's agreement in the convention that it would indemnify the owners of Portuguese vessels and cargoes detained by British cruisers between 1 June 1814 (the cut-off date for claims under the 1815 treaty) and the date on which the permanent Mixed Commissions assembled. Claims would be received and liquidated by the London Commission, but vessels which had loaded slaves north of Cape Palmas and not in Portuguese territory would not be entitled to compensation, nor would vessels detained more than six months after ratification of the 1815 treaty with slaves taken from north of the Equator.

The convention also included Special Instructions concerning the means of detention of suspected slavers. Visits and captures were to be made only by cruisers of the British and Portuguese navies, and only by those carrying copies of the Special Instructions. Only vessels actually carrying illicit slaves could be detained, and arrests south of the Equator might be made only after hot pursuit from north of the Line. Searches were to be conducted "in the most mild manner" and by officers of at least lieutenant rank. Slave cargoes were to be left on board untouched, and were not to be disembarked until arrival at the place of trial, unless a commanding officer decided that the state of health of the slaves demanded urgent disembarkation of all or part of the cargo. It was also directed that no vessel should be visited or detained while in a port or roadstead belonging to either signatory, or within cannon-shot of its batteries ashore. Instead, representation was to be made to the authorities ashore. If there were no such authorities available, vessels with slaves on board could be detained within cannon-shot of their respective territories.

The British legislation to carry this convention into effect (58 Geo. III, c. 85) was passed on 5 June 1818, and the Crown declared that the British half of proceeds from slaver sales would go to the captors. The payment of bounties, or head money, for slaves condemned by the Mixed Commissions was set at £10 per slave, as it was for condemnations under the Abolition Acts. The Portuguese Regent, for his part, issued an *Alvara* at Rio in January 1818 that officers of illegal slavers were to be banished to Mozambique for five years.

Despite its serious shortcomings, especially its failure to address the massive trade south of the Equator, the new convention did represent a step forward, and it did at least impose on Portugal a share of the responsibility for the trial of its slavers. The signing of the agreement aroused fury in Brazil, although in itself Britain's opposition to the Trade caused little concern to the merchants of Rio and Bahia who were determined to continue the traffic. Lacking secret assurances from Lisbon akin to those made to the Cuban planters by Madrid, the Brazilian planters and merchants began to contemplate independence from Portugal. They also accelerated their imports. The British Minister in Rio reported in May 1818 that 25 ships had arrived since the beginning of the year, "none of them bringing less, and many more, than 400 [slaves]".

It was time now for Castlereagh to renew his efforts against the Spanish Trade. In late 1815 the Spanish Foreign Minister had blocked a British proposal for abolition north of the Line, insisting that "discussion with England on this subject was at an end". A few months later, however, the Council of the Indies recommended to the King of Spain an immediate and total, if conditional, abolition of the slave trade. The King thereupon proposed to Britain that the Spanish Trade might be abolished immediately north of the Equator, and south of it in five years' time, on payment by Britain of £1.5 million to compensate for losses, and on agreement by Britain to take action against the Barbary pirates. Castlereagh found these conditions unacceptable, and he reminded the Spanish Minister in June 1816 that, under a *Cedule* of 1804, the Spanish trade had been permitted for only twelve more years, and was now therefore illegal under Spanish law. In passing he also complained about the piratical conduct of slavers under Spanish colours.[12] The Spanish Government ignored the matter of the *Cedule* and denied all knowledge of pirates, refusing to accept that they were Spanish subjects and leaving it to the British Government to destroy them.

Nevertheless, there was encouraging progress, which culminated in a treaty signed at Madrid on 23 September 1817 by which Spain promised to abolish the slave trade throughout her dominions on 30 May 1820. On that date the Trade would become unlawful to Spanish subjects on all parts of the coast of Africa, although a further five months would be allowed for completion of voyages begun before the deadline. An undertaking of more immediate effect was abolition of the Spanish slave trade north of the Equator on the date of ratification of the treaty, 22 November 1817. In this case a further six months of grace would be allowed for completion of voyages. The treaty included measures for establishing and regulating Mixed Commissions, and for issuing Special Instructions to cruisers and passports to legitimate slavers, an exact repetition of these parts of the Portuguese Convention except that there was no requirement for a London Commission and the permanent Mixed Commission courts would be in Havana and Freetown. Nor was there need for a clause to regulate visits and detentions in ports and roadsteads. There was, of course, a price for London to pay: a sum of £400,000 as compensation for losses already suffered by Spaniards during Britain's suppression campaign.

Admiral Home Popham at Jamaica reported that the Spanish treaty had caused an immediate rise of 25 per cent in the price of slaves there, and a rise of 100 per cent was expected for those coming from north of the Line.[13] Cuban opposition to the agreement was strong; the planters had offered the Spanish Government $2 million to be allowed to maintain the Trade, and another $5 million per year thereafter as long as the private permission continued. When this attempt had failed, Cuba's planters, merchants and officials agreed that no treaty by Spain would be allowed to interfere with their slave trade, and it seems that Madrid subsequently decided secretly to allow Cuba to break Spain's anti-slaving law. In the event, actual slaver losses in the subsequent few years amounted to only a tiny proportion of the slavers sailing from Cuban harbours, and there was no apparent reduction in the traffic: about 205 vessels between January 1816 and September 1817. Many of the Spanish-flagged vessels taken were owned and commanded by Americans, despite their bogus masters and papers. The King of Spain did not use the £400,000 to compensate the Cuban merchants; he used it to buy five warships from the Tsar, a project which partly motivated his signing of the treaty.

With the departure of *Cherub* and *Inconstant* in the spring of 1817 the continuation of the campaign at sea fell upon the lonely but able shoulders of Robert Hagan, now at last confirmed in the rank of lieutenant. During 1817 his colonial brig *Prince Regent* took six slaver prizes, and two more in January 1818, leading to the release of 464 slaves. On 29 March his boats boarded the Spanish schooner *Laberinto* after a five-day search of the Rio Pongas, and he took the Portuguese schooner *Gramachree* at the Rio Nunez on 12 April after she had groped her way through the creeks from Bissau to avoid detection. She was an American whose supercargo had obtained Spanish papers from the Governor of Havana at a cost of $2,000. Six days later, also in the Nunez, Hagan's boats found the Spanish brig *Esperanza* after a patient search of 11 days. The vessel had been largely dismantled, and her masts were wrapped with palm branches; she was so well hidden that she was given away only by the voices of her crew. On 21 October there was the rarity of a capture of an English vessel, the brig *Two Boats*. Although she was empty she was known to be on passage from a Mr Lightburne's factory on the Rio Pongas to the Scarcies, but there is no mention of her in the court records. After a chase of 30 hours on 7 December, Hagan took the Spanish brigantine *San Juan Nepomuceno* with 272 slaves, and a week later he contrived to surprise the Portuguese schooner *Linde Africano*, trading with her cargo on deck. Another English vessel fell to him on 21 January 1818. This was the schooner *Hannah*, on passage from the Rio Pongas to the Banana Islands with five slaves, and she was condemned. *Prince Regent*'s final success came on 31 January when she released 130 slaves from the Spanish schooner *Belle Machancho*, which had been destroyed by her crew to prevent capture.

Thereafter the *Prince Regent*'s efforts were constrained by the predictable failure of the Colonial Office to procure for Hagan the Special Instructions required by the Portuguese and Spanish treaties, despite repeated requests by Governor MacCarthy who complained that: "I am debarred from giving my support to the cause of humanity, and cannot without those Instructions prevent the Spanish and Portuguese Traffickers taking slaves from rivers within Ten Miles of Sierra Leone."[14]

However, the treaties clearly specified that only vessels of the signatories' navies might detain slavers, and colonial vessels were considered to be beyond the pale. So, after January 1818 *Prince Regent* was limited to revenue enforcement and logistic duties until her career was interrupted on 13 October 1818 in the River

Gambia, where she struck an uncharted pinnacle of rock and was judged to be damaged beyond local repair. Happily, that misfortune was not similarly the end of Lieutenant Hagan's anti-slaving career. He initially transferred to the colony's remaining 74-ton schooner, but MacCarthy had higher ambitions for his protégé.

There was little respite for *Cherub* after her arrival at Spithead in mid-July 1817. She was refitted for return to Africa, and sailed again on 11 October under the command of Captain George Willes. By 29 November she was off Grand Bassa, in unsuccessful chase of a vessel which successively hoisted American, British and French colours. There were further frustrating chases during the following month of patrolling along the Gold Coast and Slave Coast, and an excessively close encounter with a loaded Spanish slaver which ran on board *Cherub* off Anamabo, carrying away the cruiser's jib-booms and spritsail yard and stoving in her cutwater, but Willes had probably received a copy of the Spanish treaty by then, and would have known that his assailant was still immune from arrest.

On his arrival at Princes Island on 27 December he took action on less sure grounds. He identified a schooner at anchor up the harbour as the *Concha*, which was wanted for piracy against two English vessels off Popo, the *Diana* and the *Nimble*. *Cherub* ran almost alongside the schooner, but grounded in two fathoms, and when Willes saw the pirates preparing to open fire he directed his marines to drive them from their guns with musketry. He then boarded, killing the mate and wounding two seamen. While *Cherub* was kedging herself off the shoal, the indignant fort fired several ineffectual rounds at her, and Willes had to spend two days convincing the Governor of the *Concha*'s guilt before he was allowed to sail with her as prize. Early in January an auction was held on *Cherub*'s quarterdeck of damaged and perishable items found in the prize, but Thomas Roberts and Richard Crump had attempted to relieve the *Concha* of more tempting articles: gold dust and doubloons, and earned four dozen lashes each for theft.

When *Inconstant* arrived home in October 1817 she was in such a poor state that the Admiralty ordered her to be taken to pieces, and her officers and ship's company transferred to the frigate *Semiramis* (42). On 21 December Captain Sir James Yeo, in his new ship, sailed yet again for the African coast, rehoisting his broad pendant in the Channel. This third deployment for Yeo and his men would, however, be brief and unsatisfactory. Between her arrival at Sierra Leone on 17 January 1818 and her departure for the West Indies in late April, *Semiramis* managed to board only one slaver, a Spaniard she was not permitted to detain. Towards the end of a final cruise to leeward, the Commodore met *Cherub* at Cape Coast in mid-February.

On 8 January 1818 *Cherub* and her prize had, between them, boarded and detained two vessels off Porto Novo. Willes released one of them with some of his prisoners from the *Concha*, but the master of the other, the Portuguese brig *Difforso*, complained that he had been plundered of tobacco and rum a week earlier by armed men from a Spanish slaver called *Descubridor*. As she was cruising past Quitta Fort on 15 January, *Cherub* sighted a brig which slipped her cable on seeing the cruiser. After a chase of over two hours, and several rounds of gunfire from *Cherub*, the brig hove to and struck her colours. She was the *Descubridor*, and, on finding the stolen goods on board, Willes removed her crew and sent her to Cape Coast to await the Commodore's arrival. *Cherub*, meanwhile, returned to Porto Novo to find a witness of the piracy to give evidence in court. The only man available from the *Difforso* was an African.

On his return to Cape Coast, Willes found that the *Descubridor*'s cargo was deteriorating, and he was obliged to sell it by public auction before continuing his cruise along the Gold Coast, with the *Concha* in company. *Semiramis* arrived on 19 February, and the Commodore, agreeing with Willes's course of action, sent *Cherub* and her two prizes back to Freetown. After storing and watering at St Thomas, and towing the brig for much of the passage, *Cherub* reached Sierra Leone on 15 April to find that the Vice-Admiralty court was not empowered to try cases of piracy. The frustrated Willes landed the 460 slaves from his prizes, already ravaged by smallpox, and waited for Yeo to return. The surviving 433 slaves were condemned and released.

On parting company from *Cherub* at Accra Roads on 1 March, *Semiramis* made an uneventful passage via St Thomas to Ascension, and returned to Sierra Leone on 21 April. There Yeo and Willes agreed that the two prizes should be taken to Jamaica, where both cruisers were bound and where the Vice-Admiralty court could try the charges of piracy. *Semiramis* provided the prize crew for the *Descubridor* and an ailing Commodore sailed with her on 24 April for the West Indies, but *Cherub* had not yet finished her work on the African coast. After supplying *Prince Regent* with spare boatswain's stores, Willes sailed for the Gold Coast on 29 April and despatched the *Concha* to Jamaica.

As she passed Grand Sesters for the last time *Cherub* landed her Kroomen, then pursued a brig off the St Andrews River until the chase showed a French ensign, and, on 15 May, encountered a ship under Spanish colours off Cape Three Points. That chase lasted for nearly ten hours, and the slaver replied to *Cherub*'s fire with four long twelves and musketry. The quarry had been extensively hit in hull and rigging before she was brought-to, and the boarding party found that she was the *Josepha*, last from Puerto Rico but initially from New York. She had been the British packet *Windsor Castle*, taken by the Americans during the war, and her master and nearly all her crew were American. Although she had trade cargo to purchase 600 slaves she had only 35 on board, and Willes, not wanting to return to Sierra Leone, decided to send her to join his other prizes in Jamaica. He was also probably doubtful of achieving a condemnation at Freetown, with the court there in a state of change. A week later, after transferring 16 male slaves to *Cherub*, he detached the *Josepha* for Jamaica, while *Cherub* continued to Accra and then Ascension before heading for the West Indies herself. *Josepha*'s adventures continued; she was sent from Jamaica to Havana and then back to Sierra Leone where, a year after her capture, she quietly sank at her moorings. The Spanish Commissary Judge subsequently declined to try her case because of doubt whether *Cherub* had held the Special Instructions at the time of the arrest.

There was trouble awaiting Willes at Jamaica. Articles from the plundered merchantmen *Diana* and *Nimble* had been found in the *Concha*, and her master had been condemned and executed for piracy, but the man who had originally admitted to being the master of the *Descubridor* was now claiming to have merely been a passenger. Furthermore, the agents for the *Descubridor*'s Havana owners had induced magistrates to state that no offence had been committed, and the one witness to the alleged act of piracy, the *Difforso* crew member sent by Willes to Jamaica to give the necessary evidence, had not been permitted to do so because he was black. The Attorney-General had consequently decided that no charge could be brought, and, when *Cherub* arrived on 12 July 1818, Willes found that the *Descubridor*'s owners were claiming restitution of the vessel and cargo, as well as compensation for an excessive number of slaves said to have been waiting for embarkation. When Willes did not immediately pay he was arrested, and bail was set at £6,000, a sum beyond his means. On agreeing to give up the *Descubridor* and provide an account of the slaves landed at Sierra Leone and the perishable goods sold at Cape Coast he was released, after paying jail fees and other expenses of his arrest. The Spanish Government subsequently complained that the *Descubridor* had been illegally detained for slaving, but Willes insisted that she had been arrested for piracy, and the Foreign Office seems to have dismissed the complaint. However, the owners brought a successful case against Willes in the Court of Guildhall, and in January 1820 he was ordered to pay the enormous sum of £21,180. In November 1822 he managed to reach a compromise with the Havana slaver at a figure of £6,507. His legal costs amounted to £329.9s, against which he was able to set the money remaining from the sale of the perishable goods from the brig. The Treasury agreed to pay the balance of £62.11s.6d.! Such were the outrageous risks to which commanding officers were exposed.[15]

Semiramis sailed from Jamaica on 26 July, touched at Havana and steered for home. On 21 August, just south of the Grand Banks of Newfoundland, Commodore Yeo died, another victim of either the mosquitoes of the west coast of Africa or their cousins at Jamaica. He was only 35.

Lord Castlereagh's diplomatic efforts were further rewarded on 4 May 1818 by the signing of an Anglo-Dutch treaty at The Hague. By this it was agreed that, within eight months of the ratification of the treaty, Dutch subjects would be prohibited from taking any part in the slave trade. There would be reciprocal rights of visit and detention, except in the Mediterranean and in European waters outside the Straits of Gibraltar to the north of 37°N and to the east of 20°W. Alone among the treaties to date, this one placed a limit of 12, without special consent, on the number of cruisers on either side authorised to make such visits, and stipulated that the names of the vessels so authorised, and issued with copies of the treaty, were to be reported between the signatories. In all other aspects the agreement was virtually identical to the Spanish treaty of September 1817, although there was no provision for British payments to the Netherlands. Mixed Commissions would be set up at Freetown and Surinam. British legislation to carry the treaty into execution was passed on 31 March 1819 (59 Geo. III, c. 16).

Another, more ambitious, initiative by Castlereagh met with less success. He had secured agreement at the Congress of Vienna for a permanent conference of the European powers to collate information and institute action against the slave trade, and 14 meetings were held in London between August 1816 and the first full conference of Foreign Ministers at Aix-la-Chapelle in 1818. At this first peacetime conference to resolve their differences, France, Russia, Spain and Portugal were presented by Castlereagh with the revolutionary proposal that an international Right of Search in time of peace, essential in British eyes to effective abolition of the Trade, should be enforced by an armed international police force on the African coast. This idea was generally perceived as yet another measure to ensure British supremacy at sea, and found no support, even from Tsar Nicholas who had himself considered the formation of a "neutral institution", with an international fleet and its own court and headquarters in Africa, to combat the slave trade. Clearly there was, as yet, no alternative to a policy of negotiating bilateral anti-slaving treaties.[16]

During these years, Governor MacCarthy in Sierra Leone was giving much care and attention to the welfare of his expanding population of emancipated Africans. With 6,400 of them in the settlement by 1819, and only 115 Europeans, the challenge was appreciable, but MacCarthy was anxious to prepare the ex-slaves to earn their own livelihoods after their initial year of government support. He established new villages in the hinterland and divided the peninsula into parishes. He took boys from the villages as government apprentices to skilled tradesmen, and, with the arrival of a surveyor, he initiated a formal system for allocating land to African families. To serve the parishes curates were sent out from England, preferably married clergymen whose wives could act as parish schoolteachers. In 1816 MacCarthy gave enthusiastic support to the Church Mission Society's new school just outside Freetown for 350 children of both sexes, and he bombarded the Colonial Office with requests for schoolbooks, material for clothes-making and tools for crafts and agriculture.[17]

MacCarthy's ambitions were also directed beyond internal affairs. With the return of Senegal and Goree to the French, he was concerned that Britain would be unable to exploit the trade opportunities of the River Gambia, and in 1816 he founded a new settlement on St Mary's Island in the mouth of the river, taking the precaution of naming it Bathurst after the Colonial Secretary. The French reacted by establishing a factory at Albreda, 30 miles upriver on the north bank, and in 1819 MacCarthy feared that Colonel Schmaltz, his opposite number in Senegal, was planning to form a military post at the factory. Not only would this threaten British trade in the river, but also it might encourage an extension of the French slave trade already flourishing at Senegal and Goree. Anxious also that the French might show unwelcome interest in the Isles de Los, he garrisoned Crawford Island, at the centre of the group and with the best anchorage.

After the departure of *Cherub* in late May 1818 there was no cruiser on the Coast for some months, but an alert passer-by made an arrest at Madeira in early June 1818. Captain Dundas arrived at Funchal in the frigate *Tagus* (42) to find at anchor the brig-sloop *Sappho* (18), Commander Plumridge, and the merchant ship *Alfred* of London. Following a dispute between the master and supercargo of the *Alfred*, the British Consul-General asked Dundas to investigate, which he and Plumridge did. Dundas formed the suspicion that the *Alfred* was in fact a slaver, not least from the number of guns and quantity of powder he saw on board her, but he was unwilling to arrest her in Portuguese waters. After much delay, *Alfred* sailed on 10 June, with a warrant officer sent by Dundas ostensibly to keep the peace. *Tagus* followed her to sea, and it came as no surprise to Dundas to see guns and carriages being ditched from the merchantmen a short while later. He quickly boarded her, and *Sappho* took her to Sierra Leone, but the Vice-Admiralty court mistakenly restored her. The well-informed Zachary Macaulay of the African Institution had already warned his agent in Freetown to keep an eye open for her, and subsequent investigation in England not only showed that she had indeed been on a slaving voyage but also gave leads to sources of British interest and capital involved in the Trade.

After six months with no Royal Navy presence on the Coast, the frigate *Tartar* (42) arrived at Freetown on 2 December with Commodore Sir George Collier, Bt. Aged 44, this educated, energetic and immensely experienced officer had been almost continuously in command at sea between 1800 and the end of the American war. He had distinguished himself in the sloop *Victor* by the chase and sinking of the French corvette *La Flèche* in the Indian Ocean in 1801, an action which earned him Post rank the following year. Thereafter he served almost exclusively in frigates, culminating in *Leander*, one of the 56-gun fir-built ships designed to combat the heavy American frigates during the war of 1812. There could hardly have been a better choice for the West Africa Squadron, and in due course he would command the strongest force on the Coast for some years.

6. Commodore Sir George Collier, Bt

Collier's orders differed little from those of his predecessor, again requiring departure from the Coast before the next year's "Rainy and Sickly Season", but making mention for the first time of the use of Ascension Island for "refreshment". Rendering protection and assistance to the British forts and settlements remained the first priority, and he was glibly told to look into the "several bays and creeks" between Cape Verde and Benguela, particularly on the Gold Coast, at Whydah, in the Bight of Benin and in Angola to seize slave vessels under the authority of "the several Acts of Parliament", and "to use every other means in [his] power to prevent a continuance of the traffic in slaves". As far as Portuguese and Spanish slavers were concerned, he would be sent copies of the treaties. There was no attempt to advise him on what "other means" were within his power.[18]

After a brief stop at Sierra Leone, *Tartar* sailed on 6 December 1818 to run down to the Gold Coast settlements, meeting several slaving vessels: French, Spanish, Portuguese and American; but none with slaves embarked. At Princes, Collier heard of a piratical slaver cruising between St Thomas and Cape Lopez, and, while heading towards Annobón in search of her, he chased and boarded a Spanish schooner of only 90 tons, with 250 slaves on board. As was usually the case with officers new to the Squadron, Collier was appalled by the conditions he found in the slaver. However, he knew that she could not be adjudged until the arrival of the Spanish Commissioners in Freetown, and he regretfully released her after 24 hours. Two vessels loaded with slaves, one of them probably the pirate, had sailed from Annobón two days before *Tartar*'s arrival there, and Collier gave up the search. There had been some fever on board the frigate, but it subsided on passage to Ascension, and on 18 February 1819 the Commodore returned empty-handed but healthy to Freetown.

In Collier's absence there had been an incident which caused further diplomatic embarrassment with France. In the early hours of 9 February, the brig-sloop *Redwing* (18), Commander Frederick Hunn, one of the St Helena squadron, had sighted in bright moonlight a strange sail which had made to escape. At daylight the stranger hoisted a French ensign and shortened sail, and *Redwing* boarded her. She was *La Sylphe*, 51 days out of Bonny for Guadeloupe, and of her cargo of 388 slaves 20 had already died. Aware of French anti-slaving legislation and supposing that there was a court in Sierra Leone with a French commission, Hunn sent her into Freetown. The surviving 364 slaves and most of her crew were landed, but Hunn, finding no court to adjudge the case, took his prize onward to the Cape of Good Hope, *Redwing*'s destination. There, on the extraordinary grounds that *La Sylphe* had contravened French legislation, the Vice-Admiralty court condemned her. The Admiralty, no doubt the target of Foreign Office anger, sharply told the Commander-in-Chief at the Cape to send *La Sylphe* to the nearest French possession, and Hunn, a half-brother of George Canning, incurred Their Lordships' displeasure for this clearly illegal seizure.

Hunn was not the only commanding officer to find himself in difficulty. In October 1818 Captain Scobell, late of the *Thais*, reported to the Admiralty that he had been summoned to appear before the Court of King's Bench to answer a charge of false imprisonment and seizure of property brought by Robert Bostock and John McQueen, the two British slavers arrested by *Thais* after their factory at Mesurado had been burned in May 1813. After conviction at Freetown they had last been heard of in the Portsmouth hulks, but it was now suggested that the trial in Sierra Leone had been unauthorised. The appalled Scobell suspected that the two convicts had decided to act after hearing of the deaths of two key witnesses. He wrote:

> I cannot learn it was ever instanced to agitate the decision of a Vice Admiralty Court after so many years lapse without appeal or mention, and if such may be, what Officer is safe and who can risk the construction that the Common Law may after time put on even their Lordships' orders.[19]

He voiced an anxiety which would be felt by many of his successors. In this case the Government accepted the Admiralty's recommendation that the defence should be handled by the Treasury Solicitor, but it would be years before a judgment was reached.[*]

Tartar sailed again on 28 February 1819 to retrace her steps to leeward, visiting the roadsteads at Cape Coast and Accra before stretching out to the Bights islands. At Cape Coast Castle Collier learned that the Governor had received a threatening letter from the King of Ashanti, and he resolved to return via the Gold Coast, but was then distracted by events at Princes. On his visit to the island in January he had met with evasion and misinformation about two vessels in the harbour: the half-Portuguese, half-Spanish schooner *Armistad* and the Spanish brig *Gavilan*; and at St Thomas he had found a Spanish schooner which had been plundered of her slaves by a vessel answering the description of the *Gavilan*. On her return to Princes on 22 March *Tartar* discovered the *Gavilan* in a new paint-scheme and ready to sail, and the Portuguese authorities attempted to prevent Collier from communicating with the shore. Realising that something was afoot, Collier determined to take a look at the back of the island, although the Military

---

[*] No record has been found of the outcome of this case.

Commandant, Major Xavier, tried to dissuade him. The fort, surprisingly, fired a 21-gun salute as *Tartar* weighed on the morning of 24 March, and as she cleared Port Antonio she sighted two schooners running for the land and a mass of canoes in the offing. The schooners, for whom the salute had certainly been intended as a warning, immediately hauled their wind and stood away.

During the night *Tartar* rounded the north end of the island, and at daylight she saw the schooners closing the bay on the west coast. Collier hoped to catch them landing slaves, but they were too quick for him, partly because *Tartar*'s cutter capsized on lowering. While the crew was being rescued, *Tartar* opened fire on both schooners, and one, later found to be the *Armistad*, cut her berthing ropes to the other slaver and made all sail to the north, hoisting Portuguese colours. When the boats were clear and heading for the second schooner, still at the beach, the frigate set off in chase of the *Armistad* and caught her at 0300 on the twenty-sixth. On board were found a licence from the Governor of Princes for slaving at Cabinda, and the log of a slaving voyage to Bonny in January, as well as one slave boy in a locker, wrapped in a Spanish ensign.

The boats' crews had found the second schooner deserted but with canoes leaving her. They recovered six exhausted slaves from the beach and took the schooner, the *Princessa*, round to Port Antonio. The batteries initially tried to prevent her from entering, but she managed to anchor in the harbour. When *Tartar* arrived on 27 March Collier sent in his First Lieutenant to bring her out unless he found her to be Portuguese. There was no evidence of nationality on board, but the fort opened fire when she attempted to sail. The First Lieutenant thereupon called on the Governor to enquire whether he was detaining the schooner, but the Governor, Jose Ferrara Gomez, prevaricated, and the *Princessa* was taken out. Collier wrote to the Governor to explain the position and asked for an explanation of the activities of the schooners. He received in reply a storm of abuse and an accusation of murdering slaves. By this time the crew of the *Armistad* had admitted that the two schooners had illicitly loaded 500 slaves at Bonny and, as Collier had suspected, they were destined for the *Gavilan* at Port Antonio. Clearly Gomez and his henchmen were fully aware of the transaction. After transferring the six rescued slaves to *Tartar*, Collier sent the *Princessa* in *Armistad*'s wake to Sierra Leone, but there is no record that either vessel came to trial. In the continuing absence of Portuguese and Spanish commissioners, there was currently no court with the necessary jurisdiction.

After a very brief call at Ascension, *Tartar* returned to Sierra Leone on 1 May 1819 and sailed for Barbados and home three days later. Collier just missed the arrival of the first of his squadron. On 7 June the ship-sloop *Pheasant* (22), Commander Benedictus Kelly, entered the Sierra Leone River after calling at Cadiz. She brought with her the first members of the Mixed Commission courts: Thomas Gregory, the British Judge, together with his nephew Edward; Don Francisco Le Fer, to be the Spanish Commissary Judge; and Don Juan Campo, the Spanish Commissary Arbitrator. Edward Fitzgerald, who had been Judge of the Vice-Admiralty court since November 1817, was to be the British Arbitrator, and Molloy Hamilton, King's Advocate of the Admiralty court, was appointed Registrar. The Spanish Commission was closely followed by the Dutch Judge and the Arbitrator, Dow van Sirtema and Mr Bonnouvrié, who arrived in the Dutch sloop-of-war *Komet* on 28 August. The British officials were appointed to all three courts, but there was still no sign of the Portuguese commissioners to form the third of them. Despite the deadline of one year after ratification set by the Portuguese Convention for the formation of the Commissions, the Rio court did not open until December 1819, and the Portuguese commissioners for Freetown were not appointed until February 1820. The excuse offered by the Portuguese was that good men could not be found to undertake such an unprofitable task in a bad climate.

The Admiralty's curtailment of cruising on the slave coast during the rainy season apparently applied only to Fifth Rates. *Pheasant* sailed for the Bights on 9 June, and on 12 June the *Morgiana* (18), Commander Charles Strong, arrived at Sierra Leone. *Morgiana* was one of the flush-decked ship-sloops built of cedar at Bermuda, more spacious than the quarterdecked sloops and probably the most suitable vessel for the task yet to be sent to West Africa. Only a day astern of her had been the ship-sloop *Erne*, but in passing the Cape Verde Islands in darkness *Erne* had failed to give a wide enough berth to the rocks at the eastern end of the island of Sal and was totally wrecked, happily without loss of life. Her Commanding Officer, Commander Timothy Scriven, attributed the disaster partly to "that island being laid down in the Admiralty chart very incorrect". The two sloops were soon to be joined by three more cruisers, and it is clear that the appointment of the Spanish and Dutch commissioners had stimulated renewed effort from the Admiralty after some years of negligible interest in the suppression campaign. With what seemed to be a sound legal basis now established, and international courts in place, there was motivation for deploying more men-of-war to the Coast.

The first victim of this new phase of the campaign fell to *Pheasant*, cruising in the Bight of Biafra. On 30 July, off the mouth of the River Campo, the sloop intercepted the Portuguese schooner *Novo Felicidade*, a vessel of only 11 tons, with

71 slaves on board. Seventeen men, shackled in pairs, and 20 boys were crammed, one on top of another, in a space 17 feet in length, 7 feet 3 inches in breadth, and 1 foot 8 inches high. Beneath them were stowed the yams for their food. One man was in the last stages of dysentery, creating, as Commander Kelly put it, "an effluvion too dreadful for description" among the yams. Thirty-four women were stowed in a space 9 feet 4 inches long, 4 feet 8 inches wide, and 2 feet 7 inches high. Unlike the men, they were not shackled, and had probably been allowed on deck at times. Although the master claimed that he had loaded his cargo south of the Line at Cabinda, he had no passport for his voyage, and the vessel's position aroused sufficient suspicion for Kelly to arrest her. He shifted 22 of the male slaves into *Pheasant*, took his prize in tow and, cutting short his cruise, headed for Freetown to save the lives of the slaves.

Apart from one death, the care of the sloop's Surgeon had achieved some improvement in the condition of the Africans by the time they reached Sierra Leone on 17 August, but Judge Gregory took the abnormal step of ordering their removal ashore before adjudication. By this time the slaver's master and crew had admitted that the slaves had been bought in Old Calabar and that the vessel belonged to Gomez, the Governor of Princes Island, but the continuing absence of Portuguese commissioners appeared to prevent the vessel coming to trial. In November, however, in a rare gesture of cooperation, the Portuguese Ambassador in London agreed that the case might be heard by the British Judge alone, and the *Novo Felicidade* was condemned.

Morgiana returned from her first cruise on the day following *Pheasant*'s arrival at Freetown. She brought with her the Spanish schooner *Nuestra Señora de Regla* of Havana, taken off Petty Bassa on 10 August. It had come on to blow and the schooner had been obliged to slip her cable with only one slave of her cargo embarked and leaving her master ashore. Thereupon *Morgiana* snapped her up to become the first slaver condemned by the Spanish Mixed Commission court, and, although she measured only 50 tons, her sale produced handsome net proceeds of £759.11.8.

A less satisfactory encounter came on 13 August, when *Morgiana* chased a large schooner for 30 miles in thick and dirty weather. In attempting to make her escape inside a ledge of rock off Grand Bassa, the schooner drove ashore in heavy surf and was lost. A boat from *Morgiana* tried to intercept canoes lifting slaves from the wreck but was fired on by the schooner, and some slaves were undoubtedly drowned when the vessel broke up within half an hour. Commander Strong felt, however, that the incident would have a salutary effect along the Windward Coast, and *Morgiana* continued to make her presence felt as far as

Accra. The sloop showed herself fast enough to speak with every vessel she chased, apart from one brig which just escaped after a pursuit of a hundred miles, and Strong reported with satisfaction that "We have sailed well enough to convince them (for every vessel I have spoken has been slaving) that there are Men-of-War upon the Coast with whose sailing they cannot trifle." On return to Freetown on 18 August the ship's company was remarkably healthy despite "the rains falling in Torrents for days together".[20]

At the beginning of August 1819 two gun-brigs, *Snapper* (12) and *Thistle* (12), sailed from Spithead to join the Squadron, the first of their type on the Coast since *Daring* had been scuttled in 1813. Of their several limitations for the Station, their inability to provide adequate prize crews had at least been appreciated, and the Admiralty ordered them to hire 12 Kroomen each, additional to their complements of 50. Returning to Sierra Leone in *Thistle* was Lieutenant Robert Hagan. Governor MacCarthy had sent him home in December 1818, after *Prince Regent*'s grounding, to report to Lord Bathurst at the Colonial Office and carrying a strong recommendation that he should be given command of a suitable vessel in which to return. Commodore Collier helpfully provided his opinion of what would constitute a "suitable vessel":

> should combine fast sailing with strength, to be about 350 tons burthen and capable of mounting 14 twenty-four pound Carronades with two long nines for chase Guns it would be advisable that she should be coppered to the Bends and to be fitted with Iron Tanks for holding her water. It would be absolutely necessary that she should be of sufficient height out of the water to admit air skuttles (*sic*) being cut in order to give a free ventilation on the lower Deck, which in the wet season would be the means of keeping the Ships Company in health.[21]

Thistle fell well short of this ideal, but at least Lord Bathurst, realising that colonial vessels no longer had a part to play against the slave trade, had passed on MacCarthy's recommendation to the Admiralty, and Their Lordships had been wise enough to invest in the experience and talent of Lieutenant Hagan.

It was probably on Hagan's initiative that *Thistle* delayed for a fortnight at the Cape Verde Islands to salvage stores from the already plundered wreck of the *Erne*. Then, after a pause for water and provisions at Sierra Leone, she headed down the Grain Coast. Off Trade Town on 9 October she detained the Dutch schooner *Eliza* of St Eustatius, but, by the time *Thistle*'s boat could board her, the slaver had managed to land all but one of her cargo. On the following day a second Netherlands schooner, the *Virginie*, became a more satisfactory victim, with 32 slaves on board but abandoned by all save one of her crew. With his ship's company now depleted by the absence of two prize crews, Hagan was obliged to abandon his intention to make for the Bight of Biafra, and returned to Freetown. There his plan was again thwarted, this time by his patron, the Governor. MacCarthy was anxious to visit his fledgling settlement at Bathurst and asked for passage in *Thistle*, but a controversy had brewed with the court over one of Hagan's Dutch prizes, and it was not until 19 December that she at last sailed for the Gambia.

Lieutenant James Henderson in *Snapper* had arrived at Sierra Leone on 4 September 1819 to discover that the colony had suffered appallingly during a severe wet season. Of the 118 Europeans in Freetown at the beginning of the year 58 had died, and two of the foreign commissioners were just recovering from fever. It was perhaps not surprising that the Portuguese were having difficulty in filling their appointments. Henderson also learned from the Governor that the French, American and Spanish slavers were very active in the Rio Pongas and Rio Nunez, and that the Spaniards were now sometimes using French colours. With his wood and water completed, Henderson was probably glad to leave Freetown for the Bights to seek *Pheasant* for orders. On 30 September, by chance in passing, he found the 150-ton Spanish schooner *Juanita* off the River Costa on the Quaqua Coast, not generally a worthwhile hunting ground. She had only nine slaves on board, probably in the midst of loading, and Henderson sent her back to Sierra Leone, where she was rapidly condemned.

Snapper finally fell in with *Pheasant* off Cape St Paul's on 3 October, and Kelly, currently the senior officer on the Coast, sent her to victual at Cape Coast Castle and then to rejoin *Pheasant* at Princes Island. *Pheasant* herself had sailed from Sierra Leone in early September, bound for the Bights. She found the Windward Coast swarming with fast slaving schooners and was in almost continual (but unsuccessful) chase for nearly a fortnight. *Pheasant* boarded three French slavers, provisioned at Cape Coast, and then stationed herself west of Cape Formoso on the longitude of Lagos, in the optimum position to intercept Portuguese slavers heading south after loading illegally on the Slave Coast. As even the Foreign Office had appreciated, vessels bound for Brazil from the Portuguese possessions south of the Equator steered directly across the Atlantic, carried by the South-East Trade Wind and the South Sub-Tropical Current, and it was a reasonable presumption that any Portuguese slaver found north of the Line was trading to or from a prohibited area. *Pheasant* suffered almost incessant rain throughout her cruise, and Kelly's officers were badly hit by fever. By late October he had lost his Gunner, Surgeon, Master and Acting Pilot. However, Kelly's tactics met with success on 6 October when he arrested the Portuguese brig *Vulcano do Sud*, bound from Lagos to St Salvador with 270 slaves. Leaving the Portuguese master, the boatswain, one white sailor and the African cook on board, he despatched her to Freetown, under the command of Midshipman Castles, with a prize crew of four British seamen, two Kroomen and two Cape Coast natives. The sequel was tragedy.

The first inkling of what had happened came in 1821 from the boasting of a Portuguese seaman in another *Pheasant* prize that he had taken part in the murder of the *Vulcano do Sud* prize crew. It was only in the following year that the full story came to light, when one of the Cape Coast natives of the prize crew, Quashie Sam, appeared aboard *Morgiana* at Bahia. One afternoon, a week into the *Vulcano*'s passage to Freetown, he had been feeding the slaves while Castles was fishing in the chains; one British seaman was at the wheel, two were in the foretop and another was on the forecastle. Quashie Sam heard a shot and saw the Portuguese master cutting down Castles with a cutlass. The midshipman fell overboard, bleeding, and the quartermaster already lay dead at the wheel. The master then fired at the two men in the foretop and they both fell into the sea. The Portuguese sailor and the cook ran to the forecastle and killed the fourth British seaman before throwing him over the side. The two Kroomen jumped overboard and were drowned. The Cape Coast natives were driven below and told that they would be sold on arrival in Brazil, but if the brig was to meet a British warship they would be killed. At Bahia the slaves were transferred to another vessel and the *Vulcano do Sud* was scuttled. The two Cape Coast men were enslaved, and Quashie Sam passed through the hands of several masters before he courageously made good his escape. Castles and his crew had paid the price of inattention, inexperience and, in all probability, fatigue.[22]

Kelly, happily ignorant of the fate of his men, called at Accra at the end of October 1819, checked the anchorage at Whydah, and took advantage of an improvement in the weather to make a survey of part of the Slave Coast. After several weeks in the Bight of Benin he noted a great reduction in the slave trade between Cape St Paul's and Cape Formoso, and he was almost certainly justified in attributing it to the presence of a cruiser; his sloop had provided the first protracted watch on the Slave Coast for some years. *Pheasant* found *Snapper* as planned at Princes late in December, and sent her to cruise to the south-east of Whydah and keep an eye on the Whydah anchorage, a likely port of call for Portuguese treaty-breakers. After a week of refitting rigging at Princes,

Pheasant called at Cape Coast for provisions before heading for Sierra Leone in mid-January 1820.

There was still no Portuguese Commission in Sierra Leone, but on 4 November 1819 the London Slave Trade Commission, established by the Additional Convention of June 1817, opened to judge claims for compensation on Portuguese seizures made since 1 June 1814 and condemned by the Vice-Admiralty court under the treaties of 1810 and 1815. The process would be a dispiriting one for Britain. Amongst its apparent errors, the Vice-Admiralty court at Freetown had misinterpreted the wording of the 1810 agreement with Portugal which had prohibited the Portuguese slave trade on any part of the Coast of Africa "which did not form part of the Prince Regent's dominions and in which the trade had been abandoned by the European powers which formerly traded there". Britain had tended to ignore the final phrase of this description, and a case in point was the Old Calabar River; it was not a Portuguese possession, but it certainly had not been abandoned by other European slavers. Furthermore, it was highly questionable whether Britain had been entitled to seize Portuguese slavers under the 1810 treaty at all.

Alexander Marsden and Justinian Casamajor sat as the British Judge and Arbitrator, and their Portuguese partners were Antonio Julião da Costa and John Jorge. Disagreements between Marsden and da Costa were frequent, and began with matters of principle. Da Costa demanded indemnity which would give owners at least 69 per cent profit on capital exclusive of interest, and that profit should be allowed not only on a slaving voyage but also on an intended return voyage. He gave way on this final point, but the other issues were decided by the method used by all the Mixed Commissions for resolving disagreements between the judges on individual cases: one of the two Arbitrators would be selected, by drawn lots, to act as a third judge. It appears that in every such instance the final judgment was made in line with the nationality of the successful Arbitrator.

Two or three clearly invalid claims were dismissed by the Commission, but, between January 1820 and March 1823, claims were upheld in 23 cases, a high proportion of the detentions made. *Comus, Inconstant, Princess Charlotte, Prince Regent, Ulysses, Brisk* and *Bann* were all declared to have made illegal seizures, although no provision was made for their erstwhile commanding officers to argue their cases. Disagreement between the judges tended to concentrate on the sums to be allowed, and, with the Portuguese commissioners obviously bent on extracting the maximum amount of money, some excessive awards were almost certainly made. It seems, however, that decisions on illegality of capture were generally fair, with the exception of two clear miscarriages of justice in the cases of the *General Silveira*, taken by *Brisk* in 1815, and the *Santa Antonio Milagroso*, detained by *Bann* in 1816. Both vessels were undoubtedly guilty of illegal slaving, but became beneficiaries of the Arbitrator lottery. The latter judgment drew a virulent written protest from Marsden.

After lengthy birth pangs, the Anglo-Portuguese court in Rio de Janeiro opened in mid-December 1819 with Henry Hayne and Alexander Cunningham as British Judge and Arbitrator, and Silvestre Pinheiro Ferreira and Jose Silvestre Ribello as the Portuguese commissioners. It still had no permanent home, and it would be nearly two years before it would be presented with a case. British commissioners were by now also established in Havana; Henry Kilbee had been appointed as Judge, and a Mr Jameson as Arbitrator as early as January 1819, but there was nothing for them to do. In November 1820 a list would be promulgated by the Admiralty of four cruisers on the Leeward Islands Station and nine on the Jamaica Station holding copies of the Instructions from the three treaties, but it would be some years beyond that before the first detained slaver would be carried into Havana. Since the end of the American war in 1814, the Royal Navy in the West Indies had concentrated its efforts against the menace of piracy, and there had been negligible success against the slavers in recent years. Two French vessels, the brig *Hermione* and the ship *La Belle*, had been taken into Antigua with 723 slaves by HM Ships *Barbadoes*, *Columbia* and *Chanticleer* at the end of the French war, and they had been condemned as prizes, rather late in the day, in October 1815. In September 1816 HM Brig *Bermuda*, shortly before she was wrecked south of Tampico Bar in November of that year, had taken an unidentified vessel with 221 slaves before the Vice-Admiralty court in the Bahamas, but without success. Lieutenant Pakenham, however, had appealed to the High Court of Admiralty and had won his case. Finally, in January 1817, the St Vincent court had condemned another unidentified vessel with 152 slaves detained by the frigate *Orpheus*, Captain Piggott. It was not much of a haul considering the volume of slave traffic making for Cuba, but, the distraction of piracy apart, the West Indies courts may have been more wary than their equivalent in Sierra Leone about trying American slavers under Spanish colours.

On 5 October 1819 a new addition to the West Africa Squadron sailed for the Coast from Gibraltar. She was the ship-sloop *Myrmidon* (20), Commander Henry Leeke, and she soon drew attention to her arrival. Hearing at Freetown about a number of slavers at the Gallinas and the neighbouring Mana and Sagary rivers, Leeke sailed to find them. At sunset on 10 December *Myrmidon* was close-in to the north of the Gallinas, and Lieutenant Nash, the First Lieutenant, set off inshore with the pinnace and the cutter, wisely carrying an extract of the treaties in his pocket.

The boats found a number of anchored schooners and got alongside the largest of them. After slight resistance from her crew of 25, Nash captured her and found that she was the 150-ton Spanish *Voladora*, although she was later condemned under the name of *Nuestra Señora de los Nieves*. She held a passport from Havana for slaving south of the Line, but her part cargo of 122 slaves had been loaded in the Gallinas. Nash also found that she was the only schooner with slaves embarked; the others had landed their cargoes on detecting *Myrmidon* that morning. However, the other vessels did not intend to let the British get away easily with their prize. When they saw that the *Voladora* had been taken they fired their broadsides into her, and five of them hove up in a line to give her three or four broadsides apiece as she slipped her cable and headed out of the anchorage. Fortunately, the Spanish gunnery was ineffective, and the boarders suffered only one man wounded. Leeke expressed his "high approbation" of Nash's conduct.[23]

At daylight on 15 December off Cape Mount *Myrmidon* arrested the Spanish schooner *Virgien* on suspicion of piracy, and Leeke put Lieutenant Belcher on board with a large prize crew of 25. At that moment a fleet of schooners hove into sight and Leeke made the signal to chase. By 1100 *Myrmidon* had boarded three vessels, one of them with 140 slaves on board but under French colours, although papers on board showed that she was operating for Spaniards. Then, after a three-hour chase, she drove a schooner ashore. The vessel was totally destroyed, but the cargo of slaves was landed by the crew, apart from nine who were saved from drowning by the sloop. Belcher in the *Virgien*, meanwhile, with *Myrmidon* out of sight, made sail in pursuit of a large brig. When she was within gun-shot the brig hoisted Spanish colours, fired a broadside into the schooner, bore up and ran ashore. The slaves jumped overboard, and the brig was wrecked.*

Much more satisfactory was an episode of daring and initiative by two of Leeke's warrant officers a few days later. The boats had been sent inshore round Cape St Ann on Sherbro Island to gain intelligence, and on the morning of 23 December they landed John Baker, master's mate, and John Evans, clerk. These two contrived a meeting with a slave trader named John Kearney and pretended to be from a New York schooner trading for slaves but driven from the Gallinas by a man-of war. Kearney, who had been an officer in the Royal African Corps,

---

* As a suspected pirate, the *Virgien* was likely to have been dealt with by an Admiralty court, but no record of the case has been found.

boasted that he embarked almost every slave bought between Cape St Ann and the Gallinas. He employed a black agent to buy his slaves, and supposed that no one suspected him because, he claimed, he was empowered by the Governor of Sierra Leone to seize slavers. He owned a small schooner ostensibly for legitimate trade, but bought nothing but slaves. He had information on every man-of-war sailing from Freetown, and passed it to his slaver friends so that they could evade capture. The appearance of the *Myrmidon*, he said, had "completely alarmed" the slaver masters, who had trusted in the superior sailing of their schooners but had found that the *Myrmidon* could come up with most of them. He added, in a strong Irish brogue: "By Jasus, the *Myrmidon* had given the slavers a breakfast one morning and returned the evening of the next day and gave them a ball and supper!"* Kearney said he would see the "Americans'" trade goods the following morning and provide 300 slaves in two or three days' time. Baker and Evans returned to their boat after a good morning's work, and Kearney became a target for the cruisers.[24]

Morgiana returned to her patrol on the Grain Coast after a short respite in late August, and on 8 September she again encountered the phenomenon, not apparently seen before the arrival of *Myrmidon*, of a slaver running herself ashore to avoid arrest. A suspicious schooner had been sighted at daybreak, and had tried every means of escape during an eight-hour chase. She then drove ashore in the surf near Trade Town, and within a quarter of an hour she was on her beam ends with the sea breaking over her. Commander Strong believed her to be full of slaves. The next chase came on 17 September. *Morgiana* and her boats pursued another schooner off Cape Palmas throughout the day, and sighted her again at dawn the following day. The sloop came up fast until it fell calm, and Lieutenant Head then took three boats to board the schooner. She was the *Fabiana* of Havana, with 13 slaves on board, and her master was ashore at Trade Town purchasing the rest of her intended cargo of 300.

Knowing that slavers lying at anchor would weigh the instant they sighted a square-rigged vessel, Strong sent his boats inshore during the night of 25 October, under the command of Lieutenant Ryves, hoping to catch some of them by surprise. At 2000, after a very long pull, Ryves boarded an English ship under Cape Mount, and was told that three or four other vessels, certainly slavers, were lying 12 miles further north. After a short rest the boats set out again, and at

---

* Bearing in mind that *Myrmidon* had only just arrived on the Coast, and recalling *Morgiana*'s chases in August, it seems likely that Kearney was talking about *Morgiana*, a better sailer than *Myrmidon*. The recollections of the conversation by Baker and Evans may well have been distorted by pride in their own ship.

7. Capture of the Spanish schooner *Esperanza* by the boats of HM Sloop *Morgiana* on 10 December 1819

last found themselves close to four schooners. They quickly boarded the first, a Portuguese vessel, to prevent her getting underway, but she had no slaves, and so they moved rapidly on to the next. She too was Portuguese, the *Cintra*, newly built in America, a vessel of 137 tons with 15 men and two guns. The vessel had embarked slaves at sunset that evening, but only 26 of the 200 she intended to take to Trinidad de Cuba. By now the other two schooners, a Spaniard and a Frenchman, had weighed and run, but it was learned that neither had slaves on board. *Cintra* was sent to Freetown, where she was satisfactorily condemned, and *Morgiana*, her men reduced to two-thirds allowance for bread and rum, set off to find provisions at Cape Coast.

Events on 10 December dictated a change of plan. That morning the ship was standing along the shore to the south of Little Bassa when she sighted two schooners heading out from the land on her weather bow. *Morgiana* quickly made ground on them under all sail, but at 1000 it fell calm, and the cutter and gig were sent away under Lieutenant Head's command to examine them. Before the boats reached her at 1130 the first schooner hoisted French colours, but Head went alongside to give his crews a 15-minute rest before putting off for the other vessel. For a while this second schooner pulled away from the boats, but the gap was then reduced to a quarter of a mile, at which point she hoisted Spanish colours and appeared to prepare for action. When the boats had closed to a cable's length she gave them two rounds of grape and canister from her midship gun. The boats boarded regardless, one on each side.

As they came alongside, shot was thrown down into them, hitting Head and one of his men, and the bowman of the cutter was badly wounded as he hooked-on. However, the boats' crews were quickly on deck, and within a couple of minutes had driven the defenders below, killing two of them and wounding the schooner's master and three others. Midshipman Mansell, who commanded the gig, was first on board, closely supported by Marine Lord. Mansell was immediately attacked by the master and two men, but was saved by William Harris, one of his gig's crew, who cut down the master. The schooner was the *Esperanza* of 137 tons, with a crew of 21. She was bound for Puerto Rico, with 41 slaves of her expected cargo of 325. Four seamen received cutlass wounds in this gallant little action, but all recovered. *Morgiana*, now running short of men as well as provisions, headed for Freetown with her prize. She arrived at the end of the year, and lay in the river until 20 January 1820, seeing *Esperanza* quickly condemned but finding no court to judge the Portuguese *Cintra*.

By now the Commodore had rejoined his squadron, and the team which would operate on the Coast with little change for the next two years was now formed.

Sir George Collier's orders gave him considerable latitude for operations, but they again required *Tartar* to leave Africa before the start of the next rainy season. This time, however, he was required to leave England with 20 supernumeraries to fill empty billets in the Squadron, and on sailing for home he was to embark any officers and men invalided, replacing them with supernumeraries, volunteers from *Tartar*, or, if necessary, authorising vessels to bear a higher proportion of natives.

Emulating Hagan in the *Thistle*, Collier paused at the island of Sal on 26 November 1819 to recover a few more stores from *Erne*, although the wreck was by then in a bad state. Calling at Porto Praya on 2 January 1820 he discovered that a new Governor had arrived, a Portuguese naval officer "of talent and respectability" who appeared to be sincere in wanting to stop the illicit slave trade, but, as Collier pointed out, it was easy to elude the Governor's good intentions in small craft from Senegal and Goree. From the Cape Verde Islands *Tartar* made for Goree, where Collier heard that Governor MacCarthy was in the Gambia with *Thistle*. The Commodore arrived at Bathurst on 7 January, removed the Governor into the relative comfort of the frigate, and, with *Thistle* in company, sailed for Sierra Leone.

At Freetown on 13 January 1820 Collier found *Morgiana* and *Myrmidon* awaiting his arrival and showing "a considerable degree of debility apparent from the countenance of [their] crews generally" after their cruises on the Grain Coast. His solution was to send them to sea and "give their crews the benefit of the healthy Trade Winds for a few days which cannot fail of being extremely strengthening", and he was probably not far wrong. The wastage of Surgeons in the Squadron during his absence, and the difficulty of finding replacements, became a concern to him. He wrote disparagingly to the Admiralty in March 1820: "It appears to me that fear of the Climate has operated as strongly on the minds of the Medical Gentlemen in this Squadron as Disease itself", but his critical opinion was in contrast to a report by Commander Strong of *Morgiana* on his return to Freetown on 24 December:

> It is my duty to say that since I joined the Navy I never met a more zealous, kind and careful Medical Man than I have found in Mr Forrester; his attention to our Sick which has seldom been less than sixteen during the rains, when so ill himself that he could scarce walk forward to visit them, calls for my warmest acknowledgements.[25]

This commendation, reflecting a gratitude frequently apparent in the Squadron, was one of the last letters written by Charles Strong from *Morgiana*. In January

1820 he was deservedly promoted to captain, and was replaced by Commander Alexander Sandilands.

Health was not the only concern of the moment for the Commodore. On meeting Lieutenant Hagan he heard that Judge van Sirtema in the Anglo-Dutch court had initially refused in November to condemn *Thistle*'s prize *Eliza*, arguing that, as the treaty specified slaves in the plural, one slave on board was insufficient for condemnation.[26] Hagan, as usual, did not mince his words. He found the conduct of the court "arbitrary, unjust and unprecedented". He was, he wrote, allowed neither to employ anyone to carry on the prosecution nor to do it himself. He had been given no access to papers, all proceedings being in secret, and he had heard only by chance of the obstruction of van Sirtema who, Hagan insisted, "has been an advocate of the slave dealers and not an impartial judge". The court had perhaps been surprised by Hagan's outspoken persistence, quoting its own regulations in support of his demand to be allowed to appoint a proper person to act for him as captor, and when the *Eliza* case went to the Arbitrator she was condemned. Collier was probably even more concerned to hear of an objection raised in another case that it was not consistent with the spirit of the treaty that a ship's boat should make a capture out of gun-shot of her parent man-of-war.* If that principle had become established it would have been devastating to the Navy's operations.

It was not long before Collier had his own difficulty with the Netherlands court. Having despatched *Myrmidon* and *Morgiana* to the Grain Coast in company on 20 January 1820, he headed north for the Rio Pongas and Rio Nunez with *Tartar* and *Thistle* five days later. Hagan heard at the Isles de Los that the Frenchman who had commanded the *Louis* was now in a slaving brig at the Pongas and issuing threats against *Thistle*. Undeterred, the shallow-draught *Thistle* went over the northern bar and into the river in support of three of *Tartar*'s boats carrying seamen and marines and commanded by Lieutenant Marsh. All emerged on 2 February with a fine Spanish hermaphrodite schooner of 180 tons, the *Francisco*, and the Dutch brig *Marie* of similar size. Both had been captured on 30 January while loading intended cargoes of 300 slaves each, but had on board only 69 and two respectively. Hagan's knowledge of the river and the practices of the slavers had enabled the captures to be made quickly, virtually by surprise and almost bloodlessly. The slavers, described by Collier as "Renegadoes, or the refuse of every Country", had received the boarders with small-arms fire, but there was time for only one gun to be fired in return. After this success the boats had gone

---

* Collier does not indicate which case he is referring to here, but it was probably either *Fabiana* or *Esperanza*.

ten miles upriver to Kissing, where they met with irregular fire from the jungle, and then onward to Bangalang, where the slave trader John Ormond had his town and factory. At Kissing there was an American schooner awaiting slaves, and there were two more at Bangalang.

Tartar was back at Freetown on 5 February, followed a few days later by *Thistle*, and the obviously efficient Anglo-Spanish court condemned the *Francisco* on the eighth. The Dutch *Marie* was a different matter. It was over a week before she came to trial, and Collier was impatient to sail. Then van Sirtema refused to allow Collier as Commanding Officer of the captor vessel to be present in the court or to employ any form of counsel. This was the treatment that Hagan had received, and Collier was equally outraged. In his opinion "the circumstance of the court being a secret one" rendered it "not only quite useless" but also threw "an unnecessary obstacle in the pursuit of justice". Then, to his fury, he found that van Sirtema had refused to condemn ten of the *Marie*'s twelve slaves on the grounds that they were crew members.

It was difficult enough to catch the slavers; as the Commodore said, "it is only by great cunning / or great accident / they can be surprised with slaves onboard." To be obstructed, and indeed threatened, by a Mixed Commission court was intolerable, and Collier complained bitterly to the Admiralty. He protested, first, about the in camera proceedings:

> so that here I believe I may venture to say the most extraordinary of all Courts of Justice I ever heard, or read of / different from that of the barbarous Nations of Africa who are known never to Condemn, until the palaver as they term it, is fairly talked, and in open Court too / in a Court of Justice with Judges.

He demanded that the court should be open, at least to the interested parties during the examination of witnesses, and that those parties should be allowed to cross-examine witnesses as well as to plead their own cases openly in court. He was deeply concerned that a captor might, after such a closed trial, be arrested and perhaps be unable to pay the damages awarded against him "by the secret sentence of the court from which there is no appeal". He demanded that the officers of his squadron, bound by their orders to seize illicit slavers, at the risk of heavy damages if judgment should go against them, should be permitted to employ proper persons to represent them, especially if their duties took them away from Sierra Leone.[27] The Commodore's complaints were surely justified, and Their Lordships passed his letter to Lord Castlereagh, but there would be no easy resolution of this major new difficulty.

* * *

With the ending of the Napoleonic War in 1815 there had come a period of aimlessness in the suppression campaign. The Navy's presence on the coast of West Africa had declined to a mere token, entirely ineffectual against the slavers, whose numbers had burgeoned with the return of the French. In the West Indies the Jamaica and Leeward Islands Squadrons had become preoccupied with the piracy plague, and it would be the best part of a decade before they turned their attention to the slave trade. In consequence, the slavers, increasingly using vessels which could outrun almost any cruiser, enjoyed four years of scant interference in the western Atlantic and the Caribbean.

Underlying the faltering of this ill-directed campaign at sea and on the Coast was the inadequacy of the treaty provisions upon which the operations of the Navy were necessarily founded. In a period of naval retrenchment the Admiralty was understandably reluctant to commit cruisers to the Coast on such a flimsy basis. The illusion had persisted for some time that Britain was entitled to take action against foreign slavers flouting their own nations' domestic legislation, particularly in the case of American slavers. This mistaken notion had, however, been sharply dispelled after a brief spate of incidents with French vessels. Treaties signed with the Portuguese in 1810 and 1815 were confusingly worded, and led to a number of well-intentioned but illegitimate condemnations. With the lifting of wartime restrictions on trade; the unwillingness of the USA, Portugal and France to enforce their anti-slaving legislation; and the availability of the flag of Spain as a disguise, the slave trade flourished. The colonial officials of France and Portugal were frequently embroiled in the Trade themselves, and a new curse had emerged in the shape of piratical slavers.

Throughout this period the few cruisers deployed tried hard to meet their remit and, well supported by locally manned colonial vessels, they had their successes, not infrequently in the teeth of determined opposition. Consequently there was a steady trickle of emancipated slaves into Sierra Leone where they, and the Navy, had a good friend in Governor MacCarthy. Slaves brought to Freetown were released whether or not their vessels were condemned, even though commanding officers were liable to be sued for damages by the owners of vessels deemed to have been detained illegally.

Noble efforts by Lord Castlereagh to achieve international abolition and suppression of the Trade, focused initially on the Congress of Vienna, came to naught, and he had redirected his attention to negotiating bilateral treaties with the slaving nations. The resulting agreements with Portugal, Spain and the

Netherlands in 1817 and 1818 represented a major step forward, but, although these treaties were undoubtedly the best that could be achieved at the time, they contained serious flaws which would be exploited to the full by the slavers.

The Admiralty had responded to the treaties by expanding the West Africa Squadron to a strength of six vessels. This was still sadly inadequate for the suppression task, but, with the Portuguese and Brazilian trade south of the Equator forbidden them, these few cruisers could maintain a fairly regular, if very thin, presence in the principal slaving regions north of the Line. Still generally lacking the speed of their adversaries in open water, they were most likely to achieve success against slavers caught while loading their cargoes, and the fresh intelligence usually needed to make such seizures seems to have flowed fairly freely, despite the absence of any organisation to gather it. Intercepted vessels would sometimes put up a fight, and the faster cruisers were now occasionally finding that a chase would run herself ashore to evade capture.

The innovative Mixed Commission courts, introduced by the new treaties, were slow to start, but promised a sound judicial framework for the international policing operations in which the Royal Navy was now involved. However, it had soon become apparent that the principles upon which these courts operated were rooted less in natural justice than in political expediency. The legal risks run by commanding officers in the past had at least been at the hands of British justice, although, as Commander Willes discovered, that could be grossly unfair. These officers now felt that some of the foreign Commission judges were in league with the slavers. The Navy appeared to have entered a legal minefield laid against the enemies of the slavers, a hazard which would claim numerous naval victims in the years to come, and good officers, doing their duty as they saw it, would find scant support from the Crown they served when the legal dice fell against them.

CHAPTER 7

The Most Evident Falsehoods, 1820–4

False face must hide what the false heart doth know.
WILLIAM SHAKESPEARE

AT LAST, with the stationing of a permanent squadron on the slave coast, under the command of a commodore well established in his post, there was now a possibility of patrolling the main slaving areas of the Coast with some degree of regularity. But half a dozen cruisers was still a pathetically small force to police the Squadron's enormous region of responsibility. Even though British interference in the massive Portuguese slave trade south of the Equator was firmly forbidden, there remained about 2,400 miles of coastline for the Navy to watch, much of it indented with bays and rivers. Nevertheless, the areas favoured by the slavers at that time were fairly well defined, which permitted some concentration of effort. Added to that, as the cruiser commanding officers became experienced on the Station, and as the natives and legitimate traders of the rivers and the masters of British merchantmen on the Coast became more familiar with the men-of-war, a flow of intelligence sometimes enabled intervention to be fairly precisely targeted.

To windward of Sierra Leone, the most northerly area requiring regular attention centred on the rivers Pongas and Nunez, already well known to Lieutenant Hagan of *Thistle*. The relative proximity of these rivers to the Squadron's base at Freetown allowed a degree of reliance on intelligence to trigger reactive operations in that area. Occasional visits were needed even further to the north at the Rio Grande and the Portuguese settlements at Bissau and Cacheu, but, beyond that, the Squadron was powerless against the slave traffic from the French settlements at Goree and Senegal, illegal though it was. To leeward of Sierra Leone, the Trade was active along much of the coast from Sherbro Island to Cape Palmas, operating primarily from the rivers, especially the Gallinas, and at Cape Mount, Cape Mesurado, Trade Town, Grand Bassa and Little Bassa. Along this coast the Commodore tried to ensure a frequent presence.

To the east of Cape Palmas there was little slaving on the long stretch of the Ivory Coast, the Quaqua Coast and the Gold Coast, and the next region of high intensity began at Cape St Paul's: the Slave Coast. Here the northern shore of the Bight of Benin was dotted with slaving posts, the greatest of them being Whydah and Lagos. Less prominent were Little Popo, Grand Popo, Porto Novo and Badagry, but all contributed extensively to the slave trade, despite the difficulties of their open, surf-pounded beaches. Further around the Bight of Benin and into the Bight of Biafra, several of the many river mouths of the Niger Delta harboured slavers, but the principal trade of the Delta lay at its eastern extremity, in the rivers of Bonny and New Calabar. A little further east, far up a long estuary, lay another great slaving centre in the Old Calabar River. Finally, there was an intermittent traffic from the Cameroons River, where the Bights shore curves to the south. The Commodore made every effort to keep as constant a watch as possible on the whole of this Bights coast, from Cape St Paul's to the Cameroons.

Occasional visits to the Cape Verde Islands and to the Bights islands of Princes and St Thomas were also necessary to keep an eye on the use made of them by the Portuguese as staging posts for their illegal slave trade. The Commodore was also required to provide an occasional supportive presence at the new British settlement of Bathurst on the River Gambia, and at the African Company forts on the Gold Coast. Fortunately, some of these visits could be conveniently combined with calls for stores and provisions. Wood, fresh water and fresh produce were readily available at the Bights islands, and, by means of occasional visits by supply ships from England and by stocks brought out by cruisers newly arrived from home, stores depots had been established for the Squadron at the Gold Coast castles of Cape Coast and Anamabo as well as at Freetown. A little further east, Accra, with its dilapidated fort, was the Squadron's primary source of cattle and grain. Ascension, too, had become an important port of call, both for reprovisioning of the garrison and for recuperation of cruiser ships' companies away from the heat and fevers of the African coast.

Despite the efforts of Lord Castlereagh at the Foreign Office in London, and of the cruisers on the Coast, there had been no reduction in the slave trade. All major nations still involved had enacted legislation against the traffic, but that had scarcely hampered the slavers. Washington was at last initiating some naval enforcement of its anti-slaving laws, but it was no inconvenience to American slavers to hoist false colours, and there was no sign that Madrid, Lisbon or The Hague might be inclined to follow Washington's example. The Madrid Government, as expected, made numerous demands to delay the approaching deadline of 30 May 1820 for prohibition of the Spanish slave trade, agreed by the 1817 treaty. A new date of 31 October was accepted by Britain, but not until 10 December did the

Captain-General of Cuba acknowledge that he had received orders to enforce the agreement. Thereafter, it was apparent that the mother country was giving covert encouragement to the island's planters and slave merchants to ignore this inconvenient legislation.

As far as Portugal was concerned, even the colonial officials at the Bights islands and Bissau were prominent players in the illegal traffic north of the Line, supplementing the massive, and still legal, southern hemisphere traffic to Brazil. Perhaps most galling of all to the Royal Navy's cruisers was the French trade. Like the United States, France refused to consider allowing foreign men-of-war, least of all British, to interfere with its slavers. Nor, despite its abolition legislation, had it yet taken any significant preventative action itself; "on the contrary", wrote Commodore Collier, "she gives to the Trade all countenance short of public avowal." Not only had the French slave traffic expanded enormously since the end of the Napoleonic War, but also the flag of France had become an effective protection against British cruisers for the slavers and pirates of all nations on the west coast of Africa. In Collier's estimation, not less than 60,000 slaves had been taken from north of the Equator in the year 1819–20, most of them to Martinique, Guadeloupe and Cuba, and when *Tartar* called at Havana in July 1820 the Commodore counted no fewer than 40 vessels fitting-out "avowedly for the Slave Trade, protected equally by the Flags and Papers of France and Spain".

Among the native chiefs and the European outcasts and mixed-race traders who supplied the slaver masters with their cargoes, a handful had by now earned particular notoriety as leaders in their trade. Pre-eminent among the mostly semi-European or American slavers on the rivers Pongas and Nunez was John Ormond, known as "Mongo John".* His father had probably been a Liverpool Irishman and his mother a chief's daughter, and as a boy he had been stranded in Africa by his father's death. He had returned after some years at sea to reclaim his father's property from his mother and her relations, and now lived like a native chief among his slave barracoons at Bangalang on the Pongas, enjoying the company of the girls of his harem, mostly gifts from neighbouring chiefs who were anxious to remain in his favour.

To the south of Sierra Leone, John Kearney, the Irishman encountered by *Myrmidon*'s two warrant officers in late 1819, had achieved some success on the River Sherbro, but he was small fry by comparison with his neighbour on the Gallinas, Don Pedro Blanco. This haughty Spanish don, once a master mariner in Malaga, had built his factory in 1813 amid the desolate marshes and low-lying islands of the river, and for 15 years he never as much as crossed the Gallinas

---
* "Mongo" was an African term meaning "Chief of the River".

bar. He lived in state, surrounded by every luxury his success could provide, but unlike most of his kind he never succumbed to drink and drugs. Each of the favourite women of his seraglio had her own island, and Blanco himself lived on another small island with his sister. To warn him of the approach of cruisers, he built 100-foot watchtowers on outlying islands. His arrogance and cruelty, and his surprising generosity, were well known, and he had become a figure of power and influence along the Windward Coast.

Whydah was the headquarters of a trader of almost regal status, Da Souza, or "Cha-Cha", as he was known along the Coast.* He was probably a mixed-race Brazilian from Rio de Janeiro, and claimed to have been an officer in Dom Pedro's Imperial Guard. His first appearance in the Trade was at Havana, and he probably reached Whydah in a slaver. There he had won the favour of the local chiefs, become slave-broker to the King of Dahomey, and built a palatial mansion. Collier found him living "in prodigious splendour", and lamented that he "assumes the rights and privileges of a person in authority, granting papers and licences to the slave traders in all the form and confidence of one empowered to do so by the Portuguese Government."[1]

Further to the east, along the shores of the Niger Delta and the Bight of Biafra, there were no intermediary merchants between the chiefs of the coast and the slaver masters, and at Duke Town, on the Old Calabar River, the Trade was in the hands of a renowned native potentate known as Duke Ephraim. He was a brutal despot who delighted in human sacrifice, and those in his power who displeased him were liable to public beheading. Masters of ships trading for slaves or palm oil in the Old Calabar needed to tread warily and kowtow assiduously.

* * *

Affairs on the slave coast certainly offered no evident grounds for optimism among the abolitionists, and, although the West Africa Squadron was not disheartened as it continued its campaign in early 1820, the Commodore and others were angry and concerned at what they saw as obstruction and bias against them on the part of the new Mixed Commission courts. Obstruction there certainly was in these early days, and it came initially from the Netherlands and Spanish judges, Colonel van Sirtema and Don Francisco Le Fer. It was van Sirtema who had demanded that proctors (advocates) should be excluded from the Anglo-Dutch court and who had opposed condemnation of the *Eliza* on the grounds that she had only one slave on board. Le Fer had declined, on a flimsy excuse, to try *Cherub*'s case

---

* "Cha-Cha" meant "Bustle".

against the undoubtedly illegal *Josepha*, and lost no time in echoing van Sirtema on the matter of proctors.

The British commissioners resisted this outrageous objection to professional representation of captors, but were mindful of the Foreign Secretary's directive to "cultivate a spirit of conciliation and harmony", and finally acquiesced. They pleaded with the Foreign Office to reach agreement with the partner nations on rules for the Mixed Commissions, but it would be another two years before the Foreign Office ordered that the courts must be open to all when the commissioners were exercising their judicial functions. Commissioners Gregory and Fitzgerald could not have supposed that their difficulties would be eased by the much-delayed arrival of their Portuguese counterparts in February 1820, but, after it opened in May, with João Altavilla and J. César de la Figaniere e Morão as Portuguese Judge and Arbitrator, the affairs of the new Anglo-Portuguese court ran smoothly.

On 20 January 1820, a few days after *Tartar* and *Thistle* had sailed for the Rio Pongas, *Myrmidon* and *Morgiana* left Freetown to cruise between Cape St Ann and Cape Mount, with *Morgiana* now under the command of Commander Alexander Sandilands. Arriving off the Gallinas on 25 January they sent away their boats before daylight, and dawn revealed six schooners at anchor. Five were Spaniards with no slaves embarked, and the sixth, the *La Marie*, hoisted French colours. Nevertheless, *Morgiana*'s boats boarded *La Marie* and found 106 slaves but few Frenchmen. They also discovered documents which convinced Leeke and Sandilands that the slaves had been shipped by John Kearney, and the rash decision was made to detain her, despite her alleged nationality.

Later that day, Leeke found himself on apparently firmer legal ground when *Myrmidon* was running along the coast to the south of the Gallinas, still in company with *Morgiana*. There were several schooners at anchor at the mouth of the River Mana, and one of them, under Portuguese colours, was seen to be hurrying Africans into a boat. As Lieutenant Smith approached the schooner in *Myrmidon*'s cutter, a boat pushed off and pulled "with great eagerness" for the shore. Smith managed to intercept the boat, and found one slave, apparently the last to be removed from the schooner to join another 200 ashore. Leeke recognised the slaver as one he had encountered during *Myrmidon*'s previous cruise, and twice seen landing her cargo before he could board her. She was the 270-ton *San Salvador*, with a crew of 33, a third of them Americans, and a hefty armament of eight 18-pounders, bound for Havana. She had left her offloading just too late on this occasion, but it remained to be seen whether she could be condemned when the Portuguese Commission finally began work.

On the evening of 30 January, as the two sloops lay becalmed about five miles north-west of Cape Mount, they detected three sail and a great deal of firing off the mouth of the River Manna. Lieutenant Nash of *Myrmidon* was despatched with the boats of both ships to investigate. By 2300 he was alongside the largest of the three vessels, a brigantine of 240 tons, and captured her after a sharp action. She was *L'Arrogante*, a "Patriot Privateer" flying the flag of Artigas[*] but fitted-out at Baltimore, and most of her crew of 40, and some of her officers, were British subjects. It was believed that she had robbed a French schooner a few days earlier and temporarily imprisoned the master and crew, and that, under Spanish colours, she had fired a volley of musketry into an American schooner without warning. It seemed that the Artigas colours were being used to give the false cover of a Letter of Marque to acts of piracy, and it was for piracy that Leeke arrested her. He also took the two Spanish slavers which *L'Arrogante* had just seized, the *Anna Marie* and *El Carmen*, although neither had yet embarked the slaves waiting for them at the Gallinas.

Myrmidon returned to Sierra Leone with her five prizes on 8 February, but the cases were to meet with difficulty and delay. It was decided that *La Marie*'s slaves should be emancipated at Freetown, but that the vessel would be taken by *Myrmidon* to Senegal or Goree and handed over to the French authorities. The Vice-Admiralty court judged that *La Marie*'s slaves, although supplied by Kearney, had been purchased by the French vessel's master before capture and were not therefore liable to confiscation,[†] which meant no head money for *Myrmidon* and *Morgiana*. Leeke held the British subjects from *L'Arrogante* under arrest in *Myrmidon*, although it seemed unlikely that a charge of piracy could be proved against them. The *Anna Marie* and *El Carmen*, being empty of slaves, were released. The case of the *San Salvador* had to await the arrival of the Portuguese commissioners and was not dealt with until June 1820. Ultimately it was decided on the opinion of the Portuguese Arbitrator that the slave in the boat did not constitute a slave on board, and, the letter of the treaty taking precedence over its spirit, the slaver was restored. Castlereagh felt that the court had made a decision "in opposition to the Design of the Treaty", and he asked all the British commissioners to press their foreign opposite numbers to adopt a more liberal interpretation of the relevant clause. A disappointed Leeke headed north for the River Nunez and then Boa Vista.

---

[*] "Artigas", or General Artigas, seems to have been the name used at the time for Simon Bolivar, and Governor MacCarthy described *L'Arrogante*'s colours as Venezuelan. The use of this flag was apparently a device for giving a veneer of legality to what was essentially piracy.
[†] However, the slaves from the French vessel *La Marie* were settled in Sierra Leone – at the new village of Waterloo!

Morgiana had bade farewell to *Myrmidon* off Cape Mount, and was making for Cape Coast when, on the morning of 3 February, she boarded the schooner *Prince of Orange* at anchor off Little Bassa.* Commanded by Job Northrup, this vessel, like *L'Arrogante*, had been an Artigas privateer, and she had recently captured a Spanish schooner called *L'Invincible* while loading slaves at the Rio Pongas. Taking command of the *L'Invincible* and renaming her *L'Invincible the Second*, Northrup had then seized the American schooner *Swift* and shifted his flag and commission into her. *Swift*'s master had been ashore when his vessel was taken, and Northrup had made a sham sale of the *Prince of Orange* to the mate of the *Swift* who, with some of *Swift*'s crew, was found on board *Prince of Orange* by Sandilands. Manned entirely by Americans and Englishmen, the *Prince of Orange* had sailed last from Baltimore under the name of *La Constantia*.†

Having heard this convoluted tale, Sandilands decided to send the *Prince of Orange* to Sierra Leone to be tried for piracy, and continued his passage to leeward. The next day he boarded the French schooner *La Jeune Estelle* and learned that on 3 February she had been plundered of 73 slaves, probably by the *Swift* and *L'Invincible the Second*. Sandilands was anxious to find the two pirates, and he was in luck. At daybreak the next morning he found close at hand two schooners answering the description of the pair he wanted. Both hoisted Artigas colours and tried to escape. In a light breeze and using sweeps, the *Swift* got clear, but Lieutenant Head in *Morgiana*'s boats managed to come up with the second vessel. She was, as expected, *L'Invincible the Second*. There were 28 slaves on board, some of them from *La Jeune Estelle*, but none of the original slaver crew. Keeping his prisoners safely in custody in *Morgiana*, Sandilands sent his prize to Freetown. The British commissioners were not averse to trying *L'Invincible* as a Spanish slaver, but the Spanish judge was opposed to the idea, and the vessel was released.

On her arrival at Cape Coast on 15 February, *Morgiana* became involved in another curious episode. The British ship *Prince of Brazil* was making her way out of the anchorage on her way to Fernando Po, supposedly to establish a settlement. She had on board a considerable number of Africans who were said to have received bounties and an advance of wages for taking part in the project, but they claimed that they were there against their will and had been given nothing. They said that the chiefs on the Windward Coast who had sent them on board had been paid with bars of iron, and they begged to be taken home. Two who had

---

* The native word "Piccaninny" was sometimes substituted for "Little" in place names.
† The British commissioners' report to the Foreign Office compounds the confusion by referring to the *Prince of Orange* as *L'Invincible* and the Spanish slaver as *Industria*. The fate of the *Prince of Orange* is not known.

escaped to Elmina had been sent back to the ship in irons. There was no record in the log of the Africans coming aboard, and the master and surgeon complained that the man in charge of the expedition, a Mr Robertson, had kept them in ignorance of its purpose. Robertson had purchased two tenders at Sierra Leone, and one of them, the *Jane Nicholl*, was returning from Whydah, where she had landed about 40 Africans to form another "establishment".[2] The business smelled of slavery to Sandilands, and he sent the ship and the *Jane Nicholl* to Sierra Leone, but there is no evidence that charges were sustained in the Vice-Admiralty court.

With the Commodore's court cases over, *Tartar* sailed on 21 February 1820, also initially for the Gallinas, and was joined there a week later by *Thistle*. During the passage south, Collier had sent his boats inshore in an unsuccessful attempt to gain information about the slaver Kearney, and the boats also boarded the French schooner *La Catherine* in the belief that she was Spanish. Finding a part cargo of 50 slaves on board, Collier decided to send her back to Freetown to be taken north to one of the French settlements. She failed to arrive at Goree, and it was later heard that she had been recaptured by her crew, who had taken *Tartar*'s prize crew to the West Indies.

To both the Governor and the Admiralty Collier expressed his growing frustration and anger at the brazen behaviour of the French slavers, declaring his view that: "The many vessels under the French flag openly slaving upon this Coast renders it absolutely necessary that these violations of the laws of their Nation should be sent to answer for the same before the proper French authorities."

He predicted that the flags of France and America "will probably in another year cover the whole windward coast" unless Their Lordships countenanced his practice of detaining those with slaves on board, and he unwisely cited the case of *Redwing* and *La Sylphe*. MacCarthy, concerned about French slaving almost on the shores of Sierra Leone, was sympathetic, but the Admiralty was not. Collier was firmly told to desist. *La Sylphe* was not to be used as a precedent, and the Commodore was reminded that Commander Hunn of *Redwing* had earned Their Lordships' displeasure for that illegal seizure.[3]

While *Thistle* was taking *La Catherine* to Freetown, *Tartar* boarded a pair of Spanish schooners at sea, the *Esperanza* and the *Anita*, both of which had landed part of their trade goods to Kearney but had not yet loaded slaves. As a novel means of disrupting their plans, Collier decided to take them with him to leeward beyond Cape Palmas. Temporarily abandoning her two unwilling consorts soon after dawn on 2 March off Grand Bassa, *Tartar* bore up in chase of the American-built schooner *Gazetta*, under Spanish colours, and caught her at sunset. The 82 slaves were released from their shackles, and Collier wrote that "the gratitude of these poor beings for this kindness is beyond description."[4] Concerned for the safety of his prize crew, Collier took on board *Tartar* 16 of the motley bunch of ruffians crewing the slaver before despatching her to Sierra Leone, and then made sail to rejoin the two schooners.

The Commodore had brought with him a tender under the command of *Tartar*'s Lieutenant Finlaison,* and on 4 March Finlaison boarded two slaving schooners under French colours but of uncertain nationality, the *Jeune Estelle* and the *Joseph*, both of which had been unsuccessfully pursued by *Tartar* on 2 March. The former had been visited in February by *Morgiana*, and it was alleged that she had taken revenge on Northrup's *Swift* by seizing part of the *Swift*'s consignment of slaves at Trade Town. During the search of the *Jeune Estelle*, Finlaison's suspicion was roused by a cask closed at the bunghole by canvas. The hoops of the cask were knocked off and two African girls were found inside, almost suffocated, apparently two of the slaves originally destined for the *Swift*. Although he was not legally entitled to do so, Collier had the two children removed to *Tartar*, and on their arrival in Sierra Leone they were taken into the care of missionaries. By 1824 one was happily married and the other was chief monitor at the church school at Leopold. The angry and frustrated Commodore was obliged to release the *Jeune Estelle* and the *Joseph* with their remaining slaves after endorsing their papers.[5]

After releasing the two Spanish schooners south of Cape Palmas, to beat their way back to the Gallinas, *Tartar* continued to Cape Coast. There, on 14 March, she found a dejected *Snapper*, her Commanding Officer, Lieutenant James Henderson, and her Assistant Surgeon both dead of fever. The Commodore sent the gun-brig back to Sierra Leone, appointing Lieutenant R. J. Nash of *Myrmidon* to command her, and putting one of his *Tartar* midshipmen into *Myrmidon* as Second Lieutenant.

The drain on manpower caused mainly by disease was a constant concern for the Commodore. Fortunately there was in the cruisers a supply of excellent lieutenants ready and anxious to seize the opportunities for command offered by death or invaliding, and there were also, in the larger vessels, midshipmen and master's mates passed for lieutenant who could be provisionally promoted by the Commodore as necessary. A greater worry than the loss of commissioned officers was the wastage of specialist warrant officers, particularly Surgeons and Masters. There were very few qualified assistants available on station to fill vacancies, and Collier bemoaned the "great mortality among Surgeons of the squadron".

---

* The identity of this tender is not recorded.

Morgiana had left Cape Coast shortly before *Tartar*'s arrival, and when the Commodore moved along the coast to Accra he found that *Pheasant* had just arrived there from her Bights patrol. Sending *Pheasant* to Cape Coast for provisions and then to Ascension to caulk, Collier paid a short visit to Princes Island. He found no sign of any slaver in the Bights except for a small vessel which had landed a cargo of slaves at the house of Acting Governor Gomez just before *Tartar* arrived at Port Antonio. As the frigate headed back to Cape Coast, fever made an appearance, and the Commodore was anxious to sail without delay for the healthy climate of Ascension. However, he reluctantly became involved in a dispute developing between the Governor of Cape Coast Castle and the King of the Ashantis.

The natives of Cape Coast had become wealthy through trade with British merchants under the protection of the castle, and the Ashantis, a powerful nation to the north, clearly entertained ambitions to control the coastal region. A British consul, Mr Dupuis, had been sent to the Ashanti capital of Kumasi to negotiate an agreement, but he had weakly acquiesced to a treaty which was distinctly disadvantageous to the British merchants and the Cape Coast natives. It was probably as well for Mr and Mrs Dupuis that Collier took them on board for passage home when *Tartar* sailed for Ascension on 16 April.

The fever abated in *Tartar* within a week of her leaving the Coast, and she reached Ascension on 29 April in relatively good health to find that *Pheasant* had arrived there in a debilitated state. The Commodore therefore ordered Commander Kelly to remain at the island for three weeks for the recovery of his ship's company and to refit, and then to head for the Bights islands initially for water and then to cruise in the vicinity, keeping in touch with Cape Coast Castle. After taking in a supply of turtles and pumpkins, *Tartar* sailed on 2 May for an eight-day passage to Sierra Leone. Collier had ordered a concentration of the Squadron at Freetown so that he could make arrangements for the rainy season before *Tartar*'s departure, and he was pleased to find all except *Pheasant* in harbour. He was rather less pleased to learn that preparations were in an advanced stage for an expedition to the Rio Pongas.

An unfortunate train of events had been set in motion by Lieutenant Hagan while watching the northern rivers in *Thistle*. Hearing that a merchant vessel belonging to a British merchant on the Isles de Los had been seized by one of the Rio Pongas slavers, a maroon called Curtis, Hagan, never slow to take action, had unwisely sent a boat into the river on 4 May to demand release of the vessel. The boat had been fired on, possibly because Curtis thought the hated Hagan himself was in it, and Midshipman Robert Inman, a particularly promising youngster, was killed. In *Thistle* it was believed that others of the boat's crew had been killed or wounded, that the wounded were in the hands of Curtis, and that the bodies of the dead had been mutilated. Hagan, his brig short of provisions, had brought the news back to Freetown, where the senior officer was Commander Leeke of *Myrmidon*. Although hesitant to embark on a major operation in the absence of the Commodore, Leeke and the Governor agreed that a rescue attempt should not be delayed, and MacCarthy placed Captain Chisholm with four officers and 150 soldiers under Leeke's command. At this juncture Collier arrived, and, although he "lamented very much" Hagan's initial lack of judgment, he gave full support to the decision by the Governor and Leeke. Producing a model Operation Order, he despatched the force in *Myrmidon*, *Morgiana*, *Snapper* and *Thistle* on 12 May, particularly instructing it to avoid damage to the property of innocent people.

On arrival off the river on 15 May the landing force of soldiers and ships' marines was transferred to the gun-brigs which crossed the bar the next day, and, in the hope of depriving Curtis of support, the local chiefs were assured that the objectives were simply to recover *Thistle*'s men and boat, and to arrest Curtis. Two days later the brigs were taken as close as possible to Curtis Town, and armed boats under flags of truce approached the mud fortification which Curtis had manned with the help of neighbouring chiefs. Although the fort also hoisted a white flag, the boats were heavily engaged as they reached the shore, but opposition was quickly overcome, the town was burned, and the force moved into the surrounding villages in an unsuccessful search for the captured boat's crew. After a night back on board, and armed with intelligence from a local trader, the soldiers marched several miles inland to search the village of Mungo Brama, who had played a leading role in the attack on *Thistle*'s boat. Overcoming some resistance, the force destroyed the village and a large quantity of Curtis's slaving trade goods. Thanks to the intervention of King Yando Coney, two seamen were then released, and a further four who had taken refuge with Mongo John at Bangalang were returned on 19 May. Curtis remained at large. The cost to the landing force of this largely successful but avoidable episode was three soldiers wounded and one marine corporal dead of heat exhaustion.

When he sailed for the West Indies and home on 4 June, the Commodore was obliged to leave behind *Tartar*'s Lieutenant William Finlaison to replace Commander Sandilands in *Morgiana*, who had to be invalided. By November, Lieutenant Nash of the unfortunate *Snapper* had followed Sandilands, and, in the absence of the Commodore, Commander Kelly promoted Midshipman James Pratt of *Pheasant* to temporary command of the gun-brig. *Myrmidon*'s Surgeon had died too, and the Commodore had made a plea to the Admiralty for Assistant Surgeons to be appointed to all vessels of the Squadron to allow for the inevitable wastage of these essential warrant officers.[6] A further concern for him was the extent of rot in the running rigging of his ships, and, lacking adequate Navy Board stocks on the Station, he had to buy a large quantity of expensive rope in Sierra Leone. Finally he deployed his ships to avoid the worst effects of the rains: *Myrmidon* north to patrol the coast to windward, *Thistle* to one of the northern islands to caulk and to keep an eye on the rivers in the north, *Morgiana* to the Gold Coast with stores for *Pheasant* and then to Ascension, and *Snapper* to Ascension to caulk and recover the health of her ship's company.

To ease one of the Governor's constant problems, *Tartar* took with her to the West Indies a number of Spaniards arrested in slaving vessels. There were still no proper arrangements for captured crews, who were mostly Creoles or Europeans with families in foreign colonies, and they generally wanted passage to the West Indies. The Colonial Office was currently paying for lodgings, rations and passage, but the King's Advocate at Doctors' Commons saw this as an undeserved bounty and was advising the Foreign Office that countries of origin should meet the costs. While the arguments continued, Governor MacCarthy had difficulty in making adequate provision from his scant resources.

Pheasant left Ascension on 16 May, cutting short the period of recuperation specified by the Commodore and fuelling a growing tension in the relationship between Collier and Commander Kelly. She called at Dixcove to collect one of *Erne*'s masts to replace a crippled topmast, and then called on the new Governor of Princes who appeared to be taking firm action against the slave trade. On 9 July she was coasting north from Gabon to Corisco Island in the hope of rescuing survivors from the wrecked British merchantman *Liverpool* when she sighted a small schooner at anchor close inshore. Not far from the schooner, one of *Pheasant*'s boats taking soundings along the shore was fired on by natives on the beach. The remaining boats were armed, manned and sent under the command of Lieutenant Jellicoe to bring the schooner off or, if necessary, to destroy her. Meanwhile, the schooner was hauled onshore, and, as the boats' crews attempted to board her, a large crowd of natives rushed out of the bush and fired a heavy volley of musketry, killing one seaman and wounding Jellicoe, five seamen and a Krooman. Jellicoe prudently retreated, and Kelly tried to destroy the schooner with gunfire, but was prevented by shoal water from getting into effective range. Poor Jellicoe had lost a hand as a midshipman, and now, in a nugatory skirmish, a piece of langridge had carried away much of his throat. Collier later asked the Admiralty that the wounded Krooman should be given the equivalent pension of a British seaman, but his request probably fell on deaf ears.

After rescuing five members of the *Liverpool*'s crew held as slaves by natives near the River St Benito, *Pheasant* boarded the French slaver *La Prothée* ten days out of Bonny. She had over 320 slaves on board, mostly children aged between six and twelve, and already 30 had died of dysentery. Otherwise she found little or no slaving between the Cameroons and the Equator. On reaching Accra on 25 August Kelly discovered that a Spanish schooner had landed 50 slaves to await the arrival of a full cargo. They were being held by native traders close to the undefined boundary between British and Dutch Accra, and Kelly demanded their release to him, threatening otherwise to bombard the town. The traders did not comply, and Kelly was as good as his word. The town had been evacuated and no one was hurt, but an hour's gunfire produced the required result.[7] This unconventional measure was welcomed by the British merchants in Accra and approved of by Collier and the Governor of Elmina. Strangely, there was no protest from either the Freetown court or the Foreign Office.

At the end of July 1820, Governor MacCarthy had at last left Sierra Leone for his first home leave since his arrival in Africa in 1812, having served in Canada for some years directly before that. When he sailed in late July, initially for the Gambia in *Myrmidon*, he appointed Major Grant at Bathurst to be Lieutenant-Governor in his stead, and Grant had not long taken up the reins in Freetown when he heard the alarming news that the Gambia settlement was under threat of attack by two neighbouring kings on both banks of the river. *Morgiana* was lying at Freetown, and Grant asked Commander Finlaison to investigate. With "the zeal so peculiar to Captain Finlaison, and so characteristic of the British Navy", *Morgiana* sailed on 4 October and anchored off Bathurst on the twenty-sixth. The following day, with the Commandant of Bathurst on board, the ship headed upriver as far as Fort St James. There Finlaison held a conference with the kings, who denied meditating any attack. He then invited them on board, presented them with gifts, demonstrated the firing of a few guns, and assured his guests that *Morgiana* or another ship would visit frequently. The kings were apparently struck with awe, and the Bathurst merchants were highly gratified and reassured.[8]

Grant's next challenge came from the southern flank of the colony. The threat of French colonial expansion on the Coast had long caused Governor MacCarthy some uneasiness, but in May 1818 he had expressed a new disquiet about the potential effect on his settlement's trade of a proposed American protectorate on the River Sherbro. Early in 1820 his anxiety was sharpened by the arrival of the United States merchant vessel *Elizabeth* with 85 black people and stores to start an "asylum" for liberated slaves on the Sherbro. Not only was this too close to Sierra Leone for comfort, but also MacCarthy was convinced that the site was dangerously unhealthy. In this latter concern he was right, and within months the settlers were in serious trouble. This project did, however, bring a welcome development.

Escorting the expedition was the Sixth Rate *Cyane*, Captain Trenchard, of the United States Navy, and she was authorised to detain American slavers and land their slaves at the new settlement. Trenchard showed encouraging zeal in his abolition duties, and by late May he had taken four slave vessels and put any British crew he found in them on board Royal Navy cruisers. In late October 1820, Grant received a letter from Trenchard reporting that most of the Sherbro settlers were dead, and that there was no longer any authority to whom American cruisers could deliver freed slaves. He asked that Grant would accept them in Sierra Leone for the time being, and Grant agreed to do so pending further arrangements between the two Governments, promising that they would be treated in all respects as were slaves freed by British cruisers.

The final British captures of 1820 fell to *Thistle*. The gun-brig had been badly hit by fever, and in mid-July she had, with some difficulty, reached Goree with all her officers and men sick. However, returning from the Cape Verde Islands, the irrepressible Lieutenant Hagan arrived off the Rio Pongas on 12 September and seized the English-owned sloop *Two Sisters* with 15 slaves. Pausing at Freetown for water and provisions, he headed for Trade Town, where he heard of a Spanish slaver in Little Cape Mount River. On 16 October, under cover of a shower of rain, *Thistle*'s Master took the boats into the river and surprised the schooner *Nuestra Señora de Montserrate* of Havana with 85 slaves, some of whom had been taken from the French schooner *Industry* when she was wrecked on the bar of the river. With no officer to put on board his prize, Hagan was obliged to take her back to Sierra Leone, where she and the *Two Sisters* were condemned, the latter in the Vice-Admiralty court. Possibly as a result of this river expedition, *Thistle*'s Master died in November, and Hagan was left with a midshipman as the only officer beside himself. Lieutenant Pratt in *Snapper* was in identical straits, and by February 1821 he needed to be invalided himself, to be replaced in command by Lieutenant Thomas Evans. The rainy season took its usual toll on the other ships' companies of the Squadron, and, among others, a lieutenant and the Surgeon were sent home sick from *Pheasant*, and the First Lieutenant from *Morgiana*.

On 31 January 1821, Commodore Collier once again brought *Tartar* to anchor off the Sierra Leone River. On board was Lieutenant-Colonel Burke, sent out as Governor in MacCarthy's absence, but his tenure of office was very brief. Within a week he was sick and on his way back to England, leaving Grant to reassume the government. The Commodore found *Thistle* in harbour, and was there to meet *Myrmidon* and *Snapper* when they arrived the following day. *Pheasant* and *Morgiana* were away in the Bights. Robert Hagan found himself again high in the Commodore's favour as the only commanding officer to make a capture during Collier's absence, but it was Commander Finlaison of *Morgiana* who was to draw the first blood of 1821.

While still well to seaward in the Bight of Benin on 14 February, and making for Cape Coast to meet the Commodore, *Morgiana* encountered the Portuguese schooner *Emilia* out of Lagos, and took her after a short chase in latitude 3° 50´ N. The decks of the 158-ton vessel were crowded with 398 slaves, among whom "fearful disease" was breaking out, and Finlaison took 100 of the Africans into *Morgiana*. As was usually the case with Portuguese slavers caught north of the Line, the *Emilia*'s master swore that he had shipped his cargo well to the south, in this case at Cabinda, but the slaves stated otherwise, and the freshness of their brands and the vessel's low expenditure of water supported their evidence.

Tartar, meanwhile, had been making her way eastward, examining the Gallinas, Cape Mount and the forts at Dixcove and Secondee, before reaching Cape Coast on 2 March to find that *Morgiana* and her prize had arrived the previous day. Advising Finlaison to take the southern route, Collier sent him on his way to Freetown. It was the last the Commodore would see of *Morgiana*. The sloop was unable to make Ascension, and she bore away for the coast of Brazil, arriving at Bahia on 20 May with one day's provisions left after a passage of two months. She took her prize onward to Rio de Janeiro, arriving on 7 July, presenting the Rio Mixed Commission court with its first case.

A saga of prevarication and obstruction ensued, to the immense irritation of Henry Hayne and Alexander Cunningham, the British Judge and Arbitrator. Their Portuguese opposite numbers, Silvestre Pinheiro Ferreira and Jose Silvestre Ribello, were cooperative, but the Minister for Foreign Affairs, Vieira, wasted no opportunity to thwart the Commission. The *Emilia* was condemned on 28 August, but it was not until the end of February 1822 that the sale of the vessel, improperly taken out of the hands of the Commission by the Portuguese authorities, was finally made. The officers of the *Emilia*, having escaped prosecution, had their property restored to them, and Haynes had little faith that the arrangements made for the emancipation of the slaves would be honestly carried through.[9] Although *Morgiana* was now free to sail, she was in a poor material state, largely thanks to rot, and Finlaison decided to make good the defects affecting ship safety while the resources of Rio were available to him. It was not until 22 December that the sloop left the coast of Brazil for Ascension.

The first few months of 1821 were scarcely less frustrating for *Myrmidon*. After meeting the Commodore in Freetown, she sailed on 10 February for a cruise to leeward, and on the morning of the sixteenth she was lying at anchor under Cape Mount when a schooner rounded the point. On making out the *Myrmidon* the schooner bore up and made all sail in flight. The chase lasted until noon of the following day, when Commander Leeke found his quarry to be the Spanish *Carlotta*, with neither papers nor slaves. Her master and her trade goods had been landed at the Gallinas to buy slaves, and Leeke suspected her of cruising to find more slaves by piracy, a not uncommon practice by the Cuban slavers. Encountering *Tartar* off Cape Coast, Leeke handed the *Carlotta* over to the Commodore, who had her taken a few degrees to the south and turned loose. It was later heard that, after beating back to the Gallinas, she had sailed at the beginning of June with a cargo of 270 slaves and had capsized in a squall. The slaves in irons had, of course, drowned, but the handful of survivors from the crew, including the master, had reached Sierra Leone. Collier hoped to see them there!

Tartar and *Thistle*, after calling at Accra for provisions, sailed in company for the Bight of Biafra, with *Thistle* some distance ahead disguised as a slaver. Governor Smith at Cape Coast Castle had reported that a great many Portuguese vessels had passed into the Bights during the past few months, and Commander Kelly in *Pheasant* had corroborated Smith's complaint that Elmina was providing these Portuguese with canoes for use on the Slave Coast. As Collier wrote, "it is a Positive fact, no description of country produce is ever shipped in Portuguese Vessels from this part of the Coast, yet hundreds of Vessels under that Flag anchor here during the year".[10] The two cruisers boarded only one of these Portuguese vessels, off Whydah. She was empty, but clearly awaiting a slave cargo. Her passport from Brazil, as usual, was for a slaving voyage to Cabinda, but there supply could not cope with demand, and she was one of many such vessels seeking illegal cargoes north of the Line. Collier recorded his opinion of her on the face of the passport.

On 23 March the two men-of-war arrived off the River Bonny, and *Tartar* was preparing her boats to enter the river by the east channel with *Thistle* when a schooner was sighted at anchor in the channel. *Thistle* closed the entrance under Spanish colours, and a canoe approached her under the impression that she was a slaver. As the canoe came alongside the gun-brig its crew took alarm and were starting to paddle away when Hagan, with, as Collier put it, "that presence of mind and activity I have so long known him to possess", leapt into the canoe, followed by a seaman and a Krooman. With the evening closing in, and anxious not to delay while waiting for *Tartar*'s boats, Hagan hid his Acting Master, the excellent Admiralty Midshipman Charles Lyons,* in the bottom of the canoe with 30 men, and made the native crew paddle for the schooner. When the schooner hailed, the men in the canoe replied that all was well, and the reassured slaver crew left their stern chaser and returned to the cabin. Surprise was complete.

As the boarding party gained the schooner's deck, the Spaniards began shooting from the cabin, and Lyons, a seaman and a marine were slightly wounded, but the vessel was quickly taken. In the noise and confusion, however, about 50 terrified female slaves jumped overboard and were killed by sharks. The canoe, meanwhile, escaped and spread the alarm upriver. The prize was the *Anna Maria* from Cuba, measuring less than 180 tons but originally with 491 slaves on board. Her slave rooms were filthy, intensely hot and only 2 feet 11 inches in height, and Collier recorded that the occupants were "linked in shackles by the leg in pairs, some of them bound with cords; and several had their arms so lacerated by the tightness that the flesh was completely eaten through". Most of the slaves were suffering from dysentery, and those who could do so were "clinging to the gratings to inhale a mouthful of fresh air, and fighting with each other for a taste of water, showing their parched tongues, and pointing to their reduced stomachs". The supercargo, and effectively the master, was either American or English but claimed to be naturalised Spanish, a wild villain who credibly asserted that he would have blown up the vessel if he had been able to reach the magazine in time.

Tartar's boats pushed on, under the command of Lieutenants Marsh and Graham, to find a Portuguese slaver said to be lying upriver. After a tedious row they found her on 24 March, and initially faced a fire of grape and musketry as they approached. As the boarders came up the ship's side the slaver crew ran below, and later claimed that they had ceased fire as soon as they realised the identity of the boats. The vessel was the ship *Donna Eugenia*, armed with six 12-pounders, bearing a passport to buy slaves at Cabinda and Malembo, and with a part cargo of 83 Africans loaded at Bonny. Before Collier could send the two prizes to Sierra Leone, he had to shift more than 200 slaves out of the *Anna Maria* to allow room for the schooner to be worked. Nearly 100, including 40 sick, were put into *Tartar*, of whom twelve died on passage, and 125 were taken to the *Donna Eugenia*, of whom 38 died. Of the 272 left in the *Anna Maria*, 34 were subsequently lost. In the melee of the capture, the crew of the *Anna Maria* had managed to cut every rope within reach, and the Prize Master, Lieutenant

---

* Although ships' commanding officers continued the age-old practice of entering promising youngsters of their own acquaintance, or the relatives of friends and fellow officers as midshipmen in their commands, the Admiralty was now also entering and appointing midshipmen. These young men were known as Admiralty Midshipmen, and many of those in the West Africa Squadron seem to have been of outstanding quality.

Knight, felt himself justified in confining them in irons on the slave deck for the passage to Sierra Leone. Having despatched the *Anna Maria*, the Commodore took *Tartar*, *Thistle* and *Donna Eugenia* to Fernando Po for water and yams. The prize then departed for Freetown, and, on 5 April, the two cruisers headed for the Old Calabar River.

Tartar sent her boats into the estuary on 6 April, escorted by *Thistle* for as far as Hagan thought it prudent to take the brig. Having grounded once and refloated, *Thistle* anchored, and Hagan took his boats to join *Tartar*'s Lieutenant Marsh. Off Duke's Town on 9 April the force found two Portuguese slavers. The first to be boarded was the *Constantia*, a 73-ton brigantine from Princes Island. She had 250 slaves on board, which the master swore he had loaded at Cabinda, and a false log supported his ludicrous claim that he was at Old Calabar purely for provisions. As the boats moved on to the second slaver they boarded an English merchant vessel in error. Unfortunately this gave a warning to the Portuguese brig *Gavião*, which was then found to have on board eight Africans whom the Commodore believed to be slaves, and Duke Ephraim himself wrote to Collier that he had sold three men to the master, of whom two had been landed in haste on the approach of the British boats. The boarding party had found a Portuguese crew member in the hold attempting to pull trousers onto an African who was thought to be the third of these slaves.

When the two prizes reached *Tartar* on 13 April, 60 or 70 miles from Duke Town, Collier boarded the *Constantia* and found her the most crowded slaver he had ever seen. A dead woman was almost buried under the living, and he decided to move all the male slaves into *Tartar* to give the 77 women adequate space. One man died before reaching the frigate. The four vessels then made for Fernando Po to redistribute the slaves. Although he had no hesitation in sending the *Constantia* for adjudication, Collier was doubtful about achieving condemnation of the *Gavião*, despite receiving a certificate from the British merchantmen masters in the river declaring that she was slaving. It was all too apparent to him: "The interest every slaver has in disproving the charges against his vessel, tempts them to swear to the most evident falsehoods."[11] Finally he was swayed by the need to use her to take slaves to Freetown. Keeping about 70 slaves in *Tartar*, many of them sick with fever, dysentery and a skin eruption called craw-craw, Collier distributed the remainder between *Constantia*, *Gavião* and *Thistle* before sending the prizes on their way under the care of the gun-brig. Thirty of the slaves would die on the passage to Freetown.

Tartar returned to Cape Coast, where she was joined by a sickly *Pheasant*, and, after ordering Commander Kelly to take his ship to Ascension to recover, the Commodore sailed on to Accra. Here he discovered that the condemned Dutch slaver brig *Marie* had been sold by the Mixed Commission to a British owner who, in turn, had sold her on to the slaver Da Souza at Whydah. The transaction illustrated how easy it was, despite the Commodore's warnings, for condemned slaving vessels to find their way back into the Trade. It was also apparent that British merchantmen sold in the Bights often fell into the hands of slavers at Princes Island.

The Commodore then followed in *Pheasant*'s wake to Ascension, where he told Kelly to remain for at least a month to allow his debilitated ship's company to recuperate before returning to the Gold Coast. *Pheasant* had been reduced to the parlous state of having only one anchor, and Collier hoped that Kelly, during his stay, might be able to beg another from a passing merchantman. The Squadron had lost and damaged an excessive number of anchors on the Coast, and the Commodore had asked the Admiralty to supply an extra bower anchor to each sloop and gun-brig sent to the Station.

After a nine-day passage from Ascension, *Tartar* reached Sierra Leone for the last time on 5 June. She found *Snapper* in harbour after visiting the northern rivers and giving support to Gambia merchants launching the gum trade at Portendic. *Myrmidon* too had returned from the north, where she had been trying to deter the Portuguese slave trade between the Rio Grande and the Cape Verde Islands. Before that she had been watching the Grain Coast, for which Collier commended Leeke's "zeal and perseverance". Neither vessel had made an arrest. *Thistle* and three of the prizes had arrived, but Hagan had been obliged to leave the *Gavião* on the edge of the South-East Trade to make her own way.

Indeed the *Anna Maria*, *Donna Eugenia* and *Constantia* had already been condemned, and the *Anna Maria* had been not only condemned but also sold and refitted, all in unusual haste. The Commodore was furious that the Spanish schooner had been dealt with before his return with her master, partly because he had intended to charge her with piracy, and partly because he believed he could show that she was English-owned. Proof of English ownership would have brought her under the Abolition Act, and head money would have been doubled; Collier had resolved not to make any pecuniary gain himself from the emancipation of slaves, but he had every intention of doing his best for his officers and men.

The commissioners had condemned the *Anna Maria* in the absence not only of the master and the captor, which was not generally their practice, but also of the one remaining Spanish Commissioner, Le Fer, who had retreated to the

Gambia for the rainy season.* Collier believed they had done so because they did not wish to uncover an English connection or any link between English and Cuban merchants. He particularly suspected the motives of Fitzgerald, Chief Justice of the colony and Judge of the Vice-Admiralty court as well as British Arbitrator. Thomas Gregory had resigned as Commissary Judge, and had been replaced in mid-1820 by his nephew, Edward Gregory, a young man who, Collier considered, was subjected to an "over-ruling authority assumed by [Fitzgerald]". The obvious ill feeling between the Commodore and Fitzgerald extended, in Collier's view, to animosity between Fitzgerald and several of Sierra Leone's principal officers.[12] However, all useful evidence against the *Anna Maria* was now destroyed, and all the Commodore could do was to make his farewells. On 17 June *Tartar* sailed for Barbados and home. As she cleared the River Sierra Leone, her prize *Gavião* entered harbour.

When the *Gavião* came to court, Collier's doubts about the case were realised. As usual, no representation by the captor was allowed,† and, in denial of the evidence of Duke Ephraim's letter to Collier, the commissioners stated that it had been clearly proved that no slaves had been taken on board at the Old Calabar. The verdict centred on the status of seven slaves embarked at St Thomas and Princes Islands. They had been shipped, allegedly to help work the brig, with the consent of the Governors of the two islands, at least one of them a slaver, and the *Gavião*'s master described them as free men because, he said, he intended to give them their freedom and employ them. Gregory believed that the *Gavião* should be condemned, but Altavilla considered that not only had she not been engaged in the slave trade but also that the owners were entitled to restitution. Fitzgerald, as Arbitrator, sided with Altavilla against condemnation, but accepted that there had been intent of slave trading, and consequently believed that damages should not be allowed. The *Gavião* was therefore released without restitution, but some weeks later the case was reopened to consider a claim for damages. The captor was given three days in which to rebut the charge, despite the fact that he was far away, and, entirely unbeknown to Collier, an award of £1,520 was made against him. This was undoubtedly a gross miscarriage of justice on several counts, but Collier's subsequent protestations to the Admiralty probably availed him nothing.

---

* Le Fer specifically forbade the trial of Spanish vessels in his absence. He promised to return to Freetown if required, and expected to be able to do so within the two months allowed for completion of cases.

† Collier claimed to have been ejected from the court on one occasion, and his Agent, Walsh, was summoned by the court to appear before the Governor for interfering in the case of the *Donna Eugenia*.

This was a disappointing ending to this fine officer's three years in command, and he could find no compensating satisfaction in any reduction in the slave trade as a result of his efforts. A British supercargo, widely experienced on the Coast, estimated that, between June 1820 and April 1821, 3,700 slaves had been taken from the Grain Coast, 1,200 from Accra, and 22,600 from the Bights, mostly Bonny and Calabar. During this same period the British cruisers freed about 1,200. Thankless though his task had been, Collier's professional dedication was matched by his care for the men of the Squadron and for captured slaves, and his loathing of the slavers was almost equalled by his contempt for the lawyers he believed to be obstructing justice. In the opinion of *Marshall's Naval Biography*, "No officer of his standing in the Service was more generally known or higher in estimation as a brave, experienced, clever seaman, and most generous, warm-hearted, friendly man."[13]

His period on the Coast had worn him out, and, having been unjustly accused of cowardice during the war of 1812, he committed suicide in 1824 at the age of 50.

Commander Kelly in *Pheasant* was now temporarily the senior officer on the Coast, and, the health of his ship's company somewhat improved, he sailed from Ascension on 23 June for Cape Coast. Here he was joined by *Myrmidon* and, at Accra a few days later, by *Snapper*. Then, instead of returning to Sierra Leone as previously ordered by the Commodore, he made for Fernando Po for wood and water, a disobedience which drew a complaint to the Admiralty from Collier in England. While waiting off Cape Formoso for *Snapper* to rejoin, *Pheasant* and *Myrmidon* chased a Portuguese schooner for seven hours. *Myrmidon* brought her to and found her to be the *Adelaide*, with 232 slaves on board. She had been boarded only five days previously off Badagry, where the master had said that he had landed his trading cargo and was, entirely implausibly, going to Malembo for slaves. On board the *Adelaide* was discovered a seaman who had unwisely boasted to his shipmates that he had helped to murder *Pheasant*'s prize crew in the *Vulcano do Sud*.

It was probably from the *Adelaide* that a boy of 11 was emancipated, with his mother and two sisters, to be educated in Sierra Leone and, in 1864, to be consecrated the first Bishop of the Niger Territories. He was Bishop Samuel Crowther, who is largely credited with spreading Christianity in West Africa.

As they headed eastward on 27 July, still without *Snapper*, the two cruisers tried to send their boats into the Bonny River, but were thwarted by the weather. *Myrmidon*'s cutter capsized on the bar with the loss of five lives, and *Pheasant*'s jolly boat failed to return. By now Kelly was incapacitated by fever, and, on sailing from Fernando Po on 7 August, *Pheasant* headed for Freetown while *Myrmidon* returned for another attempt at the Bonny. Leeke anchored off the river mouth on 9 August and sent the boats over the bar early the following morning under the command of Acting Lieutenant Bingham. The boats found a shortcut to the slaver anchorage off Bonny Town, which allowed them to catch the vessels there by surprise at daybreak. Finding only French vessels, they boarded the first and were told of two Spanish slavers in a side creek, full of slaves and, it was said, with their crews ashore in a state of mutiny. Taking the pinnace and the gig, Bingham headed into the creek. On approaching the two vessels, the boats fired a couple of muskets to induce the Spaniards to hoist their colours, but, as the range closed to pistol-shot, the slavers hauled up their port-lids and opened fire with grape and musketry. The first round hit the sternsheets of the gig, severely wounding Bingham, Midshipman Deschamps, the Sergeant of Marines and a seaman. Bingham appeared to be dying, and Deschamps took charge of the boats, withdrew out of range and sent word to *Myrmidon*.

Leeke immediately despatched reinforcements, but it became apparent that the two vessels had barricaded themselves with iron bars and nailed-down awnings to an extent that made boarding impossible. Undeterred, Leeke decided to take the ship into the river, and settled down to sound the bars, lay buoys and wait for a spring tide and a favourable wind. On 31 August *Myrmidon* squeezed over the bar in three and a half fathoms of water and anchored in Bonny Roads just after sunset. On seeing her, the Spaniards fled ashore and sent a letter to acknowledge their wrongdoing and plead for their lives. The slaves had been landed, and Leeke delayed taking possession until the King of Bonny gave them up again. When boarders took the two Spanish vessels they found 153 slaves in *La Caridad*, a fine brig of 254 tons, 12 guns and a crew of 45, and 140 in the schooner *El Neuve Virgen*. Leeke also took the opportunity to dissuade the King from his habitual ill-treatment of British merchants, and he was able to recover *Pheasant*'s jolly boat and the survivors of the earlier attempt who had been very kindly looked after by British merchantmen in the river. To complete a most satisfactory episode, both slavers were condemned and the wounded all recovered.

A year after the grounding of Lieutenant Hagan's colonial brig *Prince Regent* in 1818, Governor MacCarthy had bought a replacement schooner of about 200 tons, and continued his confusing naming policy by calling her *Prince Regent*. At the end of 1820, Lieutenant John Kingdom was granted 18 months' leave to command her, but in September 1821 he died, and MacCarthy appointed in his stead a Mr McCoy. McCoy had been Master of *Erne* when she was stranded on the Cape Verde Islands, and he had been dismissed from the Service for her loss, but the Governor showed a generous faith in his new Commanding Officer.

Snapper, her temporary Commanding Officer relieved by Lieutenant Christopher Knight early in June, had made her way to the Old Calabar River. On 1 August Knight had boarded a departing English merchantman outside the river bar and obliged her pilot to take *Snapper* into the river. He learned from the English vessel's master that there were three slavers off Duke Town taking cargoes from Duke Ephraim, and, on the first night in the river, he sent his boats ahead under the command of Mr Cowie, his Acting Master. Following gingerly in the brig, Knight arrived off the town to find that the boats had captured the Portuguese schooner *Conceição* with a part cargo of 56 slaves. She was the property of Donna Maria da Cruz of Princes, one of the aristocratic landowners of that island who were making a good profit from supplying cargoes at Princes to slaving vessels which did not dare to risk trading on the mainland themselves. Reporting this success to the Admiralty, in the absence of a commodore, Knight reminded Their Lordships of the distance from the Bights to Sierra Leone, and of the consequent inevitability of deaths among slaves during the passage, as well as the risk to prize crews in loathsome conditions with no medical assistance possible.[14]

This well-deserved seizure, *Snapper*'s first for two years, was the Squadron's last of 1821. In mid-November the four remaining cruisers were at Sierra Leone: *Myrmidon* and *Snapper* back from the Bights, *Thistle* from the north, after an unusually quiet few months for Lieutenant Hagan, and *Pheasant* from a cruise to the Gallinas and the Gambia. In late September, before this cruise, Commander Kelly had been relieved in command of *Pheasant* by Commander D. C. Clavering, but Kelly appears to have remained in Sierra Leone as Senior Officer to await the arrival of the new Commodore. From there he despatched one more expedition before the end of the year.

Governor MacCarthy, returned from leave at the end of November, had heard of a Portuguese slaver in the Rio Pongas, and so *Snapper* and *Thistle*, with boats from *Pheasant* and *Myrmidon*, were sent to investigate. On Christmas Eve, Knight took the boats into the river and found the schooner *Rosalía* of Havana landing rum and tobacco at Mongo John's factory. Ormond and a British citizen named Lightburne had agreed to supply her with 200 slaves, but as yet there was no sign of this cargo. *Snapper* returned to Sierra Leone, but Hagan, doubtless anxious not to miss an increasingly rare opportunity at his favourite hunting ground, remained with *Thistle* off the river mouth.

Despite the Squadron's efforts, the year had seen an increase in the Trade. Activity had been extensive at all the recognised slaving areas, and the Bights coast, in particular, was swarming with French, Spanish and Portuguese slavers.

The Chiefs of Bonny and Calabar had begun to keep registers of slave vessels to help in charging duty, and the figures reaching Sierra Leone had been reckoned to be exaggerated, but when *Myrmidon* visited those rivers in October Commander Leeke found, on good authority, that between July 1820 and October 1821 the Bonny had exported 190 slave cargoes and the Calabar 162.

On 16 January 1822 *Thistle* entered the Sierra Leone River, with the *Rosalía*. Hagan had taken a second look at his quarry on 7 January, and found that, although her master and seven of her crew had died, she was ready for loading and many of her slaves were waiting ashore. Anxious to move matters along, he gathered together the local chiefs and persuaded them to have the slaves put on board. This was achieved by 11 January, whereupon Hagan seized the vessel and took her to Freetown. In the absence of a Spanish Commissioner, the British Judge and Arbitrator accepted that the *Rosalía* was a Spaniard, although she had no papers, and condemned her and 59 slaves. They admitted, however, that if there had been any representative of the vessel to claim her they would have restored her.

By the time that the details of the case reached the Foreign Office, Lord Castlereagh, that great champion of abolition, was dead by suicide, and his place as Foreign Secretary had been taken by George Canning. Canning was a disciple of Pitt, already experienced at the Foreign Office, and a strong advocate of abolition throughout his political career. His initial optimism about the prospects of ending the Trade was soon dashed. At the Congress of Verona in 1822 his recommendation to Austria, France, Russia and Prussia that the slave trade should be denounced as piracy under the Law of Nations found no agreement, and his proposal that produce of countries still engaged in the slave trade should be boycotted met with laughter from his foreign colleagues. Continental statesmen intimated that his suggested refusal to admit Brazilian sugar was evidence that self-interest lay at the root of Britain's abolition policy. Canning was consequently soon persuaded that his predecessor's policy of bilateral negotiations with the slaving nations was the only viable option.[15]

The new Foreign Secretary was not pleased with what he read about the *Rosalía*. The slaves had been embarked for the purpose of being freed, not for the slave trade, as required by the terms of the treaty. Furthermore, some of Hagan's men had been on board the slaver for two days before seizure, so the vessel had been partially under his control, and his boat had assisted in loading the slaves. Canning chastised the commissioners for condemning the slaver against the evidence, a decision which might "establish a precedent, encouraging British Officers to repeat a conduct so reprehensible, as that pursued in this instance by the Captor of the *Rosalia*." He asked the Treasury whether Hagan might be prevented from gaining any financial advantage from the capture, and the Treasury, of course, found no difficulty in making that so. Their Lordships expressed their "decided disapprobation" of Hagan's conduct, and the Commodore was told to prevent any repetition. Hagan was convinced that the outcome would have been different if he had been represented in court, and later pointed out, with some satisfaction, that the arrest of the *Rosalía* had put an end to slaving from the Rio Pongas for the remainder of 1822.

Nevertheless, intelligence reached Commander Kelly on 25 January that another schooner was slaving in the Pongas, and he sent *Snapper* to investigate. Lieutenant Knight returned to Sierra Leone to report that the vessel was the American *Dolphin* of Charleston, information which led to an admirable instance of Anglo-American cooperation. In May 1820 a new anti-slaving bill had been introduced into the United States Congress, not only prescribing the death penalty for direct involvement in slave trading, which it classified as piracy, but also empowering the President to employ armed cruisers, as he saw fit, to seize American slaving vessels. USS *Cyane* had already sailed in January with the expedition to settle the Sherbro, and President Monroe deployed several more warships to the African coast during 1820 and 1821. The sloop *Hornet* sailed in June 1820, the schooner *Alligator* in April 1821, the Sixth Rate *John Adams* in July 1821, and the schooner *Shark* in August 1821.[*] Between them they made 11 captures of American slavers trading to Havana, and American revenue cutters did worthwhile damage to the Trade in West Indies waters.

The Royal Navy was instructed to "cooperate" with these new allies in the suppression campaign, and was indeed anxious to do so, but only after repeated requests did the Foreign Office define what it meant by "cooperate": the "American Government having refused to enter into any more intimate concert" with the British Government for the suppression of the slave trade, British cruisers were to be instructed to: "give such general assistance [...] towards the attainment of this common object, as was consistent with the existing Treaties and Rights of both Nations, and with the friendly relations and perfect Amity subsisting between them."

In May 1820, Lieutenant Nash in *Snapper* had searched the Rio Pongas in the company of USS *John Adams*, Captain Wadsworth, and since then there had been various encouraging contacts between the cruisers of the two navies. In his annual report for 1821, Commodore Collier remarked that:

---

[*] The Commanding Officer of *Shark* was Lieutenant Matthew Perry, future instigator of US trade with Japan.

the American Officers in command of the Vessels of War employed by their Government in the suppression of the Slave Trade on the Coast of Africa, had upon all occasions acted with the utmost zeal in the object, and it was extremely gratifying to me to observe, that the most perfect unanimity prevailed between the Officers of His Majesty's Squadron, and those of the American Vessels of War engaged in the same views.[16]

If all subsequent commodores and commanders-in-chief on the West African coast had been able to report a similar warmth of cooperation the later history of the Atlantic slave trade would have been a happier one.

On her departure from the Coast at the beginning of 1822, the schooner USS *Alligator* left behind a tender under the command of her Master, Mr Harry Hunter, and when *Snapper* returned to Sierra Leone with news of the American slaver *Dolphin* in the Pongas, Kelly passed the information to Hunter. Although the tender was short-handed, Hunter leapt at the opportunity, but asked Kelly for assistance. Kelly willingly lent him Lieutenant James and 12 men, and on 16 February the tender returned to the Sierra Leone River with the *Dolphin* at heel. James spoke highly of "the cordiality which subsisted between himself and Mr Hunter", and of that between their men.[17] London probably, and Washington certainly, would not have approved of such an irregular, but eminently sensible, arrangement.

An unusual visitor arrived at Sierra Leone on 1 February 1822: the French brig of war *Huron*, with Commodore Du Plessis, who commanded the squadron of small vessels employed between Senegal and Goree. *Huron* had already cruised as far as Grand Bassa and boarded a few French slavers, but had arrested none because they had no slaves embarked. Governor MacCarthy had heard from a French minister that the French cruisers were operating under the same rules as the British, but they were certainly not taking action against foreign traffic, and it appeared that they were treating their own slavers as the British were the Portuguese, Spanish and Dutch. This visit was perhaps the first sign of a willingness of some French officials to attempt suppression of their nation's slave trade, and the Governor was happy to provide Du Plessis with water and stores to allow him to prolong his cruise to the Gallinas. The only other French cruiser to be seen at Sierra Leone during the year was the corvette *Diane*, which appeared off the river in mid-May but did not stop or communicate.

On 2 February 1822 the frigate *Iphigenia* (36) arrived at the Gambia under the command of the new Commodore, Sir Robert Mends, a distinguished veteran with command experience stretching back to 1796, and a Post Captain since 1800.

During the American Revolutionary War he had lost his right arm at the age of 13, and had been repeatedly wounded during the French wars. A splinter wound to his head while commanding *Arethusa* in 1809 had affected his eyesight, and he felt its effects for the rest of his life.

The Commodore's orders from the Admiralty emphasised his tasks of support to the British settlements and protection of commerce in addition to suppression of the slave trade, and he was told that Ascension, now considered part of the West African Station, was to be given necessary support. His attention was drawn to the accompanying treaties in his treatment of slavers, it was emphasised that all Spanish slaving was now illegal, and he was directed that the "several Bays and Creeks", particularly those in the Bights, "must be diligently watched and frequently examined".

It seems that Their Lordships had learned little from the 15 years of the campaign. As if he had not already enough to occupy him, Mends was also to take with him Captain Sabine of the Royal Society and Mr Don of the Horticultural Society, and "afford them every facility" for their studies. At the beginning of the rainy season he was to head for the mouth of the Amazon, for Sabine to make observations, and then return to England via the West Indies. He was to take home with him any officers and men invalided, replacing them with supernumeraries or volunteers from *Iphigenia*, or by authorising ships to bear a larger proportion of Kroomen.

Iphigenia began her deployment with a flourish. As she coasted south towards Sierra Leone she sent her boats into the Bissau anchorage on the night of 21 February, under the command of Lieutenant George St John Mildmay, and they brought out the Portuguese brigantine *Conde de Ville Flor* with 171 slaves. The slaver had, however, been lying at anchor close to the Portuguese fort, and an arrest within gun-shot of such a fort contravened the Anglo-Portuguese treaty. However, bearing in mind that some of the slaves in the prize were being shipped on the account of the Governor of Bissau, the court resorted to the treaty clause which allowed that if there were "no Local Authorities to whom recourse could be had [...] vessels so visited may be brought before the Mixed Commission", and it condemned the slaver. Nevertheless, Judge Altavilla was anxious that this should not be seen as a precedent.

A merchant brig had arrived at Sierra Leone from Virginia in March 1821 with an agent and a new batch of families for the (by then failed) American settlement on the Sherbro. Governor MacCarthy persuaded them of the unsuitability of the Sherbro and allowed them to live in Freetown until they could find a more promising site elsewhere on the Grain Coast. They soon settled near Cape

Mesurado at what was to become the city of Monrovia.* The American cruisers naturally showed an interest in this new attempt, and, on leaving the Coast at the beginning of 1822, USS *Shark* put a midshipman and a few men into a small schooner called *Augusta* to support the infant settlement.

On *Iphigenia*'s arrival at Sierra Leone on 22 February, Commodore Mends encountered the *Augusta*, and, seeing an opportunity for cooperative action while *Iphigenia* lay at Freetown, he put Lieutenant Clarkson and a small detachment of the frigate's men into the schooner for a local cruise. This curious, if happy, arrangement led to an unusual seizure. In the Gallinas on 26 February, the *Augusta* found the schooner *Joseph*, clearly a slaver but with no slaves on board, and under Swedish colours. Clarkson detained her on suspicion that she was British-owned. The two British commissioners forming the Anglo-Spanish court finally concluded that the vessel, although registered at St Bartholomew, was partly owned by an Englishman in Cuba. She was condemned, but Canning later wrote that the means by which the *Joseph* had been detained was not altogether to be encouraged.

The entire Squadron had been waiting at Sierra Leone for the Commodore's arrival, even including *Morgiana* which had returned on 8 February from Ascension after her protracted visit to Brazil. Before *Iphigenia* sailed for the Gold Coast with the Governor on 11 March, Mends sent *Myrmidon* and *Thistle* to leeward and despatched *Pheasant* on a short cruise to the northern rivers in company with *Snapper*.

Pheasant's brief visit to the Rio Nunez in March resulted in a complaint to Canning by the French Chargé d'Affaires in London. Commander Clavering had sent one of *Snapper*'s boats into the river, with a crew from both vessels and commanded by Lieutenant Helby, to find a Portuguese slaver reported to be loading there. He had employed, at his own expense, a small schooner for pilotage, and in the afternoon the boat had sighted a vessel showing no colours. Having no intelligence of any other vessel in the river, Helby assumed that she was the Portuguese. When he boarded her that night, Helby discovered that she was French, and quickly left her. The French Chargé d'Affaires, however, levelled accusations of an intentional assumption of a Right of Search, plunder and ill-usage. Canning seems to have dismissed these absurd allegations, but it would appear that the French had adopted a policy of diplomatic attack as the best defence for their own toleration of massive illegal slaving by their own people.[18]

Scarcely more successful was *Morgiana*'s first interception since her return to the Coast. Christopher Knight had handed over *Snapper* to Lieutenant T. H.

---

* The city was appropriately named after the abolitionist President Monroe.

Rothery and had replaced Commander Finlaison in *Morgiana*, becoming the sloop's fourth commanding officer during her deployment,* and he reported the arrest of the Spanish schooner *Dichesa Estrella* off Trade Town on 17 March. There were no slaves on board at the time, but her cargo was ready ashore, and Knight, taking a leaf out of Lieutenant Hagan's book, arranged with the local chiefs for 34 of the waiting slaves to be loaded before he detained the vessel. Midshipman Maclean was put into her as Prize Master, and he was obliged to anchor off Cape Mesurado for water before making for Freetown. There the slaver was wrecked, apparently because natives had cut her cable during the night. The prize crew and slaves all managed to reach shore, but five of the slaves were seized by the natives who, Knight wrote, behaved throughout "in the most barbarous manner". The survivors suffered considerable privations, but were helped and protected by the American settlers at Mesurado, at some risk to themselves. Eventually, Maclean, his crew and the remaining 29 slaves were taken to Sierra Leone by the US Schooner *Augusta*. Maclean died on board, leaving a partially written letter in which he spoke highly of the conduct of the French schooner *Adolphe*.† Knight expressed his gratitude to Lieutenant Clarkson of *Iphigenia* who, presumably, was still in the *Augusta*. Having not yet seen Canning's comments on the *Rosalía* case, the commissioners decided to condemn and release the slaves from the *Dichesa Estrella*, and Knight escaped without the "disapprobation" heaped on Lieutenant Hagan.[19]

During the last few days of March the Commodore helped Governor MacCarthy to inspect the forts at Dixcove, Cape Coast, Anamabo and Accra, all of which had become part of the Governor's responsibility when Parliament transferred the possessions of the African Company to the Crown in July 1821. Mends noted the potential of the forts for storage of provisions and naval stores for the Squadron, and expressed particular interest in converting the Cape Coast slave room into a water tank. He then refocused his attention on the Portuguese slavers on the Slave Coast, and during the first week of April he took three of them.

Off Appam on 1 April he found the brig *Dies de Feverio*, which had landed her cargo of slaves for security, but there were still ten Africans on board and receipts were found which later proved to the satisfaction of the Mixed Commission court that the ten were slaves. The second victim was the schooner *Nympha del Mar* off Whydah on 6 April. She had only three slaves on board, but another 250 were due to arrive in a few days' time. Mends would have demanded them from

---

* The *Navy List* indicates that Knight took command on 3 June 1822, but it is clear that he was already in command at the time of the *Dichesa Estrella* episode.
† The part played by the *Adolphe* is not known.

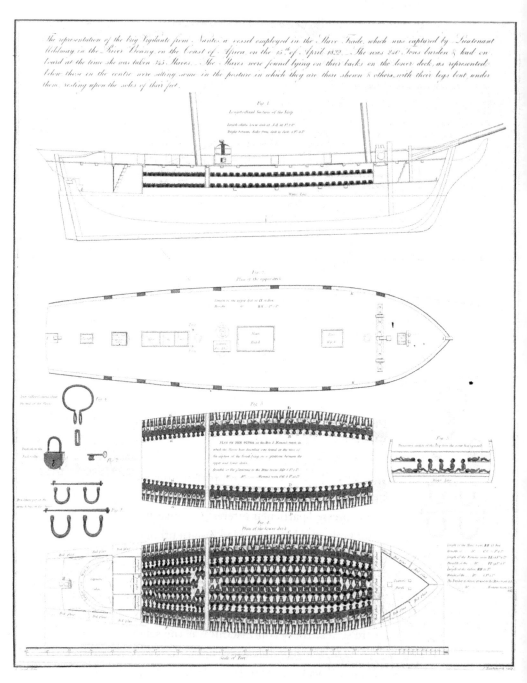

8. Slave stowage plan of the French brig *Vigilante* arrested by the boats of HMS *Iphigenia* and HMS *Myrmidon* on 15 April 1822

the King had the surf not been too high for the frigate's boats, and no native canoe would approach *Iphigenia*. The three Africans had been kept naked and chained, but when the frigate appeared they were released and dressed. Lieutenant Mildmay took command of the schooner with instructions to meet *Iphigenia* off Lagos, where there were believed to be more Portuguese slavers, and by the time Mends arrived off Lagos, Mildmay had added to the tally by detaining the polacca *Esperança Felix* in the anchorage on 7 April with 187 slaves.

By contrast with these easy arrests, on the morning of 15 April a severe encounter took place in the Bonny River. The boats from *Iphigenia* and *Myrmidon*, again under Lieutenant Mildmay's command, were sent in to examine the Bonny and the New Calabar, and, on entering, they sighted two large schooners, three brigs and a brigantine anchored in line about six miles north of the bar. Shortly after 0800 the schooners, followed by two of the brigs and the brigantine, all using springs on their cables to bring their guns to bear, opened fire on the boats at a range of about two miles. Mildmay and his men ran the gauntlet for 20 minutes, replying with the boats' guns at a range of three-quarters of a mile, and enduring a fire of grape and musketry as well as round shot as they pulled closer. Two boats lagged behind, and the others lay on their oars to wait for them. Mends, watching from one of *Iphigenia*'s tops beyond the bar, was struck by the "vivacity and continuance" of the action, and later gave testimony to Mildmay's "gallantry and decision". By now the brigs and brigantine had hoisted French ensigns, but the schooners showed no colours. With a loss of three killed and seven wounded,* the boats' crews succeeded in boarding and taking all six vessels.

The schooners were found to be Spanish: the *Icanam* of 306 tons and 55 men, with 380 slaves, and the *Vecua* of 180 tons and 45 men, with 300 slaves. Both Spanish slavers mounted eight long 18-pounders and one or two long 9-pounders. One of the brigs had apparently not resisted, and was not further molested. Between them, the other three French vessels mounted four 12-pounder carronades, all on the engaged side, and eight 9-pounder carronades, and carried 82 crew and a total of 808 slaves. The Spaniards had given muskets to their slaves, making them fire up through the hatchways, and some of these Africans were inadvertently killed by the boarders. In the *Vecua* a lighted match was found hanging over the open magazine hatchway, left by the crew as they fled.

A menacing aspect of this affair was that the resistance was premeditated and planned. The two Spanish vessels had agreed to stay together on sailing from

---

* These are the figures given by Mends. The British Commissioners reported two killed and five wounded, but they may have missed *Myrmidon*'s share of the casualties.

Havana, and the crews had sworn blind obedience to their masters, agreeing to forego all benefits from the voyage if the vessels were taken. It was said that they later bragged that they would have killed every British seaman who fell into their hands had they won the fight.

The drama and tragedy of the episode was not yet over. The *Icanam*, *Vecua* and *Esperança Felix* were put in *Myrmidon*'s charge for the voyage to Sierra Leone, and Commander Leeke decided to scuttle the Portuguese polacca because she was laggardly and her slaves sick, distributing the slaves between the faster *Icanam* and *Vecua*. On the night of 10 June, failing to shorten sail in time for a tornado, the *Icanam* capsized with the loss of two midshipmen, 16 seamen and nearly 400 slaves. Seven seamen escaped in *Iphigenia*'s pinnace which was towing astern of the schooner, and were fortuitously picked up by *Myrmidon* five days later. The *Nympha del Mar* was wrecked as well, just to the west of the Bonny, but probably without loss of life.[20]

The Spanish and Portuguese slavers were all condemned, and Mends was determined that the French slavers should pay for their criminal conduct. The brigs *Vigilante* and *Petite Betsy*, and the brigantine *Ursula*, after landing their slaves at Sierra Leone, were taken to England by Lieutenant Mildmay for trial in the Admiralty courts. In the event, the Government decided that they should be dealt with by the French authorities, and Mildmay took them onward to Cherbourg. During his five weeks in England, Mildmay incurred considerable duty travel costs and asked the Admiralty for reimbursement of one guinea per day. Their Lordships felt that they could not comply with this request. Thus did the Crown reward its naval heroes!

Myrmidon had achieved one more success before leaving the Bights and taking her charges to Freetown. Commander Leeke had sent his boats into the Old Calabar on 27 April, and the following day Lieutenant Elliot encountered the Portuguese schooner *Defensora da Patrie* heading to sea with 100 slaves. She was owned by one of the Princes Island traders and, armed with a passport for Cabinda, she had made straight for the Old Calabar. Her excuse that she had been carried north by the current found no sympathy from Leeke, but she was certainly a very sluggish sailer, and he decided to scuttle her. The visit to the river elicited from English palm-oilers the encouraging information that there had been no other slavers in the Old Calabar for four months, and in the Cameroons for five or six months. It seemed that the slavers were becoming wary of being caught by the cruisers' boats in the confines of rivers and creeks, and were shunning the Bight of Biafra in favour of the open shores of the Slave Coast, where they could detect the approach of men-of-war.

Ignorant of the misfortunes of the *Dichesa Estrella* and her prize crew, *Morgiana* had headed for the Bights, and on 15 April her luck changed for the better with the arrest of the Portuguese brig *Esperança Placido* at the mouth of the Lagos River. The master and crew had deserted, leaving on board 149 slaves and a Royal Passport from Bahia for slaving at Malembo, with permission to call for provisions at St Thomas and Princes, a common excuse for visiting slaving stations north of the Line. She was readily condemned.

This was *Morgiana*'s swansong. On 14 May she arrived at Ascension to deliver provisions and fuel before briefly returning to Sierra Leone, no doubt partly to collect her prize crew. She was there on 11 June when *Iphigenia* returned with Governor MacCarthy from the Gold Coast, as was *Snapper*, back from the Gambia. *Myrmidon* arrived too, with her remaining prize from the Bights. However, there was now a parting of the ways. *Morgiana* retraced her steps to Ascension, where Knight gave the garrison as much of his victuals and fuel as he could spare before sailing for Brazil and home after a three-year absence. When *Morgiana* arrived at Bahia, a city under siege by revolutionaries, she found *Pheasant* lying there. After an uneventful cruise in the Bights, *Pheasant* had fallen in with *Iphigenia* at St Thomas in mid-May, and the Commodore had taken the opportunity to rid himself of the astronomer Sabine. So *Pheasant*, after having to fish her foremast which was rotten as well as sprung, finally left the Coast, also after three years on station, and headed via Ascension for the mouth of the Amazon, for Sabine's benefit, and then England. *Myrmidon* made for the Cape Verde Islands to refit, and while she was at St Vincent in August she received orders to sail for Spithead, only a little under three years since she had left it. On 26 June, with the rains beginning, Mends and *Iphigenia* sailed for Barbados and home.

The two gun-brigs, *Snapper* and *Thistle*, remained on station, and, before the Commodore's departure, they were joined by two fresh ship-sloops: *Bann* (20), Commander Charles Phillips,[*] and *Driver* (18), Commander Thomas Wolrige. They were followed in mid-July by the ship-sloop *Cyrene* (20), Commander Percy Grace. Still under the orders of their absent Commodore, and with the inexperienced Wolrige as the senior officer on station, the four vessels at Sierra Leone sailed for their various cruising grounds: *Cyrene* for the northern rivers and the Cape Verdes, *Driver* and *Snapper* for the Bights, and *Bann* for Ascension to deliver stores brought by *Driver* and herself from England. *Thistle*, meanwhile, had been making two captures in the Bight of Biafra.

---

[*] *Bann* had been rerated from Sixth Rate to ship-sloop since her previous deployment to the Coast.

Thistle had been cruising the Bights for more than three months, carrying out numerous boardings but making no arrests. In June, Hagan made a check on the Bonny River, finding that there had been no slavers there since Lieutenant Mildmay's eventful visit in April, but learning from an English merchantman that there was a small Portuguese schooner with slaves in the Old Calabar. On 23 June *Thistle* found the *San José Xalaça*, a mere boat of 7 tons, and revealed a grim story. Like the *Conceição*, taken by *Snapper* in 1821, the schooner was owned by Donna Maria da Cruz, daughter of the ex-Governor of Princes, and had sailed from the Old Calabar with 30 slaves stowed between her water casks and the deck beams, a space 17 inches in height.[21] She failed to make Princes, her water and provisions ran out, ten slaves died of starvation, and she returned to the Old Calabar after six weeks. The surviving slaves had been landed when *Thistle* found the vessel, but they were given up by the custodian, probably Duke Ephraim. Hagan scuttled the schooner, and took the slaves into *Thistle*, where three more died on passage.

Armed with further intelligence from English palm-oilers that there were two slavers in the Benin River, Hagan headed west, and off Cape Formoso on 29 June, after a "smart chase", he caught the Portuguese brig *Estrella* of Bahia with 298 slaves loaded at Lagos or Badagry. *Thistle* had boarded her while empty at Cape Coast and endorsed her papers, but now she had no anchors or cables, and there was smallpox among her slaves. Manning his captures had made Hagan very short-handed, and he was obliged to ignore two more loaded slavers he found in the River Benin because he could not provide prize crews. After a passage of only nine days from off Princes, *Thistle* arrived at Freetown on 19 July and landed 23 smallpox cases, all slaves, having lost only one at sea. Thanks to a programme of vaccinations in Sierra Leone, 3,000 since April, the disease was confined to the landed slaves.

Having seen her prize condemned, *Thistle* spent the remainder of the year to the north of Sierra Leone or at Freetown, and made no further captures. At Cacheu she was told that a brig had sailed for the Cape Verdes with 500 slaves in August 1821, and the Portuguese officials, who made no attempt to conceal the Trade, regretted that *Thistle* had not arrived in time for a contest with two slavers mounting 16 guns and 11 guns.

The first of the newcomers to draw blood was *Driver*. Commander Wolrige had endorsed the papers of the Spanish schooner *Josepha* (alias *Maracayera*) at Bonny on 21 July, but, undaunted by the warning, the slaver finally loaded and sailed on 18 August. She badly mistimed her move. *Driver* was just returning from a visit to Princes, and sighted the *Josepha* as she put to sea for Havana. After "an interesting chase on all points for twenty hours", *Driver* brought her to and detained her with 215 slaves. To have hunted down a freely manoeuvring Spanish slaver in open water was a creditable achievement, the first such occurrence for 18 months. That success was followed on 7 September by a more routine seizure. Wolrige sent Lieutenant King with two boats into the Cameroons River that morning, and they found the Portuguese brig *Commerciante* rapidly landing slaves. The native chief from whom they had been bought immediately and voluntarily reloaded them, and King took the prize and 179 slaves out to *Driver*. Both slavers were condemned, but 32 of *Josepha*'s slaves had died on passage to Freetown.

There was to be a gallant but tragic sequel to these successes. After returning to Freetown from the Bights, *Driver* was required to sail for Cape Coast in mid-November with the Governor, who had recently been rewarded with a knighthood. The *Commerciante* had not yet arrived at Sierra Leone, and the sloop was therefore unable to recover her prize crew. However, the colonial schooner *Prince Regent* was due to follow to leeward a few days later, and so she embarked *Driver*'s two midshipmen, eleven seamen and marines, and four Kroomen. Calling at Cape Mesurado on 1 December they found that the American settlement was under serious attack by the native chiefs, who had sold the land to the United States. The 50 or so settlers were badly outnumbered, exhausted and running out of ammunition, and the Agent, Mr Ashman, begged *Prince Regent* for help.

Midshipman Samuel Gordon and all eleven of *Driver*'s seamen and marines volunteered to land and defend the settlement, and officers taking passage in the schooner then negotiated an agreement with the chiefs. On 17 December, Gordon fell ill with fever, and on Christmas Day he died. By a week into January 1823, eight more of *Driver*'s men were dead. Wolrige sent *Snapper* to recover the survivors, and she found only three convalescents. The settlers were deeply saddened, but grateful for their deliverance. In writing to Lieutenant Rothery of *Snapper*, Ashman acknowledged "having derived the most important benefits from the repeated interposition of the kind offices of HBM's Naval Officers", and signed his letter "With Sentiments of respect and Gratitude".[22]

Bann headed north from Ascension in August, bringing the Squadron's strength in the Bights to three. She made her first arrest off Whydah on the twenty-seventh, and it proved to be an unwise one. The Spanish schooner *San Raphael* was lingering off Whydah, waiting for her cargo and with her master ashore, but instead of allowing *Bann* to board her and find her empty, she attempted to evade. In light airs, Commander Phillips eventually brought her under grape and musket fire, driving the Spaniards from their sweeps, and arrested her. When she eventually came before the Anglo-Spanish court in January 1823 it predictably judged that she had been illegally detained, but there was no claimant for her. She therefore remained Phillips's responsibility, and he used her as a tender for some time before the court agreed to sell her.

Bann's next encounter was more successful, but it raised the question of the legality of Portuguese slaving south of the Equator from territory not claimed by the Portuguese Crown. On 29 September 1822 she detained the Portuguese schooner *Magdalena da Praca* between St Thomas and the mainland, and at only one degree north of the Line, with 33 slaves destined for St Thomas. The slaves had been loaded at Cape Lopez, at 1°S, and the Commission not only condemned the schooner but also gave its opinion that, by the terms of the treaty, Portuguese slaving was legal only between 5°12´S and 18°S.[23]

Returning to the Slave Coast, *Bann*, luckier than cruisers earlier in the year, found the brigantine *San Antonio de Lisboa* off Whydah on 5 October. This Portuguese slaver had previously been boarded while empty by *Driver*, *Snapper* and *Bann* herself, but now had 335 slaves on board. On 31 October, with immaculate timing, *Bann* caught another Portuguese slaver, the schooner *Juliana da Praca*, which had embarked her cargo at Porto Novo the previous day and was trying to clear the coast. Since mid-April, this vessel had been boarded on no fewer than seven occasions by six different cruisers before she was finally found with 114 slaves. *Bann*'s spree was still not over. Beginning her passage back to Sierra Leone from the Bights, she fell in with the Portuguese schooner *Conceição* on 13 November, with 207 slaves on board, and captured her with the boats about nine miles off St Thomas. Phillips wisely communicated with the Acting Governor of the island, who raised no objection that the capture came within the limits of protection of the Portuguese fortresses. Deciding to leave the *Magdalena da Praca* and her crew at St Thomas, probably because he had insufficient manpower to provide six prize crews, and also landing the crews of the *San Raphael* and the *Conceição*, Phillips put some of the slaves into the *San Raphael*, and herded his clutch of prizes towards Sierra Leone where, the *San Raphael* apart, they were easily condemned. The court did, however, caution him to bring back slaver crews as witnesses in future.

Bann was to have yet another encounter before reaching Freetown. Having heard from the *Magdalena da Praca* in September that there was another Portuguese slaver at Cape Lopez, Phillips was hoping to find her on his southward passage for Sierra Leone. On 3 December, by chance but not to his advantage, he fell in with his intended victim, the brigantine *Sinceridade*, 600 miles west of St Thomas, but, unfortunately, seven miles south of the Equator. Confused by the terms of the treaty, Phillips took her back to Sierra Leone, but although the

123 slaves had been bought illegally at Cape Lopez, her arrest south of the Line, not following hot pursuit from the north, was also illegal, and she was restored. Phillips complained bitterly that the slavers were perjurous and were "protected by a cunning Mulatto Advocate who receives […] the kind assistance of the British Arbitrator, Mr Fitzgerald." The case of the slaves was stood over for a decision at government level on whether compensation to the owner for their loss should be allowed, and Phillips was perhaps lucky to escape with a bill for only £100 for all other loss and damage. Nevertheless, it was clear that Fitzgerald had won no friends among the naval newcomers.

Meanwhile, *Snapper*, never one of the more fortunate vessels, had failed to replicate the successes of the other two Bights cruisers. Her one arrest, on 14 October at Mina Piccaninny to the west of Little Popo, was a mistake. Lieutenant Rothery detained the Portuguese brigantine *Nova Sorte* on the grounds that she had embarked slaves four days before arrival at Mina Piccaninny and had concealed them when *Snapper* first visited on 12 October, landing them that night. Rothery had then successfully demanded the re-embarkation of the 122 slaves and had seized the vessel. The Mixed Commission judges disagreed on the case, and Fitzgerald arbitrated. After questioning the slaves, he decided that there had been no earlier embarkation and relanding, and the vessel was restored. The decision caused some discontent in Freetown, where it was felt that the business of the courts was to condemn slavers, but the commissioners reiterated their duty to act "faithfully, impartially, fairly and without preference or favour". This was one of a number of cases in which it was found that the authorities in Brazil were stating false tonnages on slavers' passports to evade Portuguese limitations on slave loading. The *Nova Sorte*'s tonnage, for example, was given as 141.5, which allowed her 353 slaves. Her measured tonnage, however, was a little over 83, which should have permitted her a maximum of 211 slaves.

Cyrene had returned to Sierra Leone in early October from windward, and had heard that there was slaving activity at the Gallinas. On the morning of 23 October, just north of the river, she sighted two schooners which, on making her out, tacked away to the south. The westerly wind prevented the schooners from weathering Cape Mount, and *Cyrene* caught the weathermost, the Dutch *Aurora*, as darkness fell, and the leeward vessel, the French *L'Hypolite*, at about midnight. Both had slave cargoes waiting at the Gallinas factories, and Commander Grace decided to send to King Siaca to request liberation of these slaves.

On the morning of 25 October, Lieutenant Courteney took the boats into the river, through a tremendous surf on the bar, and was promptly met by musketry from the jungle on both banks. As the boats progressed against a strong ebb tide they encountered raking fire from a battery of two long 18-pounders and an 8-inch howitzer on Lower Factory Island. Landing against grape and musketry, the crews took the guns and turned them on their attackers. During their assault they saw slaves being thrown into war canoes and taken upriver. With his ammunition running low, Courteney burned the factories on that and a neighbouring island, spiked the guns, and returned to the ship with one man dead and three slightly wounded. King Siaca, a professed friend of the British, had been absent in the interior, and it became clear on his return that the masters of the prize schooners, who had been ashore at the time of the captures, had led the resistance, giving weapons and rum to the natives. Siaca blamed the French for the destruction of the factories, and he handed over 180 slaves intended for the prizes. The *Aurora* was condemned, but the case of the French *Hypolite* was, of course, a different matter.

Cyrene continued southward, and, while watering at Grand Bassa on 10 November, Grace heard of a slaver schooner at anchor off Little Bassa. She had been seen on the Coast for some while, using either French or Spanish colours. After a chase of four hours, *Cyrene* caught her and found that she was *La Caroline*, under a French ensign and claiming to be empty with a cargo waiting at Trade Town, but a search revealed five hidden slaves, Spanish, Dutch and Portuguese colours in her lockers, and instructions from a Dutch merchant in Surinam. Grace, assuming that the French ensign was fraudulent, made the foolhardy decision to send her to Sierra Leone if he could extract her waiting slaves. On returning to Little Bassa, he found a message from King Wise of Trade Town offering to give up the slaves if Grace promised not to destroy the town. Grace agreed, and 80 slaves, as well as *La Caroline*'s supercargo, were sent out to him.

La Caroline sailed for Freetown under the command of Mr William Hunter, Second Master of *Cyrene*, and an extraordinary drama unfolded. Remaining in the schooner was her master, a Monsieur Baron, a "worthless, drunken character, and so determined a villain", who apparently tried to have Hunter and his prize crew poisoned. Hunter put him in irons, but it was alleged that an accomplice had freed him and that he had caused a small explosion in an attempt to destroy the vessel. Hunter had him brought on deck for interrogation, but Baron's behaviour was so abusive and threatening that Hunter shot him dead. Hunter, apparently a mild-mannered man, was obviously rattled by the situation, probably tired and almost certainly ill. He died shortly after arriving at Freetown; otherwise he would have been tried for murder. As it was, when the news reached England, Canning was deeply displeased at the arrest of two more French vessels, his displeasure exacerbated by the killing of Baron, and directed that the two slavers should be released to the French authorities. The Admiralty ordered the immediate return of *Cyrene* to England.

Before the end of 1822, there were two international developments significant to the suppression campaign. The first occurred in Rio, where, on 1 December, the Prince Regent of Portugal was crowned Emperor of Brazil. Since 1815 Brazil had been a coequal of Portugal within the United Kingdom of Portugal, Brazil and the Algarves. Consequently she had been bound by Britain's anti-slaving treaty with Portugal. King John VI had ruled the United Kingdom from Rio, but revolutionary movements in Portugal had obliged him to return to Lisbon in 1821, and the Portuguese Cortes then wanted to return Brazil to colonial status. However, the Prince Regent, Dom Pedro, had stayed in Rio, and in September 1822 he declared Brazil independent and expelled Portuguese troops. By the end of 1823 independence was, in effect, complete, and in 1825 Portugal signed a Treaty of Recognition. Thenceforth Brazil was free of Portugal's treaty obligations to Britain, but in 1826 she signed a convention with Britain in which she agreed to abide by the terms of the Anglo-Portuguese treaty of 1815 and to prohibit the importation of slaves to Brazil after 1829. This left Brazil free to trade slaves south of the Equator for the next four years, but there was a view that Portuguese exports to Brazil were no longer legal because they no longer constituted an internal trade. Brazil also undertook to pass a law treating slave trading as piracy, but showed little inclination to fulfil that promise.

The second development was the first fruit of the Foreign Office's efforts to prevent slavers' evasion of arrest by landing slaves on sighting approaching cruisers. At Madrid on 10 December 1822, Britain and Spain signed articles explanatory and additional to the treaty of September 1817 which acknowledged that such evasion was "contrary to the true object and spirit of the Treaty". They agreed that "if there shall be clear and undeniable proof that a Slave or Slaves has or have been put onboard a vessel for the purpose of illegal traffic, in the particular voyage on which a vessel shall be captured", then such a vessel should be detained and condemned.

At Brussels on 31 December, Britain and the Netherlands made an identical addition to the Hague Treaty of May 1818. With the signing of a similar agreement between Britain and Portugal at Lisbon in March 1823, a major loophole in the anti-slaving treaties was closed.

Of even greater importance in the long term was a further Additional Article signed by Britain and the Netherlands on 25 January 1823. It stipulated that any vessel subject to examination under the treaty of May 1818, and detained at sea or at anchor on the coast of Africa in the area defined by the treaty, should be condemned as a slaver if her outfit and equipment fell within one or more of a number of categories, regardless of whether or not she was carrying slaves. The categories encompassed hatches with open gratings (for ventilation); additional subdivisions of the hold; planks for a slave deck, fitted or not; slave irons; excessive quantities of water, provisions or water casks; and excessive numbers of mess tubs or cooking-boiler capacity.* This became known as "the Equipment Clause", and its introduction represented a major step forward in the campaign and a significant diplomatic success. The vessels of nations accepting it would no longer be inviolate as they headed into the slave coast or waited for their slave cargoes, and a central source of frustration for the cruisers would be removed. However, the Netherlands was the only nation to agree to it so far, and the Dutch slave trade on the African coast was all but finished. Not until Portugal and Spain could be persuaded to sign such a clause would real progress be made.[24]

In contrast with the busy last few months of 1822, the opening months of 1823 were largely uneventful on the Coast. *Cyrene*, unaware of her disgrace, arrived at Ascension in mid-January after a cruise which took her from Trade Town to Lagos and Princes, calling at every likely slaving station but without making a single arrest. At Ascension she provided the garrison with sugar, soap and firewood, receiving turtles and vegetables in return, and leaving with requests from the Governor, the enterprising Major John Campbell, for green paint, a great variety of fruit plants in pots, more firewood, and young trees of fast growth to provide shade. Meanwhile, the other four vessels of the Squadron were either at Sierra Leone or cruising ineffectively in the difficult conditions of the rainy season.

Iphigenia had reached Spithead in October 1822, having suffered a serious outbreak of yellow fever after calling at Havana, and her officers and ship's company were turned over to the frigate *Owen Glendower* (42). Sir Robert Mends sailed with this new command from Spithead on 20 January 1823, carrying orders unchanged from the previous year except that he was no longer required to return to England for the Rains, and for the first time slave-trade suppression was mentioned before protection of British commerce.

On his way south, off Bissau on 4 March, he boarded a French vessel called *L'Africain*, an unremarkable incident except that the French later claimed that the boarders had offered violence and had plundered her of ivory, tobacco, preserved fruits and cooking utensils. The frigate's officers protested that, in fact, the master had been asked civilly for his papers, of which there were virtually

---

* See Appendix E for a full description of the Equipment Clause.

none, and the boarding party had left immediately when it was discovered that the vessel was French.

On his arrival at Sierra Leone on 20 March, the Commodore found the whole Squadron assembled and waiting for his orders. *Bann* and *Driver* he despatched to Ascension with stores, and he told *Bann* to carry on to Bahia to repair a damaged rudder. Still ignorant of the Admiralty's order regarding *Cyrene*, he sent her to Cape Coast to collect Sir Charles MacCarthy and then to cruise between Cape Palmas and the Gambia. *Snapper* was to sail for Portendic to protect the gum trade, and then return to England at the end of June. *Owen Glendower* prepared to sail for the Bights, and for *Thistle* and Lieutenant Hagan it was time to head for home.

When *Thistle* sailed for Spithead at the end of March 1823 she had been continuously on the Coast for three and a half years. For that period, her ship's company of 50 had lived and worked in their 84-foot gun-brig without any shore leave to speak of, enduring tropical heat and humidity exacerbated by a deckhead height of only five feet, attacked not infrequently by fever, and suffering the frustration of innumerable slaver boardings resulting in only 13 arrests. To maintain morale and enthusiasm in such circumstances required leadership of a high order. Lieutenant Robert Hagan, by now aged 28, provided that leadership. In all, he had been on the Coast for nearly seven years (his "total Servitude", as Mends put it): initially as First Lieutenant of *Porcupine*, then in command of a succession of colonial vessels, and finally in *Thistle*. Since his arrival to command *Princess Charlotte*, he had been home for only seven months in six and a half years, and he had been instrumental in the seizure of 33 slavers and nearly 3,100 slaves.* This brave, decisive, determined and uncompromising young officer, a consummate seaman, had inspired admiration and trust in three successive commodores, and, outspoken though he was, had earned the respect even of the British and foreign commissioners. On his departure, the Sierra Leone Council and the foreign judges addressed a letter of thanks to him and gave him a piece of plate, and the maritime community of the colony presented him with a similar address and a hundred-guinea sword. Mends wrote to the Admiralty that Hagan "takes with him the respect and good will of the whole Colony".[25] After paying off the worn-out *Thistle* to be broken up in May 1823, he was promoted to commander but not employed again until 1829. He was then appointed as Inspecting Commander of the Irish Coast Guard until 1843, when he was promoted to captain and placed on half pay.

---

* John Marshall in *Marshall's Naval Biography* and William R. O'Byrne in *O'Byrne's Naval Biography* (J. Murray: London, 1849) credit Hagan with 40 slavers and 4,000 slaves.

When Mends arrived at Cape Coast on 10 April he found the Gold Coast in a state of high tension. The Ashantis had been terrorising the coastal tribes and threatening the British forts for some time, and the murder of a sergeant of the Royal African Colonial Light Infantry near Cape Coast Castle had brought matters to the boil. Convinced of the value of a man-of war as a deterrent to the Ashantis and as a support to the forts, MacCarthy asked the Commodore to station a cruiser on the Gold Coast until peace was restored. Mends promised that *Bann*, on her return from Ascension, would remain there until relieved. For the next 15 months the Ashanti War would pose a serious distraction from the anti-slaving campaign.

As it happened, *Bann* was unable to return to the mainland until September. On 25 April she arrived at Ascension in a very sickly state, and by late May she had lost 33 dead from her complement of 110. Commander Phillips was invalided, and was temporarily relieved by Lieutenant Thomas Saumarez of *Driver*. When the sloop sailed for Bahia on 2 June to recuperate and to repair her rudder, she seemed to have left her infection, probably yellow fever, to ravage Ascension. From Bahia she continued south to Rio, perhaps in the hope of recruiting some seamen, and finally sailed for the Gold Coast in early August, arriving at Cape Coast on 5 September. *Driver* was in no better state. She arrived at Ascension on 5 May, and by the end of July, when she returned to Cape Coast to support the garrison, she had lost 31 of her 100 men to fever.

By early June, orders for *Cyrene* to return to England were on their way to the Commodore, but it would be some months yet before Mends' consequent instructions would reach Commander Grace. When Sir Charles MacCarthy had made his dispositions on the Gold Coast, *Cyrene* embarked him at Cape Coast and delivered him to Bathurst on 11 June. She then cruised in the vicinity until he was ready to return to Sierra Leone at the end of the month, and, having performed that task, she sailed for the Cape Verdes in August. The Governor returned to Freetown after his four-month absence to find that Edward Fitzgerald, the British Arbitrator, had died at the beginning of June, just one of the victims of a season of sickness in the colony described by Edward Gregory as calamitous. As temporary replacement Arbitrator, MacCarthy appointed Daniel Molloy Hamilton, the Registrar.

The only cruiser to achieve any success against the slavers during the first seven months of 1823 was *Owen Glendower*. After leaving the Governor at Accra in mid-April, Mends took the frigate south as far as Malembo. South of the Line he could not, of course, touch the Portuguese or Brazilian Trade, but Spanish vessels with slaves were liable to arrest. The only Spaniard he encountered in southern waters, however, was a schooner which managed to outsail his boats. He had more luck in the Bights in June. When the crew of the Spanish schooner *Maria la Luz* saw the British boats approaching her in the New Calabar River on 10 June they abandoned the slaver, taking most of their slaves with them, but the Chief of New Calabar, when he saw that the schooner had been taken, sent the slaves back on board. Thanks to the incompetence of the pilot, the schooner was lost on the bar when leaving the river, but the slaves were safely transferred to *Owen Glendower*, although they were so sick that Mends decided to land them at Cape Coast. The Spaniards had hoisted a French ensign to avoid being searched, but the vessel was clearly Spanish property, and was subsequently condemned as such.

A few days later, hearing that there was another Spanish slaver in the Old Calabar, the Commodore sent Lieutenant Clarkson upriver with the boats. After a long search they found her on 16 June up a side creek with the mangroves within 20 feet on either side. Many people, mostly black, were seen fleeing into the bush, and the vessel, the *Conchita*, was found fully fitted for slaving but deserted. Clarkson took her out into the river, and a canoe immediately came alongside with three slaves allegedly bought by the master. Next day Duke Ephraim sent off 55 more slaves, and declared the schooner to be Spanish. She had been flying Danish colours, and her log, in Spanish, stated that she was Danish, but there was ample evidence to substantiate Ephraim's information. The slave deck was filthy, the "poo-poo tubs" had recently been used, and it was clear that slaves had been on board until very shortly before capture; but the court, for some reason, could not later conclude that there was "clear and undeniable proof" that there had been slaves on board for illegal traffic on that voyage, as required by the newly amended treaty. The court sold the schooner, but postponed the case. It was not until April 1835 that Lord Palmerston at the Foreign Office told the Mixed Commission that, although the King's Advocate regarded the captor's conduct as not strictly regular, condemnation of the *Conchita* would have been justified, and he desired the commissioners to conclude the case. The slaver was duly condemned in July.[26]

When *Owen Glendower* anchored at Cape Coast on 31 July the Commodore reported, perhaps rather optimistically, that he had examined with his boats every river and creek from Senegal to the Congo during the five months since his return to the Coast. He had found very few slaving vessels anywhere, and was of the opinion that the Trade was in decline. The native chiefs had complained to him that the British had deprived them of the slave trade without giving them anything in return. However, his mind now turned to the Ashanti War, and he decided that the frigate must remain at Cape Coast in support of the British forts. Having landed the 230 debilitated slaves recovered from the *Maria La Luz*, he promised the castle's Commandant, Major Chisholm, all assistance his ships could afford. He ordered *Driver* to remain at Cape Coast, and sent instructions to *Cyrene* to join him immediately with whatever cruisers were at Sierra Leone, *Snapper* excepted, as she was ordered home with despatches and invalids.

When she sailed for England on 1 September, *Snapper* took another passenger, Mr Altavilla, the Portuguese Judge. Altavilla, who had earned wide respect during his time in office, was returning to Europe on leave, the latest in a series of departures among the Freetown commissioners. In 1821 there had been three casualties: Moraõ, the Portuguese Arbitrator, had returned home for his health, as had the Spanish Arbitrator, Campo, another Commissioner held in high esteem in Freetown. Le Fer, the Spanish Judge, had first retreated to the Gambia and then Goree to avoid the rains, before deciding that he was safer in Spain. In November 1822 the Dutch Judge, van Sirtema, was relieved by Major I. A. de Marée, and, with Fitzgerald's death and Altavilla's departure, the only survivor of the original Freetown commissioners was the Dutch Arbitrator, Bonnouvrié. The Spanish and Portuguese commissioners would not be replaced in any haste, and it was MacCarthy's opinion that those claiming ill health had returned home simply to suit themselves.

By the beginning of September, the *Driver*, thanks to the "zeal and attention of that excellent Officer", Commander Charles Bowen, had done much to prepare the ill-maintained guns in the forts at Cape Coast against the Ashanti threat, and her First Lieutenant, John King, was in command of a large force of Fantees 12 miles from the castle. Also serving ashore was Pringle Stokes, the First Lieutenant of *Owen Glendower*, who had garrisoned the Martello Tower outside Cape Coast Castle. In the roadstead, however, the Commodore had been suffering for two days from an illness described as "cholera morbus". On 4 September he was apparently recovering, and was walking from his cabin to the *Owen Glendower*'s quarterdeck with the elder of his two sons, who were midshipmen in the frigate, when he "was seized by an apoplectic fit" and died. Lieutenant Stokes returned on board to take command, assuming the rank of Acting Captain, and when *Bann* arrived from Rio on the following day Stokes immediately ordered her to Ascension with despatches, reporting the Commodore's death, for onward carriage to England.

Meanwhile, two of *Owen Glendower*'s boats, left behind by Mends in the Bight of Biafra under the command of Lieutenant Gray, were making a capture in the Bonny River. On 14 September they boarded the Spanish schooner *Fabiana*, finding that she had just landed most of her slaves and that all but two of her crew were ashore. Gray seized her, and two days later King Pepple of Bonny delivered on board the 120 slaves landed by the schooner. The prize and 170 slaves were taken to Sierra Leone by Mr Batt, Second Master of the frigate, but he found on his arrival that the Anglo-Spanish court would not try the case. Commodore Mends, the commissioners said, had not specifically authorised the seizure of the *Fabiana*. He had given only general orders to Gray. That contravened the terms of the treaty, and the seizure was therefore illegal. Stokes could not be allowed to correct the omission because he had not been in command of *Owen Glendower* at the time of the capture. It seemed that only the dead Commodore had the authority to do that! Not until the following summer was this impasse resolved by a letter from Canning. The Admiralty was perfectly content that responsibility for conducting the prosecution of *Fabiana* should devolve on Gray, and, with barely disguised irritation, Canning declared that the detention had been made by a competent officer, and, although the circumstances were irregular, they were not sufficiently so to preclude adjudication. He pointed out that not every deviation from the treaty necessarily invalidated a capture, and that the court should have proceeded in this case. He told it to delay no more. The commissioners quibbled on in self-justification, but finally condemned the *Fabiana* in October 1824.[27]

Stokes's period of command was brief. On 15 October 1823, the gun-brig *Swinger* (12), Lieutenant John Scott, with a ship's company turned over from *Thistle*, arrived at Cape Coast from England. She brought with her Commander John Filmore to take command of *Bann* from her temporary commanding officer, Saumarez.[*] *Bann* had just returned from Ascension, and Filmore, having taken command of her, immediately appointed himself, as the senior officer present, to command *Owen Glendower*. That was justifiable on the premise that the frigate was the most suitable ship from which to command the Squadron, although it did smack of self-aggrandisement. But Filmore then stepped well over the mark by hoisting a broad pendant, on the grounds that the Ashantis should see that the dead Commodore had been replaced. To fill the consequent command vacancy in *Bann*, he appointed Lieutenant George Courtenay of *Cyrene*, an officer already nominated for promotion to commander. On 16 October, *Driver* returned to Cape Coast from a short cruise to the southward, during which Commander Bowen had found the slave trade much diminished. *Driver*, however, was in a poor state, and Filmore reluctantly decided to send her to Ascension with supplies, and onward to Brazil for repairs. She remained at Rio until the end of January 1824.

Filmore found himself with a number of concerns: the perennial shortage of Surgeons in the Squadron, exacerbated by the invaliding of the excellent Surgeon

---

* Filmore had performed particularly well as Columbine's First Lieutenant in *Crocodile* in 1810–11.

Stewart of *Bann*; a leak in the frigate's magazine, probably thanks to rot; and the critical condition of Commodore Mends' elder son following a very painful operation. His most pressing worry, however, was the shortage of provisions at Cape Coast, and he was finally obliged to sail for Sierra Leone before Sir Charles MacCarthy had arrived at Cape Coast to take command of a field force against the Ashantis, and before the arrival of *Cyrene*. He sent Lieutenant Stokes to England with despatches in a departing transport, and left long-delayed orders for *Cyrene* to proceed immediately to Spithead. Finally, he told *Bann* to remain in support of Cape Coast Castle, unless Major Chisholm decided that her presence was no longer necessary, in which event she was to take a brief run around the Bights. Acting Commander Courtenay made the most of that opening. On his arrival at Cape Coast in the *Prince Regent* on 28 November, MacCarthy released *Bann*, and Courtenay sailed on 14 December for the Bights, called at Accra for stores and cruised the Bights until mid-February 1824.

On 14 December, *Owen Glendower* arrived at Sierra Leone on quarter rations of bread, having been without any at all for two days until she obtained a supply from an American vessel she had happened to meet. At Freetown, Filmore might have been cheered briefly by the arrival from England of two Assistant Surgeons, but any idea he might have formed that the Trade was in decline, from the several reports of reduced slaving activity in the Bights, was dashed by the news in Freetown that the traffic was heavy between Sierra Leone and Cape Palmas. On the passage, the frigate's bowsprit was found to be damaged by rot; it was fished during the stay at Sierra Leone, but the magazine leak remained, and Midshipman Mends had died at sea on 5 December.[28]

Bann took a look at the Lagos anchorage on her passage to the Bight of Biafra, and sighted a large brig which slipped her cable and made all sail to escape. After a ten-hour chase and a dozen rounds of gunfire, Courtenay boarded what he found to be the *Cerquiera* of Bahia, under "the insurgent flag of Brazil" and empty, but with a passport from the Emperor of Brazil for 761 slaves from Malembo. Her guns were loaded and primed, and her sides were greased to obstruct boarding. Courtenay was tempted to send her to Sierra Leone, but, with 25 men sick, he was too short-handed. Leaving the Slave Coast, *Bann* made for the River Formoso where, on Christmas Day, her boats found another Brazilian brig, flying the ensign of the Provisional Government of Pernambuco. She too was empty, and Courtenay decide to return at the new moon in the hope of catching her crossing the bar on the spring tide with a cargo of slaves. After visiting the Bonny and New Calabar, which had seen no slaver since late October, Fernando Po for water and yams, and the River Cameroons, Courtenay concluded that the Trade in the

Bight of Biafra was much in decline, and he returned to the River Formoso on 28 January 1824. However, his bird had flown. *Swinger* had found the brig too, and Lieutenant Scott had left a boat and crew with the British merchantman *Fletcher* to intercept her. In despair, the slaver had sailed without a cargo, and had struck heavily on the bar while leaving.

Bann returned to Lagos on 30 January, and discovered the *Cerquiera* still there, now joined by the ship *Minerva* and the schooner *Creola*, all under the novel green and yellow colours of Brazil. Believing that the Portuguese treaty still applied to Brazilian vessels, Courtenay boarded all three without the expected opposition, intending to bring on board *Bann* the crew of *Cerquiera*, and of the others if they were found to be slaving. The brig and schooner were ready to sail, except that they had not loaded slaves, and the three masters were ashore assembling nearly completed slave cargoes. The slavers' crews were taken to *Bann*, the *Cerquiera*'s forcibly so, and, after waiting 24 hours for the masters to return, Courtenay sent Lieutenants Armsink and Wilson into the river on 31 January with the pinnace, yawl and gig to extract the three masters and to attempt to bring off the waiting slaves. The slavers refused to leave, but the chief holding the slaves finally agreed to send off, on the following day, the 700 who had been paid for.

After a night in the boats, Armsink and his British merchant interpreter returned to the chief at dawn. The chief tried to detain them, but they escaped to the boats, followed by the Brazilians and a crowd of two or three thousand natives. As the boats left the beach, the crowd attacked them with muskets and bows and arrows, and three gun-batteries opened fire. The pinnace and yawl returned fire with grape and canister, and, having driven the enemy from a particularly threatening battery of nine guns, Armsink landed and spiked the guns. The boats returned to *Bann* with one seaman killed, a marine mortally wounded, Lieutenant Wilson severely wounded and five others slightly wounded.[29]

With 49 men on his sick-list and 100 prisoners to guard, Courtenay decided to make for Cape Coast, sending the three prizes to Freetown. The *Creola* was boarded on passage by a Spanish schooner which claimed that she had been chartered to take some of the cargoes of the three Brazilian slavers to Bahia, and the Spaniards plundered Midshipman Miller and his prize crew of everything they possessed, including the only quadrant on board. Miller navigated to Freetown on dead reckoning alone. The Anglo-Portuguese Mixed Commission declined to proceed against any of the three prizes, and released them; it was later heard that they had returned to Lagos to embark their slave cargoes.

On his arrival at Cape Coast on 16 February, Courtenay learned that *Owen Glendower* was away embarking troops, and that Acting Captain Filmore had been taken sick and left the frigate to return to England, handing over command to Lieutenant Woollcombe. A great deal more shocking was the news that the field force commanded by the Governor had been virtually destroyed by the Ashantis on 21 January 1824, and that, in leading a final bayonet charge with ammunition expended, Sir Charles MacCarthy had been killed.

* * *

Over the past couple of years there had certainly been a downward trend in the slave trade north of the Equator, although whether or not this was merely temporary remained to be seen. This welcome decline was evident not only in the reports from commanding officers in the Bights, and the poor haul of arrests by the British cruisers in 1823, but also in the recorded numbers of slavers arriving in Cuba: at least 26 in 1820 and 1821, reducing to ten and four in 1822 and 1823. The reasons were undoubtedly economic rather than the result of the Royal Navy's operations, although some degree of deterrence was perhaps achieved against the Spanish and Portuguese trades, both now illegal since 1820.

However, the majority of the traffic north of the Equator was now in the hands of French traders, and 24 slavers were reported fitting-out at Nantes alone in the six-month period after July 1823. They were working in large numbers between Senegal and the Gambia, on the Grain Coast and in the Bight of Biafra, scarcely troubled by their own nation's authorities and invulnerable to the British cruisers. It was their impotence against this illegal traffic, and against the insolence of the French slavers, which gave the British officers their greatest frustration. Probably it was this frustration as much as suspicion of false identity which led to the occasional illegal arrest of French slavers, to the understandable annoyance of the Foreign Office. The British cruisers were being taunted, and, to make matters worse, many of the French slave vessels were commanded by officers of the French Navy, who "delight[ed] in appearing in their Naval Uniforms when visited by the English". Almost equally infuriating was the continuing involvement of the Portuguese colonial authorities, at Bissau and on the Bights islands, in slave smuggling north of the Line.

The presence of a few American cruisers, and the cordial and mutually supportive relationship between the two navies, had been a most encouraging aspect of the past few years. US Navy deployments had tailed off, however; there was no American man-of-war on the Coast between the departure of *Alligator* in January 1822 and the return of *Cyane*, Captain Spence, in March 1823. Furthermore, negotiations between London and Washington had stalled. Repeated attempts by British ministers to induce the American administration to agree to some limited form of mutual Right of Search had come to naught. A recommendation for such an arrangement had been adopted by the House of Representatives in 1821, but rejected by the Senate, and the Secretary of State, John Quincy Adams, found constitutional objections to the idea. Adams, however, sent a draft treaty to England in January 1824, proposing to recognise slaving as piracy under the Law of Nations on condition that, although slavers might be seized by the authorities of any nation, they could be tried only by their own nations' courts. Britain agreed on every point, but, despite President Monroe's support, the convention was then so much amended by the Senate that Britain could not sign it. Meanwhile the slave trade under the American flag had dwindled, largely thanks to the US Navy's presence, but American slaving under foreign colours and with foreign papers, mostly Spanish, still thrived.[30]

Apart from one or two cases of legal hair-splitting, sometimes proving unjustly costly to captors, the Mixed Commission courts in Freetown had become more efficient and reasonable in their adjudications. The additions to the treaties gave useful expansion in the powers of the courts, but not until the Equipment Clause was accepted by Portugal and Spain would a satisfactory legal basis be achieved. Even then, the Mixed Commission courts could condemn only vessels and cargoes, not the officers and crews of slavers. These criminals could be tried only by the courts, and according to the laws, of their own countries. Experience showed that, British citizens apart, they had little reason to fear punishment.

The employment of the British cruisers on the African coast, never more than six of them, had been increasingly efficient in the hands of three successive commodores, but control had been regularly disrupted by the Admiralty's irrational requirement for the commodore to return home for the rainy season. Tactics had not significantly developed, although the flow of intelligence, largely through British trading vessels on the Coast and British consuls in foreign ports, seems to have improved. Of the 69 vessels seized in the past five years, only 26 had been captured in open water, and several of those had initially been detected inshore; the rivers and roadsteads were where the slavers could be most readily found and cornered, and boat-work remained a crucial element of the cruisers' tactics. Of the 69 seizures, 34 had been made solely or primarily by boats, and only 13 had resulted from open-water chases by the cruisers. Resistance from slavers, particularly the heavily armed and well-manned Spaniards, was becoming a common feature of the campaign, and unprovoked hostility from French ships, which knew perfectly well that they had nothing to fear from the British, caused great anger to the British officers.

Disease continued to drain the Squadron's manpower, and epidemics had occasionally rendered cruisers non-operational. Officers and men seem to have accepted with resigned patience the constant losses by death or invaliding, and it is a testament to the quality, resilience and professionalism of their crews that even the smaller vessels could absorb the loss of key officers and warrant officers to fever, hepatitis, ulcers, dysentery and other illnesses, and yet continue their work. If there was a threat to morale, it was not the conditions of service on the Coast, or even the shortage of cruisers to meet the impossible task given by a seemingly uninterested Admiralty; it was a feeling of impotence under the constraints imposed by inadequate treaties, and against the duplicity of European powers which had legislated against slaving but which were giving barely covert support to their slavers. As Commodore Mends saw the situation in June 1822: "the Traffic has not decreased; nor do I see how it can, whilst it is supported by European protection in the most open and avowed manner and defended by force of Arms." It was his opinion, albeit over-optimistic, that if the British cruisers were allowed free rein against the Trade, "it would in a short time be so cut up and harassed, as not to make it worth the risk, trouble, and disappointment".[31]

The Squadron also supposed, not entirely unjustifiably, that Ministers and their minions in London, and lawyers in Freetown, failed to comprehend the nature of the traffic that the cruisers were fighting. "It is necessary", wrote Mends, "to visit a Slave ship to know what the Trade is."

CHAPTER 8

A Lonely Furrow: Eastern Seas, 1824–8

Defarge: "It is possible — that it may not come, during our lives [...]
We shall not see the triumph."
"We shall have helped it," returned madame.

CHARLES DICKENS

IN A SHORT SPACE of time the West Africa Squadron had lost two valuable allies. The death of Sir Charles MacCarthy had deprived Sierra Leone of a Governor of outstanding ability, breadth of vision and devotion, and the Navy of a sterling friend. The Governor and the officers of the Squadron had held each other in admiration and respect, and had consistently afforded each other every possible support. Commodore Sir George Collier expressed the Navy's opinion in a letter to MacCarthy in 1820: "Your Excellency, who, I am proud to say, has upon all occasions afforded lasting and gratifying proofs to the Navy of your anxiety to anticipate every wish, and to relieve every want",[1] and, in one of several despatches of similar tone, the Governor reported to Lord Bathurst in 1823 that: "the assistance so readily given by Commodore Sir Robert Mends, by Captain Bowen RN and the Officers and Men under their Command, demands my gratitude – they are true Characteristic traits of their Profession."[2] It would be many a year after MacCarthy's death before this happy relationship between Squadron and Sierra Leone Governor was restored.

Of deeper significance was the loss of the help of the United States Navy in the suppression of slaving on the African coast. With the breakdown of anti-slaving treaty negotiations between Washington and London in 1824, to the dismay of President Monroe and thanks primarily to the refusal of the US Senate to accept even a limited Right of Search of American vessels, the USN cruisers on the west coast of Africa, apart from an occasional visitor to the Cape Mesurado settlement, were withdrawn. There was to be no progress in Anglo-American cooperation on the Coast for the next 20 years.

Attitudes remained no more encouraging among the other major slaving nations, despite public assertions of a determination to destroy the slave trade. France maintained only a small squadron for part of the year at Senegal, purely against the French slave trade, but showed no inclination to extend its activities into the Bights, by far the busiest French slaving region. Portugal's concern was to preserve the invulnerability of the Brazil Trade south of the Equator, and its colonial officials continued to connive in illegal slaving to the north of the Line. Spain's resolve to allow a vital supply of slaves to the plantations of Cuba, in defiance of its treaty obligations, was reflected in the determined obstruction in that island of all British efforts to interrupt the Trade, and in an absence of any Spanish anti-slaving cruisers. No other nation was willing to follow the example of the Netherlands in signing an Equipment Clause, and even the Dutch had no naval presence off West Africa, although they generally kept a cruiser or two in the Dutch Antilles. The only recent diplomatic success had been a Right of Search agreement with Sweden in 1824. Britain, in effect, now stood alone against the slavers.

* * *

Following the death of Commodore Mends, the West Africa Squadron had begun the year 1824 leaderless, and, with *Driver* refitting at Rio and the departed *Snapper* not yet replaced, its already inadequate strength was much depleted. Lieutenant Woollcombe, temporarily in command of *Owen Glendower*, took the frigate back to the Gold Coast from Sierra Leone in early February to join *Swinger* in support of the British forts against the Ashanti threat. There they were joined by *Bann*, returning from her Bights cruise in mid-February, and Commander Courtenay, as senior officer on the Station, exchanged commands with Woollcombe. This confusing (and inevitably unsettling) chopping and changing of commanding officers continued with the arrival of *Snapper*'s replacement, the teak-built ship-sloop *Victor* (18), another of the numerous Cruizer-class. Her first task was to convey to his post the new Commandant of the Ascension garrison, Lieutenant-Colonel Nicholls of the Royal Marines, and on finding *Owen Glendower* at Cape Coast at the beginning of April, *Victor*'s Commanding Officer, Commander Thomas Prickett, perceiving that he was senior to Courtenay, appointed himself to command the frigate. Lieutenant John Scott, who had performed with great credit in support of the Army during the past month, was taken out of *Swinger* to command *Victor*, and was replaced in the gun-brig by the elderly Lieutenant Herd, First Lieutenant of *Victor*.

Courtenay, displaced from *Owen Glendower*, was unable to reclaim *Bann*, which had been despatched to the Bights islands to water and thence to Sierra Leone to refit. To add to Courtenay's frustration, his ship under Woollcombe's command took a slaver on 10 March while on passage well to seaward in the Bight of Benin. She was the brig *Bom Caminho*, under Brazilian colours with a passport from the Provisional Government of Bahia for a slaving voyage to Malembo, and with 334 slaves embarked from Badagry. By coincidence, Woollcombe had boarded her from *Owen Glendower* in January off Elmina and had warned her against loading slaves on that coast. She was condemned to *Bann* without difficulty under the Portuguese treaty.

Victor too had escaped from the grip of the Ashanti War on 17 April to replenish the Ascension garrison. Five days later, off Lagos, she encountered a large brig flying a South American flag with a smaller brig and a lugger in company. All three cut their cables on sighting her, but *Victor*, after a chase, came up with the smaller brig, *El Vencador* of Bahia. Lieutenant Scott thereupon learned that she, with the lugger, had been seized by the larger vessel, the *Romano*, fitted-out by a company of Havana merchants with sixteen 18-pounders and a crew of 120, essentially as a pirate. *Romano* had since transferred part of *El Vencador*'s cargo into the lugger to trade it for slaves. Returning to Lagos to catch the lugger, Scott saw a boat full of men leave his quarry, which then weighed and ran ashore in heavy surf, immediately going to pieces. With the Brazilian brig in company, *Victor* headed for Princes, where Scott's only other commissioned officer, Lieutenant Turner, died.

While lying at Port Antonio, Scott made a second capture. On 8 May a small schooner-boat was seen behaving suspiciously on approaching the harbour, and the crew of *Victor*'s boat found 17 slaves on board her, loaded at the River Gabon. This little schooner, the *Maria Piquena*, owned in Princes probably by either the Governor or the odious Donna Maria da Cruz, was in no condition to make the voyage to Sierra Leone, so Scott put her slaves and papers into *El Vencador* before despatching the prize brig to Freetown. There the slaves were adjudged by the Mixed Commission and the vessel, regarded as not falling under any slave treaty, was dealt with by the Vice-Admiralty court.

Firm control of the Squadron was re-established in May 1824 on the arrival of the frigate *Maidstone* (42) with Commodore Charles Bullen CB. Bullen had gone to sea at the age of ten in 1779, had served in the *Ramillies* at the Glorious First of June, had narrowly escaped being murdered by the Nore mutineers while First Lieutenant of *Monmouth*, and had so distinguished himself at the subsequent Battle of Camperdown that he was promoted to commander in 1798. While in command of the sloop *Wasp* he had been instrumental in saving the infant settlement of Sierra Leone from attacks by natives in 1801–2, and for that service he was made Post. At Trafalgar he was Flag Captain to the Earl of Northesk in *Britannia*, and he was then continuously in command at sea, mostly in frigates, until 1817. He had joined *Maidstone* from half pay. Reputedly one of the most popular officers in the Navy, and an old friend of Robert Mends, the vastly experienced Bullen was an ideal choice for the arduous command on the African coast.

The new Commodore's orders directed him to give such support to British settlements and protection to British commerce as was required, and to prevent slaving by British subjects and by the subjects of the Kings of Spain, Portugal and the Netherlands. He was to send his cruisers to Ascension when necessary to "recruit and refresh the people", and he was to remain on station until relieved, unless sickness, shortage of stores, etc. made it necessary to return to England in *Maidstone*. As with all the cruisers, he was to complete the frigate to war complement with Kroomen.[3] Arriving at Cape Coast on 24 May, in company with the refitted *Bann*, his first action, in accordance with his orders, was to send *Owen Glendower* home via Ascension. She was in a poor material state, and probably demoralised by frequent changes of commanding officer. He also dispatched *Driver* to England. She had achieved little since her return from Rio, other than to carry Lieutenant-Colonel Sutherland from Freetown to Cape Coast as Commandant and Deputy Governor on the Gold Coast, together with reinforcements to resist the Ashantis, and she too was in a defective state.

Commander Courtenay then rejoined *Bann* to take her north in support of the Gambia settlement and the Portendic gum trade, continuing responsibilities which regularly drew a cruiser away from interdiction of the slave trade.* Woollcombe shifted to *Swinger* in place of Herd to continue the gun-brig's work with the Cape Coast land forces, but only until early July when John Scott returned to Cape Coast in *Victor*. Scott and Woollcombe then exchanged commands, and Scott found himself back in his own brig. *Victor* sailed on 6 July for Accra to meet the Commodore and then to cruise the Bights, while *Swinger* faithfully maintained her station between Cape Coast and Elmina, to guard the Army's seaward flank while the crisis of the Ashanti War passed and the threat to the forts receded.

On 4 July, after an unsuccessful chase of a slaver on passage, the frigate *Thetis* (46), Captain Sir John Phillimore,† had arrived off the Castle in a flurry of energy, enthusiasm and goodwill, bearing reinforcements for the Royal African Corps and taking the diminutive *Swinger* under her wing. The two men-of-war anchored within grape range of the beach to defend the east and west approaches to the

---

* The anchorage of Portendic, through which gum arabic was exported, lay 225 miles north-north-east of Cape Verde. The town of Portendic became Nouakchott, the capital of Mauritania.
† *Thetis* was a sister ship of *Trincomalee*, now renovated and lying at Hartlepool.

castle, armed and manned boats, and landed men to crew the guns of the castle and its outlying tower. In this they were joined by the merchantman *Woodbridge*, most of whose crew had volunteered their services. The Ashantis were defeated in a general engagement at some distance from the Castle on 11 July and then retreated, allowing re-embarkation of the sailors four days later. Many of the returning men were suffering from fever, including all ten from the tower, of whom a midshipman and two seamen died. On 22 July, having bestowed all the largesse in her power, *Thetis* sailed for England. On the Commodore's arrival at Cape Coast two days earlier, Lieutenant-Colonel Sutherland gratefully commended Captain Phillimore's cool decisiveness during the crisis.

Maidstone had returned from six weeks in the Bights, having cruised as far as the River Cameroons and visited the Slave Coast, Bonny and Fernando Po. Fernando Po had caught Bullen's eye as a potential base for the Squadron, and he reported with enthusiasm on its strategic position, natural resources and sheltered anchorage. But at Bonny there had been tragedy: while the boats were crossing the bar a roller had capsized the barge, with the loss of ten men. The survivors, rescued by the gallantry of master's mate Thomas Thompson, had been treated with "the greatest kindness, attention and hospitality" by the Liverpool merchantmen trading for palm oil in the river.[4]

The cruise had brought the frigate no seizures, and Bullen, adding the views of the British palm-oiler masters to his own observations, concluded that the brisk slave trade in the Bight of Biafra was being carried out almost entirely under the French flag, although he was under no illusion that the mere hoisting of the French white ensign was proof of nationality. He proposed to the Admiralty that any British cruiser finding a loaded French slaver should, without entitlement to reward, take her to Goree for trial, but there was no hope that France would agree to such an idea. All that could realistically be achieved was a mild visit to ascertain nationality. On the Slave Coast the Commodore boarded several Brazilian and Spanish slavers waiting for cargoes, and lamented that "I am forced to become little more than an idle spectator, & am obliged to allow them to triumph in their villainy."

On his return, Bullen found a serious shortage of provisions for his ships at Cape Coast, thanks to the delayed arrival of the routine Navy Board Transport, and he complained of the inadequacy and poor condition of the Squadron's store in the Castle, previously a slave cell. He recommended that it should be replaced with a hulk moored in the roadstead, but Their Lordships decided to delay a decision on this sensible idea, and no improvement was made. The Commodore's concern on the matter of stores was not allayed by his discovery on arrival at

Sierra Leone on 20 August that the Squadron's depot of precious naval stores and provisions had been raided. HMS *Brazen*, passing through Freetown, had helped herself to rum and bread, and the half-pay Lieutenant Austin, appointed by the Acting Governor to command the new colonial brig *Prince Regent*, had fitted-out and stored his vessel at the Squadron's expense.* Bullen's anger was then exacerbated by chastisement from the Admiralty following a complaint from Lieutenant-Colonel Nicholls at Ascension that the Commodore was neglecting the island's needs, an accusation which Bullen strongly refuted.[5]

Amid the competing operational demands and logistical obstacles, the campaign against the slavers was suffering, and *Victor* alone had been patrolling the Bights through August and early September. Her first success of the cruise came on 11 August. At 0700 she had found the Brazilian brigantine *Dianna* and had chased her in the eye of the wind until sunset, when she finally boarded her 100 miles west of Princes. She had 143 slaves on board from the Rio Formosa, having already lost nine to smallpox, and in her log Malembo had acquired a fictitious latitude of 4° 30′ N instead of 5° S. Woollcombe wrote that "of all vessels I was ever onboard of this was in the most deplorable condition", and most of her crew were drunk by the time she was seized. He kept the prize close alongside all night, and in the morning cleaned and whitewashed her before sending her to Freetown, where she was condemned.† Care was taken that no one went on board who had not already had smallpox.

On 18 September, after another five weeks of fruitless cruising, *Victor* took her fourth prize of the year, 70 miles south-west of Princes. This was another Brazilian brigantine, the *Dos Amigos Brazilieros*. Bullen later went on board her at St Thomas, probably the first loaded slaver he had seen, and, like all his predecessors, this tough veteran was appalled:

> the filthy and horrid state I found her in beggars all description. Many females were in an advanced state of pregnancy & several had infants of from Four to Twelve months of age. All were crowded together in one mass of living corruption.[6]

In fact, the 260 slaves from Badagry, although close-packed, were relatively healthy. The same could not be said for *Victor*, or indeed for *Maidstone*, which had returned

---

* The previous colonial schooner had been condemned as unseaworthy in early July and the brig had been purchased shortly thereafter. The colony showed its usual lack of imagination in naming her.
† Whitewashing the interior of a vessel was believed to inhibit disease.

to the Bights after checking that all was quiet at Cape Coast. Both ships had been attacked by fever, and although the crews were recovering, they had lost between them a lieutenant, three midshipmen and five men. Two more lieutenants, "mere skeletons", had to be invalided, and Woollcombe was convalescing.

These two ships were not the only ones to suffer from fever contracted at Cape Coast. *Swinger*, finally released from her Gold Coast vigil of more than seven months, sailed on 11 September for Freetown to refit and to check her bottom at Bance Island. She arrived without her Commanding Officer. On 26 September, to Bullen's "deepest regret and sorrow", her "active, zealous and valuable Commander", Lieutenant John Scott, had died at sea of Cape Coast fever.

Victor was sent to cruise south of the Line until the end of October, touching at Ascension, to recover her health fully. Although she was powerless against the Portuguese and Brazilian traffic in southern latitudes, she could watch for any Spanish slaving which had, by the 1817 treaty, been illegal since May 1820. She was then to call at Sierra Leone before heading north for the Gambia and the Cape Verde Islands, examining the northern slave rivers en route. *Bann* had left that northern sector in July, having discovered incidentally that all but two of the French squadron had retreated to France for the rains, and had returned, via Sierra Leone, to the Bights in late August. She had to wait until 23 October before taking the only prize of her four-month cruise. A hundred miles or so north-west of Princes she intercepted the Brazilian brig *Bella Eliza* with 371 slaves. The slaver had a passport to take a cargo from Malembo, and her master said that he had loaded at "Northern Malembo". Asked to point out this port on his chart, he put his finger on Lagos. At Freetown the brig was satisfactorily condemned and the slaves emancipated.

After sending *Victor* south, the Commodore remained in the Bights, and very soon made his own capture, *Maidstone*'s first and only arrest of 1824. On 26 September, just to the south of Princes, he chased and caught the Brazilian brig *Aviso* with 465 slaves taken from Badagry, consort of *Victor*'s prize *Dos Amigos Brazilieros*. Bullen kept the slaver with him to provision her at Princes, and then towed her to Annobón. There he found another loaded Brazilian slave brig, but, being south of the Equator, had to let her go. On 14 October he cast off the *Aviso* for Sierra Leone, where she was condemned, but by the time the slaver reached Freetown she had lost 34 slaves to dysentery and the remainder were in a wretched state. *Maidstone* then made for Ascension, where Lieutenant-Colonel Nicholls was distressed to learn that his letter to the Admiralty about the inadequacy of supply arrangements for the island had been misinterpreted to Bullen's cost, and friendly relations between Commandant and Commodore were re-established. The frigate landed a bull and a cow as earlier requested by the garrison, and Bullen was impressed by the availability of meat from wild goats and guinea fowl and by the quantity of vegetables being grown on Green Mountain.

Following Sir Charles MacCarthy's death, Daniel Molloy Hamilton, as senior member of the Sierra Leone Council, had assumed temporary government of the colony, and after meeting Bullen in May he had told the Colonial Office that the Commodore had "in the most polite and gracious manner offered to co-operate with me in any measures that might be requisite for, or prove beneficial to His Majesty's possessions on this Coast". A new Governor, Major-General Charles Turner, was appointed in June, but it appeared that Turner was in no haste to take up his post. The months passed as he arranged his affairs in London, recruited several hundred more men to his newly inherited regiment, the Royal African Corps, demanded a steamboat for the colony, and argued with the Colonial Office about his pay and allowances, his uniform and the unsuitability of Government House. The Commodore was anxious to be at Freetown for Governor Turner's arrival, and he returned there at the beginning of November to refit and replenish *Maidstone* and to wait. A month later there was still no sign of the Governor, and Bullen decided to delay no longer. It was a wise decision; Turner did not appear for another two months.

The diplomatic campaign against the Trade took a further small step forward on 6 November 1824 with the signing in Stockholm of a treaty between Britain and the Kings of Norway and Sweden. It agreed a mutual Right of Search within certain geographical limits which included the Atlantic Ocean south of the latitude of Cape St Vincent. Mixed Courts of Justice were to be formed, one at Freetown and the other on the Caribbean island of St Bartholomew. Each nation was to be allowed a maximum of 12 authorised cruisers. Vessels might be seized and condemned if there was clear and undeniable proof that one or more slaves had been embarked on the voyage in question for the purpose of the Traffic, and if she was found within one degree westward of the coast of Africa between 20° N and 20° S. It was a pact of no great or immediate significance to the struggle at sea, but it was worthwhile in denying potential false identities for slavers.

Bullen was concerned at a resurgence of the slave trade in the Rio Pongas and Rio Nunez to the northward, conducted by "very fine, fast sailing" schooners from Havana, armed and manned for piracy as well as slaving and enabled by their superior sailing qualities to evade detection. He had sent *Victor* north in December to visit the Gambia, the Cape Verdes and the northern rivers, where she found a number of empty Spanish slavers. *Maidstone*, however, headed south again for the Bights, which Bullen considered to be the Squadron's highest priority, although he resolved to station a cruiser off the northern rivers whenever he could. He paused to check the Sherbro and Gallinas rivers in passing, finding several French slavers and disturbing some Spaniards. All of the Spaniards found by *Victor* and one of those now in the Gallinas had already been reported to Bullen by the Admiralty, thanks to the flow of information from the British Judge in Havana.[7]

Swinger, now commanded by Lieutenant Edward Clerkson from *Maidstone*, had preceded the Commodore by a fortnight in returning to the Bights, losing her Master to fever shortly after leaving Sierra Leone. On 14 January 1825, 100 miles north-west of Princes, she captured the Brazilian schooner *Bom Fim* with 149 slaves. A number of private individuals had made small investments in this cargo, including four ladies in Brazil who had "particularly pointed out the qualities they wished their respective Negroes to possess". *Bom Fim* was condemned without difficulty, but Clerkson's next encounter was less satisfactory.

At dusk on 19 February *Swinger* anchored among four unidentified vessels lying off Lagos. No sooner had she come to than a large brig hailed her and then fired a shot over her. Suspecting that she was one of the Spanish pirates infesting the Coast, Clerkson replied with four double-shotted guns, and when the brig's boat came alongside full of men he took them prisoner. At daybreak he closed the brig, which was under Spanish colours and clearly ready to resist an attack, and, after a warning, fired four broadsides into her and boarded without opposition. She was the *Alerto* of Havana, about which the British traders in the Bights had been complaining for some months, as Bullen had already told the Admiralty, and which had taunted *Swinger* when Clerkson had boarded her in December. The *Alerto* was armed with sixteen long 12-pounders and numerous small arms, had a crew of 86 and was ready to load slaves. Clerkson threw overboard all her guns and other weapons, but had to release her.

Their Lordships were not pleased. Perhaps in conjunction with the Foreign Office, they remained in stubborn denial of piracy on the Coast, in the face of reports from successive commodores. They "could not but think" that Clerkson was "guilty of unjustifiable aggression" accompanied, and this was probably the crux of their concern, by "circumstances of illegal damages & destruction of property". The Commodore leapt robustly to Clerkson's defence, giving his opinion that "his prompt and decisive mode of acting will be attended with the greatest benefit to the commerce on this Coast, which has lately been disturbed by these lawless marauders".[8]

In Britain the abolitionists, now campaigning against slavery itself, had gained a further hardening of legislation against the slave trade, and on 1 January 1825 there came into force an Act of Parliament passed in March 1824. It declared that British subjects convicted of slaving would be deemed "guilty of Piracy, Felony and Robbery" and would "suffer Death without benefit of Clergy and Loss of Lands, Goods and Chattels as Pirates, Felons and Robbers on the Seas ought to suffer". In June 1824 all the earlier Abolition Acts had been repealed and replaced by a single Act which redefined the various offences associated with slave trading and their related penalties. As well as the more direct crimes of purchasing, carrying away, etc., subsidiary activities such as those involving trade goods, fitting-out slave vessels and providing insurance were included. No prosecution was ever brought under the March "piracy law", and it certainly had no bearing on the Navy's efforts, but the second Act was of greater interest to the cruisers. In addition to its main purpose, this Act realigned the bounty allowed under British law with those agreed in the various treaties, decreeing a payment of £10 for each man, woman or child emancipated, to be distributed as for prize money.* In addition, a Commodore First Class became entitled to an admiral's share of prize money.

With *Driver*'s departure from Africa in May 1824 the Squadron had been reduced to four cruisers, and her replacement, the Sixth Rate *Atholl* (28), Captain James Murray, was a welcome reinforcement when she reached the Coast in late January 1825. Her arrival was reported to Bullen by *Victor* on her return to the Bights on 25 February, together with news from Sierra Leone of General Turner's arrival on 5 February. With *Bann* again cruising to the northward, and *Atholl* required by the Admiralty to call at Cape Coast and Freetown before making for Ascension with stores, the Commodore left *Victor* and *Swinger* to cover the Bights while he sailed in haste for Sierra Leone in *Maidstone* to call on the Governor.

His arrival happened to coincide with that of *Atholl*, returning from her visit to Cape Coast. On passing the Gallinas on 7 May she had opened her campaign by capturing the Spanish schooner *Española* with 270 slaves. Even more satisfactory to the Commodore was finding in harbour two newly joined cruisers: the ship-sloop *Esk* (22), Commander William Purchas, and the gun-brig *Conflict* (12), Lieutenant John Chrystie. This brought the Squadron briefly to what appears to have been its intended strength of a frigate, four Sixth Rates or sloops, and two

---

* See Appendix I for comparative values of sterling.

gun-brigs. However, Bullen was acutely disappointed to discover that Governor Turner had sailed for Cape Coast four days earlier.

Wasting no more time, the Commodore took *Maidstone*, *Esk* and *Conflict* on a short cruise to the north, and when he returned to Sierra Leone in early May he left *Conflict* to support the Portendic gum trade and the Gambia settlement. The Governor was again in residence at Freetown, but Bullen, to his astonishment, was initially refused access to him, and Turner's first letter to the Commodore accused him of a "Suspension of my Authority by you" over the control of Navy Board Transports visiting the Coast. On Bullen's general instructions, the agent in the transport *Cato* had declined a demand that he should take to England a detachment of troops in his already fully laden vessel, and the Governor not only declared that this refusal constituted "an insult to the entire British Army" but also claimed, quite wrongly, that the transports were his to control. The Commodore was not a man to be bullied by the likes of General Turner, and his response, although mild and measured, was firm. When Turner complained to the Colonial Office, Bullen was unusually well supported by the Admiralty. He also defused a potentially damaging squabble about the Squadron's watering place at Freetown, which had involved the Governor.

Others could not defend themselves so readily. At Cape Coast and at Sierra Leone Turner found fault with almost every aspect of affairs. He harshly criticised the past and present performance of most of his subordinates, ruthlessly dismissed the Acting Commandant of Cape Coast, the long-serving but now fever-debilitated Lieutenant-Colonel Grant, and he was even disparaging of the achievements of Sir Charles MacCarthy. He was also foolish enough to arrest and clap in irons the master, mate and agent (a lieutenant of the Royal Navy) of a Navy Board transport lying off Cape Coast with whom the Castle officers had had a disagreement.* At the same time he gathered around him a small loyal coterie, including Lieutenant Austin of the *Prince Regent*, whom he wanted to be appointed to the steam vessel he had repeatedly demanded to replace the colonial brig. Although not without ability, Turner was an arrogant and self-serving man, intensely jealous of his own status, and he took care to impress on the Colonial Office that, having found his new bailiwick in a deplorable state, he alone was setting matters to rights. Cooperation of the sort which Bullen had offered to Hamilton seemed far from the new Governor's mind.

---

* This brought forth a very starchy letter from the Admiralty to Lord Bathurst at the Colonial Office: the Governor had "so far forgot himself, and was so little informed as to the extent of his authority", and Their Lordships "entertained no doubt that a prosecution [...] may be instituted and that damages would be recovered in a Court of Law". They advised that HM Government should take steps to prevent a recurrence.

At a loss as to why his efforts to establish a mutually supportive relationship with Turner had been abortive, the Commodore sailed on 12 May, with *Esk* in company, to return to the Bights. Off the Gallinas the two ships and their boats disturbed a horde of French slavers, chased the Havana schooner *Altrevida* for three hours before discovering that she was empty, and seized a schooner flying a Dutch ensign. This was the *Bey*, which had been a Virginia pilot boat and was manned mostly by Americans, and her Netherlands colours had been fraudulently obtained to cover the property of other nations. Under the Anglo-Netherlands Equipment Clause, Bullen sent her into Freetown where she was condemned. The remainder of the Grain Coast yielded only a few French and empty Spaniards, and Bullen found that the French brig-of-war *Dragon* had recently arrested three French slavers in the area, only for them to be released by the Senegal court and return to the Trade even though *Dragon* was still patrolling.

Maidstone found *Atholl* completing provisions at Cape Coast when she arrived on 7 June, and intended to replenish with bread, sugar and cocoa herself, but discovered that all of these provisions at the Castle, provided for the sole use of the Squadron, had, on the orders of the Governor, been issued to troops in transports lying in the roadstead. With his ship on two-thirds allowance of bread, Bullen headed immediately for Accra, concerned that he might be obliged to return to Sierra Leone for replenishment.[9] In the Accra anchorage he found *Victor* and *Swinger*. Their news brought no encouragement. *Atholl* and *Victor* had both visited Ascension with stores, but otherwise the three cruisers had been patrolling the Bights without any success since *Maidstone* had left at the end of February. The French slave trade was extensive and increasing, there was sickness in *Atholl* from which two lieutenants and the Master had to be invalided, and Lieutenant Clerkson of *Swinger* had died of fever off Princes on 2 April.

Victor, after an unusually short period of 16 months on station, was ordered home, and Bullen, after putting Lieutenant Poingdestre of *Maidstone* in command of *Swinger*, sent the gun-brig to Sierra Leone for refit and victuals before making her way north for the Gambia, Portendic and the Cape Verdes to replace *Conflict*. On 19 June the Commodore sailed from Accra for the Bights with *Esk* and *Atholl*, stationing them in what seemed to be the most advantageous positions for interceptions, while *Maidstone* scoured the anchorages. At Lagos he almost caught the pirate *Alerto*, now rearmed by the Governor of Princes after her encounter with *Swinger*, but it was not until well into July that the next capture was made. On 17 July *Esk* took her first prize about 100 miles west-south-west of Cape Formoso: the Brazilian sumaca *Bom Jesus dos Navigantes*. She had the usual passport from Bahia to Malembo for slaves, and, as was increasingly the case with Brazilian slavers found in these waters north of the Line, her passport also gave her permission to call at the Gold Coast, a permission for which there was no legitimate justification. Her master claimed not to understand navigation, and thought that Malembo was close to Benin. He had certainly picked up his 280 slaves not at Malembo but on the Slave Coast, and his vessel, a dull sailer like most of the Brazilians, was condemned.

On 24 July *Conflict*, relieved at the Gambia by *Swinger*, joined the Commodore near St Thomas, and a week later *Maidstone* detained the Dutch brig *Z* off the River Sombrero, just to the west of Bonny. Bullen had boarded her on the evening of 29 July but, uncertain of the meaning of the longitude limit given in the Dutch treaty, he kept company with her until she made the land and then seized her two days later. He had initially hailed her after dark in French, which had tricked her into admitting Dutch nationality, but Bullen was convinced that she also had false French papers. He found not a Dutchman on board and her papers indicated that her trade goods were owned by British merchants in the West Indies.[10] Although she was empty she was fitted for slaving, and was condemned, subsequently to be bought by the Governor as a colonial brig. The prize crew was obliged to remain ashore in fever-riven Freetown for an appreciable time after the adjudication, with a result which accounted for most of the eight seamen, three marines and an Admiralty Mate lost by *Maidstone* during that rainy season.*

Bann had sailed for England in April 1825 after very nearly three years on the Coast, and her replacement joined the Commodore at Accra on 25 August. The newcomer was the sloop *Redwing* (18), whose only previous experience against the slavers was in the embarrassing affair of *La Sylphe* in 1819. She was now rigged as a ship rather than a brig, and commanded by Commander D. C. Clavering, who had previously been on the Coast in command of *Pheasant* in 1822. The first of her misfortunes on this deployment was in striking a coral reef off Boa Vista on her passage south, losing her false keel and starting a leak. It was not to be her last.

Atholl was the next to make an arrest, taking the Dutch schooner *La Venus* just off Cape Formoso on 1 September. She had no slaves, but fell under the Equipment Clause and was sent to Freetown. To Bullen's satisfaction, Murray's search of her had revealed a set of French papers for the Dutch *Z* which that

---

* Mates, or, more properly, master's mates, were junior warrant officers of two different categories. There were those who were assistants to the Master and aspired to become masters themselves, and there were those who were midshipmen who had passed the examination for Lieutenant and were awaiting their commissions, often for many years. The former category continued to be called master's mates, but to distinguish the latter category the "master's" was omitted. Admiralty Mates belonged to the latter category, having been entered as Admiralty nominees in the Royal Naval College rather than as Captains' nominees either at the College or, more likely, at sea.

vessel had left behind in the West Indies. A week later the prize was intercepted by two strangers: a brig and a schooner. The schooner showed Spanish colours, but the brig hoisted a red ensign and a pendant,* and each fired a shotted gun at *La Venus* which, having no British colours on board, wore a Dutch ensign. *Atholl*'s Prize Master was detained and maltreated for two hours before being taken back to *La Venus* and ordered to remain in station between these two pirates, but he managed to slip clear during the night. The brig was probably the piratical slaver *Don Pedro*, heavily manned and armed, and believed to be carrying 600 slaves.

Maidstone chased a slaver for 11 hours off Princes on 4 September, finally bringing her to with gunfire, before discovering to her frustration that she was the French ship *Orphée*, with 698 slaves from Old Calabar. The slaver was about to call at Princes for water and farina, further evidence of the support the Portuguese were giving to the Trade. The next legitimate seizure was made jointly by *Esk*, *Atholl* and *Redwing* on 9 September 130 miles south-west of Cape Formoso. Their prize was the Brazilian schooner *União*, carrying the usual passport for Malembo and 361 Africans from the Slave Coast. She was badly overcrowded, and there was an appalling loss of 112 slaves during her passage to Sierra Leone. The three cruisers then joined *Maidstone* and *Conflict* at Princes before the Commodore sent them on their various ways: *Atholl* to Sierra Leone to replenish, thanks to the Army's misappropriation of the Squadron's stores at Cape Coast; *Esk* and *Conflict* to cruise the Bights for the remainder of the year; *Redwing*, a poor sailer since her grounding, to look at the Bonny, Old Calabar and Cameroons rivers; and *Maidstone* to the Gabon and back to the Bight of Benin.

After a two-month dearth, *Maidstone* made a commendable capture on 29 September off Lagos. The Spanish schooners could usually show a clean pair of heels to a frigate, but Bullen, after "a most determined and hard run" of nine hours, and every exertion and manoeuvre to escape by his prey, especially after dark, caught the *Segunda Gallega* with 285 slaves. *Redwing* seemed also to be in luck. Her boats went into the Old Calabar River on 5 October, and on the following dawn they found two Spanish slavers working downriver. The Spaniards opened fire with grape and musketry, and the boats replied for 15 minutes as they closed. Lieutenant Card then boarded the first vessel, sword in hand, and rapidly carried her at a cost of two men wounded. The crew of the second vessel took to their boats and escaped into the bush. The prizes were the *Teresa* and the *Isabella* from Santiago de Cuba, with 248 and 273 slaves respectively. Meanwhile, Commander Clavering had heard of another slaver about to sail from the Cameroons and, after despatching *Isabella* to Sierra Leone, he set off to find her, with *Teresa* in company.

---

* A pendant at the main masthead indicated a man-of-war.

Card and the boats successfully flushed their new quarry on 11 October, and she was caught after a chase of several hours by the *Teresa*, commanded by Lieutenant Wilson. This third prize was the brigantine *Ana*, also from Santiago de Cuba, with 106 slaves, but satisfaction was short-lived. At 0230 on 19 October, *Teresa* was blown onto her beam ends by a heavy squall and sank with the loss of 193 slaves, a Spaniard and four of the prize crew. *Redwing* was unaware of the disaster until the following morning, when she rescued Lieutenant Wilson and a handful of survivors from floating spars. Fortunately, 50 of her overcrowded slaves had been shifted into the *Ana* on the day before the tragedy, and *Ana* reached Freetown in safety to be condemned, having, however, lost 68 slaves to disease on passage.

Redwing returned to Freetown for refit in December with 45 slaves, probably transferred from the *Isabella* before parting company, but found no sign of *Isabella* herself. It was not until April 1827 that the story of the slaver's disappearance emerged, related by the master of another slaver prize. On passage to Sierra Leone, the *Isabella*, commanded by Mr Jackson, master's mate, had encountered the Brazilian slaver *Disuniao* and captured her. Jackson split his prize crew of another master's mate, an Assistant Surgeon, ten seamen and marines, and two Kroomen between the two vessels, and neither was able to defend herself when they were intercepted by the large Spanish piratical brigantine *Gavilina*,* flying French colours. Both were taken after a sharp action, and the *Disuniao* was plundered of slaves and everything of use before she was allowed to make for Rio, where she arrived with only five crew left alive, all mutilated. The *Isabella* reached Cuba on 29 November under the name of *Juanita*. All of the British prize crew were either killed in the action or subsequently murdered.

British vessels had not for many years been involved in the shipping of slaves, but in November 1825 Mr Pennell, the British Consul at Bahia, expressed to Canning his concern that British merchantmen were being chartered to carry trade goods for the purchase of slaves who were to be loaded in foreign vessels. The brig *George and James* of London, for example, had sailed in March from Bahia with a British master, bound for the coast of Africa north of the Line. She was laden with tobacco, which Pennell strongly suspected was destined to purchase a slave cargo for the smack *Caridade*, whose owner had chartered the brig. The master of another British brig, the *Grecian*, had consulted Pennell about a similar charter and, on the Consul's advice, the master had rejected the offer. It

---

* This appears the most likely of various versions of the name given by several sources, but the vessel appears to have operated under a variety of identities.

was Pennell's correct understanding that such a charter was illegal under British law, as long as the master knew that the proposed cargo was destined for the slave trade, but that if that knowledge was denied it was virtually impossible to prove a case against the master. In fact, unbeknown to Pennell, the *George and James* was condemned in the Vice-Admiralty court at Sierra Leone on 17 October 1825, having been arrested off Whydah by *Atholl*. She certainly would not have been the only culprit.[11]

A replacement for *Bann* had by now arrived: the Sixth Rate *Brazen* (18), Captain George Willes. It was a third deployment to the Coast for Willes; he had last been there in *Cherub* in 1818, and, probably still smarting from the disgraceful case of the *Descubridor* in that year, he fervently hoped it would be his last. Passing the Gambia in mid-October, *Brazen* had fallen in with a badly debilitated *Swinger*. The gun-brig was short of provisions and medicines, most of her people were suffering from fever and dysentery, and, while she had been in the river, fever had claimed six men and Lieutenant Poingdestre, the third *Swinger* Commanding Officer to die in 12 months. Finding *Atholl*, and much sickness, at Freetown, Willes ordered her to rejoin the Commodore at the Gold Coast or in the Bights, taking the several redundant prize crews there back to their parent ships and away from the dangers of the town. *Brazen* sailed with *Swinger* on 27 October in similar search of *Maidstone*, and 50 miles south of Cape Mesurado on 4 November she seized "by chance" the 50-ton Spanish schooner *Clara* (or *Clarita*) with 36 slaves. Claiming her to be "the fastest and finest Schooner Brig out of Havannah", Willes bought her on her condemnation and manned her as a tender, naming her *Black Nymph*.*

At Cape Coast *Brazen* found that sickness had created havoc among the troops there and had delayed movement of the Ashanti army, bringing the still extant Ashanti War to a state of quiescence. At Accra she just missed Bullen, and Willes was concerned to discover that the slave trade was now being conducted within half-gunshot of the dilapidated British and Dutch forts. When *Brazen* captured the Spanish brigantine *Ninfa Habanera* off the port with only five slaves on 17 November her master admitted that he had already loaded 50 slaves at Dutch Accra and then added them to a cargo he had waiting at Little Popo. Willes later arranged the loading of this cargo of 231 slaves into the schooner, but, because they had not been on board at the time of capture, they could not be condemned by the Mixed Commission court, and therefore brought no bounty to *Brazen*. They were, of course, freed, and the vessel was condemned on the

---

* There is some doubt about the naming of this vessel. Willes may initially have called her *Brazano*.

admission that during the course of the voyage she had illegally taken on board the 50 Accra slaves.[12]

After landing the explorer Clapperton at Badagry, *Brazen* made for Princes where, on 19 December, she encountered the English ship *Malta*. Willes found that this vessel had been in the River Danger in November and had sold to a Spanish slave trader the four female Africans she was holding on board as hostages for a cargo she had landed. With the permission of the Portuguese authorities, Willes arrested the *Malta* and sent her to Sierra Leone, where she was condemned by the Vice-Admiralty court. The vessel's master, Captain Thomas Young, was taken to England for trial, but the Admiralty court acquitted him of slaving.

Brazen, still without finding the Commodore, returned to cruise the Slave Coast, and was rewarded on 27 December by the capture of another Spanish schooner, after a 48-hour chase: the *Iberia*, with 422 slaves. Another month or more of cruising the Bight of Benin brought no further success for *Brazen*, but on 22 January 1826 the *Black Nymph* began to earn her keep for her parent ship. Under the command of Lieutenant Baldwin Wake Walker, she was on passage from Sierra Leone with dispatches for the Commodore when she caught the schooner *Vogel*, under Dutch colours. The slaver was empty, and was carrying French papers (showing the name *L'Oiseau*) as well as Dutch, but the Anglo-Dutch court condemned her to *Brazen*, *Black Nymph* being regarded as "*Brazen*'s boat". Wake Walker then engaged the Dutch vessel *Van Tromp* for 20 hours off Cape Palmas, trying to dismast her because, after manning the *Vogel*, he had too few men left for a boarding, but *Van Tromp* escaped by shipping her sweeps.

Unlike *Brazen*, the sickly *Swinger* had found the Commodore at Cape Coast, and Bullen sent her to Ascension to recover her health. Before she left the Bights, however, and still under the temporary command of Acting Lieutenant J. C. Giles, she took the Brazilian brig *Paqueta de Bahia* off Whydah on 22 November 1825. The slaver's master confessed that he had loaded his 386 Africans at Whydah, and the brig was condemned. *Swinger* then sailed for Ascension with a new Commanding Officer, Lieutenant George Matson, who had been First Lieutenant of *Esk*. At the island her men were cared for with great kindness by Lieutenant-Colonel Nicholls and his people. While the ship's company was accommodated ashore in tents and reinvigorated with fresh food, the garrison built rafts of casks and spars which enabled Matson to heave the gun-brig down to her keel and scrub her bottom in the open roadstead, an evolution that Bullen admitted he would have deemed impracticable.[13] When she sailed in mid-January 1826 for a cruise off the northern rivers, *Swinger* was again in fine fettle.

Finding no further success to leeward *Maidstone* returned to Sierra Leone in mid-December 1825 for a fortnight's refit, and was joined there by an equally empty-handed *Esk*. *Conflict* had been scarcely better rewarded. All three had boarded numerous French slavers in the Bights, and Bullen was convinced that two-thirds of those were also carrying Dutch papers. The Squadron's officers were naturally anxious to avoid being fooled by Dutch vessels with dual papers, but they were under threat of personal penalty, or even serious international incident, if they infringed the restrictions regarding French vessels. Finally, on 19 December, one of these Dutch slavers was caught out in the Old Calabar. Mr Deschamps, Admiralty Mate of *Conflict*, had boarded what purported to be the French brig *Eugene* with 265 slaves, but in boarding four other French slavers he learned that she was carrying two sets of papers. Returning to her, he found Dutch papers concealed in the master's desk identifying her as the *Charles* of St Eustatius, and she was condemned with her surviving 243 slaves.

Atholl had been more successful. After sailing from Sierra Leone in October to find the Commodore, on Captain Willes's orders, she had cruised the Bights through November and December, and had made two captures. While she was lying off Cape Coast, and no doubt to Captain Murray's surprise, the Acting Governor of Elmina had written to ask *Atholl* to investigate a suspicious schooner which had arrived off the Dutch fortress. Lieutenant Calger boarded and detained the *Aimable Claudina* under Dutch colours on 12 November, and she was condemned, but a cargo of 34 slaves which had been embarked later near Popo at the instigation of the captor, although freed, was not emancipated.

Atholl's second prize, on 28 November, was the Brazilian brigantine *São João Segunda Rosália*, taken 130 miles south of Cape St Paul's with 258 slaves bought at Lagos. Owing to his shortage of officers, Murray put in command of her two first-voyage midshipmen (Kirby and Pipon) and despatched her for Freetown. These inexperienced youngsters had no knowledge of the currents and winds of the Coast, and were faced with a passage of about 1,200 miles in the calm season. They were at sea for 65 days, and for three weeks the prize crew and slaves each lived on a handful of farina and black beans and half a pint of water per day. A Columbian privateer commanded by an Englishman, a few days before arrival at Freetown, found the prize lost and heading out into the Atlantic. With provisions and a new course given by the privateer, she made harbour on 9 February 1826 after losing the Brazilian master and 72 slaves, mostly through starvation. Murray seems not to have been high in Bullen's regard, and he had caused Their Lordships to express "regret" at the number of floggings he had inflicted in *Atholl*. He was sharply criticised by the Commodore for the *Segunda Rosália* episode.

By now Governor Turner had initiated his private campaign against the slave trade in the rivers adjacent to Sierra Leone. In July 1825 he had reported to the Colonial Secretary an "alarming increase" of slave dealing in the neighbourhood, and the "total inadequacy of the Ships of War on this Station". He then claimed, with breathtaking arrogance, that he would

> undertake at little or no expense, without the aid of the Navy, without compromising the Government, and without Risque (*sic*) of failure, to complete in six months such Arrangements as will prevent any Vessel of any Nation to carry away a Cargo of Slaves from Western Africa.

With the steam vessel he had ordered he would

> maintain our Sovereignty over the various Rivers from Senegal to the Gold Coast, a Sovereignty which I will procure from the Natives, if approved, at a small expense […] which will cause them to be considered British waters, and give us the power to exclude all Nations from them.

He required that the whole project must be under his control, and that the steamboat should have a commission to seize slavers.[14] Turner's lack of grasp of the realities of slaving, including the fact that the vast majority of the traffic was to the east and south of the Gold Coast, casts some doubt on his mental state; but his notion did, in fact, bear the seed of a development which, many years hence, and nurtured by much wiser men, would bear fruit.

Turner had bought two French ex-slaver brigs, *Eleanor* and *Susan*, and he began his campaign in late September 1825 by taking 80 soldiers to the Sherbro. He aimed to persuade the chiefs and kings of the coastal districts to abandon slaving in exchange for British protection, and to accept British sovereignty over their territories. Treaties were duly signed, and Turner ordered French slave vessels in the river to leave what he now regarded as British waters. He sent messengers inland offering similar terms to more remote chiefs, who seemed to accept them, and he mounted another expedition up the Sierra Leone River, predictably losing many of his white troops to fever. The slavers displaced from the Sherbro allied themselves with those in the Gallinas and the French dealers in the neighbourhood to re-establish the Sherbro trade by force. Not to be thwarted, the Governor determined to attack their base up the Bolm River, a tributary of the Sherbro, and his means to do so came to hand in January 1826 with the arrival of Captain William Fitzwilliam Owen and his surveying squadron.

Owen was returning home from his epic survey of the east coast of Africa, and had been ordered to chart the west coast from the Congo to the Gambia en route. Leaving Lieutenant Alexander Vidal to leeward with the ten-gun "coffin brig" *Barracouta*, he headed for the Shoals of St Ann to survey the Windward Coast himself with the ship-sloop *Leven* and the schooner *Albatross*. He and Bullen had been told not to interfere with each other's tasks, but the Commodore had been encouraged to give Owen such aid as he could to finish the survey before the rains. So Bullen lent *Leven* a pair of midshipmen from *Maidstone* and told *Conflict* to join Owen on her return from the Bights. Ignoring the urgency of the survey, Owen readily agreed to join the Governor's adventure, and, with *Albatross*, Turner's cutter *Swift* and *Leven*'s boats, he and Turner set off up the Bolm on 16 February. They routed the slavers, merely a temporary setback to the local Trade, but at a cost. As well as three seamen who were severely wounded, Owen lost three officers and six seamen to fever, including, to the Commodore's fury and distress, the two *Maidstone* midshipmen: Hutchinson and Bullen.[*] Victim of his own foolhardiness, Turner died of fever on 7 March.[15]

Conflict did not return from leeward to join Owen until late March 1826, and Bullen ordered *Swinger* to supplement the surveying squadron temporarily on her return from windward in mid-February. On joining, however, she promptly grounded between Sherbro Island and the mainland, and, by the time she was hauled off, undamaged, poor visibility had intervened to prevent any surveying before *Conflict*'s arrival.

On Governor Turner's demise, the senior member of the Sierra Leone council, Kenneth Macaulay, temporarily assumed the government, and that early part of 1826 also brought the latest in numerous changes to the Mixed Commission courts. In January 1825 Edward Gregory, British Judge since 1820, had died, and the Arbitrator since February 1824, Daniel Molloy Hamilton, had been promoted to Judge, with the Colonial Secretary, Joseph Reffell, as Registrar. Shortly afterwards the Dutch Judge, de Marée, died and was replaced by his Arbitrator, Bonnouvrié, and in June 1825 Altavilla, the Portuguese Judge, returned from leave, only to be removed in December by his Government, whom he had evidently displeased. After Turner's return from the Gold Coast in May 1825, Hamilton had taken sick leave in England, shortly to be followed by Reffell, who was replaced by Registrar Smith. The Governor, justifiably complaining of the absence of legal expertise in the colony, had become Judge himself. At last, at the end of January 1826, John Tasker Williams, a professional, Foreign Office-appointed Judge, reached Freetown to lead the British Commission.

---

[*] It seems reasonable to assume that Midshipman Bullen was related to the Commodore.

There was an unfortunate swansong from *Atholl*. She had been ordered to India and sailed from Sierra Leone in mid-January 1826, but on 1 February, well south of the Equator, she intercepted the Brazilian brig *Activo*, boarded by Murray off Elmina in November while empty, but now carrying 165 slaves. The master readily admitted that, ignoring Captain Murray's warning, he had loaded his slaves at Lagos, but felt himself invulnerable south of the Line. Murray unwisely regarded his interception to be as good as hot pursuit, and sent the slaver to Freetown. Predictably, and despite the absence of Portuguese commissioners, the Anglo-Portuguese court restored her and her cargo, and a sum of £258 was decreed against Murray for costs and damages, in addition to his own costs of £155. Nevertheless, the court was strongly disinclined to reward the slaver for his obvious breach of the treaty, and decided that payment by Murray of a further £11,100 awarded in compensation should be conditional on a final decision in London. The matter was complicated by the slaves escaping ashore while the brig was awaiting adjudication, and the refusal of Acting Governor Macaulay to give them up, thereby making Murray liable for payment of the value of the slaves. The discretionary award seems finally to have been remitted, but the episode gave a sharp reminder of the financial risks being run by cruiser commanding officers.*

At the end of December 1825 Bullen had sent *Maidstone*'s boats under Lieutenant William Gray to take a look at the northern part of the Grain Coast, and on 3 January they celebrated the New Year of 1826 by seizing the empty schooner *Hoop*, under Dutch colours, while she was hovering 20 miles off the Gallinas. The owners were believed to be Americans who had become naturalised Dutchmen at St Eustatius, a swift and simple process, but thereby their slaver fell foul of the Equipment Clause. This was to prove a fortunate capture for Bullen. On his own account he bought the brig from the court when she was condemned on 23 January, ostensibly for the "shelter and comfort of officers and crews of boats detached on distant service", but her role soon became much less passive. He renamed her *Hope*, and gave her to Lieutenant William Tucker with an Admiralty Mate, an Assistant Surgeon, 18 seamen, five marines and eight Kroomen from *Maidstone*.[16] Her purchase may be seen as the beginning of a new phase in the Squadron's campaign.†

---

* The monthly salary for a captain of a Sixth Rate was £26.17. Comparative values of sterling can be found at Appendix I.
† There is no precise evidence that *Hoop* became *Hope*, and Basil Lubbock writes in *Cruisers, Corsairs and Slavers* (Brown, Son & Ferguson: Glasgow, 1993) that Bullen's tender was the *Segunda Gallega*, taken by *Maidstone* on 29 September 1825 and condemned on 23 November. However, Bullen reported that he had made the purchase in January 1826, the time of *Hoop*'s condemnation. This and the similarity of names strongly indicate that he bought *Hoop*, not *Segunda Gallega*.

It was a pity that *Hope*, with her American-built slaver's speed and her traversing long 12-pounder and four 18-pounder Govers,* had not joined Gray and his boats when, at dawn on 1 February, they sighted a brigantine under French colours off Cape Mount. When Gray approached to board her for information she fired a shot, hoisted Spanish colours and made all sail to escape. The boats chased her until 1730, when Gray crossed her at half pistol-shot and received a full broadside of grape and musketry, fortunately without much damage. The boats and the brigantine continued crossing on opposite tacks until 1900, when the chase escaped in darkness. While *Maidstone* paid a visit to Ascension in February with a party of liberated Africans to work with the garrison, the boats, with *Hope* in company, remained on the Coast before rejoining the frigate in April.

Redwing, meanwhile, was continuing her run of misfortune. She had returned to the Bights from Sierra Leone, via Cape Coast, in January, and on the twenty-eighth she detained the empty Brazilian slaver *Pilar* (or *Pylades*) at anchor off Whydah. Clavering kept the prize with him throughout February and held her papers in *Redwing*. When he seized another vessel on 8 March he removed half of the *Pilar*'s prize crew to man the new prize, including Lieutenant Wilson, who took Clavering's orders with him. Master's mate Samuel Falconer was left as Prize Master in *Pilar*, with 11 hands. Two nights later *Pilar* and *Redwing* were separated in a squall, and the following morning *Pilar* was boarded by the Brazilian imperial brigantine *Imprehendedor* (or *Emprendedor*). According to Falconer, who had neither ship's papers nor prize order, the Brazilians stole his charts, navigation instruments and money, and told him to make for Princes. Later in the day the Brazilian vessel returned, plundered *Pilar* of most of her provisions and everything except the clothes on the prize crew's backs, removed the Kroomen and one seaman, and then took her to Bahia. On arrival, Falconer and his remaining men were ill-treated and imprisoned, and two months later the vessel and prisoners were taken to Rio, where proceedings were instituted against *Pilar* in the prize court. While the British Chargé d'Affaires defended the court case on Clavering's behalf, the Commander-in-Chief for South America tried in vain to procure the release of Falconer and his men.

Meanwhile, *Maidstone*'s boats and Lieutenant Gray, now in Bullen's small schooner *Ellen*, encountered the *Imprehendedor* while they were watching the Bonny and Calabar rivers. The Commanding Officer of the Brazilian brigantine reported that he had boarded the *Pilar*, found no Prize Master or papers, and the crew of four drunk. Suspecting that she might have been retaken he had, he said,

---

* This weapon was a compromise between the long gun and the carronade, designed by a Captain Gover.

sent her to the British Consul in Rio. He failed to mention that he was holding four Kroomen on board, and he subsequently tried, unsuccessfully, to sell them to a slaver at the Old Calabar. The Kroomen in the palm-oilers lying there created such a furore that he was obliged to discharge his captives to the palm-oilers, and they were later recovered by *Hope*. The *Pilar* was ultimately sold at Rio to the benefit of the British Treasury, and poor Falconer, a "decided drunkard" in Clavering's opinion, returned home insane in April 1827 and was sent to Haslar Hospital having attempted suicide.

The prize to which Commander Clavering had sent Lieutenant Wilson on 8 March was the *Cantabre*, taken by *Redwing* under French colours on suspicion of being Spanish with two sets of papers. It was also claimed that the master had been murdered and that the mate had begged Clavering's protection. The prize was carrying spirits in her forepeak and, despite Wilson's precautions, two of his men broke down a bulkhead to get at them. One of the "beastly drunk" culprits then tried to persuade the slaver crew to rise against their captors. He was awarded six dozen lashes when he rejoined *Redwing*, and his co-conspirator was given three dozen. The *Cantabre* then fell in with the frigate *La Flore*, flagship of the French squadron. Commodore Massieu removed Wilson and his crew and replaced them with his own men to take the prize to Goree, but, before the two vessels parted company, *Brazen* appeared on the scene. She had sailed on 5 April from Freetown to rejoin *Maidstone* in the Bights, and Captain Willes paused to pay his respects to the French Commodore. He was astonished to see the frigate's boat approaching *Brazen* with Wilson and his men in it, and he protested to Massieu at his seizure of the prize, but to no avail. When Bullen heard the story of this, as he saw it, further French protection of the use of its flag by foreign slavers, he wrote a forthright letter to Massieu. The reply was courteous but equally firm, pointing out that there was no evidence that *Cantabre* was anything but French, and that *Redwing*'s action had been illegal.[17]

The passage to Sierra Leone was a hazardous one for prizes taken in the Bights, but in this respect as in others, *Esk* was more fortunate than *Redwing*. She had left Freetown in company with the Commodore on 29 January, and on 4 March she made two captures in the Benin River. The first was the Brazilian sloop *Esperanza* of only 40 tons, and the second the 75-ton Brazilian brigantine *Netuno*. On the approach of *Esk*'s boats, under the command of Mr Richard Burrough Crawford, Admiralty Mate, both vessels tried to land their slaves, but Crawford managed to intercept two of their boats. He took four slaves back to the then empty *Esperanza* and returned 20 to the *Netuno*, bringing her total to 92. Both vessels were sent to Sierra Leone, the *Netuno* under Crawford's command,

and both were condemned. *Netuno*'s arrival, however, brought news of one of the most celebrated actions of the entire campaign.

Crawford's crew for the passage consisted of a 16-year-old Master's Assistant, five seamen and a boy of 17. Also on board were the Brazilian master and three of his crew. The *Netuno* was armed with two 6-pounder carronades, a pair of pistols and six cutlasses. At 1500 on 20 March, about 270 miles south of Cape St Paul's, the prize found herself being overhauled by a brig which, despite her French colours, gave, as she drew close, every appearance of being a pirate. After a shot passed between his masts Crawford hove to, but he declined an order to board the brig on the excuse of having no boat. A boat then came across from the brig with the master at the tiller and five other men, one of them an Irishman acting as interpreter. Speaking in Spanish and broken English the master ordered Crawford into the boat. Pointing to the British ensign, Crawford refused, and the Spaniard replied menacingly that his vessel was not a French man-of-war but the Havana brig *Caroline*. Certain now that he was dealing with pirates, Crawford ordered them at their peril not to board, and, promising to fetch his papers, he went to his cabin and returned to the deck with his pistols hidden behind his back. The pirate master repeated his demand and, when he tried to leap on board, Crawford shot him dead. The bow-man, attempting to follow, met the same fate, and Crawford then made the other three boat's crew jump overboard.

When the pirate brig then opened fire with a broadside of five guns and two swivels most of the prize crew ran below, leaving Crawford with Master's Assistant Olivine, the Brazilian master and one seaman, Frost, on deck. Frost took the helm, the master brought ammunition, and Crawford and Olivine manned one of the 6-pounders. The *Netuno* contrived to keep within effective range for her little gun, about 30–50 yards, and the pirate fortunately not only failed to close and board but also fired her broadsides high. The one-sided action continued from 1730 until 1915, with *Netuno*'s round shot continually hulling the pirate and her canister inflicting casualties on deck. One shot from the brig hit *Netuno*'s side, killing an African woman, wounding another and partially scalping Crawford with a splinter. Frost aimed the carronade for what proved to be the last shot of the engagement, and the pirate brig filled her sails and bore away.

The prize had about 50 shot holes in her sails, her main gaff was damaged, and her trysail gaff shot away. She had only four cartridges left. As the action ended Crawford collapsed from exhaustion, but he wrote his report to his Commanding Officer that evening.[18] He was given his commission as a lieutenant in reward for this courageous defence, but was invalided home with fever in September

1826. A little over five years later he would be back on the Coast in command of a gun-brig.

Esk herself continued to cruise both Bights, occasionally visiting Princes and Cape Coast for water and provisions, and in May the Commodore ordered her to patrol specifically between Princes and Cape St Paul's. *Redwing* remained in the Bights until early April, when she sailed for Sierra Leone to replenish and recover prize crews before cruising between the Sherbro and Cape Mount. *Conflict* worked with Captain Owen's squadron until the survey finished in June and Owen sailed for England with *Leven*, *Barracouta* and *Albatross*. By then Owen had angered Bullen yet again, this time for sending home *Atholl*'s prize crews although he knew full well that every cruiser on the Coast was short of seamen, and he was sharply told not to meddle again with any vessel under Bullen's command. Returning to slave duties, *Conflict* made for the Bights to rejoin the Commodore, although Bullen held no hope that a gun-brig would ever catch a slaver in chase. After being released by Owen, *Swinger* too had returned to the Bight of Benin to cruise between Cape St Paul's and Cape Formoso, but she left again for Sierra Leone in mid-April, making the passage in only nine days. By 27 April she was at the Gambia for a short visit to the settlement at St Mary's, before returning briefly to Freetown.

On 8 May, after a few months short of three years on station, *Swinger* sailed for home. Life on the Coast was always tough for the slow and cramped old Confounder-class gun-brigs, generally known as "*Pelter* brigs" in the Service, and they were the laughing stock of the slaver clippers, but *Swinger* had suffered more than her share of hardship. Soon after her arrival in 1824 she had lost eight of her 50 men to a yellow fever outbreak, and was later hit by the deaths of three commanding officers in the space of a year, including that of the excellent John Scott. However, she had always pulled her weight, and latterly under the command of Lieutenant George Matson she was clearly a happy ship. Matson, a relatively elderly officer of the old school, was known as "Old Rough and Ready", and when the brig was paid off in July 1826 his ship's company presented him with a dress sword, belt and epaulette in token of their very great esteem.[19]

After her encounter with the French Commodore, *Brazen* met *Maidstone* at St Thomas and then sailed to cover the Bights. On 17 May she had the satisfaction of taking the schooner *La Fortunée*, under Dutch colours, to the west of Princes. This slaver had repeatedly been boarded under French colours in the Bonny River, and she again tried to pass herself off as French. As *Brazen*'s boat reached her she threw her Dutch papers overboard, hoisted French colours and presented incomplete French papers to the boarding officer. Her 245 slaves were in a very crowded and sickly state, and, despite every possible care, 46 of them died on the passage to Freetown. Another 73 were lost during the wait for adjudication. The Dutch Judge considered the prize to be French on the grounds that none of the witnesses to the ditching of the Dutch papers could read, but when lots were drawn to settle the adjudication the British Arbitrator decided to condemn her.

Brazen's arrest of the Brazilian ship *San Benedicto* off Popo on 11 June was less satisfactory. The court accepted the slaver's explanation that the 25 Africans on board were canoemen, and that he was making for Malembo for slaves, having been north of the Line only for provisions. The vessel was restored, and costs of £32.10s were allowed, but the court showed its doubts by granting no demurrage. In fact the Africans were Fantees forcibly put on board by the chiefs of Accra.* When landed at Freetown they were ignored by the government and nearly starved. Bullen had these wronged friends of Britain returned home to Accra by *Conflict*.

Maidstone had been in the Bights since her return from Ascension in March, visiting Cape Coast, Accra and the islands, paying particular attention to the Slave Coast, while the Commodore directed the disposition of his other cruisers. On 18 April Bullen boarded the 212-ton Brazilian brig *Perpetuo Defensor* off Annobón, south of the Line. There was ample evidence that the 424 slaves had been loaded at Badagry, rather than Malembo as claimed by the slaver master, but Bullen was anxious not to fall into the same trap as had Murray with the *Activo*. He found that one of the slaves was a British citizen, captured by the Ashantis and sent to Elmina to be sold, and he sent the brig to Sierra Leone to be tried by the Vice-Admiralty court on a charge of enslaving a Briton. However, the prize was passed to the Mixed Commission, which, predictably, released her. Costs of £351.10s were awarded against Bullen, but, as in the case of *Activo*, the award for compensation was made conditional on a decision by the two governments involved, and Bullen apparently escaped what would have been a grossly unjust penalty.

While the prize was awaiting adjudication Bullen complained of the state of her many sick slaves, and he wanted them landed. The Governor's vessel *Susan* was moved alongside to take the healthy slaves without breaching the quarantine regulations, and, despite an application by a member of the Council for a writ of habeas corpus, the slaves were returned to the custody of the Brazilian claimant. The Governor believed that slaves reaching a British harbour automatically became free, but the decision of the colony's Chief Justice to the contrary was supported by London.

---

* The Fantees were the local people around Cape Coast, and were loyal allies of the British.

Maidstone returned to Sierra Leone in mid-May and then visited the Gambia, Portendic, Goree and Porto Praya in June and July before heading back to Sierra Leone in early August. *Hope*, meanwhile, was cruising the Bights on behalf of her parent frigate. She made her first capture off Whydah after a chase of several hours on 20 May, taking the Spanish schooner *Nicanor*, bound from Little Popo to Havana with 174 slaves. The Mixed Commission condemned the schooner, but asked the Foreign Office for guidance on dealing with future cases involving *Hope*. The judges accepted for the present that she was "a *Maidstone* boat", but felt that her status needed to be clarified. By 6 August the tender had carried out 32 fruitless boardings, but, as *Maidstone* was making for Ascension from Sierra Leone, *Hope* stepped dramatically into the limelight.

On 2 August *Hope* had run down to the Whydah roadstead and found the Brazilian brig *Principe de Guiné* ready to load slaves. This beautiful 280-ton brig, only ten months off the stocks at Philadelphia, was regarded as one of the "crack" slaver on the Coast. She was armed with a traversing long 24-pounder, four long 9-pounders, two long 6-pounders and two swivels, and carried a crew of 72. A few days earlier her men had cheered *Esk*'s boat, after a search, boasting that no sloop-of-war would ever take her with slaves. Lieutenant Tucker, hearing from the master of the *Thomas* of London that the slaver was about to load her cargo, cleared out of sight and cruised across her expected track.

Early on 5 August Tucker sighted the *Principe de Guiné* standing out under all sail, and a hard chase of 28 hours began. When *Hope* came up with the slaver the following morning she fired several shots to bring her to, but the brig returned fire, and an action of two hours and 40 minutes followed, the final hour and ten minutes at close quarters. Badly wounded in the foot at the beginning of the engagement, and now with three guns disabled, Tucker finally ran his bowsprit across the slaver's quarter in order to board. Admiralty Mate Robert Pengelly, admirably supported by Assistant Surgeon George Williams, led the boarders with great gallantry, and in a few minutes, although shot in the side, he had gained possession of the brig. The slavers had lost 11 killed and 15 wounded, and *Hope*, with less than half the strength in men, three wounded. It was a remarkable success, in both chase and action. Of the 608 slaves initially on board, two were killed in the fight, 12 were wounded, and 16 drowned in trying to swim to *Hope*. On the passage to Sierra Leone, where the prize was condemned, 12 more slaves died and one was born.[20]

A new Governor, Major-General Sir Neil Campbell, reached Sierra Leone on 22 August 1826 in the frigate *Lively* (38), Captain William Elliot. He found on his arrival that the Acting Governor had been pursuing Governor Turner's anti-slaving policy in the Sherbro. In this his key asset had been the colonial steamboat *African*, which had at last arrived on 1 April under the command of Lieutenant Austin and carrying back from sick leave Mr Hamilton, the Chief Justice, and Mr Reffell, the Colonial Secretary.* The Government had given partial approval of Turner's earlier actions, but refused to ratify his treaties with local chiefs. It was content with his commercial arrangements designed to discourage the slave trade, but would not sanction "proceedings which interfere with the rights of other nations or [...] might be construed as a desire for territorial aggrandisement". Furthermore, it would not allow military posts at the mouths of rivers, but would permit a small floating force.[21]

Armed with this directive, Acting Governor Macaulay had sent a punitive force to the Sherbro with the *African* in late April against a slave trader named Tucker, who had seized the property of colonists in the area and threatened the natives who had signed treaties with Governor Turner. The force, in boats commanded by Austin, had burned Tucker's town and stockades and had recovered most of the stolen property, although Tucker had escaped. The chiefs had subsequently renewed their undertaking to suppress the Trade, and the colonial vessel *Eleanor* was left in the river by Tucker with a detachment of native soldiers to prevent a resurgence. A small vessel, the *Revenge*, was purchased to blockade the Gallinas, and she was soon in action. On the morning of 27 April she was sent by Austin to warn off a brig sighted heading inshore at the Sherbro. Mr Murray in *Revenge* finally brought her to, but after hoisting Portuguese colours she opened fire, and after a 45-minute action Murray was obliged to withdraw.

Austin later heard that the brig was on the point of loading slaves in the Gallinas, and he took the *African* to find her. On 3 May he saw her standing out of Cape Mount Bay and he caught her after a chase of two and a half hours. She was the *Relámpago*, empty but fitted for slaving, and with a crew mostly of Spaniards and Frenchmen. Austin sent her to the Vice-Admiralty court to be tried for firing on a vessel flying His Majesty's ensign within the limits of "a British possession". The previous day, he had had another encounter when he saw a brig heading in towards the Mana River, east of the Gallinas. On boarding her he found that she was the French slaver *Atalanta*, previously warned off by Lieutenant Jeayes in the colonial brig *Susan*. She had landed her trade goods, and Austin sent her to Sierra Leone for breach of the blockade; a blockade which, incidentally, the American settlement at Mesurado, now called Liberia, had agreed to respect. It subsequently became apparent that Lord Bathurst at the Colonial Office was not entirely pleased with Macaulay's conduct as Governor.

---

* Macaulay and Austin now encountered a novel difficulty: the coal contracted for had not arrived.

Shortly before Governor Campbell's arrival the Dutch judge, Bonnouvrié, had returned to his duties, but on 10 August Judge John Tasker Williams succumbed to fever. This setback and other affairs delayed Campbell in Sierra Leone, anxious though he was to reach Cape Coast, where the Ashantis were again threatening. While the frigate *Lively* was waiting to give him a passage, news arrived that a pirate had plundered two American vessels at Cape Mesurado, and on 28 August Elliot sailed in search of her, with the *African* in company. He ran down to Cape Mount, boarding an empty Brazilian slaver en route, but it was not until he was returning to Sierra Leone on the evening of 5 September that he sighted what appeared to be the pirate. He disguised the frigate as much as possible and closed to six miles before the chase took to her heels and eventually escaped in rain and darkness. *Lively* sailed with the Governor and troops four days later, but found on arrival at Cape Coast that on 7 August the British force had, at last, won a decisive victory over the Ashantis. In the Ashanti camp had been found the carefully preserved head of Sir Charles MacCarthy, clearly an object of veneration.

As *Maidstone* was leaving Sierra Leone on 3 August she had met the replacement for *Atholl*, the Sixth Rate *North Star* (28), Captain Septimus Arabin. Bullen ordered her to Sierra Leone, Cape Coast and Princes, and she remained in the Bights for the remainder of the year, apart from a visit to Ascension in December for refit and provisions, perhaps the first occasion on which a cruiser had gone to the island specifically to collect provisions rather than to deliver them.

Bullen had cut short his call at Ascension on hearing of the Ashanti advance on Accra, and, on finding that *Maidstone*'s support was no longer needed on the Gold Coast, he had cruised mostly between Princes and Cape St Paul's until arriving at Cape Coast on 30 September to meet the new Governor, shortly after *Lively*'s departure. Here he delayed for several days to comply with Campbell's request to evacuate and dismantle the fort at Dixcove. Then he investigated the Slave Coast anchorages, finding nine empty Brazilians and Spaniards off Whydah, before flushing the Brazilian brigantine *Hiroina*, one of four vessels at Badagry, on 17 October. She gave Bullen a chase, but he finally caught her only to find that she was empty and carrying an Imperial passport for slaving south of the Line. Nevertheless, he risked sending her to Sierra Leone.

The Commodore had by now received from the Admiralty copies of a correspondence between Foreign Secretary Canning, the Brazilian Government and the British Consul at Rio de Janeiro, in which it was agreed that Brazilian vessels trafficking slaves north of the Equator might legally be detained. Apparently it mattered not whether they were loaded or empty, and Bullen decided that *Hiroina* should provide the test case. The Anglo-Portuguese court, with Governor

Campbell as British Judge, found that the vessel, which had landed part of her cargo of trade goods between Whydah and Lagos, had no legitimate reason for lying at anchor on the Slave Coast and was deliberately flouting her passport. It condemned her, and an important precedent was set.[22]

In mid-September *Maidstone* had found the entire Squadron gathered at Princes, a rare occurrence. Most had been in a satisfactory state, but *Redwing* and *Conflict* had been very sickly. Their dead included two Surgeons, and among those invalided were another Surgeon, a lieutenant and John Chrystie, Commanding Officer of *Conflict*. When the six cruisers parted company, *Conflict*, with Lieutenant Arthur Wakefield of *Brazen* in command, headed for the Bonny, Calabar and Cameroons rivers, but found nothing but French slavers before leaving in mid-November for Cape Coast and a cruise between the Sherbro and Cape Mesurado. *Redwing* sailed to investigate the reported plundering of a British merchantman in the River Benin, but her leak had worsened, and the Commodore decided on her return that the only sensible course of action was to send her home. She sailed for England from St Thomas on 27 November. *Brazen* had preceded her by a month, but after more success.

When, on 28 September, she captured the Dutch brigantine *De Snelheid* and 23 slaves, *Brazen* discovered a strange chain of events. The brigantine had been seized and plundered by the powerfully armed Spanish ship *Atalanta* at anchor off Trade Town, and her crew had been landed and replaced by men from *Atalanta*. When the two vessels were separated a fortnight later, *De Snelheid*'s prize crew took to piracy, using the plundered cargo of the American schooner *Cassandra* to buy slaves at the River Nazareth. When *De Snelheid*'s boatswain heard that the local king was about to arrest the vessel for piracy he sailed, leaving the master ashore, and two days later off St Thomas he gave chase to what he supposed to be a Portuguese slaver. He discovered, too late, that his intended victim was *Brazen*. When Willes had brought the slaver to with gunfire he found that she had no fewer than three sets of papers: Dutch, French and Spanish. Sitting in judgment on her, Reffell and Bonnouvrié felt competent to try the vessel only on her use prior to her being seized by the *Atalanta*, for which they condemned her. They turned over the slaves and crew to the Vice-Admiralty court, the slaves to be emancipated and the crew to be tried for piracy and murder, the boatswain having apparently flogged a slave to death.

Shortly before this capture, *Brazen* had been detached by the Commodore to return home, after only a year on the Coast. She sailed from Sierra Leone on 22 October, taking the invalided Admiralty Mate Richard Crawford of *Esk* with her. Crawford's ship had herself recently achieved some success. On 10 October,

50 miles north-west of Princes, she had intercepted the Spanish schooner *Intrepida* of only 100 tons, although mounting five 18-pounders and with 290 slaves crowded on board. Twenty slaves had already died and 55 more were lost on passage to Sierra Leone. A larger prize fell to *Esk* on 21 December, when her boats found the Brazilian ship *Invincival* at anchor in the Cameroons River with 440 slaves just embarked. The slaver parted company from *Esk* five days later, with Lieutenant Tollevey in command. She did not reach Sierra Leone until 20 February, having twice been struck by lightning which carried away her main topmast and shivered her mainmast and mizzen-topmast. In these strikes four men were killed and over 20 injured. To add to Tollevey's troubles he had light and variable winds for the entire passage and constant sickness among the slaves and the prize crew. Disease claimed 178 slaves before arrival and another eight afterwards, and Tollevey had to be invalided.

On Bullen's behalf, *Esk* had been keeping an eye on *Maidstone*'s boats, left behind in the Bights with *Hope* while the frigate returned to Sierra Leone for refit. Lieutenant Tucker and *Hope* again demonstrated their worth on 6 December 1826 by taking the Spanish schooner *Paulita* off Lagos after a chase of several hours. This slaver was one of those which, on sailing from Havana, had been reported by Judge Kilbee of the British Commission there, and Commander Clavering of *Redwing* believed she had committed acts of piracy against French vessels in the Benin River, where she had loaded her 221 slaves. *Paulita* was condemned to *Maidstone*, but the Admiralty was not comfortable with the activities of "a vessel called *Hope*, said to be a tender to *Maidstone*" of which it knew nothing.

Their Lordships, responding to the capture of *Nicanor* earlier in the year, warned Bullen that they "could not authorise proceedings of any private vessel", that "no vessel other than one of His Majesty's Ships, duly authorised and having a copy of the treaties, could act in the search or capture" of foreign slavers, and that "the *bona fide* nature" of vessels so employed was of particular importance. Bullen protested that he had never referred to *Hope* as a tender, but as a "boat of *Maidstone*", and that she had been privately purchased as a headquarters vessel for his boats on detached duty. He pointed out that he was only following the practice of Collier and Mends, and emphasised that Tucker had been given copies of the treaties. However, he went on to admit that *Hope*'s potential for catching the fast slavers had indeed been a factor in his decision to buy her, a potential "strongly verified" by the cases of *Nicanor* and *Principe de Guiné*, and he argued that he could not fulfil Their Lordships' directives regarding the slave trade solely with the vessels they had placed at his disposal. As a parting shot he offered[23] as a precedent the case reported by

Judge Kilbee of a capture in the West Indies by the schooner *Lion* acting as a tender to the sloop *Carnation*.*

The new year of 1827 began with an extraordinary run of success by *Esk* and *North Star*. Cruising through the Slave Coast anchorages, *North Star* had first found the empty Brazilian schooner *Eclipse* at Whydah on 6 January, and, having ascertained that she had landed her trade goods at Popo and Whydah, Arabin arrested her, relying on the precedent of the *Hiroina*. At Bonny on 31 January he sent his boats into the river and seized the 90-ton Spanish schooner *Emilia* crowded with 282 slaves, only 177 of whom survived the five-week passage to Sierra Leone. A week later the boats were in the Old Calabar and found another Spanish schooner, this one of only 31 tons and with 100 slaves on board. The inexperienced Arabin made the mistake of releasing the master and crew of this slaver, the *Fama de Cuba*, to the next Spanish vessel he encountered, contrary to the requirements of the treaties, but the court accepted that he had been motivated by humanity.

All three of these prizes were condemned to *North Star*, but the master of the *Emilia* accused Arabin before the Mixed Commission of breaking bulk in the slaver.† Arabin responded that he had removed three casks of spirits from the slave deck to prevent the prize crew and remaining Spaniards from getting drunk, and that he had, with the permission of the master, taken a spare sail from the schooner for the Spaniards transferred to *North Star* to sleep on. Less wise had been the acceptance by Lieutenant Cory of the slaver master's gift of a telescope. Although no further action was taken on the accusation, it reminded the Squadron of another pitfall to be avoided. *Emilia* was to serve Arabin well; he bought her from the court, manned her and named her, appropriately, *Little Bear*.

North Star's excellent work continued with the capture of the *Conceição de Marie*, a Brazilian brigantine, off Whydah on 4 March with 232 slaves loaded only two hours before her detention. She had the usual Brazilian passport to take a cargo south of the Line. So too did the Brazilian brig *Silveirinha*, taken at the entrance of the Old Calabar on 12 March. *Maidstone* and *Esk* had boarded this brig in November and December 1826 while she was awaiting her cargo, but *North Star* was lucky to catch her on the day after she had shipped her 266 slaves. A passage of 11 weeks to Sierra Leone cost the lives of 57 of these slaves to dysentery.

Not to be outdone, *Esk*, cruising further to the south than *North Star*, arrested the brig *Lynx* under Dutch colours off Princes on 9 January with 265 slaves on board. The slaver had French as well as Dutch papers, and the Mixed Commission

---
* See Chapter 10.
† Prize Law had long forbidden the "breaking of bulk" until a prize cargo had been condemned, a precaution against pillage, and the anti-slaving treaties had repeated this prohibition.

thought she was probably French property, but her decision to sail under Dutch nationality condemned her. On 6 February, at about 200 miles south-south-west of Cape Formoso, *Esk* added to her bag the Brazilian schooner *Venus* with 191 slaves, loaded at Whydah in defiance of her passport for Malembo, and to the south-east of Princes only two days later she intercepted another Brazilian schooner, the *Dos Amigos*, with a similar passport and 317 slaves from Badagry. No longer having any officer on board beside himself, Commander Purchas was in a quandary as to who could command this latest prize, but his Purser volunteered to take her in. At this point *Esk* returned to Sierra Leone herself, recovered the five prize crews awaiting her there, and sailed north on 10 March for a call on the gum-traders at Portendic.

By the beginning of March *Maidstone* was on her way back to the Gold Coast and the Bights after visiting the Cape Verde Islands and the Bathurst settlement in January and February. Calling at Cape Coast, Bullen received a letter from the Governor of Elmina asking for his help in discouraging slaving at Accra, a request authorised by the Dutch Government. The letter lamented that the people of Dutch Accra were openly flouting the slave-trade laws and claimed that British and Danish Accra were also involved, but the Commodore was not willing to be diverted in this way and agreed to comply only if it was necessary for the protection of British Accra. After calling at Accra he followed in *North Star*'s wake along the Slave Coast, and cut a swathe through the Brazilian slavers in the anchorages as he went.

In the space of four days *Maidstone* took five prizes: on 13 March the brigantine *Trajano* off Whydah, on the following day the brig *Venturosa* and the schooner *Carlotta* off Badagry and the schooner *Tenterdora* near Porto Novo, and on 16 March off Lagos the brigantine *Providencia*. All were empty; all had previously been boarded by *North Star*; some, but not all, had landed part of their trade goods on the Slave Coast; all had Brazilian passports for slaving south of the Line, four of them with notations permitting calls north of the Equator; and all, despite excuses about repairs and provisions, had sufficiently demonstrated intent to take slave cargoes from the Bight of Benin. They were all condemned on the *Hiroina* precedent, as was the sloop *Conceição Paquete do Rio*, taken in similar circumstances while hovering off the mouth of the Benin River on 22 March. She protested that she was entering the river to plug a leak, but *Maidstone*'s Carpenter found her watertight.

Conflict had returned to Freetown in February from a lengthy cruise off the Gallinas, Cape Mount and in the vicinity of Sierra Leone, during which Lieutenant Wakefield had been asked by Governor Campbell to investigate reports of slavers in the northern rivers, unless he had other pressing instructions from the Commodore. Wakefield declined because, he said, *Maidstone* was already in the vicinity, Bullen probably had information on vessels in the rivers, and, anyway, information of this sort was received so frequently that, if it was always acted on instantly, the Commodore's deployments would be rendered useless. This tactless response fuelled the Governor's annoyance that the Squadron's coverage of the coast between Sierra Leone and the Gambia was so sparse, and he repeatedly represented to Lord Bathurst at the Colonial Office that a cruiser should be stationed permanently in that area. He wrote that he had no hope of persuading the Commodore to do so unless higher authority intervened, and he complained of a general want of naval cooperation.[24]

This did not worry *Conflict* as she sailed for the Bights to join in the relative feast of captures in the first three months of 1827. Calling at Accra on 28 February she found the Brazilian schooner *Independencia* in the roadstead, bearing the usual Malembo passport, but, although empty of slaves, having landed part of her cargo of trade goods and carrying a large stock of provisions. Lieutenant Wakefield sent her to Freetown, where the Anglo-Portuguese court, in the absence as yet of the intended Anglo-Brazilian court, paid scant attention to a tired story of a leak and her claim that the arrest had been illegal because she was lying under the guns of the Dutch fort at Accra. After cruising the Slave and Gold Coasts for a month, on 3 April *Conflict* found another Brazilian vessel fitted for slaving, the brig *Bahia*, lying at Awey, ten miles east of Cape St Paul's. The brig's circumstances were identical to those of the *Independencia*, except that her excuse for visiting the Slave Coast was want of water and provisions. The Freetown court was fully aware that the passage from the Bights to Malembo was against wind and current, and that the only conceivable reason for an empty slaver to visit the Bights coast was to embark slaves.

Another gun-brig, *Clinker* (12), had arrived to replace *Swinger*, and, despite *Swinger*'s three years on the Coast, an unfeeling Admiralty had ordered her officers and ship's company to be turned over to *Clinker*. Before arriving at Sierra Leone in mid-February, Lieutenant George Matson had taken his new command for a brief call at Bathurst, and she was then ordered to visit the Rio Nunez and Ascension before beginning a Bights cruise. When Governor Campbell expressed regret that *Clinker*'s boats had not also searched the Rio Pongas, Matson replied that his orders had not given him the choice. On the Guinea Coast, without the speed for a successful chase, *Clinker* had to persevere for a month or more before she reaped her one and only reward of the year. On 15 May she found the Brazilian brigantine *Copioba* at anchor near Awey without slaves but fitted for slaving, with derrick rigged, trade goods partly discharged, master ashore and boat in the water. She was condemned without difficulty. *Clinker*'s cruise continued without event until she sailed again for Ascension in August for provisions, and her fellow gun-brig, *Conflict*, similarly patrolled the Bights until September, with only one break at Ascension in June. It was gruelling and thankless work for these cramped little vessels.

After a promising beginning, 1827 had become a tedious year for *Esk* too. After returning from several months at Portendic, she arrived at Ascension in July with a number of invalids, sent to benefit from the sick quarters which Lieutenant-Colonel Nicholls had been told to build for the use of the Squadron as a more convenient (and probably more effective and agreeable) alternative to invaliding to England. She then made for the Bight of Biafra for a cruise which lasted until September. French slavers there were aplenty to visit and report, but few Spaniards and Brazilians for the time being, and she made no further arrests.

North Star had found more to interest her. Early in April Bullen sent her back to Freetown from Princes to collect her prize crews, and her tender *Little Bear*, under the command of Lieutenant Thomas Crofton, made an unusual capture off Cape Sierra Leone. The Brazilian schooner *Tres Amigos* had put into Freetown for water and provisions en route to the Cape Verde Islands to load salt for Rio after landing a cargo at Loando. Shortly after she had sailed on the morning of 17 April the Governor heard that she had slaves concealed on board, and he asked Crofton to go after her. *Little Bear* caught the schooner on the night of the nineteenth, and eventually found the slaves, three young women, hidden in a space one foot high and four feet square under a fireplace, and in imminent danger of suffocation. *Tres Amigos* had embarked them at Loando without a passport, and she was condemned to *North Star*.

Heading northward during May, *North Star* sent her boats 30 miles upstream in the Pongas and an extraordinary 90 miles up the Nunez without any sign of slaving. She then had a dead beat for St Jago, where Captain Arabin gathered information from the Governor-General but drew a blank on the still active Portuguese slave trade in the Cape Verde Islands. Her next call was on the settlement at Bathurst, which she reached on 13 June. There she found the colonial steam vessel *African* with Sir Neil Campbell embarked, and the two sailed in company to coast southward and examine the rivers in passing. As they approached the northern end of the channel between the Bisagos Islands and the mainland on 18 June they fell in with the Portuguese schooner *Toninha*, and the *African*, which had on board an officer and boat's crew from *North Star*, chased and caught her. Finding 65 slaves and a baby on board, *North Star*'s officer detained her. A Portuguese army officer in the slaver claimed that the Africans were his "domestics", producing a passport from the Governor of Bissau. So Arabin took the slaver and *North Star* into Bissau to remonstrate with the Governor, who predictably denied all knowledge of this illegal traffic. On passage to Sierra Leone the *African* attempted to take the *Toninha* in tow, but, in a heavy swell, she rammed the slaver with her paddle-box and could not extricate herself. Finally, after the slaves and crew had been rescued, the schooner had to be scuttled in order to free the *African*.[25]

The court condemned and emancipated the surviving 58 slaves, but noted that the *Toninha* had been a previously condemned slave vessel, bought from the court by Commodore Bullen and sold to a Colonel Martinez, apparently the Portuguese Government's agent in the Cape Verdes, who had sold her on to return to her original trade.* The commissioners also observed that the *Principe de Guiné* had followed the same route to the same owner. Although there was no suggestion of impropriety on Bullen's part, these examples fuelled a general concern at the ease with which condemned vessels could, and often did, return to slaving. To that was added the frustration that although the nations involved in the slave trade had all by now specified severe penalties for illegal slaving, the owners, officers and crews of condemned vessels were never brought to book by their national authorities.

By then *Maidstone* and Bullen had made their final contribution to the campaign. On 10 April the frigate's boats had been sent into the Old Calabar, and while the ship was lying offshore she sighted from the masthead a suspicious vessel between herself and Fernando Po. She lost the chase in light winds and the onset of night, but at 2200 saw her again by moonlight at a distance of seven or eight miles. Lieutenant Morton, the First Lieutenant, took away the cutter and gig, and, after a hard pull, he brought-to the 86-ton Brazilian brigantine *Creola* with 308 slaves taken from the Old Calabar and crammed into a slave deck with a deckhead height of three feet. Sending the prize to meet him at Princes, Bullen collected his boats, learned that there were no slavers in the Cameroons, and made for Accra for provisions. He particularly wanted these "refreshments" for the men who had spent a great deal of time away in the boats during the cruise, recently in weather more intensely hot and with rain heavier and more constant than *Maidstone* had previously experienced. On return to Princes, the frigate took the *Creola* in tow and sailed for Sierra Leone, losing 19 slaves on passage.

Two days after her arrival on 22 May, *Maidstone* was joined by the frigate *Sybille* (48) under the command of Commodore Francis Augustus Collier CB, reputed

---

* It is not known what use Bullen made of *Toninha*. Martinez had been given a contract by the Admiralty to salvage the wreck of *Erne*.

9. Commodore Francis Augustus Collier CB

to be one of the keenest and tautest officers in the Service. The new Commodore had been one of Nelson's midshipmen in *Vanguard*, had been given command of a brig in 1806, and, after being made Post in 1808, had been in command at sea throughout the Napoleonic War. Thereafter, in combined operations with Major-General Sir William Keir, he had been instrumental in crushing the Persian Gulf pirates while in the frigate *Liverpool*, and the two commanders were reported to have worked "in the most perfect harmony". Collier inherited Bullen's orders, and the Admiralty re-emphasised his triple tasks of assistance to British settlements, protection of British commerce and prevention of the slave trade. The ludicrous phrase "the several bays and creeks" failed, for the first time, to appear in the Commodore's orders.

Bullen handed over command of the Station on 1 June 1827, and the two frigates sailed in company a few days later for Cape Coast, finally parting on 26 June. *Maidstone* called at Accra to refit her rigging and repair the fishing of her bowsprit, which had been sprung for most of her deployment, and then made for Princes and Ascension, finally sailing for England on 21 July carrying the Squadron's invalids and with Bullen recovering from a dangerous bout of "coast fever". By the time of *Maidstone*'s departure, relations between Governor Campbell and Bullen were strained. Campbell had made several complaints to the Colonial Office about a general lack of naval cooperation, although there was no reciprocal word of criticism from the Commodore. It is probable that Bullen's refusal to station a cruiser permanently off the northern rivers was attributed by the Governor to wilful antagonism by the Commodore rather than a calculated decision on the strategic deployment of woefully inadequate resources. This seems to have developed into an assumption by General Campbell, as with Turner, of a naval antipathy towards the Army, a notion refuted by the record of earlier amicable cooperation on the Station between the Navy and Army. As things stood, Bullen found as he made for Ascension that Campbell had presented the Squadron with a logistical headache.

In order to pay for provisions for slaves on passage to Sierra Leone, the practice had been for the cruisers' Pursers to present bills to the Superintendent of the Captured Negro Department at Freetown, and this procedure had been incorporated in Commodore Mends' orders to the Squadron. In May 1827 the Superintendent, on Campbell's orders, began to refuse payment, and threatened the Purser of *Esk* with arrest if he did not pay the amount of a bill from Cape Coast for the victualling of the *Intrepida*. The cruisers had no public money on board with which to pay for these provisions, and Bullen had been obliged to reimburse, from his own pocket, the supplier of the *Principe de Guiné*, who had been refused payment on *Maidstone*'s bills. The system had worked well under General Turner, but Campbell regarded payments made by his predecessors as unauthorised by the Government, and referred Bullen to papers laid before Parliament in 1824 of which Bullen was entirely ignorant. The Governor expressed "extreme surprise" that the Commodore should have presented such claims, and having neglected to discuss the problem with Bullen, he asserted that the cost should be paid for out of the proceeds of prizes and proposed no alternative procedure. The Admiralty referred the matter to the Treasury, and in December 1827 the British commissioners were told that ships should still present bills to the Superintendent of Captured Negroes, but that the cost should now be defrayed from the value of condemned vessels and cargoes.[26] Governor Campbell never heard of this decision; on 14 August 1827 he died of fever at Freetown.

Assisted by *Hope*, *Maidstone* had captured 17 prizes and 2,485 slaves during her three and a quarter years on the Station, and the Squadron under Bullen's command had brought another 41 slave vessels to trial. Bullen, after a period as Superintendent of Pembroke Dockyard and Captain of the yacht *Royal Sovereign*, was knighted in 1835, gained his flag in 1837 and was ultimately promoted to Admiral in 1852.

Within weeks of his taking command of the Station, Collier had become apprehensive for the safety of *Redwing*. On 13 June 1827 that habitually unfortunate sloop had returned to Sierra Leone from her refit in England, still under the command of Commander D. C. Clavering, and three days later had sailed for Accra. She was neither seen nor heard of by the Squadron thereafter. About three months later wreckage marked *Redwing* was washed ashore at Matacong near the Isles de Los, confirming her loss, but the circumstances of her fate remained a mystery until 1841. A Spanish slaver master finally revealed to the Governor of St Vincent in the Cape Verdes that he had seen *Redwing* in chase of two slavers as a tornado approached. None of the three had made the necessary drastic reduction in sail, and all had capsized in the squall, leaving no survivors.

Conflict had gone to Ascension for provisions in June, and, when she returned to the Bights in July, her sister *Clinker* left her Bights patrol and made first for St Helena and then Ascension. The regular requirement to reprovision the latter island had been something of a burden to the Squadron for the past decade, but development under the direction of Lieutenant-Colonel Nicholls had by now enabled the garrison to repay the debt, and Ascension had become a refuge from disease, a haven for recuperation and a source of plentiful fresh food and spring water. The Commodore had little hope of either of the two gun-brigs ever catching a slaver; they were, he said, forever chasing and losing.[27] Nevertheless, he kept them in the Bights, *Clinker* for the remainder of the year and *Conflict* until the end of September when, with her mainmast badly sprung, she was sent first to Sierra Leone and then home.

Esk, the other recent visitor to Ascension, had returned to the Bights in August after delivering invalids to the island, and joined the Commodore, who was cruising to the north-west of Princes in *Sybille*. *North Star* remained at Portendic until the gum trade was over for the year, and in August she returned south to cover the rivers between the Gambia and Cape Mount. While cruising off the River Sesters, Captain Arabin heard that a Havana brig, which had robbed an American vessel on her previous voyage, had landed money at the Gallinas to buy slaves. He waited off the river, out of sight of land, and sighted a suspicious brig on 31 October. The breeze was too light for the ship to chase, and so he sent away the boats. They were unable to gain on the brig, but forced her offshore. After losing contact during the night, the pinnace, commanded by Lieutenant Boultbee, came up with the chase the following day. The brig had run-out and depressed her guns, but Boultbee boarded over the bows and, although met with musket fire, quickly gained possession. *North Star* took the brig, the *El Gallo*, to Mesurado in the hope of an identification by American settlers involved in the earlier robbery, but without success, and Arabin had to release her.

The new Commodore was much impressed by *Hope*'s record, but the matter of tenders remained contentious. George Canning at the Foreign Office had reassured the British commissioners that they had acted correctly in the cases involving *Hope*, *Black Nymph* and *Little Bear*, basing their decisions on the principle that tenders should not be considered as distinct from parent ships. In an apparent softening of its attitude, the Admiralty then directed that officers commanding tenders should be given signed copies of the treaty Instructions. However, Governor Campbell had, not unreasonably, been concerned that condemned slavers purchased initially as tenders had subsequently been resold without papers at the Cape Verdes and then found their way back into the Trade.* Consequently, late in 1827, the Admiralty directed the Commodore to discourage officers from buying tenders, and if such vessels were purchased they were not to be resold except into the King's Service. Collier had pre-empted this instruction in August by asking the Admiralty for permission to purchase a tender if he had the means to do so, and he was told that if he could find a suitable vessel at a moderate price he was at liberty to buy her, and, on his reporting her details, papers authorising her to act against slavers would be sent to him.[28]

Collier's opportunity came on 6 September, when *Sybille* was cruising off Lagos. At a little after midnight the frigate detected a brig reaching off the land on her lee bow, and, in a fresh breeze, she crowded on sail in chase. The brig's master, overconfident after six very successful slaving voyages, made the mistake of trying to cross *Sybille*'s bows to get to windward of her, and she came within range of the frigate's bow chasers. After several well-aimed rounds, the slaver, for all her speed, was obliged to surrender. She was the 257-ton, American-built *Henriquetta*, under Brazilian colours, about 91 feet in length on her gun deck, with three guns, a crew of 38, and 569 slaves loaded only a few hours earlier at Lagos. By the time she reached Freetown she had lost 27 of her slaves, and in their 12 days of living ashore before *Sybille*'s arrival, two of the prize crew contracted fever. One died, but the other, an apparently hopeless case, was saved by the frigate's Surgeon, Robert McKinnel, using large doses of sulphate of quinine.

Sybille remained in the Bights until the end of November, taking one more prize, on 12 October, the Brazilian schooner *Diana*, at about 180 miles west of Princes. During the chase the slaver had thrown some of her papers overboard,

---

* Bullen does not mention how he disposed of *Hope*, but he probably sold her before leaving the Coast.

and there was consequently some doubt about where she had embarked her 87 slaves, but the court finally accepted that the Africans were from Benin, and she was condemned.

However, the prize which particularly interested Collier was the brig *Henriquetta*, renowned as the fastest and most successful slaver on the Coast, and a rare capture. He arrived at Sierra Leone on 12 December, shortly after she had been condemned, and on 31 December 1827 he was able to report to the Admiralty that he had bought for £900 a very fast vessel, mounting two short 12-pounders and one long 18-pounder on a swivel amidships, which "appears well adapted for the purpose intended".²⁹ He placed her under the command of one of his lieutenants, William Turner, manned her with a crew of 55 from the frigate, gave her one of his boats, and renamed her *Black Joke*. Under the command of a succession of outstanding officers, she was destined to become the most successful of the Squadron's tenders and the most famous cruiser of the campaign.*

In December the Squadron was joined by a relief for *Esk* and a replacement for *Conflict*. They were the ship-sloop *Primrose* (18), Commander Thomas Griffinhoofe, and the gun-brig *Plumper* (12), Lieutenant Edward Medley. *Plumper* was not only of a type which Collier, and Bullen before him, regarded as virtually useless for catching the current generation of slave vessels, but also she and her boats arrived in a disgracefully defective state, and she sailed "most wretchedly".

There had been another new arrival on the Coast at the beginning of September, one who was to become something of a thorn in Commodore Collier's side; although, on balance, an asset to the suppression campaign. This was Captain Owen, who had so angered Commodore Bullen, sent back to West Africa in the Sixth Rate *Eden* (26) with the task of establishing a settlement at Fernando Po. On several occasions previous commodores had recommended the use of Fernando Po as the base for the West Africa Squadron, and in 1826 Commodore Bullen had been asked by Lord Bathurst at the Colonial Office to investigate its potential. It was ideally placed for observation and interdiction of the slave traffic of the Bight of Biafra, there was an excellent anchorage in Maidstone Bay at the north-west extremity, it offered plentiful fresh provisions and water, and it was relatively healthy. Furthermore, if (as Bathurst proposed) the Mixed Commission courts were to move there from Freetown, slavers detained in the Bights and their captors would be spared the long voyage to Sierra Leone against wind and

---

* She became the third small vessel to appear in the *Navy List* with the name *Black Joke*. According to Lubbock in *Cruisers, Corsairs and Slavers*, this curious name may have come from an eighteenth-century term for a dance similar to an Irish jig, or, less likely or appropriate, from the north-country word "joke", or "jook", for a duck.

current, with its associated loss of slaves. The island did, admittedly, belong to Spain, but that was not apparently seen as an obstacle.[30]

Owen had been ordered by the Admiralty to proceed with *Eden* and the transport *Diadem* to Sierra Leone after collecting artificers at Plymouth. At Freetown he was to embark more artificers, labourers and a detachment of troops, and confer with the Acting Governor, Lieutenant-Colonel Lumley, on arrangements for the intended establishment at Fernando Po. He was to find the most suitable site on the island for a settlement, probably Maidstone Bay, build accommodation for the Mixed Commission courts, make arrangements for the move of the courts, and prepare facilities for the resettlement of liberated Africans. Commodore Collier was directed not to interfere with *Eden* except in case of necessity, and Owen was told not to interfere with the other vessels on the Station. Owen was also specifically ordered not to deviate from his task to look for slavers. However, Owen had become accustomed to independent command, and was liable to make his own rules and to circumvent instructions which did not suit him. He was directed to report to the Governor and to the Commodore, but it is not clear whether or not he had a responsibility to the Colonial Office as well as the Admiralty, and it was probably equally unclear at the time. The water was muddied further by the issue to *Eden* in June 1828 of a copy of the treaty Instructions, which authorised her to take slavers. This mixture of contradictory arrangements and confused lines of accountability, combined with the contrasting personalities of Collier and Owen, was a recipe for trouble, as should have been apparent even to the Admiralty.[31]

At Sierra Leone, Owen had purchased a small schooner, the *Horatio*, and Acting Governor Lumley had given him the steam vessel *African*, by then in a very run-down state. While *Eden* proceeded to Fernando Po with the *Diadem* and the *Horatio*, via Cape Coast and Accra, the *African* remained on the Gold Coast under the command of Lieutenant Badgley of *Eden*. She may have been left behind as a provisions transport, but Owen had put *Eden*'s jolly boat on board her as a device to give the *African* authority, albeit questionable, to detain slavers. The somewhat eccentric Owen had given names to his ship's boats, and this jolly boat was called the *Hay*. On 23 October, Badgley in the *Hay* detained the empty Brazilian sumaca *Sao João Voador* off Quitta, and on the following day he took the Brazilian schooner *Vencedora*, also empty, while at anchor off Whydah. Both were fitted for slaving, but the inexperienced Badgley neglected to gather evidence that they intended to load slaves north of the Line. The two vessels were subsequently released by the Anglo-Portuguese court, and costs, but not demurrage, were awarded against Owen.

Taking his prizes initially to Fernando Po, Badgley arrived on 31 October to find that Owen had anchored his vessels in Maidstone Bay four days previously, bestowed royal names on the bay's principal features, and chosen Point William, one of the arms enclosing Clarence Cove, as the site for his settlement. Within days Owen was complaining of a shortage of seamen and labourers, the beginning of a steady flow of grievances about a lack of resources, and by the end of the year he had irritated the Commodore on two counts. The first was by a request that the ships of the Squadron should run supplies to him from Sierra Leone. Collier replied that his ships were there to suppress the slave trade, not as transports, and reminded Owen of the orders regarding mutual non-interference. Owen, he said, already had a tender, and he should use her to supply him. Nevertheless, *North Star* and *Esk* would make deliveries in passing.

The second annoyance arose from a call on the Commodore in Freetown by Lieutenant Robinson of *Eden*, who had arrived in one of the detained vessels. He was wearing a beard, and Collier told him to leave the frigate instantly and "not to presume again to disgrace the uniform of the Service and his Country by appearing so unlike a British officer". Owen's response was to regret that Robinson's beard "should have drawn from you such an extraordinary order"; the lieutenant "may possibly in wearing his beard have been influenced by the example of his Captain who has worn his these six years".[32]

The beginning of 1828 saw *Esk*, leaky from worm damage to her bottom, being lightened and caulked at Sierra Leone, prior to a call at Fernando Po in company with the Commodore in early February. Taking passage via Ascension with cows and ewes for the garrison, she then made for home after a creditable deployment of three years, which included nine prizes (one of them shared) and the capture of 2,240 slaves, and in which Commander Purchas had earned Commodore Bullen's praise as "this highly meritorious, active and zealous officer". Collier noted also that *Esk* had lost or invalided fewer men than any other sloop on the Coast over a similar period. *North Star*, meanwhile, arrived at Fernando Po on 25 January with stores from Sierra Leone, and then began a cruise in the Bights. *Primrose*, after cruising the Grain Coast in January, followed *North Star* and *Esk* to Fernando Po in February and also began a Bights patrol. *Plumper* paid a visit to the Gambia settlement in late January, and returned to Sierra Leone on 7 February, in no fit state to make any useful contribution to the campaign. *Clinker* was still in the Bights, and *Sybille* sailed from Freetown in mid-January to visit Fernando Po in company with *Esk*, no doubt to allow the Commodore to establish a more amicable working relationship with Captain Owen.

On 12 January 1828, a mere fortnight after hoisting British colours for the first time, *Black Joke* took her first prize.

* * *

For the past four years, once the distraction of the Ashanti War was removed, the West Africa Squadron had been able to concentrate almost entirely on its anti-slaving task, although the Portendic gum trade still required periodic attention. Since the departure of the United States Navy and the death of Governor MacCarthy, the Navy had, however, suffered a dearth of friends on the Station. Not only had it continued to find obstruction rather than help from other nations who were supposedly committed to suppressing the slave trade, but also the hitherto warm relationship between the Commodore and the Governor of Sierra Leone had become unsupportive and even antagonistic. This deterioration seems inconsistent with the past record and known personalities of Commodores Bullen and Collier, and it is difficult to see how they might have contributed to it. The attitude of Governors Turner and Campbell, on the other hand, might well have been coloured by earlier bad experiences of antipathy between the Navy and the Army, but they appear also to have misunderstood and resented the independence of the Commodore.

This resentment was particularly apparent in their complaints of lack of naval cooperation in their attempts to suppress the Trade in the vicinity of Sierra Leone, but the Commodore had a very broad remit and grossly inadequate resources with which to fulfil it. The earlier policy of dispersing the cruisers intermittently along the entire slave coast north of the Equator had produced disappointing results, and it was apparent that a greater concentration of effort was necessary. Bullen had identified the Bights as his top priority, and Collier had subsequently shown no sign of disagreeing with him. Of the five or six cruisers generally on station, there had, in the past three years, rarely been fewer than three, and often four or five vessels positioned in the Bights. The northern rivers and the Grain Coast had not been entirely neglected, but both areas, the former especially, had received little coverage.

The results indicated that Bullen's judgment had been sound. By comparison with the 50 vessels and 6,200 slaves taken in the years 1820–3, the Squadron had captured 67 slavers and 10,750 slaves in 1824–7 with similar force levels. Four of the prizes were either lost or retaken, but only six were restored or released by the courts, against twelve in 1820–3, a figure which indicates a greater understanding and sympathy among commissioners and commanding officers for each other's difficulties, and a greater willingness by the courts to apply the spirit rather than merely the letter of the treaties.

There could be no illusion, however, that the Squadron's work was making decisive inroads on the slave traffic north of the Line, most of which was by now in the hands of French slavers who remained invulnerable to the British cruisers and scarcely discomfited by those of their own nation. The South Atlantic was still the legitimate preserve of the Portuguese and, increasingly, the Brazilian slavers, and the Brazilians had been straying liberally and illegally into the Bights, despite their recent liability to seizure while empty. Unlike the Spaniards with their Baltimore clippers on the extensive Cuba trade, these Brazilians had mostly been using non-specialised and relatively sluggish vessels, which largely explained why 37 of the Squadron's 67 prizes were Brazilians; but, as the capture of those two fine brigs *Principe de Guiné* and *Henriquetta* showed, that might be changing. The illegal Portuguese traffic north of the Equator had retreated solely to the Portuguese possessions north of the Gambia. The Dutch, discouraged by the Equipment Clause in their treaty, had not appeared in large numbers, and nothing had been seen of the Stars and Stripes.

No amendment of British strategy was apparent. Still the only element of the slave traffic under attack was the sea passage; the shore traders sold their slaves without constraint, and, scandalously, even the slaver officers could feel safe from personal penalty, despite the laws passed by their nations in accordance with the anti-slaving treaties. Governor Turner had produced the only novel policy idea, that of restraining the slave-trading coastal chiefs by treaty, but he had lacked the wisdom to bring it to worthwhile fruition.

Nor had the Squadron's tactics on the Coast changed appreciably, although much less reliance had been placed on boat work in the rivers, largely because of the speed with which the slavers now loaded and sailed once their cargoes were ready. The exposed Slave Coast anchorages and other slaver waiting areas had been regularly visited, and had produced a third of the captures as well as information on likely times of departure of laden vessels, but half of the prizes had been taken at sea, not infrequently after a chase. Many of these successes had resulted from judicious patrolling across the departure routes of the slavers in the Bights, and, as will be recounted in Chapter 10, that had also been the case in the West Indies, where Admiral Halsted's schooners had shown their paces against the Cuba slavers.

Similarly swift vessels under British colours were now beginning to make their mark on the African coast, thanks to the initiative of Commodore Bullen, and their appearance heralded a new phase in the campaign.

CHAPTER 9

Tenders and Tablecloths: Eastern Seas, 1828–31

Surely we should find it both touching and inspiriting, that in a field from which success is banished, our race should not cease to labour.

ROBERT LOUIS STEVENSON

NOT ONLY HAD THERE been no progress in strategy or tactics during the first 20 years of the suppression campaign on the slave coast, but also there had been not the slightest advance in the quality of men-of-war deployed to the West Coast of Africa Station. The improved system of warship construction introduced by Sir Robert Seppings, Surveyor of the Navy, had become common practice by 1815, but since then the Admiralty had invested little in further design improvement, or, indeed, in building programmes. A number of sailing trials had taken place in the early 1820s involving 18-gun Sixth Rates and 28-gun frigates by several designers, including Seppings, but they accomplished little or nothing, and certainly brought no advantage to the West Africa Squadron. Of the vessels on the Coast at the beginning of 1828, *Sybille* had been launched at Toulon in 1791 and taken from the French in 1794, *Primrose* was of 1810 vintage, *Conflict* had left her slip in 1812, to be followed into the water in 1813 by *Esk*, *Clinker* and *Plumper*. Only *North Star* was relatively new, having been launched in 1824, but to an 1816 design.

None of the cruisers joining the Squadron during the past 20 years was truly suitable for addressing the challenge of the West Africa campaign. Above all they lacked the sailing qualities to compete with the swift and agile slaving schooners on the Cuba trade, and their habitability fell grossly short of what was necessary for health and good morale in the heat and humidity of the African coast. Particular criticism has been levelled in the past at the deployment of frigates to the Coast, on the grounds that their size led to their premature detection by slavers. In fact, the sighting of *any* ship-rigged vessel would alarm the slavers because, in those waters, she would almost certainly be a warship, but there is no evidence that Fifth Rate frigates were at a greater disadvantage in this than their lesser sisters.

A vessel hull-down was most likely to be sighted on account of her sails, not her masts, and, if she avoided setting her upper sails until they were needed for a chase, a frigate's height of mast would give her the sighting advantage against a slaver carrying the habitual cloud of canvas. By comparison with smaller cruisers, a frigate had the additional benefits of greater endurance, a good outfit of boats, better habitability and a more numerous ship's company, which could sustain the depletion inflicted by the despatch of prize crews and wastage incurred through disease.

A relatively big ship's company and large complement of officers and warrant officers conferred another benefit, which was beginning to bring a marked improvement to the Squadron's results: a frigate could afford to man one or more tenders. The Mixed Commission courts, in accordance with the treaties, continued to sell condemned slavers on the open market, and most of them found their way back into the slave trade, but the occasional arrest of a Baltimore clipper slaver meant that there were vessels of excellent quality and advanced design available for purchase by the Navy. In the past, several commodores and commanding officers, using private money, had purchased small vessels from the courts or elsewhere for a variety of support tasks, but Commodore Bullen began a new trend with *Hope*. He purchased her as "a boat of *Maidstone*", but used her not only as a headquarters vessel for the frigate's boats, her purported role, but also, to the irritation of the Admiralty, as an additional cruiser. The reason for Whitehall's objection to Bullen's worthy initiative is unclear, but it may simply have been that Their Lordships regarded the Commodore's acquisition as implying criticism of Admiralty deployment policy. Be that as it may, Commodore Collier had now overcome opposition in London, and, with the blessing of the Admiralty, had bought the *Henriquetta* with the avowed intent of using her to arrest slavers. This admirable development made a start on the agonisingly long process of rectifying the material shortcomings of the Squadron, and, as *Henriquetta* was launched on her career as HMS *Black Joke*, it heralded one of the more dramatic phases of the suppression campaign.

* * *

Two factors set *Black Joke* apart from previous tenders on the Coast. The first was that she was entered in the *Navy List*, and the second, of crucial significance, was that she was issued with her own set of the anti-slaving Instructions defined in the various treaties. She was manned and administered by *Sybille*, and on the advice of the commissioners at Freetown to Commodore Collier, she carried one of *Sybille*'s boats to ensure that her prizes would be condemned to her parent frigate, but she was authorised by those Instructions to cruise independently. With this authority en route, if not yet actually in his pocket, Lieutenant William

Turner opened his new command's account on 12 January 1828. After a chase which may have surprised and shocked his victim, he made the otherwise unremarkable capture of the Spanish schooner *Gertrudis* off the Gallinas River. Of the slave cargo, embarked only two days earlier in the river, all 155 survived to be emancipated, and the slaver was condemned.

After an uneventful cruise into the Bights, the tender's next encounter, on 2 April, presented a sterner test. At 0200, about 40 miles north-east of St Thomas, *Black Joke* sighted a brig to the south-west and gave chase. At daylight Turner hoisted his colours, and the brig responded by showing a Red Ensign and running down to within hail on *Black Joke*'s weather beam. She then shifted her colours to Spanish and gave her name as *Providentia*. Master's mate Harvey and a boat's crew crossed to the supposed Spaniard to investigate her, and were immediately taken captive. An officer from the Spaniard then returned in *Black Joke*'s boat to demand Turner's papers and the tender's identity, claiming to suspect that she was a Colombian privateer. Turner produced his commission and the Commodore's orders, but this failed to satisfy the other brig, and Turner was ordered to send across 15 men, presumably as hostages, and to take his vessel to Princes; otherwise the *Providentia* would open fire.

Turner naturally declined the proposition, and detained the brig's officer and his boat's crew; whereupon the *Providentia* fulfilled her threat with round shot and grape, and *Black Joke* returned fire. Having just one 18-pounder to his opponent's fourteen guns, 12- and 24-pounders, Turner skilfully manoeuvred to lie on the *Providentia*'s bow where, with his single gun on a sweep mounting, he could engage effectively while remaining largely immune to the returning broadside fire. The action continued for two hours, *Black Joke* firing high to prevent the other brig from closing. The *Providentia*, her sails and rigging cut to shreds, then hoisted a flag of truce and sent back Harvey and his men. Harvey reported that the vessel was indeed a Spanish privateer, and that she had already been boarded by *Sybille*. Turner returned his detainees to the privateer and left her to lick her wounds.[1]

It was only two days earlier that *Sybille* had boarded the *Providentia*, and the Commodore had formed the opinion that she was a pirate. He was almost certainly correct, and he was shortly to be proved right in his suspicion of another case of false identity in an incident in March. After *Sybille*'s visit to Fernando Po, which she reached on 9 February 1828 with the leaky *Esk*, she had cruised the Bights and, on 19 March, had intercepted the French-flagged *La Fanny* 115 miles south of Cape Formoso. This schooner was three days out of the Calabar River with 266 slaves, and Collier knew that a French cruiser had recently been into the Calabar and arrested all slavers under the White Flag of France. As *La Fanny* had not been apprehended, Collier concluded that she must have produced Dutch papers. Then *Sybille*'s men found a Dutch ensign and pendant under the master's bunk, and *La Fanny* was sent to Sierra Leone. The Mixed Commission could find no further evidence of other than French nationality, and it was decided to despatch her to Senegal. This was unwelcome news to *La Fanny*'s crew, who reckoned that Anglo-Dutch justice in Freetown was preferable to the French alternative in Senegal, and they admitted that the vessel's Dutch papers had been destroyed during the chase by *Sybille*. Satisfied that Collier had been right, the Mixed Commission condemned the schooner, and the surviving 238 slaves were emancipated.

There was a further success for *Sybille* on 13 April when she took the empty Portuguese schooner *Esperanza*. The vessel was carrying a passport for Cabinda to load slaves, and claimed to believe that she was on a direct course for that port, although she had not seen land. Her master had been in command of the *Trajano* when she was arrested by *Maidstone* in March 1827 for precisely the same offence of slaving north of the Equator, and having been captured within sight of Lagos, nearly 900 miles north-north-west of her claimed destination, *Esperanza* was similarly condemned.

After a largely frustrating four-month cruise, the Commodore returned to Freetown on 13 May to find *Primrose*, *Clinker* and *Plumper* all lying in harbour and unable to proceed to sea because the Squadron store was empty of provisions. *North Star* arrived on the twenty-seventh and the entire Squadron, *Black Joke* apart, found itself in the same predicament. This unfortunate hiatus did at least give the Commodore an opportunity to review the achievements of the year to date, and they were sadly sparse. *Plumper*, with her defective copper, her hull clearly needing urgent survey and her boats in poor condition, was virtually useless and should never have been deployed to the Station. *Primrose* and *Clinker* had both been in the Bights but had made no arrests, eight French slavers had been boarded by the Squadron between 1 January and 14 May, and the sole prize, other than those of *Sybille* and *Black Joke*, had fallen to *North Star*.

Having made her delivery of provisions to Captain Owen's base at Fernando Po in late January, *North Star* had cruised the Bights to no avail until 20 April. A small compensation for her three months of tedium then fell into her lap in the shape of the unladen Brazilian brig *Terceira Rosália*, lying at anchor off Popo. She was yet another of the many Brazilian and Portuguese vessels with passports issued in Brazil for loading slaves at Cabinda, and, having no excuse for her presence off the forbidden Bights coast, she was condemned without hesitation. The Foreign

Office continued to review such cases, however, and it was not until the end of December that the Earl of Aberdeen, who had superseded the Earl of Dudley in May, conveyed to the commissioners his agreement with the condemnation.

Since the death of John Tasker Williams in 1826, the posts of British Commissary Judge and Freetown Vice-Admiralty Judge had been filled by a succession of governors, acting governors and even the British Commissary Registrar, Joseph Reffell, none of them with any legal training. At last, in January 1828, the Foreign Secretary appointed George Jackson to be the Judge, although he would not arrive until late August, reappointed William Smith as Arbitrator, and Reffell, whose primary post was Secretary of the Colony, as Registrar. In early April, however, John William Bannister arrived to be the Chief Justice and, in the absence at the Gold Coast of the lieutenant-governor, Lieutenant-Colonel Lumley, Bannister assumed the duties of the Judge. Two weeks later, Lumley returned to Sierra Leone and relieved Bannister as Judge, but a mere fortnight after that Lieutenant-Colonel Dixon Denham, who had achieved fame as a Niger explorer, arrived to assume the governorship of the colony and consequently to supersede Lumley as Judge. This convoluted saga was not yet at an end; on 9 June, after nine days of fever, Denham died, and Lumley again found himself as Lieutenant-Governor and Judge. Poor Lumley lasted only until 2 August, when fever also claimed him, and, there now being no Field Officer in the colony, Samuel Smart, King's Advocate, succeeded him. George Jackson finally arrived in HMS *Medina* on 26 August to put the British Commission onto an even keel, and Smart reverted to the post of Arbitrator while Smith was on leave. The British judiciary had re-established stability, but at the time of Jackson's arrival there was not a single foreign commissioner in Sierra Leone. It was as well that the West Africa Squadron had not been heavily burdening the Commissions with business.

In early July Captain Owen arrived with *Eden* at Sierra Leone, shortly preceded by the Spanish schooner *Emprendedor*, which he had taken 60 miles to the east of Princes on 11 June. The prize had been carrying three Africans who had gone aboard voluntarily at Little Popo on a promise that they would be landed further along the coast, but Owen found them under a nailed-down hatch in the fore-hold, 450 miles from the promised landing place. Schooner and Africans were condemned. Owen also brought unwelcome news to Sierra Leone. Writing from *Eden* in the harbour, he informed Lieutenant-Governor Lumley that the new establishment at Clarence Cove on Fernando Po was now ready to receive the Mixed Commissions, and that *Eden* was standing by to convey them thither.

It was abundantly clear that, uncongenial though conditions in Sierra Leone certainly were, the commissioners had not the slightest desire to exchange them for an outpost in the Bight of Biafra, a further 1,400 miles from home. In great haste, Reffell wrote to Lumley that "as no orders or arrangements of His Majesty's Government have yet been received […] regarding their removal […] it is quite impossible for the measure to be carried into effect." There the matter rested, and *Eden* sailed on 21 July to resume what, despite the intentions of the Admiralty and the Colonial Office, seems to have become Owen's preoccupation: hunting slavers in the region of the Calabar and Cameroons rivers.[2]

The shelving of the proposed transfer of the Mixed Commissions to Fernando Po was probably not altogether unfortunate. The move would have saved the prizes taken in the Bights from the long, often dangerous and (for the slaves) too frequently fatal passage against wind and current to Freetown, and it would have enabled the cruisers to recover their prize crews comparatively rapidly. Against that, Fernando Po lacked the considerable infrastructure developed in Sierra Leone for the reception, support and integration of emancipated slaves; it also lacked the civil administrative officials available in Freetown, few though they were, temporarily to fill the regular gaps appearing in the ranks of the British Commission. Added to that, foreign governments, already having considerable difficulty in providing willing commissioners for Sierra Leone, would have found it nigh on impossible for Fernando Po. Furthermore, for a small squadron based so far to leeward it would hardly have been feasible to give even minimal attention to the slave rivers north-west of Cape Palmas, let alone those north of Freetown.

Even before he had chosen the site for the new establishment at Clarence Cove in October 1827, Captain Owen had deployed *Eden*'s tender, the merchant vessel *Africa*, in company with the ship's jolly boat under the command of Lieutenant Badgley to hunt slavers on the mainland coast adjacent to Fernando Po. It was not part of his remit to do so, and he clearly had only a vague understanding of the contents of the treaties. For Owen, however, these matters seem not to have constituted obstacles to his pursuing what appeared to offer a lucrative sideline, and in late January 1828 he again despatched Badgley on a foray into the Calabar rivers.

Lieutenant Badgley was given command of *Royal Admiral*, *Eden*'s pretentiously named pinnace, and again had in company the merchant schooner *Africa*, re-engaged as *Eden*'s tender for the expedition. He was furnished with intelligence from the merchantman *Margaret* of Liverpool that three slavers were waiting for cargoes in the New Calabar, and his Commanding Officer had told him to detain every Spanish, Portuguese and Dutch vessel he found either with slaves embarked or fitted for slaving. In addition he was instructed to find sources of livestock and fresh provisions, and to make arrangements for their collection. Finally he was to seek "free labourers" for the construction work at Clarence Cove.

Making his way initially up the Old Calabar to Duke Town, Badgley found several slavers with multiple false papers, and discovered that the master of one of them, apparently a Spaniard under French colours, had shot the mate of the British palm-oiler *Kent*. Finding that he was sharing the river with *North Star*'s boats, and that *North Star* herself had anchored outside the bar, Badgley wisely left both the attempted murder and the slavers to Captain Arabin. Then, acting on information proffered by the Liverpool merchantman *Neptune*, off the Bonny bar on 1 February, Badgley went in search of a Spanish slaver lying in the St John River with, it was said, 150 slaves embarked.[*] Having anchored the *Africa* six miles from the river mouth, he took the pinnace into the river on 3 February and found his quarry anchored three miles upstream. Although resistance was expected, the slaver crew ran below as Badgley boarded, and then, leaving only two or three dying seamen on board, officers and men abandoned their vessel and refused to have anything more to do with her. There were only two slaves on board, the remainder of the cargo being held ashore. The prize was the schooner *Felis Victoria* of Havana, and the following day Badgley sent her to Fernando Po where Owen kept her, using her at least twice for visits to the Calabar, before despatching her to Sierra Leone on 13 May. The Mixed Commission, clearly irritated by the delay, condemned the vessel and emancipated the two slaves.

One of those visits was by Owen himself in mid-March, and, finding two Spaniards in the river waiting for slaves, he stationed Lieutenants Robinson and Badgley with the pinnace, cutter and 50 men to prevent their escape. On 14 April, after what must have been an exceedingly tedious and uncomfortable vigil, the two boats, by now with Acting Lieutenant Mercer in command, boarded the schooner *Musquito* inside the river mouth. She was carrying 126 slaves, and it was discovered that her master/supercargo was still ashore with another 125. There was clear evidence that the vessel was Spanish, but her Spanish papers had been destroyed, and at the time of the boarding she was under French colours and with French papers. Lieutenant Robinson, rejoining in the *Felis Victoria* during the boarding, seized her and took her to Fernando Po, where Owen landed all the surviving slaves, two having jumped overboard on passage. His excuse for this irregular (and indeed illegal) action was to avoid the likely mortality on the voyage to Sierra Leone, but he admitted that it was his intention to employ the

---

[*] The St John River was one of the minor Delta rivers, otherwise known as The Second River of Brass, flowing into the Bight of Biafra between the New Calabar and Brass.

slaves at Clarence Cove. Having then loaded a number of slave children, Owen despatched the schooner to Freetown, where she was condemned.

Before returning to Fernando Po, *Eden*'s two boats visited the Cameroons River and, on 18 April, were fortunate to find the Brazilian schooner *Voadora* at the river mouth with 234 slaves. She too was sent initially to Fernando Po, where Owen landed all but 47 of the slaves together with the slaver's medicine chest and a quantity of provisions. His justification on this occasion was overcrowding and sickness, a far from uncommon condition of captured slavers. In condemning the *Voadora* and her 45 surviving slaves the Mixed Commission expressed some sympathy with Owen, but considered that such landing was "subject to great objection". The Admiralty, however, when it heard about the *Musquito* episode, was a good deal more forthright, and its sharp response was accompanied by copies of the treaties to forestall any future claim of ignorance from Owen.

Throughout these months there was a steady trickle of self-pitying complaint from Owen to the Admiralty. Among his grievances was his perception (probably justified) of an increasing prejudice, particularly in the Mixed Commissions, against the Fernando Po settlement; a prejudice which, he felt, extended to Owen himself. Added to that, he resented the supposed unfriendliness of Commodore Collier towards him, and he grumbled that shortage of officers prevented him from taking *Eden* to interdict the slave traffic from the Bonny, Old Calabar and Cameroons rivers, a shortage resulting, he said, from the need to employ them on detached duty. This last complaint stung the Admiralty into a retort in August that his lack of officers resulted from his sending them off hunting slavers, which was not the task given to him. Apparently not in the least abashed, Owen replied that he had been told to seize any slaver he fell in with, and that he had never lost sight of his principal aim.

Owen had no one but himself to blame; he was undoubtedly taking useful action against the slavers, but he was operating well beyond his instructions, and he clearly had no compunction about treading on the Commodore's toes. He insisted that Collier's squadron should not make use of the north coast of Fernando Po without consulting him, and nor should it communicate with the natives of the island. With admirable patience, Collier willingly agreed and assured Owen that the Squadron's services were at his disposal, other than in ferrying provisions. Outstanding surveyor though he was, Owen – perhaps because of his extended periods of independent command – had become arrogant and virtually insubordinate, intolerant of constraints on his freedom of action, and a selfish and uncooperative colleague. He deserved credit for his achievement in establishing the settlement on Fernando Po, pointless though it ultimately became, but he was an unnecessary irritant to the Commodore and the Mixed Commissions.³

Partly as a result of the absence of stores in Sierra Leone, the months of May and June saw no success for any of the Squadron other than *Sybille*'s nimble tender. On 16 May *Black Joke* was cruising the Bight of Benin, 70 miles or so south-east of Cape St Paul's, when she intercepted the Brazilian brig *Vengador* on passage from Lagos to Rio with a massive cargo of 645 slaves. Although she mounted eight guns, the slaver offered no resistance, and the boarders discovered that she was none other than the renamed *Principe de Guiné*, gallantly taken by *Maidstone*'s tender *Hope* two years earlier. Having been bought at public auction by Commodore Bullen and employed as a tender, she had been sold to a foreign buyer and then transferred to a new owner in Bahia to resume the work for which she had been built. The sale of captured vessels by the courts was naturally popular with the Navy because of the prize money it generated, but this recycling of slavers, a common practice, was a scandal which could not be allowed to continue. Lieutenant Turner returned to Sierra Leone with his prize, losing 21 slaves on passage, and the brig was condemned for the second time in her career.

After disposing of her prize, *Black Joke* returned to her patrol of the Bights, as did *Sybille*, and it was the parent frigate which made the next capture, but not until early July. On 29 June she had chased a topsail schooner under Spanish colours until the quarry had got close in to the shore near Badagry where, with no escape possible, the Spaniard had run on shore and shortly after gone to pieces. She appeared to be full of slaves, but the surf was so heavy that *Sybille*'s boats could not approach her to save lives or take possession. On 4 July, however, she successfully intercepted the 63-ton Brazilian schooner *Josephina* about 120 miles south-west of Cape Formoso with 79 slaves and two logbooks, one of them belonging to the schooner *Voador*. The master claimed that he had embarked his slaves in Cabinda, but they had clearly been loaded in the Cameroons River, and the prize was condemned with her surviving 77 slaves.

The *Voador* herself was shortly to fall into the clutches of *Clinker* as Lieutenant Matson's second prize in August after an otherwise fallow year. His first success came on the fifth, when he cornered the Brazilian brig *Clementina* in the Cameroons River with 271 slaves. The brig had a passport for Cabinda and Malembo, but her master admitted to loading in the Cameroons. On the slaver's passage to Freetown for condemnation she suffered a horrific death toll of 115 of her cargo. On 20 August Matson then found the *Voador*, a Brazilian, at anchor in the River Bimbia, adjacent to the Cameroons. The boarding party was welcomed with a round shot and a volley of musketry, to which they replied with small arms, and there were numerous wounded on both sides as well as one or two deaths. Among the killed was the slaver's master. Although no slaves were embarked, the schooner was fitted for slaving, and she was subsequently condemned for violation of her passport for Cabinda and the River Zaire, a principle for which there was Foreign Office approval and numerous precedents. Brazil subsequently claimed that the arrest had been illegal, and made a complaint against Matson for cruel conduct. After an investigation, the Foreign Office responded in January 1831 that the complaint was groundless, and stated that the affray had been entirely attributable to the "violent and premeditated attack made upon the British boats by the *Voador*'s people [...] and to their subsequent temerity, after the boarding had been effected, in twice attacking the British Officers and Seamen".[4]

Kroomen formed a part of every ship's company in the Squadron, and the Commodore regarded their value as incalculable. As he told the Admiralty, it was impossible for Europeans to endure the sun, rain and fatigue that the Kroomen tolerated, and he voiced the general view of the Squadron's commanding officers in representing that the two-thirds rations allowed for Kroomen was inadequate for men working as hard as they did. He considered, however, that the reduced spirit ration was quite sufficient, and explained that precautions had to be taken to prevent Kroomen from selling their rum to the seamen. With unaccustomed generosity the Admiralty authorised Collier to issue such increased allowance as he considered necessary.

The Navy Board was less cooperative when the Commodore applied for an additional supply of paint for hammock cloths. These cloths protected the lashed hammocks stowed in the nettings along the ships' rails during the day, and, in the persistent heavy rain of the African coast, the weatherproofing paint was being worn off more quickly than usual. It was often necessary to pipe down hammocks during the day, and the men frequently had wet bedding. The Board, who had presumably never experienced wet bedding, felt unable to comply.[5]

Primrose had not met with any success during 1828 until she encountered the Brazilian schooner *Nova Virgen* at sea off Lagos on 28 July. The slaver master swore that he had loaded his 354 captives at Malembo, in accordance with his passport, but the evidence of the slaves and the low consumption of water contradicted him. It was obvious that he had taken his cargo illegally on the Bights coast within the past three days, and the vessel was condemned in November, by which time 34 of her slaves had died. She was the first vessel to be tried by the new Anglo-Brazilian Mixed Commission, her case having been opened in the Anglo-Portuguese court and then withdrawn and begun afresh. In reporting this trial, Judge Jackson complained to the Foreign Office that the Brazilian

Government was now including in its slaver passports permission to touch at Princes, St Thomas or even Elmina, thereby providing a pretext for the presence of these vessels in the Bights, a violation of the spirit of the convention with Brazil.

Mr Joseph de Paiva was appointed the Brazilian Commissary Judge in September, and, although not yet at Freetown, he lodged a protest against all actions by the Anglo-Brazilian court prior to the arrival of the Brazilian commissioners. Judge Jackson pointed out that the parties to the treaty of 1826 had undertaken to establish their Mixed Commission court within one year of ratification, and Brazil had only itself to blame for the absence of its commissioners for cases heard to date. Furthermore, it had been agreed between Britain and Portugal in March 1823 that, in the absence of the Portuguese commissioners, cases could be judged by the remaining commissioners. In the event of vacancies not being filled within six months, Portuguese vessels taken into Freetown after the expiry of that period no longer had right of appeal to the Anglo-Portuguese Commission in Rio.

Black Joke, under orders to deliver despatches to *Primrose* off Lagos and to take a look at Whydah, was running down the Slave Coast at 1600 on 27 August when she saw a brig and two schooners weigh and stand out from Whydah Roads. Taking them to be slavers, Lieutenant Turner hoisted Brazilian colours at his main topgallant masthead as a disguise and closed one of the vessels, a two-topsail schooner, which was seen to be signalling the other two while beating to windward. By 1700 *Black Joke* was at half-gunshot from the schooner which shortened sail and hove to, but still displayed no colours. Turner then hauled down the false colours and set his main course and boom mainsail. At this, the schooner made sail, showed French colours and tacked inshore. Turner tacked after her, and at 1830, lying on *Black Joke*'s lee quarter, she fired a shot and hoisted Spanish colours. Turner set his topgallants, hoisted colours and pendant, and went for her.

The two-topsail schooner fired another round and the other schooner, followed by the brig, opened fire. At half pistol-shot from the larger schooner, Turner hailed her three times, but was answered with the schooner's larboard broadside. Returning fire, but hoping to draw the two-topsail schooner, clearly a superior sailer, away from the other two, *Black Joke* stood off the land. At 2330 the bigger schooner was sighted on *Black Joke*'s weather quarter and, when the tender rounded on her, she bore up and made all sail. The two consorts were then seen to be coming down with all sail set. The chase continued until 0420 when the schooner rounded to and fired her broadside. *Black Joke* replied, and then, loading her guns almost to the muzzle with round shot, grape and canister, she ran alongside and fired them together with a volley of musketry.

Although both vessels were still sailing at six knots, Turner immediately boarded with all hands except the man on the wheel and the two boys, and after a short struggle, he took possession with the loss of his gunner's mate. Having found the schooner's signal book, he made the private signal to the brig to close, and when she did so he boarded and seized her too. The other schooner was out of sight, but Turner was in no position to chase her with his hands full of prisoners, double the number of his own men.[6]

The prize schooner was the "Buenos Aires privateer" *Presidente*, effectively a pirate, fitted-out at St Bartholomew as were many of her ilk, and the other two vessels, both Brazilian, had been seized and plundered by her off Whydah. The pirate master and most of his officers and crew were English, and the remainder were Americans. When told by one of his Brazilian prisoners of the identity of *Black Joke* when she was first sighted, the pirate had replied: "She is a damned fine brig, has but one gun, will do very well for us, and I must have her." He and five of his men died in the attempt to take her, and 20 of his crew of 95 were wounded. In fact the Commodore had given *Black Joke* a second gun only a few days earlier, the 12-pounder carronade from the frigate's launch, mounted on a traversing carriage. The pirate mounted six 12-pounder carronades and a long 12-pounder.

There were no slaves in either of the prizes, and the *Presidente* was wrecked off Sherbro Island on passage to Sierra Leone without loss of life, but *Black Joke* was awarded salvage for the recapture of the Brazilian brig *Hosse*. The survivors from the *Presidente* were sent home for trial on charges of piracy, robbery and plunder on the high seas, but were acquitted for lack of evidence.

At the opposite end of the Squadron's effectiveness scale from *Black Joke* was *Plumper*. Added to the gun-brig's barely seaworthy state, with serious rot in major timbers found by a survey in September 1828, were the shortcomings of Lieutenant Medley. Not mincing his words, Collier wrote to the Admiralty, along with the survey report, that: "From the first moment of this vessel's arrival on this Station her commander has been full of complaint and difficulties, and [*Plumper*] has been the most useless and inefficient vessel I have ever met in His Majesty's Service." Two days later he forwarded a letter of complaint from Medley of the Commodore's conduct towards him, and he drove home his opinion that *Plumper* was a slovenly vessel and that Medley had been negligent in several respects: watering, gunnery practice, and not bringing sick men to medical survey. Collier regarded him as unfit to command a warship, and it was revealed that the Admiralty had made a note in 1815 that Medley "should not be again taken from half-pay on any pretext".[7] The Commodore may have been more tetchy than usual, suffering as he was from another bout of fever "of the tertial type", but recovering under the treatment of Surgeon McKinnel with sulphate of quinine. Nevertheless, it is inexplicable that Medley had been given a command when so many high-quality half-pay lieutenants were begging the Admiralty for sea appointments.

Having spent most of the year nursing her defects in either Freetown or Fernando Po, *Plumper* did manage to hobble into the Bight of Biafra in October, and on the seventeenth she redeemed herself to some extent by taking the Brazilian sloop *Minerva da Conceição*, about 250 miles south-west of Cape Formoso, with 105 slaves loaded at Lagos. The vessel's master had undergone a similar experience at the hands of *Maidstone* when in command of the slaver *Conceição Paquete do Rio* in March 1827, but was apparently unchastened by it. It may have been this new condemnation which, in November, stung Judge Jackson into protesting to the Foreign Secretary that the laws of all nations party to anti-slaving treaties with Britain specified punishments for convicted slavers, but that these laws appeared never to have been acted upon. He expressed his concern at the impunity with which slaver officers conducted the Trade, and at the encouragement given to the Trade by this lack of legal action.

Since taking the *Nova Virgen* in late July, *Primrose* had cruised in vain until she intercepted the Brazilian schooner *Zepherina* 30 miles south of Lagos on 14 September. *Black Joke* was also in the vicinity and joined in the chase, but although the tender was present when the slaver was brought-to, it was *Primrose* which made the arrest and discovered 218 slaves on board. The schooner held a passport for Cabinda and, in contravention of the spirit of the Anglo-Brazilian treaty, clearance for Elmina, Princes and St Thomas en route.

The *Zepherina* encounter was Lieutenant Turner's last in command of *Black Joke*. His exploits had earned him not only promotion to commander but also the admiration of his shipmates in *Sybille*, who presented him with a two hundred-guinea sword bearing the inscription: "A token of respect and regard from Commodore Collier, the Captain, officers, and ship's company of H.M.S. SYBILLE, to Capt. Wm. Turner, for his zeal and gallantry while Lieutenant-commanding the BLACK JOKE tender."

Having handed over his command on 14 November to Lieutenant Henry Downes, another *Sybille* officer, Commander Turner, together with the *Presidente* prisoners, sailed for Spithead in *Plumper*.

The gap in the Squadron's ranks left by the departure of *Esk* in February was at last filled with the arrival on station of the ship-sloop *Medina* (22), Commander W. B. Suckling. She had been ordered by the Admiralty to make for Ascension via Fernando Po, and while still to the north of the Equator on 3 October she fell in with the Brazilian schooner *Penha da Franca* and found 184 slaves on board her. The owner later claimed that her cargo had been loaded at Ambriz and that adverse winds had driven her north of the Line, but this implausible tale was disproved by papers on board showing that the slaves were from Lagos. The inexperienced Suckling neglected to issue the prize with the certificate required by the treaty, but the court decided that this failure did not invalidate the capture.

Medina's promising start led into a fruitful November. Heading north from Ascension for the Niger Delta she encountered the Spanish brig *El Juan* about 180 miles west-south-west of Princes Island on the twelfth, and arrested her with 407 slaves taken from the Bonny River. During the chase the slaver heeled so far that the Africans made a rush for the main hatchway grating, and the crew, fearing a revolt, fired through the grating. After the capture, the boarding officer declared that he had never seen such carnage as he found in *El Juan*'s hold, although 378 slaves lived to be emancipated.

On 23 November the sloop made her third capture. This prize was the Brazilian schooner *Triumpho*, carrying 127 slaves from Benin, and she had been furnished with one of the duplicitous Imperial passports for slaving at Cabinda which allowed her to call at "any of the Ports on the western Coast of Africa". *Medina* found her about 90 miles south-south-east of Cape Formoso. Both prizes and their slaves were satisfactorily condemned. Less satisfactory was the boarding of the schooner *Allega Gallega* of Havana on 4 December. She was found to have no slaves on board, but six months later Suckling's successor received an affidavit from one of the slaver's seamen revealing that an African woman brought on board on the Grain Coast had been thrown overboard weighted with an anchor ring and stock on the approach of *Medina*'s boats.

Once the Squadron's stores crisis at Sierra Leone in May had been overcome, *North Star* headed northward, apparently the only foray in that direction during 1828. Having spent a week at Bathurst, more to encourage the merchant community than to discourage slavers, she sailed on a cruise in mid-June, and was back in the Bights at the beginning of August. On the eighth, about 100 miles north-west of Princes, she boarded the Brazilian schooner *Sociedade* bearing a slaving passport for Malembo but, although still empty of slaves, the schooner had already landed her considerable cargo of trade goods on the Bights coast. Captain Arabin arrested her, and she was condemned at Freetown.

By now, serious rot had been found in *North Star*'s fore- and mainmasts, and in late August she put into Fernando Po for repairs. Taking this challenge in their stride, the ship's company and Captain Owen's men cut and prepared new masts, lifted out the old, stepped and rigged the new, and the ship was at sea again by 27 September, when she took *L'Aigle* with 464 slaves from the Old Calabar. Captain Arabin despatched his prize for Freetown under the command of Lieutenant Blythe, with a Master's Assistant, 12 seamen and 12 Kroomen. She was never heard of again. There was a happier outcome from the seizure on 17 October of the *Santa Effigenia* 50 miles south of Badagry. The master of this Brazilian schooner simplified the arrest, and subsequent condemnation, by admitting that he had loaded his 218 slaves at Badagry.

North Star's tender *Little Bear* had not been granted the same degree of independence that *Black Joke* enjoyed, and neither had she achieved comparable success, but on 30 October she made a double strike in the Cameroons River.* The first of her two Brazilian victims was the *Estrella do Mar*, empty of slaves and claiming that she was in the river for repairs. However, her passport was irregular, as was that of the second culprit, the schooner *Arcenia*, which was loaded with 448 slaves. Both were condemned to *North Star*. Only two days later this short burst of success for *North Star* concluded with the seizure of the small Spanish schooner *Campeadora*, taken after a chase when she grounded on the Bonny bar. Into her slave deck were crammed 381 captives, loaded in the Bonny River, but a large part of her provisions had been thrown overboard during the chase, and it was clear to Captain Arabin that slaves and prize crew would face starvation if despatched directly to Sierra Leone. He therefore diverted to Fernando Po to ask Captain Owen if he might land 200 slaves to ease overcrowding as well as the victualling crisis. Owen, himself short of provisions, agreed to take 150. *Campeadora* arrived safely at Freetown, and was condemned in mid-December.

Having had his offer of conveyance to Fernando Po declined by the Mixed Commissions in July, the disgruntled Owen returned in *Eden* to his settlement. Just to the north of the island on 11 August, the day before his arrival, he fell in with a suspicious brig under French colours, and on boarding her he found that she was the *Henrietta*, two days out of Old Calabar with 425 slaves and a crew of mixed nationality. Fictitious French and Spanish papers were found on board, but, as was later admitted, her true Dutch papers had been ditched overboard as *Eden*'s boat approached. Undeterred by earlier criticism, Owen landed 59 of the slaves at Fernando Po, because of their "sickly, crowded and miserable state", before sending the prize to Freetown in company with his tender *Horatio*. The brig was condemned, but 60 slaves died on passage, and a further 15 succumbed before emancipation.

---

* The identity of the Commanding Officer of *Little Bear* is not recorded.

By mid-1828 doubts had arisen in London about the future of the Fernando Po settlement, and investment in its development was suspended. Owen, it was decided, should be removed from command of *Eden*, thereby losing his claim on the slave head money she earned, and should become an official of the Colonial Office as Superintendent of Fernando Po. Naturally this plan failed to find favour with Owen, and in September he declined the post. The matter was not pressed by London, but Owen did decide to base himself ashore (although not relinquishing command of *Eden*) and to keep the ship at sea as much as possible, under the temporary charge of Acting Captain Harrison, both for the health of the ship's company and to optimise her potential for interdiction of the slave traffic. He also purchased, at his own expense, the schooner *Cornelia* as another tender, and armed her with one long 7-pounder swivel gun and two 6-pounders.[8]

His tenders served Owen well, as did his intelligence sources in the Old Calabar, but *Cornelia*'s first encounter with a slaver had a tragic outcome. On a run to the mainland for livestock on 4 October, she fell in with *Les Deux Amis* with, it was believed, 360 slaves on board. The tender gave chase and was coming up with her quarry at 0400 on 5 October when she lost sight of the slaver in a tornado. When the weather cleared there was no sign of *Les Deux Amis*, and there was little doubt that she had been overwhelmed with the loss of all on board.

Cornelia's next supply run to the Old Calabar, in company with *Horatio*, was a little more productive. The pair returned to Fernando Po with a small schooner called the *Portia*, which had fallen into the hands of Duke Ephraim after she had been seized by a "Buenos Ayres pirate", and he was happy to turn her over to Owen for £50, the price of his expenses. On 16 October *Cornelia* was once again in the Old Calabar and found the pirate vessel there, but she was under French colours and holding French papers, so Lieutenant Kellet wisely took no action. Owen, however, on hearing that the pirate was again in the river, ordered her to be arrested when next encountered.

The pirate, a ship of 300 tons, 11 guns and 58 men, was well known to Owen. She was the *Venus*, holding French papers, apparently from a vessel called *Coquette*, but preying on other slavers on the Coast under Buenos Aires colours. Having landed trade goods for 500 slaves at Duke Town in late August she was expected to return for her cargo in three months' time, and on 16 November she arrived a mere half-hour before *Cornelia* returned to the anchorage, enough time for Duke Ephraim to warn her that Captain Owen had ordered her arrest. The tender was now commanded by Acting Lieutenant Mercer, a midshipman not yet even passed for lieutenant, who, without hesitation, ran *Cornelia* close alongside the *Venus*, "shot ahead of her and at her hawse with a spring on his cable, put 20 men on board and had possession in less than three minutes." The pirate crew fled below the moment *Cornelia*'s boats boarded, but the master and some of his officers escaped ashore. At Mercer's request, however, Ephraim hunted them down and returned them. The fate of the *Venus* is not known, but 17 of her crew, together with an African prosecution witness, were shipped to England for trial. The court judged that the ship was neither a slaver nor piratical, that the ten Africans found on board had been embarked at Cape Palmas to help work the ship, that she may have had authority from France to suppress the slave trade, that she may have had a privateering commission from Buenos Aires, and that there were no grounds for prosecution in a British court. Owen, and anyone with knowledge of the Coast, would have considered the judgment naive!

Some confusion with the seizure of the pirate ship *Venus* was introduced on 13 November by Owen's arrest off Fernando Po of the schooner *La Coquette* with 220 slaves, under French colours but with the Dutch alias of *Venus*. Seamen who had seen the vessel at St Thomas reported that whenever a French cruiser appeared, a frequent occurrence, *La Coquette* hoisted Dutch colours.* Owen already knew her, and her master had attempted to bribe Owen with 500 dollars to let him pass. At Sierra Leone, the slaver's master, having failed in an attempt to retake the schooner on passage, then tried to pass himself off as French. After six weeks in harbour with the Africans still on board, the Anglo-Dutch Mixed Commission tried the case, taking the precaution of having the French Commodore in attendance. In the nick of time, however, the tender *Horatio* arrived with the schooner's Dutch papers, which had been reported ditched prior to capture, showing an authenticated bill of sale from a French to a Dutch owner at St Martin. Her mate had tried to sell them to Owen after the prize had sailed, but Owen had relieved him of them and sent them to Freetown. *La Coquette* and the surviving 185 slaves were condemned.

Cornelia had not yet finished her valuable service for 1828. In mid-November she brought news to Clarence Cove that the brig *Neirsee*, believed to be Dutch, was about to sail from the Calabar River with slaves.† Her master had offered Owen 1,000 guineas to let the brig pass, but Lieutenant Badgley was sent in *Cornelia* on 22 November to await her departure on the river bar, and two days later the tender reappeared with *Neirsee* and 280 slaves in company. The slaver produced French papers, which Owen believed to be fictitious, and she was sent to Freetown with

---

* It may be assumed that this is the Caribbean island of St Thomas.
† It is not clear whether this was the Old or the New Calabar River.

the tender *Horatio*. However, the two vessels were separated in a tornado and, on 24 December, the slaver was retaken. A month later, at Guadeloupe, flying Dutch colours and showing Dutch papers, she landed the slaves, five Kroomen of the prize crew and three free British African passengers. She then sailed for Martinique, where apparently she belonged under the name *Estafette*, and when in sight of Dominica she put the Prize Master and the remaining eight of his crew into an open boat to make their way to the island. The brig finally headed via St Eustatius to St Bartholomew, where she was sold, and hoisted Swedish colours. The Governor of Guadeloupe refused to give up either the slaves or the Kroomen and free British Africans, although two seamen and seven black men from *Eden* who had been in *Neirsee* eventually reached Portsmouth from the Caribbean in September 1829.

The schooner *Portia*, the other member of Captain Owen's brood of tenders, was not to be outdone by her companions. Sent to collect bullocks from the Old Calabar, under the command of Mr Simmons, *Eden*'s Master, she was crossing the river bar on 18 December when she saw a brig outward bound. The brig bore up and tried to get back into the river, but was obliged by the tide to anchor. Simmons knew that a Spanish slaver brig called *Bolivar* was due to sail from the Calabar under French colours and falsely named *Duc de Bordeaux*, and he boarded this suspicious vessel in *Eden*'s galley with 12 men. He found her eight carriage guns loaded with grape, small arms prepared, a crew of 46, half of whom were Spanish, and 426 slaves. He also learned that she had turned back on account of a mistaken report of a man-of-war to seaward of the bar. He was convinced that she was the *Bolivar*, and he immediately sent her master to *Portia* and secured the remainder of the crew below. Owen was equally confident of the slaver's identity, but instead of sending her to Sierra Leone he kept her at Clarence, made arrangements for the return of the officers and crew, and landed the slaves until he might find it convenient to take them to Freetown. However, he was able to report on 19 December that all slaves that he had previously detained at Fernando Po had now been condemned by the Mixed Commissions.

Eden herself boarded a suspicious vessel, a large brig of eight or ten guns, in the Bight of Biafra in November, and found that she was the *Guadeloupienne*, commanded by a man named Troubriant, without slaves but with a remarkably large crew of between 120 and 160. It seemed that another vessel belonging to Troubriant had collected a large cargo of slaves and left them in the hands of King Pepple in Bonny to be delivered to yet another Troubriant slaver. Pepple, however, had sold them to a different customer, and, when intercepted by

Owen, Troubriant was on his way to destroy Pepple's town. Having sent two of his officers to *Eden* with *Guadeloupienne*'s papers, Troubriant slipped away in the darkness, taking *Eden*'s Gunner with him. Despite appearances, Owen was sure that *Guadeloupienne* was not French, but he gave her papers to the French Commodore, Vilaret de Joyeuse.

At about this time an anonymous letter was received by the Admiralty alleging misconduct by Owen on a number of counts, ranging from seizures in contravention of the treaties and mishandling of prizes, through illegal landing of slaves to the flogging of black soldiers without court martial. There is no doubt that Owen had brought these accusations upon himself by flouting regulations and by regularly antagonising the authorities in Sierra Leone, but he made a spirited defence, generally claiming that his intentions and actions had been misinterpreted, and also that there were those in Freetown who wished to discredit him in order to avoid a move of the Mixed Commissions to Fernando Po. He was convinced that the Mixed Commissions were the source of the information on which the letter was based, and he strongly suspected that the author was Reffell, Colonial Secretary and Registrar of the British Commission.

Meanwhile, the British Commission had reported the irregularities of Owen's behaviour to the Colonial Secretary, and Aberdeen had in turn drawn the attention of the Lords Commissioners of the Admiralty to their officer's conduct, asking them to require an explanation of Owen. By then Owen had written to Secretary Croker at the Admiralty refuting or excusing the allegations of the anonymous letter, and although Owen was invited to explain himself and to have the slaves he had illegally landed at Fernando Po properly condemned when the Mixed Commissions were established on the island, the matter appears not to have been carried further.[9]

While activity had been relatively intense in the Bight of Biafra during the closing months of 1828, *Sybille* and the Commodore had been absent from the Station at Ascension and St Helena since mid-September, returning to Freetown on 30 November before cruising the Bights for the remainder of the year. Collier explained this absence as necessary for refitting the frigate and refreshing her ship's company, but, as he reported to the Admiralty in January 1829, he had been dangerously ill, and his own debilitation and need for recuperation may explain this retreat by the frigate to a healthier climate.

The year 1829 was to be remembered in the Squadron chiefly for tragedy, but it began promisingly enough, and it was a tender which brought the first success of the new year. On 6 January Lieutenant Badgley in the *Cornelia* seized the brig *Jules* and the schooner *La Jeune Eugenie* off the Old Calabar bar.

Both were apparently Dutch, the former with 220 slaves and the latter with 50. Undeterred by earlier criticism, Owen kept the prizes at Fernando Po for a week before sending them to Freetown under the command of Prize Masters who were unable to bear witness to the arrests or to the authenticity of the papers produced, and with neither evidence of the vessels' nationality nor any form of declaration by himself. In Sierra Leone there was no patience left with Owen's flouting of the treaty regulations, and the Commissioners found that there were too many irregularities for the cases to come to court, although it is probable that the slaves were released. This result would have come as a disappointment to the Commodore, who had inspected *Jules* in Fernando Po and, finding her fast and new, had manned her from *Sybille* for the passage to Freetown and wanted to buy her. The King's Advocate subsequently disagreed with the Mixed Commission, declaring that the flying of Netherlands colours at the time of seizure was sufficient to justify proceedings.

Sybille was not the only visitor at Clarence Cove in early January. *Clinker* arrived on the third having lost her mainmast, and Owen's carpenters set to work once again to cut a replacement from the forest. Thanks to their efforts, George Matson and his gun-brig were at sea off the Gabon by 24 January, while the Commodore was patrolling off Princes Island.[10]

On the day after *Cornelia*'s captures, *Medina*, cruising 80 miles or so southwest of Cape Formoso, boarded the Brazilian schooner *Bella Eliza* and found 232 slaves loaded, as her master admitted, in Lagos, and the familiar Imperial passport permitting a slaving voyage to Cabinda, touching illicitly at the Guinea Coast, Princes and St Thomas. Also on board was the mate of *Minerva da Conceição*, taking passage from Lagos, where he had been stranded when his sloop was arrested with slaves supposedly loaded at Cabinda.

In Sierra Leone Major H. J. Ricketts had returned from evacuating the Gold Coast forts to find that Governor Lumley had died and that Smart, an officer junior to himself, was acting as Governor. He therefore, as senior Field Officer of the Royal African Corps in the colony, assumed the duties of Lieutenant-Governor and British Commission Arbitrator on 11 November 1828, and Smart returned home on sick leave.

At the time there was concern that the Portuguese slave trade, largely squeezed out of the Bights by the Brazilians, was breaking out afresh to the north, a concern fuelled by the seizure of two Portuguese vessels by *Primrose* at the River Cacheu on 15 January. First, the brig *Vingador* was caught with 223 slaves loaded in the Cacheu on 11 January, of whom 220 survived to be emancipated, and then, on the same day, the galliot *Aurelia*, surprised while taking slaves on board by the

10. Capture of the Spanish brig *El Almirante* by HM Brig *Black Joke* on 1 February 1829

sloop's pinnace, commanded by Lieutenant Parrey.* The court had some difficulty in distinguishing between slaves and "bona fide domestics" in this second case, but the vessel was condemned and 29 slaves were emancipated.

Heading back to Freetown from her cruise to windward, *Primrose* diverted to take a look into the River Nunez where, 120 miles upriver on 31 January, her boats boarded the schooner *Favorite* of Havana, ready and waiting to embark a cargo of slaves from the factory ashore; but, with no Equipment Clause yet agreed with Spain, she could not be touched. During this cruise *Primrose* also had an encounter with five schooners she found at anchor off the Gallinas River, probably awaiting the gathering of their slave cargoes. Commander Griffinhoofe managed to catch one of them, but only after hitting her with several shots.

Primrose was not the only cruiser to find success on the Windward Coast at this juncture. The commissioners reported to Aberdeen at the Foreign Office in February that the French Commodore, then at Sierra Leone, had sent one of his vessels to the River Pongas, and she had returned to Freetown with three slaver prizes, presumably French.

For four months *Black Joke* had found no action to relieve the tedium of patrolling the Bights, but when she arrived off Lagos on 15 January 1829 she found five apparently Brazilian vessels lying in the anchorage. Hearing that the slave cargo of one of them, a brig, was almost complete at the factory, Lieutenant Downes watched her carefully from the offing, and at daybreak on 31 January it was reported to him from the masthead that a brig was in sight on a bearing of east-south-east, standing to the south under a press of canvas. In a light south-westerly breeze *Black Joke* made all sail in chase, but by 0930 the wind had died and Downes ordered: "Out sweeps." By 1400 the chase could be identified as the "Brazilian" slaver, and, with another three hours of gruelling work at the sweeps necessary to bring the quarry within gun-shot, *Black Joke*'s crew of 47 was undoubtedly glad of an earlier temporary reinforcement by a mate and seven hands from *Medina*. At 1745 the slaver shortened sail, hoisted Spanish colours and wore twice to give the tender both broadsides, without effect. It was then sunset and Downes decided to postpone action until daylight, keeping the chase close at hand. She tried every manoeuvre to escape during the night, but at dawn the two brigs were within one and a half miles of each other and becalmed.

---

* The Mixed Commission report in the State Papers records the arrest of *Aurelia* as taking place outside the river. However, Basil Lubbock writes in *Cruisers, Corsairs and Slavers* (Brown, Son & Ferguson: Glasgow, 1993), in some detail, of capture by *Primrose*'s pinnace inside the river, and this carries conviction. Lubbock also writes that *Aurelia* had been an English yacht which had often accompanied King George IV in the yacht *Royal George*.

At 1230 a light westerly breeze sprang up, and two hours later *Black Joke* had closed to within grape range on the slaver's weather quarter. The chase then wore and engaged with her larboard broadside. The tender answered with three cheers and a rapid and accurate fire with her two guns. At 1515, having appreciated that the slaver mounted 14 guns against his own pair, Downes resolved to board and stood directly for the Spaniard, but the wind fell light and the slaver again managed to bring her larboard guns to bear, keeping up a steady fire of round shot and grape. A light breeze returned briefly at 1530 and, when the slaver tried to wear, the two brigs came into close contact. *Black Joke* found herself in a commanding position on the Spaniard's larboard quarter, and for 20 minutes, with her two guns and small arms, she raked the slaver fore and aft. At 1550, her fire completely silenced, the Spaniard hailed to say that she had struck, and at that moment there came sufficient wind for *Black Joke* to run alongside and take possession.

The prize, a slaver which had previously been boarded without success by several cruisers, was the 360-ton *El Almirante* of Havana, with a crew of 80 men and a cargo of 466 slaves. Her broadsides consisted of four long nines and ten Gover 18-pounders, massively outweighing *Black Joke*'s armament. The tender's casualties, however, amounted to two mates and four seamen wounded, of whom one of the latter died later, against 15 killed and 13 wounded, one mortally, in the slaver. The third mate was the only Spanish officer to survive. Both brigs were considerably damaged in hull, yards, sails and rigging, and it was found that the slaver's starboard main shrouds had been cleanly sliced away, apparently by a two-fathom length of chain loaded into the pivot-gun by *Black Joke*'s native cook, determined to strike a personal blow as a free African. When the Commodore subsequently saw the prize he remarked that: "I never in my life witnessed a more beautiful specimen of good Gunnery than the Stern and quarter of the Spaniard exhibits."

Lieutenant Downes was later promoted to commander for this gallant action. Two of his mates, Butterfield and Slade, had already been selected for promotion, and a third, Hardy, who had been wounded, was made lieutenant for his part in the engagement. The prize was condemned, and of the slaves 11 were killed in the action, 39 died on passage and 416 were emancipated.[11]

By now, *Sybille* had been away from England for 22 months, and Collier received warning that it was the Admiralty's intention to leave her on the Coast for four years. He replied that the ship's bottom was sorely in need of cleaning, general recaulking was necessary, and her bowsprit required replacement. He might also have added that four years in that theatre would be a grossly inhumane imposition on his ship's company. His recommendation was that she should be ordered home. This supposed martinet also took time to draw to the attention of Secretary Croker at the Admiralty a request for discharge from the Service of his marine servant. Collier had been "three times dangerously ill", and believed he owed his recovery largely to the "kindness and unremitting attention shown by my servant John Farrant, Private Marine 3rd Class". He begged Their Lordships to approve the request and, if not, to permit Collier to purchase Farrant's discharge. The Admiralty agreed to the purchase on Farrant's return.[12]

Sybille opened her 1829 account with two arrests in February. On the sixth she intercepted the 173-ton Brazilian brigantine *União* about 90 miles south of Cape Formoso, with 405 slaves crammed into her. She was the most crowded slaver Collier had ever seen, and even her Imperial passport from Bahia to Cabinda limited her to 370. The slaver admitted to the Mixed Commission that she had loaded in Lagos, and the surviving 366 slaves were emancipated. *União* had been seen ditching papers during the chase, and many of them were recovered. Some were letters in cypher which, together with similar letters found in *El Almirante* by *Black Joke*, revealed the routes taken by slavers making for Havana, and Collier sent copies to the Commander-in-Chief in the West Indies. The letters also warned the addressees in Cuba and Brazil of the difficulty in evading the British squadron and advised sending only fast and heavily armed slaving vessels to the Guinea coast.

On 19 February *Sybille* seized the Brazilian brig *Andorinha* at anchor off Lagos while discharging her trade cargo, but without slaves. This was a notorious vessel of 343 tons, a crew of 43 and seven guns, better known on the Coast as *Black Nymph*. She held the usual Imperial passport for Cabinda, touching at the Gold Coast, St Thomas and Princes, which the Mixed Commission had no difficulty in declaring illegal. In the Lagos anchorage Collier had also found two French vessels and a Swede discharging cargo. They were not slavers themselves, but had conveyed trade goods to purchase cargoes in advance of the arrival of the slavers, which would consequently need to remain at anchor for only a few hours.

Encouraged by the successes of *Black Joke* and Captain Owen's tenders, the Commodore decided to add two more ex-slavers to his squadron. The first was the schooner *Arcenia*, taken by *Little Bear* in October 1828, which he renamed *Paul Pry*,* and the second was *Black Nymph* which, in Collier's opinion, sailed better than any slaver he had yet encountered. She became the *Dallas*, named after a naval friend, but the lower deck unfairly (and inevitably) preferred "*Dull Ass*".

---

* It is not known why *Paul Pry* was so renamed by Commodore Collier. Lubbock records in *Cruisers, Corsairs and Slavers* that she was known on the lower deck as "*Peeping Tom*", or "*Little Inquisitive*".

It was fortuitous that *Paul Pry* was in harbour under the command of Lieutenant Edward Harvey when *Andorinha/Black Nymph* arrived at Freetown. The Prize Master was Mr Browne, the Acting Master of *Clinker*, a supernumerary in *Sybille*, whom Collier had selected for the task for want of a midshipman. On passage on 15 March *Andorinha* fell in with the Brazilian brig *Donna Barbara*, four days out of Lagos with 376 slaves, with a crew of 18 and four 12-pounders. Mr Browne, with admirable initiative and courage, detained the Brazilian, outnumbered though his prize crew was. This was, however, as the Mixed Commission put it, "mistaken zeal in the service", because he had no authority under the Anglo-Brazilian treaty to make such an arrest, and the court would have been obliged to declare the seizure illegal. On seeing the two prizes entering the Sierra Leone River, Lieutenant Harvey realised, with commendable alertness, what must have happened, and he boarded the *Donna Barbara* and made her prize, being an officer in the rank of lieutenant and having from the Commodore the necessary written Instructions. The court had no patience with a claim that the Brazilian's slaves had been loaded in Cabinda, and the *Donna Barbara* was condemned to *Sybille* and 351 slaves were emancipated.

However, that was not the end of the story. On reviewing the case in mid-1831 the Judge in the High Court of Admiralty, Sir Charles Robinson, declared that *Sybille* should not receive any of the benefits of the capture, contending that *Paul Pry* was 1,500 miles away from *Sybille* at the time of arrest and could not be considered part of the squadron which the Commodore commanded. This flew in the face of all custom and practice regarding tenders, as well as precedent in the current campaign. Collier appealed, and his Counsel argued that:

> The practice of allowing boats belonging to ships of war to go in quest of slave vessels had been productive of the best results. During the seven years prior to the adoption of this plan the whole squadron only captured 9679 slaves, while in the three years since the adoption of the plan 12470 slaves had been seized and emancipated.[13]

Sybille won her case.

February 1829 saw appreciable activity off Fernando Po, beginning with the confusing case of the *Adeline*. Apparently she was arrested under Dutch colours by *Eden* in Clarence Cove on the ninth, having been detained in December for firing into *Eden*'s boat and wounding a man. By the time she arrived at Sierra Leone the master had died, there was inadequate evidence that the vessel was Dutch, and disagreement was found between the statements of Captain Owen and Lieutenant Badgley, the boarding officer. The case was not then allowed into court, but it seems that further evidence subsequently came to light, particularly of the *Adeline*'s involvement in providing slavers at St Thomas (West Indies), Martinique and Guadeloupe with fraudulent French papers, and she was condemned for being fitted for slaving. No doubt the circumstances had become clear to the Mixed Commission, if to no one else.

The seizure of the Brazilian schooner *Mensageira* on the Bonny bar by *Eden*'s tender *Cornelia* on 15 February was more straightforward. As usual in Owen's cases, the prize was taken to Fernando Po and, perhaps justifiably on this occasion, all of the 353 slaves were landed because of the unhealthy state of the vessel. When the schooner sailed for Freetown 127 slaves were kept on the island, and only 177 of the remainder survived to be emancipated. It seems that of those, 66 then died ashore. This may have been the first case heard by Mr de Paiva as Brazilian Commissary Judge, and he fully agreed on the condemnation with the British Arbitrator, Governor Ricketts. He had, however, protested at his first sitting of the court against all judgments and sentences of the Anglo-Brazilian Commission prior to his grossly delayed arrival. Ricketts quite properly dismissed the complaint.

Eden herself made a capture on 26 February, at the mouth of the Calabar River, of the Netherlands schooner *Hirondelle* with fraudulent French papers. She was carrying 113 slaves and cargo to purchase 200 more. Owen had the vessel surveyed at Fernando Po to assess her fitness for the passage to Sierra Leone, landed 34 of the slaves and, with his habitual disregard of prize law, also landed the cargo and sold it at an estimate of its value. The schooner was condemned at Freetown, but 24 of her slaves had died. Owen later explained that he had been obliged to detain the last two slavers he had captured in order to obtain provisions for the settlement, an implied complaint that Freetown had failed to supply him, but only a month earlier he had reported that circumstances at Fernando Po enabled him to dispense with the tender *Horatio*, part of his own supply chain. In July the Mixed Commission emancipated by decree the slaves landed at Fernando Po from *Mensageira* and *Hirondelle*.

On 8 March *Eden*'s tender *Cornelia* was hailed by a corvette claiming to be the French man-of-war *Amphritite*, but Lieutenant Henry Kellett, suspecting that she was a pirate, kept company for two days until the corvette anchored off the Bonny River and sent in her boats. Having received word of this from Kellett, Captain Owen sent *Eden*, under the command of Acting Commander Badgley,* to arrest her. When Badgley found her, the corvette was under Spanish

---

* It is not clear whether Owen had any authority for this promotion of Lieutenant Badgley.

royal colours. She weighed rapidly, but *Cornelia* brought her to and *Eden* took possession. At Fernando Po she was found to be the *Diana* of Cadiz, previously *La Fama*, with a crew of 140 and mounting eighteen long 12-pounders, which had already taken almost 600 slaves from Whydah that year. She had a Letter of Marque from Havana and had landed her master and cargo at Bonny, but she had no slaves. Owen was convinced that she was a slaver and a pirate, but he had to let her go. That, it transpired, was *Eden*'s final effort against the slavers.

North Star left the Coast on 11 February, after 30 months on station, to head for Brazil via St Helena and Ascension. The Squadron was thereby reduced to *Sybille*, *Medina*, *Primrose*, *Clinker* and *Sybille*'s tenders, and Collier, leaving the Bight of Biafra to the attentions of *Eden* and her tenders, continued to concentrate his efforts on the Bight of Benin and on the approaches to Princes Island and the Gabon River, with one or other of his cruisers occasionally showing herself, without apparent result, off the Windward Coast. The only successes of the next few months fell to *Sybille* and her prime tender.

With, unusually for her, an undramatic capture, *Black Joke* returned to the limelight on 6 March. Her victim was the Brazilian brigantine *Carolina*, taken about 85 miles south-east of Lagos with 420 slaves. The master originally told Lieutenant Downes that he was bound from Lagos to Bahia, but then swore that he had loaded his cargo at Malembo. Finally he abandoned this ludicrous claim and admitted embarking the slaves at Onim only 19 hours before capture.* He held a commercial passport for a voyage to Onim, revealing a new practice of Brazilian slavers being issued not with the usual Imperial passports for slaving south of the Line but with commercial passports for the Guinea coast, a ruse to protect them from arrest if they were caught without slaves. This was of no help to *Carolina* which, with her slaves, was condemned.

Also on 6 March, *Sybille*, cruising off Princes Island at daybreak, sighted a large brigantine on her weather bow and made sail in pursuit. By 1300 the chase was hull-up and was clearly a slaver with a cargo capacity of 400 or so. The frigate was then hit by a "tremendous tornado" which carried away her main yard, and when the weather cleared the slaver, last sighted under all sail, was no longer to be seen. There was little doubt that she had capsized and sunk. A French frigate to leeward had seen nothing of the slaver, but closed the damaged *Sybille* to offer help. She was wearing the broad pendant of the French Senior Officer on the Coast, Commodore Vilaret de Joyeuse, who, as Collier reported to the Admiralty, "behaved in the kindest and most handsome manner, offering every assistance and", as Collier had no suitable spare yard, "supplying *Sybille* with a European

---

* Onim is an alternative name for Lagos Island.

spar of 60 feet by 14 inches." This was greatly superior to anything available on the Coast, and to part with such a valuable item so far from home was a noble gesture on the French Commodore's part, reflecting not only his own generous personality but also a thawing of Anglo-French relations in those waters.

On 23 March *Sybille* was off Whydah where, at dawn, she seized the Portuguese brig *Hosse* with 182 slaves loaded at Whydah the previous day. This vessel's recent career had been eventful, and indicative of the post-condemnation adventures of many captured slavers. She had been the *Trajano*, a Portuguese slaver condemned to *Maidstone* in 1827, sold at public auction on the orders of the Mixed Commission, and resold twice, reaching the hands of the infamous slaver Cha-Cha at Whydah. He procured colonial papers for her from the Governor of Princes, under the name of *Hosse*, but she was seized by the alleged pirate *Presidente*. The *Presidente* had then made the fatal error of attempting to capture *Black Joke*, and the tender was awarded salvage on the recaptured brig. *Hosse* had then shipped a cargo of merchandise at Freetown and sailed for Whydah to return to her slaving career. After this latest arrest she and her surviving 166 slaves were condemned, but it is scarcely surprising that this and many similar tales should cause intense frustration to Collier and his men.

Medina, now under the command of Commander Edward Webb, described by the British commissioners in Freetown as "an indefatigable officer", left the Bights for a cruise off the Gallinas in April. Over the course of several weeks she boarded five Spanish vessels fitted for slaving and ready to receive cargoes, and one, which clearly had slaves on board, she chased for 24 hours before losing her during the night. The absence of an Equipment Clause in the Spanish treaty continued as another cause of frustration.

It was left to *Sybille* to make the final arrest of the spring, and it was one which the Commodore may later have regretted. In the Bight of Biafra on 29 April she intercepted the Spanish schooner *Panchita* with 292 slaves from the Calabar, bound for Havana. The prize arrived at Freetown on 18 May and her diseased slaves were landed, but the Liberated Africans Department was crowded with the sick and it was another 11 days before the healthy slaves disembarked. The British commissioners were both ill, and it was not until 24 May that the vessel and surviving 259 slaves were condemned. By then the entire prize crew was dead of the fever epidemic sweeping Sierra Leone.

There was, of course, a general concern at the level of losses of slaves on passage after capture and before emancipation, but there was no evident solution to this distressing problem. The Africans had to remain in the slavers en route to Freetown, and there was a severe limit on how far the humanity of the tiny prize crews could alleviate conditions in the slave decks. Doctor Boyle, Surgeon to the Mixed Commissions, proposed, with little thought, that captors should provide medicines to prizes before dispatching them to Sierra Leone. The Commodore pointed out the impossibility of doing so unless the cruisers were supplied with at least ten times the quantity of medical stores currently allowed. He forwarded a suggestion from his Surgeon, Robert McKinnel, that warm clothing, red woollen shirts in particular, would be of more use to the sick in captured vessels.[14]

Another proposal had irritated Collier. The Victualling Board, at the instigation of Captain Owen, suggested that the Squadron's provisions store, which had originally been at Cape Coast Castle but was now at Ascension Island, should be moved to Fernando Po. The Commodore was unequivocally opposed to the idea. Clearly Fernando Po was closer to hand, but he was determined that his ships should retire to the healthy and invigorating climate of Ascension at least annually for refit and recuperation, and the passage to and from Ascension was relatively easy from his primary operating area in the Bights. He also saw great advantage in visits to St Helena for water and cattle, and there the men could go ashore and enjoy themselves. That was impossible on the Coast, where "the women will not come off to them". The other alternative was the Cape Verdes, occasionally used in the past, but Collier saw them as remote and as unhealthy and unsuitable for refitting and repair as Freetown and Fernando Po.[15]

The Commodore was merely annoyed by the Victualling Board, but in early May he was outraged, as he told Secretary Croker at the Admiralty, at a letter from Judge Jackson in Freetown accusing Lieutenant Turner, lately of *Black Joke*, of plundering a prize, and, furthermore, at Jackson's sweeping and unsubstantiated statement: "I have some reason to believe that the practice complained of is of too frequent occurrence." The Judge had received his information from the Registrar, who had it from "elsewhere". Collier remarked that the Mixed Commission should be aware of what value to place on the oath of a Brazilian slaver master, and he pointed out that prizes were regularly plundered not by the cruisers but by the men put on board by the Marshal of the Court after condemnation. He had seen them at it in a *Sybille* prize. Collier then attacked the Mixed Commission's recommendation that a survey and inventory should be carried out on each slaver prize immediately after capture. The idea was entirely impracticable, he said; it was difficult enough simply to count the slaves. Nevertheless, a typically unsympathetic Admiralty agreed that a survey of stores was to be made at the moment of capture.

Several other personnel matters were occupying the Commodore's mind. One was the shortage of midshipmen in the Squadron, and particularly in *Sybille*, causing especial difficulty in finding Prize Masters. The Admiralty, claiming to know better than its commander at sea, unhelpfully told him that the Squadron was up to complement. Collier was also concerned that no arrangements had been made for payment of prize money to Kroomen, invaluable members of the Squadron, and the Admiralty did at least invite the Treasurer of the Navy to suggest how that might be achieved. Then the Commodore brought to Their Lordships' attention the fact that several smugglers, sent by the courts to *Sybille* as an alternative to jail, were serving in *Black Joke*, where they had "conducted themselves with the greatest gallantry". Their behaviour had been exemplary, and Collier wrote that he would consider it a favour to himself if they were allowed to return to England when *Sybille* was ordered home. Sadly, these concerns were soon to be submerged by tragedy.

The loss of the *Panchita* prize crew to disease in Sierra Leone was the harbinger of much worse to come. *Eden* had suffered badly from fever in 1828, losing 41 of her ship's company, 23 of them ashore at Fernando Po, before retreating to Ascension to recover in October. However, her tribulations were far from over. The yellow fever epidemic which claimed *Sybille*'s men in *Panchita* began in Sierra Leone at the end of April 1829, and, by unhappy chance, *Eden* arrived there on 1 May. While she was in Freetown, Doctor Boyle noted that she was less clean than the other men-of-war, and that her men were often to be seen in a drunken state in the streets by day and night. The ship was a prime target for infection, and by the end of May, after a passage of "incessant rains and frequent tornadoes accompanied with much thunder and lightning", with "the main deck [...] crowded with sick and constantly wet", 27 of her men were dead. She reached Fernando Po on 11 June with all of her officers, apart from the First Lieutenant and the Gunner, dead or incapacitated. Among them, she had lost the estimable Lieutenant Badgley, the Surgeon and two Assistant Surgeons.

When the ship was released from quarantine in mid-June she was emptied of her stores and tanks so that she could be cleaned, fumigated and whitewashed. She sailed on 9 July, once again under Owen's command, with 58 able-bodied Europeans and 23 convalescents on board, calling briefly at Princes, and then (on the orders of the Commodore who, at the direction of the Admiralty, took her under his orders on 13 July) she headed for St Helena. On passage the fever continued its ravages. The ship's main deck was crowded with the hammocks of the sick, and when the new Surgeon embarked at Fernando Po became incapacitated, his "intellect deranged", Captain Owen, who had been ill himself, took over the care of his men. He had a horror of phlebotomy, having witnessed the frequently fatal results of such "energetic treatment", and, with a great deal more sense than much of the medical profession, he favoured mild measures, "brisk purgatives", and applying quinine until recovery.* *Eden* anchored at St Helena on 23 August having lost a further 31 men in June, 32 in July and 7 in August. Of 30 men left in hospital at Fernando Po only 19 remained alive by December. The ship's total death toll between 1 May and 1 December was 110, 13 of them African natives, and this from a nominal ship's company of 160.† [16]

The suffering was not confined to *Eden*. In mid-June HMS *Champion*, carrying supernumeraries for Fernando Po, called at Freetown, contracted the disease and conveyed it to Clarence. *Sybille* anchored there on 21 June, but Commodore Collier forbade any contact between the frigate and the shore or with *Eden*. Collier had always been strict concerning the health of his people: he warned his men of the dangers of drunkenness and punished it severely; he was particular about cleanliness and ventilation in the ship and insisted on alleviating any humidity on the lower deck;‡ he avoided men becoming wet unnecessarily; night watch-keepers were required to wear a "blanket dress"; wooding and watering were mostly done by Kroomen, who were supposedly immune to yellow fever; and seamen employed ashore were always given Peruvian bark and wine after breakfast. However, on 22 June *Sybille* received on board eight marines from *Eden* and one from shore who had arrived in *Champion*. It was enough. The frigate's first case of fever appeared on 26 June, and at Princes on 2 July the epidemic began in earnest. It was soon of a "most malignant character", claiming the lives of 22 of the 69 men attacked, but, as the ship headed south, it ceased suddenly on 28 August, and she arrived at St Helena on 12 September, three days after the last death, without a man on the sick list.

Sybille suffered even more severely through *Black Joke*. The frigate's men in the tender contracted the fever at Freetown, and by the time the epidemic was exhausted they had lost 23 of her ship's company of 45. These, together with the losses incurred by prize crews, brought *Sybille's* death toll to 57. *Plumper*, now under the command of Lieutenant John Greer, rejoined the Squadron at Sierra Leone on 1 July, and, although she appears to have escaped the fever initially, in the closing months of the year she had to land 36 of her 50 men to the military hospital in Freetown, where 24 of them died and 11 were invalided.

The year 1829 was indeed a dreadful one, as can be judged by comparison with 1826, which may be considered average, although it cost *Maidstone* 29 deaths to disease. In that year the Squadron, with a total strength of 1,043 men, lost 57 to

---

* Owen had seen one of his Surgeons bleed himself to death during *Leven's* survey of the east coast of Africa.

† As the African natives seemed to have a degree of immunity to yellow fever, it is probable that malaria or another fever was also present in the ship.

‡ This was done with Brodie's Stoves, coal-fired heaters patented by Alexander Brodie in 1780.

disease and six to accidents. In 1829, from a strength of 792, 204 men died of disease and two from accidents.

After a brief call at Freetown in February, *Primrose* had been at sea almost continuously until mid-August, mostly in the Bights, and apart from the loss of an officer left in Sierra Leone as a Prize Master, she escaped the epidemic. This may have owed something to Commander Griffinhoofe's insistence that whenever there was risk of contact with the shore his men should be dosed night and morning with bark mixed with rum.* *Clinker*'s fruitless cruising, broken only by a respite at St Helena in May followed by a visit to Ascension to collect medical stores for the Squadron, had similarly protected her.

In June, before the distraction of the fever attack, Commodore Collier had expressed his concern at the increased number of Spanish slavers in the Bight of Benin. He was aware of eight of them awaiting cargoes at Whydah, one of them a formidable frigate-built ship pierced for 26 guns with a picked crew of 170, the *Veloz Passagera*. There were another four between Whydah and Cape St Paul's, and in April there had been eight at the Bonny River. One of these Bonny slavers was another sizeable ship, the 24-gun corvette *Fama de Cadiz*, which had left the river in the company of a five-gun schooner and a brig of 14 guns and, as the "privateer" *Providentia*, had been given a bloody nose by *Black Joke* after she misguidedly attacked the latter in April 1828. On departure this trio was chased by *Primrose* and initially showed an inclination to engage, but spoke to each other and decided to separate and run for it. *Primrose* then lost them during the night. It was believed that the ship was carrying 900 slaves and the others 700 between them.

It had been Collier's intention not to leave the Bights to refit and refresh *Sybille* at St Helena until the *Veloz Passagera* had been brought to book, but the fever epidemic left him no option but to retreat south, and in his absence he left *Black Joke* to cover Lagos and stationed *Medina* between Cape St Paul's and Whydah. Commander Webb had no luck on that hunting ground and, no doubt to his particular disappointment, the *Veloz Passagera* failed to make an appearance. Further to the south-east, however, on 7 August he picked up the Brazilian schooner *Santo Jago* with 209 slaves, and ten days later, off the Gabon, he arrested the Spanish schooner *Clarita* from Bonny with 261.† The *Clarita* had smallpox on board when seized, and only 201 slaves survived to be condemned with the vessel. Webb managed on this occasion to achieve a survey of the slaver's stores on capture, as demanded by the Mixed Commissions.

---

* Bark and wine was a more common concoction, but *Primrose* had no wine on board.
† There is no mention of this case in the Mixed Commission records, and available details are sparse, but there is no reason to suppose that the vessel and the surviving slaves were not condemned.

August was a surprisingly successful month for *Plumper*, too. Little was expected in the way of seizures by the two gun-brigs, and the Admiralty displayed remarkable pig-headedness in continuing to deploy these unsuitable vessels to the Coast in defiance of the recommendation of every West Africa Commodore. Early in the month, however, while cruising 90 miles to the east of Princes, *Plumper* caught the Brazilian schooner *Ceres* with 279 slaves. The slaver had loaded her cargo in the Cameroons River, and simultaneously and with moronic stupidity, her master had replenished with water from the river alongside. Too late the prize crew discovered that the water was brackish and dirty, and dysentery was added to the woes of the slaves, most of whom were already suffering from worms. Nearly 60 per cent of them died before condemnation on 22 September. This was the schooner's second appearance before a Mixed Commission; she had been condemned as the *Gertrudis* in February 1828 and sold.

Venturing into the Cameroons River on 22 August the gun-brig then cornered the Brazilian schooner *Restamador* with 277 slaves, her final success before being struck disastrously by yellow fever.* *Clinker* was unable to emulate her sister, making her sole capture of the year on 31 October when she intercepted the Brazilian brig *Emilia* 80 miles north of Princes, carrying 157 slaves shipped at the Bonny River and an irregular passport from Pernambuco. Owing to the disruption caused in Sierra Leone by the fever, the brig and her surviving 148 slaves were not condemned until 1 May 1830.

Another Brazilian *Emilia* had already fallen victim to *Sybille*'s tender *Dallas*, patrolling the Bight of Benin under the command of Lieutenant Harvey. She seized this slaver schooner off the mouth of the Benin River on 16 August with 486 slaves, and they were condemned a mere five weeks later. On 1 November, 50 miles further west, she made another sizeable capture, the Brazilian schooner *Tentadora*, with 432 slaves. This second slaver held a passport for Malembo from Bahia, but had made straight for Lagos. Her adjudication too was delayed until 1 May 1830, by which time a grim total of 112 of the slaves had succumbed.

Dallas's fellow tender, *Paul Pry*, had met with no success since the *Donna Barbara* incident in May, and her career came to an unfortunate end during the autumn. She had been chasing a suspected slaver since dawn, and shortly after 1600 she was within range for a warning shot from her single 6-pounder. Small arms had been prepared for boarding and piled under a tarpaulin on the quarterdeck. No sooner had the shot been fired than Lieutenant Browne was obliged to shorten sail under the imminent threat of a tornado. The schooner quickly recovered from the initial blow of the wind, but amid the lashing rain and bolts of

---

* Details are sparse on this case.

lightning the charges in the piled weapons began to explode. The mate fell with both legs shattered below the knee by musket balls, and several other men were wounded. There was no Surgeon on board, but it was apparent that the mate's legs required urgent amputation. The French cook undertook this ghastly task with his carving knife in appalling conditions. Browne made for Fernando Po where, luckily, he found one of the cruisers. The mate, the 19-year-old Mr Allen, Acting Second Master of *Sybille*, survived the ordeal, but, with an Admiralty pension of only two shillings and sixpence a day, his prospects were poor. After this incident the Commodore sold *Paul Pry*, and in October he wrote to beg Their Lordships to employ Allen as a clerk in a public office.[17]

With his ship recuperating in the South Atlantic, Captain Owen was able to return to his prolific correspondence to Secretary Croker, and, on 2 August, he sought to justify his boarding French-flagged slaving vessels in the Bights. On *Eden* coming under the Commodore's orders, Collier had sent Owen extracts of his secret instructions from the Admiralty prohibiting interference with slavers under French colours, or any others not authorised by the anti-slaving treaty Instructions. Owen suggested that his conduct would have been different if he had seen those orders earlier, but claimed that the right of examination remained and was necessary. He was undoubtedly reflecting the views and experience of the Squadron's commanding officers in pointing out that most of the vessels wearing the French flag on the Coast were in fact Dutch, Spanish or Portuguese, and he urged that the prohibition orders should specify "vessels under the French flag or others, those flags being legally assumed and worn, and the vessels being legitimately employed according to their passports", which, he was sure, was their true intention. Slavers with false French colours and papers were frequently being allowed to proceed, and he mentioned one he had encountered, the *Coquette*, which had hoisted her master's tablecloth in lieu of French colours.* [18]

Medina had joined *Sybille* at St Helena on 18 September for refit, and the two sailed in company on 25 October. *Sybille* had been caulked, refitted, her copper repaired, but still had her fore- and foretopsail yards sprung. *Medina* headed for Sierra Leone to collect stores for the Squadron and to pick up despatches for the Commodore, now addressed to "The Senior Officer, West Africa". *Sybille* called at Ascension, where the Agent Victualler had on charge "a very large quantity of provisions for Squadron and island",† and on 2 November, provisioned for five months, she sailed for the Bights. To Secretary Croker, Collier once again argued

---

* The replacement of the White Flag of Bourbon by the Tricolore in August 1830 would, in time, prevent such a simple abuse.

† These would have been dry and preserved provisions.

the case for his cruisers to visit St Helena in rotation during the rainy season as the most amenable refuge for general refit and as a source of fresh provisions of the best quality at a moderate price. Soft bread, fresh meat and oxen, vegetables and beer were all readily procured there, as were plentiful wood and water, not available at Ascension, and, unlike the ports of call north of the Line, it was safe there for the men to go ashore.*

After a brief call at Accra, *Sybille* returned on 19 November to Fernando Po, where Collier found a letter from the Admiralty (which clearly had crossed with his recent commendation of St Helena) requiring the Squadron in future to refit and refresh at Ascension and forbidding him from going to St Helena. No doubt Their Lordships were concerned about distance from the Guinea coast, but Collier pointed out again the want of adequate water and fresh provisions at Ascension. He had already instructed his commanding officers to make for St Helena if struck by sickness, but undertook to rescind that order, and acknowledged that Sierra Leone, with all its disadvantages, would often have to be used for refits.

En route to Fernando Po, the frigate had carried out her final boarding of 1829 when she encountered the *Eliza*, carrying 400 slaves but under French colours. The slaver was nominally registered at Martinique, but her last port of call had been the Dutch island of St Bartholomew, where she had shipped two-thirds of her crew. Collier was convinced that her French papers had been falsely obtained there too, to disguise Dutch nationality, but he had unequivocal instructions to desist from interfering with her like.

Returning to the Bights from her task at Sierra Leone, *Medina* achieved her last success of the year on 10 December, seizing the Brazilian schooner *Não Lendia* 110 miles south-south-west of Bonny. The schooner held the usual passport from Bahia to Cabinda, but her 184 slaves had been loaded at Lagos. Her case was not judged until the following May, and 25 of the slaves had died before they could be condemned.

On 4 October a relief for *North Star* arrived at Freetown. She was the Sixth Rate *Atholl* (28), absent from the Coast since March 1826, now commanded by Captain Alexander Gordon. *Atholl* brought with her *La Laure*, a vessel under French colours, which she had seized on 1 October 130 miles west of the Isles de Los with 249 slaves who were part of a cargo recovered from the Spanish schooner *Manuel*, wrecked at the River Shebar. The slaves were landed on arrival at

---

* Prices were: bread – three pence per pound; meat – sixteen pence per pound; vegetables – four pence per pound. Collier managed to have the price of strong beer from the island's brewery reduced from 2 shillings per gallon to 1 shilling and sixpence for HM Ships, it being recommended for the sick!

Freetown, and it seems that *Atholl* then sailed to take *La Laure* to Goree, but then brought her back. Reports on subsequent action are contradictory, and obviously there was indecision on how to deal with the case. Apparently the Ordnance Storekeeper and Collector of Customs arrested her on return, for reasons unclear and without authority, and even put soldiers on board until the Governor told him to remove them. As Spanish property, the slaves were finally condemned by the Anglo-Spanish Mixed Commission, but when Gordon's agent was unable to establish that the vessel was Spanish property, the Collector instituted proceedings in the Vice-Admiralty court. In May 1830 the Colonial Office reported to the Admiralty that the vessel had been condemned.*

Atholl, leaving the *La Laure* confusion behind her, sailed on 16 October for the Squadron rendezvous in the Bights, and on 9 December she achieved a satisfactorily clear-cut arrest when, 50 miles south of Lagos, she encountered the Brazilian brigantine *Emilia*, the third slaver of that name to be seized within four months. The prize was carrying 187 slaves, of whom 59 were added to the dreadful death toll of slaves in the long wait until May 1830 for condemnation.

By October *Black Joke* had largely recovered from the debilitation of the fever epidemic, and command had passed from Henry Downes, promoted for the *El Almirante* action, to Lieutenant Edward Iggulden Parrey of *Primrose*. A grateful Commodore had presented Downes with a polished oak vase bearing silver-gilt ornamentation as "a tribute of admiration and respect from Commodore Collier to Lieut. Henry Downes, for his gallant conduct when in command of H. M. tender BLACK JOKE."

Shortly after the capture of the *Carolina*, however, Downes had to be invalided home in poor health.

Parrey's first opportunity came on 10 November, when he took the Spanish brigantine *Cristina* with 348 slaves. *Black Joke* found the slaver stranded on the Scarcies Bank, off the river of that name, onto which the current had carried her in calms and poor visibility. Despite every effort to refloat her she filled with water and was lost, but Parrey rescued the slaves and crew, and salvaged some sails. The death toll before condemnation was dreadful, despite the proximity of Sierra Leone: 132 were lost, and 75 of the 216 emancipated were children. Walter Lewis presided as Judge in the case, in the absence of Ricketts through sickness, and there was still no Spanish Commissioner.

---

* Presumably this was the result of the trial in the Vice-Admiralty court, but the grounds for condemnation are not apparent. The report of *La Laure*'s condemnation was, however, contradicted in a report from the British Commissioners to Lord Palmerston in September 1836 which noted that she had been restored as French.

Eden, rigging refitted and health restored, sailed from St Helena on 7 September and called at Ascension, where Owen sold his tender *Cornelia* to Commander Boteler of the survey vessel *Hecla*. The British commissioners had summoned Owen to Sierra Leone to explain some of his flouting of the anti-slaving treaties, and *Eden* reached Freetown on 3 October. As Owen saw it, the "hostile feeling of our judicial courts" had diverted him from his proper place of duty, the Bight of Biafra, which had allowed several large slave cargoes to be successfully shipped, despite the presence of *Primrose*, which they "beat in sailing". These included the *Nueva Diana* which, he heard, had taken off 1,500 slaves from Bonny.

On return to Fernando Po in late November, Owen sent Lieutenant Mercer with a pair of boats to search the Cameroons River, and was rewarded with the capture, by Acting Lieutenant Roberts, of the Brazilian brigantine *Ismenia* at King Bell's Town on the twenty-eighth. Although she was empty of slaves she still had a cargo on board; and, in a further illegal act, later reluctantly approved by the Foreign Office, Owen sold part of it at Fernando Po. By now Owen was aware that he was soon to take *Eden* to Brazil, and he lamented that he would have to buy the *Ismenia* to enable his prize crew to rejoin at Rio, but the prize was not available for purchase. It was not until June 1831 that, despite a protest by the Brazilian Judge Mr de Paiva, the brigantine and cargo were finally condemned on account of an irregular slaving passport from Rio. The two judges were in disagreement, and, in the absence of a Brazilian Arbitrator, the decision fell to Mr Smith, the British Arbitrator.[19]

The 1829 yellow fever epidemic, or fear of it, had cut a swathe through the Mixed Commissions in Sierra Leone. Judge Jackson fell ill and returned home in June, leaving the lieutenant-governor, Major Ricketts, as temporary Judge; Joseph Reffell, the Registrar, died in July, and was replaced by Thomas Cole; Samuel Smart, the acting Arbitrator, was also obliged to depart on sick leave, but, fortunately, William Smith had returned and was able to resume the post. By mid-July Mr de Paiva was the only foreign Commissioner remaining in the colony. There was further disruption at the end of the year when the Foreign Office appointed W. W. Lewis as Registrar, and then Ricketts, who had been unwell for 18 months, returned home, having appointed Captain Evans R. A. C. to act as Governor and Judge in his absence. A Captain Fraser arrived in the colony in early January 1830, found he was senior to Evans, and claimed both posts. In February news arrived at Freetown that Lieutenant-Colonel Findlay had been appointed Lieutenant-Governor of the Gambia, but when London became aware of the unsatisfactory situation in Freetown it took the easy option of shifting Findlay, and he assumed governorship of Sierra Leone on 26 April 1830.

The beginning of the year 1830 saw a brief improvement in *Sybille*'s fortunes. On 7 January Lieutenant Edward Harvey in *Dallas* arrested the Brazilian schooner *Nossa Senora da Guia* just off Lagos, where her 310 slaves had been loaded the previous day. Unhappily, 58 were lost to smallpox on passage and another 14 died at Freetown after arrival on 19 February and before disgracefully delayed condemnation on 13 May. There were further repercussions from this arrest: stores had been removed from the prize, and Judge Findlay appears to have been more concerned about the apparent loss of 120 fathoms of four-inch cable and a new mainsail than about the deaths of the slaves.

Slave losses in the next prize were even more appalling. She was the Brazilian schooner *Umbelina*, taken on 15 January by *Sybille* herself 150 miles south of Porto Novo. Collier despatched her immediately for Freetown with the 377 slaves she had loaded at Lagos, but before arrival on 13 March she had lost 194 of them, and, in the two-month wait for condemnation, another 20 died in harbour. The Commodore was successful again on 23 January, seizing another Brazilian 60 miles south-west of Lagos, the brigantine *Primeira Rosália*, with another cargo from Lagos. Before she was condemned, also on 13 May, the brigantine had lost 40 of her 282 slaves.

Judge Findlay reported stores deficiencies in all of these prizes, and *Medina* was the next to fall foul of his mission to eradicate pilfering. On 2 February she took the Brazilian brigantine *Nova Resolução* just below Cape St Paul's, and intercepted 200 miles south of her departure point with an unusually small cargo of 43 slaves taken from Awey. On being condemned on 13 May, a busy day for the Mixed Commission, the brigantine was reported to be missing livestock, rope, muskets and powder. She had, however, lost only one slave. It should have come as no surprise to the court that the livestock had been consumed during the passage, and it was explained that the weapons and powder had been taken into safe custody and replaced once the vessel had entered harbour.

Further discrepancies were later found, and the Foreign Office requested an investigation by the Admiralty. Commander Webb was invited to explain himself, and he, in turn, sought the comments of the Prize Master. Mr Pearne reported that the rope had been used to fish the brigantine's foremast after a lightning strike; tobacco had been issued to the slaves; flannel had also been given to the slaves to make themselves decent; and a gaff-topsail had been cut up for the same purpose; a lead had been lost while sounding; a studdingsail had been used to repair the foretopsail; eight muskets had been exchanged for provisions at Kroo Sesters; two sweeps had succumbed as firewood; there had been no fore-royal sail on board as claimed; and all liquor cases had been landed at Sierra Leone, except for three cases misidentified at the initial muster. The matter of the liquor might have warranted further investigation, but otherwise there was no evidence here of criminal activity. It seems that neither the Freetown commissioners nor the London bureaucrats, amid a general lack of understanding of life at sea, could comprehend the difficulties encountered by a prize crew on boarding a slaver, and expected a faultless inventory. However, Findlay was hardly unreasonable in demanding that Prize Masters should deliver accounts of stores necessarily expended on passage when entering ships' papers at court.[20]

There was some astonishment in Freetown on the evening of 23 January when the slaver brig *La Louise*, brought in two days previously by Lieutenant Ramsay of the *Atholl*, suddenly put to sea again. It transpired that *Atholl* had chased and boarded her 30 miles off the River Gabon late in 1829 while she was under French colours, but Captain Gordon had been assured by Owen, who had examined her before loading in the Old Calabar, that she had Dutch as well as French papers. The master and mate admitted to Gordon under oath that the vessel was indeed Dutch, her Dutch papers and colours having been ditched during the chase, and he despatched her for Sierra Leone. On passage, however, Ramsay began to suspect that the brig really was French, and after arrival at Freetown her officers admitted that their statements to Gordon had been false. To avoid embarrassment or worse to his Commanding Officer, he therefore took her to sea and handed her back to her master, together with the cargo of 226 slaves. The Collector of Customs protested, but Ramsay, who had made the usual report to the Marshal of the Mixed Commission court, responded robustly. As was to be expected, a protest was received from the Senior Officer of the French squadron, on the matter of *La Laure* as well as *La Louise*, but both Ramsay and Captain Gordon had undoubtedly acted correctly.

Captain Owen and *Eden* said farewell to West Africa on 23 February, but 30 hard months on the Station had not earned them a rest. They sailed from Ascension for Brazil, where the prospect of more cruising against the slavers awaited them. Only six months later it became clear that there was no future for the settlement at which they had laboured so hard. In September the Foreign Secretary, admitting that no progress had been made in negotiations with Spain to permit a move to the island by the Mixed Commissions, enquired of the Admiralty whether it attached importance to Fernando Po as a naval station. Their Lordships replied dismissively that it was handy as a rendezvous, but had no use beyond slave-trade suppression. This would have brought acute disappointment to Owen who, although he had been a troublesome colleague and subordinate during his period at Fernando Po, was a man of ability and intense dedication. That dedication may have been tinged with self-interest, but his personal campaign against the slavers in the Bight of Biafra, which had initially been a sideline and had then increasingly become his central (if unofficial) objective, had brought appreciable benefit.

It seems that *Atholl* caused irritation to Acting Governor Fraser in two instances during March and April. Gordon, heading north to give protection to the gum trade vessels at Portendic, undertook to carry 150 liberated male slaves from Sierra Leone to the Gambia, but declined to take 50 women. This seems to have been a wise decision, but Fraser probably had no notion of what conditions would be like in a man-of-war 114 feet in length crowded with 200 people in addition to her complement of 175. There was then a complaint that an *Atholl* boat's crew had damaged a drain in Freetown and that sailors had committed an assault at the watering place, not the first alleged contretemps there involving Squadron seamen. Gordon dismissed the accusations, but there remained a suspicion of friction between the Navy and the people of Freetown.

Primrose and *Clinker* achieved some success at the end of March. Acting Commander Edward Parrey had been promoted out of *Black Joke* to command *Primrose* after Thomas Griffinhoofe died at Ascension, and on the twenty-fourth he captured the Spanish schooner *Maria de la Conception* at the River Pongas. The prize, with her 79 slaves intact, was quickly condemned. On the same day *Primrose*'s boats, under the command of Lieutenant Butterfield, took the Spanish schooner *Conchita*, also in the Pongas and also with 79 slaves.

While these rare Windward Coast captures were taking place, *Clinker* was cruising off Cape Formoso, and on 27 March she seized another Spanish schooner, the *Altimara*, with 249 slaves shipped from the River Brass. As was now all too common with prizes taken in the Bights, her losses in slaves were severe: 51 before condemnation on 11 May. This was *Clinker*'s final contribution of note. She took 50 liberated slaves from Sierra Leone to the Gambia in May, but by June she was in a very poor state, with (among other defects) serious rot in hull planking and both masts, and her mainmast sprung in two places. Furthermore, her ship's company was very run-down, and at the end of the month, once essential repairs had been completed at Princes, she was sent home.

Black Joke had recorded no captures since her shocking losses to fever in 1829, but on 1 April, now under the temporary command of her senior mate, William Coyde, she added to the list of Spanish prizes with the arrest of the brigantine *Manzanares* on (for her) an unfamiliar cruising ground 200 miles west of Cape Mesurado. With three guns and 42 men, the brigantine was apparently a match for *Black Joke*, but the capture was uneventful, and *Manzanares*, with the surviving 349 of her 354 slaves shipped from Cape Mount, was condemned.

Since her return to the Bights in November, *Sybille* had generally kept well offshore, but she met *Black Joke* at Princes in early January 1830. At about the same time, she apparently embarked a boy from the *Tyne*, which was passing, and which was said to be healthy. The boy may have been the source of infection, but whatever the case, the frigate was again struck by yellow fever while cruising off Cape Formoso later in the month. By early February it had become a very alarming outbreak, causing "most dreadful havoc among all classes on board".

By all accounts *Sybille* was not only a tautly disciplined ship, thanks to Commodore Collier, but also a happy one. However, the blow of this second epidemic dispirited the ship's company, and they were particularly downcast at the death of the revered old Master, Tom Collins. Morale was seriously endangered by the men's conviction that the disease was contagious, and the Surgeon, Doctor McKinnel, determined to demonstrate his belief that it was not. He directed his Assistant Surgeon, Alexander McKechnie,[*] to collect a pint of black vomit from the next man to be attacked.[†] On hearing that this had been done, McKinnel took a wine glass from the gunroom and returned to the half-deck while the men were at their dinner. There he encountered Lieutenant Green, who was going below after his watch. McKinnel then, in the words of McKechnie,

> called [Green] over, and filling a glassful of the black vomit, asked him if he would like to have some of it; being answered in the negative, he then said, "Very well, here is your health, Green," and drank it off. Dr. McKinnel immediately afterwards went to the quarter-deck, and walked until two o'clock to prevent its being supposed that he had resorted to any means of counteracting its effects [...] It is almost unnecessary to add that it did not impair his appetite for dinner, nor did he suffer any inconvenience from it afterwards.

There can have been few instances of such a revolting but selfless and heroic act.

Collier decided to head once again for the haven of St Helena, and, by the time the ship arrived on 13 March, she had lost 26 men from her 87 cases of fever. However, the island did not provide the respite expected, and a further outbreak beginning on 22 March brought another six deaths from 22 cases. Leaving her boom boats behind to allow better ventilation, the ship put to sea on the twenty-ninth to find cooler weather, reaching 36° S before the fever

---

[*] Commodore Collier writes "McKenzie", but the *Navy List*, supported by Lloyd and Coulter in *Medicine and the Navy*, shows "McKechnie". It is assumed that the Commodore's memory was at fault.

[†] The patient, named Riley, died about two hours later.

entirely disappeared. The Commodore paid warm tributes to his Surgeons: "The unremitting attention of Dr McKinnel to the sick surpassed anything I had ever experienced", and, of McKechnie, who had been one of the first with fever, but soon recovered, "his conduct has been beyond all praise".[21]

The Admiralty had told Collier that it intended to leave *Sybille* on station until 1831, but that he should return home earlier if he felt it necessary, leaving his senior captain in command. Before sailing from St Helena, he had decided that neither his ship nor his ship's company was in a fit state to remain on the Coast, and he had sent to Sierra Leone to summon Captain Gordon in *Atholl* to Ascension, there to meet *Sybille* and take over command of the Squadron. However, the St Helena schooner carrying the Commodore's letters had been intercepted by a pirate felucca called *Desperado*, some of her crew had been murdered, she had been damaged and plundered, and the letters had been taken. Hearing of this, Gordon sent *Primrose*, in company with *Black Joke*, which had become too defective to be left on her own, to hunt the pirate, and, guessing correctly, sailed himself for Ascension in the hope of finding *Sybille* there. *Medina* was to remain in the vicinity of Sierra Leone until late May, recover the Squadron's prize crews, and then make for Princes.

There was one final success for Collier. Shortly before his departure from the Coast, his tender *Dallas* took the Spanish schooner *Madre de Dios* with 360 slaves after an epic chase in the Bight of Benin. Having sighted the slaver becalmed at dawn, the short-handed tender's crew spent from about 0900 until dusk at the sweeps, and, with the return of a westerly breeze, Lieutenant Harvey skilfully managed to keep the chase inshore of him during the night. At daybreak the schooner was again seen at a distance of seven miles, and, running eastward before a light wind, *Dallas* closed the gap to two miles during the day. The Spaniard was running out of sea room, and the following morning she was still inshore of the tender which, in a moderate breeze all day, continued steadily to close her. When darkness fell it appeared that the slaver intended to run herself ashore, but, in a freshening wind, she was shortly seen to cross *Dallas*'s bows. Rounding on the same tack, the tender brought her under gun and musket fire until she surrendered. This dogged pursuit had lasted for over 60 hours. The fate of this prize is, however, not known. There is no record of her appearance before the Mixed Commission court, and she may have been lost on passage to Sierra Leone.*

---

* The only account found of this incident is in Lubbock's *Cruisers, Corsairs and Slavers*, which is generally reliable. There is no reason to doubt its authenticity, but there is no confirming reference from either Commodore Collier or the Mixed Commission court, and *Dallas*'s log has been lost. The date of the seizure is not known.

Dallas rejoined *Sybille* at St Helena, and there the Commodore sold her to a merchant of the island. The frigate, refitted for her voyage home, then made for Ascension, where *Atholl* had arrived on 20 May. The handover of command to Captain Gordon completed, *Sybille* and Francis Augustus Collier sailed for Spithead after three gruelling years on station.

Gordon's first concern during the six months of his interregnum was the deteriorating material state of *Black Joke*, which *Atholl* had inherited from *Sybille*. Parrey had the tender hauled up ashore at Princes by the time *Atholl* arrived on 14 June, to find *Medina* and *Primrose* also present. Webb and Parrey had ascertained that *Veloz Passagera* and nine other slavers were awaiting cargoes at Whydah, and also had reason to believe that the pirate *Desperado* had returned to her home port of Barcelona. Armed with this intelligence, Gordon ordered *Medina* to cruise in the vicinity of Whydah, and, having collected materials for *Black Joke*'s refit from Fernando Po, he stationed himself in Collier's favoured Squadron rendezvous, 150 miles north-west of Princes, to intercept traffic leaving the Bights. *Primrose* was despatched to Ascension with sick and invalids, and to take a look at the Brazilian slave trade south of the Equator. *Clinker* and *Plumper* were to rejoin at the rendezvous, *Clinker* to refit in preparation for her passage home.

On arrival at Princes, Gordon had found news from Lieutenant John Adams, who had relieved Lieutenant Greer in command of *Plumper*, that on 12 May, about 25 miles off Trade Town, his boats had captured the Spanish schooner *Loreto* (alias *Corunera*) with 186 slaves from Little Bassa. Three slaves were lost overboard in a tornado on passage to Freetown, but the schooner and surviving slaves were condemned. *Plumper* herself reached Princes shortly afterwards, needing a refit and with Adams unwell, and Gordon sent her to Ascension.

Any hope that there might have been for a more capable replacement for *Clinker* were dashed by the arrival at Sierra Leone on 11 July of her sister gun-brig *Conflict*, last on the Station at the end of 1827, and now commanded by Lieutenant George Smithers. When she made her way to Princes in early September, Gordon decided that she would be best employed in embarking a party of *Sybille* men, stranded on the Coast by the frigate's premature departure, and taking them to Ascension to await passage home. From there she sailed for the Gambia.

Atholl, meanwhile, cruising off the Bonny River on 3 August, seized the Spanish schooner *Santiago*, bound for Santiago de Cuba with 162 slaves taken from Bonny two days earlier; 153 of the slaves survived to be emancipated. Less satisfactory was *Medina*'s capture on the eighteenth of the same month of another Spanish schooner, the *Atafa Prima*, 80 miles south-west of Cape Formoso. Commander

Webb reported that the master had seized six Africans from Grand Bassa and confined them in irons, but the Africans later told the Mixed Commission that they had gone aboard the schooner of their own free will, and not only were the vessel and Africans restored, but also poor Webb was charged £134.5s damages.

In mid-1830 the Treasury decided that the £10 bounty on captured slaves had become excessively expensive, and that the time was ripe to economise. On 16 July, therefore, an Act (1 Will. IV, c. 55) was passed to reduce the bounty. It declared that "whereas it is expedient to reduce the different rates of Bounties" payable under earlier acts, for all slaves seized from and after the fifth day of October 1830 the existing rates were repealed. In lieu, a bounty of £5 would be paid upon every man, woman and child slave seized and condemned. It is probable that, when this news eventually reached the African coast, the men of the Squadron would have doubted the "expediency" of the measure, and would have drawn discouraging conclusions concerning the value placed upon their endeavours and hardships.

During her cruise south of the Line, *Primrose* had found that the Brazilian slave trade was at a relatively low ebb, and the King of Loango had admitted that he had been obliged to kill about a hundred of his captives because there were no slave vessels on hand and he could not afford to feed them. On 3 September Acting Commander Parrey, having brought *Primrose* back to Princes Island, was relieved by Commander William Broughton and returned home. Parrey must have been bitterly disappointed when he heard that, a mere three days after his departure, *Primrose* had intercepted the long-awaited Spanish slave ship *Veloz Passagera* and fought one of the most renowned actions of the campaign.

Primrose was heading for Badagry when she sighted a sail at 1730 on 6 September, and crowded on canvas in pursuit. She lost sight of the chase in poor visibility, but at 2330 she was again seen, close at hand and standing close-hauled for *Primrose*. Broughton's hail was not answered, and he fired a warning gun to leeward. This too produced no response, and he tacked, ranged up on the weather quarter of the stranger, fired two shotted muskets over her and ordered her to heave to. She did so, and was seen to be a ship somewhat larger than *Atholl*. Sent across to examine her was the First Lieutenant, Edward Butterfield, only lately confirmed in rank but greatly experienced on the Coast, having served in *Atholl* and *Sybille*, and having been a mate in *Black Joke* during the *El Almirante* action. He had also boarded the *Veloz Passagera* at Whydah and, of course, had immediately recognised her. Initially he was told that the ship's master was ill, but, when he

11. Capture of the Spanish ship *Veloz Passagera* by HM Sloop *Primrose* on 7 September 1830

looked down into the main deck, Butterfield saw not only the crew at quarters, armed and keeping the broadside guns trained on *Primrose*, but also, at the foot of the companionway, the master, who produced the ship's papers, volunteered that he was bound for Princes for wood and water but repeatedly refused Butterfield's request to see round the ship. However, Butterfield had seen and smelled enough to tell him the ship was loaded with slaves. On Butterfield's return, Broughton demanded to send another boat, but the slaver refused, claiming that he did not know in the darkness that the sloop was not a pirate. Broughton then wisely decided to stay by the Spaniard through the night and board her in the morning.

At 0545 on 7 September, with ensign at the peak and Union Flag at the foremast-head, Broughton bore down on the slaver, hailed several times through a Spanish interpreter that if she did not heave to and permit a boat to board within five minutes, he would fire into her. As the time expired and with the vessels almost touching, *Primrose* fired her broadside, and *Veloz Passagera*, which mounted 20 guns of various calibres from 9-pounder to 18-pounder, immediately replied in kind. On firing his second broadside, Broughton ordered his helmsman to lay the sloop alongside the Spaniard's starboard bow, and he led his boarders up onto the slaver's rail, well above *Primrose*'s forecastle. This disparity of height, and the movement between the vessels in the swell, made boarding difficult and dangerous, and the slaver crew fought with great resolution. As he reached the Spaniard's rail, Broughton was piked in the abdomen and fell into *Primrose*'s fore-chains, but Butterfield led the boarders onward, with Acting Lieutenant Foley, the Acting Master and all the midshipmen right behind him. The slavers disputed possession right down to the orlop deck, leaving a bloodbath behind them, but in about 12 minutes the ship was taken.

Of her mixed-nationality crew of 150, *Veloz Passagera* had lost 46 killed and 20 wounded, of whom six later died. By contrast, *Primrose*'s casualties were astonishingly light: three killed and twelve wounded, and five of the Spaniard's 556 slaves were killed by the British broadsides. The sloop's Acting Surgeon Lane, although sick with fever, dragged himself from his berth to tend wounds from grapeshot, cutlass, musket ball, pike and bayonet, and it seems that he saved all of *Primrose*'s casualties. The prize was severely cut up in the hull, but her rigging and spars were uninjured. The sloop suffered considerable damage to rigging, boats and upper works, but not to the hull.

The severely wounded Broughton was warm in his commendation of Butterfield, to whom he entrusted the prize; of Foley, who was given temporary command of *Primrose*; of Mr Fraser, the Acting Master; of Mr Williamson, the

Acting Purser; and of Midshipman Bentham.* He praised his ship's company not only for their steadiness in action but also for their exertions in refitting the two vessels afterwards. Governor Findlay shared Broughton's anger at the unjustified resistance of the slaver crew, and he insisted that, on her sailing for England, *Primrose* should take 24 of them to be tried for the murder of the sloop's men. The Foreign Secretary, Lord Palmerston, later told the commissioners in Freetown that a charge of piracy was not considered to be appropriate and that, as British courts had no jurisdiction, the 24 men would be sent to Spain for trial, as indeed they were in the spring of the following year.† Of the *Veloz Passagera*'s 551 surviving slaves, 21 died before the prize was condemned and their fellows were emancipated.[22]

On her return from Ascension in early October, *Atholl* remained on the Windward Coast, and on the seventeenth, 210 miles west of Cape Mesurado, she intercepted the Spanish schooner *Nueva Isabelita* with 141 slaves taken from Little and Grand Bassa. Captain Gordon was expecting the arrival of a new Commodore at any moment, and on taking his prize to Freetown, where she and 139 slaves were condemned, he waited in harbour for a week at the end of October in the hope of handing over command. However, *Medina* was holding the fort in the Bights while *Primrose* refitted in Ascension, and would have to head there herself for provisions by the end of the month. Gordon therefore abandoned his wait in order for *Atholl* to relieve *Medina*, and left instructions for *Primrose* to remain in the vicinity of Freetown on her return.

In early October, before departing for Ascension, *Medina* sent her boats into the entrance of the Old Calabar River, where they captured the Spanish brigantine *Pajarito* with 239 slaves. There was a less happy story from her pinnace and cutter sent to examine the rivers between Cape Formoso and Bonny. On 6 October there was a mutiny in the cutter, commanded by Mr George Herbert, Mate. When Herbert ordered the crew to pull against the tide for a Liverpool brig in one of the rivers the Coxswain, James Pitt, said he would "be damned and buggered if he would steer". Thereupon, Herbert hit Pitt with a piece of wood, and then William Calvert, seaman, came aft, collared Herbert and said, "You bugger, I will throw you overboard." In self-defence, Herbert then stabbed Calvert. This sad affair, probably the result of fatigue, poor leadership and possibly liquor, broke Herbert, who shot himself during *Medina*'s passage to Ascension. It is perhaps surprising that the strain of cruising the slave coast did not give rise to more such outbursts.

---

* Most of the sloop's officers and warrant officers were "Acting", testament to her losses while on station.

† Lubbock records in *Cruisers, Corsairs and Slavers* that 28 slavers, rather than 24, were taken to England, tried for piracy under a special commission, acquitted and sent to Havana.

For some time there had been concern in Sierra Leone at the growth of the slave trade in the vicinity, and at Commodore Collier's insistence on concentrating the Squadron's efforts in the Bights with only occasional forays to windward. In early October Governor Findlay complained to the Colonial Office of the Squadron's "inefficiency", and his stinging attack was conveyed by Downing Street to the Admiralty. Africans were being carried off from close to the borders of the colony, he said; the trade of Sierra Leone and the Gambia was being damaged; and Africans would not attend to legitimate production when they could so easily sell their fellow creatures. He believed that the Squadron cared nothing for this, and was interested only in making head money and in "increasing the expense of the British Nation by bringing many thousands of slaves to this colony". He thought that head money should be done away with.[23]

Nevertheless, when Captain Gordon heard that a Sierra Leone citizen had sent a vessel to the Rio Pongas to sell guns and ammunition to a slaver, he sent *Plumper* north to investigate. Lieutenant Adams had only just brought the gun-brig back to Freetown after a fortnight searching the northern rivers, and he had left his boats in the Pongas. Arriving at the river bar on 5 November, he was warned by the boats' crews that the slave dealers were planning to put 400 men into canoes to murder *Plumper*'s men and destroy the boats. Memories of *Thistle*'s experience in the same river ten years earlier were still fresh. In darkness, the gun-brig entered the river at 2000 and anchored. At 0200 the next morning a schooner was detected sweeping downriver close under the trees. She was hailed and told to anchor, and it was found that she was the *Admiral Owen*, belonging to a Mr Smith. Her master admitted that he had sold guns to a French slaver, and she was despatched to explain herself to Governor Findlay. Later that morning, *Plumper* forged onward to Hell's Gates, 30 miles upriver, to cover the boats, and on the morning of 7 November three boats went even further to examine a number of vessels. One of them was a Portuguese schooner, the *Maria*, with 35 slaves. She was brought down to the brig which, the following morning, worked her way out of the river and sailed for Sierra Leone, where the prize and all 35 slaves were condemned. *Plumper*, however, paid heavily for her temerity in the Pongas; she subsequently lost 27 men to fever.*

Black Joke returned to the limelight in early November, by which time she had become tender to *Atholl* under the command of Lieutenant William Ramsay, Gordon's First Lieutenant. She had been suffering the frustration, increasingly common in the Squadron, of watching a burgeoning French slave trade while

---

* Commodore Hayes initially reported that 20 had died, but he subsequently amended the figure to 27 out of the 38 exposed.

strictly forbidden from taking any measure to discourage it. Between 5 and 9 October, on her cruising ground off the Old Calabar, she had boarded French slavers carrying 1,642 slaves, and, on departing the area, she had left ten French-flagged vessels in the river.

On 9 November, however, she had the satisfaction of taking a particularly fine Spanish brigantine, the *Dos Amigos*, which, it transpired, was a singularly appropriate name: the two vessels were to become famous comrades-in-arms. While *Black Joke* lay unobtrusively offshore, Ramsay sent his two boats into the Cameroons River, and, on the morning ebb tide, the gig, commanded by Mr Jenkins, Mate, encountered the Spaniard running out of the river. The brigantine turned on her heel and struggled back to King Bell's Town, ten miles upriver, chased by the gig and cutter and followed by *Black Joke*. Jenkins was too late to prevent the Spaniard's 563 slaves being offloaded and herded into the bush, but he seized the vessel. On *Black Joke*'s arrival next morning, Ramsay demanded the return of the slaves, threatening to fire the town. He was offered a token number, but then he found that the slaver's coppers had been ditched, and there remained no means of feeding them. He sent the brigantine, under Jenkins's command, to Freetown where she was condemned for having had slaves on board during the voyage on which she was captured. The commissioners called upon Ramsay to explain certain deficiencies in her stores, and the Treasury subsequently charged Gordon a sum of £4.14s for the losses.

Having sailed from the Gambia for Sierra Leone in company with *Primrose*, *Conflict* opened her account with a particularly gallant boat action 100 miles west of the River Pongas. In the early afternoon of 24 November she sighted a schooner to the south-east, but it soon fell calm, and at 1600 Lieutenant Smithers sent away his gig with Mr Hyne and six hands, commanded by the Master, Mr Rose. As the gig approached the schooner it was greeted by heavy fire of cannon and musketry, and the chase's deck was crowded with men. Rose lay on his oars and signalled *Conflict* for help. *Primrose* was in sight but too far away to assist, and Smithers sent his pinnace to join the fray. At 2000 the two boats, with 17 seamen and six marines, were pulling hard to overtake the schooner with her sweeps out, but at 2030 the chase swung round to present her broadside and opened a heavy fire. A big charge of grape shot away five of the pinnace's oars, hulled the boat between wind and water, and severely wounded several men. Undeterred, Rose ordered the gig to board to port and the almost-waterlogged pinnace to starboard. Resistance was desperate, but within five minutes the schooner was taken. She was the Portuguese *Nympha* with a crew of 40, of whom 13 were killed and seven wounded. The boarders, who had braved the fire of a pivoted long

9-pounder and a 12-pounder carronade, lost four dangerously wounded and ten slightly wounded. The slaver and her 167 slaves were condemned, and the wounded *Rose* was offered a lieutenant's commission. Four of the Portuguese were sent to England to answer for firing on *Conflict*'s boats, but, probably fruitlessly, they were later sent to Lisbon for trial.

On 30 November the new Senior Officer West Africa, Commodore John Hayes CB, arrived at Freetown in the frigate *Dryad* (36). Throughout the Service he was renowned as "Magnificent" Hayes in honour of his brilliant seamanship in saving his ship, *Magnificent* (74), during a gale in the Basque Roads on 12 December 1812, an episode widely regarded as the supreme example of seamanship in the days of sail. As well as being a great seaman, Hayes was also a skilled ship designer, and the ship-sloop *Champion* (18), which took part in the 1825 experimental trials, was among the vessels built by the Admiralty to his design. Shortly before sailing for Africa he had completed the cutter *Seaflower*, and he had been permitted to take her with him as a tender. *Dryad*, by contrast, did nothing to reduce the average age of the Squadron, having been launched in 1795. She did, however, have some excellent officers, and her second-in-command was a commander, William Turner, recently promoted for service in command of *Black Joke*.

On her passage out, *Dryad* had diverted for a fruitless pirate-hunt, and at Boa Vista in the Cape Verde Islands she had embarked the guns and an anchor from HMS *Erne*, wrecked in 1819. They had been salvaged by the Portuguese Government Agent, Colonel Martinez, and the Admiralty had presented him with a silver salver in gratitude, presumably not realising that he was probably the leading dealer in slaves on the Windward Coast of Africa. Hayes had also boarded a Brazilian brig which proved to be a "trade ship", one of the increasing number of vessels carrying trade goods to the Coast for slave purchases, thereby allowing the slave vessels themselves to load and sail without delay and at minimum risk. This one, having delivered her trade goods, was in ballast from Angola and, as was commonly the case, bound for the Cape Verdes to load salt for her return to Brazil.

Plumper had been sent to keep a lookout for *Dryad*, and when Hayes first sighted her on 29 November she was in pursuit of a brig. The chase could easily have outpaced *Plumper*, but she hove to and hoisted the White Flag of France, readily admitting to her ownership and her cargo of nearly 300 slaves, confident that the British could not touch her. As well as illustrating for the new Commodore one of the Squadron's particular difficulties, the encounter provided an opportunity for a briefing by Lieutenant Adams on the current state of affairs on the Windward Coast and in the northern rivers.

It is probable that Hayes had also met Collier before leaving England at the end of September, and he was well primed to make a number of decisions on arrival at Freetown, and to launch two complaints at the Admiralty.* The first of these concerned the disgraceful non-payment of bounties to Kroomen, who, having "done the deadly work of the Squadron" and thereby "preserved many valuable European lives", had not received fair reward for their services. The second, a matter upon which Hayes clearly already had strong feelings, probably reinforced by the incident he had observed between *Plumper* and the French brig *Felicité*, was a protest at the Squadron's impotence in the face of French slaving. There could be no end to the slave trade, he said, while the French slavers were permitted their current freedom, and he lamented that "the number of French flags is increasing daily".[24]

The Commodore immediately despatched *Primrose* for home without waiting for her replacement; he issued an emphatic order to the Squadron, under threat of dire penalties for disobedience, prohibiting improper use of stores found in prizes or their removal; he allocated *Dryad*'s launch to the Agent Victualler to be coppered and used as a stores lighter in order to remove the need for the cruisers' men to go ashore in Sierra Leone, which he was determined to prevent for the sake of their health; he directed that work should finish at 1600 in vessels at Freetown, and that the Kroomen, "those valuable people", should not "be unnecessarily harassed"; he gave "most positive orders" that commanding officers must "on no account whatever" send boats up rivers, or even in chase if there was a probability of their losing sight of the parent vessel; and he issued a general order, at the urging of the Admiralty, that every captured vessel should be given a medical officer, when practicable. He also decided upon a novel role and operating pattern for *Conflict* and *Plumper*.[25]

Lieutenant Smithers in *Conflict* was the first to receive the Commodore's new orders for the gun-brigs, which he was to pass on to the absent *Plumper*. He was directed to remain at anchor in the Sierra Leone River for three weeks to receive prize crews, so that none should go ashore, and, when relieved by *Plumper* or another vessel, he was to seek information from the Governor on the state of the slave trade in the vicinity. Then, for three weeks, he was to cruise off the Gallinas, Sherbro, Pongas and Nunez rivers as appeared to give the best probability of intercepting slavers. *Conflict* and *Plumper* were to continue to alternate, always ensuring that one of them was at Sierra Leone. This admirable arrangement made good use of the sadly limited gun-brigs, it might mollify the Governor, who was

---

* These were addressed to The Right Hon. J. W. Croker Esq, Secretary at the Admiralty, but would have been received by his replacement, Captain The Hon. George Elliot.

constantly critical of the Squadron's apparent lack of interest in the colony and its neighbouring rivers, and it would reduce the losses to disease among prize crews in Freetown.

Plumper was absent from Freetown because she had again been sent to the Pongas River. The Governor had learned that a British citizen named Josiffe (or Joseffe), a former servant of Governor Turner, was carrying on the slave trade on the Pongas. He had been kidnapping emancipated Africans from Sierra Leone, taking them north and selling them to Portuguese and Spanish slavers. Thirty of his victims had been found in the *Nympha*, as had an incriminating letter he had signed. Findlay asked Hayes to provide a vessel to take the police magistrate and Ensign Findlay, the Governor's son, with a colonial government boat, to the mouth of the Pongas to demand that the chiefs of the area should hand Josiffe over. The Commodore willingly complied, and sent *Plumper*. Over a year later, however, Hayes learned from the Admiralty that he had been accused of failing to cooperate with Findlay. He protested that he had complied precisely as requested, but that when the initial plan failed he had declined to take part in a subsequent expedition up the Pongas to find Josiffe. Not only did he believe that it had little chance of success, but also he had no intention of exposing his people to disease in that river after the appalling losses that *Plumper* had suffered there the previous month.

Following this wild goose chase, *Plumper* was sent on another: to find *Black Joke*, which had headed for Ascension after the *Dos Amigos* action and had not returned to Freetown as expected. On Boxing Day, not long into her passage, the gun-brig was fortunate to encounter the Spanish schooner *Maria* about 60 miles south-west of Cape Mesurado, and Adams took her, apparently without a fight, on finding her cargo of 505 slaves. After losing eight of them on passage, the schooner was condemned with the survivors.* *Black Joke* later made her way safely to Freetown, but in her increasingly decrepit state she needed further repairs during January 1831. *Plumper* returned north, this time to the Nunez River.

Conflict, instead of immediately taking up her station in the Sierra Leone River, had been sent north to convey Chief Justice Jeffcott to the Gambia. On passage she fell in with the schooner *Caroline* loaded with 51 slaves, carrying French papers, and, for an ensign, flying a tablecloth at her peak. As the boarding officer was about to leave the slaver, he heard some of the Africans calling in English to claim protection. They had been abducted from Sierra Leone, and Jeffcott advised

---

* There is little information on this capture, an extraordinary one for a gun-brig, and it must be assumed that the schooner was either becalmed or dismasted!

Lieutenant Smithers to take in the schooner. Four kidnapped British subjects were subsequently landed, and the schooner was escorted by *Medina*, recently returned from Ascension, to face French justice in Goree, together with a letter of remonstration from Findlay. The episode provided evidence to support the Commodore's contention that hundreds of liberated Africans had been carried away by slavers protected by French colours, for which tablecloths would suffice, and by French papers which could be easily procured. He predicted that the problem would only grow worse.

In the middle of December, only a few days after *Primrose* had sailed, her replacement arrived at Freetown, under the command of Commander Joseph Harrison. She was the ship-sloop *Favourite* (18), bearing the name of a West Africa cruiser last seen on the Coast in 1814, but for once a newly built vessel, launched in 1829.* New she was, but otherwise she offered no design advance on her fellows in the Squadron. As soon as she had completed with stores and water, Hayes sent her initially to Cape Palmas and then to cruise between Cape Mesurado and the Shoals of St Ann until the end of February.

The new Commodore's initial dispositions were giving greater emphasis to the Windward Coast and the northern rivers than Collier had favoured, but he wished to have more information on the state of the Trade before making further deployment decisions. He therefore directed Captain Gordon to take *Atholl* down to leeward to make a survey of general trade and of the slave trade, to make enquiries at Grand Sesters into murders allegedly committed there in two British merchantmen, to investigate opportunities for cruisers to take freights of gold dust and ivory, and to assess what measures could be taken to make legitimate trade more successful. Having visited the rivers and settlements as far as Cape St Paul's, he was to proceed directly (which Hayes emphasised) to Princes and cruise between 6 and 45 miles to the west of the island until the Commodore joined him. *Medina* was on passage to Cape Formoso, and if Gordon should encounter her to the west of Cape Palmas he was to send her to the Commodore at Freetown, or, if he found her to the east, he was to order her to cruise to the east of Princes, again between 6 and 45 miles from the island. The execution of these orders was to suffer a brief delay. On 29 December *Atholl* succeeded where Captain Owen's survey had failed in finding Coley's Rock off Cape Palmas. Having struck it, she hung there for a short time with ten fathoms of water on both sides of her, before sliding off, sustaining damage to her keel which degraded her sailing and steering.

---

* Correspondence continued to spell her name "*Favorite*", but the *Navy List* had adopted the modern spelling of "*Favourite*".

Atholl's sole action against the slave trade for the next four months was to discover at Popo in late January the Spanish schooner *Mannelita*, a slaver which, Gordon had heard, had returned to the coast having previously taken a slave cargo and acquired gold and ivory by piracy. The master was out of Gordon's reach ashore, and the remainder of the schooner's crew had not been involved in the previous voyage. Gordon held her for three days, but, finally, all he could do was to throw overboard her three guns, one of them a long 24-pounder.

Lieutenant-Governor Findlay did not find that Hayes's dispositions met the needs of his colony, and he made the extraordinary request to the Foreign Office that a small man-of-war, carrying the anti-slaving treaty Instructions, should be placed under his command. On hearing of this, Hayes retorted to Elliot at the Admiralty that it was "a question too absurd for Their Lordships to entertain for a moment", and he pointed out that Sierra Leone would always have at hand one of his brigs, instructed to pay due regard to any request from the Governor which did not infringe the Commodore's instructions. The Admiralty replied to the Foreign Office that Their Lordships could not comply with the request; it was "contrary to the Customs of the Service to place any of His Majesty's Ships under a Colonial Government", and the Senior Officer had been instructed to grant all protection in his power.[26]

By the end of January 1831, Hayes's new tender, the cutter *Seaflower*, was cruising between Cape Formoso and the southern end of Fernando Po. In command of her was *Dryad*'s senior lieutenant, Henry Vere Huntley, with a complement of three midshipmen, an Assistant Surgeon, a boatswain's mate, a carpenter's mate and 38 seamen, marines and Kroomen. She was armed with an 18-pounder carronade on an elevated slide and a brass 6-pounder. *Black Joke*, meanwhile, was undergoing another refit at Freetown, but by mid-February she was back at work off the Gallinas River. In command was Lieutenant William Castle, who had commanded the schooner *Speedwell* in the West Indies in the early 1820s and had been First Lieutenant of the frigates *Isis*, *Sybille* and *Dryad*. Hayes was giving him a chance to win promotion on a cruise in *Black Joke*.

Very soon after he had exchanged roles with Ramsay, Castle's opportunity came. At 0800 on 22 February she sighted a suspicious schooner 30 miles southwest of Cape Mount and settled down to a long chase. When darkness came the schooner, after losing sight of *Black Joke*, bore up on the assumption that the tender would hold her course until morning, but Castle managed to keep the chase in sight using his night-glass and cut her off. Still she tried to escape, and Castle eventually had to fire into her to bring her to, killing the cook and two slave children, and wounding two other children. The schooner was found to be the Spaniard *Primero*, of only 130 tons burthen but carrying 311 slaves from the Gallinas, including 155 children, four of them babies. Her slave deck was a mere 26 inches under the beams, and the slaves were nearly stifled.

Hayes's most immediate concern regarding the slave trade was the extent of French involvement, and the inability of the Squadron to counter it. At the end of January 1831 he wrote an impassioned letter to the Admiralty on the subject, describing the sufferings inflicted on the slaves, venting his frustration, and concluding, "Gracious God! is this unparalleled Cruelty to last for ever?" As yet, there was no indication that it would not.[27]

* * *

Commodore Collier's justifiable concentration of his scant resources in the Bights, combined with Captain Owen's private campaign based on Fernando Po, had undoubtedly caused difficulty for the slavers in that sector, primarily the illegal Brazilian trade north of the Equator, and, but for the hammer blows of the fever epidemics, might have achieved more concrete success. The fact remained, however, that for all the improved management of effort and for all the tragic sacrifice of lives, the slave markets of the Western Hemisphere remained well supplied, and no perceptible progress had been made against the Trade as a whole.

A degree of concentration had been necessitated by the continuing chronic lack of cruisers. Not only the intensity of slaving, but also the availability of Fernando Po as a forward operating base and the relative proximity of Ascension, increasingly popular as a welcoming and invigorating refuge, had dictated that the Bights should receive the lion's share of attention.

Nevertheless, as the redistribution by Commodore Hayes of cruisers to the northern rivers was demonstrating, the Spanish trade in that region was again thriving, and although Dutch and Portuguese activity was generally much diminished, the Portuguese were still shipping cargoes from the Rio Pongas to the staging post of the Cape Verde Islands. The protestations by the Governor of Sierra Leone about Collier's strategic neglect of the Windward Coast had been highly parochial but increasingly valid.

The use of a mixed bag of ex-slaver tenders had usefully extended the cover of the Squadron, accounting for nearly a third of the seizures, although none had replicated the success of *Black Joke*. That famous brig's achievements had been based on her ability to operate virtually as an independent cruiser, her sailing qualities, and, primarily, the excellence of her succession of commanding officers. However, she was now almost worn out. The tender concept had been well proved in the eyes of the officers on station, but doubts remained that the Admiralty was convinced.

Apart from shortage of cruisers, there were three pressing concerns for the authorities in-theatre. The first was the continuing inability of the courts, and the reluctance of parent nations, to punish the officers of detained slavers, most of whom returned to the Trade unrepentant and undeterred. The second was the long-standing requirement of the Mixed Commission courts to sell captured vessels on the open market, thereby ensuring that those of any quality were quickly recycled as slavers. The third was the oft-deplored impotence of the Squadron against a blatant and expanding illegal French slave trade, particularly in the Bights. There was no sign of resolution on any of these concerns.

CHAPTER 10

Gallant Pinpricks: Western Seas, 1824–31

Their force is wonderful great and strong;
and yet we pluck their feathers little by little.

SIR FRANCIS DRAKE

THE ENDING IN 1815 of the many years of worldwide conflict had released onto the seas great numbers of redundant privateersmen, paid-off naval seamen and adventurers who could no longer find an outlet for their energies, or earn a living, in legitimate warfare. Many of them turned to piracy, and nowhere in the world's oceans did they congregate more thickly than they did in and about the Caribbean. They followed in the long tradition of the buccaneers of the Spanish Main, preying upon some of the richest trade routes in the world, and it is scarcely surprising that Great Britain, with its acute interest in the security of those trade routes and of its island possessions in the region, should have ensured a continuing strong naval presence in the Caribbean after the ending of the Napoleonic War.

It is equally unsurprising that the suppression of piracy should have taken a markedly higher priority among the Royal Navy's roles in the West Indies than did the interdiction of the Cuban slave trade. The Admiralty's orders in November 1817 to Rear-Admiral Sir Charles Rowley, the Commander-in-Chief on the Jamaica Station, made it clear that his first duty was the protection of the island of Jamaica, and his second was "to protect the British Flag and Commerce from any Pirates that may infest the Seas within your Station" and to bring such pirates to justice. Their Lordships did not see fit to mention the slave trade.

However, with nearly a thousand sugar plantations during the 1820s and a rapidly increasing acreage growing coffee, as well as the rich merchants in the great city of Havana requiring domestic servants, Cuba was the northern hemisphere's greatest consumer of West African slaves. It has been estimated that the island's slave population in 1817 was 200,000, and the average annual number imported in the five years prior to the expected Spanish abolition of the Trade in 1820, promised in the Anglo-Spanish Treaty of 1817, was over 25,000. The resulting glut, together with the realisation that Spanish abolition would not, in reality, amount to a constriction in the flow of slaves, led to a market-driven reduction in the Cuban Trade to an annual average of about 4,000 in the early 1820s. Late in the decade, however, the number increased to 15,000 per year, despite the efforts of the West Africa Squadron.

The vast majority of voyages to bring slaves into Cuba originated at that island's harbours: 192 out of a total of 247 in the period 1821 to 1830, and, as was unashamedly announced in the city's press, as many as 40 slavers sailed from Havana in the eight months from June 1824 to January 1825. Although there was a clear need to interdict the transatlantic traffic at the market end of the Middle Passage as well as at the supply end off West Africa, a very much easier and more effective option would have been to arrest these vessels as they emerged from Havana and Santiago de Cuba on their outward-bound voyages to the slave coast, but the absence of an Equipment Clause in the treaty of 1817 prevented such action. The Commander-in-Chief on the Jamaica Station had therefore, in addition to his other operational tasks, to attempt to intercept an average of two laden slavers per month inward-bound for Cuba. There was no possibility of prior intelligence on times of arrival to improve the chances of interception, and, although Havana was likely to be the preferred arrival port for these slavers, they had numerous alternatives for disembarking their cargoes. Eliminating piracy was largely proactive work, but, although the two tasks were not entirely incompatible, action against the slavers was a matter either of patient waiting and watching or of pure chance.

Brazil was the other great slave market, but as yet the traffic into Rio de Janeiro, Bahia and its lesser Brazilian slave harbours was free of interruption by the Royal Navy. That traffic was entirely under the Portuguese flag, and as long as Brazil remained a Portuguese colony its Trade from the Portuguese territories on the African coast remained legal under the Anglo-Portuguese Treaty of 1817. However, Brazil was moving towards independence, and it was difficult to predict how that event might change matters south of the Equator.

* * *

Added to the Navy's difficulties in intercepting Cuban-bound slavers was the obstructive attitude, indeed the enmity, of its supposed allies in the anti-slaving campaign, the island's Spanish authorities, led by the corrupt and virulently Anglophobe Captain-General Vives. As Britain's indefatigable Mixed Commission Judge in Havana, Henry Kilbee, wrote to George Canning in 1825: "It is universally believed that abolition was a measure which Great Britain, under the cloak of philanthropy, but really influenced by jealousy of the prosperity of this island, forced upon Spain by threats or other means."[1]

Ever since the Anglo-Spanish Treaty of 1817 had established the Mixed Commission in Havana, Kilbee had been obliged to salve his frustration by reporting to the Admiralty the arrivals and departures of slavers, and by complaining, fruitlessly, to the Captain-General at every landing of a slave cargo of which he became aware. However, in the closing days of 1824, the Havana court was at last furnished with its first case.

There had been three Stations to which the vessels of the Royal Navy operating against the Caribbean pirates and the Cuban slavers might belong: North America, Leeward Islands and Jamaica, but in 1823 the two West Indies squadrons were amalgamated into the Jamaica Station, with Vice-Admiral Lawrence Halsted as Commander-in-Chief. By 1824 the struggle by the Royal Navy and the United States Navy against the pirates had achieved sufficient success to allow Halsted occasionally to reallocate one or two of his vessels to interdiction of the slave trade.

In this, as in his anti-piracy work, the Admiral's handful of small schooners provided his most useful cruisers. They were a mixed bag, ranging in size between 70 and 200 tons. Four were captured pirates: *Lion* (ex-*La Gata*), *Renegade* (ex-*Zaragozana*), *Assiduous* (ex-*Jackal*) and *Union* (ex-merchantman *City of Kingston*), and one was a purchased merchant vessel: *Speedwell* (probably ex-*Royal George*). They variously mounted three or four guns, and were manned by between 31 and 35 men. All had been designed to cope with the hazardous shoal waters common in the Caribbean and particularly those surrounding Cuba; *Union*, for example, with a length of 80 feet and a beam of 24 feet, was fitted with two pivoted centreboards which permitted a draught of only six feet six inches.[2]

Detection of slavers at sea was considerably more difficult for Halsted's cruisers than it was for the West Africa Squadron. The volume of general traffic in the approaches to Cuba was much heavier here than off the African slave coast where any strange sail sighted was likely to be slaving, and here it needed keen eyes to spot the occasional slaver amidst the legitimate shipping. This difficulty, combined with the infrequency with which cruisers were stationed to intercept the slave traffic in these early days, resulted in a very low risk to the slavers.

The Spanish schooner *Relámpago* may therefore have considered herself most unfortunate to have been caught by *Lion*, Lieutenant Francis Liardet, on 14 December 1824 at the junction of the Nicholas and Santaren Channels, 30 miles off the north Cuban coast. As Commodore Bullen on the African coast had heard, *Lion* was acting as tender to the brig-sloop *Carnation* (18), Commander Rawdon Maclean, but, unlike Bullen's *Hope*, *Lion* was a commissioned man-of-war. She was not carrying treaty Instructions, but Maclean presented his copy to the court, and the *Relámpago* and her 159 slaves were condemned only nine days after her capture.

Kilbee was justifiably worried about the subsequent safety of the emancipated Africans, for whom no adequate provision had been made. The Captain-General placed them with private employers and public establishments under strict conditions, but there was clearly a severe risk of their being reduced again to slavery. There was general concern on the island at the presence of freed slaves, and representations were made that they should be removed. The problem of safe and humane resettlement of emancipated slaves would continue to be a thorn in the flesh for the Havana court. The *Relámpago* was sold, and in July 1825 Kilbee reported that she had sailed once more for Africa under a new name.

Even before *Relámpago* was condemned, *Carnation* was hunting another slaver, this time with *Union* and *Assiduous* in company. They chased a suspect vessel into Cardinas (or Cabanas) on 22 December 1824, and when one of Maclean's officers applied to the Spanish brigantine-of-war *Bellona*, also in the harbour, for permission to search the vessel his application was refused and he was told that she was the *Magico* from Sisal in ballast. The British officer persisted, and when he boarded the *Magico*, accompanied by a Spanish officer, it was clear to him that the vessel had just landed slaves. Maclean reported the incident to Kilbee, who then challenged Captain-General Vives with not only *Magico*'s breach of the treaty but also *Bellona*'s aid to that violation.

When Vives asserted that no vessel could be detained unless she had slaves on board, Kilbee pointed out that the Additional Article signed at Madrid in December 1822 allowed arrest if it could be shown that slaves had been illegally put on board during the vessel's current voyage. Vives denied all knowledge of the Article, and was angry that the Spanish commissioners had acknowledged it, but he undertook to investigate the episode on the basis of the original treaty. The result of this investigation, a denial of the obvious truth, was that the *Magico* had entered Cardinas to escape "insurgent privateers", not to land slaves. Poor Kilbee complained to the Foreign Office that his practice of reporting slave landings to the Captain-General had become a "disgraceful farce", but he was urged to persist. Vives, with blatant hypocrisy, told Kilbee that there was "nobody more ready than I to observe the Laws, and cause them to be obeyed".[3]

In the first half of 1825, Cuba despatched 19 slavers for Africa and 20 (five French and 15 Spanish) landed cargoes in the island. To give the lie to Vives's declaration, 12 of the outbound vessels had been escorted clear of the island by Spanish warships in two convoys.

There was no further arrest until October 1825, although in April the Sixth Rate *Valorous* (26), Captain the Earl of Huntingdon, chased and drove ashore on the Yucatan coast the Spanish brig *Victoria* after she had landed her cargo of 250 slaves on the south coast of Cuba.

On 5 October *Lion*, now commanded by Lieutenant Edward Smith, captured the Spanish brigantine *Isabel*. This slaver, one of the largest in the Cuba trade and carrying 12 guns, had probably been stolen by her master while the owner was ashore buying slaves at the Gallinas, and had sailed with only 50 slaves. In the vicinity of Nuevitas she was chased by *Lion* and ran herself ashore. The brigantine's master escaped with several of the crew and most of the slaves, but Smith hauled off the vessel and took her, together with the remaining ten slaves, into Havana where Kilbee, with his temporary Spanish colleague Don Andres de Jawegiu, condemned them. Kilbee spoke highly of Jawegiu, who was replaced after this case by the permanent appointee, Judge Pinillos, on his return from leave in Spain.

A year after her escape from *Carnation*, the Spanish brigantine *Magico* met her match when she was sighted by *Union*, Lieutenant A. B. Lowe, on 20 January 1826 off the Great Bahama Bank. On the following day Lowe brought her to action on the Bank, and they exchanged shots for over half an hour before *Magico* made off under all sail, occasionally firing her stern chasers. *Union* kept up the pursuit until the slaver ran herself ashore near Manatí two days later. The crew, helped by local inhabitants, landed about 200 slaves and then deserted. About 30 slaves, driven into the sea, had been drowned in the escape, and 179 remained alive on board, many of them wounded by the Spaniards. On taking possession, Lowe immediately had the slaver's magazine searched, and a lighted slow-match was found leading to a barrel of powder. The salvaged *Magico* was condemned with the surviving slaves.[4]

The next victim fell to the "coffin brig" *Ferret* (10), Lieutenant William Hobson, a cruiser most unlikely to catch a Spanish schooner in chase, but, on 3 February 1826, she was fortunate to find one aground near Salt Cay on the eastern edge of the Great Bahama Bank. This was the *Fingal*, and her 58 slaves, with some of the crew, had been landed on a small uninhabited cay close at hand, from which *Ferret* rescued them. Only a month later, *Speedwell* had a similar experience when she found the Spanish brigantine *Orestes* aground near Grass-Cut Cays in the Gulf of Providence. Unlike Hobson, who refloated the *Fingal* and had her condemned, Lieutenant James Bennett of *Speedwell* was unable to get his schooner close to the stranded slaver, and had to content himself with rescuing her slaves and most of her crew. The *Orestes* had been chased by two unidentified schooners she thought British but probably pirates, until she had grounded on 28 February, and the crew had then escaped to a cay. When he boarded the wreck on 5 March, Bennett found 238 slaves, as well as several bodies, and he ferried them first to a cay and then to *Speedwell* some distance away. Four were drowned in the surf during the rescue, and others died after the capture or in *Speedwell*, but 212 were safely handed over to the Mixed Commission for emancipation. Bennett found space for the master and mate of the slaver on board his grossly overloaded schooner, but had to leave the remainder of her crew on their cay after giving them provisions and water.

This flurry of activity in early 1826 concluded with a two-day chase by *Union* of the Spanish brigantine *Palowna* through the shoals of the Great Bahama Bank. The slaver had the edge on *Union* in the strong breeze, but for a time the two were within gun-range and exchanged fire which killed three Spaniards. Finally losing his quarry in darkness on 28 April, Lieutenant Lowe, now amongst a maze of reefs at the south-western edge of the bank, wisely anchored. The slaver charged onwards, holed herself on a coral head and sank in deep water. Of her crew of 29 men and her 165 slaves only her master and one man survived.

There followed a fruitless three months, but the interval saw the arrival of two more Royal Navy schooners. The larger of the two was *Nimble* of 170 tons, previously the *Bolivar*, probably a pirate, and bought into the Service. She was commissioned by Lieutenant Edward Holland, given a complement of 41, and armed with a pivoted long 18-pounder amidships and four broadside 18-pounder carronades. The smaller of the two was the first of a pair that Admiral Halsted had ordered to be built from the lines of *Assiduous*. She was the 70-ton *Magpie*, built in Jamaica, armed with two long 9-pounders and two 18-pounder carronades, and manned with a complement of 35. Her tragically short career, under the command of Lieutenant Edward Smith, began with a confrontation in Havana which starkly illustrated the mendacity of the island's authorities.

On 16 August 1826, Commander George Jackson in the new ship-sloop *Pylades* (18), with *Magpie* in company, chased into Havana the Spanish schooner *Minerva*, which he was certain was a slaver. Jackson sent in Lieutenant Nott to report his suspicions and to investigate further, but Nott not only found no one prepared to listen to his report but also discovered that the schooner was already occupied by a Spanish officer and guard. Finally an officer at Government House agreed to examine the schooner, but would not allow Nott to accompany him on board. After enduring further prevarication, Nott returned to *Pylades*, and Jackson sent *Magpie* into harbour. By now the British Commission had entered the lists; not Judge Kilbee, who was on leave, but William Sharp Macleay, the Arbitrator and Acting Judge. He had heard that there were 200 slaves in the *Minerva*, and became aware that, between Nott's departure and Smith's arrival, two boatloads of Africans had been landed from her on the shore opposite the city. Smith then set a watch on the slaver by a boat commanded by Nott, who saw six boatloads of Africans being landed during the night at the Havana public wharf. The following day the Captain-General refused to believe Nott's report, but Smith obtained an order from the Spanish Commodore allowing him to visit the schooner, and found plentiful evidence that she had just landed slaves. The Captain-General, displeased at the Commodore's acquiescence, remained unmoved, and the *Minerva* evaded justice.⁵

There was a sequel to this disgraceful incident. Commander Jackson heard that some of the *Minerva*'s slaves were to be taken from Havana to Matanzas clandestinely in the steam vessel *Mexicano*, and *Pylades* lay in wait for her. On 21 August she took the steamer back into Havana, but although Judge Macleay considered the arrest legal, the Spanish commissioners disagreed, presumably on the grounds that the vessel was engaged in merely a legitimate internal transaction, and she was liberated.

On 27 August, *Magpie* was cruising alone off the Arrecifes de los Colorados on the north-west coast of Cuba, and lying becalmed. When a lurid black cloud was seen, heralding a "Norther", Lieutenant Smith reduced sail in the expectation of a heavy squall. A rapidly approaching line of spray heralded the wind, and it hit *Magpie* with extreme violence, taking her aback. Fine seaman though he was, Smith could do nothing to save his schooner, and she capsized, with the loss of all but two of her ship's company. Most of her people were taken by sharks, and it was two days before the survivors were rescued.

The final capture of 1826, that of the brigantine *Nuevo Campeador*, fell not to a schooner but to the frigate *Aurora* (46), Captain Charles Austen, off Santiago de Cuba on 29 August. The slaver, pointlessly flying false Dutch colours, was carrying Spanish and Dutch papers, and the master admitted that she was owned in Santiago de Cuba. Her 263 slaves were in a very poor state, and Austen felt it necessary to land 36 of them at Santiago. While in harbour he gave permission for the slaver master to visit his supposedly sick wife, on the Spaniard's solemn promise to return, but the man escaped into the interior, as did most of his crew. The *Nuevo Campeador* and her surviving 217 slaves were condemned.*

Admiral Halsted's cruisers, still primarily concerned with piracy, did not trouble the Havana Mixed Commission court again for over a year, but *Nimble*

---

* This must have been the incident recorded by Midshipman Harry Keppel (later Admiral Sir Henry Keppel) in his journal as having occurred on 6 August, involving the Sixth Rate *Tweed* and the brig-sloop *Harlequin* as well as *Aurora*, all three of whom appear to have shared the bounty. Keppel, in *Tweed*, remarked that "as [the slaves'] heads were shaved and greased, they looked, before we got close, like so many thirty-two pound round shot".

was unfortunate not to bring in a prize in December 1827. On the nineteenth she encountered the Spanish slaver brig *Guerrero* off Orange Cay on the western edge of the Great Bahama Bank and gave chase, bringing her to action after nightfall. Half an hour later the brig signalled her surrender, but, as Lieutenant Holland was lowering a boat, the slaver again filled her sails and ran for the Florida coast. *Nimble* was quickly after her, but both vessels then ran onto the Florida Reef. The following morning, three American wreckers came to their assistance, and *Nimble* was refloated. The *Guerrero*, however, was fast aground, and her 520 slaves and most of her crew were transferred to the wreckers. On the way to Key West, the Spaniards managed to take control of two of the wreckers, escape the rudderless *Nimble*, and land 400 slaves in Cuba. On arrival at Key West with the third wrecker, *Nimble* was obliged to hand over the remaining 120 slaves to the American authorities. The *Guerrero*, mounting 14 guns and with a crew of 95, was probably the largest slaver on the Havana Trade and considered to be the best sailer, and it was a fine achievement by *Nimble* (5) to catch her. Holland subsequently made an optimistic claim for £10 head money on each of the 120 slaves he had been obliged to surrender to the Americans and a further bounty payment on those recovered by the Spaniards. The Treasury, with uncommon generosity, granted him £5 each for the 120.

In mid-1827 Vice-Admiral Halsted was relieved by Vice-Admiral The Hon. Charles Fleeming, who thereupon reported to the Havana Mixed Commission the names of vessels under his command holding slave treaty Instructions. They were the Fourth Rate frigate (cut-down two-decker) *Barham* (50), the flagship; two 46-gun frigates: *Aurora* and *Druid*; the Sixth Rate *Valorous* (20); five ship-sloops, all of 18 guns: *Arachne*, *Espiegle*, *Harlequin*, *Scylla* and *Pylades*; four 10-gun "coffin brigs": *Fairy*, *Beaver*, *Ferret* and *Bustard*; and, most valuable of all for the suppression of piracy and slaving, seven schooners. *Nimble*, *Speedwell* and *Union*[*] remained of the original group, *Lion*, *Assiduous* and *Renegade* having been worn out and sold out of the Service, and to them had been added four new vessels, built to Admiral Halsted's orders in Bermuda. The first of these was *Monkey*, sister of the ill-fated *Magpie*, and the subsequent class consisted of *Pickle*, *Pincher* and *Skipjack*. This number was soon to be reduced by the loss of *Union*, Lieutenant C. Madden, wrecked near to New Providence in the Bahamas on 17 March 1828,[†] fortunately without loss of life.

---

[*] Lubbock in *Cruisers, Corsairs and Slavers* thought it likely that the original *Union* had, by this time, been replaced with another vessel of the same name, but this conjecture is not supported by David Lyon in *The Sailing Navy List* (Conway Maritime Press: London, 1993).
[†] This is the date given by Lubbock in *Cruisers, Corsairs and Slavers*. Lyon, in *The Sailing Navy List*, gives, rather doubtfully, 17 May.

In October 1827 the Hydrographical Office reported to the Admiralty that a work of considerable value to the West Indies cruisers had been completed. Master Anthony de Mayne in the brig *Kangaroo* had finished his survey of that part of the Great Bahama Bank forming the north side of the Old Bahama Channel. The Spanish slavers, who tended to keep to the Bank until their way into harbour appeared clear, did not know the shoals well, just certain safe channels, and, once charts had been drawn from the survey, the cruisers would have a marked advantage in pursuits in those hazardous waters.[6] Only a year later, *Kangaroo* fell victim to her dangerous work, being wrecked on Hogsty Reef off Cuba.[*]

The question of whether or not an arresting commanding officer was required to be present in court for judgment of an alleged slaver continued to be debated in the Havana Mixed Commission. The matter had been resolved years earlier in Sierra Leone, and one wonders why the Foreign Office had not, at that stage, given a clear directive to the British commissioners in Cuba that the presence merely of the Prize Master was the normal requirement. As it was, Judge Kilbee contended that the commanding officer's presence was not necessary unless a declaration from him under oath was required. Arbitrator Macleay perversely argued that a case could not proceed without the arresting officer, and, not surprisingly, the Spanish commissioners agreed. Consequently, arresting cruisers continued to be constrained in their movements, entirely unnecessarily, until adjudication of their cases had been completed. It was not until September 1828 that HM Advocate-General communicated his opinion in support of Kilbee, pointing out the obvious conclusion that for a cruiser to be obliged to leave her station for every capture "would endanger the very object of the treaty". The Spanish commissioners, however, declared in February 1829 that they had been directed by the Spanish Crown to continue the current practice, and refused to discuss the matter further.[7]

At the end of April 1828, Kilbee reported that he was taking leave, and took the opportunity of his letter to explain to the Foreign Office the standard procedure in Havana on his becoming aware of the landing of a slave cargo. The British commissioners would report to the Captain-General that a cargo had been landed; the Captain-General would reply that he had directed the Naval Commander-in-Chief to investigate; the Commander-in-Chief would reply that he had examined the vessel's log which showed no evidence that she had been slaving; the Captain-General would then tell the commissioners that the vessel was innocent.

---

[*] Lubbock writes in *Cruisers, Corsairs and Slavers*, probably correctly, that the wreck was on Hogsty Reef, although Lyon records in *The Sailing Navy List* that it occurred in the Bahamas. The entire ship's company was saved.

A newcomer to the Station was the next to make an arrest. This was *Grasshopper* (18), a ship-sloop commanded by Commander Abraham Crawford. She had been sent by Admiral Fleeming to examine the reefs of Los Colorados off the north-west coast of Cuba, reputed to be a pirate haunt, and on 26 June 1828 she fell in with the Spanish schooner *Xerxes*, nearing her destination with 405 slaves. After a 26-hour chase, which took the pair almost to the middle of the Gulf of Mexico, Crawford made his arrest. The slaver was subsequently condemned, and the slaves were emancipated without further loss.

Shortly after seeing *Xerxes* condemned, *Grasshopper* was involved in another episode of perfidy on the part of the Cuban authorities. She sighted the schooner *Esperanza*, a suspected slaver, and chased her into Havana on 9 July. There the schooner was found to be in ballast, having undoubtedly landed a slave cargo elsewhere on the island. To Acting Judge Macleay's surprise, the Spanish authorities indicated their intention to prosecute the vessel and to cast him in the role of prosecutor, presumably to embarrass him. Macleay wisely declined the invitation on the grounds that it was incompatible with his position as a British Commissioner, but passed all his evidence to the Captain-General. The *Esperanza* was acquitted in a farce of a trial in the Spanish Court of Admiralty which was based, entirely improperly, on the Anglo-Spanish Treaty rather than the Spanish *Cedule* of December 1817 which had banned Spanish slaving. The court's logic was that, in the absence of an Equipment Clause to the treaty, a vessel could not be seized by a cruiser unless she had slaves on board, and, as no slaves had been found in *Esperanza*, she must be acquitted. This amounted to an admission that, whatever the stipulations of Spanish law, no slaver would be convicted in the Cuban Court of Admiralty.[8]

At 0600 on 2 August *Skipjack*, Lieutenant James Pulling, sighted a suspicious brigantine off Haiti on a course for Cuba and gave chase. At 0830 Pulling hoisted his colours, and at 1000 he fired two round shot. The brigantine responded by hoisting Spanish Royal Colours* and replying with two rounds. The two exchanged sporadic fire until about 1400, when *Skipjack* came within grape range. The chase then shortened sail and hauled down her colours. It transpired that the brigantine, which (to excuse her failure to surrender immediately) claimed to have believed that *Skipjack* was a Columbian privateer, was the Spanish slaver *Intrepido*, of 151 tons and seven guns. She had sailed from the coast of Africa with 343 slaves and a crew of 43, but at her arrest only 27 of her crew and 153 slaves remained alive. Pulling described the latter as being in "dreadful confinement", and a further 18 died before the vessel's condemnation.

---

* Warship colours as a disguise.

Over three months passed before the next arrest, and this was another success for *Grasshopper*. The encounter took place on 22 November, beginning as a suspicious brigantine emerged from the passage between Martinique and St Lucia, unusually distant from Cuba for a slaver interception, and a matter of pure chance. In a strong south-westerly wind, which favoured *Grasshopper*, Commander Crawford, while making his preparations, feigned a lack of interest in the stranger, encouraging her to come on. With his quarry on the beam, he went about, rapidly shook out a reef, set topgallants and gave chase. Initially the sloop gained steadily, but a calm in the early afternoon gave the advantage to the slaver. With wind freshening, slackening and then freshening again as the day wore on, the two vessels alternately made and lost ground. As darkness fell it seemed likely that the chase would be lost to sight and escape, but then, as the moon rose, there came a heavy squall. The cruiser shortened sail in time, but the blow was too much for the slaver, and, her sails torn and rigging much damaged, she was obliged to surrender. The prize was the Spaniard *Firme* with a crew of 43, four passengers and 487 slaves, including a babe in arms, embarked at Popo on 18 October. By contrast with the *Intrepido*, she was very healthy, having lost only five slaves on passage, although three more were to die before the vessel was condemned.

On arrival in Havana, Commander Crawford was moved to commend the slaver master, Guiseppe Fornaro, for his humane care of his slaves. Fornaro, for whom this was the second capture in 19 voyages, was a most unusual example of his profession, as revealed by the extraordinary scene witnessed by Crawford at his arrival on board the prize on the morning after her arrest:

> I found all the slaves, men, women and children, a remarkably fine, healthy race – fat, sleek, good humoured – and, to all appearances, happy and contented. It was quite a pleasure to witness their joy when the captain of the *Firme* stood upon the deck. They all crowded round him, and in every way they could testified the pleasure they felt at his presence, showing by their action that they looked upon him as their friend and father. Not one moody or sullen-looking black did I observe among the whole.[9]

This remarkable slaver had graphically demonstrated that this essentially inhumane traffic could be conducted without much of its usual cruelty, and he had undoubtedly profited in his career thereby.

In his report to the Foreign Office, Acting Judge Macleay gave an analysis of the financing of *Firme*'s voyage to illustrate the profits that might be made. Her cargo of trade goods cost $28,000 and consisted of gold, silver, raw spirit, handkerchiefs, printed cotton and gunpowder; crew pay, which was to be forfeited in the event of shipwreck or capture, amounted to $13,400; and the cost of all stores was $10,600. Mandingo Africans were selling at $300 each, and so the potential value of her slave cargo was $145,200. The expected profit from the voyage was therefore $93,200. Fornaro and his officers were shareholders in the venture.[10]

Officers of the port of Havana were among the other shareholders in the two slaving enterprises terminated by *Grasshopper*, and the sloop found herself exceedingly unpopular when she arrived with her prize. A hostile and jeering mob crowded the landing place when Crawford went ashore to call on the Captain-General, and there was a strong rumour that an attempt was to be made in darkness to recapture the *Firme* and even to set fire to *Grasshopper*. Additional marines were stationed in the prize and boats rowed guard around the two vessels at night. Crawford himself half-expected to feel a knife in his back when ashore on business.

The following week, an attempt by Lieutenant Pulling of *Skipjack* to earn similar unpopularity met with frustration. On the morning of 29 November he sighted a suspicious schooner inshore on the north-western coast of Cuba. He fired a gun to bring her to, but there was no response, and he gave chase. As *Skipjack* came within range, the chase fired on her and then ran herself ashore about nine miles to the west of Havana. There she managed to land her cargo of slaves with the assistance of a number of boats from shore which kept up a brisk fire on *Skipjack*. With no intimate knowledge of that dangerous part of the coast, Pulling had some difficulty in coming to anchor in a position from which he could cover his boats as they pulled for the chase, and before they could board her the slaver blew up. On deserting her, the crew had set fire to her clearly with the intention of killing the boarders and any remaining slaves. In fact one sick African remained on board, unharmed, and some papers were recovered identifying the vessel as the *Maria*, a Spaniard. Her log was found to have been annotated by Captain Owen in the Old Calabar River. Following this attempt, the British commissioners expressed their sense of Pulling's "activity and enterprise" during his short time on the Coast. He had, they wrote, "greatly alarmed the slave traders".

As Macleay reported to London in January 1829, the previous year had seen an appreciable increase in the Cuban slave traffic. In 1827 a total of 27 slaving voyages had cleared from the island and ten had arrived. In 1828 the figures rose to 63 and 28, and three vessels had landed two cargoes each. The increase resulted from the demands of new land being brought under sugar cultivation, thanks to sugar bringing very high prices in the past year, and the consequent large profits to be made from slaving. These profits were enormous by comparison with the risk of loss, which was slight, and that of punishment, which was nil thanks to the certainty of protection for the traffic by the Cuban Government. Adding to the demand for slaves to work new sugar plantations was the fact that many coffee planters were facing ruin after three years of low prices on the European markets, and were turning to sugar production. The number of slaves needed to work sugar was three times that for a similar acreage of coffee.

In response to the flow of complaints from the British commissioners in Havana, the Earl of Aberdeen at the Foreign Office assured Kilbee and Macleay in March that orders had been sent from Madrid to the Captain-General of Cuba to exert measures better calculated for suppression of the "disgraceful traffic, against which the efforts of His Majesty have been so long and so anxiously directed". Aberdeen must have been extraordinarily optimistic if he imagined that such orders, even if actually despatched, would achieve the slightest benefit when received by the official who was currently demonstrating his intentions by providing naval escorts for convoys of outward-bound slavers along the north coast of Cuba. A Royal Order by His Catholic Majesty did indeed materialise, but, as expected, and as reported by Macleay in June, it had little influence on the Government of the island, and the slave trade continued to be carried on as openly as ever.[11]

On 19 February 1829 the new schooner *Pickle*, Lieutenant John Bunch Bonnemaison McHardy,* joined the fray. She had been given a cruising ground off the north-east coast of Cuba, between the Windward Passage and the Old Bahama Channel, and was off Puerto de los Navanjos when she sighted a strange sail, a schooner, and gave chase. The schooner ran herself ashore near Puerto del Padre, 20 miles east of Puerto Manatí, and the crew fled with their cargo of slaves into the interior. One sick and exhausted African was left on the beach, and he was rescued by *Pickle*'s boats as they seized the slaver, although he died two days later. The prize was found to be the 70-ton *Golondrina*, which had sailed from Little Bassa on the Grain Coast at the beginning of January with about 78 slaves, and she was heaved off the beach without damage. McHardy was anxious to return to his original task of investigating an act of piracy, and, although it was the carnival season, the Mixed Commission court was immediately summoned. Despite there having been no slave actually on board the vessel at seizure, the *Golondrina* was condemned.

The smallest of the Admiral Halsted brood of Bermuda schooners, *Monkey*, Lieutenant Joseph Sherer, drew her first slaver blood on 7 April. Her victim, four times her tonnage, was the Spanish schooner *Josepha*, which she detained near the

---

* "Bunch" is spelt variously "Bunce" and "Bunche" in different sources.

Berry Islands in the Bahamas with 206 slaves. *Monkey* had been lying there in a small anchorage formed by the Stirrup Cays and known as Slaughter Harbour, a favourite place for the Royal Navy schooners to lie in ambush for slavers attempting a clandestine approach to Cuba through the Bahamas. After the prize was brought into Havana on 11 April, one slave died and one was born. While the two schooners lay in harbour awaiting adjudication, there was a rumour that the slavers would try to cut out the *Josepha*. However, it was clear that *Monkey* was well prepared for such an attempt, and the slavers thought better of it. After condemnation of vessel and slaves, Acting Judge Macleay was able to report that he had received every cooperation from the Spanish commissioners in this case, and that the Captain-General had given every assistance to *Monkey*.

After the frustrating encounter with the *Golondrina*, McHardy and *Pickle* suffered another disappointment on 16 April. The schooner was lying at anchor in the small north-eastern Cuban harbour of Gibara when a strange brig appeared. McHardy immediately suspected that she was a slaver, and he was confirmed in his suspicion when she passed *Pickle* under Spanish colours and with Africans on deck. The schooner's boats boarded the slaver before she anchored, but the master refused to divulge any information. McHardy therefore took possession of the vessel until he had consulted the shore authorities, whereupon the Spaniard became more cooperative and produced the brig's documents. The slaver was the *Bolador* from Puerto Rico, licensed to carry the number of slaves she had on board, and her previous voyage had been from Africa with slaves for Puerto Rico. It soon became apparent that the Cuban authorities intended to take no action, and the Anglo-Spanish Treaty forbade British cruisers from making arrests in Spanish harbours. McHardy was obliged to release the brig.

The month of June 1829 saw two of the more notable actions of the campaign in Cuban waters. In the first of these, *Pickle* and McHardy, after their two largely abortive slaver encounters, were able to demonstrate their true mettle. The schooner was again cruising off Puerto de los Navanjos, disguised and behaving as a West Indian coaster, with her main topmast and flying jib-boom on deck. At daybreak on 5 June she sighted a suspicious topsail schooner to the eastward, heading along the coast towards her. Her disguise allowed *Pickle* to place herself between the stranger and the shore without causing alarm, and when his quarry was directly in the eye of the north-easterly wind, McHardy swiftly sent up his topmast, rigged out his jib-boom and set all sail in chase. The stranger immediately headed off close-hauled on the starboard tack. This was *Pickle*'s best point of sailing, and she gradually made ground on the chase until, at 1620, she came within extreme gun-range. She hoisted her colours and fired a shotted gun,

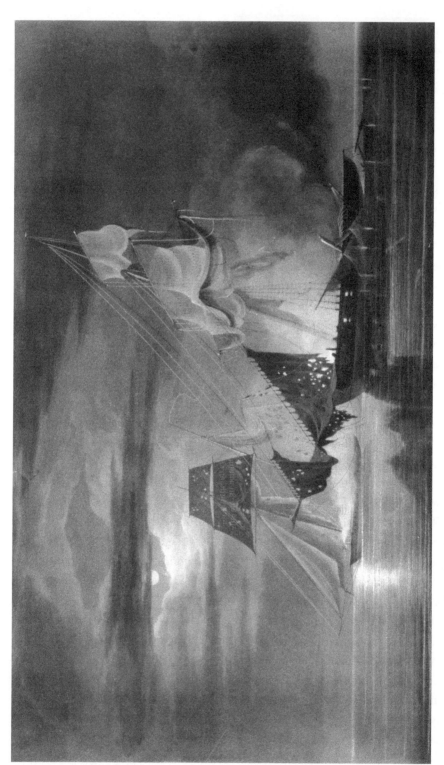

12. Capture of the Spanish schooner *Voladora* by HM Schooner *Pickle* on 6 June 1829

but there was no response, and McHardy held his fire for a further two hours. Then, with darkness coming on, he tried several more rounds, still to no effect. The night was dark and overcast, and *Pickle* had great difficulty in keeping sight of the chase, but, while gradually overhauling the schooner, McHardy skilfully manoeuvred to maintain his leeward and shoreward position.

By 2300 McHardy was able to fire a shot over the chase and she, seeing that she could not escape and realising that she was the more powerful vessel, bore down on *Pickle*. There then began an 80-minute action within pistol-shot, in pitch darkness apart from the flash of gunfire. By the end of it, *Pickle* had shot away her adversary's mainmast, crippled her foremast and reduced her almost to a wreck. The fight had been by no means one-sided, and *Pickle* had been subjected to very well-directed musketry as well as round shot. On her calling that she had surrendered, the chase was ordered to send a boat to *Pickle*, but she replied that she had none that would float. So McHardy told her to stay close at hand until daylight, and he manned *Pickle*'s sweeps to keep his conquest in sight.

At dawn *Pickle*'s mate, Mr W. N. Fowell, took possession of the prize and found her deck covered with wounded and drunken men, as well as a tangle of debris. He sent her wounded master to the schooner, and McHardy learned that his prize was the 240-ton Spanish schooner *Voladora*,* carrying 335 of the 357 slaves she had embarked at Popo before sailing on 29 April. During her wait for her cargo the slaver's log had been annotated by both *Sybille* and *Black Joke*. She was armed with two long 18-pounders and two long 12-pounders, as well as 40 muskets, 20 pairs of pistols and an array of swords, boarding pikes and boarding knives. Against this armament, *Pickle* had one long 18-pounder on a pivot and two 18-pounder carronades which became unserviceable early in the action. *Voladora* had probably begun the engagement with 66 men, of whom at least eight were killed, and, of the 14 wounded, two subsequently died.† *Pickle*'s complement amounted to 39, of whom five were boys. Her casualties, all inflicted by musketry, were one killed, one mortally wounded and eight wounded.

Pickle suffered only superficial damage, and, with her rigging repaired and new sails bent, she was able to tow her prize, of twice her tonnage, to Gibara, arriving

---

* McHardy reported the name of the prize as *Boladora*, but the British Commissioners, in their despatch to the Foreign Office, give her name as *Voladora* (FO 84/92). In this account the names reported by the Commissioners have generally been accepted because, unlike the captor, they had all obtainable documentation and witnesses to hand, as well as time for deliberation. *Voladora* is therefore believed to be correct, although Lubbock and others have preferred *Boladora*.

† McHardy was doubtful about the *Voladora*'s crew and casualty numbers. He found 52 men on board, but estimated that there had originally been 66. He was sure that more had been killed than the eight bodies found, and clearly suspected that some had gone over the side.

13. Capture of the Spanish brig *Midas* by HM Schooner *Monkey* on 27 June 1829

the following day. There, in a very hostile atmosphere, she jury-rigged the slaver and laid in provisions for the voyage to Havana. The *Voladora* was condemned, and the surviving 331 slaves were emancipated.[12]

An action involving an even greater disparity of force took place only three weeks later. *Monkey*, probably the smallest commissioned man-of-war in the Navy at the time, gladly escaping Havana after the condemnation of the *Josepha*, had headed through the Straits of Florida for her station at the northern end of the Great Bahama Bank. On rounding Bimini Island on the morning of 27 June, she sighted, at a distance of six or seven miles, a large brig at anchor on the Bank. On the appearance of *Monkey*, the brig weighed anchor and stood away under easy sail. In a very light breeze, *Monkey* closed to three miles and the brig then re-anchored, putting a spring on her cable in order to control the direction of her broadside, a clear statement of her intention to resist arrest. As the schooner closed in the brig fired a wild broadside, but Lieutenant Sherer waited until, in only a faint breeze, he had worked in to point-blank range across the brig's stern before opening a rapid fire from his single gun,* shooting high to avoid injuring the slaves. The spring on the brig's cable clearly failed in its purpose, because she never brought effective fire to bear on the schooner.

After 35 minutes of this punishment there was a hail from the after end of the brig to say she had surrendered, but musket fire continued from the forecastle and another round or two of canister and grape were necessary to achieve total submission. The prize was the Havana slaver *Midas* of 360 tons, mounting four long 18-pounders and four 12-pounders, and carrying a crew of 53 and a cargo of 400 slaves.† Of *Monkey*'s 26 officers and men, not one had been hurt in the engagement, and the slaver had lost only one killed and three wounded. The dead man was the mate and supercargo, and, it was claimed, the owner. Mortality among the slaves had been dreadful, however. *Midas* had sailed from the Bonny with 562 Africans and, in the long delay before condemnation, owing to the contrary winds which kept the two vessels at anchor on the Bank for several days, a further 88 were added to the death toll to disease on the Middle Passage. To make matters worse, 31 threw themselves overboard before arrival at Havana, and the court was able to emancipate only 281. A small bonus to *Monkey* was the discovery of $1,360 in the brig, and this was condemned with the vessel on 14 July.[13]

---

* In *The Sailing Navy List* Lyon shows that *Monkey* mounted two 12-pounders and a five-and-a-half-inch howitzer, but accounts of this engagement indicate that she had only one gun at the time, presumably a 12-pounder and probably on a pivot.
† The *Midas* had been the slaver *Providencia*, apparently the Brazilian brigantine taken by *Maidstone* off Lagos on 16 March 1827, condemned at Freetown and sold by the court.

The arrival in Havana of three valuable prizes in the space of only a little over two months, and especially the two most recent actions, had a most salutary effect on the slave merchants, and it was particularly galling for them to see the minuscule size of the latest captor, *Monkey*'s length equalling that of the main boom of her prize brig. However, those most concerned by the turn of events were probably the five British citizens found in the *Midas*. Initially they were sent to Jamaica, but Judge Tucketts of the Vice-Admiralty court there reported that the Attorney-General had cast doubt on the validity of the commission sent to the island for the prosecution of slaving cases. The five were therefore returned to England and brought to trial in the Admiralty Court at the Old Bailey for "feloniously aiding and assisting in carrying away divers (*sic*) persons from the Coast of Africa to be dealt with as slaves". There was insufficient evidence of British citizenship in the case of two of the accused, but the others were sentenced to death on 25 October 1830.[14] However, it seems that the sentences were commuted to transportation for life. On conclusion of this case, which had necessitated the return to England of witnesses from *Monkey*, it was clear that repetition of such a costly procedure should be avoided, and new commissions were issued to the Vice-Admiralty courts at Jamaica and St Kitts to try offences against the slave trade laws.

The convicted seamen from *Midas* had apparently been survivors from a Liverpool merchantman wrecked on the Cuban coast, but Acting Judge Macleay expressed his concern in November 1829 that men were being enticed to desert from HM Ships at Havana and then inveigled into the slave trade. He was probably reflecting the experience of Captain Napier of the frigate *Galatea* (42) who reported in December, after a visit to the port, that not only had he seen slave vessels "of all descriptions" openly fitting-out, but also that the slavers were offering bounties of as much as $100 a man for British seamen, and that he had lost six men by desertion.

In September, Admiral Fleeming reported to the Admiralty that he had sent the sloop *Icarus*, Commander Mayne, to St Thomas to enquire after the slaver *Neirsee*, taken by *Eden*'s tender *Cornelia* in November 1828 but retaken on passage to Freetown. Mayne discovered that, as the West Africa Squadron officers were all too well aware, the frequent changes of colours, names and masters made identification extremely difficult. However, he concluded that the *Neirsee* had fitted-out at Martinique or Guadeloupe under French colours, and had then obtained Dutch, Swedish and Spanish colours at islands belonging to those nations. She had returned initially to one of the Cuban harbours under the name *Martin* or *San Martin*, and there had been correspondence concerning her between the Captain-General and the French Consul. Then she had sailed for one of the French islands with the name of *Estafette*. Fleeming remarked on the open manner in which the slave trade was carried on between the French possessions in the West Indies and the coast of Africa, with vessels usually American-built, fitted-out at either Martinique or Guadeloupe (like the *Neirsee*), and with guns obtained in Cuba. They made use of a depot of seamen of all nations at St Bartholomew, and having Danish, Dutch and Swedish subjects on board as supercargoes, surgeons or stewards, they had ready-made "masters" on board for whatever colours they needed to display to avoid arrest.

The year 1829 saw something of a downturn in the Cuban slave trade. There was an increase from 28 to 33 in the number of slavers arriving safely, but a fall from 63 to 45 in the number sailing for Africa. There had been a comparative failure of the voyages which had departed from the island in 1828 owing mostly to high slave mortality, but the relative success of the cruisers on both sides of the Atlantic had also been a significant factor. Furthermore, the ruin of most of the small coffee estates had flooded the island's market with slaves. Many Havana shopkeepers had been shareholders in slaving voyages and had been bankrupted. To make matters worse, there was concern that the number of liberated slaves on the island might lead to an insurrection. Admiral Fleeming heard that, in response to a request from the Cuban authorities that these *emancipados* should be removed, the Spanish Government had directed that they should be shipped to Spain to be employed on works at Ceuta or on growing tobacco near Seville. He also heard, although he could hardly credit it, that slaves who had been liberated after completing their period of servitude, and who had settled on the island with wives and families, were to be included in the scheme. He remarked that a representation by Macleay had "prevented execution of this barbarity".[15]

In June 1828 there had been a suspicious death in *Nimble*, and her Commanding Officer, Lieutenant Holland, was subsequently relieved of his command and sent to the schooner *Firefly*, which was building at Bermuda.* He was replaced by Lieutenant Sherer of *Monkey*, who consequently had the satisfaction of making the final slaver capture of 1829. On the evening of 16 November, in the channel between the southern tip of Great Abaco and the Berry Islands in the Bahamas, *Nimble* fell in with a suspicious schooner, made sail in chase and opened fire. During a pursuit of an hour and a half, the chase replied with round shot and grape, but when the head of her mainmast had been shot away and her rigging had been shredded, Lieutenant Sherer was able to take possession of her. She was

---

* Whatever blame was attached to Holland for this incident seems not to have seriously damaged his career. In 1838 he took command of the brig *Dolphin* on the west coast of Africa.

the *Gallito* of St Thomas, under Spanish colours, with a crew of 16, a cargo of 136 slaves, and armed with a long 9-pounder pivot gun. This capture brought Sherer's total of slaves liberated to a creditable 743, and his Admiral referred to him as "this active and fortunate officer", no doubt intending this as a favourable comment.

It was not until 9 April that the first arrest of 1830 took place, and, unusually, the captor was not one of the schooners. The ship-sloop *Sparrowhawk* (18), Commander Thomas Gill, was on passage to Jamaica to collect the Admiral's despatches for New Providence when she fell in with the 43-ton Spanish schooner *Santiago* 60 miles off Santiago de Cuba, and, finding 108 slaves on board, detained her. Gill, an elderly commander who had lost a limb in the Service, felt himself unjustified in diverting to Havana and decided to send in the slaver with Mr Robert Miller, Mate, with four hands, together with the *Santiago*'s master and an African sailor. The remainder of the Spanish prisoners, mostly sick, were landed at Santiago de Cuba. The prize had three and a half feet of water in the hold at capture, and the leak gained during the voyage along the southern coast of Cuba until, on 23 April, with six feet of flooding and two of his men sick, Miller made to run the schooner ashore on Cape San Antonio.

By good fortune, the ship-sloop *Slaney* (20), Commander Charles Parker, hove in sight at that moment, and Parker transferred slaves, prisoners and crew from the sinking prize to the sloop, landing them safely at Havana on 27 April. The case was a difficult one for Macleay. It was the first instance in which the captor commanding officer had not accompanied his prize, and the Spanish commissioners persisted in demanding his presence despite the clear instructions to the contrary from the Foreign Office to Macleay. There was also a shortage of evidence, and the court felt that it needed dispositions on oath from the master of the detained vessel and at least two or three of the principal members of her crew. All that could be achieved, pending the appearance of Gill, was a provisional condemnation of the slaves, but they were eventually condemned to *Sparrowhawk*. The commissioners in Sierra Leone would have found no such obstacles to finalising the case.

It was of concern to Macleay that the documentation presented to the commissioners by captors was often inadequate, and he urged that cruisers should, in addition to copies of the complete treaties, be supplied with the forms of affidavit and certificate approved by the British and Spanish Governments, with blanks to be filled in by commanding officers. He was giving forms he had produced himself to cruisers visiting Havana, but, considering the frequent changeover of ships, he recommended that those joining the Station should be issued with approved versions before leaving England. The Foreign Office also asked the

Admiralty to provide the ships with copies of the Government's guidance to the Mixed Commissions agreed with Madrid.[16]

In 1823 the Leeward Islands Station had merged with the Jamaica Station, and in 1830 there was a further amalgamation. In early April, Vice-Admiral E. G. Colpoys, in the Fourth Rate *Winchester* (52), accompanied the Jamaica Station flagship *Barham* to Havana to take over command from Admiral Fleeming. Then, while lying at Nassau on the twenty-eighth, he received orders to proceed to Bermuda to relieve Rear-Admiral Sir Charles Ogle in command of the North America Station. As Commander-in-Chief of the resulting West Indies, Halifax and Newfoundland Station, Colpoys would inevitably spend much of his year in or around Nova Scotia and Bermuda, delegating the direction of affairs in the Caribbean to one of his captains, and, although the schooners continued their conscientious vigil, it may be that the infrequency of the Commander-in-Chief's visits to the Cuba theatre led to a decline in the vigour with which the frustrating campaign against the slave trade was pursued.

The only other arrest of 1830 was achieved by the sloop *Victor* (18), Commander Richard Keane, on 11 June. While cruising off Punta Escondida on the south coast of Cuba, not far from Santiago de Cuba, she encountered the Spanish brigantine *Emilio*, bound for Santiago, and seized her with 192 slaves loaded at the New Calabar, 20 having been lost on passage. Keane also found three British subjects among the crew of 20. One of them died before arrival at Havana, and another, a Bermudan who had deserted the slaver but been forced on board again, was subsequently released by Colpoys, but the third, Robert King, a Dubliner who had been invalided from HMS *Blanche* on the South America Station a few years earlier, was kept in *Victor* for return to England. The slaver was condemned at Havana without difficulty, and the slaves were emancipated.

Robert King was held prisoner in the flagship at Portsmouth until, at the end of July 1831, he was released on the recommendation of the Treasury Solicitor because there was no officer from *Victor* to give evidence against him. Only a week later it was learned in the Treasury Chambers that Commander Keane had arrived in England, and the Solicitor, believing that a conviction might now be procured, asked the Admiralty to hold King, but it was too late.[17]

It had long been the practice of the British commissioners in Havana to send information they had gleaned on the movements of Cuban slaving vessels to the Foreign Office for onward despatch to the West Africa Squadron via the Admiralty. The intelligence invariably reached the Commodore on the slave coast far too late for any useful reaction, and it eventually occurred to the Foreign Office that it would be advantageous if Havana were to communicate directly with the

Commodore. Late in 1830 Macleay was told to do so, but, this clearly not being a novel idea to him, he replied that he had no means at his disposal to achieve direct communication because legitimate traders between Cuba and the coast of Africa were virtually non-existent.

Only two arrests in 1830 seemed a poor return for the Navy's efforts in Cuban waters, but the reason for it was a marked decline in the slave trade. In 1829 the number of slaver sailings for Africa had been 45, and 33 slave cargoes had been safely landed in Cuba. The figures for 1830 were 29 sailings and five cargoes landed. Sickness, shipwreck, piracy and capture had all contributed to the reduction, but the main factor was the low price of sugar on the European markets. There had been no hindrance to the Trade from either Madrid or the Cuban Government, and the Royal Order of March 1830 urging more effective measures by the Cuban authorities to eliminate the slave traffic had, as the British commissioners had anticipated, proved futile. Indeed, it had never been promulgated in Cuba.

The Havana Mixed Commission court was not troubled at all in the first six months of 1831, but the absence of a second British Commissioner was at last addressed in London. Mr Charles Mackenzie accepted the appointment of Arbitrator, but was proving reluctant to journey to his post. This was much to the annoyance of the new Foreign Secretary, Viscount Palmerston, who had replaced the Earl of Aberdeen in late 1830. In August 1831, to Macleay's regret, the Spanish Arbitrator de Quesada resigned, being replaced by Brigadier-General Don Juan Montalvo y O'Farrill.

Although there had been no adjudications, there had been one arrest. Off the Stirrup Cays among the Berry Islands at 0630 on 18 June, *Pickle*, with a new commanding officer, Lieutenant Taplen, encountered a schooner running down towards Cuba. On seeing *Pickle*'s colours the stranger hauled to the wind on the starboard tack, and Taplen made all sail in chase. At 1000 *Pickle* fired a shot, but the schooner still failed to show any colours, and the pursuit continued until, at 1100, with the man-of-war gaining fast, the chase hove to and hoisted a Portuguese ensign. Portuguese slavers were unusual in these waters during this period, and this one, the 95-ton *Roza* with a crew of 12 and 157 slaves, although carrying a passport from Lisbon for a commercial voyage to the Cape Verde Islands, had loaded slaves at the Cacheo River and sailed for Matanzas.

Any hope that Lieutenant Taplen might have entertained that he would be relieved of his prize at Havana was of course dashed; since the independence of Brazil, the only court with jurisdiction over Portuguese slavers was at Freetown. Acting Judge Macleay, anxious to be helpful, suggested that the slaves might be landed at Nassau, although he advised Taplen to check in advance on the local laws there while remaining offshore. Having successfully followed this recommendation, a not inconvenient course of action because Nassau was the victualling base for the schooners, Taplen dispatched the empty slaver for Sierra Leone under the command of his mate, Mr Fowell. Following her arrival on 26 September she was satisfactorily condemned, and her 157 slaves were emancipated *in absentia*, one having died and another having been born before arrival at Nassau.

Mr Fowell, while waiting at Freetown for an opportunity to rejoin *Pickle*, volunteered to command Lieutenant-Governor Findlay's colonial schooner in defence of the Gambia settlement against the aggression of the King of Barra. Seizing this opportunity with both hands, he earned a high accolade from the commander of the Royal African Corps troops in the campaign, Lieutenant-Colonel Kingston: "to the Zeal and Bravery which Mr Fowell has already exemplified in the Barra War he has shown an Interest and activity deserving of the Warmest praise."

Lieutenant-Governor Rendall of the Gambia added to this his own expression of very warm approbation, and Rear-Admiral Warren, newly arrived as the Commander-in-Chief on the Cape Station, forwarded both letters to the Admiralty.[18]

Taplen had found two British subjects in the *Roza*, Lawrence Lindsay and Henry Hadden, and, although consideration was given to trying them at Jamaica, they were landed at Bermuda and taken to Spithead in HMS *Falcon*. It became clear, however, that they had been victims of misfortune rather than voluntary slavers. They had been shipwrecked in a Liverpool packet on the island of Boa Vista in the Cape Verdes, and the *Roza* being the only vessel in the area, they (perhaps unwisely) had boarded her for passage to Havana for return to England.

On 21 May 1831 the little *Monkey* was wrecked on the treacherous bar at Tampico on the Gulf coast of Mexico, happily without loss of life. This made a sad reduction to the schooner force, but there had been three reinforcements in recent years. *Firefly* had been acquired in Bermuda in 1828; *Kangaroo*, adopting the name of the wrecked survey vessel, had been purchased in Jamaica in 1829; and *Minx*, newly built in Bermuda, had joined early in 1830. *Firefly* and *Kangaroo* mounted what was becoming a standard armament for the schooners: two long 12-pounders and a five-and-a-half-inch mortar, but *Minx* was armed with 12-pounder carronades instead of the long guns; not, it would appear, a satisfactory arrangement given the poor record of carronades in schooner actions.

The eight schooners: *Nimble, Pickle, Pincher, Skipjack, Speedwell, Firefly, Kangaroo* and *Minx* remained the front line in the struggle against the slavers in the seas of the West Indies, searching the cays, combing the approaches to

Cuba, and endeavouring to spot slaving vessels among the myriad of legitimate merchantmen, many of them from Baltimore, Charleston and the Chesapeake, with the same rakish lines as the American-built Cuban slavers. Gallant though their actions had been in the past few years, they had scarcely dented the slave traffic into Cuba.

South of the Equator, slave trading under the flag of Brazil had remained legal until 13 March 1830, and the merchants of Rio, Bahia and Pernambuco had been making the most of the final months of undisturbed trafficking. In the last half of 1829 the ports of Brazil had seen the landing of 35,301 live slaves, and the number had risen to 40,409 in the first half of 1830, almost all from the Congo and the coasts of Loango, Angola and Benguela. In November 1829 the British Minister in Rio de Janeiro received notification from the Brazilian Minister for Foreign Affairs that the Brazilian Government accepted that legal slave trading by Brazilian citizens would cease on 13 March of the following year, in accordance with the Anglo-Brazilian Convention of 1826, but pointed out that it had been agreed with the Earl of Aberdeen, the British Foreign Secretary, that slaving vessels still at sea after that date would be allowed to complete their voyages as long as they had cleared the coast of Africa by 13 March. Imports fell dramatically in the first half of 1830. Live landings at Rio in the last six months of 1829 had been 30,839, but dropped to 1,300 in the following half-year.

The evasions then began. At the end of March 1830 there were still vessels in Bahia which had been given passports for legal slaving, and their intention to continue trading was clear. Others had recently sailed under new names. Well-founded rumours were rife in Rio that the merchants intended henceforth to import Africans under the pretext that they were colonists or servants. It also appeared likely that slaving into Brazil would continue under the Argentine flag. The British commissioners in Sierra Leone were no more confident than were the diplomats in Rio that the Brazilian slave trade would cease simply because the government in Rio had agreed with Britain that it should. Their suspicion was that Brazilian slavers would purchase Portuguese papers, readily available for a few dollars at the Cape Verde Islands, Princes or St Thomas. As things stood, such papers and Portuguese colours were likely to give them a clear run south of the Equator.

As was argued by Commissioner Smith at Freetown, the Additional Convention to the 1815 Anglo-Brazilian Treaty, signed in July 1817, had permitted the Portuguese to continue slaving south of the Line only to its own colonial possessions. With its declaration of independence in 1823 Brazil was no longer a Portuguese possession, and, having acknowledged that independence in 1825, Portugal could not deny the fact. However, the Lisbon government had neglected to declare an end to the Portuguese slave trade in the southern hemisphere. That trade had dwindled as Brazilian merchants increased their share of the traffic into their ports, and Britain had not pressed the point, but, with the end of legal slaving by Brazil, the matter of Portugal's standing on her abolition treaty with Britain had become a matter requiring urgent resolution.

The difficulty was brought into focus by an arrest on the morning of 2 December 1830. The frigate *Druid* (46), Captain G. W. Hamilton, had been lying at anchor off Bahia, keeping a weather eye on the general state of affairs in the town, when she sighted to windward a schooner which initially approached the harbour and then stood to sea again. Her suspicions aroused, *Druid* weighed and gave chase. As the range closed, Hamilton fired a gun, which induced the schooner to hoist Portuguese colours and brought her down to the frigate. A boarding party found that she was the *Destimida*, bound from St Thomas and Princes to Bahia, with irregular papers and with five slaves whom the master pretended were members of the crew. She was in a leaky state, and Hamilton took her the ten miles into Bahia before a more thorough search revealed a further 50 slaves concealed in the bottom of the schooner. The Africans, who declared that they had been loaded at Whydah, were removed to *Druid* to allow the schooner to be careened, and the majority of them were retained in the frigate for their safety on the subsequent voyage to Rio.

On 22 January 1831 the long-unemployed Anglo-Brazilian Mixed Commission court at Rio, of which the British commissioners were now Alexander Cunningham and Frederic Grigg, completed its investigation of the case. It concluded that the *Destimida* was indeed Portuguese and therefore not subject to the court's jurisdiction. There no longer being an Anglo-Portuguese court in Rio, and the option of sending the slaver to Freetown having apparently not been considered, the schooner was restored to her owner. The court did, however, feel itself empowered to liberate the slaves. To the intense displeasure of the Commander-in-Chief on the South America Station, Rear-Admiral Baker, the slaves were not removed from the schooner, and from *Druid*'s care, for a further month, thereby effectively immobilising the frigate.[19]

If the Portuguese intended to continue slaving into Brazil there was clearly little in the present arrangements to deter or prevent them, and intend to continue they certainly did.

* * *

At the beginning of the 1830s the Cuban slave trade was operating at a lower volume than had been the case in recent years, and to some there seemed to be hope that the Trade was in terminal decline. Admiral Colpoys' schooners, and even, on a few occasions, his larger vessels, had indeed achieved notable successes against Cuba-bound slavers, but, although these seizures angered the Havana merchants, they had not significantly damaged the Cuba trade, and had certainly not deterred it. Spain's economy continued to rely on Cuban sugar, and the Cuban sugar plantations continued to depend on imported slave labour. The authorities in Havana, particularly a succession of powerful, even tyrannical, Anglophobe Captains-General, gave their support to the Trade while lining their pockets, and a weak Spanish Government, while paying lip service to its treaty with Britain, acquiesced in the illegal trade. The downturn in the traffic was the result merely of a temporary glut in the slave supply, and the fact was that the level of the slave trade into Cuba was still, despite the best efforts of Admiral Colpoys' brave little schooners, driven solely by market forces.

South of the Line, the slave traffic into the ports of Brazil, primarily Rio de Janeiro and Bahia, although observed by the cruisers of the South America squadron, had remained invulnerable to their attentions until March 1830. At the ending of legal slaving, that immunity for Brazilians had apparently ceased, but, with slave trading reclassified at that point by Brazil as piracy, the future jurisdiction of the Anglo-Brazilian Mixed Commission was brought into question. The probable continuation of an illegal Brazilian slave trade was not the only concern in the western Atlantic. Portugal had persisted after Brazil's independence in running slaves into its former colony, in contravention of the Anglo-Portuguese Treaty of 1817, although Britain had neglected to challenge this clearly illegal practice.[*] Portuguese slavers were now also joining the traffic into Cuba, heretofore a virtual Spanish monopoly. In the light of this continuing Portuguese activity north and south of the Equator, the absence of Anglo-Portuguese Mixed Commission courts in the West Indies and South America was certain to present a difficulty which had already manifested itself.

A quarter of a century after the Royal Navy had first been deployed against the slave trade, its task in the western Atlantic had expanded massively. At the same time the judicial infrastructure supporting that task was looking disconcertingly fragile.

---

[*] The 1817 agreement prohibited slaving by Portuguese vessels bound for any port not in Portuguese dominions.

CHAPTER II

To the Cape Station: Eastern Seas, 1831–5

The triumph of hope over experience.
DOCTOR SAMUEL JOHNSON

BY EARLY 1831 the legal and judicial system for dealing with suspected slavers under the flags of Portugal and Brazil was beset with uncertainty. Although riddled with frustrating loopholes, the Anglo-Portuguese Treaty of 1815 and its Additional Convention of 1817, to which Brazil was later added as a signatory, had provided a workable basis for action, but these agreements now seemed to have been overtaken by events.

The Anglo-Brazilian Treaty of 1826, ratified by Brazil on 13 March 1827, had agreed abolition of the Brazilian slave trade north of the Equator forthwith, and of the Trade south of the Equator three years after ratification. The Brazilian Trade from the Bights had certainly not stopped immediately in accordance with the treaty, but the intensity of that illegal traffic in recent years appeared to indicate that efforts were being made to build up the number of slaves in the country before a halt was called. The total number of slaves imported into Rio de Janeiro had risen from about 30,000 in each of the years from 1826–8 to 45,000 in 1829 and nearly 60,000 in 1830.

The assessment by *Primrose* in the summer of 1830 that there had been a vast reduction in the Trade south of the Line, for some years a Brazilian monopoly, indicated either a glut in the market or a move towards abolition. A further hopeful sign was the absence of a single Brazilian slaver prize since February 1830. Late in 1829 the Brazilian Government had declared its acceptance that slave trading by subjects of Brazil would become illegal on 13 March the following year, but that was a far cry from effective abolition or from a commitment to suppress a future illegal traffic.

A clear indication had reached the Foreign Office that the slave merchants of Brazil fully intended to continue importing Africans after abolition, protesting that they were "neither colonists nor servants".[1] The British commissioners and the Chargé d'Affaires in Rio were instructed by London to deal with such conduct as illegal slaving, but it was not certain that the Brazilian Government would continue to recognise the Mixed Commission courts after 13 March 1830. At that point it might argue that Brazil had fulfilled the terms of its agreement with Britain and claim that the treaty had automatically terminated. A further complication was that the Brazilians had declared illegal slaving to be piracy, a crime with which the Mixed Commission courts were not empowered to deal.

Even if Brazil wished to enforce abolition it would undoubtedly need considerable help in doing so, and if Brazil were not so inclined Britain would probably feel authorised and obliged, on the strength of the 1826 Treaty, to take unilateral action against the Trade from Loango, the Congo, Angola and Benguela as well as from the slave coast north of the Equator. Either way, the implications for the Royal Navy were considerable. The West Africa Squadron was already grossly overloaded by its responsibilities north of the Equator, and was quite incapable of extending its remit to cover a further 1,000 miles of coastline to the south, at least not to any useful effect. Not only would a much larger force be required, but also base facilities and logistical support would need to be rearranged. Freetown, at the northern end of 3,000 miles of slave coast, would necessarily have a reduced role, the case for developing Fernando Po would be strengthened, both Ascension and St Helena would have bigger parts to play, and the Cape would probably become involved.

The command structure would also have to be reviewed. The entire West Africa slave coast, from 15° N to 15° S latitude, had the dimensions of a major naval station, and an operation extending along its whole length would encroach appreciably on the existing Cape Station. Instead of controlling the campaign itself, through a squadron commodore, it would appear to be necessary for the Admiralty to delegate responsibility for such an extended affair to the commander-in-chief of a newly established Station encompassing the whole slave coast.

The judicial infrastructure too was faced with difficulty. Even if the Anglo-Brazilian Mixed Commissions were to survive, a single court on the African seaboard in Sierra Leone would be too remote for seizures in the Eastern Atlantic south of the Equator, and the Rio de Janeiro court was no more helpfully placed. In the event of Brazil's withdrawal from that joint system the British Vice-Admiralty courts would probably have to resume jurisdiction, but on what legal basis that might be remained to be seen.

The second major quandary for Great Britain in the early 1830s was presented by the persistent failure of Portugal to legislate against its slave trade south of the Equator. The freedom granted to its slavers by the agreements of 1815 and 1817 to continue trading south of the Line specifically excluded any traffic to ports not in the dominions of Portugal, and the achievement of independence by Brazil clearly removed that country as a legitimate destination. Portugal ignored this change in status, and, perhaps partly because the volume of the Portuguese trade to Brazil was not great in the late 1820s, Britain did not press the point at the time. However, lest there remained any doubt on the matter, the Advocate-General gave his opinion in February 1831 that, as the Portuguese trade had not been altogether abolished as stipulated by the treaty of 1815, vessels slaving under the Portuguese flag were liable to arrest both to the north and the south of the Equator. Furthermore, until Portugal declared that the slave trade had universally ceased and was prohibited throughout its entire dominions, the Convention of July 1817 would, he said, continue in force. The question for Britain was what action it should take, or could afford to take, particularly at sea.

From the point of view of the commanding officers of the West Africa Squadron, whose attention was still concentrated primarily on the Bights, the matter of most pressing concern was what might be done to curtail the apparently invulnerable French slave trade on those coasts. They continued to be intensely frustrated too by their inability to take action against slave vessels not actually carrying slaves, except in the case of the Netherlands, the only nation subject to an Equipment Clause in its treaty with Britain. The value of that measure had been demonstrated by the virtual extermination of the Dutch slave trade.

Meanwhile, the Lieutenant-Governor of Sierra Leone was not worried by matters in the Bights; his concern lay in the increased Portuguese slaving from the northern rivers, some of it on his own borders, and repeated appeals to the Colonial Office had resulted in no significant action by the Navy.

Then there was still indecision on the future of the British settlement on the Spanish island of Fernando Po.

It was difficult to discern what real progress had yet been achieved against the slave trade, but, still hampered by utterly inadequate resources, the Navy remained loyal to its long-pursued and single-themed strategy of interception of slavers at sea.

* * *

Although no one since the departure of Captain Owen, either on the African coast or in London, had shown any enthusiasm for retaining the settlement at Fernando Po, there had yet to be a final decision on its future, and the Foreign Office continued negotiations with Spain to exchange Crab Island in the West Indies for Fernando Po. A new superintendent had been appointed to relieve Owen: Lieutenant-Colonel Edward Nicholls of the Royal Marines, who was reputed to have fought in over a hundred actions during the Napoleonic wars, and who had been markedly successful in command of Ascension Island. Then in July 1831 the West Africa Squadron's victualling depot was moved from Freetown to Fernando Po, effectively making the island the Squadron's regular rendezvous.[2]

Commodore Hayes had sent Captain Gordon in *Atholl* along the coast to leeward to assess, among other factors, the state of the slave trade, and, pending receipt of Gordon's report at a planned meeting of *Dryad* and *Atholl* at Fernando Po, he turned his mind to the matter of his tenders.

His immediate concern was the need to acquire a replacement for the virtually worn-out *Black Joke*, but uncertainty hanging over the entire future for tenders as anti-slaving cruisers was a worry to him. He had been only partially reassured by the arrival in January 1831 of eight sets of Instructions for tenders, leaving blank spaces for the names to be filled in. They had been signed by Lord Melville and Sir George Cockburn, and Hayes was concerned that the recent change of Board of Admiralty would render them worthless. Furthermore, there was no indication of the extent, if any, to which he might make further purchases. No one on the slave coast was in the least doubt of the effectiveness of Baltimore clipper ex-slavers as cruisers, but it seemed that the Admiralty was not yet persuaded, despite the fact that they were also cheap. As Hayes pointed out to Secretary Elliot after reiterating the operational case for ex-slaver tenders, "very superior vessels" could occasionally be bought at Sierra Leone for a sixth of their value, and £2,000 would provide all the tenders the Squadron might need for three years, while a suitable vessel could not be built, fitted-out and sent out to the Coast for less than £4,000. His apparently irrefutable argument fell on deaf ears, and Their Lordships replied that they did not wish any further purchase of tenders to take place.[3]

However, the Commodore, very wisely, had already reported his intention to replace *Black Joke* with one of the two prize vessels shortly to become available at Freetown, and before the Admiralty's embargo arrived he bought *Atholl*'s prize, the splendid brigantine *Dos Amigos*, renaming her *Fair Rosamond*. On the excuse that he had been doubtful that he would be able to acquire the brigantine, he recaulked *Black Joke* and endeavoured, or so he told Secretary Elliot, to make her seaworthy for smooth waters in the vicinity of Sierra Leone, intending to run her on for as long as she would last. These were decisions that Hayes would not regret, and, although *Seaflower*, which he tended to keep on a short rein, was something of a disappointment, *Black Joke* and *Fair Rosamond* were to serve him well.

There had been an encouraging encounter on 26 February 1831 off Cape Mesurado when *Black Joke*, fresh from her capture of the *Primero*, while in chase of an unidentified schooner sighted a strange frigate also in pursuit of the schooner. The chase, having come under fire from both men-of-war, turned towards the frigate, which had hoisted United States colours and was identified as USS *Java*. The frigate boarded the schooner, which *Black Joke* then passed within hail for Lieutenant Castle to ask permission also to board. The American boarding officer refused, but Castle then went aboard *Java* to call on Captain Kennedy. He was received with every civility and offers of assistance and information, and Kennedy expressed a hope that if Castle should encounter any American slavers he would proceed against them as pirates, as he would himself if he found a British slaver.

Two days later *Black Joke* fell in with *Java* again, and Kennedy asked how far Castle intended to remain cruising close inshore, observing that the waters were dangerous, the currents strong and the charts bad. Castle replied that he intended to continue for a further 40 miles and offered to remain ahead of the frigate to take soundings. Kennedy declined the offer, but said that if he happened to find a slaver he, having signalled Castle, would detain her until *Black Joke* arrived.[4] Slaving under the Stars and Stripes had not been a matter of concern for some years, but the renewed presence of the US Navy on the African coast, coupled with the friendly and cooperative attitude of Captain Kennedy, shone a hopeful ray of light onto a generally bleak scene.

This was the final incident in Castle's brief period in command of *Black Joke*. With the brig recaulked and all thoughts of "smooth waters in the vicinity of Sierra Leone" forgotten, he headed for the Bights, rejoining the Commodore at Princes Island in April. There he was appointed in temporary command of *Medina* to replace Commander Webb, and in October he was promoted. Ramsay returned to *Black Joke*.

Intelligence had reached Hayes that there were some heavily armed and strongly manned slaver brigs in the Bonny and Old Calabar rivers, and as *Black Joke* sailed from Princes for a Bight of Biafra cruise, he warned Ramsay to be prepared for battle. At 1100 on 25 April, about 40 miles south of the mouth of the Old Calabar River, the tender surprised a suspicious brig and gave chase on a south-easterly course in a fresh breeze from the south-west. The stranger was an excellent sailer, and it was not until 2100 that she came within range of *Black Joke*'s long gun. A shot ahead to bring her to received instead the response of three broadside guns. The wind fell light as the two vessels came under the lee of Fernando Po and, with both resorting to sweeps, a running fight continued until 0130 on the twenty-sixth, when a light breeze sprang up.

It was clear to the chase that she could not escape, and she shortened sail to defend herself. As *Black Joke* came close up it was found that the chase had five guns to the tender's two, and Ramsay realised that he would have to board quickly. As the two vessels came together, Ramsay, Mr Bosanquet, the mate, and

Mr Hinde, a First Class Volunteer,* along with 15 men, leapt aboard the chase, but then the tender cannoned off her adversary, isolating a badly outnumbered boarding party. Those remaining in *Black Joke* quickly manned the sweeps and brought her back alongside, then lashed the vessels together. Once reinforced, Ramsay's men were soon in complete control of what they found to be the 300-ton Spanish brig *Marinerito* and her 496 slaves. She mounted a long 18-pounder and four short 18-pounders, and, with the clear intention of fighting if intercepted, she was manned by a crew of 12 officers and 65 men.

Thirteen Spaniards had been killed and 15 wounded, four dangerously so, and *Black Joke* had lost one killed and four wounded, the latter including Ramsay and Bosanquet. Midshipman Pierce had survived three narrow shaves: his hat was removed by a musket ball, he was pushed overboard by a sabre thrust which punctured his jacket and shirt, and he then evaded drowning by catching hold of the slaver's trailing foresheet and reboarding. Assistant Surgeon Douglas earned Ramsay's particular praise for being "involved in every department", and James Cooper distinguished himself in steering the tender and manning the long gun. Cooper was subsequently advanced to First Class Petty Officer; a boy, Fail, was entered on the ship's books in consequence of the death of his father, the one man killed; Douglas was promoted to Surgeon; Bosanquet was given his lieutenant's commission; and Ramsay was soon to be made commander.

Panic among the overcrowded slaves had caused the death of 27, and the remainder were desperate for water. They overturned a tub when it was first filled, and when it was refilled so many of them tried to drink that their heads wedged in the tub. On reaching Fernando Po, Ramsay was obliged to land 107 desperately ill Africans before despatching the slaver for Sierra Leone under Bosanquet's command. Of the 496 slaves at capture, 373 were emancipated on arrival at Freetown in *Marinerito*, 62 arrived safely later in *Plumper* and *Seaflower*, 4 remained sick at Fernando Po and were emancipated, and 57 died at Fernando Po or on passage to Freetown.†5

---

* The source of this account is Ramsay's report to Hayes two days after the event, but Basil Lubbock writes in *Cruisers, Corsairs and Slavers* (Brown, Son & Ferguson: Glasgow, 1993) that Hinde, "a tiny mid, not fifteen years of age" remained on board *Black Joke* and took charge of getting the tender back alongside and lashed. If Lubbock is right it would be an injustice to Mr Hinde, who was undoubtedly a First Class Volunteer rather than a "mid", not to record this version of events. A First Class Volunteer (a title introduced in 1794 to replace the misleading "Captain's Servant") had to serve for two years before becoming eligible for promotion to the Warrant rank of Midshipman.

† Different sources record varying figures, but those given are a best estimate based on the incomplete figures in the Mixed Commission's annual report to the Foreign Office, and on the court's minutes of the trial (which disagree!).

There are few records of the reaction of slaves to their release in captured slavers, but it was observed in a report on the *Marinerito* incident that:

> All slaves appeared to be fully sensible of their deliverance and upon being released from their irons expressed their gratitude in the most forcible and pleasing manner [...] The poor creatures took every opportunity of singing a song, testifying their thankfulness to the English,* and by their willingness to obey and assist, rendered the passage to Sierra Leone easy and pleasant to the officers and men who had them in charge.[6]

The slaver was condemned, as were 31 gold doubloons found in the cabin chest of her late master, a welcome bonus for the captors. The Commodore lamented that he had no authority to buy such a well-found vessel, which he expected to be sold for less than £1,000. Had he been allowed to do so, he would have transferred the ship's company of one of the gun-brigs into her and sold the gun-brig to defray the cost. The exchange would have rid him of what he regarded as a useless cruiser and replaced her with one "ten times her worth".

Black Joke, already in a precarious material state, had suffered considerable shot damage aloft and alow, and her larboard quarter and larboard forward bulwark had been stoved in. Nevertheless, she was ready for sea by 7 May, and after provisioning and watering from *Dryad*, which was at Fernando Po, she was sent out again to cruise the Bight of Biafra, and Ramsay was given freedom to station himself anywhere between Cape Formoso and the eastern shore as intelligence and his judgment should dictate.

Despite the high incidence of sickness, it had not been possible to put a Surgeon into the *Marinerito* for her passage to Sierra Leone as required by the Admiralty. Echoing the complaint of predecessors, Hayes had pointed out that the Squadron had barely enough medical officers for its own purposes. If Their Lordships' instructions were to be obeyed, he wrote to Elliot, there was a requirement for four Assistant Surgeons in *Dryad*, two in *Atholl* and two in each of the sloops and gun-brigs. The Admiralty at last ordered them to be provided.

While *Dryad* was lying at Clarence Cove at the beginning of May, the Commodore had *Atholl* surveyed to ascertain the extent of damage inflicted by Coley's Rock. He found that there was no immediate need for her to go home, and he sent her to Princes to complete her water, to Annobón to land the *Marinerito*'s Spanish crew, and to Ascension to replenish with bread, which was in short supply at Fernando Po. He instructed her thereafter to examine

---

* Clearly they failed to appreciate that their saviour, Ramsay, was a Scot!

the African coast between the Equator and 10° S to discover the extent of the slave trade, to ascertain the principles of general trade, and to assess whether any openings existed for British trade. Meanwhile, *Dryad* sailed to visit Cape Coast and Accra. In the last week of April, however, while the two ships were in company at Princes, there had erupted a clash between the Commodore and Captain Gordon of *Atholl*.

In late January, Hayes had sent an amendment to Gordon's orders via *Medina* which had returned to Freetown from a fruitless Bights cruise. *Atholl* was, without delay, to take a consignment of clothing to Fernando Po for the Royal African Corps contingent, and to offer assistance to Lieutenant-Colonel Nicholls. If *Medina* managed to deliver her message before *Atholl*'s planned arrival at Princes, then she was no longer to make for Princes. This seems to have sowed some confusion in Gordon's mind, and his interpretation was that he was required to hang around pointlessly in the vicinity of Fernando Po instead of cruising against the slavers. He believed furthermore that, on account of the Commodore's orders, which allowed him no discretion, his protracted period at Fernando Po had resulted in the deaths from fever of the Master and a boy in *Medina*, and the "more than probable" death of the Purser of *Atholl*. He had earlier been annoyed by (as he saw it) cavalier treatment by the new Commodore at the handover of command, and also by Hayes having taken, without Gordon's agreement, the First Lieutenant of *Atholl*, Lieutenant Ramsay, and a number of her men to man *Dryad*'s tenders,[*] a matter on which he subsequently complained to the Admiralty. On meeting Hayes at Princes in late April, he expressed his concerns in a private note which, although implicitly critical of the Commodore's instructions in the phrase "your last order I have obeyed so rigidly that valuable lives have been lost which I shall never cease to regret", was essentially a pleading of his case and was couched in courteous and friendly terms. It also contained an offer by Gordon to take Hayes's letters home and to call on Mrs Hayes.

Hayes was furious and challenged Gordon to name the men who had allegedly died in consequence of his orders. Gordon, taken aback by this response, replied that he had meant no criticism, was actually blaming himself for remaining overlong at Fernando Po, was concerned about the damage to his ship, and was suffering from an ulcer and a "highly irritable state of Stomach and Bowells (*sic*)". He wrote that, in consequence of his indisposition, he was obliged to give up command of his ship and return to England, and he asked to be excused further

---

[*] Ramsay supported the Commodore's contention that this had been done with Gordon's agreement, but Ramsay had benefited greatly from the arrangement, and would have been unwise to contradict Hayes.

correspondence on the matter. Now was the time for Hayes to pull back from open confrontation with no harm done, but he was not to be placated and, having replaced Gordon with Commander Webb of *Medina* and put Lieutenant Castle into *Medina*, he ordered that an enquiry into Gordon's allegation be conducted by these two officers and Commander Turner of *Dryad*. The enquiry concluded that no deaths had resulted from the Commodore's orders, and Hayes applied for Gordon to be tried by court martial.[7]

It seems that this unpleasant escalation resulted from overreaction by Hayes to an unwise, if understandable, indiscretion by Gordon, and one wonders whether there might have been a history of ill feeling between the two men. When Gordon came to trial in September 1832, before two flag officers and the commanding officers of the post ships at Portsmouth and Spithead, including Sir Francis Collier, he faced five charges involving "Disrespect", "Negligent Performance of Duty", "Un-Officerlike Conduct" and a fraud concerning bounties. Three were found to be not proved, and the court decided that it was not authorised to try the charge relating to a letter of complaint written to the Admiralty by Gordon while he was on half pay. The remaining charge, relating to the private note, was judged to be partially proved; the court found that Gordon had been unguarded and had used an improper expression, but had apologised on realising the offence caused, and he was admonished to be more circumspect in future. This negligible sentence implied heavier censure of Hayes than of Gordon for an unnecessary altercation.

Favourite had returned to the Bights after her uneventful Windward Coast cruise earlier in the year, and on 23 July the Commodore despatched her from Fernando Po on a foray south of the Line similar to that of *Atholl* in May. Commander Harrison found it a struggle to make southing against a strong northerly set inshore until, having investigated some way into the Congo, he crossed the stream of the river, after which he experienced little difficulty. He also found that word of his coming travelled down the coast more rapidly than *Favourite* could sail to windward. On 9 September the sloop reached St Paul de Loando, where she replenished with water and bullocks and found 15 vessels at anchor, all apparently slaving.

In his communications with all the governors and captains-general of the Portuguese settlements on the coast Harrison was readily informed of recent visits by Spanish slavers, an intrusion into what had been purely a Portuguese and Brazilian trade until that year. He learned too that the Spaniards were powerful vessels; the *Montezuma*, for instance, which Hayes had told him to watch for under the name of *Fama de Cadiz*, was a vessel of 600 tons armed with sixteen long 12-pounders and four long nines. Trade other than slaving was, he discovered, trifling, and there seemed to be no worthwhile openings for British merchants; there was a government monopoly on ivory and there was no gold.

On her way north on 14 September, *Favourite* boarded a Portuguese schooner on passage from Rio to Ambriz and found slave irons ready for use, but, there being no slaves on board, she could not be detained. After calling at Fernando Po with his report in late September, Harrison set off for a Bights patrol between latitudes 2° and 4° N, and longitudes 4° and 8° E.

It had been the Commodore's intention earlier in the year to send home both *Atholl* and *Medina*, and *Medina* had indeed departed, but *Atholl*, even after her southern hemisphere cruise, remained on station. The reason for the change of plan is not apparent, but it may have been that, having got rid of Gordon and put Edward Webb into the ship in the acting rank of captain, the Commodore was happier to have her under his command. Continuing in her investigatory role, she sailed from Fernando Po in early July for the coast of the Bight of Benin and the Gold and Ivory Coasts to enquire into the state of the slave trade and cases of piracy, to find out whether there was any requirement for men-of-war to carry gold and ivory to England as freight,* and to try to pinpoint Coley's Rock and another off-lying rock reported between Cape Palmas and the St Andrews River. Having touched at all the regular slaving posts and forts between Lagos and Cape Palmas, Webb was able to report on 2 October that he had encountered only four vessels slaving, all of them Spanish, and that the Trade appeared to have decreased greatly in the Bight of Benin and ceased entirely on the rest of the coast.

While their larger sisters had been making their enquiries along the coasts to the south and west, the three tenders had continued their watch for slavers in the Bights. *Seaflower*'s maiden cruise under the command of Lieutenant Henry Huntley had brought only one chase, exciting but ultimately unsuccessful, when a schooner she sighted at dawn off Whydah managed to beat back and land her slaves before the tender, gaining fast, could catch her. On rejoining the Commodore and his two other tenders at Princes Island in April, Huntley transferred to the newly commissioned *Fair Rosamond* when Ramsay relieved Castle in *Black Joke*. Unlike his comrade-in-arms, Huntley had to wait until July for his first victim.

As night was falling on 20 July, *Fair Rosamond* was heading into the Lagos anchorage when she passed an outward-bound Spanish schooner. Under the illusion that the tender was a fellow slaver, as Huntley hoped, the Spaniard held her southerly course, but as soon as the schooner was lost to sight in the darkness

---

* This means of giving secure carriage for high-value, low-volume consignments was welcomed by merchants and provided a lucrative sideline for commanding officers.

Fair Rosamond hauled round in her wake with all sail set. After an anxious night of fast sailing on a broad reach, Huntley and his men were relieved at daybreak to sight the schooner seven miles on the lee beam. After a six-hour chase the schooner came within range of the tender's single long gun, a nine-pounder on a pivot, and Huntley fired a warning shot over her. By now the schooner had jettisoned all moveable weight, and finally she cut away her anchors. By wetting her sails she increased the distance a little as the wind dropped, but a freshening breeze and experiments in trim brought *Fair Rosamond* back into range, and a second shot cut down the Spaniard's studdingsails on one side. Several more rounds into her rigging had no effect on the chase, but then a shot across her deck killed her helmsman. The Spaniard hauled down her colours and hove to, but with all sail still set. Huntley realised that she meant to fill again as soon as the tender's boarding boat was in the water, and he ranged up within 50 yards on the schooner's beam before hailing her master to lower his sails and come aboard. The Spaniard feigned not to understand either English or French, so Huntley, by all accounts a determined and impatient man, ordered his marines to fire at the schooner's jib halliard block until the halliard was cut. As the jib came down they shifted their aim to the peak halliard block on her fore gaff. At this the Spaniard hailed in good English that he would come aboard. The schooner was the *Potosi* with 192 slaves, nine of whom died before condemnation at Freetown.[8]

Fair Rosamond's next chase finished less satisfactorily. As she was running along the land one evening within sight of the Lagos anchorage, she sighted a large brigantine getting underway and standing out to sea. Huntley immediately came round after her, and tried a shot at her in the gathering darkness, but the range was too great. He managed to keep her in sight until about 0200, when rain obscured her. At dawn she was resighted only four miles away, but the breeze dropped, and the sweeps had to be manned. By 0800 the brigantine was within gun range, and she lowered her sails. The tender came up within hail and Huntley told the Spanish master to come on board with his papers, only to discover that the brigantine had no slaves and that her cargo was still in the barracoons. When the Spaniard was asked why he had made a chase of it, he replied that he just wanted to compare *Fair Rosamond*'s sailing with that of his brigantine. At this, the exasperated Huntley told the slaver that the information would have to be paid for, and he sent his midshipman and ten hands to unbend the brigantine's sails; unreeve her running rigging; throw overboard her guns, powder and all other weapons; and run out her bower anchor cables to their clenches. Warning the Spaniard not to try any more tricks with a British man-of-war, he sent him back to his chaotic vessel. Unfortunately the slaver later managed to slip out undetected with about 400 slaves.

It does not seem to have occurred to the British commissioners at the Freetown courts that, following expiry of the 13 March 1830 deadline, there might now be some difficulty in trying Brazilian slavers, and it was left to the Commodore to point out that, if these Brazilians were now to be regarded in law as pirates, the Mixed Commission would have no jurisdiction over them. The Admiralty, reminding the Foreign Office that there was no "court of piracy" at any of the British settlements in Africa, sought instructions. The British Commissioners, Findlay and Smith, somewhat tardily, voiced their concerns to Viscount Palmerston, who had recently replaced the Earl of Aberdeen as Foreign Secretary. Raising the diplomatic temperature, the Brazilian Chargé d'Affaires in London then demanded the immediate dissolution of the Anglo-Brazilian Mixed Commission courts, and De Paiva, the Brazilian Judge in Freetown, announced instructions from his Government that, as the Brazilian slave trade was finished, the Mixed Commission was no longer necessary. He had been told to complete the long-delayed trial of the *Ismenia* by 30 June 1831, bring the affairs of the Commission to a conclusion, and return home.[9] In August, after a number of diplomatic exchanges on the subject, Lord Palmerston, not a man to stand any nonsense from foreigners, put his foot down. He declared that HM Advocate-General's view of the matter differed from that of the commissioners, and that the Anglo-Brazilian Mixed Commission should continue to apply the provisions of the treaty concerning condemnation of slave vessels and emancipation of slaves. De Paiva reported at the end of July that he had been ordered to remain in post.

After her visits to Cape Coast and Accra in May and a cruise in the Bights, *Dryad* had made for Ascension, where she watered and reprovisioned before sailing for Princes and Fernando Po on 9 September. A week before *Dryad*'s departure, *Conflict* arrived at the island. She had been struck by fever; her Commanding Officer, Lieutenant Smithers, was sick and one man had died on passage, but, as usual, Captain Bate RM and his garrison provided every possible support and care, including clearing the gun-brig's hold and landing her iron tanks, suspected sources of infection. Once restored to health, and loaded with provisions and water, *Conflict* sailed for Accra and Princes.*

Meanwhile the tenders *Black Joke* and *Fair Rosamond* maintained their watch for slavers, the former in the Bight of Biafra and the latter in the Bight of Benin. During the first few days of September, Lieutenant Huntley was refitting his brigantine at Clarence Cove when he heard from a British palm-oiler that two large and powerful Spanish brigs, the *Regulo* and the *Rapido*, were lying in the

---

* *Dryad* reprovisioned with 2,782 lb of fresh meat, 1,930 lb of vegetables and four live sheep, and took 15 tuns of water. *Conflict* embarked 397 lb of fresh meat, 590 lb of vegetables and seven tuns of water.

Bonny River with slaves, intending to sail in company, but their draught prevented them from doing so until the next spring tide. This gave *Fair Rosamond* three days to finish her refit and three to reach the Bonny. In fact she made good time, arriving on the eighth and anchoring to seaward of the shoals, where her lookout could command both channels of the river mouth. Huntley was hoping that *Black Joke* would make her appearance, this being her cruising ground, and early the following afternoon she hove into sight. Ramsay had gleaned from a French slaver information similar to that reported to Huntley, and although Huntley apologised for poaching on *Black Joke*'s station, it was clear that the two needed to act in concert to deal with two slavers believed to be mounting between them fourteen long 12-pounders, four 18-pounder carronades and a long 18-pounder on a pivot.

As expected, at high water on the morning of 10 September the two brigs were sighted, emerging under clouds of sail. The two tenders delayed showing any canvas until their bare spars had been detected; and, in any case, they were becalmed. At 0930 the slavers were obviously alerted, just as a light breeze reached the tenders, which weighed and set all sail. The slavers, by contrast, were suddenly becalmed and anchored just inside the outer bar. *Black Joke* had taken longer to weigh, and in her poor condition gradually dropped astern of *Fair Rosamond* as the two skirted the Baleur Bank and ghosted inshore. At 1300 the sea breeze reached the slavers, which weighed and stood to seaward under fighting canvas as if to engage the tenders. When the Spaniards were at about three or four miles from *Fair Rosamond*, the *Rapido* lost her nerve and bore away for the river, setting all sail. The *Regulo* was obliged to follow suit. Unlike the slavers, Huntley had no pilot, but fortunately he had a new chart by Captain Owen, and he headed, with studdingsails set and a freshening wind, into the maze of shoals and broken water obstructing the river mouth. In the rough water *Black Joke* gained ground, and she was only two cables astern when the two tenders emerged unscathed into the deeper water of the river. On crossing the bar, *Fair Rosamond* had closed to about a mile and a half from the *Regulo* which, in turn, was the same distance astern of the *Rapido*.

As the tenders passed through the anchorage five miles upriver, the crews of the British palm-oilers cheered, and one master managed to board *Fair Rosamond* by boat as she passed, bringing eight musket-armed volunteers to reinforce the tender's men. They were most welcome, although Huntley wanted cutlasses rather than muskets. With a prize crew away and ten men sick, he was very short-handed, having, beside himself, a mate, an Assistant Surgeon, ten seamen, four marines and six Kroomen.

About two miles above the anchorage the two slavers turned into a tributary entering the Bonny from the east, but their movements remained visible over the intervening low jungle from *Fair Rosamond*'s masthead, and it could be seen that they were desperately trying to land their slaves. As the tender entered the tributary, taking in her studdingsails, she found the *Regulo* hard aground in the mud and surrounded by canoes into which a mass of Africans were being crowded. Huntley fired a shot over her, sending the canoes paddling for the shore, and the slaver's colours came down. Leaving *Black Joke* to take possession of the *Regulo*, *Fair Rosamond* then passed on to deal with the *Rapido*, which was throwing her slaves overboard still shackled in pairs. As the tender approached, two Africans were sighted in the water ahead, and they were hauled aboard by catching their shared leg-chain with a boathook.

Having landed or drowned her cargo of 450 slaves, the *Rapido* appeared prepared to resist capture, but a round shot over her deck, two volleys from the marines, and the sight of Huntley and his mate, Mr Robinson, four marines and two seamen leaping aboard with bared cutlasses, sent the Spaniards scurrying below, leaving their master alone on deck with two dead men. The crew was found to number 61, of whom three were wounded.

Meanwhile, *Black Joke* had gone alongside the *Regulo*, boarding and arresting her without resistance. Most of her crew of 56 had jumped overboard. Only eight of her expected fourteen long 12-pounders were mounted, but they should have been more than adequate to deal with the two tenders. Of a cargo of about 460 slaves, 207 remained on board, but Ramsay decided to land five found to be infected with smallpox, an action later judged by the Foreign Office (at a safe distance) to have been "unprecedented and improper".

As the tide ebbed, all four vessels grounded in filthy mud, and the men of the two tenders had to endure nearly a week of heat, rain and back-breaking labour before they had lightened the *Regulo* and kedged and warped the prizes and their own vessels into the main river. Meanwhile, the number of fever cases mounted in *Fair Rosamond*. In the work of extrication the tenders received immense assistance throughout from two palm-oilers: the Liverpool ship *Huskinson*, Captain Thomas Clegg, and the London barque *Rolla*, Captain James Ballerney, who loaned men and boats, laid out a bower anchor and chain, and provided pilotage advice. Without their help the four vessels would have been unable to leave before the next spring tides. As it was, they sailed on 16 September, although *Black Joke* did not clear the shoals until the following day. The Admiralty, sending copies of Huntley's report to the owners of the two merchantmen, acknowledged "the meritorious conduct of the Masters".

The two prizes were condemned, the *Rapido* only on the evidence of the two slaves, fortuitously rescued, who provided the necessary proof that the brig had been slaving. Most of the remainder of the *Rapido*'s cargo had drowned or fallen victim to sharks or crocodiles; a month later a palm-oiler counted 100 mangled and shackled bodies lying on the riverbank. Of *Regulo*'s cargo, 28 died after capture and 164 were emancipated. Ramsay could not account for ten missing slaves.

The Commodore expressed his "perfect approbation of the conduct and exertions" of all involved, and he promoted *Fair Rosamond*'s mate, Mr Robinson, to command *Plumper* in place of Lieutenant James Sulivan, who had died only four months after relieving Lieutenant John Adams. By contrast, it seems that the principal concern of the Mixed Commission and the Foreign Office after this dramatic affair was the fate of an anchor and cable missing from the *Regulo*. The simple explanation, provided by Huntley on *Fair Rosamond*'s return to Sierra Leone in October, was that *Black Joke* had weighed the anchor during the salvage operations and had provided the slaver with a temporary replacement pending an opportunity to make an exchange. That a department of state should involve itself in such pettiness was hardly likely to generate respect among the officers of the West Africa Squadron for those directing affairs in London.[10]

Their Lordships declined to confirm Robinson's appointment, and *Plumper* was under the command of Lieutenant Creser when, at the behest of Lieutenant-Governor Findlay, she sailed for the Gambia on 5 September with a military force from Sierra Leone. The Gambia was at open war with the Kingdom of Barra, and at a critical time its only support had come from the French brig-of-war *La Bordelaise*. Troops were then sent from the French colonies of Senegal and Goree, and the corvette *La Bayonaise*, calling at Freetown shortly after *Plumper*'s departure, immediately sailed for the Gambia at Findlay's request. *Seaflower* followed her with a small reinforcement, arriving on 7 December to find a serious situation. Hayes received Findlay's plea for help via *Fair Rosamond* at Fernando Po on 4 December, and he concluded that *Favourite*, believed to be on passage from Ascension to Sierra Leone, was best placed to respond. *Atholl* was heading for Ascension to replenish with bread and flour, not available at Fernando Po, to enable her to sail for England, and Hayes despatched *Fair Rosamond* to intercept her at Ascension and send her to the Gambia without delay. On arriving there, Captain Webb was to take charge of operations until the arrival of the Commodore, or until the state of affairs allowed him to sail for home and leave Commander Harrison of *Favourite* in charge. *Conflict* was either at Ascension or en route for Sierra Leone, and *Dryad* was about to head for Ascension for bread and flour but would make haste for the Gambia. As it happened, *Favourite* returned to the Bights and *Atholl*, remaining in ignorance of the crisis, made directly for home. This failure was unimportant in the event because peace had been restored by 7 January, but it demonstrated the difficulty, indeed the impossibility, of rapid redeployment of such a wide-ranging squadron.

Despite satisfactory cooperation in the Barra conflict, December 1831 saw a reignition of the ill feeling of the previous year between Governor Findlay and the Navy. The British slaver Josiffe had returned to the River Pongas for another cargo, and Findlay, assured by the Colonial Office that the Commodore had been directed by the Admiralty to aid the destruction of the British slave trade in the Pongas, asked Commander Harrison to take action to seize the slaver. In a rather surly reply, Harrison wrote that he was under orders not to send his boats into any river for any reason and, even if he was allowed to do so, he doubted whether he could achieve anything. *Favourite*'s draught, he said, made taking her into the river neither prudent nor practical. The Admiralty later sought an explanation from Harrison, who repeated the reasons he had given to Findlay, quoting the emphatic orders from Hayes concerning boats in rivers, and there appears to have been no further action.[11]

This unfortunate exchange coincided with the publication of its report by a commission appointed to enquire into the slave trade on the rivers Pongas, Nunez and Gallinas, and Lord Goderich at the Colonial Office, clearly not a man conversant with conditions in the rivers of west Africa, asked of the Admiralty whether it might be proper to employ "several armed boats or schooners" on those rivers to visit slave factories and arrest canoes, etc. carrying on the slave trade. Their Lordships replied dryly that they relied on every exertion being made by the Commodore.

Not only had the French moved generously to the defence of the Gambia, but also there had been a marked decline in their slaving activities in the Bights. *Atholl* had found no French slavers during her summer survey of the coast, and, unusually, there were none in the Bonny at the time of the *Regulo* and *Rapido* action. It is estimated that about 500 slaving voyages from French or French colonial ports had taken place between French abolition of the slave trade in 1818 and 1831, but the abolition movement in France had been gaining strength during the 1820s and the French Navy's West Africa patrols against French slavers had become gradually more extensive and effective. In 1830 the bourgeois King Louis-Philippe, an Anglophile and a member of the Society of Christian Morals, came to the throne. He was inclined towards abolition and gave support to a recommendation from Captain Alexis Vilaret de Joyeuse that trading in slaves should be declared a crime. Vilaret de Joyeuse was fresh from command of the French West Africa squadron, the Commodore who had so generously supplied the disabled *Sybille* with a replacement main yard. A new abolition bill passed into French law with only minimal opposition, ending what the historian Serge Daget called "seventeen years of tautology, bad faith, good reasons and countertruths". Slaving or attempting to trade in slaves would henceforth be heavily punished with imprisonment, and freed slaves would be liberated in the colony for which they had been intended.

Early in November 1831, France opened discussion with Britain on a bilateral convention for suppressing the slave trade at sea, a matter on which the Admiralty and Foreign Office had been pestered for years by the West Africa commodores and West Indies commanders-in-chief. London initially proposed a format directly similar to the Spanish and Portuguese treaties, but the French responded that their law precluded a Mixed Commission. Other main points were rapidly agreed, and the convention was signed in Paris on 30 November. Thenceforth there existed a mutual Right of Search by vessels of war, commanded by lieutenants or above, of merchant vessels suspected of slaving. Specific men-of-war were to be authorised to make such visits, and were to be issued with written instructions agreed by the two nations. The numbers of vessels to be so authorised were to be fixed annually by special convention, and the number of authorised cruisers of neither nation was to exceed double that of the other. Lists of these vessels and their commanding officers were to be exchanged. Private flag signals were to be used by the authorised cruisers of the two nations to identify each other, and mutual assistance was encouraged. The requirement for mildness in the manner of visits was emphasised, and vessels found to be slaving or fitted-out for slaving were to be handed into the jurisdiction of the nation to which they belonged.

The geographical limits within which this mutual right might be exercised was bounded by the west coast of Africa, by the meridian of 30° W reckoned from the Meridian of Paris (27° 40´ W from Greenwich), and by the parallels of latitude of Cape Verde and 10° S. It was also to be exercised all around the islands of Cuba, Puerto Rico and Madagascar to a distance of 20 leagues, and out to 20 leagues from the coast of Brazil.* Hot pursuit was to be allowed outside these limits of vessels sighted, suspected and chased within the limits.

Discussion continued well into the following year on which ports should be used for adjudications under the convention. Britain proposed that Fernando Po and Rio de Janeiro would be suitable for both nations, that British vessels captured off Cuba should go to Jamaica, and that French vessels arrested off Puerto Rico and Madagascar should be taken to Martinique and the island of Bourbon

---

* The English league equalled three nautical miles, or one twentieth of a degree of latitude. However, there was no internationally agreed measure of a league.

respectively. The French naturally objected that Fernando Po belonged to neither Britain nor France, and insisted that French vessels captured on the African coast should be taken to Goree, but the Admiralty expressed a deep concern that such a long passage from the Bights would incur appalling loss of life among the slaves. There the matter rested for a time. Meanwhile, under the convention, the Admiralty nominated all of its cruisers on the West Coast, apart from the tenders, and the French Ministry of Marine nominated four in African waters.[12]

The British list of nominated cruisers reflected changes in the West Africa Squadron in the New Year of 1832; *Atholl* had returned to England and three fresh cruisers were on passage to join. A more profound change was afoot, however. The Admiralty had decided that the West African Station needed a commander-in-chief, but, conscious of the developing requirement for anti-slaving operations south of the Equator, and perhaps lulled by the recent low level of slaving activity in the Gulf of Guinea, the expectation that Brazil would declare abolition and the signing of the Anglo-French Convention, Their Lordships opted for the economical solution of incorporating the West Coast into the Cape of Good Hope Station. The current Commander-in-Chief at the Cape was Commodore Schomberg, although in mid-1831 the Commander-in-Chief South America, Rear-Admiral Sir Thomas Baker, had been appointed to the Cape Station, apparently in addition to his existing command. The critical state of affairs in Rio Janeiro had, however, prevented Baker from leaving for South Africa and this extraordinary additional appointment was cancelled. The new nominee to relieve Schomberg, and also to replace Commodore Hayes, was Rear-Admiral Frederick Warren.

Warren had sailed for Africa in the Fourth Rate frigate *Isis* (58), designated as his flagship at the Cape, with two additions to the West Africa Squadron: *Brisk* and *Charybdis*. These two were nominally of the ten-gun "coffin brig" class, but the Constructor of the Navy, Seppings, had responded to the many complaints about these vessels by modifying a number of them for chasing slavers. Consequently, *Brisk* and the brand-new *Charybdis* were rigged as brigantines, with the addition of light square sails to be set on the mainmast; their upper works had been lowered; their lower rigging was secured to bulwark stanchions, instead of projecting channels and chains, in order to reduce drag when the vessel heeled; and their armament was reduced to a long-gun on a pivot amidships and two carronades. These much-improved cruisers were commanded by two officers already renowned on the slave coast: *Charybdis* by Lieutenant R. B. Crawford, who had defended the prize *Netuno* so gallantly in 1826, and *Brisk* by Lieutenant E. H. Butterfield, First

Lieutenant of *Primrose* at the capture of the *Veloz Passagera*. Lagging somewhat astern of Warren's small flotilla was an entirely novel addition to the Squadron, the paddle steam-vessel *Pluto*, commanded by Lieutenant George Buchanan.

The newly commissioned *Pluto*, 135 feet in length, 24 feet in beam and with a draught of only six and a half feet, had a complement of 80 and was armed with two 18-pounders mounted on pivots forward and aft. Her two-cylinder side-lever engine delivered 100 horsepower, giving a speed of about seven and a half knots. Naturally the Admiralty was concerned about the expense of supplying fuel for her as far away as Fernando Po, to which the Navy Board had been directed to deliver "600 Chaldrons of Coals", and Warren was instructed to give Lieutenant Buchanan "the most positive direction never to consume any for raising steam and keeping it up unless for the purpose of chasing suspicious vessels". Otherwise he was to use the sails on his two masts, and his paddle-boards were removable to enable him to do so.* Buchanan was to find a convenient place to procure wood for his furnace, and he was to report the quantities of coal burned, the occasions on which it was used and the length of time steam was kept up on each occasion. The Commodore was also to report the location at which wood was to be procured and a comparative statement on the costs of burning coal and wood. Bureaucracy was alive and kicking! Warren did not expect much coal consumption, and estimated that there was sufficient at Fernando Po for three months. He pointed out that the Government was paying £3 per chaldron of coal, but a Sierra Leone merchant was offering it at 30 shillings, and he expected the Navy Board to be able to contract for less than that. He asked for supplies to be sent to Sierra Leone, not Fernando Po, and he suggested that steam vessels should be furnished with steam saws for cutting logs.[13]

Pluto had been ordered to make directly for the Gambia, as had *Brisk*, which had gone on ahead, while, on 5 January 1832, *Isis* and *Charybdis* called at Tenerife. The Admiralty had required Warren, while at the island, to form a judgment on the common practice of "ships touching at the Wine Islands making purchases of that Article without being apparently regulated by any specific rule". As a result of Warren's recommendation, made, it would seem, at the Cape of Good Hope, the Admiralty decreed that wine should not be purchased for the public service at Madeira, Tenerife or any other of the Wine Islands because it could be bought more cheaply at the Cape. This embargo was all very well for those heading for the Cape, but it could hardly have been as popular in vessels joining the West Africa Squadron. Warren had perhaps not made the best of starts in the eyes of the cruisers' officers.

---

* It is not clear whether *Pluto* was rigged as a brig or a brigantine.

After a brief visit to Goree, *Isis* and *Charybdis* arrived at Bathurst, where Warren found *Plumper* in the river as well as *Brisk* and the transport *Parmela*. He was disgusted to discover that the fortifications of Bathurst Island were in a dilapidated condition and that, the present gathering apart, there had been no visit by a British warship since May 1831. It seemed that it was only the arrival of the Governor of Senegal with a brig-of-war and 200 native soldiers that had saved the settlement from destruction at the hands of the King of Barra. Having sent *Charybdis* to Sierra Leone with orders to Hayes to remain there to await the arrival of the flagship, Warren held a palaver with the troublesome King and other chiefs and told them that he would station a force on the coast to keep the peace. Leaving *Plumper* with *Brisk*'s marines in the river, the Admiral sailed for Sierra Leone with *Brisk* in company, arriving on 27 January to find *Charybdis*, *Seaflower* and the French frigate *Hermione* in harbour. While heading into the anchorage, the two new arrivals had encountered the heartening sight of two French brigs-of-war passing with a Spanish slaver apparently taken off the Gallinas.

The Commodore arrived at Freetown in *Dryad* on 28 January 1832 to be met by Warren with a stark choice: either to haul down his broad pendant and remain on station under the Admiral's command, or to return home immediately with the pendant flying. Not surprisingly, Hayes chose the latter, giving his decision by a rather curt letter in which he protested that he had been given no intimation of Their Lordships' intentions and that the turn of events had taken him entirely by surprise. He told the Admiral that *Dryad* had landed stores and spare sails at Fernando Po and Ascension in order to make space and improve ventilation on board, and Warren gave him permission to touch at both places before returning home.[14]

Black Joke's future lay in the balance. She was worn out, and the Admiralty had ordered that she should be destroyed, but Hayes reported to Warren on 1 February that when the tender had been emptied at Fernando Po during the autumn of 1831, in preparation for breaking her up, it had been discovered that she was not in as poor a state as had been supposed. *Dryad*'s Carpenter, Mr Roberts, had repaired her "at trifling expense" using timber cut at Fernando Po, and, in the opinion of the Commodore, she was "now as strong as when first built". The refit seems to have been a more extensive matter than Hayes was prepared to admit, but, armed with this report, Warren sought Admiralty permission to transfer the brig to Lieutenant-Governor Findlay as a colonial vessel. Findlay had been given Colonial Office approval to purchase a vessel suitable to maintain communication with the Gambia at all seasons of the year, but had been unable to find one. *Black Joke* seemed to offer the ideal solution.

Charybdis departed on 31 January to relieve *Black Joke* on her cruising station between Cape St Paul's and Fernando Po, to allow the tender to return to Sierra Leone to pay off. As the gun-brig sailed, a further addition to the Squadron joined from England. This was *Pelorus*, an 18-gun brig-sloop, now in her twenty-fourth year and under the command of Commander Richard Meredith. Two days later the flagship, with this new arrival and *Brisk* in company, sailed for the Admiral to call at all the settlements on the coast and to visit Fernando Po and Princes Island before heading for Ascension. That same day *Pluto* finally joined. Laggardly she may have been, but she made the passage from Gibraltar to the Gambia entirely under sail and logged over eight knots at times. Meanwhile, *Seaflower* was sent with stores to the Gambia to replenish *Plumper* before making for Ascension to report to the Admiral on developments at Bathurst and to rejoin Hayes.

Warren was not impressed by what he found at Fernando Po on his arrival on 21 February. Provisions had been damaged because the victualling store was so unsatisfactory, and Boatswain's and Carpenter's stores left there had likewise been damaged by the damp climate. He compared the unhealthy Fernando Po most unfavourably with Ascension as a depot and refitting location for the Squadron, and he recommended removal of all stores to Ascension, the total shutting-down of the Fernando Po base, and abandonment of the proposed move of the Mixed Commission courts. Anchoring at West Bay on 28 February, he found Princes Island much more to his liking, praising the convenience and excellence of the watering place and the plentiful stock and fruit. He reported to the Admiralty his intention to direct the Squadron to repair there. He was similarly impressed by Ascension during his 14-day visit. He put in train some further improvements to the facilities and defences, and he regarded the island, with its good potential for production of foodstuffs, as admirably suited to support the Squadron, although ships would have to be regulated to avoid having to draw on the island's poor supply of water.

The brevity of his progress along the northern slave coast and through the Gulf of Guinea did not deter Warren from reaching conclusions on a variety of affairs on the Coast. While acknowledging the Admiralty's anxiety to reduce the cost of "the Coast of Guinea Squadron", he believed that he could not hope to suppress the slave trade and protect commerce with fewer than five small vessels and either a sloop-of-war or a small frigate to take charge during his absence at the Cape. He had experienced the Coast while the slave traffic was at a relatively low ebb, but it is still difficult to comprehend how he could have underestimated the necessary force to such a degree. He asked for an additional two cut-down brigs, rigged to carry larger topsails than *Brisk* and *Charybdis*, because "quantity of sail is everything in this fine weather country". He had not set eyes on a slaver at sea, let alone experienced a chase, at this juncture. At least he held fire on reporting a comparison of the performances of *Pluto* and his brigs.

Reaching well beyond his remit, but probably conveying the views of Lieutenant-Governor Rendall at Bathurst, Warren then recommended that the Gambia River should be settled by emancipated slaves, and he implied that the Mixed Commissions might be moved to Bathurst. In August, Viscount Goderich at the Colonial Office objected "decidedly" to both of these suggestions. He still believed that Fernando Po was the place for the courts, and if Spain objected then they should stay in Freetown. A few days later, however, he had come to the conclusion that negotiations with Spain to exchange Crab Island in the West Indies for Fernando Po were unlikely to succeed, and, in the light of the Admiralty's decision to break up the naval establishment on the island, that the Fernando Po settlement should be withdrawn. As far as the Gambia suggestion was concerned, Governor Findlay at Sierra Leone was free to transfer liberated Africans to the Gambia if he so wished, as long as each individual African gave his consent to the move, and Rendall was to report details of his proposals for further settlement on the river.[15]

There had been no slaver capture since the *Regulo* and *Rapido* episode in September 1831, and, in the absence of *Fair Rosamond* on passage to Ascension, it inevitably fell to *Black Joke* to break the fast. As the brig's refit was reaching its completion in the late autumn, news had been received of Lieutenant Ramsay's promotion to commander, and, on his departure, Lieutenant Huntley had persuaded the Commodore to let him have *Black Joke*. Mr Robinson, mate of *Fair Rosamond*, whose appointment to *Plumper* had been denied by the Admiralty, was made Acting Lieutenant and given command of the brigantine to take her to Ascension. So it was Huntley who, about 50 miles south of the Bonny River on 15 February 1832, snapped up *Black Joke*'s final victim, the Spanish schooner *Frasquita*, with 290 slaves. Unlike many intercepted Spanish slavers, the *Frasquita* succumbed quietly, and she was despatched to Freetown for condemnation, losing 62 of her slaves on the passage.

Ten days later *Black Joke* was lying in wait off the Bonny for a Spanish slaver, known to be loading in the river, when *Isis*, *Brisk* and *Charybdis* hove over the horizon, having sailed from Fernando Po on 23 February after the Admiral's inspection. Warren ordered Huntley to try the tender's rate of sailing against the gun-brigs, but it seems that *Black Joke* ran rings around *Brisk* and *Charybdis*. The unwelcome result of the competition was apparently not reported by Warren. The Admiral then sent his Flag Captain and *Isis*'s Master and two carpenter's mates to survey the tender, and their report was very much at variance with that from Commodore Hayes. It enclosed "tasters" of crumbling timber from a number of sites, and these Warren sent to the Admiralty without further comment.* *Black Joke*'s fate was sealed. She was ordered to Sierra Leone, and the Admiral directed Hayes to return to Freetown from Ascension, instead of making straight for England, to attend to the breaking-up of the brig, the sale of her gear and the discharge of her Kroomen. It was undoubtedly time for the tender to be paid off, but the decision to destroy her still seems tainted by a whiff of spite. Findlay might have made good use of her, and, as Hayes wrote later, such was the brig's celebrity that she could have been sold for £1,200. She should also have been replaced when a suitable condemned slaver became available, and the wait would not have been long.

On 19 March, *Pelorus* opened her account with the arrest of the Spanish brig *Segunda Teresa* in the Bight of Benin, about 140 miles south of Badagry. Like the *Frasquita*, she was carrying a passport for Princes and St Thomas, but was loaded with 459 slaves from the Slave Coast. The surviving 445 slaves were emancipated, but the capture was soured on two counts. During the passage to Sierra Leone the Prize Master, Lieutenant de Saumarez, had a great deal of trouble with one of his seamen, and on arrival at Freetown, after consultation with Lieutenant Huntley in *Black Joke*, he had the man flogged. Commander Meredith, however, was an opponent of flogging, and had banned it in *Pelorus*. When de Saumarez rejoined he was placed under arrest, and remained so for 18 months before he was sent home for court martial. The court, sitting in June 1834, considered his action was not in direct opposition to Meredith's orders, that there had been a "spirit of insubordination" among the prize crew, and that de Saumarez had been justified in "recourse to such punishment". He was consequently acquitted.[16] Judging from his odious (and probably drunken) behaviour towards a sentry at Simon's Town during his period under arrest, de Saumarez was probably not a pleasant man, and he may have contributed to insubordination among his prize crew. Moreover, the master of the *Segunda Teresa* claimed that de Saumarez had taken his gold watch, chronometer and sextant, but that accusation does not appear to have been pursued.

Segunda Teresa was a particularly beautiful, Philadelphia-built clipper on her first voyage, and she would have made an ideal successor to *Black Joke*. As it was, she was subsequently bought by Captain The Hon. R. F. Greville, a member of the Royal Yacht Squadron, and renamed *Xarifa*. An article in the United Services Journal of 1833 describes her: "This beautiful vessel forms the perfect *beau ideal* of naval architectural symmetry. The long low rakish look and airy tracery of Cooper's fanciful Water Witch appear realised in the *Xarifa*."

---

* These "tasters" still exist, in Admiralty (ADM) 1/74 at the National Archives.

However, the Admiralty had finally turned its face against the use of tenders, and an opportunity was lost.

Following his initial assessment of affairs on the slave coast, Admiral Warren made one wise decision. Before *Isis* sailed from Fernando Po, he left orders for Commander Harrison of *Favourite* that, on *Conflict*'s arrival at Sierra Leone, he was to order *Plumper* to Freetown, de-store her, put her out of commission, reduce her to a hulk, whitewash her thoroughly inside and make her comfortable accommodation for "the reception of Crews of Prize Vessels to the Squadron". He was to place her in the care of four Kroomen, a petty officer and three landsmen, borne on the books of *Favourite* but under the charge of the Lieutenant-Governor. *Conflict* was then to take on board Lieutenant Creser and his *Plumper* ship's company, and proceed to Spithead. As it transpired, *Conflict*, which had clearly received garbled word of the Admiral's intentions, was already well advanced in the task of reducing herself to a hulk by the time Harrison found her at Freetown, while *Plumper* was still employed at the Gambia. The roles of the two vessels was therefore reversed, but the objective was achieved. The campaign would be little disadvantaged by the loss of these two ineffective gun-brigs, and the health of prize crews kicking their heels at Freetown would be preserved by keeping them clear of the town.

On departing the Bights for Ascension and the Cape, Warren placed "the small vessels on the Coast of Africa" under the command of Commander Harrison with orders to suppress the slave trade and protect commerce in the absence of the Admiral.

Two of *Fair Rosamond*'s men had been ill with smallpox when she arrived at Ascension, and Captain Bate RM, the Commandant, placed her in quarantine. Acting Lieutenant Robinson, apparently not a man to squander time, acquired some copper sheets, put his ship's company under canvas ashore, and managed to heave down the tender and copper her before she was released to return to Sierra Leone with Hayes. *Dryad* and *Fair Rosamond* arrived at Freetown on 29 April, as did *Seaflower*, returning from a visit to the Gambia. They found, with some relief, that *Conflict* was in harbour. She had made a bad passage to Ascension, and nothing had been heard of her since she had sailed from there in the last week of February. It transpired that on her passage south she had got several degrees to leeward of the island, and on her intended return to Accra was six or seven degrees out in her reckoning. Hayes was concerned at Lieutenant Smithers' health, and concluded that the state of his mind had been affected by the climate and "ebriety".[17]

Hayes had diverted to Sierra Leone, on Warren's instructions, to oversee the sad task of destroying *Black Joke*. The tender had become something of an icon on the Coast, greatly admired and widely held in affection. It was said that Hayes had received a petition from the emancipated slaves in Fernando Po begging him not to destroy her, and Lieutenant Leonard of *Dryad* expressed the general feeling in the Squadron:

> this favourite vessel – the terror of slave dealers and scourge of the oppressors of Africa, has done more towards putting an end to the vile traffic in slaves than all the ships on the station put together [...] Her demolition will, therefore, be hailed as the happiest piece of intelligence that has been received at the Havannah, and wherever else the slave trade is carried on, for many years.

In her career of four years and four months the brig, under the command of a succession of fine young officers and crewed by enthusiastic and gallant warrant officers, seamen, marines and Kroomen, had made 13 captures, had fought a suspected pirate into submission and had been instrumental in the seizure of 3,692 slaves. Perhaps almost as important, she had acted as a powerful deterrent to the slavers.

It had been Hayes's intention to break-up his tender, but Admiral Warren decided that not only would the process be unnecessarily time-consuming but also there was unlikely to be a demand for her timber. Therefore, under the Commodore's direction, she was hauled up on the shore of the Sierra Leone River and, on 3 May 1832, after her masts, yards, sails, rigging and all moveable articles had been removed, she was destroyed by fire. Her stores and surviving materials were then sold at auction.*

A few days later, with *Fair Rosamond* and *Seaflower* in company, *Dryad* sailed for England, and, at the request of Governor Findlay, Hayes escorted the brigantine *General Turner* on her passage to the Gambia with liberated Africans. This was Hayes's final service at sea. After hauling down his broad pendant he applied his energies to naval architecture, although he ultimately reached Flag rank.

On the day that *Black Joke* was being reduced to ashes at Sierra Leone, Edward Butterfield in *Brisk* made his first and only arrest, detaining the Spanish schooner *Prueba* off the Bonny River. The prize, bearing a passport for St Thomas, was carrying 318 slaves destined for Havana, of whom 34 died on the 23-day passage to Freetown. The schooner and surviving slaves were condemned. It was to be over three months before the next seizure, and that was to be the last one of 1832.

This paltry level of success following the withdrawal of the tenders, added to the very poor return during the past year from vessels other than the tenders, leads inevitably to the conclusion that the loss of the ex-slaver cruisers was

---

* It is regrettable that Hayes did not think to take off the lines of the brig before she was burned.

seriously damaging to the campaign. It must be borne in mind that the Brazilian and Portuguese slavers had virtually deserted the Bights, where Hayes, Warren and Harrison had continued to concentrate their efforts, but in January 1833 the British Commissioners in Sierra Leone, Judge William Smith and Arbitrator H. W. Macaulay, sworn in to their posts in February 1832, dismissed the inference that the small number of seizures in the previous year had resulted from a decline in the slave trade. They believed that the Trade had continued "with as much perseverance and success as ever" under the Spanish flag, they had heard of Portuguese vessels at Whydah intending to take slaves to Brazil, and they had information that the French were still slaving. In their view, there had been no lack of zeal on the part of the cruisers, but there were too few of them.

Also, they concluded, a new system employed by the slavers was more effective in avoiding capture. Instead of a single vessel being used to land trade goods and then load slaves, the trade merchandise was being landed in advance by a vessel which then cleared two or three hundred miles offshore, hoping to decoy a cruiser which expected her to return for slaves, before returning home empty. Meanwhile one of the fast-sailing slavers would, as soon as the African cargo was ready and the coast clear, slip in, load rapidly and sail undetected. More realistically than Admiral Warren, the commissioners called for five or six vessels in the Bights and the same number between Cape Palmas and the Gambia. They also urged, again, that an Equipment Clause should be introduced to the treaties.

In fact, Lord Palmerston had already returned to this last matter in November 1831, directing the British Ambassador in Madrid to press the Spanish Government not only to order the Cuban authorities to fulfil the requirements of the abolition treaty but also to adopt a stipulation that vessels found equipped for the slave trade within certain latitudes should be liable to condemnation. Other diplomatic activity that autumn included a report from Rio that an order had been issued by the Brazilian Government for stricter measures to prevent illegal importation of slaves. As expected, neither of these moves brought the desired results. In a flurry of decision-making in November 1832 the Foreign Office declared first that all Spanish vessels trafficking slaves, both north and south of the Equator, were liable to condemnation, and second that the British settlement on Fernando Po was to be withdrawn altogether. The following month Palmerston, in a most necessary initiative, called on the foreign parties to the slave trade treaties to agree to the destruction of condemned vessels, and the sale of the resulting materials, in order to halt the scandalous recycling of these vessels into the Trade. There was no early response.[18]

Uncertainty remained on how to deal with Portuguese and Brazilian slavers, and, responding to an Admiralty query in February 1832, Lord Palmerston reported that the Portuguese Government had been called upon to declare the slave trade abolished throughout its empire and to agree with Britain appropriate alterations to the Additional Convention of 28 July 1817. Until such changes had been made, however, British cruisers would have to adhere to the existing treaty. Nevertheless, the separation of Brazil from the mother country had, under the existing agreement, rendered illegal any Portuguese slave trading into Brazil, although the British Government had postponed any action on this point in the hope that a declaration from Lisbon would make it unnecessary. The question of the mode of proceeding against the crews of Brazilian slavers was still under consideration by the Law Officers of the Crown.

There was still smouldering resentment in Brazilian circles about the precedent set by the *Hiroina* judgment of 1826, on which many Brazilian vessels, although empty of slaves, had subsequently been condemned. In mid-1833 the Government in Rio proposed that the decisions of the Anglo-Brazilian Commissions should be subjected to arbitration by a third Power. This would have flown directly in the face of the principles accepted in the treaty of 1826, and Palmerston rejected it out of hand. A further note on the subject from the Brazilian Minister in London received a similar response.

The reports from *Atholl* and *Favourite* after their reconnaissance voyages south of the Equator in 1831 had raised concern in the Anglo-Spanish Mixed Commission about the legitimacy of seizures of Spanish slavers south of the Line. It seemed that prizes from the southern hemisphere might soon be brought into Freetown, and clarification was needed. Confusion had been sown by the Foreign Office because, although orders to Hayes dated 8 September 1830 had made it abundantly clear that all Spanish slavers, north and south, were liable to arrest, instructions to the Mixed Commission from Lord Palmerston on 22 December 1830 had included the pre-1820 restriction on seizures south of the Line. When the matter was raised by the commissioners in August 1832, the Foreign Office acknowledged its error and corrected its instructions regarding Spanish slavers, but, in contradiction of the opinion of the Advocate-General reported in February 1831, persisted in its prohibition of Portuguese arrests south of the Equator.

During this exchange, the British commissioners sent a copy of a locally produced standard format of the form of declaration required by the courts from commanding officers making arrests, in accordance with the treaties. There had been too many errors and omissions in these declarations in the past and, although the commissioners had generally been forbearing with incomplete submissions, it was high time for standardisation. The format shown in Appendix F was adopted.[19]

The latter half of 1832 was even less successful on the slave coast than the first six months had been. The Commander-in-Chief had directed that a vessel should be stationed at the Gambia for the security of the Bathurst settlement and its trade, and this guardship made an occasional foray to the northern slave rivers or the Cape Verdes. However, that distraction apart, *Favourite*, *Brisk*, *Charybdis*, *Pelorus* and *Pluto*, under the tactical command of Commander Harrison in *Favourite*, continued to concentrate their attention on the Bights. Warren had indicated an intention to visit the Coast early in 1833 but, having retreated in *Isis* to the Cape and delegated local command to the senior officer north of the Equator, he seems in this and his subsequent extended absence to have made no attempt to influence the pattern of operations of the small vessels on the Coast, despite the relative ease of communication between the Bights and Ascension and between Ascension and the Cape.

For *Charybdis* the only notable incident of 1832 was an unwelcome one. In May, her mate, Charles McDonnell, was ashore on Princes Island for some shooting and, in demonstrating the attributes of his firearm to Assistant Surgeon Naulty, he took aim at a coconut. Unfortunately two Kroomen, left behind by *Pluto* to cut wood, were in the line of fire, and McDonnell succeeded in shooting one dead and wounding the other in his private parts. When *Favourite* then met *Charybdis* and *Brisk* at sea, Harrison took all three into the Princes anchorage to conduct an enquiry into the incident. Not only was this a demoralising episode for Crawford and his ship's company, but also it became a distraction from the primary task (albeit brief) for the greater part of the West Coast force.

Favourite made the only seizure of the second half of 1832, and this long-awaited success proved to be the sole arrest of her deployment, scant reward for her patient vigil. On 15 August, about 60 miles south-west of Fernando Po, she intercepted the Spanish brig *Carolina* carrying 426 slaves. The prize was duly condemned, but only 369 slaves survived to be emancipated.

Pelorus arrived at Ascension from Sierra Leone in early June for a short break and replenishment before return to the Coast, but she was back at the island only a month later on passage to the Cape, reducing the slave coast force to four. *Brisk* followed her south to Ascension in August for stores and respite, and Butterfield, whose long-deserved promotion to commander had recently been announced, found Lieutenant Josiah Thompson waiting there to relieve him and take the brigantine back to the Bights.

Early in 1833 Lieutenant-Governor Findlay made another bid to draw more of the Navy's suppression effort towards the northern rivers. He reported to the Colonial Office that four slavers had recently left the Nunez and Pongas rivers with full cargoes, the last with 400 slaves, but he also conveyed a welcome piece of news: the notorious slaver John Ormond, "Mongo John", had shot himself a short time previously. Since then, Theodore and Joseph Canot had become the leading traders on the Pongas, Theodore buying on the river and Joseph selling in Havana. The colonial schooner *Queen Adelaide* returned to Freetown from a cruise to the northern rivers and the Isles de Los in late March with a report that a Canot schooner had sailed a few days earlier with 250 slaves and another was awaiting a cargo. A worrying traffic on the Sierra Leone River had been discovered by Judge Smith and Registrar Lewis of the Mixed Commission, who had made a most commendable expedition up the river as far as Port Logo. They had found regular slave factories, trade goods coming from the Pongas by canoe and slaves being taken to the Pongas from Sierra Leone. This bid by Findlay, like his previous attempts, brought no perceptible change of policy.

Two more of the "coffin brig" class, *Griffon* and *Forester*, had sailed from England in January 1833 to join the force on the slave coast. They, however, were two of three recently completed (rather than modified) to the configuration and armament of *Brisk* and *Charybdis*, and similarly rigged as brigantines. Unfortunately their arrival off West Africa was delayed, considerably so in *Forester*'s case. They were caught in a great south-westerly gale in the Channel approaches, and, while Lieutenant James Parlby in *Griffon* turned back to Plymouth, Lieutenant William Quin opted to seek shelter for *Forester* in St Mary's Roads in the Scilly Isles. On 13 February she parted her cables, struck on the Crown Bar and fetched up on a reef off St Martin's. She was salvaged, but it was not until the autumn that she was able to join the West Africa force under the command of Lieutenant George Miall, who had previously commanded the schooner *Minx* in the West Indies.

In mid-January 1833 the Commander-in-Chief sailed from Simon's Bay at the Cape, with his flag in the frigate *Undaunted* (46), to visit the Bights and the Windward Coast as far north as the Gambia, concerned as much for the security and trade of the settlement at Bathurst as he was for the suppression of the slavers. His visit was brief, and he was back at the Cape by the end of April, but it had served to change his opinion on the number of cruisers necessary to make an impact on the slavers. He had complacently believed that the scarcity of arrests had reflected a diminution of the slave trade rather than a shortage of men-of-war, but his revised assessment, generally coinciding with the views of the British commissioners in Freetown, was that the Trade was being carried out "with great activity", and more widely spread than for some years, from 7° S to the Cape Verde Islands. He thought that few slavers were using the Cameroons River or the Calabars, perhaps deterred by *Pluto*, which had been in the Bights continuously from March to October 1832, but that there was much trading of slaves in the northern rivers, "'tho much diminished since Sir F. Collier's time". He was convinced that slaver masters were being told to avoid areas in which previous captures had been made.[20]

Warren's conclusion was that no serious check to the Trade could be applied without eight or nine brigs or brigantines stationed off the principal slave rivers. He regarded ships and ship-sloops as of little use compared with the cut-down brigs which, in his opinion, were admirably suited to the task. He was inclined to think well of *Pluto* too, although she had made no arrest. In Fernando Po he had been told that there was a fear of the steamer on the opposite mainland, and that several slavers had awaited calms and then used sweeps close inshore to avoid her.* They were unlikely to have been aware that she had no keel and could beat to windward only with great difficulty.

It was not until 22 February 1833 that *Charybdis* was able to break her fast and redeem herself. Patrolling off the mouth of the Bonny River, she snapped up the Spanish schooner *Desengano* as she emerged with 220 slaves. The slaver produced the familiar Royal Passport from Havana for lawful commerce at St Thomas, but she had made directly for the Bonny. She and her surviving 209 slaves were condemned at Freetown.

An addition to the force arrived at Ascension from Mauritius in late March 1833, but the elderly brig-sloop *Trinculo* (16) was probably not what the Commander-in-Chief had had in mind while writing his recommendations.† Having encountered Warren at Ascension when *Undaunted* called at the island on her return voyage to the Cape, Commander James Booth took *Trinculo* onward to Fernando Po. In May, however, Booth was invalided to England, and it is not clear who initially commanded *Trinculo* thereafter; but court records show that Josiah Thompson of the *Brisk* was her Commanding Officer for an arrest in July.‡

Britomart, yet another of the "coffin brigs" and one of the original ten-gun

---

* One would have thought that these were precisely the conditions to give *Pluto* the advantage, but her main propulsion was sail, and she would not raise steam until she had made a suspicious sighting. Wise or not, the slavers had, so far, made successful escapes.

† Lyon's *The Sailing Navy List* gives *Trinculo*'s armament as two 6-pounder long guns and sixteen 32-pounder carronades, in common with the rest of her class, but the 1833 *Navy List* shows her as a 16-gun sloop.

‡ Presumably Thompson was the only Lieutenant immediately at hand suitable for temporary command of a sloop, and his First Lieutenant probably took command of *Brisk* in his absence.

variety, reached the slave coast in the early summer under the command of Lieutenant William Quin, lately of the damaged *Forester*. She performed the Bathurst guardship duty, and indeed navigated 50 miles up the Gambia River before making for the Cape in August and then sailing in September for Sierra Leone and another cruise on the northern slave coast. A further reinforcement had arrived from the east coast of Africa in company with Warren on his passage north in February. This was *Curlew*, one of the many unmodified ten-gun brigs like *Britomart*, but launched as late as 1830. Her Commanding Officer, Commander Henry Trotter, was unusually senior for such a command, but he was shortly to supersede Harrison of *Favourite* as Senior Officer of the small vessels on the Coast. At the Commander-in-Chief's direction, Harrison inspected the forts at Cape Coast and Accra before handing over to Trotter in May and taking *Favourite* home via Ascension and Sierra Leone.

By then *Curlew* had begun to make her mark. The Spanish schooner *Veloz Mariana*, bearing a similar cynical passport to that of the *Desangano*, had made directly for the Old Calabar River and loaded 290 slaves, but on 23 April Commander Trotter met her on her way to sea and sent her off to Sierra Leone. Twenty-five slaves died on passage, but the schooner and her remaining Africans were condemned. At her subsequent sale the *Veloz Mariana* proved to be the most valuable vessel of the year's haul of slavers, realising £479.0.9, of which half was, of course, the Spanish Government's moiety.

Only a fortnight later there was achieved an equally undramatic but more momentous seizure: the first by a steam-powered man-of-war. On 5 May 1833, *Pluto*, now commanded by Lieutenant Thomas Ross Sulivan, was cruising 50 miles south-east of the Bonny River when she fell in with the *Josepha*, a Spanish brig which had shipped her cargo of 278 slaves in the river on the previous day and was making her break for the open sea, bound for Santiago de Cuba. There was a dreadful death toll of 85 on the slaver's passage to Freetown, but the prize and the surviving slaves were condemned without difficulty.

The pace of arrests had accelerated appreciably, and *Favourite*, almost at the end of her 30 months on the Coast, soon added the next success. Contributing to the evacuation of the establishment at Fernando Po, she sailed from Clarence Cove on 6 May for Sierra Leone, and on the thirteenth, by chance, she encountered the Spanish sloop *Indio*, four days out of the Bonny River and homeward-bound for Santiago de Cuba with 117 slaves. She too carried a worthless Royal Passport for legitimate trade at Princes. Nine of her slaves died before her condemnation at Freetown. *Favourite* made a final call at Ascension in June, and then headed for England.

In mid-May 1833 *Pluto* escaped the Coast for the first time since her arrival on the Station, and headed for Ascension to refit her machinery. This was a novel task for the island, but, as ever, Captain Bate and his enterprising garrison gave every assistance in their power. In early August, with, as Bate reported, her machinery working perfectly, she sailed for Princes.[21]

The two original brigantines had failed to fulfil the expectations of Admiral Warren, and *Charybdis*'s deployment had been particularly disappointing. In July 1833 she arrived at the Cape for a refit, and not only had she only one slaver arrest to her credit, but also her Commanding Officer, Lieutenant Richard Crawford, was in trouble. The Commander-in-Chief may already have been ill-disposed towards Crawford, whom, on *Charybdis*'s arrival in Simon's Bay, he had challenged about purchases of provisions made, it would seem perfectly legitimately, at Sierra Leone and during a cruise to the Gambia and the Cape Verde Islands in October and November of the previous year. Now the brigantine's Second Master had made a number of serious allegations against his Commanding Officer, and Warren was left with little alternative but to order a trial by court martial.

Not having the officers available at the Cape to form a court, the Commander-in-Chief sent Crawford home in his brigantine in September to face trial on board HMS *Victory* on 9 December. He faced eight charges ranging from unauthorised disposal of stores, through falsification of the ship's log, to assault with a telescope on an able seaman causing, in modern parlance, grievous bodily harm. Four of the charges were found to be not proved; two, including the assault, were judged partially proved; and two were proved. This promising young officer was sentenced to be dismissed from the Service, although the court, "in consideration of the high testimonials" produced by Crawford, recommended him to the favourable consideration of the Admiralty. The sentence was consequently revoked, but he was left on half pay.*

The elderly sloop *Trinculo*, under the temporary command of Lieutenant Josiah Thompson, belied her years on 7 July when, on passage from Sierra Leone to the Bights, she intercepted the Spanish schooner *Segundo Socorro* about 60 miles west-south-west of Cape Mount and after a short chase arrested her with 307 slaves taken from the Gallinas River and destined for Cuba. The Spaniard and her slaves were condemned ten days later, and it was discovered that she had been the American-built slaver *Planeta*, taken off Cuba by *Speedwell*, condemned at Havana in August 1832, and sold to her present master by a Havana merchant

---

* There was a happy sequel to this episode. Crawford became involved as a volunteer in the First China War, distinguished himself with outstanding initiative, daring and gallantry, and was then given command of the experimental brig-sloop *Mutine*.

for $3,000. Her master revealed that he had previously made 13 slaving voyages unmolested, the last four to the Gallinas, and for this final passage to Africa he had been given a passport at Havana for legitimate trade at St Thomas but had headed directly for the Gallinas.

The most dramatic episode of 1833 was the relentless pursuit of a pirate vessel and her crew by Commander Trotter of *Curlew*. During a visit to Princes Island in May he had been shown a newspaper article reporting the seizure by a pirate schooner of the Salem merchantman *Mexican* in September 1832, south-west of the Azores. Trotter also heard that a schooner named *Panda*, which answered the description of the pirate vessel, was believed to be trying for a cargo of slaves in the River Nazareth. Trotter determined to bring the pirates to justice, and on the evening of 3 June *Curlew* anchored nine miles off the mouth of the Nazareth. At dawn the following day, Trotter led his three boats upriver and found the *Panda*, but was unable to board before the crew had made their escape ashore. A lighted match to the magazine was found and bravely extinguished by John Turnbull, seaman, and an open cask of gunpowder was discovered alongside the galley fire. Some papers and a suspicious array of foreign ensigns found on board gave Trotter sufficient evidence to seize the schooner.

Five recently joined members of *Panda*'s crew surrendered, and, having anchored the schooner alongside *Curlew*, Trotter opened negotiations with the local chief, King Pass-All, for the remainder of the schooner's men to be given up. After several fruitless days, Trotter took the *Panda* back into the river, there being insufficient water over the bar for *Curlew*, and anchored her just off the King's town. An ultimatum was delivered to the King, but when the schooner's long 12-pounder was fired to indicate its expiry a spark fell on some loose powder, igniting the powder casks in the magazine below. The vessel's stern was blown out, and she filled and sank. The Purser, the Gunner and a marine were fatally injured, and one Krooman was never seen again. The starboard rail remained above water and the survivors, including the injured Trotter and a badly hurt marine, clung to it until two of the sloop's boats rescued them.

The five Portuguese prisoners, none of whom had been involved in the piracy, revealed that some of the original *Panda* crew had left the vessel at various places on the coast, and in his efforts to track them down Trotter took the *Curlew* to the Gabon, St Thomas, Princes, and to Whydah where 13 slavers, waiting in the roads to load, were boarded, more in the hope of finding pirates than slaves. Finally the sloop looked into the Bonny, but by then Trotter was dangerously ill with fever, and on 17 August he was landed at Clarence Cove to be nursed back to health by Lieutenant-Colonel Nicholls, Commandant of Fernando Po.

During Trotter's convalescence, five Spanish sailors who had escaped to Fernando Po from Bimbia Island in the mouth of the Cameroons River, and a sixth who was caught on Bimbia Island, were identified by one of the Portuguese prisoners as members of the pirate crew, and one of them turned King's evidence. He admitted to the piracy of the *Mexican* and that the *Panda* had gone to the Nazareth for slaves. All were then sent to Ascension. Meanwhile, it was heard that the pirate Captain, Don Pedro Gilbert, was still at the Nazareth, and Trotter was loaned a small English barque to take there a dozen of his men in disguise, commanded by one of his mates, Henry James Matson, who in due course was to become one of the most renowned officers of the campaign.*

King Pass-All's kraal had been moved close to Cape Lopez, and on landing there Matson encountered Gilbert. Despite coming under suspicion, Matson bravely maintained his deception of trading for ivory, and even managed to secure the King's son as a hostage. When the *Curlew* then arrived, successfully disguised as a Portuguese slaver, the King was finally persuaded to surrender Gilbert and the men remaining with him. However, it was known that the pirate mate and a number of the *Panda*'s crew had arrived at St Thomas in the schooner *Esperanza*, believed to have been purchased by Gilbert from the Governor of the island. Trotter accosted the Portuguese Governor, who denied all knowledge, but the *Esperanza* was found and, on Trotter's orders, seized by Matson for aiding the pirates. The Governor then admitted that the *Panda*'s men were on the island, and, after much prevarication, gave them up.

It was found that one more pirate remained at Cape Lopez, and in mid-January 1834 Trotter sailed with the *Esperanza* in company, commanded by Matson. Trotter was again ill, but no difficulty was anticipated. The King, however, was now in the midst of a drinking bout and determined to take his revenge on Trotter. He contrived to seize three unsuspecting officers, including Matson, and their boats' crews, badly maltreating some of them, and obliged Trotter to pay a £100 ransom for their release. He then refused to return the prisoners. Two more days of bargaining between Matson and the King were brought to a halt by the arrival of *Fair Rosamond*, and the captives were released. Trotter was now determined to recover the men's clothes and his ransom money, and he sent Matson in the *Esperanza* to Princes Island to find *Trinculo*. On Matson's return with the sloop, Trotter launched an attack on the town with eleven boats. The natives were driven into the bush, but Trotter desisted from burning the town.

---

* This young master's mate should not be confused with Lieutenant George William Matson, known as "Old Rough and Ready", who commanded *Clinker* in the late 1820s. It is not known whether or not the two were related.

With a large number of his men sick, Trotter decided to sail directly for Ascension, and after a month of recuperation there *Curlew* and *Esperanza* left for England at the beginning of April. Trotter's almost obsessive pursuit of the *Panda*'s crew was barely consistent with his primary duty as Senior Officer of the force on the northern slave coast, and Warren wrote to Elliot at the Admiralty of his "extreme regret" at Trotter's "indiscreet and inexplicable" behaviour. The Law Officers of the Crown considered that seizure of the *Panda* had been justifiable because "Pirates being considered the Enemies of all mankind can have no place of legal shelter", but they doubted that the refusal of the African chief to give up the *Panda*'s crew would have justified any loss of life ensuing from Trotter's attack. In contrast with the Commander-in-Chief, Washington was delighted that the pirates had been captured and handed over to American justice, and in September 1835 the Admiralty conveyed to Trotter the thanks of the President of the United States and, "as a mark of the sense which my Lords entertain of your conduct", promoted him to the rank of captain.* [22]

The owner of the schooner *Esperanza*, seized at St Thomas for aiding the pirates, was less favourably impressed by Trotter's behaviour, and threatened legal action for the detention of the vessel and maltreatment of three of the crew, Spanish subjects. The schooner had, indeed, never been brought before an Admiralty court, and the men, although never charged with any offence, had been imprisoned in *Victory* and almost extradited to the United States. In October 1834 the Foreign Office arranged that if the Admiralty would deliver the *Esperanza* to Lisbon the Portuguese Government would return her to her owner, and the matter was then forgotten.

At the end of November 1833 *Pelorus* sailed from the Cape to return to the Bights, and on her arrival, although she did not encounter *Curlew*, Commander Richard Meredith relieved Trotter as Senior Officer.

It was of great concern to Lieutenant-Colonel Nicholls that there was an extensive trade in slaves in the vicinity of Fernando Po. He drew particular attention to a large brig lying at Princes Island under Brazilian colours, but apparently owned by a French trader, for which slaves were being brought from the Old Calabar in a smaller vessel. When the Frenchman, Gaspard, arrived in the Calabar River all legitimate commerce had ceased in the scramble to supply him with slaves. Although he was busy de-storing Clarence Cove for evacuation, Nicholls pleaded with the Colonial Office to be given charge of the merchant

---

* The pirates were returned to Boston in HMS *Savage* in August 1834, and Pedro Gilbert, a gentlemanly figure, was hanged in June 1835 with five of his men. The Mate, De Soto, was pardoned and was later on friendly terms with Matson.

steamer *Quorra* and a crew of 12 white men, in addition to his Kroomen and his "allies on the mainland coast", with orders to clear the Bights coast of slavers and to keep it so. For this he sought no pecuniary reward. Nicholls' admirable enthusiasm does not seem to have been matched by his appreciation of the magnitude of the task, or of the constraints imposed by the bilateral treaties.

It was something of a surprise that the old *Trinculo* was by far the most successful cruiser of 1833. She already had *Segundo Socorro* to her credit, and two months later she seized her second victim. By then Josiah Thompson's brief period of command had finished, and another temporary Commanding Officer, Acting Commander R. L. Warren, had joined.[*] Having taken her first prize off Cape Mount, she made her way to the Bight of Biafra, and she was about 30 miles southwest of Cape St John on 18 September when she met and detained the Spanish schooner *Caridad* on her way to Cuba with 112 slaves. While the captured schooner steered for Freetown, losing five of her slaves before emancipation, *Trinculo* headed for Ascension and the Cape. Her two prizes were not a fair measure of her work during the cruise; it was illustrative of the frustrated efforts of all of the cruisers that she had also boarded 22 suspected slavers, almost all Spanish, many of which had water and supplies loaded and slave platforms laid, but all empty of slaves.

Having shed the "acting" rank of commander on leaving *Trinculo*, Josiah Thompson resumed command of *Brisk* in time to make the arrest of the 78-ton brigantine *Virtude* on 23 October. He intercepted her just off the mouth of the Old Calabar River as she put to sea with 350 slaves. She was under Portuguese colours but had been fitted-out at Bahia and, as was increasingly the case with Portuguese slavers from the Bights and the northern rivers, she was bound for Cuba. *Virtude* and her surviving 314 slaves were condemned.

Only five days later there was a capture even more unlikely than those achieved by *Trinculo*. Since returning to the Cape in *Undaunted* in April 1833, the Commander-in-Chief had worn his flag in whichever vessel happened to be lying in Simon's Bay, and, although she had been deployed as the Commander-in-Chief's flagship, *Isis* had rarely performed the role. In July she left the Cape and headed north via St Helena and Ascension to make a rare cruise in the Bights. Although her Commanding Officer, Captain James Polkinghorne, was clearly senior to both Henry Trotter and Richard Meredith, it seems that these two officers retained, in succession, the duty of Senior Officer on the slave coast. On 28 October, in defiance of Admiral Warren's low opinion of ship-rigged cruisers against slavers, *Isis* arrested the Spanish brig *El Primo*, which had sailed from Cadiz ostensibly

---

[*] It seems likely that this was a relation of the C-in-C who had complained bitterly of not being allowed to employ patronage to replace Crawford in *Charybdis*.

for Princes Island. She was intercepted 60 miles north-west of Princes having loaded 343 slaves at the Bonny River, and was making for Santiago de Cuba. Polkinghorne sent her off to Freetown, where she was condemned with 335 slaves.

Commander Booth, restored to health, returned to the Station as a passenger in the refitted *Forester*, but the brigantine arrived at Ascension in mid-December with a sprung mainmast and was then diverted to Tristan da Cunha to investigate the loss of a merchantman, with the result that Booth was nearly two months too late to rejoin *Trinculo* before she sailed on 2 December for a novel cruising ground.

In perhaps the only significant initiative in the campaign under Admiral Warren's command, *Trinculo* was directed to cruise between Ambriz Bay and Cape St John, a 600-mile stretch of coast, almost all of it south of the Equator, in the hope of threatening the Brazil traffic from the Congo and Loango. Success, when it came at the end of December, was, however, against smaller fry. On the twenty-seventh, at the northern end of her patrol line and 85 miles east of St Thomas, the sloop continued her run of success in seizing the small Portuguese schooner *Apta* on the short passage from Cape Lopez to St Thomas with 54 slaves. Although only 30 feet long and 11 feet in beam, this 20-ton schooner had 65 people on board, and her small size and "crazy condition" persuaded Richard Warren that she was unseaworthy for a voyage to Sierra Leone. The same was true of a second schooner, the *Santissimo Rosario e Bon Jesus*, caught the following day, which was even smaller at 14 tons and with 71 people on board, 54 of them slaves. Both slavers were built and owned at Princes, and the second was found 40 miles south-east of the island on her way home. Warren decided to land slaves and stores at Fernando Po and deliver his prizes into the charge of Lieutenant-Colonel Nicholls, and it was not until mid-June, when he arrived at Freetown in *Forester* with the slavers' papers, that the two cases could come to court. The vessels were condemned, and the slaves emancipated in their absence. Despite the intention that the freed Africans landed from several slavers over the years at Fernando Po should be transferred to Sierra Leone, it was decided in 1835 that the move would not be in the interest of the people themselves, and the plan was abandoned for the time being.

In continuation of this run of better fortune, a capture of unusual significance took place on 8 January 1834. *Pluto* was lying at anchor off the River Bonny when at 1000 she sighted a sail to the west-north-west. Lieutenant Sulivan quickly recognised the stranger as the Spanish brig *Vengador*, which *Pluto* had twice boarded and once chased over the Bonny bar. Having weighed, shipped paddle-blades and raised steam by 1025, *Pluto* took up the chase of the brig, which was steering south-west under a heavy press of canvas with a fine breeze on her starboard quarter.

At 1130 Sulivan fired a shotted gun to bring the slaver to, but she immediately replied with two guns, keeping up a well-directed fire with round shot, grape and, later, canister, evidently aiming at *Pluto*'s paddle-wheels, while the gun-vessel closed in from leeward on the brig's beam and quarter. Meanwhile, the chase jettisoned her anchors and boats. At 1225, when the two vessels were 300 yards apart, the slaver lowered her colours and came to the wind on the starboard tack, but Sulivan, standing on the paddle-box, could see that her crew was still at quarters and preparing the guns for a starboard broadside. Ordering his boarders to the starboard bow, Sulivan lay *Pluto* aboard the brig's larboard quarter, at which *Vengador*'s men retreated below and left the deck to the boarders. The Spanish master, Don Pedro Badia, was notorious on the Coast, and resistance would have been stiff if he had managed to persuade his crew to obey him. The intended cargo had been 500 slaves, but 94 had died before embarkation. Despite the best efforts of the prize crew, especially the Kroomen, 17 slaves jumped overboard on passage to Freetown, and only 376 of the 406 captured were emancipated. The first successful chase of a slaver by a steam-powered British man-of-war, in a good sailing breeze, had been accomplished.[23]

There had been no casualties on either side, thanks largely to Sulivan's decision to withhold fire to preserve the lives of the Africans and to avoid crippling this "remarkably fine" vessel and jeopardising a quick passage for her to Sierra Leone: "a consideration of the utmost importance", in Sulivan's view, for the benefit of her slaves. *Vengador* had a crew of 47 and mounted two long pivot-guns on traversing sweeps amidships, an 18-pounder and a 12-pounder. This was her first slaving voyage, and she was bound from the Bonny to Havana. Her master and two men remained on board with the prize crew for the passage to Freetown, and *Pluto* landed the remainder at Princes Island.

It was ironic that just after *Forester*, designed for chasing slavers, had hoisted the Commander-in-Chief's flag at Simon's Bay, the frigate *Isis*, originally destined as the flagship, was making a second capture on the slave coast. On 16 February 1834, about 30 miles off the Forcados River in the Bight of Benin, she fell in with the 78-ton Spanish brigantine *Carolina*, with 350 slaves. The Spaniard, at the beginning of her passage from Lagos to Havana, made a run for it, but *Isis*, to her great credit, brought her to after a chase of four hours. Twenty-seven slaves died on the passage to Freetown, and 100 had to be landed to hospital before the case was adjudicated and the *Carolina* condemned. After a ten-month cruise *Isis* returned to the Cape on 2 May, and, although there had been no particular sickness on board, the Admiral found her ship's company in an emaciated state.

On the same day, *Trinculo* returned to Simon's Bay from her five months on the built *Lynx*, sister of *Griffon* and *Forester*.* They were joined by their half-sister *Charybdis*, restored to the Station after being dragged from useful employment by the wretched affair of Richard Crawford's court martial, and now under the command of Lieutenant Samuel Mercer, who had cut his slaver-chasing teeth as one of Captain Owen's midshipmen in *Eden*.

It was perhaps no surprise that the first of the newcomers to make an arrest was *Fair Rosamond*. Having arrived in the Bight of Biafra after storing and watering at Ascension in March, she seized the Spanish schooner *La Pautica* off the Old Calabar River on 26 April. There was no need for the ex-tender to demonstrate her sailing prowess; the slaver was lying vulnerably at anchor after leaving the river, and *Fair Rosamond*'s boats boarded her and discovered 317 slaves destined for Havana. Rose made the error of removing the entire crew of the slaver prior to her departure for Freetown, with the exception of the master's servant, incurring the displeasure of the Mixed Commission who, having condemned the prize and the surviving 270 slaves, pointed out the treaty requirement that the master of a prize and a part at least of her crew should appear before the court. It was then claimed that a boat and oars had been removed from the slaver by *Fair Rosamond*, but Rose later mollified the commissioners by explaining that the boat had been used to land *La Pautica*'s officers at Old Calabar instead of Fernando Po at their request, and, the boat having been returned, he sought instructions from the court on its disposal. Rose subsequently admitted to the Commander-in-Chief that he had sent only one crew member with the prize because of his apprehension that prisoners would smuggle liquor to the prize crew and then retake the vessel. The master had agreed that the servant should be sent to speak on his behalf.

The next prize, which fell to *Charybdis* on 14 June, held an unusual passport. She was the Portuguese brig *Tamega*, which had been cleared for a voyage from Gibraltar to Lisbon, then via Rio or Bahia to the coast of Africa and back to Lisbon, clearly not for slaving. From Bahia, however, she had made for Lagos and embarked 444 slaves. She was about 50 miles south-south-west of Lagos on her return to Bahia when *Charybdis* sighted her and began a four-hour chase. Jettisoning her two guns availed the slaver nothing, and she was taken without a struggle. During the adjudication at Freetown, which resulted in condemnation and the emancipation of 434 slaves, it was claimed that six hams, one cheese, some vermicelli, a new log line and a half-barrel of wine had been removed into *Charybdis* by the boarding party.

---

* Lyon, in *The Sailing Navy List*, shows that *Lynx* was built as a brigantine, but Admiral Campbell referred to her, probably mistakenly, as a brig.

The Prize Master admitted to the wine and some yams, but asserted that the wine had been given by the master of the *Tamega* to the Second Master of the brigantine. The commissioners awaited an explanation from Lieutenant Mercer when he next visited Sierra Leone. It was not until April 1836 that the commissioners were able to report that the disappearance of the articles had been satisfactorily explained. Mercer had taken the half-cask of wine to Princes at the request of the owner, the vermicelli had been eaten by the Portuguese prisoners in *Charybdis*'s gunroom, and the slaver's master was sure that the remaining items had been stolen by his own crew.[26]

Judge Smith was absent from Sierra Leone at the time of this adjudication, and his post was temporarily occupied by Lieutenant-Governor Temple, who had replaced the worn-out Findlay in December 1833. Although there were fairly frequent changes, often temporary, in the personnel of the British Commission at Freetown, the posts were always filled. The same could not be said of the foreign Commissions, whose officials tended to find the risks and hardships of Sierra Leone intolerable for any length of time. There had, however, been one marked exception: Jozé de Paiva, the Brazilian Judge. By June 1834 he had been continuously in post for nearly six years, but, on the twenty-ninth of that month, fever finally claimed his life. He was temporarily replaced by the Arbitrator, Matthew Equidio da Silveira.

Another commanding officer shortly found himself in considerably deeper hot water than had Mercer. By 30 June, *Pelorus* had been lying in wait for some days at the mouth of the Cameroons River for a Spanish schooner known to be at anchor upriver and about to load her slave cargo. At 2115 that evening, his patience finally exhausted, Commander Meredith sent his pinnace and cutter into the river, in tow by the steam vessel *Quorra*, under the command of his First Lieutenant, Barrow. Knowing well that they could not arrest the schooner, the *Pepita*, while still empty of slaves, they first landed at the barracoon where the slaver's intended cargo was incarcerated and removed three African boys. On boarding *Pepita* at 1145, Barrow's men drove the Spanish crew aft and lowered the three children into the slave deck, where the platforms were already laid. They then arrested the vessel for having slaves on board. The following morning the 176 slaves held ashore for the *Pepita* were sent on board at the captor's instigation, and the native chiefs supplying them were paid with the trade goods remaining on board for their expenses in feeding the slaves, provisioning the schooner and providing pilotage. The slaver grounded on leaving the river, was damaged and strained before being floated, and was taken initially to Fernando Po before making the passage to Freetown, followed by *Pelorus*. The seizure was southern part of the slave coast, and at the beginning of June, with Commander Booth once more in command, she set off into the Indian Ocean for a diplomatic mission to the Far East. By then the Commander-in-Chief had moved into the more spacious quarters of *Isis*, and *Forester* sailed, probably gratefully, for the River Gambia on 12 May.

While his flag was in *Forester*, Warren had taken the opportunity to question ten Kroomen serving on board. As he then reported to the Admiralty, seven of them appeared to be entitled to shares from no fewer than 52 condemned slave vessels, and money owed to some of them should have been paid six years earlier. This was yet another indication of the length of service given by these essential and brave, skilled and astonishingly loyal native seamen, and of the disgracefully cavalier treatment meted out to them by an ungrateful British Government.[24]

Commander Trotter's pirate-hunt had not been the only distraction from suppressing slaving in the Bights. The dismantlement of the Fernando Po settlement and base facilities had continued throughout most of 1833, and several cruisers had been involved in evacuating people from the island: *Favourite* to Cape Coast in February, *Pluto* to Sierra Leone in February, *Charybdis* to Sierra Leone in April, and *Favourite* again to Sierra Leone in May. The transport *William Harris* carried the stores to Ascension, arriving with her second cargo in July, just as the rollers set in to hamper unloading. It was not until late April 1834 that she returned on her third trip with the last of the Clarence Cove establishment, reporting before her departure for England that the irrepressible Lieutenant-Colonel Nicholls, released from his duties, was proceeding up the River Niger in the steamer *Quorra* to enquire into the circumstances of the fatal wounding of the explorer Richard Lander.

With the loss of the depot at Fernando Po, the Commander-in-Chief had to reconsider arrangements for logistical support of the cruisers on the West Coast. He regarded Sierra Leone as unfit for the purpose because stores deteriorated so rapidly there, and he sought a decision from the Admiralty on whether the West Coast vessels should be supplied by the yard at Simon's Town or whether a stock should be kept at Ascension. Considering the distances involved and the shortage of cruisers, a depot at Ascension would seem to have been the obvious option. Indeed, that was the decision which Warren eventually received, but in early May 1834 he was still awaiting a response, anxious to receive it before leaving the Cape for a visit to the West Coast.

An Order in Council was promulgated in March 1834 on a matter of acute interest to the officers and men of the cruisers: the distribution of prize money, bounties and head money. Henceforth the net proceeds would be distributed on a new scale. Flag officers present at a capture would receive one-sixteenth of the whole; the commanding officer one-sixth of the remainder, or one-sixth of the whole if there was no flag officer; and the remainder would be divided among the officers and ship's company according to a scale of ten "Classes". The members of each of these Classes would each be allocated a number of "shares". Senior lieutenants, for example, constituted the First Class and would each receive 55 shares. Surgeons, Pursers, Gunners, Boatswains and Carpenters were included in the Third Class with 25 shares each; midshipmen, Masters' Assistants, gunners' and boatswains' mates, and captains of tops, among others, were of the Fifth Class, allocated ten shares each; and so on down to the Tenth Class, who were (a little confusingly) boys of the second class, who earned only half a share. The rank of the commanding officer was immaterial to his share.[25]

Early 1834 saw the return to the slave coast of *Fair Rosamond*, no longer a tender but a commissioned man-of-war commanded by Lieutenant George Rose. She had been in the hands of the dockyard at Portsmouth and her sail plan had been cut down during a refit, but her fine qualities had apparently not been spoiled. By coincidence, her erstwhile Commanding Officer, Lieutenant Henry Huntley, also returned at about the same time in command of the newly

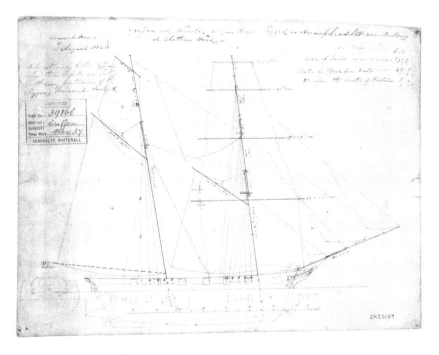

14. Profile of the ten-gun brigantine *Griffon*

of course illegal, and the commissioners quickly saw through Meredith's foolish trick. The surviving 153 Africans could not be emancipated because they had been shipped by the captor and not for the purpose of the slave trade, but they were freed nevertheless, and the schooner was restored to her owner.* It was decided that the owner was entitled to compensation, but not for the loss of the slaves, who, he claimed, were not his. Meredith admitted fraud, and crippling damages of £1,091.19s.4¼d were awarded against him.†

Fair Rosamond made her second arrest on 5 August, 75 miles south-east of Princes Island. On sighting the cruiser at 1000 the slaver took to her heels under all sail and sweeps, but she was caught only an hour later. She was the Spanish schooner *Maria Isabel*, which had been under Portuguese colours until just before sailing from Santiago de Cuba, and she was holding an unusual provisional passport from the Havana Navy Board for Princes. The schooner had made directly for the River Benin and loaded 146 slaves, purchased with English trade goods acquired at Jamaica: tobacco, gunpowder, muskets and dry goods.

After a 26-day passage to Sierra Leone in company with his prize, Lieutenant Rose discovered that the master of *Maria Isabel*, José Mauri, not only had twice appeared before the Freetown Mixed Commission, as an officer and joint owner of the *Gazetta* in 1820 and the *Atafa Primo* in 1830, but also was suspected of having been mate of the pirate vessel *Pelicano*, which had seized HMS *Redwing*'s prizes, *Isabella* and *Disuniao*, and murdered the prize crews in 1825. He had subsequently bought and commanded the *Pelicano* under various names, and had taken her to the River Nunez for slaves. While there he had drunkenly boasted of the murders of the British prize crews, and he had so irritated the natives there that they had refused him assistance in pilotage. Consequently he had grounded his vessel on departure and been obliged to remove her guns and stores and blow her up.‡ The court condemned the *Maria Isabel* and the surviving 131 slaves, but regretfully concluded that there was insufficient evidence for further action against Mauri.

The Commander-in-Chief, after a 16-month absence from the slave coast, sailed from Simon's Bay in *Isis* at the end of July 1834 to make his way north to relinquish command of the Cape Station. He called at St Helena and then Ascension, where he found that the improvements to facilities which he had instigated had been completed. In reporting these advances to the Admiralty

---

* This was a matter of legal semantics. The slaves were treated precisely as they would have been had they received formal emancipation.
† Meredith's pay as Commanding Officer of a sloop was £23.0s.4d per month.
‡ When the fate of the two prizes was initially revealed the pirate vessel was identified as the *Gavilina*, but she obviously operated under various names. She had originally been HMS *Kite*, an 18-gun brig sold out of the Service in 1815.

he complimented the Commandant, the excellent Captain Bate of the Royal Marines, in glowing terms. *Pluto*, *Lynx* and *Griffon* were all at the island at the time, and, with the exception of *Griffon*, which had another three weeks of refit to complete, they sailed soon after the departure of *Isis* on 10 September, *Lynx* for Princes and *Pluto* for home. *Isis* was making initially for the Gambia, where Warren was to meet his replacement.[27]

On 17 September 1834, *Lynx* was three days into her passage from Ascension to the Bights when she encountered a loaded Spanish slaver which, having departed the African coast safely, was, no doubt, expecting a clear run on the trade winds to the Caribbean. The slaver did everything in her power to escape but, after a chase of some hours, Lieutenant Huntley took her without resistance about 330 miles north-east of Ascension and a little over four degrees south of the Equator. She was the American-built schooner *Arrogante Mayaguesana*, bound from Loango to Puerto Rico with 350 slaves. This was a milestone case for the Freetown Anglo-Spanish Mixed Commission: the first slaver brought before it from south of the Line; and it hoped that her condemnation would encourage cruising on the southern slave coast. There was a high mortality among the slaves, the greater part of whom were children, and only 309 lived to be emancipated, although the court noted that the slaver's master had "displayed an unusual degree of humanity and attention to his slaves". The Prize Master, Master's Assistant Stephen Johns, was unable to present the court with the requisite Declaration from the Commanding Officer of *Lynx*, believing it, and the list of stores seized, to have been mislaid "through the darkness of the night and the hurry and confusion of the detention". The court, however, had every faith in the very experienced Henry Huntley and invoked the leeway granted by Lord Palmerston that not every deviation from the treaty necessarily invalidated a capture, and it proceeded to condemnation.

The new Commander-in-Chief was to be Rear-Admiral Patrick Campbell CB, famous for HMS *Dart*'s cutting out of the French frigate *Désirée* from Dunkirk Roads in July 1800, an action which James, the naval historian, rated for gallantry with the most notable feats of the Napoleonic Wars, including Cochrane's capture of the *Gamo*. He hoisted his flag in the frigate *Thalia* (46) at Spithead on 6 September, and sailed for the Gambia.

Having heard that there were many vacancies for mates and midshipmen on the coast of Africa, Campbell successfully sought permission to take six midshipmen as supernumeraries in the flagship to distribute among the cruisers. While at Tenerife, Captain Wauchope of the *Thalia* asked permission of the Admiral to purchase a small quantity of wine at the Cape for the use of the midshipmen. It was necessary, he said, to drink quantities of liquid in the tropics. It was not healthy, however, to drink water alone on the coast of Africa because of the risk of dysentery, but if the youngsters were allowed to mix spirits with their water it would "lead to habits of intemperance injurious to their future character". Apparently a commander-in-chief was not authorised to grant such a request, and the matter was referred to the Board of Admiralty. The Board refused permission.[28]

Without any diversion for investigation of affairs on the slave coast, *Isis* and Admiral Warren arrived off Bathurst on 24 September, finding there the brigantine *Buzzard*, sister of *Forester*, *Lynx* and *Griffon* and launched as recently as March that year. Her short period on the coast had brought no reward, and she was currently keeping the routine watch over the Bathurst settlement and its merchants. From those merchants Warren received a warm welcome, and they fêted him in gratitude for his interest and support. Then, *Thalia* having arrived on 10 October and the handover of command having been completed that day, *Isis* sailed for home.

The first task for Admiral Campbell was to replace the Commanding Officer of *Buzzard*. He had found that Lieutenant McNamara, an officer of long service, was worn out and broken in health, and he offered the command to the lieutenants of the two flagships. Astonishingly none of them grasped this golden opportunity, and so the Admiral ordered Clement Milward, Third Lieutenant of *Thalia*, into the brigantine. It was an order for which Milward would soon have good reason to be grateful. The new Commander-in-Chief then made for Freetown, where he found the victualling store in good repair and gave his blessing to the proposal for a new wharf. From Sierra Leone *Thalia* took him on a tour of inspection of Cape Coast Castle and the fort at Accra before heading for Princes Island, St Helena, Ascension and the Cape of Good Hope. The forts were in excellent shape, and the accounts at Cape Coast showed that the value of its exports in the latter half of 1833 had reached a healthy £80,412. The Admiral was also pleased to find that there was no sign of slaving in the vicinity of the British forts between Cape Apollonia and the River Volta.

Returning to lesser matters while at Princes, he dealt with a request from Midshipman Frederick Robinson of *Charybdis* who, at the age of 26, had despaired of getting his examination for lieutenant and begged for discharge from the brigantine and return to England. Campbell, no doubt sympathising with the dispiriting slowness of promotion for junior officers, and despite the shortage of midshipmen on the Station, approved the request.

Since her arrival on the Station, *Griffon* had found little relief from the tedium of Bights cruising, and a week at Ascension in July followed by a short refit at the Cape had provided her only break. The closing stage of her return passage north at the end of October, however, brought a burst of excitement. Having sailed from Ascension for Princes on 22 October, the brigantine was 75 miles west-south-west of St Thomas at daybreak on the thirty-first, and still 11 miles south of the Equator, when she sighted a two-topsail schooner hull-up to leeward on the larboard tack. Lieutenant Parlby ordered all sail in chase, and the stranger was seen to bear up and make all sail. *Griffon* gained fast, and by 0715 she was within gun-shot and opened fire. Just before 0800 the chase shortened sail, hove to and surrendered without resistance. She was the Spaniard *Indagadora*, bound from New Calabar for Havana with 375 slaves. It was later found that she had appeared in 1829 before the Anglo-Dutch Freetown court as the *Adeline*, had been purchased by a Sierra Leone merchant on condemnation, sent to England for sale and registered there, sold at Cadiz, fitted-out there for slaving, and loaded with trade goods at Gibraltar. On condemning her and 363 slaves, the court attributed the unusual speed with which she had purchased her Africans to the superiority of her British trade goods.

Only three days later *Griffon* found herself again in luck. At 1100 on 3 November, by now 40 miles south-east of Princes, she sighted a sail to leeward, and soon made her out to be a brigantine standing to the west. She gained fast on this chase too, and was able to open fire at 1745. As shot began to fall very close, the stranger cut away her studdingsail halyards, rounded to, and submitted without a struggle. It came as no surprise that she was another Spaniard, the *Clemente*, bound from the Bonny for Havana with 415 slaves. She had experienced disruption to her outward passage from Havana as a result of an encounter in mid-Atlantic with HMS *Pylades*, Commander Edward Blanckley, returning home from the South America Station.

Until she had identified *Pylades* as a man-of-war *Clemente*'s behaviour had been threatening, and she had been flying a warship's colours which she then rapidly shifted. She was also seen to be covering her 12-pounder long gun with a sail. Commander Blanckley initially had no intention of boarding her, it "blowing fresh with a heavy swell", but this suspicious conduct and his recognising her from Buenos Aires, where she had been fitting-out as a slaver, changed his mind. A search revealed not only the usual slaving equipment, on which account she could not, of course, be seized, but also the papers of a British merchantman, and Blanckley took her into Portsmouth on suspicion of piracy. There it was decided that there was insufficient evidence to bring her to trial, and she was released. The Advocate-General considered that some compensation might be appropriate but should not be "vindictively pressed". Subsequent events off West Africa almost certainly removed this threat to Blanckley.

At the condemnation of *Clemente* the court paid an unusual compliment to Lieutenant Parlby and to Mr Gammon and Mr Noddall, the Prize Masters of *Griffon*'s two slavers, in which the losses on passage to Freetown had been particularly light: twelve slaves in *Indagadora* and seven in the *Clemente*. "We never knew", they wrote,

> any prize to arrive here when the papers were more regularly prepared, where more attention had been paid to the comfort of the slaves and the kind treatment of the detained crews or where the clean and orderly state of the vessels was productive of happier results than in the case of the *Indagadora* and *Clemente*.[29]

Before these two successes, *Griffon* had endured many fruitless and dispiriting months in the Bights, and, in an age when expressions of appreciation from those in authority appear to have been rare, this praise must have been most welcome.

Lord Palmerston's anti-slaving diplomacy was brought temporarily to a halt in November 1834 with his replacement as Foreign Secretary by the Duke of Wellington. Palmerston's closing months in office had brought a pair of minor successes and one keen disappointment. Denmark and Sardinia had joined the Anglo-French Convention in July and August respectively, but hopes that the United States would do likewise were finally dashed in October. In 1833 Britain and France had simultaneously invited the USA to accede to the Convention, and Britain had proposed a reciprocal Right of Search within certain geographical limits and with the other restrictions specified in the agreement with France. The proposal required that the Right of Search would extend to the coasts of the United States, and Britain offered in turn to extend it to the British West Indies. Washington delayed negotiation, but in March 1834 the proposal was refused, largely on account of the demand for a Right of Search to the American coast. In July Britain agreed to waive that offending clause, and in September the French again joined in urging accession, but on 4 October Secretary of State Forsyth stated that the determination had "been definitely formed, not to make the United States a party to any Convention on the subject of the Slave Trade". It would not be long before the implications of this sad failure would begin to manifest themselves.

Mid-December 1834 marked the beginning of a relatively productive three-month period off West Africa. Richard Meredith in *Pelorus* opened the sequence of arrests when he encountered the Spanish schooner *Sutil* close off the north-east

15. Capture of the Spanish brig *Formidable* by HM Brigantine *Buzzard* on 17 December 1834

coast of Fernando Po on the seventeenth and, although the schooner did her best to escape, took her after a chase of some hours. On the slaver's first night at sea after loading her cargo in the Old Calabar River, 27 of the Africans had been suffocated in a "ferocious scramble for room". She had 307 slaves on board at the time of capture, but a further 79 died before arrival at Freetown, and 12 more before emancipation, thanks to the criminal stupidity of the Spaniards in taking the drinking water for the slaves from alongside in the river. In declaring condemnation and emancipation the commissioners emphasised that no blame was attached to the prize crew for the deaths, and regretted that the Prize Master, Mr Judd, "a promising young man", had died a few days after arrival at Sierra Leone as a result of the "miserable privations he had endured in common with the slaves".

Sutil might have been acting as tender to the next prize, the Spanish brig *Formidable*. Both had initially been to the Gallinas River, and they had shipped their slaves on the same day in the Old Calabar. They were also both seized on 17 December. *Formidable* was well known on the Coast and had been frequently boarded, most recently in the Old Calabar by *Pelorus*. She was a 200-ton former brig-of-war, armed with two long 16-pounders and six 18-pounder Govers and manned by 66 men, far outgunning and almost matching for men that day's adversary, *Buzzard*. *Formidable* had been fitted-out to deal with the three-gun brigantines which her master had apparently dismissed as "tom-tit cruisers". As she emerged from the river at 0830 *Buzzard* was waiting for her, and a chase of seven hours began.

The wind was very light, and Lieutenant Milward's men were constantly at the sweeps from 0930 until 1615. At 1130 the chase cut her longboat adrift; at 1200, with the vessels under four miles apart, she began throwing overboard her stores; and at 1600 she ditched her stern-boat, ran-out stern guns, hoisted a large Spanish flag and took in her studdingsails. At 1615 *Buzzard* opened fire with small arms at pistol-shot range. The chase immediately replied with her stern guns, and *Buzzard* then "gave her the great guns" as they could be brought to bear. The brig's colours came down at 1650 and Milward ceased fire, but his hail to discover whether the chase had indeed struck was answered by a discharge of stern guns which shot away the brigantine's flying jib-boom. Milward reopened fire, but outgunned as he was, he knew that he had to board quickly.

It was another ten minutes before *Buzzard* could put her larboard bow alongside the brig's starboard quarter, and by then the slaver's fore and main topmast stays had also been shot away, her running rigging had been cut to pieces and her sails were holed and torn. The boarders then had to slash their way through high boarding netting before they could reach the slaver's deck. They were met by not only the Spanish crew but also a number of armed Africans in red jackets; however, the struggle was short and sharp and Milward's outnumbered men soon had possession, at a cost of six wounded, two of them dangerously so. The entire engagement had lasted only 45 minutes.

Milward later described resistance as "trifling", but the cost to the slaver's crew indicated a fight of some determination: six killed, including the mate; five seriously wounded, the master amongst them; and eleven others wounded, three of them armed slaves. *Buzzard*'s forestay and foretopmast stay had been cut through, her running rigging and sails were considerably damaged, her flying jib-boom had been shot away, and her bumpkin was carried away in the boarding. The Commander-in-Chief drew Their Lordships' attention to Milward's "very gallant conduct" in this action, and for this and "subsequent meritorious conduct" recommended his promotion. This reluctant commanding officer was consequently promoted to commander.

There had been 712 slaves on board *Formidable*, and their tribulations were far from over when she was taken. During the 42-day passage to Freetown the prize was struck by lightning which killed ten slaves and two of the prize crew, six slaves jumped overboard, and the number of Africans arriving at Freetown was horribly depleted by disease to 408.* A further 18 died after landing and before emancipation. *Formidable* too had filled her water barrels from alongside in the Old Calabar River, with grim consequences, and the inexperience of the prize crew contributed to the losses. With *Buzzard* in a crippled condition, Milward had been unable to afford anyone more senior than his Acting Second Master as Prize Master. When the Surgeon to the Court, who must have been accustomed to harrowing sights, visited the brig in Freetown, he described her state as "quite appalling".[30]

The next arrest involved two chases and a battle of wits which thoroughly tested the skill and experience of Henry Huntley. On 20 December *Lynx* ran into Whydah Roads and boarded the 15 slavers lying there. The most advanced in her preparations for loading her cargo and sailing was the *Atrevido*, a fine American-built brig. Not quite so advanced was what appeared to be a sister vessel, the *Fortuna*. *Lynx* then retreated over the horizon. She stood off and on until dawn two days later when two brigs were sighted, one on the weather bow and the other on the weather quarter. Huntley decided to pursue the one further to seaward, gaining on her slowly all day until at dusk she could be seen down to

---

* It was not uncommon for slaves to jump overboard from prizes if the opportunity offered. Apparently many believed that in death they would be returned to their own countries.

her courses. The danger now was that *Lynx* would overtake the chase in darkness, but Huntley's excellent judgment was rewarded when the moon rose at 0200 and the brig was sighted in the moonlight within gun range. One shot through her boom mainsail head brought the chase to, but she was the empty *Fortuna*, sent out as a decoy, not the loaded *Atrevido*.

Lynx returned to just out of sight of Whydah to wait for the *Atrevido*, with nothing but steadying sails set, and at daylight on 24 December a brig was sighted heading directly for *Lynx*. On finally sighting the cruiser, the stranger tacked and steered close-hauled to the north-west. *Lynx* immediately set all sail, and the chase continued all day in a fresh breeze. The brig was visible down to the foot of her topsails just before darkness hid her, and Huntley now had to guess her movements during the night. He was convinced that she would alter to the south-east, and, after careful calculation, he issued orders for the night which initially put *Lynx*, under easy sail, on a course of east-south-east, directly away from the bearing on which the chase had last been seen. At dawn on Christmas Day the brig was sighted on the lee bow, heading straight for *Lynx*.

The pursuit resumed, and, despite Huntley's poor opinion of his brigantine's sailing qualities, *Lynx* had the legs of the Baltimore clipper brig, and she was almost within gun range when the two vessels were hit by a heavy tornado and were obliged to reduce sail rapidly. The chase was entirely hidden by rain for the next two hours, and Huntley's judgment was again put to the test. Confident that the brig would not continue to run to the eastward, he brought *Lynx* about on the starboard tack, heading north-west, and when the sky cleared and the sun emerged at 1300 the chase was revealed, as Huntley had expected, becalmed and well within gun-shot. One round over her brought her Portuguese colours down. She was indeed the *Atrevido*, and she was carrying 494 slaves destined for Montevideo. The chase had finished 100 miles south of Whydah, and *Lynx* took her prize to Princes, where she encountered the still empty *Fortuna*, before sending her to her condemnation. She lost 12 slaves on passage, but the Prize Master, Mr Arundel, was commended on the "superior order and cleanliness" of his vessel. The whole episode was something of a triumph for Huntley.*[31]

The new Commander-in-Chief, while at Ascension in December, submitted his initial views on affairs on the slave coast. He emphasised the "very considerable

---

* This account is distilled largely (via Lubbock in *Cruisers, Corsairs and Slavers*) from Sir Henry Huntley's book *Seven Years' Service on the Slave Coast*, which may have a natural tendency to self-congratulation, but there is no denying the skill and perception demonstrated by Huntley on this occasion.

extent" of the Spanish and Portuguese trade, and he drew the undeniable conclusion that to achieve "anything like an effective check on this Traffic on such an extensive line of Coast" he had too few cruisers. He could afford only two between Cape Verde and Cape Palmas, one of them almost permanently at the Gambia, and he should have four. Although many cargoes were being shipped from the Gallinas River and its vicinity, he wrote, the greater portion of slaves were taken from the rivers flowing into the Bights, and he needed at least six vessels constantly cruising between Cape Palmas and the Congo. There were often only three or four on the whole of that station. He submitted that he should be sent more cruisers, and requested one or two steamers. He also recommended a drastic reduction in the size of the Ascension garrison from 216 to 132, proposing to leave all 47 liberated Africans there but removing the Kroomen.

On his arrival at the Cape in January 1835, Campbell found his attention being drawn to support for the Army in its confrontation with "Kaffir" tribes, but, left to their own devices, his cruisers in the Bights continued their run of success into the New Year.

The tensions of lower-deck life had briefly surfaced in *Fair Rosamond* in October. The Kroomen had been paid prior to discharge, and Able Seaman Edward Fitzgerald had lain in wait for Jack Fryingpan, Krooman, had "maliciously and wilfully" knocked overboard the Krooman's hat which contained the whole of his pay, £12.1s.1d, and had been heard to say later that he had "only paid him what he owed him". Lieutenant Rose requested that he be allowed to draw a bill for the amount of the Krooman's wages to be charged against Fitzgerald.

Such grievances would probably have evaporated in *Fair Rosamond* on 3 January 1835. Compared with her earlier feats, the capture of the small Portuguese schooner *Maria* was a minor event, but it brought to an end a disappointing five months. The schooner was on passage from Mayumba Bay to Princes with 48 slaves and a quantity of trade goods, and Rose's boats made the arrest close to the mouth of the River Gabon, seven miles north of the Equator, after a nine-hour chase in light and variable winds. *Fair Rosamond* towed her prize to West Bay, arriving with only one day's wood and water remaining. Unable to carry sufficient provisions in either *Fair Rosamond* or *Maria* to take the slaves to Sierra Leone, Rose landed all but four of the Africans at Fernando Po and sent the prize to Sierra Leone, where she was condemned. The schooner was typical of the vessels shipping slaves from the mainland to Princes.

There was a more gripping encounter on 15 January in the Old Calabar River. Commander Meredith in *Pelorus* had heard three days previously that a Spanish slaver was on the point of sailing from Duke's Town, and he sent three boats 60 miles upriver in ambush. On the fifteenth the cutter had to be detached for more provisions, leaving the First Lieutenant, Barrow, with a First Class Volunteer and 21 seamen and marines. It had been ascertained from the master of an English brig that the slaver was loaded and was prepared for action, and a reconnaissance after sunset discovered her moving downriver under tow by several canoes. The boats lay at a narrow part of the stream, and as the slaver came abreast they pushed off. It was a calm, clear, moonlit night, and after two hails the marines opened fire to cover the approach. The reply was a continuous hail of musketry until the boats touched the vessel's side, but although several oars were shot away, no one was hit because the Spaniards could not see for smoke.

The slaver had boarding netting rigged, so the boarders climbed through the gun ports. As they did so a double-shotted 18-pounder misfired directly over the boats; the second, it was later discovered, to have done so. The slaver was quickly carried, miraculously without loss of life on either side, and Barrow immediately anchored her. He then found that she was the 300-ton polacca barque *Minerva*, carrying 650 slaves, manned by two officers and 32 men, and armed with four guns and 40 muskets. Brandy had been distributed about the decks and under the guns, presumably to stimulate resistance. It was then discovered that the master was absent, and an hour later he returned, unaware of events, with another 25 slaves in a canoe. One of the boats detained him as he came alongside, and the slaves were put on board. When the prize was later condemned, these 25 could not be emancipated with the other 444 surviving slaves because of the circumstances of their embarkation, but they were, of course, freed into Sierra Leone.[32]

From a narrow perspective the year seemed to have begun relatively well, but the wider picture was vastly less encouraging. It would later become clear that the flow of slaves across the North Atlantic was climbing to a peak in 1835. The full magnitude of the Trade could hardly have been comprehended at the time, but it was apparent to the Foreign Office, the British commissioners at Freetown and Havana, the cruisers, and even the new Commander-in-Chief, that further measures were pressingly necessary if the suppression campaign was to achieve any degree of effectiveness.

* * *

At the beginning of the 1830s, the White Flag of Bourbon, replaced in 1830 by the French Tricolore, had distinguished the prominent nationality among the slavers on the coast of Africa north of the Equator, but the Anglo-French treaty of 1831, enforced by French cruisers, had at last served virtually to sweep the

French slavers from the seas. Their place, however, was soon filled by the resurgent Portuguese who, although they did not achieve pre-eminence, began to take an increasingly significant slice of the burgeoning Cuba Trade. By the middle of the decade the Spaniards of Cuba were by a sizeable margin the leaders in the northern hemisphere traffic, and a number of factors, including cholera epidemics among the slaves on the Cuban estates and a fear among the Spanish merchants that restrictive legislation might be in the offing, had led to a great expansion in the volume of the island's slave trade.

Britain's force level on the African coast had remained generally unaltered, as had its strategy and tactics, and the one appreciable change to the command structure had, far from achieving improvement, ensured that there was no authority in the main area of activity able to develop and initiate advances to counter the expanding and increasingly sophisticated slaving operation.

Withdrawal of the ex-slaver tenders from the West Coast force was a backward step, motivated probably by a concern at the Admiralty that its control was being bypassed and that its ability to provide effective cruisers was implicitly being questioned. Nevertheless, by 1835 the force on the slave coast was of a more homogeneous constitution. There had been a deal of coming and going between the northern slave coast and the Cape, but of the nine cruisers generally on station, six were brigantines specifically configured for the anti-slaving task and, if not ideal, decidedly an improvement on most of their predecessors. Of the three others, *Fair Rosamond* compared well with the brigantines, and it was a pity that there were not more like her. It was understandable that a sloop, currently *Pelorus*, should be retained as the Senior Officer's vessel, and the marked exception to the overall improvement was the unmodified brig *Britomart*, of a class which had long been shown to be unsuitable. The experimental deployment of *Pluto* could hardly have been judged an unqualified success, but neither was it a total failure, and it could be seen that there was great potential for steamers on the Coast as the efficiency of their machinery increased.

The number of cruisers was still grossly inadequate, particularly if the slave coast south of the Equator was to be policed, as had become necessary with the supposed abolition of the Brazilian slave trade. As it was, no cruising could yet be conducted on a regular basis south of the Line by the Cape of Good Hope squadron, and an occasional presence as far south as the Congo was all that could be hoped for. Extraneous tasks, defence of the Gambia settlement and its trade in particular, continued to draw cruisers away from the suppression campaign, but trade protection remained an essential element of the Commander-in-Chief's responsibilities, and would continue to be so.

There had been a number of gallant actions during the past four years, but an inclination among the slavers to surrender without resistance was increasingly apparent. No doubt the Royal Navy's oft-reinforced reputation for courage and determination was responsible for this welcome tendency, and probably its record of ferocity and irresistibility in boarding actions was a particular deterrent. When the considerable risk of death or wounds in a fight with a cruiser was weighed against the improbability of any personal penalty at the hands of national courts following unresisting capitulation, the conclusion was clear to most.

Less easy to explain was what seems to be a lack of resolve among slavers encountering cruisers in open water to make any attempt to escape. The vessels making the transatlantic passage, particularly to Cuba, could almost all be classed as Baltimore clippers, many of them American-built, and (in theory at least) their speed should have been more than a match for most of the British cruisers on any point of sailing. Records of failed chases are not readily available, but although the tendency had been reversed in 1834, a high proportion of the arrests in recent years involved an entirely passive response from the slavers intercepted, even when the capturing vessels were the relatively sluggish *Isis*, *Pelorus* or *Trinculo*. It can only be assumed that there was an assumption among many of them that flight from a determined cruiser was hopeless.

It may well be that slaver masters, in the not unrealistic expectation of avoiding detection on clearing the Coast, mostly neglected to hone their sailing skills or to train their crews. The men-of-war ships' companies, on the other hand, were all anxious to extract optimum performance in their sailing, and they certainly had no shortage of time to practise. The officers and men of the smaller and faster of the cruisers took particular pride in the sailing qualities of their vessels, and one of their few delights was to race each other whenever opportunity offered. The theoretical speed of chase and chaser was not necessarily the deciding factor in a pursuit.

Although the red and gold of Spain had supplanted the colours of France as the predominant flag among the slavers, little else had changed on the Coast. The best efforts of the cruisers had continued to offer nothing more than inconvenience to the Trade as a whole, and not even that to the southern hemisphere slavers. Meanwhile, the transatlantic slave traffic was growing to enormous proportions. Many more, and better, cruisers were needed to achieve a degree of effectiveness in the suppression campaign, and so was some initiative to replace the current discredited strategy, but for the moment opportunity for progress lay primarily in the hands of the Foreign Office.

CHAPTER 12

An Uncertain Sound: Western Seas, 1831–5

If the trumpet give an uncertain sound, who shall prepare himself to the battle?
I CORINTHIANS

AFTER THE HIGH LEVEL of traffic in 1828 and 1829, there had been a clear decline in the Cuban slave trade in 1830, an abatement caused by temporary market forces and probably aided by a sickly season. The following year saw no recovery in the Trade, but neither was there a continuing reduction. The island's plantations still needed new slaves, and the planters knew that neither Madrid nor the colonial authorities would place any obstacle in the way of fresh supplies; indeed, the reverse was likely to be the case. The disruptive efforts of the handful of the Royal Navy's little schooners in the approaches to Cuba, aided and abetted by the British commissioners of the Mixed Commission court at Havana, and of the sorely stretched West Africa Squadron, would continue to be largely ineffectual. Certainly that ineffectiveness would persist as long as the Navy's cruisers remained hamstrung by the absence of an Equipment Clause from the Anglo-Spanish Treaty.

For a time it had appeared that coffee might become Cuba's primary crop, but sugar was regaining the ascendancy, and the two thousand or so coffee plantations would soon be ruined by hurricanes. About a thousand sugar plantations remained, each with an average of 70 slaves, although some employed several hundred. Steam power was being introduced to some of the sugar-cane mills but this had little effect on the requirement for slaves, and continuation of the illegal slave traffic would be essential to the interests of the planters, the Cuban authorities and the Spanish economy for the foreseeable future.

With the ending of legal slaving by Brazil in March 1830 there came not only an expectation of a lively contraband traffic of slaves into the country but also uncertainty over the continuing jurisdiction of the Anglo-Brazilian Mixed Commissions. Clearly the Royal Navy's squadron on the South America Station was now empowered to intercept slavers under the Brazilian flag, but it remained to be seen whether the Brazilian Government would continue to support the judicial arrangements necessary to deal with prizes.

Under the Anglo-Portuguese agreements, Portuguese slaving south of the Equator remained permissible only between Portuguese territories, and the traffic to Brazil under Portuguese colours had therefore been illegal ever since Brazil had gained independence. Portugal had still failed to fulfil its undertaking to legislate against the Trade, but that seemed to present no obstacle to enforcement by Britain of the 1815 Treaty and the Additional Convention of 1817. However, no clarification had been forthcoming from London for the Royal Navy's Commander-in-Chief at Rio de Janeiro, and there was another serious difficulty with seizures of Portuguese slavers in the western Atlantic: the only Anglo-Portuguese Mixed Commission was at Freetown.

Both Brazil and Portugal had largely abandoned trading from the Gulf of Guinea, despite the preference of the Brazilian planters for the hard-working Africans from Dahomey, Lagos, Benin and Bonny, although Portuguese slaving from the rivers north of Sierra Leone was again active. The main source of supply for the Brazil market remained the African coast south of the Line, particularly Angola and Benguela. With the approach of the March 1830 abolition date, the merchants of Rio and Bahia had taken advantage of the time supposedly left to them and had imported large numbers of slaves. As in Cuba, this led to a glut during the early 1830s, but legal niceties were most unlikely to deter the traders from renewing the flow once demand recovered, as it certainly would, to serve the plantations of cotton, sugar and, increasingly, coffee. Coffee was becoming the great new Brazilian product, and its cultivation, harvesting and processing involved as much hard manual labour by slaves as did sugar production.

* * *

In the West Indies there were no slaver arrests in 1831 after *Pickle*'s seizure of the *Roza* in June, and in the latter half of the year the only report concerning the slave trade from the British Acting Commissary Judge, Mr Macleay, followed the arrival in Havana of a Portuguese schooner in ballast on 10 December, after she had landed slaves elsewhere on the coast. It was Macleay's belief that the Portuguese were landing considerable numbers of slaves on the more remote parts of the island.

Otherwise communication from Havana concerned changes in the Mixed Commission and the Cuban administration. The Spanish Arbitrator, Colonel Don Rafael de Quesada, resigned in August and was replaced by Brigadier-General Don Juan Montalvo y O'Farrill. Macleay expressed his appreciation of de Quesada's impartiality, declaring that "in no instance whatever have I remarked this Gentleman to be actuated by any other principles than those of the strictest integrity and honour." He was equally warm in his commendation of the Spanish Judge, His Excellency the Conde de Villanueva, who was replaced in February 1832 by the Conde de Fernandina, praising his "utmost good faith and impartiality" and his "invariable kindness and attention".[1] He certainly could not write anything similarly creditable of the Anglophobe, corrupt, dishonest and self-serving Captain-General Vives, whose impending departure he reported at the end of December. On being relieved in his post by Lieutenant-General Don Mariano Ricafort in May 1832, this bane of the lives of Judge Kilbee and Judge Macleay was made Count of Cuba by a grateful Spanish Government.

It was with disappointment that Macleay reported at the beginning of 1832 on the state of the Cuban slave trade. He had hoped to detect a further drop in the level of traffic in the previous year, but the number of slaver departures from Havana had showed a small increase to 32, and arrivals had held steady at 36, a number partly sustained by the poor success rate of the Navy's schooners. It must have given Macleay a modicum of compensating satisfaction in the new year that the Spanish authorities had been told that he had, at last, been appointed Commissary Judge, and Mr Charles Mackenzie was the new Commissioner of Arbitration.

The year 1831 had ended sadly for *Pickle*. On 11 December her Commanding Officer, the admirable Lieutenant Thomas Taplen, was drowned when a boat capsized. As a temporary measure, Admiral Colpoys promoted Mr Edward Stopford of *Winchester* to the rank of lieutenant to command the schooner, although he intended to appoint a more experienced officer to her in due course.

There had been precious few captures in 1831 to encourage the schooners and, lest the Admiralty should conclude that there had been a lack of effort, the Commander-in-Chief assured Their Lordships that he was very satisfied with the "exertions and perseverance" of *Speedwell*, *Pickle* and *Pincher* in their suppression work. They had chased and boarded several French vessels, and there was "no want of zeal and activity". While on the subject of the French, Colpoys mentioned that *Pincher* had reported several French vessels landing slave cargoes in Cuba, and he recommended that in future such slavers should be arrested and delivered to the French authorities. He was firmly directed to refrain from doing so.[2]

Early in 1832 a major rebellion of slaves in Jamaica obliged the Commander-in-Chief to concentrate the whole of his disposable force at the island while *Winchester* made for Bermuda to collect stores for the ships around the Leeward Islands. Lieutenant William Warren in *Speedwell* was, however, free to maintain his patrol on the southern coast of Cuba, and at daybreak on 6 April, about 45 miles south-east of the Isle of Pines, he sighted a suspicious sail on his weather bow heading for the south-west point of that island. *Speedwell* gave chase, and the stranger hauled her wind, took to the sweeps and used all means to escape. Warren fired a lee gun and hoisted his colours, but the chase took no notice, and at 1130 *Speedwell* opened fire. Fire was returned, and an anxious chase of 11 hours continued. When the stranger finally struck her colours and hove to she was found to be the Spanish schooner *Planeta*, carrying 239 slaves from the River Cameroons and six free Africans, and armed with one long 9-pounder. The free Africans proved to be five Kroomen picked up from an open boat off Fernando Po, and a Calabar pilot. Warren asked the Captain-General for these six to be handed over to Admiral Colpoys for return home, and Vives agreed. One slave died before vessel and cargo were condemned.

The north coast of Brazil formed part of Admiral Colpoys' domain, and he occasionally sent vessels of his squadron to cruise there for a short time, mainly against pirates. *Sapphire* (28), Captain The Hon. William Wellesley, and *Pickle* had been the most recent visitors, and they returned in May from the coasts of Para and Maranhão. They reported that, before prohibition, large slaving vessels from the coast of Africa had landed their best slaves at Bahia before making for Pernambuco, Maranhão and Para to dispose of the remainder, but now schooners of 30 to 120 tons were making the passage, in the fine season, from the Cape Verdes directly to the Bay of St Marcos. There the slaves were transferred to country craft which took them upriver to the plantations. Even if Colpoys had possessed the cruisers to intercept this traffic they would have lacked the quantities of water and provisions necessary to take or send prizes to either Rio or Sierra Leone, and he suggested that one of the West Africa Squadron should be stationed off the Cape Verde Islands at the appropriate times of year. There was little hope of that with the current force level off West Africa.

Shortly after her return, *Sapphire* boarded the Spanish brig *Churruca* just before she entered Havana and found plentiful evidence that she had very recently landed a cargo of slaves on the coast. The Cuban authorities pretended to investigate Captain Wellesley's allegation, but, as usual in such cases, reported that nothing incriminating had been found in the vessel's log and dismissed the matter.

The total absence of slaver prizes brought to the Havana Mixed Commission by Spanish cruisers might suggest that there was no Spanish naval presence in Cuban waters. However, the importance of the island to Spain's economy demanded a substantial force in the vicinity, and in June 1832 Admiral Colpoys reported that lying in Havana was a squadron under the command of Rear-Admiral Laborde in a 74-gun flagship. The squadron further consisted of two frigates of 48 guns, one sloop of 20 guns, a brig of 16 guns, a brigantine of 8 guns, two 5-gun schooners and two gunboats. Any degree of effort by a force of this size would have caused great damage to the Cuban slave trade, but, of course, no such effort was forthcoming.[3]

Lieutenant Warren had concluded that the Spanish slavers were favouring a track south of the Isle of Pines, and, once *Planeta* had been condemned, he returned there with *Speedwell*. He did not have to wait long for his judgment to be rewarded. On 3 June, close to the site of *Planeta*'s arrest, he intercepted the Spanish brig *Aguila* and, thanks to *Speedwell*'s superior sailing qualities, he took up a commanding position on the brig's weather bow. There then began a "smart action" with the brig's eight 32-pounder carronades, two long twelves and 53 men matched against *Speedwell*'s four 18-pounder carronades, one long 18-pounder and crew of 46 officers and men. After an hour the brig, with her sails and rigging much cut about and two men wounded, struck her colours. *Speedwell* remained unharmed. The brig, at about 350 tons one of the largest slavers out of Havana, had a cargo from Loango of 616 Africans, of whom 12 died before emancipation. As the British commissioners later reported, *Aguila*'s seizure greatly annoyed the Cuban slave traders, and it demonstrated that, although increased in size and force, the Havana slavers were no better able to cope with the smallest class of British men-of-war. Having condemned the brig, the commissioners commented that the whole affair was "highly creditable to Lieutenant Warren's nautical skill and gallantry".

There was no resting on laurels by Warren. With *Aguila* disposed of, he sailed for Jamaica, but sure that the potential of the Isle of Pines hunting area was not yet exhausted, he took passage close to the island. Sure enough, on 25 June *Speedwell* fell in with the Spanish schooner *Indagadora*, a vessel regarded as notorious by the British commissioners. Warren brought her to just off the southern shore of the Isle of Pines after a seven-hour chase, and found 134 slaves from Lagos on board: 103 men, 12 women and 19 boys. The slaver's master had died of fever, and the mate was in command, but the vessel was clean and otherwise healthy, and there were no further deaths before she was condemned. However, the episode was not without incident. While *Indagadora* was lying at Havana awaiting trial, an attempt was made to rescue her acting master, the boatswain and the steward. A shore boat had been given permission to go alongside the prize to give "segars" (cigars) to the three prisoners, but, while the officer-in-charge was preoccupied with furling the main-deck awning, the three jumped into the boat and made sail. Reacting swiftly, the officer also managed to leap into the boat and, with the help of *Speedwell*'s boats, prevented the escape. The culprits were handed over to the Spanish Admiral.

This run of success earned for William Warren a very warm commendation from the Commander-in-Chief, who pointed out to the Admiralty that Warren had, with only short intervals, served in the West Indies for nine years and had been "indefatigable in the discharge of his duties". Colpoys declared that he had "never met an officer more deserving of advancement", and at the end of August, on the strength of this unreserved recommendation, Warren was promoted to commander "as a special mark of [Their Lordships'] approbation".[4]

Lieutenant John Potbury of *Nimble* had taken a leaf out of Warren's book, and in mid-July he too was trying his luck off the Isle of Pines. On the thirteenth, after a four-hour chase, he snapped up the large Portuguese brig *Hebe* of Bahia, 43 days out of Loando with 401 slaves. Although the slaver was armed with eight guns, 6-pounders and 9-pounders, she offered no resistance.* On the advice of the British commissioners, Potbury followed the procedure used for previous Portuguese prizes and took the Africans to the Bahamas. After calling at New Providence to seek permission of the Governor, he landed the 385 survivors on Highborne Cay, 36 miles from Nassau, before sending *Hebe*, via Bermuda, to Sierra Leone to be condemned.

The arrival at Havana for emancipation of the very large cargo of slaves in the *Aguila* brought to a head a growing concern that the city was becoming inundated with freed Africans and the impossibility of finding enough "respectable and responsible persons" to take charge of them. It was generally acknowledged that it was essential to the peace and safety of the island that the local authorities should begin to send them away, although there were British reservations that this would contravene the Anglo-Spanish Treaty. Initial proposals by Spain that these *emancipados* should be shipped to Spain or one of the Spanish possessions, or to Sierra Leone, found no favour in London. A counterproposal that they should be taken to Trinidad at Spanish expense was regarded by Madrid as too costly, and then an arrangement for employing them on public works had reduced the anxiety of the Cuban authorities. However, Lord Palmerston voiced his deep concern about the condition of the freed slaves on the island, and the Cuban population, sick of drunkenness and daylight robbery in the streets of Havana, wanted them gone at almost any cost. At last it was agreed that slaves emancipated within the last two years, and all new arrivals, should be removed to Trinidad entirely at the expense of Spain.[5]

In September, Captain Arthur Farquhar in *Blanche* (46) was despatched from Bermuda to the Leeward Islands to become Senior Officer on that region of

---

* Admiral Colpoys reported that *Hebe* had six 12-pounders and two 9-pounders, but the Mixed Commission was likely to be more accurate on this sort of detail.

the Station in the rank of Commodore. *Victor* (18) sailed for Barbados to place herself under Farquhar's command, and the 18-gun sloops *Arachne*, *Columbine* and *Sparrowhawk* were under orders for the same destination.

The Commander-in-Chief's primary preoccupation, however, was continuing unrest in Jamaica following the rebellion of 20,000 slaves at Christmas 1831, which had resulted in the deaths of 14 white people and a million pounds' worth of damage. It had been put down with great brutality, and reprisals had included 500 killings and executions. Revulsion in Britain at this slaughter led to a powerful swing of public opinion against slavery and acceleration of progress towards Abolition. Nevertheless, for Admiral Colpoys the concern was for the calm and security of British territories on his Station, and in mid-October 1832 there were eight men-of-war at Jamaica: *Pallas* (42), *Gannet* (16), *Tweed* (20), *Fly* (18), *Pearl* (20), *Nimble*, *Pickle* and *Minx*. They were soon to be augmented by *Sapphire* (28) and *Ariadne* (28). *North Star* (28) was also on passage for the island, and the flagship, *Winchester* (52), was about to sail from Halifax to join the gathering. Of the 26 vessels on the entire Station, 12 had been drawn to Jamaica, and, of particular implication to operations against the slave trade, three of them were schooners.

Consequently it was hardly remarkable that the next slaver arrest did not follow the usual pattern of an interception by a schooner in Cuban waters. On 21 November, over four months since the previous seizure, the sloop *Victor*, Commander Robert Russell, was cruising between Tobago and Grenada when she fell in with the Spanish brig *Negrito*. When the brig was brought-to 40 miles north of Tobago after a chase of some hours, *Victor*'s boarding party found 526 slaves taken from Whydah and destined for St Thomas, as well as a crew of 30 and seven passengers. In addition to the prize crew, Russell left the slaver's master and six of his men on board *Negrito* for the 20-day passage to Havana and followed with the remainder of the slaver crew. So, to fulfil the persistent and entirely unnecessary demand of the Spanish commissioners that captors should be present to give evidence at Havana adjudications, *Victor* was dragged the length of the Caribbean Sea from her proper station to ensure the condemnation of *Negrito* and emancipation of her surviving 490 slaves.

At the beginning of January 1833 the dispirited British commissioners were obliged to report no reduction in the Havana slave trade. In 1832 there had been 27 safe arrivals at the port, two of them Portuguese as in the previous year, and 31 slavers had sailed, of which 12 had already returned. The commissioners had little knowledge of traffic at the other Cuban harbours, but were aware of one arrival each at Santiago and Trinidad. There had been no improvement under the new Captain-General, and it was abundantly clear to the slave traders that not only were their activities being "winked at" by the local authorities but also that Spain was determined to avoid any further restriction being placed upon the Trade.[6]

By late March *Nimble*, now under the command of Lieutenant Charles Bolton, had escaped from Jamaica, and on the twenty-ninth she was beating up to the Windward Passage on her way to Nassau when she sighted a suspicious schooner and gave chase. After a pursuit of four hours before the wind the schooner was brought-to about 65 miles south-south-east of Santiago de Cuba, and Bolton boarded her without a struggle. She was the 100-ton *Negreta*, under Spanish colours, heading for Santiago with 196 slaves from Bonny, carrying a crew of 16 and mounting a long 6-pounder. In light winds and calms the voyage to Havana took 12 days, and when the slaver ran out of food *Nimble* was obliged to supply her with bread and any other victuals suitable for the slaves that she could find.

The two vessels arrived at Havana to find cholera raging in the city, and the authorities refused to allow Bolton to land the slaves, demanding that the prize should be taken to Sierra Leone. That was entirely unacceptable to the British commissioners, but their proposal that the slaves should be taken directly to Trinidad was agreed. However, Bolton judged that *Negreta* was not seaworthy for the voyage, and he was asked to submit his requirements for alternative means of transport. He requested a vessel of at least 100 tons, 50 days' provisions and water for the slaves and ten seamen, fuel for the voyage, a good chart of the coast, supplies for the seamen to return to Havana, and a pilot. Having taken on board essential stores and given the Mixed Commission the evidence necessary to condemn the slaver, he took *Nimble* and *Negreta* out to the cays of the Cay Sal Bank to await the transport. "A fine large brigantine", the *Carolina*, arrived, but no slave deck had been laid and the provisions had just been thrown into her. It took *Nimble* a day to put the transport in order and to transfer the Africans, and then Bolton escorted her for the first stage of her voyage to Trinidad before heading for Nassau to replenish the schooner's now virtually exhausted provisions. The Mixed Commission condemned *Negreta* and the 195 slaves despatched in the *Carolina*, but not until temporary replacements had been found for the Spanish commissioners who, together with all the landed gentry, had retreated from Havana to their estates to escape the cholera.

On 5 July *Ariadne* called at Havana on passage from Jamaica to Bermuda, and Captain Charles Phillips reported that three days earlier he had fallen in with the Spanish schooner *Segunda Gallega*, finding in her very clear signs that she had just landed slaves on the coast of Cuba. The British commissioners, knowing the schooner as a notorious slaver, felt duty-bound to denounce her to the Captain-General, and Phillips, probably new to the hypocrisy of the Cuban authorities, may have expected a useful outcome. The commissioners, however, wearily awaited, and inevitably received, the stock response that examination of the vessel's log gave no indication of involvement in the prohibited traffic in slaves.

Concern about the taking of slaves from the vicinity of Sierra Leone, a matter which had worried successive governors of the colony, led in August to a suggestion from the Colonial Office to the Admiralty that the commanding officers of cruisers on the West Indies Station should question captured slaves as to who had sold them into slavery. This might, so the officials of the Colonial Office hoped, identify slave traders on the coast of Africa, particularly those adjacent to Sierra Leone. In fact there seems generally to have been no particular care taken by slave traders to conceal their identities, and the Colonial Office would probably have gained the desired information from enquiries at Freetown. As it was, the suggestion demonstrated an extraordinary lack of comprehension of conditions to be expected in slavers captured in the western Atlantic, and of the difficulties facing captors and prize crews, who had quite enough to occupy them without interrogating Africans in the squalor of slave decks.[7]

Lord Palmerston continued to assure the British commissioners that Madrid was being "urged in the strongest manner" to accept the addition of an Equipment Clause to the 1817 Treaty, and there seems to have been growing apprehension among the slavers that suppression efforts might soon become more effective. The passing of British slave emancipation legislation in 1833 added to their fears, and it was reported to London that in Costa Rica and Cuba every available vessel was being taken up for the slave trade. British industry was benefiting from increasing orders for trade goods, and the Colonial Office heard that manufacturers in Glasgow were unable to keep pace with demand. Slave merchants who had been in the habit of purchasing their trade goods at Jamaica were now opting for the greater security of St Thomas in the Danish West Indies, and it was learned that 25 slavers had loaded trade cargoes at the island in the period 11 May to 25 July 1833, and that another ten were in the harbour in early August to do the same.

In mid-1833 Admiral Colpoys was relieved by Vice-Admiral Sir George Cockburn GCB, one of the most distinguished sea officers of the Napoleonic period. He had first gone to sea, under the patronage of Lord Hood, in 1786, and by 1793 he was in command of the sloop *Speedy*. The following year he was made Post at the age of 21, and in 1796, while under Nelson's command, he ran the frigate *Meleager* under the batteries at Larma to capture six armed vessels. Later that year he was in command of *Minerva* (38), flying Nelson's broad pendant, when, although being chased by a Spanish squadron in the Gut of Gibraltar, he famously

16. Vice-Admiral The Rt Hon. Sir George Cockburn GCB, Commander-in-Chief on the West Indies, Halifax and Newfoundland Station

stopped to lower a boat and recover a man overboard. Then in December the same year, in the presence of the Spanish fleet, *Minerva* took the *Sabina* (40) after a three-hour fight and defeated the *Matilda* (34), which hauled off after a half-hour of close action. *Minerva* was in refit in Gibraltar in November 1797 when Cockburn took command of three gunboats and went to the assistance of a convoy becalmed dangerously close to the Spanish shore. He kept 30 gunboats at bay throughout the night, and saved the convoy. In these exploits he earned the praise of Nelson and the approval of Jervis. After the frigate *Phaeton*, he commanded the 74-gun battleships *Aboukir* and *Pompée*, and in the latter, as a commodore, he was instrumental in the capture of Martinique. He was promoted to Rear-Admiral in August 1812, and, with his flag in *Marlborough*, he commanded the naval forces in Chesapeake Bay for combined operations on the American coast and the destruction of Washington in August 1813. In 1815 he was appointed Commander-in-Chief of the Cape Station and conveyed Bonaparte to St Helena in his flagship, *Northumberland*. He was given a seat on the Board of Admiralty in 1818 and reappointed to it ten years later, and he was made Vice-Admiral in 1819.

The new Commander-in-Chief found that, of his squadron of 27 men-of-war, there were only five schooners, one of them, *Pincher*, being employed as a tender. This was an entirely inadequate force of small cruisers to patrol the approaches to Cuba in addition to the myriad of other tasks for which small, fast vessels were needed. As 1833 drew towards its close, however, *Nimble* was more than pulling her weight against a traffic which showed every sign of increasing.

Under orders from Captain Walpole of the frigate *Pallas*, now the Senior Officer at Jamaica, *Nimble* had sailed from Port Royal on 12 October to cruise between Jamaica and Cuba, where she fell in with *Firefly*. On 30 October she boarded the Spanish schooner *Veira*, but found that she had landed her slaves on the Cuban shore further to windward. Seeking information, Lieutenant Bolton took *Nimble* into Santiago de Cuba on 5 November and learned that a vessel had been sent out to warn the *Veira* of the cruiser's presence, and she had consequently landed her slaves further east than was the usual practice. *Nimble* sailed again the following day and arrived off the Isle of Pines on the ninth. Clearly word that this was a dangerous stretch of sea had not yet spread among returning slavers, and on 10 November Bolton added to the success that Potbury and Warren had found there.

At daybreak he sighted a large schooner about ten miles to leeward standing in towards the land, and he made all sail in chase. *Nimble* overhauled her quarry quickly, and when the distance apart had closed to four miles or so the stranger, seeing no chance of escape, wore round, shortened sail and hove to about seven miles from the south-west point of the island. As Bolton took in his studdingsails and square sail and prepared for action, the stranger showed Spanish colours and fired a blank gun. Bolton then hoisted his colours, and, the moment he was within range, fired two muskets over the chase. The Spaniard replied with a well-aimed round shot, and *Nimble*, closing as fast as she could in a breeze which was falling very light, opened fire. By the time that the range had closed to half pistol-shot the stranger had received two 18-pound shot between wind and water and several through her upper works and sails. Her mainmast was nearly cut through, her rigging was much damaged, and her master was mortally wounded. She then struck her colours and cried for quarter. The schooner was the 101-ton *Joaquina* with 327 slaves, 51 days from the Bonny River, armed with one long 12-pounder, and her defence, which Bolton described as "desperate" and worthy of a better cause, had lasted for nearly an hour.

Two Africans had been killed in the action, and a Spanish seaman and two Africans were wounded. There were no casualties in *Nimble*. The drama was not yet played out, however. Mr Armitage, the Prize Master, found himself with not only a critically damaged vessel and Spanish prisoners in a dangerous mood but also a mutinous cargo of Africans. Indeed, several of the 21 slaves lost earlier in the voyage had been executed for mutiny. During the passage to Havana he foiled a murderous plot by the Spaniards and some of the slaves to retake the prize; and when he was caught in a severe northerly gale on a lee shore and the plugs came out of the shot holes, he and his small prize crew, by repairing, baling and beating offshore, managed, despite the despair of everyone else on board, to bring the schooner to safety in a near-sinking condition. Bolton was properly warm in his praise of Armitage, his senior mate, for this fine performance.[8] On the arrival of the two vessels at Havana on 16 November the circumstances of the plot were reported to the Captain-General in the hope of action against the culprits. One slave died of his wounds and three of dysentery before emancipation, and at the muster of slaves by the Mixed Commission two were found to be missing, leaving 321 to be emancipated when *Joaquina* was condemned on 21 November.

On the following day Bolton sailed again for the Isle of Pines, which he believed was "the great place of resort" for slavers. At 0900 on 3 December a schooner was sighted to windward from the masthead, running along the land under all sail. On identifying *Nimble* the stranger ran herself ashore inside the reef of Playa Larga. It was blowing strongly and a heavy sea was running, and it took *Nimble* until nearly 1600 to beat up to the stranded schooner, by which time crew and slaves had landed and the vessel, her masts cut away, had been blown up and burned. *Nimble* managed to recover only a piece of canvas with the name *Amistad Habanera* painted on it. The next morning, eight to ten miles further to windward, she encountered another vessel run ashore and burned, and local fishermen told Bolton that she too had landed slaves.

There was not long to wait for another encounter. At dawn on 7 December, midway along the south coast of the Isle of Pines, a sail was sighted running down towards *Nimble*. She was a two-topsail schooner, with her topsails still furled. Clearly she had mistaken the cruiser for a fellow slaver, and was nearly hull-up from *Nimble*'s deck before the truth dawned on her. She immediately hauled her wind, and there began a chase of nearly eight hours, for almost all of which *Nimble*'s men were working hard at the sweeps. When the cruiser came up to her, the schooner gave no resistance to the boarding party, who found that she was the 102-foot *Manuelita*, under Spanish colours. She had a crew of 34 and was carrying 485 slaves, 477 of whom lived to be emancipated when the slaver was condemned. Her armament had consisted of one gun, but it had been heaved overboard to aid her escape from a cruiser off the coast of Africa. Bolton was glad to receive *Manuelita*'s confirmation that the slaver he had chased ashore four days earlier had indeed been the *Amistad Habanera*. According to Bolton, the arrival of the prize at Havana caused a great sensation because she had been considered the fastest vessel sailing from the port. By now *Nimble* had only seven days' provisions remaining, and she made for Nassau to replenish before heading for Jamaica for further orders.

The Governor of Trinidad had expressed his willingness to take emancipated slaves from Havana, on condition that he was given a month's notice of their arrival, but he was not prepared to accept a disproportionate number of males, and many more men than women were being emancipated. When the *Manuelita* was purchased by the Cuban authorities to evacuate the *emancipados* from *Joaquina* and *Manuelita* to Trinidad she was therefore loaded with all the female Africans and an equal number of men, a total of 212.

Pickle was despatched from Jamaica to replace *Nimble* off the Isle of Pines, but the final arrest of the year on the North America and West Indies Station was made by a sloop in the far south-eastern reaches of the Station, 2,400 miles from Cuba. On Christmas Day the Spanish schooner *Rosa*, carrying 292 slaves, was running north-westward on the South-East Trade Wind, helped on her way by the Guiana Current, about 450 miles north-east of the Bay of St Marcos on the north coast of Brazil. She was little more than halfway through her voyage from Whydah to Santiago de Cuba when she had the misfortune to fall in with the brig-sloop *Despatch* (16), Commander George Daniell. The prize crew brought her into Havana on 24 January 1834, and the Captain-General, anxious to be rid of the slaves to Trinidad, urged immediate adjudication.* However, the Spanish commissioners, as was their policy, refused to deal with the case in the absence of the captor, and Judge Macleay was obliged to acquiesce because Commander Daniell had retained the slaver's papers.

Rosa therefore lay in quarantine until *Despatch* arrived on 12 February, when she and the 290 surviving slaves were condemned. At the adjudication it was revealed that *Rosa* had been boarded by *Trinculo* in July to the west of Accra, but before she had loaded her cargo. The master protested to the court that he had been obliged against his will to take the slaves on board, and that it had been his intention to hand them over to the Cuban authorities, but that he had been prevented from doing so by the British man-of-war. The Mixed Commission was apparently unmoved by this novel defence.

On 1 January 1834 the British commissioners in Havana made their routine report to Lord Palmerston on the previous year's slaving activity in the port. There had been 38 slaver sailings, including two Portuguese, an increase of seven on 1833. Of these, eight had already returned and landed slaves, and a total of 27, four of them Portuguese, had arrived safely during the year. These figures did

---

* The Spanish Government bore the cost of maintaining the slaves from the time of their arrival at Havana until their emancipation, but expenditure was repaid from the proceeds of the sale of the condemned prize. Spain also bore, without repayment, the cost of shipping the emancipated Africans to Trinidad and maintaining them from the moment of emancipation until they were delivered to the Governor of Trinidad.

not include the comings and goings at Santiago de Cuba and Trinidad de Cuba, but the commissioners were aware that slavers were being fitted-out at Trinidad, and they had heard from a naval officer that the Trade was being carried out at Santiago in "indecently glaring a manner". They conveyed the discouraging news that the ravages of cholera among the black population during 1833 had added impetus to the Cuban slave trade, but they expressed a hope that Spain under the "enlightened Regency of the Queen Mother may not give the constant protection of its tribunals to the slavers".

In March the British commissioners heard that the British slave trader Josiffe, upon whom successive governors of Sierra Leone had long wished to lay hands, was keeping a coffee house at Matanzas, and had an estate in the neighbourhood worked by slaves he had brought with him when he last escaped from the coast of Africa. The net had begun to close on the slave trader.[9]

The schooner *Pickle* was experiencing particular ill-fortune in the matter of her commanding officers. Edward Stopford, appointed as a temporary replacement after the drowning of Thomas Taplen, had been relieved by Lieutenant Christopher Bagot, but in April 1834 Bagot found himself facing a court martial in Port Royal. There were two charges: of causing the death of Marine Grainger, killed by the discharge of a carronade fired by Bagot, and of inflicting upon Able Seaman Phillips the excessive punishment of 48 lashes for insubordination and insolent conduct. There was no suggestion that the killing of Grainger had been deliberate, but there was little doubt that Bagot was entirely to blame. On 6 February the schooner's long gun and two carronades were being cleared away for an exercise firing, and Bagot, who, in the opinion of one witness, tended to impetuosity and intemperance of language, was "out of temper" and annoyed at the slowness with which the larboard carronade was being made ready. Before the weapon was run out, and while Grainger was stooping close to the muzzle to ship the fighting bolt, Bagot, who had not the slightest business to do so, angrily grasped the trigger lanyard and fired the unshotted gun. Grainger was instantly killed by the blast. Found guilty on both charges, Bagot was dismissed from his ship and the Service. His replacement in command of the schooner was Lieutenant Archibald Bulman.[10]

In mid-1834 the Commander-in-Chief was exercising command from *Vernon* in Bermuda, and had delegated control of the southern part of the Station to Commander Gordon of *Pearl*. Hearing in early May that the slaver *Estralia* would soon be landing her cargo at Punta Barracaos, to the east of Santiago de Cuba, Gordon ordered Lieutenant John McDonell in *Firefly* to cruise between Santiago and the expected landing site. After a week on patrol, McDonell sent a boat into Santiago on 17 May and discovered the *Estralia* lying in harbour, having landed her slaves 30 miles to the westward just as *Firefly* began her cruise. So McDonell bore up for the Isle of Pines and, on 21 May, began to cruise that happy hunting ground.

At 1430 two days later, he sighted a schooner standing along the land and he made all sail in chase, but when night fell he lost sight of her in the darkness. Luck was with him, however, and at daybreak on 25 May he found the same schooner standing in for the land. He closed her under full sail until about 1000 when the breeze fell to a dead calm. *Firefly*'s men then toiled at the sweeps for six hours without pause, and at 1545 they came up with the chase about 50 miles south of the western end of the Isle of Pines. She surrendered without resistance, and was found to be the Portuguese slaver *Desfrique* with a crew of 14, and 215 slaves taken from Angola. The prize was taken initially to Havana and then, on the advice of the British commissioners, to Nassau, where the surviving 205 Africans were landed, before making for Sierra Leone for condemnation in the Anglo-Portuguese court.

Judge Macleay reported to the Foreign Office in June that, following the demise of General Ricafort, General Tacon had been appointed as Captain-General of Cuba. Although the new incumbent was supposedly a Liberal, appointed by a Liberal government, he was clearly disposed to protect the slave trade to the best of his ability, in compliance with instructions from Madrid. Macleay also reported that Mackenzie, who had earned the disfavour of the Foreign Office, had relinquished the post of Arbitrator and was about to leave Havana for England.

Nimble returned to the fray in August when she captured the Portuguese schooner *Felicidad* off Cape Maisí, the eastern extremity of Cuba, on the eighteenth. Lieutenant Bolton made his arrest, unresisted, at about midnight after a two-hour chase, but was unable to prevent the master and officers escaping ashore by boat with the vessel's papers. The schooner had been French, but on a slaving venture to the Rio Nunez she had been prevented by the French anti-slaving squadron from leaving the river, and she had been sold there to the Portuguese Governor of Bissau, who had loaded her with slaves and despatched her for Santiago de Cuba. The experienced Bolton wasted no time in taking the prize to Havana and, although he wisely informed the British commissioners about the seizure, he proceeded directly for Nassau, where he landed the 162 survivors of the 164 slaves found at capture. He then sent the *Felicidad* to Freetown, where she was condemned.

Having returned to cruise the north-eastern coast of Cuba, *Nimble* achieved another, partial, success on 30 October when, after a three-hour pursuit, she chased ashore a Spanish schooner on Punta de la Vaca, to the west of Cape Lucrecia. The crew escaped ashore with some of the cargo of slaves, but left the sick and injured master with his servant on the beach. *Nimble* embarked these two along with 272 Africans, and sailed on 1 November to embark provisions at Nuevitas. The Spanish master died before arrival, but had revealed that the schooner was the *Carlota*, bound from the Gallinas for Havana with 350 slaves. On the morning of 3 November *Nimble* sailed for Havana with an experienced pilot on board, and Lieutenant Bolton, his vessel grossly overcrowded and short of water, decided to make his passage through the Old Bahama Channel that night.

During the night, however, the schooner found herself battling a north-easterly storm, and the wind, aided by the current, drove her towards the Cayo Verde reef on the southern side of the Channel, where, at 0100 on the morning of 4 November, she struck. She fell over on her beam ends and bilged, with the sea breaking over her. In Bolton's words, "The scene of horror that now presented itself may be imagined but cannot be described." To prevent the schooner from splitting apart immediately, Bolton ordered the foremast to be cut away, but in its fall it carried away the head of the mainmast, killing several Africans. All hands were then employed in building a raft, and, when dawn revealed Cayo Verde half a mile to the north-west, they began the dangerous work of getting women and children into the boats and landing as many as could be carried. When the boats returned, the Africans, in an uncontrollable panic, swamped the jolly boat. The gig rescued as many from the sea as it could carry, but the schooner's best boat was lost. Men swam to the gig to tell her not to approach the wreck, but to make for the mainland for assistance and to return with water and provisions. The raft now became jammed under the wreck, so the stump of the mainmast was cut away and, with other spars, made into a second raft. Only one small boat remained, and, afraid that the Africans would sink her if she came alongside, Bolton ferried his ship's company to her on the raft to be landed on the island, and the boat continued all day to disembark the Africans until darkness and exhaustion brought the work to a halt.

By daylight on 5 November the wind had moderated, and the sea had lifted the wreck inside the reef. The remainder of the Africans were landed in safety, and such provisions as could be recovered, although soaked with seawater, were salvaged. The fresh water had been entirely ruined, and the survivors drank rain-water from holes dug in the ground. The gig returned with water and provisions the following day, other boats soon arrived with supplies, and on the eighth the local Commandant of Marine appeared and remained until the end of the rescue. The Spanish schooner *Amistad* was chartered to evacuate those on the island,

and she arrived on the eleventh. Bolton, his five officers and 33 men had managed to save themselves and 197 of the slaves, but 70 Africans had been killed or drowned in the wreck and three more died before arrival in Havana, two having died earlier in *Nimble*. Carrying the survivors and all stores and equipment that could be salvaged from the wreck, *Amistad* reached Havana on 14 November.

The Mixed Commission emancipated the slaves, and having given *Nimble*'s men lodgings in the dockyard, the Spanish Commander-in-Chief sent them in one of his schooners-of-war to Nassau where they were best placed to take passage to Jamaica, although seven had deserted in Havana. Rather surprisingly, the crew of the *Carlota* and the 73 slaves with whom they had escaped were apprehended by the Cuban authorities and placed at the disposal of the Mixed Commission, but the court, quite properly, declared itself to have no jurisdiction over them. In due course, Bolton, his officers and crew faced a court martial to answer for the loss of their schooner, but all were acquitted. A factor said to have contributed to the wreck was that the uncontrollable noise of the Africans had drowned out the sound of the breakers on the reef. This sad end to the career of a fine little cruiser dealt a blow to the suppression campaign in Cuban waters, and the departure of Charles Bolton, commended by the Governor of the Bahamas as "a most active and intelligent officer", exacerbated the blow.[11]

A few months later Sir George Cockburn's force of invaluable schooners was further depleted by the loss of *Firefly*. On the night of 27 February 1835, while carrying passengers from Belize to Jamaica, she was virtually becalmed and a heavy swell carried her onto the Northern Triangles reef in the Bay of Honduras. Despite every possible effort by Lieutenant McDonell, she became a total loss. After considerable adventures the majority of her people were saved, but the gig was smashed in surf while attempting to reach the mainland and her occupants drowned. McDonell survived and was exonerated by a court martial.

At the turn of the year the British commissioners at Havana made their routine report to the Foreign Secretary, addressing it incorrectly to Palmerston, who had been superseded in November by the Duke of Wellington. By now William Macleay had been reappointed as Judge and Edward Schenley had arrived in Mackenzie's place as Arbitrator. They were obliged to reveal depressing figures. The number of slaver departures from Havana in 1833 had been 38; in 1834 it had risen dramatically to 61, of which two were Portuguese. There had been 33 arrivals during the past year. These statistics did not include the other ports of Cuba, and it was clear, if only from the destinations alleged by captured vessels, that the traffic from Santiago was considerable. The commissioners blamed the cholera epidemic for the heavy demand for replacement slaves, and lamented the poor rate of success of the cruisers and the protection given to the Trade by the Cuban authorities. On this last contributory factor they commented despairingly that never before had there been "so glaring a contempt" for the Anglo-Spanish Treaty.

The early months of 1835 saw some improvement in the capture rate, beginning on 14 January when the ship-sloop *Cruiser* (18), Acting Commander James Baker, fell in with the Spanish schooner *Maria* off Salt Cay on the northern edge of the Old Bahama Channel. After a short chase Baker detained the schooner and found 346 slaves shipped from the Bonny. The prize was placed in quarantine on reaching Havana, and was not condemned until eight days after arrival, by which time four slaves had died.

The second arrest of the year was made by another 16-gun brig-sloop, *Racer*, but she was of a different generation from *Cruizer*, having been launched as recently as July 1833. On 22 January, under the command of Commander James Hope, she captured the Spanish schooner *Julita* close to the island of Tortuga, off the northwest coast of Haiti. The prize was sent ahead of the cruiser to Havana with her cargo of 342 Whydah slaves, and the Captain-General, not wanting the slaves at the island a moment longer than necessary, urged the Spanish commissioners to try the case immediately. The commissioners, of course, insisted on the presence of the captor at the adjudication, and *Julita*'s condemnation had to await *Racer*'s arrival, by which time two slaves had died.

Racer made a second arrest on 12 March off the Anguila Islands on the Cay Sal Bank, where she intercepted the Spanish brigantine *Chubasco*, with 253 slaves from the Rio Pongo. When the prize arrived at Havana she was immediately placed in quarantine because of smallpox on board, but Judge Macleay arranged for the slaves to be landed at an estate to leeward of the city and kept isolated. The prize crew and *Chubasco*'s men remained in the brigantine and were ordered to be isolated for 40 days. *Racer* arrived two days later, but Commander Hope was under orders to proceed to Jamaica without delay and could not wait for the quarantine period to expire. He was allowed to deposit a captor's declaration, and he sailed after taking all but four of his prize crew out of *Chubasco* and consequently hoisting the quarantine flag. When the prize emerged from quarantine on 23 April she was condemned, and all 253 slaves were emancipated.

The old 18-gun ship-sloop *Arachne*, by then in her twenty-seventh year, was on passage from Veracruz and coasting along the north shore of Cuba on 27 March when she fell in with the Spanish polacca schooner *Joven Reyna* about 90 miles to the west of Havana. Commander Burney found on board 254 slaves from the River Congo and Ambriz, and he took his prize into Havana, where she was placed in quarantine. *Arachne* was carrying a large sum of money, and

Burney was anxious to leave, but he had to wait until the slaver was admitted to pratique on 6 April. She was condemned on the following day, and the 254 slaves were emancipated.

It was not until 8 April, about 20 miles north-north-west of Little Cayman, that one of the sadly reduced force of schooners made a first arrest of the year. *Skipjack*, Lieutenant Sydney Ussher, was not cruising against slavers but on passage from Belize to Jamaica when, at 0830, she sighted a brig on her weather bow. Ussher tacked to close her, and she immediately took in her studdingsails and hauled to the wind. *Skipjack* made all sail in chase, and when, at 1100, Ussher hoisted his colours, the chase showed a Spanish ensign and man-of-war pendant and fired two guns to leeward. *Skipjack* continued to close, and at 1130 the brig hauled down ensign and pendant. At 1415 the chase was within gun-shot and Ussher fired a gun ahead to bring her to, but she refused. She shortened sail, however, and *Skipjack* opened fire on her. The brig returned fire with round shot and grape until 1515 when she wore, apparently to rake the schooner. Ussher immediately wore and engaged on the other tack. At 1630 the brig wore again and made all sail to escape. The pursuit recommenced, with continuous fire from *Skipjack*'s long-gun, returned by the brig's stern chasers. At 1950 the wind fell light and *Skipjack* shipped her sweeps, but the breeze returned at 2030 and the exchange of fire continued for another two hours. That was enough for the brig. She rounded to and showed a light, and, as *Skipjack* ran under her stern, she hailed that she had surrendered.

Ussher found that his prize was the 116-foot *Marte* of Barcelona, bound for the Isle of Pines and carrying 442 slaves from Loango, all that remained of the 600 loaded on 25 February. The prize mounted six 18-pounder Congreves* and two long 12-pounders against *Skipjack*'s long 18-pounder and two 18-pounder carronades, and she carried a crew of 56 with an armoury of 50 muskets and 50 cutlasses. *Skipjack* had received damage only to sails and rigging in the action, but the slaver had suffered one seaman and seven slaves killed, and six seamen and eleven slaves wounded. Ussher spent 9 April repairing the brig's damage, and then was obliged to put a prize crew of a mate and 14 men into her, very nearly half his ship's company. He took most of the prisoners into *Skipjack*, but, having insufficient men to guard them and to work the schooner, he made for Grand Cayman to land some of the Spaniards. Then he headed for Jamaica to report to the Senior Officer, now Commodore Pell of *Forte*, and to reinforce his depleted crew before sailing for Havana. *Skipjack* arrived on 29 April to find that the prize, with smallpox and dysentery on board, had been placed in quarantine. It was

---

* These were weapons developed by Sir William Congreve, best described as elongated carronades.

not until 8 May that the slaver was condemned and the surviving 403 Africans emancipated. Pell later told the Commander-in-Chief that the *Marte*, a very fine brig, was of "formidable appearance" and had the "size and determination to fight". In commending Ussher the Admiral gave his opinion that "few instances have occurred where more gallant determination has been shewn".[12]

The British Government still insisted that equal numbers of male and female *emancipados* should be sent to Trinidad, despite the disparity in gender of the slaves arriving in prizes. A solution to this difficulty was offered by the new Governor of the Bahamas as he was passing through Havana. Some of his islands, he said, had an excess of women, and he would welcome some men being sent to him. His suggestion that emancipated Africans should settle in the Bahamas was viewed favourably by the Colonial Office, which also proposed to the Foreign Office in April that the Trinidad arrangement should be extended to Honduras. In late May the Havana commissioners reported to Wellington that cholera had reappeared in the city, and that no more Africans would be sent to Trinidad for the time being. They also, probably unaware of the Colonial Office's proposal on the subject, suggested sending *emancipados* to Honduras.

* * *

On the South America Station, with his base at Rio de Janeiro and his flag in the battleship *Warspite* (76), was the Commander-in-Chief, Rear-Admiral Thomas Baker. At the beginning of 1831, to fulfil his responsibilities on the Atlantic and Pacific coasts of South America, he had, apart from his flagship, a force of five frigates and four sloops. Those responsibilities had not, of course, included the suppression of the Portuguese and Brazilian slave trade while those two nations had been allowed to continue slaving unhindered south of the Equator, but it had seemed at least that action against the Brazilian Trade would fall to him in March 1830 with the ending of the period of grace allowed by the 1826 Treaty. Although the Emperor Dom Pedro had declared in his annual speech from the throne in May 1830 that the Brazilian slave trade would soon be declared illegal, the deadline had come and gone with no new directive to the Commander-in-Chief. Furthermore, the suppression Instructions issued to cruisers in compliance with the 1826 agreement were not replaced with updated versions reflecting the new situation. Even the very small number of cruisers stationed in areas in which they were likely to encounter slavers were unclear about what action they were permitted to take now that the Brazilian Trade was supposedly abolished. The campaign against the slave trade south of the Equator, only one of the

17. Rear-Admiral Sir Thomas Baker, Commander-in-Chief on the South America Station

Commander-in-Chief's tasks, was unlikely to be pursued with vigour in the absence of firm and unambiguous direction from London.

Nevertheless, there had been one slaver arrest off the Brazilian coast. On 11 January 1831, the Mixed Commission court in Rio de Janeiro, for years a sinecure post for its commissioners, was called to try the case of the schooner *Destimida*. The frigate *Druid* (46), Captain G. W. Hamilton, had been lying at anchor on the morning of 2 December 1830, about ten miles south-west of Bahia, when she sighted the schooner closing the coast and then standing offshore again. His suspicions aroused, Hamilton weighed in chase. Having caught what he found to be the badly leaking *Destimida*, and discovered 50 slaves embarked, he arrested her, despite her Portuguese colours. The Mixed Commission decided that she was indeed Portuguese property and, as such, having been trading south of the Line, she was "not included in the Alvara of 1818". She was restored to her owners, and her slaves emancipated and disposed of as "hired servants". The fact that the court referred to the Portuguese Decree of 1818, purely national legislation, rather than the Anglo-Portuguese treaties of 1815 and 1817, illustrates the degree of confusion suffered by suppression efforts south of the Equator at this early stage, but, in any case, the Rio Mixed Commission, solely an Anglo-Brazilian court, had no jurisdiction over Portuguese vessels.[13]

In Brazil the suppression campaign was virtually in a state of limbo, but other affairs were moving apace. Revolution erupted in early 1831 and the pro-Abolition Emperor abdicated, took refuge with the Royal Navy at Rio, and, on 13 April, sailed for Europe in the frigate *Volage* (28), Captain The Right Hon. Lord Colchester, accompanied by the French frigate *La Seine*. This left an infant on the throne and a provisional regency to rule the country. In many provinces foreigners, especially Portuguese, were denounced, and some were murdered. Not only had the abolitionist cause suffered a blow by the deposition of the Emperor, but also there was little chance of a formal declaration of abolition from Brazil in the resulting state of chaos.

Amid this turmoil, Admiral Baker received an appointment as Commander-in-Chief at the Cape of Good Hope. It is not clear whether or not the Admiralty really intended that he should command both the South America and the Cape Stations, but there was no sign of a replacement for him in South America. It defies belief that he might have been expected to shoulder responsibility for such a vast area, and it is more likely that the new appointment was an Admiralty error, but, in any case, as Baker pointed out in early June, with much of South America in a state of warfare or revolution, and with unrest in Rio threatening the safety of British subjects, it was inadvisable that he should leave for the Cape. The appointment was cancelled.

It had been the Commander-in-Chief's intention in the spring of 1831 to visit his command on the west coast of the continent, leaving Captain The Hon. William Waldegrave of *Seringapatam* (46) in charge of the vessels remaining on the east coast. He directed Waldegrave to exercise command from Rio in his absence, and to continue to use it as the base for refits, replenishment and the reception of orders, and as the principal rendezvous for the east coast squadron. As far as slave-trade suppression was concerned, he regretted that shortage of vessels had prevented him from "following up" the "system of interception" he had instigated when prohibition had supposedly begun, although he did not explain his system. He acknowledged that the trade had been "materially repressed" for a time, but he had reason to suppose that it would restart with renewed vigour whenever there was a "chance of it being conducted with impunity". Clearly he misunderstood the reasons for the temporary decline.[14]

The only recent deployment by Baker specifically against the slave trade had been to direct the frigate *Tyne* (28), Captain Charles Hope, to "keep the traffic in check". Apparently finding nothing to attract his attention in a northward cruise along the Brazilian coast, Hope continued north to the Cape Verde Islands, well beyond the boundaries of the South Atlantic Station, seeking information on slaver movements. Arriving at Porta Praya on 5 March, he was astonished to discover that the British Consul, from whom he was expecting a briefing on the state of the Portuguese slave trade in the islands, had abandoned his post the previous May and had handed over his duties to the American Consul. Whether or not this abandonment was authorised is not known, although London was (or later became) aware of it. In any case, the American gentleman showed no interest in assisting the Royal Navy with its enquiries, and Hope returned to the coast of Brazil none the wiser.

Owing to the upheavals in Rio, Baker (now Sir Thomas) delayed his departure for the Pacific coast until the beginning of August, and then delegated authority over the east coast force not to Waldegrave, who was apparently absent, but to Captain Hamilton of *Druid*. By November Hamilton had been superseded as Senior Officer by Captain The Right Hon. Lord James Townshend in the newly arrived Fourth Rate frigate *Dublin* (50). It was Townshend who, at the beginning of that month, despatched Captain Charles Paget in another newcomer, *Samarang* (28), to Bahia with orders to replenish *Tyne* to enable her to remain at Pernambuco. Townshend's intention was to keep a ship at, or in the vicinity of, each of these two slaving ports.

On 6 November, in response to news that a slaver had arrived in the vicinity of Santos, Townshend sent the ship-sloop *Pylades* (18), Commander Edward Blanckley, to the southward in search of her. Blanckley succeeded in detecting the slaver, but she had tucked herself behind an island before *Pylades*'s arrival and had landed her cargo. Then she attempted to escape to seaward, but in her haste to evade the sloop she grounded and became a total loss. On returning to Rio on 23 December, Blanckley reported that there was a considerable trade in smuggled slaves in the neighbourhood of Santos, despite the apparent desire of the local authority and inhabitants to suppress it. He had discovered that slaving vessels would approach the coast, make a private signal and then stand offshore again. Coastal craft would then sail to disembark the slaves and land them at uninhabited islands, thence to be sent to the mainland as convenient.

The Commander-in-Chief returned to Rio on 19 November, and on 31 December he sent *Pylades* to relieve *Tyne* off Pernambuco. She took with her orders for Captain Paget in *Samarang*, appointing him Senior Officer Northern Ports and directing him to employ *Samarang* and *Pylades* to protect British interests and to intercept the African slave trade along the whole coast between Cape St Roque and the Abrolhos Islands, a distance of 800 miles. By now there was general acceptance that loaded Brazilian slavers were liable to seizure south of the Line, but there was clearly adequate opportunity for the slave trade to be "conducted with impunity" along that coast if the slavers should so wish, especially if they took the precaution of keeping clear of the ports of Pernambuco, Olinda and Bahia.

In November 1831 the Brazilian Government enacted a law denying entry to all Africans and requiring re-exportation, at the expense of the shipper, of any found to have arrived clandestinely. It proposed a contract with "African Authorities" to grant returning Africans asylum in order to prevent them being re-enslaved. The notion of asylum was, of course, nonsense, and the legislation flew in the face of the Anglo-Brazilian anti-slaving convention. As the British commissioners in Rio explained to Lord Palmerston in February 1832, the measure, which Britain clearly could not condone, reflected a growing anxiety of the Brazilian Government to prevent an increase in its black population. A subsequent proposal by Brazil in May 1833 that contraband Africans should be returned to Africa and handed over to the British authorities in Sierra Leone received a similar lack of sympathy in London.[15]

In April 1832 Alexander Cunningham, the British Commissary Judge in Rio, died, and the Vice-Consul in Rio, Richard Pennell, assumed the duties of Judge until the permanent replacement, Sir George Jackson, arrived in December. It seemed, for the time being at least, that Jackson's duties, for which he would be drawing a salary of £1,200, were unlikely to be onerous.

The frigate *Thetis* had been wrecked at Cape Frio in 1830, owing, so Admiral Baker had concluded, to extraordinary currents which had not previously been appreciated, and she had been carrying a large cargo of bullion. The recovery of this treasure and other stores had become a distracting task for the force on the east coast of Brazil, involving the use of a diving bell and underwater work at the limits of current technology. Baker had delegated on-site control to Commander T. Dickinson of the sloop *Lightning*, but, although the salvage seems to have progressed satisfactorily, Dickinson had annoyed his Admiral, not least by claiming excessive credit for the success of the operation. The Commander-in-Chief expressed not only his "extreme disapprobation" at contravention of orders by *Lightning*'s Commanding Officer, but also his marked displeasure with a report from that officer in June 1832 which he found to be seriously unbalanced, selective and self-serving, and he withdrew his recommendation for Dickinson's promotion to captain. By that time 694,000 dollars had been brought to the surface, and at the end of July, with most of the treasure and stores recovered, the operation was brought to a close.

Amid the real and potential upheavals in Brazil and elsewhere in South America, Baker's ships were largely concerned with the safety of British people and interests, but Edward Blanckley in *Pylades* still kept an eye open for slavers, and at Bahia on 13 September 1832 he detained the small schooner *Friendship* under British colours. She was empty, but Blanckley suspected that she had been slaving, and he was sure that she had been navigated illegally and armed without a licence. He obtained permission from the President of the Province to take her to Rio for trial and, although it had been the intention that the Mixed Commission should try the case, in the event she was delivered over to the Brazilian authorities. She was subsequently bought by the Brazilian Government and fitted-out as a cruiser.

By this time, Captain Paget in *Samarang* had relinquished his post as Senior Officer Northern Ports and had been sent to the River Plate. *Pylades*, relieved of her prize, headed south to replace *Samarang*, releasing her to make for the Pacific and leaving Commander The Hon. John De Roos in *Algerine* (10) to watch the northern coast. Towards the end of the year, De Roos reported that all was quiet at Bahia and Pernambuco, and that he was concentrating on intercepting the slave trade as well as surveying the entrances of the small harbours on the coast.

In December 1832 Admiral Baker received orders to relinquish command of the Station and, leaving Captain Townshend in temporary charge, he sailed in

18. Rear-Admiral Sir Michael Seymour, Bt, KCB,
Commander-in-Chief on the South America Station

Warspite for Spithead. His replacement was Rear-Admiral Sir Michael Seymour Bt, KCB who arrived at Rio on 11 April 1833 in the Trafalgar veteran *Spartiate* (76), Captain Robert Tait. Seymour, who had lost an arm in *Magnificent* at the Glorious First of June, had made his name in command of *Amethyst* (36) when, in November 1808, after a long and bloody action at close quarters, he took the French *Thetis* (44). The King presented him with the naval gold medal, Lloyd's with a hundred-guinea plate, and Cork and Limerick with the freedom of the cities. The following April he captured *La Niemen* (46) and was made a baronet.

On his arrival the new Commander-in-Chief found a command consisting (his flagship apart) of one Fourth Rate frigate, three Sixth Rates, five sloops, of which *Beagle* was fully employed in surveying, two brigantines and a tender. Although this represented a recent marginal increase, it was barely adequate for the routine tasks on both coasts of South America, let alone the demands of slave trade interdiction.

Seymour also inherited continuing uncertainty on the question of Portuguese slaving south of the Line. In 1831 the Advocate-General had expressed his view that, following the independence of Brazil, Portuguese slaving south of the Equator had, in accordance with the Anglo-Portuguese treaty of 1817, become illegal. Portuguese slavers were therefore liable to seizure in both hemispheres. Not only had the Foreign Office not reflected that opinion in a new directive, but also it contrived in February 1832 to sow further confusion in response to a query from the Admiralty. It wrote that, as there was no legitimate object for which Portuguese citizens could engage in the slave trade, HM Government was justified in calling on the Portuguese to declare the slave trade abolished throughout the Portuguese Dominions and to agree with Britain appropriate alterations to the Additional Convention of 28 July 1817. But, it said, until that had been done, it would not be competent to British cruisers to seize Portuguese slavers under any other circumstances than those specified in the treaty. This entirely, and culpably, failed to answer the crucial question of whether or not British cruisers were now authorised to arrest Portuguese slavers south of the Line. Not satisfied with this obfuscation, the Foreign Office then added that the mode of proceeding against crews of Brazilian slavers was still under consideration by the Law Officers of the Crown.[16]

Shortly after his arrival, Seymour decided to dispose of the flagship's tender, the schooner *Adelaide*, an ex-slaver bought by Baker. She was publicly sold, and on 28 June she cleared from Rio under Portuguese colours. On 23 October she returned in ballast having, undoubtedly, landed a slave cargo.

The new Commander-in-Chief instructed his ships that whenever the ports at which they were stationed were tranquil they were to keep to sea and use their utmost exertions to suppress the slave trade. He also assured the Brazilian Government that he was ready to cooperate in any of its measures to prevent importation of slaves. Brazil was showing increased enthusiasm for taking action itself against the Trade, and one of its cruisers had brought in a slaver in February; although it was said that the officer first offered command of the schooner-of-war in question had refused it because of the odium, and even personal risk, he would incur if he arrested a slaver. There was even a report that a schooner and a brig-of-war were being commissioned to cruise against the Trade on the coast of Africa.[17] Nevertheless, Brazil resented the jurisdiction of the Mixed Commission, and early in 1833 it proposed that the sentences of the Anglo-Brazilian court should undergo revision by being submitted to the arbitration of a third power. Palmerston predictably rejected the idea out of hand, reminding the Brazilian Government that its treaty with Britain specifically prohibited any appeal.

At about the same time, Lord Palmerston returned to the charge on the subject of an Equipment Clause, proposing that an appropriate Additional Article should be inserted in the 1826 Convention. The Brazilian Minister for Foreign Affairs replied in July that his Government was generally in favour, but that the populace tended to want to defeat the good intentions of the Government, and it was likely that the parliament would reject the measure. A proposal that condemned slavers should be broken-up received no reply, but the British Ambassador, Mr Ouseley, warned Palmerston not to expect agreement. The Brazilian Minister had remarked that "wanton" destruction of property which might be turned to useful purpose would meet much resistance, and it was probable that a counterproposal would be made to the effect that the vessels should be bought by one or other of the two governments and armed and commissioned as cruisers. Ouseley also reported that the Brazilian Government was proposing to the General Legislative Assembly that "a sufficient number" of small vessels should be armed to form "a sort of *cordon sanitaire*" on the coast. Two schooners had already been bought and fitted-out for slave-trade suppression, but there was no indication of what "a sufficient number" was considered to be.

In May 1833 Admiral Seymour reported that he understood that there were several vessels belonging to inhabitants of the Argentine Republic and the Banda Oriental currently on the coast of Africa, and he enquired how he should deal with them if they were encountered.* He was told, in circuitous terms, that, as Britain had no treaty with either of the nations he mentioned, his cruisers were not entitled to detain slavers under their flags. The Consul-General in Montevideo, Mr Hoad, revealed a similar misunderstanding when he reported to Lord Palmerston that vessels were being fitted-out at that port for slaving, and that, in an interview with the Minister for Foreign Affairs, he had been told that clearance had been given for importation of 2,000 African "colonists". Hoad had remonstrated with the Minister and suggested that Montevideo vessels found by the Royal Navy to be slaving would be liable to arrest. He earned only reproof from Palmerston for his pains, and it was pointed out to him that, as Britain had no treaty with Montevideo, the Royal Navy had no right to intercept vessels under its flag, let alone bring them to trial.[18]

Before receiving via the Admiralty the Foreign Office reply about Argentine slavers, Seymour, having learned that in 1824 the Buenos Aires Government had declared the slave trade to be piracy, told the British Ambassador in Rio that he supposed that the Buenos Aires flag should not be permitted to be used as a shield for illegal traffic. He therefore sent the newly arrived brig-sloop *Snake* (16), Commander William Robertson, to cruise off the River Plate, and ordered her to seize slavers under the Buenos Aires flag. The Commander-in-Chief presumably withdrew her on receipt of the Admiralty's letter, and by mid-October she was cruising to the south-east of Cape Frio. It is clear that Seymour, despite the imprecision of instructions from London on the subject, was placing increased emphasis on disrupting the slave trade. He reported to the Admiralty in December that "no opportunity was lost of sending every disposable ship on that service", but that "the great extent of the Coast of Brazil, along the whole of which the Trade is carried on, renders our attempts comparatively insignificant."

Nevertheless, there had been a modicum of encouragement. The ship-sloop *Satellite* (18), Commander Robert Smart, was lying just off Rio on the morning of 15 November. The weather was foggy but clearing when she sighted floating in the sea a number of buckets and other utensils probably thrown overboard from a slaver. She shaped her course to the southward, and soon came across a brig in the act of ditching her cooking coppers. Smart immediately attempted to recover the coppers, but they eluded the grasp of his boat's crew. Nevertheless, it was apparent that the brig, the *Paqueta do Sul* under Portuguese colours, had very recently landed a cargo of slaves, and Smart arrested her on the grounds of

---

* The Banda Oriental became Uruguay.

the March 1823 Additional Article to the Anglo-Portuguese Treaty.* A prize crew took her into Rio, where the Mixed Commission accepted that she had indeed carried a slave cargo and was partially Brazilian owned; indeed, she had previously made five slaving voyages under Brazilian colours, and it "partially condemned" her for the proportion of the value that the Brazilian part-owner had in her.

To Seymour, who in *Satellite*'s absence had represented the captor himself, this was "a most extraordinary sentence", and there had been a sequel to the arrest which had already caused him disgust. Showing not only persistence but also an extraordinary personal involvement in the case, he had discovered the whereabouts of the *Paqueta do Sul*'s slaves and had pointed out to the Brazilian authorities the building in which they were being held. However, the subsequent search by the authorities was then botched, deliberately so in Seymour's opinion.

There was one final arrest in 1833, and it gave rise to one of the more shocking episodes of the entire campaign. On 25 November, *Snake* seized the 300-ton barque *Maria da Gloria* about 70 miles south of Rio. The barque was bound from St Paul de Loando to Rio with a cargo of 423 slaves, but she was under Portuguese colours. Nevertheless, she was brought before the Mixed Commission in the hope of proving Brazilian ownership. The Commander-in-Chief again represented the captor, and he pointed out that the vessel was Brazilian-built, had never left Brazil until the voyage in question, had sailed from Rio and was returning there (by the master's admission) intending to land her slaves clandestinely on the Brazilian coast, and that her owner had long been resident in Brazil and was a notorious slave dealer. His forceful argument that the prize was Brazilian rather than Portuguese, and therefore should be condemned, failed to persuade the court, which concluded that the owner was Portuguese. Consequently it declared itself incompetent to judge the case. The easy course of action then would have been to hand over vessel and slaves to the Brazilian Government, but Seymour was determined to bring the barque to condemnation if he could, and, after consulting the British Ambassador, he opted to send her to Sierra Leone.[19]

Having been surveyed and stored, the *Maria da Gloria* sailed on 4 January 1834 under the command of Lieutenant The Hon. Joseph Denman of *Snake* with a midshipman, an Assistant Surgeon, 11 men and a boy as prize crew.† By then her cargo had been reduced by sickness to 390. She arrived at Freetown on 19

---

* This was the Article stating that if there was clear and undeniable proof that a vessel had been carrying slaves during the voyage on which she was currently engaged she could, although empty of slaves at the time of capture, still be condemned.

† *Snake* would not see her prize crew again for a very long time, and Seymour filled the vacancies from *Spartiate*.

February, having lost 55 slaves during the 46-day passage, despite, as Denman put it, the "zealous and unremitting care of the Assistant-Surgeon" and the "most anxious attention" given by the prize crew "to every circumstance which could possibly tend to mitigate suffering". Adjudication by the Anglo-Portuguese Commission then took over three weeks, and, as Denman had feared, the vessel was restored, although without compensation. The judgment was based on the existing Anglo-Portuguese Treaty under which, the court believed, detention of Portuguese slavers was permitted only north of the Equator, and the commissioners quoted the Instructions to cruisers authorised by that treaty: "No Portuguese merchant-man or slave-ship shall, on any pretext whatsoever, be detained, which shall be found anywhere on the high seas south of the equator, unless after a chase which shall have commenced north of the equator."*

It would appear that neither the Ambassador in Rio, Admiral Seymour, nor Commander Robertson in *Snake* was aware that this unequivocal directive remained extant, or else the prize would not have been sent to Freetown. By the terms of the 1817 Treaty the constraint on southern hemisphere arrests of Portuguese vessels slaving to Brazil had been rendered obsolete by the independence of Brazil, but the Foreign Office, instead of removing residual uncertainty, had contrived to generate confusion. The Admiralty was equally blameworthy for its failure to extract clarification from the Foreign Office and issue up-to-date and unambiguous instructions to its cruisers. Both departments bore a heavy responsibility for the unfolding tragedy of the *Maria da Gloria*.

By the time that the barque was released into the hands of her master, 309 slaves remained alive. Before she was allowed to sail again for Brazil, however, the compassionate Lieutenant-Governor Temple intervened to order that 64 of the sick should be landed and freed, and it seems that she finally left Freetown on 11 April. A Brazilian corvette found her again on the coast of Brazil on 13 May, making signals off Bahia. By the time the slaves were landed on 23 May, the 430 loaded at St Paul de Loando on 26 October had been reduced to 225, and that would not have been the end of the death toll.† It is perhaps surprising that as many as that survived seven months and three Atlantic crossings in the hold of a slaver. The Brazilian naval officer who boarded the barque on 13 May later wrote that "such a spectacle of misery can hardly be conceived as the 230 living skeletons presented".

---

* It is difficult to comprehend how this case took so long to reach a conclusion, but the court took satisfaction in having finished within four weeks. Its report to the Foreign Office is contained in 98 pages of verbiage.
† It is not known how many of the 64 slaves landed in Sierra Leone died as a result of their privations.

Lieutenant Denman, meanwhile, had achieved the first stage of a challenging journey from Freetown to rejoin the *Snake* by agreeing with the owner of a recently purchased condemned slaver brig to take the new acquisition to Liverpool. He and his crew arrived in England on 4 May having lost one marine on the passage. His experience in the *Maria da Gloria* had a profound effect on Denman, colouring his attitude to the slave trade and strongly influencing his development into one of the leading figures of the suppression campaign.

The Commander-in-Chief had lamented at the beginning of 1834 that the Trade was much increased, that the measures adopted to screen the participants were so notorious that it was impossible to believe that the Brazilian Government was ignorant of the facts, and that the ease with which Brazilian vessels could be nominally sold to Portuguese owners had put almost the entire Brazilian slave trade under the Portuguese flag. Shortly afterwards, however, the Rio Mixed Commission noted that when the slavers heard that the *Maria da Gloria* had been despatched to Freetown many Portuguese-flagged vessels switched to Brazilian colours, judging that Brazil offered many means of escaping punishment whereas there was "certainty of irremediable loss and condemnation" if taken to Sierra Leone. News of *Maria da Gloria*'s restoration naturally led to a reversal of this flow.

In October 1834, Lord Palmerston criticised the Rio Mixed Commission for not trying the case of the *Maria da Gloria*. The owner, although possibly Portuguese by birth, was, he wrote, resident at Rio and carrying on his business in that city. He asserted that, by the Law of Nations, the national character of a merchant was taken from the place of his residence and his mercantile establishment, and not from the place of his birth. It was a pity that this crucial criterion had not been established for the Mixed Commissions many years earlier. He also informed the Commission that he was proposing to the Portuguese and Brazilians the annexation of Additional Articles to the existing treaties which "experience has shown necessary to prevent the citizens of either nation to engage in the slave trade".[20]

The Foreign Office went on to suggest that the Commander-in-Chief should be directed to modify instructions to cruisers, ordering them not to detain vessels under Portuguese colours "except they should present the clearest evidence of that Flag being fraudulently assumed". This smacks of hand-washing, and an attempt to shift blame for the *Maria da Gloria* debacle onto Seymour. Instructions to cruisers stemming from the suppression treaties were the responsibility of the Foreign Office and the Admiralty, and not even the Commander-in-Chief had any discretion in their wording. In any case, when *Maria da Gloria* had returned to Brazil five months earlier, an angry Admiral Seymour had come to a conclusion identical in content to the Foreign Office suggestion. Even if that prosecution had been successful, the South America squadron would have found it impractical to sustain a flow of Portuguese prizes to Freetown because of the inevitable prolonged absence of prize crews.

The Portuguese question was not the only source of exasperation for Seymour. He felt that he could place no reliance on the Brazilian authorities in any matter connected with the slave trade, and complained of "extraordinary delays" that had taken place "which amount almost to denial of justice". The particular delay concerning him was that of the *Paqueta do Sul*. Two months after her trial she was still lying in the harbour, rapidly decaying, and there was no sign that the sentence of the Mixed Commission was being carried into execution. The Commission had explained that it had no authority to interfere with the sale which, it said, was the responsibility of the Municipal Judge. In fact, under the terms of the suppression treaties, the sale of condemned slavers was a task for the Mixed Commissions, and that function had long been performed most efficiently by the Freetown courts.[21]

This was not the only example of the reluctance of the British commissioners in Rio to assert their authority. When, in June 1834, a Brazilian cruiser brought in the small slaver *Dois de Marco*, of disputed nationality, the British commissioners initially accepted the intention of the Brazilians to proceed against her in the national court under Brazilian law, in contravention of the Anglo-Brazilian Treaty, an intention supported by the Brazilian Commissary Judge. Sir George Jackson, the British Judge, made a belated representation to the British Ambassador, who then extracted from the Brazilian Department of Justice an instruction that the prizes of Brazilian cruisers, including the *Dois de Marco*, should be adjudged by the Mixed Commission.

Commander Robert Smart in *Satellite* took a chance on 15 June with another arrest of a Portuguese-flagged slaver. At daybreak he sighted a schooner heading in towards Isla Grande, but she, on seeing the sloop, hauled off and made a run for it. After a chase of 30 miles or so, the schooner lost her foreyard and *Satellite* was able to bring her to. She was found to be the *Duquesa de Braganza*, which had cleared from Rio under Brazilian colours on 23 January ostensibly for Montevideo, but she was carrying 270 slaves from Loando. While on the coast of Africa she had adopted Portuguese colours, but the Mixed Commission discovered that she had no bill of sale to support a change of ownership and nationality, and she was condemned as Brazilian. Furthermore, the Brazilian authorities stated their intention to try the master, owner and pilot in the national courts on a charge of piracy.

Sir Michael Seymour was a sick man, worn down by the climate and worry about the slave trade. In April he was taken ashore with a fever to the house of the British Consul, Mr Hesketh, and, although his state of health fluctuated, there was a general decline over the following months. On 5 July he was able to make his report to the Admiralty on the *Duquesa de Braganza* case, but four days later, at the age of 65, he died. It is evident that he was held in high regard and affection by his officers and men, and, despite the paucity of his resources, the breadth of his Station's responsibilities, and the frustrations and obstacles he had met in countering the slave trade, he had shown a deep personal commitment to the suppression campaign.

Sir George Jackson introduced a sour and unedifying note into arrangements for the Admiral's funeral at Rio. Apparently holding an inflated notion of his own status as Judge in the Mixed Commission, he refused the invitation from the Flag Captain to the ceremony in protest at the position he had been allocated in the cortege. The embarrassed Commissioner of Arbitration, Frederic Grigg, feeling obliged to follow the lead of his superior, declined the official invitation but indicated his intention to attend in a private capacity. He assured Captain Tait that no disrespect was intended either to Admiral Seymour or to the Naval Service.[22]

Appointed to replace Seymour was Rear-Admiral Sir Graham Eden Hamond KCB, who had been a 14-year-old midshipman in *Queen Charlotte* at the Glorious First of June, had been made Post just before his nineteenth birthday, and had been commended by Nelson for his conduct in command of *Blanche* (36) at Copenhagen. Off the south coast of Portugal in October 1804, while in command of *Lively* (38), he was involved in the capture of three Spanish frigates laden with treasure, and the destruction of a fourth. Two months later he took a further Spanish treasure ship off Cape St Vincent and, in company with *Polyphemus* (64), captured a 36-gun frigate laden with specie and valuable merchandise. These prizes were disposed of as Droits of Admiralty and the captors were awarded only a small proportion of the proceeds. Furthermore, when Hamond carried home in *Lively* the enormous sum of five million dollars accrued from these captures he found that the arrangements for freight money had been temporarily suspended, and he received no remuneration.

The new Commander-in-Chief reached Rio just two days before the arrival of the one further prize of the year, the product of a most unusual arrest. She was the brig *Rio de la Plata*, flying Montevideo colours, encountered by pure chance in mid-Atlantic on 28 November by the ship-sloop *Raleigh* (18), Commander Michael Quin. The brig was carrying 521 slaves from Benguela and Loando, supposedly for Montevideo, but Quin claimed that she was Brazilian. The brig was despatched on the 1,500-mile passage to Rio, while *Raleigh* continued on her way to the Cape, but Quin, with astonishing neglect, sent away the prize crew with only the clothes they were wearing and without bags or bedding. The only member of the crew not to become sick was the Prize Master, Mr Julian, Mate, whose conduct earned Admiral Hamond's praise. *Raleigh* was on passage to the East Indies Station, and it seems most unlikely that she was carrying the requisite Instructions authorising the detention of slavers, but no doubts on that score were raised by the Mixed Commission.

Pending adjudication, the Brazilian Government gave permission for the surviving slaves, the majority of them sick, to be landed at Rio under the care of the *Raleigh* prize crew and a party from the gun-brig *Rapid*: two mates and eight men all told. On 6 January, the Sixth Rate *North Star* arrived, and, on the following day, her Commanding Officer, Captain Octavius Harcourt, visited the slaves. He found the sailors vigilant, but a guard promised by the Brazilians had not materialised, and that night the refuge was attacked by an armed gang. One marine was severely wounded and 189 slaves were kidnapped. Following Harcourt's remonstrations the next day, a small guard was provided by the Brazilian authorities, and Harcourt stationed a guard boat in support.

It was no surprise that the nationality of the *Rio de la Plata* was a matter of disagreement in the Mixed Commission. The Brazilian Judge took the view that the vessel was sailing legitimately under the Montevideo flag, carrying the requisite papers, and authorised by the Government of the Oriental Republic of Uruguay to introduce Africans as "colonists", an activity breaking no law and which no other government had a right to impede. Jackson argued that the offence was committed by a Brazilian subject, and the term "colonists" was merely a subterfuge. The lot fell to Grigg, who agreed with his Judge, and the slaver and remaining 221 slaves were condemned. Since capture 111 of them had died. The Brazilian Judge later remarked wryly that if the nationality of the vessel's owner had been judged from his place of residence, as decreed by the Foreign Office in the case of the *Maria da Gloria*, then the *Rio de la Plata* would have been released.

During February and March 1835 two slavers were taken into Rio by Brazilian cruisers. Both had been taken under Portuguese colours but were claimed by the captors to be Brazilian. The Brazilian Judge was replaced by the Arbitrator M. Pereira de Souza during the first case, and he disagreed with both his predecessor and Jackson. Grigg was called upon to arbitrate, and he supported de Souza rather than Jackson. The slaver was consequently released to the Brazilian Government.

The second case produced the same result. The Royal Navy was not alone in its frustrations at the hands of the Mixed Commission.

The British commissioners were by no means unsympathetic. In their March 1835 report to the Foreign Secretary, addressed to Wellington but received by Palmerston, they deplored the constantly increasing importation of slaves into Brazil and emphasised the necessity of obtaining from Portugal an absolute renunciation of the slave trade, without limitation of latitude, and of arranging the means of jurisdiction of Portuguese prizes by means either of an Anglo-Portuguese Mixed Commission at Rio or of a separate court. In 1834, they wrote, 32 vessels had sailed from Rio under the Portuguese flag, in all probability owned wholly or partially by Brazilians. The same applied to vessels under the colours of Argentina and the "Montevideo Republics". They also complained of the venality of the Brazilian local authorities on the coast and regretted the Government's lack of physical power to stop the Trade.[23]

* * *

It was at the market end of the transatlantic slave trade that the ineffectiveness of the Royal Navy's efforts was most apparent. Twentieth-century research by David Eltis shows that imports by Brazil increased from 3,500 in 1831 to 21,500 in 1834, and into Cuba from 16,100 in 1831, through a drop to 13,600 in 1832 and 13,800 in 1833, back to 16,700 in 1834. These figures were not accurately known at the time, but the trend was clear, and, dispiriting though it was, much worse was to come. By mid-1835 it was obvious from the mounting number of slaver sailings in 1834 and early in 1835 that a massive increase in the traffic was taking place. So it transpired. Subsequent analysis shows that in 1835 Brazilian imports almost doubled to 40,900 and the figure for Cuba leapt to 25,700. Reaching the United States, Puerto Rico and Montevideo were a further 9,300, a number which included slaves shipwrecked in the Bahamas and those released in the British Americas from slavers captured by the West Indies cruisers. This proved to be the high-water mark for the Cuban Trade, but the figures for Brazil would continue to rise.

The reasons for the increase in the Brazil slave trade seem more complex than those influencing the rise in Cuban imports. The demand of an expanding economy was the primary factor, particularly in coffee production, and demand was exacerbated by the general policy of Brazilian plantation owners to work their slaves to death. Added to that was improved efficiency in the traffic; the Brazilian Government might condemn the Trade, but governors, magistrates and other minor officials oiled its wheels. Furthermore, there was now a realisation that the Royal Navy posed little threat to slavers approaching the Brazilian coast, and virtually none to vessels flying Portuguese colours. As far as Cuba was concerned, the situation was more clear-cut. First, the large numbers of slaves killed by recent outbreaks of cholera needed to be replaced, and, second, the slave merchants believed that time might be running out for them. As the Governor of the Bahamas told the Colonial Office at the end of March 1835, the people of Cuba and Puerto Rico were "making extraordinary efforts" to procure slaves in expectation of a further Anglo-Spanish treaty.

Diplomatic negotiations continued, but otherwise there had been no development of the British strategy of slaver interception at sea as the sole means of stopping the Trade. Apart from the obvious difficulties presented by large sea areas, long coastlines and paucity of resources, the commanders-in-chief in the western Atlantic were hamstrung in pursuing this strategy by glaring inadequacies in the anti-slaving treaties. In Cuban waters the continuing absence of an Equipment Clause in the Anglo-Spanish Treaty, combined with the determination of the Cuban authorities to evade Spain's treaty obligations, was seen as the paramount obstacle.

Off the shores of Brazil the lack of an Equipment Clause in the Anglo-Brazilian and Anglo-Portuguese treaties was a less pressing concern. Here the primary difficulty was the availability, south of the Equator, of Portuguese colours and papers as a means of evasion, and it seems extraordinary that Britain had allowed this gaping loophole to remain open. The 1817 Additional Convention to the 1815 Anglo-Portuguese Treaty had made it quite clear that that Portuguese slaving to any territory other than a Portuguese dominion was illegal. By no stretch of the imagination could Brazil still be considered a dominion of Portugal, and in 1831 the Advocate-General had expressed his opinion that, following the independence of Brazil, Portuguese slavers were liable to seizure in both hemispheres. The British Government had failed to endorse that legal opinion and declined to enforce the treaty, to the great detriment of the suppression campaign. That this inaction should have been allowed under the auspices of a Foreign Secretary as forceful as Lord Palmerston is particularly puzzling. It can only be assumed that the Foreign Office was hoping that, given gentle encouragement, Portugal would honour its commitment to abolish the slave trade of its own accord, and that seizures off Brazil would defeat this objective. If so, there was little indication that this hope was other than forlorn.

As it was, the efforts of the commanders-in-chief on the South American Station to interdict the slave traffic had met with virtually total failure. Although the landing areas were concentrated around Santos, Rio de Janeiro, Bahia, Pernambuco and Maranhão, the scope for slaving vessels to avoid the attentions of the tiny number of cruisers available was immense. A total of 14 or 15 men-of-war to attend to British interests on both sides of South America was barely adequate even without the requirements of slave-trade prevention. Not only was it necessary for far more cruisers to be dedicated to that task, but also they needed to be types of vessel suited to the operation: schooners, brigantines or modern sloops. In early 1835 the South America squadron possessed only four vessels in this category.

The Commander-in-Chief on the West Indies and North America Station was scarcely better placed. There were innumerable landing places available to the Cuba slavers on the island's shores, and a very large number of small men-of-war would have been necessary to cover an adequate proportion of the coast and the island's approaches. Moreover, the Station included much of the north coast of South America, another region of interest to the slavers. The schooners had performed creditably in the hazardous waters around Cuba, but with the wrecking of *Monkey*, *Nimble* and *Firefly*, and the sale of *Minx*, the number of schooners on the Station had been reduced to three: *Pincher*, until recently employed as a tender to the flagship, *Pickle* and *Skipjack*. Of Admiral Cockburn's other 22 vessels, only two were modern Symondite sloops.*

The centre of gravity of the slave trade on the western side of the Atlantic, at least as far as it concerned the Royal Navy, had shifted to the southward, driven by market forces rather than any action by Britain. The Cuban traffic had been no more than marginally inconvenienced by the naval action, and the Navy's efforts off Brazil had been even more ineffectual. The suppression campaign was at a low ebb in the western Atlantic, and, unless major diplomatic advances were forthcoming and the numbers and quality of cruisers markedly improved, the campaign would be condemned to a continuing state of drift.

---

* Vessels designed under the reign of Captain (later Sir William) Symonds as Surveyor of the Navy were known as Symondites, or Symondite (see Chapter 13).

CHAPTER 13

The Spanish Equipment Clause: Eastern Seas, 1835–8

*It was that fatal and perfidious bark
Built in th'eclipse, and rigged with curses dark,
That sunk so low that sacred head of thine.*

JOHN MILTON

THE WEST AFRICA CRUISERS had achieved a steady attrition of the slave traffic in the Eastern Atlantic since the beginning of 1830, taking 47 Spanish prizes and 8 Portuguese, but their endeavours had imposed no perceptible constraint on the flow of slaves reaching the markets in the Americas. Indeed the influx to Cuba had, after a slight dip in 1832 and 1833, risen to a new high in 1834, and showed signs of an even greater rise in 1835. The captures represented only a tiny proportion of the number of slaving voyages, and although it was now free to attack the Brazilian trade south of the Equator, the squadron on the Cape Station had made no inroads at all on the traffic leaving Loango, the Congo, Angola and Benguela. For all the determination and skill of the men at sea, the campaign, which had seen neither new initiative nor beneficial innovation for many years, was meandering hopelessly. Diplomatic, strategic and material change and improvement were desperately needed if it was to remain Britain's intention to suppress the Trade.

Some improvement was already in the wind on the material front at least, in terms of quality if not in numbers. There was widespread belief in the Navy that the design of its ships was inferior to that of other navies, and in its quest for improvement the Admiralty had initiated in 1821 a long series of competitive trials between vessels, generally sloops and small frigates, by a number of designers. *Pylades* by Seppings, the Surveyor of the Navy, *Orestes* by the School of Naval Architecture under the direction of Professor Inman, and *Champion* by Captain Hayes had performed well in the 1824–5 trials, but little of direct value had been learned. However, a new name in the field, that of Commander William Symonds, was gaining a high reputation. In the early 1820s his designs for several yachts had earned the admiration of some influential customers who had drawn the attention of Lord Melville, First Lord of the Admiralty, to his work. Melville consequently invited Symonds to design the 18-gun sloop *Columbine*, which competed successfully in the 1828 trials, was praised by the *United Service Journal* as "one of the prettiest and fastest vessels of her size that ever swam", and was highly regarded by her officers and men.*

Symonds's reputation was further enhanced when the Duke of Portland, who much admired *Columbine*, commissioned him to design the brig *Pantaloon* with the intention of offering her to the Admiralty "at a moderate price, provided, on trial, it beat a great majority (seven out of ten) of their own vessels of the same class, which might be tried with it". Trials were decisive, and *Pantaloon* was purchased by the Admiralty in December 1831, a few months after the new First Lord, Sir James Graham, had asked Symonds to design the 50-gun frigate *Vernon*.† In 1832 Graham not only closed the School of Naval Architecture but also appointed Captain Symonds as Surveyor of the Navy, the first non-professional officer to hold the post,‡ two decisions which caused long and bitter controversy.

Symonds's formula for building fast vessels consisted of a broad beam, maximum beam some way above the waterline, and a wedge-shaped (rather than rounded) underwater form. Symonds described it as "great beam and extraordinary sharpness". This produced vessels which, although generally fast, rolled with a jerky and uneasy motion which rendered them poor gun-platforms. As far as cruising against slavers was concerned, this fault was of little consequence when set against the essential advantage of speed, but, by 1835, no Symondite had been allowed to show her paces on the west coast of Africa. However, the 16-gun brig-sloops *Racer* and *Snake*, belonging to what was probably the most useful all-round cruiser class designed by Symonds, had achieved success against slavers on the Brazilian coast. It would not be long before *Columbine*, *Pantaloon* and their Symondite successors would have their chance on the coast of Africa.

Symonds, however, was not alone in designing fast yachts for eminent members of the Royal Yacht Club, and initially his closest competitor was Joseph White of Cowes. White was commissioned by the Earl of Belfast to design and build a brig which would be the superior of *Pantaloon* in speed and in all the qualities appropriate to a man-of-war, and he produced the beautiful *Waterwitch*. In 1834 she was bought into the Navy, and two years later joined the Cape Station shortly in advance of Symonds's first two vessels.

Unbeknown to those at sea, or in the Courts of Mixed Commission, there had also been valuable progress in diplomatic negotiations with Spain, negotiations

---

* *Columbine* was initially rigged, most unusually, as a barque.
† *Vernon* was so named, not after Admiral Vernon, but, at Symonds's behest, in gratitude to Lord George Vernon, the member of the Royal Yacht Club who had been primarily responsible for persuading Melville to commission Symonds to design *Columbine*.
‡ That is, not a Master Shipwright.

which were on the brink of bearing fruit long sought by the officers of the West Coast cruisers and the British commissioners in Freetown and Havana.

Of strategic change there was still no sign.

* * *

As the Sierra Leone commissioners had observed with satisfaction in their report to the Foreign Secretary at the start of 1835, the French slave trade, so active at the beginning of the decade, had all but disappeared from West Africa. Credit for this was due to the French squadron on the Coast, and, they believed, to a "dread of punishment" among the French slavers.[1] Unfortunately, there was no cause for a similar fear of punishment for the Portuguese, who gratefully replaced the French.

The commissioners' opinion was that the attention Rear-Admiral Sir Patrick Campbell's cruisers were continuing to give to the Bights was inducing the Portuguese and the expanding Spanish trade to resort to the southern ports and to the rivers between Sierra Leone and the Gambia. Once again they complained that the northern rivers were seldom visited by men-of-war, although, as they failed to acknowledge, Campbell was keeping a guardship at the Gambia. Their assessment was undoubtedly fairly accurate as far as it went, but the slave trade was still very considerable in the Bights, and the cruisers were as thinly spread as was possible without their becoming entirely ineffective. Although suppression of slaving and support of British commerce on the west coast of Africa were his primary concerns, the Commander-in-Chief's responsibilities extended eastward into the Indian Ocean as far as Mauritius. His force to cover this enormous area amounted in early 1835 to one frigate, three sloops, six brigs, three brigantines and a schooner.

It was one of the extraneous tasks liable to distract cruisers from their anti-slaving duties which fell to the ageing ten-gun brig *Britomart* in February 1835. The brigantine *Griffon* had called at Freetown before sailing for the Gambia on 5 January, and had delivered to Lieutenant William Quin of *Britomart* the Commander-in-Chief's orders to sail for the Bights and place himself under the orders of Commander Meredith of *Pelorus*, currently Senior Officer on the slave coast. Before leaving Sierra Leone on 4 January, Quin received a request from the Acting Governor to take despatches to Mr President Maclean at Cape Coast Castle, but on his arrival at Cape Coast he was urged to go to the immediate assistance of Maclean, who was beleaguered at Axim with the remains of a force defeated by the King of Apollonia. This monarch had been plundering native British subjects and the vessels landing at his town, and was guilty of "other cruelties and villainies", but Maclean's punitive expedition from Axim to Apollonia had come to grief.

Britomart reached Axim two days after leaving Cape Coast, and on 1 February, with her support at hand, Maclean recommenced his march along the beach with the 400 men he had gathered. He was then attacked from the bush by an overwhelming force, and most of his men deserted, leaving him with only 112. Lieutenant Quin, however, in a freshening breeze, was able to take his brig close enough to the shore to cover Maclean with his guns, and he cleared the beach of the advancing enemy. Many attacks were similarly repelled during the day as the march progressed. When Maclean camped for the night, *Britomart* anchored close inshore and, the deserters having relieved the force of its ammunition, provisions and water, Quin resupplied it with everything it needed, including 3,500 rounds of ball cartridge. This was no mean feat, the surf being very severe and the deserters having removed all but one of Maclean's cargo canoes. This sole remaining communication with the beach capsized several times during the night, necessitating repairs which further interrupted the work.

When the march restarted the following morning Quin put his cutter into the water with a party of marines to give Maclean musketry support, and he kept *Britomart* close to the surf in order to command the shore. At 1000 he saw several hundred of the enemy gathering at the place of Maclean's earlier defeat, so he landed all his marines and a small-arms party of seamen, and he opened fire with his starboard broadside. The enemy fled, and Maclean was able to "march triumphantly" to the British fort at Apollonia, having burned 17 towns on the way. The brig took Maclean's 15 wounded into the care of Assistant Surgeon Donovan, and remained off Apollonia to enable Maclean to enforce his demands on the King. She then headed for Cape Coast to replenish her stores, to Accra for ball cartridge, and to Princes to join Commander Meredith. This fine piece of work had avoided a defeat which would have caused great damage to British trade on that coast, and Quin, not surprisingly, received effusive thanks from President Maclean, the Council and merchants of Cape Coast, and the merchants of Accra.

Buzzard, meanwhile, was achieving a more conventional success. At 1940 on 2 February, after a chase of nearly four hours, she took the Spanish schooner *Iberia* about 30 miles off the Bonny. The slaver was hit by a round shot, but she surrendered only when Lieutenant Milward's men were in the act of boarding. On board were 313 slaves loaded earlier that day. As the Spanish prisoners, many of them drunk, were being removed to *Buzzard*, the Spaniards suddenly shifted to one side of the gig and capsized her. Two were drowned, and a search for the boat failed in the darkness. The *Formidable* prize crew had not returned to *Buzzard* and, already four men below complement, Milward was very short-handed. He therefore escorted his prize to Freetown, arriving on 3 March. The schooner was condemned, and 305 slaves were emancipated.

Three weeks later, *Forester*, which had been on the Station since late 1833, at last made her first arrest, and, as if to compensate for her long famine, she and *Buzzard* were the only two vessels to register captures during the ensuing five months. This maiden success for Lieutenant George Gover Miall and his brigantine followed a sighting at 0600 on 23 February and a seven-hour chase which culminated about 160 miles south-west of Cape Formoso. The prize was the Spanish brigantine *El Manuel*, five days out of the Bonny with 387 slaves. Her master had been in command of *La Pautica*, taken by *Fair Rosamond* in April 1834. The surviving 375 slaves were emancipated at Freetown, and *El Manuel* was condemned.

Forester's next victim was the *Legitimo Africano*, a Portuguese schooner of only 45 tons but carrying 200 slaves, shipped by the notorious Da Souza at Whydah. She made no attempt to escape, and Miall boarded her a few miles to the south-west of Cape St Paul's on 20 March. The master made the novel assertion that his Africans were not slaves but were instead going to Bahia for education. The 186 who arrived safely at Freetown were, nevertheless, emancipated. Ten were unaccounted for, perhaps because of a miscount at capture, and one had fallen overboard. In condemning this diminutive transatlantic slaver, the British commissioners commended the Prize Master for bringing his extremely crowded command through a 42-day passage in the tornado season.

Rapidly evening the score for the year, Lieutenant Milward, still in temporary command of *Buzzard*, encountered on 28 March the Spanish schooner *Bienvenida*, with 430 slaves. *Buzzard*, returning to her station from Sierra Leone, made her arrest of the American-built slaver about 420 miles to the west of Princes where the Spaniards, who were 13 days out of the River Gabon for Havana, might reasonably have felt that they had made their escape. The schooner was carrying, in addition to her slaves, 25 passengers, crews of condemned vessels, whose return to Cuba was thus interrupted for a second time. Before her condemnation, the *Bienvenida* lost 63 of her slaves, 18 of them after landing.*

Meanwhile, a number of changes had been taking place in the Squadron. Commander Booth of *Trinculo* had been granted permission by the Commander-in-Chief to return home on important private business, and the First Lieutenant

---

* Lieutenant Milward reported to the C-in-C that the slaves numbered 432. The Mixed Commission figure of 430 has, however, been taken (in this case as in others producing a disparity) as being more likely to be correct. The captor's count was inevitably made in difficult conditions, and the court officials in Freetown, sticklers for detail, had the time to make a more careful record.

of *Thalia*, Henry Puget, was made acting commander to relieve him. The brig-sloop *Pelican* (18), Commander Brunswick Popham, and the more youthful "coffin brig" *Curlew* (10), Lieutenant The Hon. Joseph Denman, had been ordered to the Station in January to replace *Pelorus* and *Brisk*, and *Curlew* had made her appearance at Freetown in early March before making for the Bights. A clearly irritated Admiral Campbell received no notification of these impending arrivals until May, and he asked the Admiralty in future to send relieving vessels directly to the Cape to enable him to make arrangements for their employment and for sending home the relieved cruisers from the Coast. The inefficiency, wastefulness and frustration involved in delegating command of operations on the slave coast to an officer so remotely stationed was rarely more apparent. By the time Campbell made this plea, another newcomer had arrived at Sierra Leone on 30 April, the ten-gun brig *Rolla*, sister to *Curlew*, commanded by Lieutenant Frederick Glasse.[2]

There had been all too frequent changes in the government of Sierra Leone and the British Commission at Freetown. On the demise of Governor Temple in August 1834, Thomas Cole, Secretary of the Colony, became Acting Governor, and held the post until the arrival, in mid-February 1835, of Major Henry Dundas Campbell as Lieutenant-Governor. The Commissary Judge, William Smith, had been on leave at the time of Temple's death, and Cole immediately inherited the duty of Acting Judge until replaced by the new Governor in February. Judge Smith retired during his absence on leave, and Arbitrator H. W. Macaulay was promoted to relieve him in April 1835, being replaced as Commissary Arbitrator by the Registrar, Walter Lewis. Lewis was replaced, in turn, by M. L. Melville. This general rearrangement lasted only until 5 July, when sickness obliged Macaulay to retreat to Ascension in *Brisk* and the Governor returned temporarily to the Judge's post. Macaulay reappeared in late August, but by that time "climatorial illness" had taken Arbitrator Lewis away to England until November. This constant disruption was scarcely conducive to efficiency and consistent judgment in the Mixed Commissions, and, when both Macaulay and Lewis requested permission to leave the colony during the unhealthy months of June, July and August each year, the Foreign Secretary made it clear that he would not allow them to absent themselves simultaneously.

In the early part of the year the Squadron had suffered a spate of troubles involving warrant officers. In March *Lynx* lost Mr John Taylor, her Gunner, in a stupid accident at Ascension. The verdict of the subsequent Board of Inquiry was "That deceased came to his death by drowning in the attempt to obtain a Man of War Bird, shot by Corporal Wilson." Then, in May, Lieutenant Miall of *Forester* was obliged to report that the drunkenness of his Assistant Surgeon,

Mr John Rees, had probably contributed to the death from fever of one of his Gunner's crew. Rees had been drunk for two or three successive days and nights during the early part of the man's sickness, and Miall proposed not only a charge of neglect but also that Rees should be removed from medical charge of *Forester*. Rees, who acknowledged his alcoholism, was aged 55 and, after 23 years' service, had advanced no further than the lowly post of Surgeon in a brigantine. He was perhaps deserving of sympathy, and he was invalided. Less worthy of sympathy was Mr John Weston, Second Master of *Charybdis*. On the brig's arrival at Simon's Bay in mid-May, the Commander-in-Chief, investigating a complaint by Lieutenant Mercer, concluded that Weston had neglected his duty and had been disrespectful to his Captain. Campbell admonished him and wisely exchanged him with the Second Master of the Sixth Rate *Imogene*, which happened to call at the Cape. Mercer himself did not escape censure: he had directed Weston to continue keeping the warrant officers' account while under arrest and not permitted to attend issues of stores.

Admiral Campbell was concerned that commanding officers were too frequently resorting to the maximum punishments allowed by the regulations, and in May he indicated to them his desire that they should not do so in future except in extreme cases. He particularly directed that corporal punishment should not be inflicted on Kroomen other than in instances of gross insubordination or of specific serious crimes listed in the Articles of War.[*] Ill-behaved Kroomen could always be discharged and replaced.[3]

Assistant Surgeon Stevenson of *Pelorus* had been sent by Commander Meredith to Freetown in the slaver prize *Minerva*, perhaps the only instance in which a commanding officer had so diminished the medical cover of his own vessel in order to benefit a cargo of slaves. *Pelorus* did, however, have two Surgeons, the only cruiser to be so fortunate, despite the experience of nearly thirty years of the campaign and the repeated begging of Senior Officers on the Coast for more medical officers. During her voyage from the Old Calabar the *Minerva* lost 206 of her cargo of 675 slaves, and on his arrival at Freetown Stevenson presented the British commissioners with a report on the care of slaves in vessels sent in for adjudication. The report, together with the comments of Dr Ferguson, Surgeon to the Courts, was forwarded to the Foreign Secretary.

Having recommended that all cruisers should carry two Surgeons, the report listed the common diseases found in slavers: dysentery, diarrhoea, ophthalmia, ulcers and itch, pointing out that over half of the sick suffered from one or other

---

[*] The 36 Articles of War, originally drawn up in the 1650s, provided the legal basis for naval discipline.

of the first two. It also listed the medicines of which additional supplies were required: castor oil, laudanum, lunar caustic (silver nitrate), salts, compound sulphur ointment and some other common ointments, "basilicon", for example, but it was acknowledged that stowage presented difficulty. It recommended that blankets should be provided to cover the sick, but it explained that eight or ten per patient would be necessary because of fouling, and that the washing of blankets was generally impossible. Stowage, again, was a great problem. Finally it emphasised the importance of separating the sick from the healthy, a measure usually impracticable because of overcrowding. Of course, the tragic fact remained that the problem of containing the diseases assailing captured slavers was too massive for the medical skills of the day and the material resources that could be provided, particularly in the ghastly conditions of the slave decks. It seems that the only action taken by the Admiralty was to dispatch a supply of blankets, all other suggestions being considered impracticable. Nevertheless, there was no shortage of sympathy, goodwill and effort among the captors, and Dr Ferguson commented on the report that he had "always witnessed among Prize Masters and others in captured slavers a strong and praiseworthy desire to meliorate the condition and contribute to the comfort of the slaves by all means in their power".[4]

Forester completed her run of success on 15 June with the arrest of the American-built Spanish schooner *Numero Dos* with a cargo of 154 slaves, 13 of whom died of dysentery before emancipation. Both *Forester* and *Britomart* were cruising off the Bonny, and the slaver, which had sailed for Santiago de Cuba, was intercepted by Miall as soon as she gained the open sea. She tried to run back into the river, but offered no resistance. The boarding party found her slaves horribly cramped, with space between decks of barely two feet. On reaching Freetown, the Prize Master told the Mixed Commission, who condemned the prize, that he had seen ten more Spanish vessels in the Bonny when he left.

Lieutenant Jeremiah McNamara, whose invaliding from *Buzzard* had given Milward his golden opportunity, had resumed command of the brig, and Milward, no doubt sadly, rejoined *Thalia* at Ascension in August. McNamara, however, an elderly lieutenant worn out by long service, finally succumbed to the climate in the Bights at the end of July, and the Admiral appointed his Flag Lieutenant and nephew, Patrick Campbell, to command *Buzzard*. Acting Commander Puget of *Trinculo*, who had reported McNamara's death from Princes, was, of course, not to know of the Commander-in-Chief's arrangements for a considerable time, and he appointed Lieutenant Thomas Roberts, his First Lieutenant, to temporary command of the brig. Roberts was apparently known as "Old Four-Eyes" and, probably because little escaped his attention in *Trinculo*, it was believed that he slept wearing his spectacles.

Pelican, the most recently arrived of the veteran sloops which the Admiralty still persisted in deploying to the Station, had been launched in 1812, and it was hardly surprising that when she reached the Cape in early July from the Coast and Ascension she needed a refit. Returning from the Cape, *Charybdis* joined *Thalia* and the Commander-in-Chief at St Helena on 22 July before sailing three days later to stretch across to Benguela and work her way up the coast to Princes, looking into the rivers as she went. *Thalia* arrived at Ascension on 13 August and found there *Brisk* and *Britomart*. *Brisk*, which had deposited the sick Judge Macaulay at the island, was about to sail for Sierra Leone to discharge her Kroomen and then make for Spithead. She was a month astern of *Pelorus*, which had found orders for England on arrival at Ascension, and had sailed for home on 18 July. A delay to *Pelorus*'s planned departure for England had been caused by news in April that the French had instituted a blockade of Portendic, ostensibly to prevent arms-running to the interior, and Commander Meredith had felt it necessary to visit the Gambia before heading for Ascension and then home. As it was, a small squadron under Captain Lockyer had been sent from the Tagus to deal with the Portendic difficulty. *Thalia*, meanwhile, after only a couple of days at Ascension, set sail for return to Simon's Bay.

By then, *Fair Rosamond* had broken the six-month arrest monopoly of *Forester* and *Buzzard*. On 29 July, about 65 miles north-east of Princes Island and after a very hard chase of nearly 90 miles, she had taken the Spanish brigantine *Volador* with 487 slaves, one day out of the Bonny, bound for Havana. The master of the slaver, who claimed to have been chased frequently and unsuccessfully by cruisers on both sides of the Atlantic, sighted *Fair Rosamond* at about 1000 and, while maintaining his course, made all sail to escape. He had been instructed to give battle if molested, but failed to do so as *Fair Rosamond* closed in; only, in Lieutenant Rose's opinion, because he could not induce his crew to support him. On taking possession at 2000, the boarders found *Volador*'s pivoted long 18-pounder loaded with shot and a length of chain cable, and there were several similar pieces of chain and double-headed shot on her deck. The slaves had been confined for some time before embarkation, and 59 died before condemnation and emancipation.

Lieutenant Roberts in *Buzzard* opened his score on 2 September when he took the Spanish schooner *Semiramis*, with 477 slaves. The schooner, a fine new vessel bought at Baltimore at the end of February, tried to escape by re-entering the Bonny River, from which she had emerged that day, but there was insufficient depth of water and she was captured in the channel without a struggle. There were cases of smallpox, ophthalmia, dysentery and ulcers on board when the slaver reached Freetown, and 51 slaves died before emancipation. The Mixed Commission helpfully achieved adjudication within 24 hours of the schooner's arrival so that the prize crew could sail in *Rolla* on 13 October to rejoin *Buzzard*.

Having parted company from the Commander-in-Chief at St Helena in late July, *Charybdis* made the mainland coast south of St Paul de Loando and found there 32 slavers, all under Portuguese colours but almost certainly Brazilian. At Ambriz she boarded a Spanish brigantine waiting to load 500 slaves, and Lieutenant Mercer, to his mortification, learned from a British merchantman that a Spanish brig had sailed two days earlier with 450. *Charybdis* then struggled her way 20 miles up the Congo against a stream of four or five knots, anchoring on 11 August and finding Portuguese and Spanish vessels waiting for slaves. Onward she went to Loango, and, on boarding vessels in the bay on the sixteenth, discovered that a Spanish brig with 500 slaves had sailed the previous forenoon. Mercer set out in pursuit, but after a fruitless 48 hours he hauled in for the land. Again he was told by an English merchantman that two Spaniards were trading for slaves at Mayumba, but that they were currently making for Annobón for farina. On 23 August he anchored in the River Gabon, and, as well as three French brigs and an English schooner trading for ivory and camwood, he found there two empty Spanish slaving schooners. After calling at Fernando Po for fresh meat and vegetables, *Charybdis* worked her way along the shore to the west of Bonny before stretching over to Princes to find the Senior Officer, arriving on 6 September. Although frustrated to have made no captures, Mercer had gained useful information on slaving south of the Line, and he assessed that the annual trade from the coast between Camma Bay and St Paul de Loando amounted to about 130 vessels carrying upwards of 47,000 slaves.

Charybdis had to wait until 11 October to make an arrest, and she had to work hard for it. She first sighted her quarry, a brig, at 1440 on the eighth, but the chase was lost to sight at sunset. Mercer persevered, and he found her again at 1640 two days later, but again darkness intervened. At 0510 on the following morning the brig was again sighted about six miles to leeward, and at 0745, about 30 miles north-west of Cape Lopez, she struck her colours. She was the Spanish *Argos* of Havana, carrying 429 of the 500 slaves with whom she had left the Bonny on 3 October. The slaver master initially claimed that, the Bonny River being very sickly, the missing 71 had simply died, but he later admitted that on the final day of the pursuit he had thrown overboard ten who appeared likely to die. His crew said that most of the others had gone the same way. Forty-six more died on passage to Freetown, and another 17 before emancipation. However, the two matters of particular interest to the Mixed Commission were, first, that Mercer had required the slaver master to sign a declaration that nothing had been removed from his vessel, a "highly improper" action according to the court, and second, that items had, in fact, been taken. *Charybdis* was short of oil for her binnacle, and Mercer apparently told his clerk to purchase the items, four jars of "sweet oil" and several dozen sperm candles, but it seemed that the Prize Master took them without payment "for the use of His Majesty's service". At Freetown the officer who had purloined them offered to pay for the articles. That offer was refused, but the captor's agent was allowed to replace them before the condemned brig was sold.[5]

A few days later, *Britomart* made her first capture of the year. At dawn on 16 October she sighted a schooner off the Niger Delta and chased her resolutely until catching her about 35 miles south-east of the Bonny two days later at first light and taking her without resistance. She was found to be the Spanish slaver *Conde de los Andes* with a cargo of 282, cleared for Princes and St Thomas (as was frequently the case) but actually three days out of the River Brass and bound for Havana. Lieutenant Quin took her initially to Fernando Po before despatching her to Freetown. At only 87 tons she was very crowded, but the loss of slaves on passage was only 13, for which the commissioners commended the Prize Master, Mr Burslem. A second arrest for Quin followed on 16 November, when he took the Portuguese schooner *Theresa* bound for Montevideo. He intercepted her, and boarded without resistance, close to the coast about 30 miles west of Lagos where the 214 slaves had been loaded. She was condemned and 202 surviving slaves were emancipated.

Acting Commander Puget, still the Senior Officer on the Coast in *Trinculo*, had been told in late October by the master of a slaver that it was known in Havana that the slave trade would soon be stopped, and that not only was every available vessel being sent for slaves but also that more were being ordered from America. The Spaniard had said "perhaps two hundred will come here soon". Puget wrote to the Commander-in-Chief that to the southward the slavers were "literally swarming", and he particularly mentioned a very fine brig called *Christina*, well armed and capable of carrying 1,000 slaves. However, he said, to do any execution "we should have at least six more vessels, and fast-sailing too". He followed that with a letter in mid-November reporting that his cruisers had recently failed to catch three slavers, one of which had been chased by *Buzzard* for six days, and

Forester had pushed another very hard into the Bonny, where she had relanded 600 slaves just before the cruiser's boat got to her. The letter continued:

> The whole Coast is in such a stew, they know not what to do, and I am obliged to shift my plans very often to compete with their manoeuvres – was I in a Steamer and two or three more Vessels under my command, many should not escape – however we do all we can to stop them, but *Trinculo* is not half fast enough for a Senior Officer as he must be moving.[6]

Clearly something momentous was afoot, but neither had warning of it been given to the Commander-in-Chief or the Senior Officer on the slave coast, nor had the Admiralty taken any measures to cope with it.

The rumour rife in Havana was well founded. On 28 June a new Anglo-Spanish Treaty, including an Equipment Article (or Equipment Clause, as it became known), long desired by the Navy and the British commissioners and long-sought by the Foreign Office, had been signed at Madrid and ratified on 27 August.

The treaty granted a mutual Right of Search and arrest of vessels engaged in the slave trade and, crucially, of vessels fitted for the slave trade. The several "equipment" criteria demonstrating "fitted for the Slave Trade" were those used previously in the Anglo-Netherlands Treaty (Appendix E), only one of which needed to be present. The Mediterranean was excluded from the treaty provisions, as were European waters bounded to the south by latitude 37° N (the latitude of Cape St Vincent) and to the west by longitude 20° W of Greenwich. The two Anglo-Spanish Mixed Commissions were to be replaced by Mixed Courts of Justice, identical in purpose and constitution to their predecessors, and the existing courts would retain jurisdiction until their replacements were established. As in the past, the commanding officers of nominated cruisers were to be issued with individually addressed copies of agreed Instructions. A most important, and long overdue, new provision was that condemned vessels were to be broken-up and the components sold.[7]

The architect of this key achievement was Viscount Palmerston, who had returned to the Foreign Office to replace the Duke of Wellington in April. Considering the determination with which he had pursued this objective, it was surprising that he did not despatch copies of the treaty to the Sierra Leone Mixed Commission, with directions to act accordingly, until 29 October. Equally extraordinary was his failure, until mid-November, to point out to the commissioners that British vessels could not be brought before the courts until legislation had been passed in London, but that Spanish vessels could be judged immediately. These communications did not reach Freetown until the New Year of 1836, and the delay generated much difficulty and frustration.

Lieutenant Denman returned to Freetown in *Curlew* on 23 November, and he was followed three days later by three Spanish vessels he had arrested, the schooners *Victorina* and *Josepha* off Cape Mesurado on 17 November, and the brigantine *General Manso* off the Gallinas the day after. All were devoid of slaves but were fitted for slaving. The incautious Denman admitted that he was in possession of neither a copy of the treaty nor any new Instructions, but had been told by a merchantman that a treaty which included an Equipment Clause had been signed. The commissioners denied knowledge of such a treaty let alone any instructions to act upon it, and declined to express "an opinion upon the propriety of the detention of any vessels which might possibly come before us judicially". Denman, realising that he had overstepped the mark, made a deal with the Spaniards that he would guarantee their freedom from molestation for long enough to regain the positions of their arrest in exchange for an undertaking not to press any claim against him regarding their detention. This arrangement was undermined by the arrival from England of the ship-sloop *Pylades* (18) under the command of Commander William Castle, who had been one of *Black Joke*'s commanding officers. He was carrying notification of Denman's promotion, as well as Denman's replacement in *Curlew*, Lieutenant Edmund Norcott. *Pylades* sailed on 28 November to cruise on the Windward Coast, and on 3 December *Curlew*'s sister, the brig *Leveret* (10), Lieutenant Charles Bosanquet, arrived from England. Both brigs now held copies of the treaty and the Instructions, *Curlew*'s having been dated 29 September.

In accordance with Denman's agreement, the slavers *General Manso* and *Victorina* sailed on the morning of 4 December in the charge of two of *Curlew*'s officers. Shortly afterwards *Leveret* sailed in pursuit, recaptured them and returned with them to the anchorage. Norcott was in *Leveret* at the time, and *Curlew* was in sight, so Norcott and Bosanquet attempted to commence proceedings as joint captors. The British commissioners, however, dismissed the case, endorsing the Proctor's two petitions: "Not granted, as we have received no authority from His Majesty's Government to carry into effect the Treaty under which the vessel has been seized, nor have we even been informed that such a Treaty is now in force." Denman, who must have been mortified at this turn of events, assured the Spaniards that this second arrest had taken place without his knowledge or concurrence. Certainly the conduct of Norcott and Bosanquet had been distasteful, if not dishonourable.

On Castle's orders, *Leveret* sailed for Ascension and the Bights on 8 December, leaving the three slavers in the anchorage under the control of *Curlew*'s Prize Masters. In a further attempt to stimulate action by the commissioners, Norcott sent copies of the treaty and the Instructions to the Mixed Commission, but was dismissively told that it had already been shown both documents by Denman. To justify their inaction, the British commissioners pointed out to the Foreign Secretary that there had been nothing which "would technically prove the genuineness of [Denman's] copy of the Treaty and Instructions"; there had been no accompanying letter from the Admiralty, and there had been no signature on the treaty. The Instructions had, admittedly, been signed by the Lords Commissioners of the Admiralty, but the copy had not been addressed to *Curlew* and might therefore have served for any of HM Ships. Furthermore, there was nothing to show that the treaty had been ratified. Added to that there appeared to have been no Act of Parliament to carry the treaty into execution, and, as Parliament had been prorogued on 10 September, such an Act would have to wait for the next session of Parliament. In short, they believed they had no choice but to sit on their hands.

The next turn of events was the arrival on 10 December at the Mixed Commission of a petition from the master of the *General Manso* explaining the agreement made with Denman, which was supposed to have given him five days' grace after leaving Freetown, and claiming that a sum of £2,900 had been taken from his vessel. He asked that his brigantine should be restored and the money returned. The commissioners responded curtly that the petition was rejected because the vessel was not before the court. There the affair rested until such time as the commissioners were convinced that they had the authority to proceed to adjudication.

The next arrest demanded little soul-searching by the Mixed Commission. At 1630 on 27 November the American-built Spanish schooner *Norma* had cleared from the Bonny River and had set her course for Havana when she was sighted by *Buzzard* at a distance of about nine miles. She immediately tacked and stood for the safety of the Bonny, but *Buzzard* caught her at 2100 just off the mouth of the river. There were 249 slaves on board, of whom 11 died of disease on the passage to Sierra Leone and four jumped overboard. Slaves from the Bonny had been found to be particularly prone to self-destruction, and, after these four had drowned, the Prize Master took the precaution of putting the male slaves in irons. The prize arrived at Freetown on 28 December, and the Mixed Commission began proceedings under the 1817 Treaty. Some uncertainty was caused by the arrival, at last, of the Commission's copies of the 1835 Treaty, but it was decided to conclude the adjudication under the earlier treaty, and the prize was condemned on 6 January, together with the 234 surviving slaves.

For the time being, that was the end of the orderly process of adjudication at Sierra Leone. The cruiser commanding officers now had their hands on the eagerly desired Instructions authorising seizure of Spanish vessels equipped for slaving, and the result was, at least by comparison with earlier arrest rates, a relative bloodbath. Unfortunately, the enthusiasm of the cruisers was matched by the British commissioners' determination not to enter into adjudication of vessels taken under the Equipment Clause; not, at least, until they had received formal notification from the Foreign Office that the legislation to bring the treaty into force (which they mistakenly believed necessary for adjudication of Spanish vessels under the treaty) had been passed by Parliament. As Lord Palmerston had written to the Mixed Commission on 16 November, legislation would indeed be necessary to allow British vessels to be brought before the court, but there was no obstacle to the adjudication of Spanish vessels.[8] That letter, however, took several months to reach Freetown. This combination of factors caused a logjam of empty Spanish slavers under arrest in the Sierra Leone River, each under the control of a deeply frustrated prize crew.

The first to be added to *Curlew*'s original threesome was the *Tres Tomasas*, taken on 19 December about 130 miles south-west of Freetown, also by *Curlew*. The slaver was on passage to the Cape Verde Islands, ostensibly for repairs to her masts, but almost certainly to obtain Portuguese colours and papers. Proceedings against her on behalf of Lieutenant Norcott were begun on 4 January 1836, but were then suspended until mid-July, when she was condemned.

Late December was also a heartening period for the arrest of loaded slavers, with three Spaniards taken in quick succession. On 22 December *Trinculo*'s boats seized the brigantine *Isabella Segunda* of Santiago de Cuba as she lay at anchor on the Bonny bar. They found 347 slaves on board, and it seemed that a further 27 had been landed immediately before the vessel was boarded, apparently to pay for the subsistence ashore of those members of the crew who were expected to be landed by the British after the imminent arrest. Thirteen of those captured died of dysentery before emancipation. The slaver master complained to a dubious court that 6,000 cigars and three pairs of razors had been stolen from the brigantine before she met *Trinculo* outside the river, but Acting Commander Puget subsequently asserted that every boat returning from the prize had been searched. Puget declared that, on being asked whether anything had been stolen, the master replied that: "no, they are a fine set of honourable fellows". Puget adds that, in the words of the master, "his own rascally crew had not only robbed him of everything but threatened and would have murdered him had not His Majesty's Sloop's boats come up with her as quick as they did."⁹ Still ignoring the terms of the 1835 Treaty, the Mixed Commission sold the brigantine after condemnation instead of destroying her.

The 64-ton New York-built schooner *Ligera*, bound for Cuba, was taken by *Buzzard* on 24 December, about 40 miles south-east of Cape Formoso, carrying 198 slaves shipped the previous day from the River Nun. *Buzzard* sighted the slaver at 1500 and boarded her, without resistance, after a chase lasting over six hours. Lieutenant Roberts, still not replaced by the Commander-in-Chief's nominee, judged his prize to be unseaworthy and took her to Princes, where he intended to caulk her. However, he found that she was making over three feet of water an hour at anchor in smooth water, and, after taking the precaution of a survey of the schooner by *Fair Rosamond*'s officers, he transferred her slaves to *Buzzard* and left her in the custody of the Portuguese authorities. After what must have been a most uncomfortable passage, the surviving 192 slaves were emancipated and the *Ligera* was condemned in her absence.* The court gave Roberts a blank Commission of Appraisement and Sale to be delivered to "some respectable and trustworthy person" at Princes to conduct the sale, and he was directed to destroy the vessel's slave coppers and most of her water casks. From their experience of "Lieutenant Roberts' high character for intelligence and correctness" the commissioners had no doubt that their directions would be strictly complied with.

The third of this trio of captures was achieved by *Fair Rosamond* on 28 December when she intercepted the schooner *Segunda Iberia* about 160 miles south-south-east of Cape Formoso. The schooner, probably built in Baltimore and bound for Santiago de Cuba, had also been to the River Nun for her 260 slaves.† She declined to make use of her pivot-mounted long 9-pounder, and was taken without difficulty. Between capture and emancipation 22 slaves died, 16 of them on passage and six after landing. On arrival at Freetown, the slaver master alleged that his private property had been plundered by the cruiser's crew. Not for the first time in such circumstances, Lieutenant Rose was called upon for comment, and the case was delayed for his response. Rose had persuaded the master to sign an exculpatory statement, a measure which the court regarded as not only useless but also improper and suspicious. He had also caused every man's bag and mess to be searched, fruitlessly, but a number of items had been recovered from the

---

* It had become the practice in Freetown to land slaves before adjudication if the Surgeon of the Court recommended it, as he usually did. Some of the deaths between capture and emancipation were therefore likely to occur ashore.
† The initial count had been 267, but the court accepted that on the crowded slave deck of a detained vessel a correct count could rarely be made.

boat carrying the slaver's crew to *Fair Rosamond*. Although the slaver's cabin boy claimed to have seen a trunk being broken open by a man-of-war's man, there was evidence that the bulkhead of the master's cabin had been stove in prior to capture. A pair of pistols, claimed to have been lost, had actually been given to the Prize Master for safekeeping, and had been returned to the master at Sierra Leone, as promised. Rose later soured his relationship with the commissioners, and weakened his position, by ignoring a letter from them on the subject of the alleged theft. Although it was delivered when *Fair Rosamond* arrived at Freetown on 19 June 1836, he failed even to acknowledge the letter before sailing on the twenty-second. Although the slaver was condemned, there was, as usual with these accusations, no satisfactory resolution of the theft case.

What appeared to be the next candidate for a long wait at Sierra Leone reached the river on 27 December. *Pylades* had sailed from Freetown on 28 November after her arrival from England, but she had remained on the Windward Coast, and on 25 December, 100 miles west of Freetown, she seized the Spanish brig *Tersicore* under the Equipment Clause. Rather strangely, a month later the Captor's Proctor asked for the case to be withdrawn, having apparently concluded that there was insufficient evidence that the brig had been fitted for the slave trade. The vessel was returned to her master.

The rate of Spanish Equipment Clause arrests gathered pace early in 1836, beginning with *Curlew*'s capture of the brigantine *Rosarito* at anchor in Accra Roads on 2 January. The vessel's master claimed that the slave irons on board were part of a cargo for sale, but when the case was eventually brought to adjudication in July the court declared that under no circumstances could it consider shackles, bolts and handcuffs to be articles of lawful commerce.

The next two victims, however, were loaded slavers, both Spaniards. At noon on 13 January, *Pylades*, still cruising the Windward Coast, sighted the schooner *Gaceta* about 100 miles south-west of Cape Mount and began a chase which lasted until 1730 the following day. The prize, another New York-built vessel, bound for Matanzas, was taken without resistance and was found to be carrying 225 slaves taken from New Sesters and Sanguin eight days earlier. On 21 January, a few miles to seaward of the Bonny River, *Lynx* took the brig *Vandolero*, with 377 slaves shipped the previous day in the New Calabar River. The slaver was sighted as she left the river at 0600 and set her course for Cuba, but she was not boarded until 1500, having thrown overboard her guns, anchors and cables, boats and stock during the chase. Although well armed, she offered no resistance. *Gaceta* reached the Sierra Leone River on 24 January, and *Vandolero* on the 29 February, but, for some extraordinary and unexplained reason, the Mixed Commission chose to deal with the latter under the 1817 Treaty and the former under the 1835 Treaty. The result was that *Gaceta* had to wait in the anchorage until her condemnation on 5 July, by which time her slaves, who had been landed on arrival, had been reduced from 225 to 169, thanks to the smallpox epidemic which had been raging in the colony for 12 months. *Vandolero*, meanwhile, had been condemned on 8 March and sold, and 343 slaves had been emancipated. Thirty-four had died before adjudication, and a baby born on passage to Freetown had died shortly afterwards.

The flow of prizes arrested under the Equipment Clause then began in earnest. *Leveret* took two schooners, the *Atafa Primo* and *Zema*, on 25 January, close inshore about 20 miles south-east of Sanguin on the Grain Coast. The *Atafa Primo*, after a long wait, made her second appearance before the Mixed Commission, having been prosecuted by Commander Webb of *Medina* in 1830 but later restored. This time she was condemned, as was *Zema*. An even bigger haul was achieved by Acting Commander Puget in *Trinculo*, a particularly well-deserved success. He had sent Lieutenant Robert Tryon into the Bonny River with the sloop's boats on the evening of 22 January, and, on the twenty-eighth, Tryon arrested four Spaniards he had found at anchor waiting for slave cargoes, and landed their crews.

King Pepple of Bonny immediately reacted by stopping trade with British merchantmen in the river, and the masters of these vessels persuaded Tryon to land with them and confer with the King. On stepping ashore they were surrounded by the Spanish crews and natives, seized, stripped naked, put in irons, chained by their necks to a log of wood, and otherwise ill-used. Tryon, however, managed to send a message to his Commanding Offer by bribing a canoe-man, and Puget resolved to take *Trinculo* into the river. His only option was to use the South Channel, which had only 15 feet of water at the top of the tide, although the sloop was drawing nearly 18 feet.* However, by transferring spare spars and other gear, provisions, cables and all but four rounds of shot into *Lynx* (which had, fortuitously, just arrived from Ascension), *Trinculo*'s draught was reduced to 16 feet. There was a heavy breaking surf on the bar, but Puget determined to run the sloop at it and try to "bump" her over into the deep water of the river. Three times she struck the bottom hard, but after each strike she was lifted onwards by a big comber. Once in the river, Puget anchored close off the town and fired four blank rounds. King Pepple released his captives at once. On 30 January the sloop sailed, via the Portuguese Channel, with the four prizes in her wake, and while *Trinculo* made for Princes, the Spaniards headed for Sierra Leone under escort by *Lynx*.[10]

---

* The alternative route was by the Eastern or Portuguese Channel, but that, although relatively deep, was too narrow to navigate with other than a wind astern, which was not forthcoming for Puget.

Puget's prizes, so courageously and skilfully won, were the schooners *Feliz Vascongada* and *Eliza*, the brigantine *Diligensia* and the brig *Maria Manuela*. The first of these had been condemned in 1834 as the Portuguese slaver *Despique* and then sold, with foreseeable consequences. The *Eliza* was carrying a Royal Passport for a voyage of lawful commerce and a Customs House clearance from Santiago de Cuba for a case of 120 leg irons as part of her cargo, indicative of the cynicism of the Cuban authorities as well as the slavers towards the Anglo-Spanish treaties. Lord Palmerston asked for a copy of the clearance in order to lay it before the Spanish Government with a request for censure of the issuing officer and orders to prevent a repetition, surely a wasted effort. The *Maria Manuela* held a Customs House clearance for "fifty pipes of rum" to which had been crudely added "and water", a pathetic attempt, probably after capture, to justify the excessive number of water casks on board. The four Spaniards had to wait until September for condemnation.

There was one further Equipment Clause seizure in January, and the first week of February saw three more. Lieutenant Rose in *Fair Rosamond* arrested the brig *El Esplorador* 120 miles south-east of Cape St Paul's on 29 January, and she was condemned in August. Then on 5 February *Charybdis* took the schooner *Matilde* on the Equator close to the island of St Thomas, and found among her slaving gear a cooking stove of novel design which appeared no larger than was normal for a legitimate merchantman, but which could be expanded to twice the size for feeding slaves. *Matilde* was condemned in July. The schooner *Mosca* was taken in Whydah Roads on the sixth by the boats of *Britomart*, but when she was finally brought before the court the master protested that no proper authority had been shown by the seizing officer, and that no certificate, as prescribed by the treaty, had been produced. Although the absence of Lieutenant Quin prevented investigation of this allegation, the court took the view that, as advised by HM Government, such trifling deviations from treaty stipulations were insufficient to invalidate capture, and *Mosca* was condemned in early September.

There came a further allegation of misconduct in the seizure of *Mosca*. The claim was made to the Mixed Commission that two anchors and cables had been unnecessarily slipped and lost, and that a dozen swabs had been stolen. Mr Henry Cox, Mate, of *Britomart*, who was in charge of the two boats which had taken the slaver, explained that shortly after the capture he had seen two suspicious vessels standing towards him, and as *Britomart* was some distance in the offing and, furthermore, had no boats in which to send assistance, he judged it prudent to join the gun-brig "with all possible expedition". There was a heavy swell in the roadstead which would delay weighing, and he had ordered the prize's cables to be slipped. Lieutenant Quin admitted that a few swabs had been taken for the use of His Majesty's Service. The court appears to have decided against further action.[11]

The Commander-in-Chief had sailed from Simon's Bay in late November 1835 to visit the West Coast, with his flag in *Thalia*. Having touched at St Helena, he arrived at Ascension on 12 December to find *Lynx* and *Forester* there. He then headed north for a two-day call on the Gambia before sailing for Sierra Leone where, on his arrival on 6 January, he learned that the trouble off Portendic had been resolved by a treaty between the French and the natives of the interior. The Admiral found *Pylades* at Freetown, and heard that both *Pylades* and *Leveret* had called at the Gambia on passage from England, that *Rolla* had relieved *Griffon* as the Gambia guardship, and that Commander Castle had sent *Curlew* south with copies of the new treaty and Instructions. *Pylades* then sailed for the Bights to replace *Trinculo* and for Castle to relieve Puget as Senior Officer, while *Thalia* made for Accra, where she found *Leveret* and the newly arrived *Waterwitch*. The Commander-in-Chief ordered *Leveret* to the Cape with the despatches brought from England by *Waterwitch*, and onward to Mauritius with those received from *Pylades*.

Joseph White's much-admired brig *Waterwitch*, now armed with two 6-pounders and eight 18-pounder carronades and commanded by Lieutenant John Adams, had been ordered from England to the Cape with despatches and to follow Admiral Campbell's orders. However, she encountered *Griffon* off the Gambia on 29 January, and, learning that the Admiral was in *Thalia* on the Coast, Adams wisely made for Accra, arriving a day before the flagship. *Waterwitch* then began her first Bights cruise in time to account for the third victim of the first week of February, the brig *El Casador Santurzano*, which she caught on the sixth off Whydah. Being empty of slaves, the prize had to wait until late August to be condemned.

It was not until 27 February that Admiral Campbell was able to write to the Admiralty to acknowledge its letter of 1 October 1835 telling him of the signing of the Spanish Treaty and forwarding the Printed Instructions for his cruisers. His formal notification of this crucial development came even after that from the Foreign Office to the Freetown commissioners who, on 2 January, had received 12 copies of the treaty and "directions pending further appointment and definitive establishment of the Mixed Court of Justice". The Admiral, in response, reported to Their Lordships that he had issued copies of the Instructions to *Thalia, Griffon, Rolla, Trinculo, Britomart, Forester, Lynx, Charybdis, Fair Rosamond, Curlew, Buzzard* and *Pelican*. *Waterwitch* would undoubtedly have been issued with her Instructions before leaving England.[12]

By then, *Thalia* had announced her arrival in the Bights with the arrest of a loaded slaver. At dawn on 8 February she sighted a westward-bound brigantine to seaward of the Quaqua Coast, and chased her until 1730, bringing her to about 140 miles east-south-east of Cape Palmas. The prize, which neither altered course nor added sail during the pursuit, was the Spanish slaver *Seis Hermanos*, bound for Cuba from New Calabar with 189 slaves. Her master had died two days before the capture, and there was a further loss on the passage to Sierra Leone of 12 slaves from dysentery and of four who threw themselves overboard. A baby was born and died. One case of smallpox surprisingly survived, and the disease did not spread, thanks to the "judicious arrangements of the prize officers". Smallpox was, however, rife in the Lower Hospital at Kissy, where sick slaves were generally received at Freetown, and the invalids from *Seis Hermanos* were landed to the Upper Hospital. Two more died ashore, and 171 were finally emancipated. Admiral Campbell reported the episode as a joint capture by *Thalia* and *Waterwitch*, and so the brig was presumably in sight at the time, but the slaver was condemned solely to the frigate.

After this relatively rare seizure of a laden slaver, the flow of arrests under the Equipment Clause recommenced, but from south of the Line instead of the Bights. Returning from Ascension, *Forester* made for Loango Bay, once the principal slaving area for the Spaniards and for some time regarded as safe from the British cruisers. There on 9 February Lieutenant Miall snapped up two prizes, the schooner *Golondrina* and the brig *Luisa*. The latter had probably been the *Jules*, condemned by the Anglo-Dutch court in 1829, and in the same year her master had been brought into Sierra Leone in *La Laure*, which had been restored as French.* Both prizes lay waiting until August for adjudication and condemnation. Ten days after George Miall had caused surprise and consternation in Loango Bay, *Charybdis* also paid a visit and was rewarded with the detention of the empty brig *Tridente*. This American-built vessel was nearly new and had recently been sold to her Spanish master for $4,000.† She too became a fixture in the Sierra Leone River until her condemnation in August.

During his short visit to the Bights, the Commander-in-Chief ordered *Trinculo* to proceed to Spithead in accordance with Their Lordships' direction, and he conveyed to the Admiralty his high opinion of the sloop's temporary Commanding Officer. Acting Commander Henry Puget had been in command of *Trinculo* for

---

* There is some uncertainty about the fate of *La Laure*. She was certainly not condemned by the Mixed Commission, but the Commissioners may have been mistaken in mentioning that she had been restored in 1829. A Colonial Office report of 1830 indicated that she had been condemned as a result of a case brought by the Collector of Customs, presumably in the Vice-Admiralty court.
† According to the British Commissioners, this $4,000 equated at the time to £1,000.

twelve months, and for seven of those he had been Senior Officer on the West Coast, in which role, wrote his Admiral, he had made wise dispositions for the protection of British commerce as well as for maximum effectiveness of slave-trade suppression. He had also made judicious arrangements for replenishment of supplies, and to this the Admiral attributed, in great measure, the healthy state of the cruisers. The "activity, intelligence and zeal" with which Puget had conducted his duties met with Campbell's approval, and he was recommended to Their Lordships' favourable notice. As a result Puget received the strong approbation of the Board of Admiralty, and was subsequently promoted to commander.

For the past year the British commissioners in Freetown had been worrying about the slaves landed from various vessels at Fernando Po, most recently the *Apta* and the *St Rosario e Bon Jesus*, and emancipated in their absence. They found it unsatisfactory that these people had been abandoned on Spanish territory, and wished them to be taken to Sierra Leone. Governor Campbell had agreed with them and, clearly with scant understanding of what was involved, had "requisitioned" Lieutenant Rose of *Fair Rosamond* to provide "speedy conveyance" to Sierra Leone, an impracticable demand upon which Rose had contrived to avoid taking action. However, a letter from the Admiralty to the Senior Officer at Sierra Leone requiring him to send a cruiser to Fernando Po to recover the Africans reached Admiral Campbell in late January, and he sent Commander Castle in *Pylades* to evaluate the situation. On arrival on 2 February, Castle conferred with Mr Beecroft, who was in charge of the residual establishment on the island, and he learned that the surviving ex-slaves had settled with the natives in the interior, in many cases having married. They had no wish to be moved, and, Beecroft said, it would be impossible to collect them. Indeed, they would abscond if they learned of an intention to do so. The Commander-in-Chief reported in these terms to the Admiralty, and there the matter rested for a time.

On 15 February, *Thalia* and Admiral Campbell sailed from Princes Island for Ascension on the first leg of their return voyage to Simon's Bay. The frigate had on board a 15-day stock of spirits, adequate for what was generally a passage of 12 days to Ascension. After six days of light and baffling winds the flagship had, however, made good only 170 miles of the 1,200 or so to her destination, and the Commander-in-Chief instructed Captain Wauchope to reduce the daily rum issue to two-thirds of the ration. When the grog was mixed that day some of the petty officers represented to the Captain that the ship's company wished to have the full allowance and accepted that they would have to do without if the frigate ran out of rum. Wauchope replied that the Admiral had ordered the reduction. When grog was then piped no one mustered for it. After waiting for ten minutes, Wauchope passed the order for the petty officers to come instantly for their grog. None did so. Campbell then ordered aft all petty officers and "expostulated with them on the impropriety of their conduct". He directed that grog should be piped again and required a report of anyone refusing to receive it. All took their grog. Subsequent investigation showed that, as has ever been the case in such disturbances, the trouble had originated with very few men, and they were punished. One in particular, Thomas Carter, who had gone about the messes fomenting the protest, was discharged with disgrace. Campbell enquired whether the men had any complaint against their Captain and officers, but there was none, and it was the Admiral's view that the officers of *Thalia* were most attentive to the welfare of their men. He subsequently received two letters of apology from the petty officers and NCOs of Marines, and he believed that they were heartily sorry and ashamed of their behaviour. The Admiral had committed the sin of interfering in the internal affairs of the flagship, thereby undermining the authority of her Commanding Officer, and he did so on a matter of particular sensitivity to the lower deck. Once this crassness had generated a predictable response, the trouble, which certainly amounted to a mutiny, seems to have been sensibly handled. Discipline was restored, and the subsequent remorse of the petty officers and NCOs rings true.[*][13]

While at Ascension, the Commander-in-Chief had another unpleasantness to report. A letter from Lieutenant Milward in *Buzzard* informed him that rumours had been circulating that Milward's conduct during the capture of the slaver *Formidable* had been less than courageous. It alleged that one of *Buzzard*'s warrant officers, Mr Bickford, had been present when the rumour was being discussed, and had failed to contradict the slur against his Commanding Officer. An enquiry ordered by the Commander-in-Chief and conducted by Captain Wauchope and Acting Commander Puget found that the rumour, undoubtedly prejudicial to the reputation of Milward, was utterly false, and had originated with two drunken men who were subsequently discharged from the Service. It also found that Bickford's conduct had been highly reprehensible, and he was dismissed from *Buzzard* and sent home in *Trinculo*.

Charybdis cruised south past the mouth of the Congo after her success at Loango Bay, and on 25 February she was running out of Ambriz Bay when she sighted three vessels at anchor: an American ship, a Portuguese brig and a schooner which was showing no colours. When *Charybdis* was three miles from

---

[*] The source of this account is Admiral Campbell's report to the Admiralty, but there is no reason to doubt the veracity of his version of events. As it happened, *Thalia*'s passage to Ascension took 15 days.

the anchorage the schooner weighed and made sail. Lieutenant Mercer took up the chase and fired his pivot gun to bring her to, but the schooner made every effort to escape. The attempt was in vain, and she grounded in stays.* *Charybdis* anchored and lowered her boats, but by the time they reached the schooner she was abandoned and bilged. The boarders found that she was fitted for slaving and discovered a Spanish ensign in the hold, but were then obliged by the heavy surf to leave. It was later learned at St Paul de Loando that the slaver had been the *Ligera* of Havana, and that when *Charybdis* found her the master and mate had been ashore at the slave factory, leaving the boatswain in charge.

The flow of prizes taken under the Spanish Equipment Clause in the Bights resumed in March, beginning with two captured in the River Nun by *Britomart*'s boats on the sixth. They were the Havana schooner *General Mina* and the schooner *Dos Hermanos* of Santiago de Cuba. In both cases Lieutenant Quin had failed to forward with the prize the necessary declaration on the circumstances of the arrest, but the pair were not brought to adjudication until the beginning of October, which gave time for the omission to be rectified on *Britomart*'s next call at Freetown.

On 10 March *Fair Rosamond*'s boats seized the schooner *La Mariposa* off Duke's Town in the Old Calabar River. The master had landed his sails and most of his cargo on the previous night, fearing arrest, and it had been his intention to clear the vessel and destroy her on the approach of British boats. As it was, he was taken by surprise, and sufficient evidence remained to condemn the schooner. When she arrived at Freetown on 18 April she was in a very leaky condition, and by July the Mixed Commission allowed her remaining cargo and stores to be landed and sold. The vessel herself was hauled ashore pending adjudication, but began to fall to pieces after a month. She was then sold by public auction, and what remained of her was condemned by the court at the end of September.

A few days later, the initial rush of Equipment Clause arrests came to an end with a final clutch of three. *Waterwitch* took two of them, both schooners: the *Galanta Josepha* off Little Popo on 13 March, and the *Joven Maria* about 25 miles off Whydah the following day. *Galanta Josepha* had been waiting seven months at Whydah for a slave cargo, and had gone to Popo for provisions. *Joven Maria* had first landed part of her cargo at Whydah and then called at two other slaving places before making for Princes Island. There her master had declared that he had found the Trade so indifferent elsewhere that he was returning to Whydah. Both were condemned in late September. *Charybdis* brought the phase to a close with another capture south of the Line, the brig *El Mismo* off Ambriz, also on

---

* That is, she had her head to wind while attempting to go about from one tack to the other.

14 March. The brig had completed three successful slaving voyages between October 1833 and October 1835, but her condemnation in October 1836 brought her career to an end.

When the last of these prizes arrived in the Sierra Leone River there were 25 vessels in the anchorage awaiting adjudication, each with a languishing prize crew and bearing stark testimony to the abysmal mismanagement by the Foreign Office of the introduction of the 1835 Anglo-Spanish Treaty measures. There they remained for months while the British commissioners awaited the specific written authority from the Foreign Secretary which they believed necessary before they could begin adjudications. On 20 March Judge Macaulay left Freetown for sick leave in London.

By April 1836 there were few potential prizes in the Bights. The majority of the year's Spanish slavers were lying in the Sierra Leone River, and those still free were considering their options for evading the greatly increased dangers introduced by the new treaty. The Portuguese and Brazilians, meanwhile, continued taking their slaves from the lower-risk areas south of the Equator.

As the pace slackened, the Commander-in-Chief arrived at the Cape on 23 March and shifted his flag into *Badger*, the mooring vessel at Simon's Bay, before despatching *Thalia* once again for the Bights on 1 April. *Pelican*, despite holding a copy of the anti-slaving Instructions, was dividing her time between supporting military operations at Algoa Bay and cruising the east coast. *Leveret*, also with Instructions, which would have been issued to her in England, departed the Atlantic on 22 March, bound for Mauritius. A newcomer, the ship-sloop *Scout* (18), Commander Robert Craigie, arrived at the Cape on 24 May from England via the Gambia, Sierra Leone and Ascension. She was relatively new, having been launched as recently as 1832, but was built to a familiar Seppings design. With *Scout* taken under command, the Commander-in-Chief gave orders for *Britomart* to proceed to Plymouth, and for *Scout*, once refitted, to sail for the West Coast, which she did on 15 June.

The first Portuguese prize of the year on the West Coast was made on 4 May, just off the south-eastern point of Fernando Po, by *Buzzard*, now at last under the command of Admiral Campbell's nominee, Lieutenant Patrick Campbell. She was the brigantine *Mindello*, sighted in calm weather that morning, 12 hours out of the Cameroons River, and taken by the boats in the afternoon, having made no attempt to escape or resist arrest. The brigantine had 267 slaves on board at capture, but lost four of them on passage to Freetown, where she arrived on 2 June and was brought immediately to adjudication. Evidence from her crew was contradictory and generally fabricated, and she had been subjected to convoluted sale arrangements, but she was certainly American-built and believed to be Brazilian property. It transpired that she bore similarities to the notorious *Maria da Gloria*. Both had been equipped for the Trade at Rio, and had been to St Paul de Loando, where they had received new passports for the destination of Montevideo. As the commissioners noted, the Customs House authorities in both Rio and St Paul were either exceedingly neglectful or participants in illegal acts. On 14 June she was condemned, and the surviving 257 slaves were emancipated.

Captain Wauchope assumed the duties of Senior Officer on the West Coast when *Thalia* arrived in the Bights, and he ordered *Forester* to remain at Sierra Leone to take charge of the prizes detained in the river. Lieutenant Miall gave Governor Campbell and Mr Lewis, currently the British commissioners, a copy of the recent Act of Parliament to bring the 1835 Treaty into force, and begged them to begin adjudication of the prizes. He was most anxious that the prize crews should be removed from the dangers of disease at Freetown in the approaching rains, and was concerned at the manpower shortages imposed on the cruisers by the prolonged absence of prize crews. He also pointed out the impossibility of preventing the plunder of the prizes by the Spanish crews still living in them. The commissioners, encouraged in their obstinacy by a letter received in May from Lord Palmerston which approved of their refusal to adjudge the first five vessels arrested under the Equipment Clause, declined Miall's request because they had still not received formal notification of the legislation.[14]

The problem of plunder was illustrated by the case of the *Matilde*, taken by *Charybdis* in February. The original Prize Master had departed after handing over the vessel to the Prize Master of the *Tridente*, Mr Rowlatt. He had not lived on board *Matilde*, and the British seaman left in charge of her had fallen sick. The ship-keeper for a considerable period had been a liberated African, and the two Spaniards who remained on board had been caught thieving by Rowlatt. The deficiencies subsequently discovered included six sails, spare shrouds, a compass, barrels of vinegar, a lower boom, bags of musket shot and a 4-inch hawser. This was hardly petty pilfering.

At last, on 5 July 1836, the new Anglo-Spanish Mixed Court of Justice, established under the 1835 Treaty, opened in Freetown. Copies of the Act of Parliament had arrived, as had the appointments of Macaulay as Judge, Lewis as Arbitrator and Melville as Registrar. With Macaulay absent on sick leave, Lewis took the oaths of Arbitrator and Judge, and Lieutenant-Governor Campbell temporarily assumed the duties of Arbitrator.

Lord Palmerston gave personal directions that "Instructions […] be given to take care that vessels condemned for Slave Trade are so completely & effectively broken-up that the Pieces shall not be capable of being easily put together again". The instructions given by the commissioners to the Marshal and Commissioner of Appraisement and Sale regarding this breaking-up required him to land all merchandise, stores, masts, spars, rigging, boats and all moveable equipment and parts of the condemned vessel, and then to ground her high on the sand at Thompson's Bay near Freetown. There she was to be cut in half athwartships, and the halves were to be sold in two separate and distinct lots. Each purchaser was to give a bond of £150 that his part would not be used for reconstruction of the vessel.[15] The commissioners acknowledged that breaking-up could be made more complete, but the remains would be proportionately reduced in value and nearly unsaleable. The first Spanish vessel to be so destroyed was the small schooner *Gaceta*, and the costs involved amounted to £9.8s.6d, at the rate of 4s.7¼d per Spanish ton. Future expense would be less, the Court having negotiated a contract with the Clerk of Ordnance Works for a fee of four shillings per Spanish ton for the first 50 tons of a vessel and two shillings for each ton above that. It was believed likely that the materials sold would be used for building small coasting craft, but the Collector of Customs had indicated that he would not be able to grant British registration for such vessels, and the commissioners recommended that registration should be allowed in order to encourage sales.

July was a good month for *Buzzard*. Lieutenant Campbell had received information of a slaver in the Bonny River about to sail, and on 30 June he despatched Mr Benjamin Fox, Mate, with two boats, manned and armed, to wait off the bar. On 2 July Fox was returning to *Buzzard*, and he sent a man to the masthead to watch for her. However, the lookout sighted instead a vessel standing out of the river. The boats immediately set off in chase of the stranger, and "with great difficulty and exertion" succeeded in catching her at 1100, despite having already pulled more than twenty miles that morning. The prize was the Spanish brigantine *Felicia*, bound for Havana with 401 slaves after lying for 32 days in the Bonny, and she was condemned with her remaining 355 slaves. Campbell expressed high praise of Fox for having brought the brigantine downriver and over the bar without a pilot, and, indeed, in the absence of her master, who had deserted in the pilot's canoe with 24 slaves and the most important ship's papers.

Fox later reported that the slaver had taken precautions against earlier capture while empty. She had been fitted with conventional hatches, but carried, as part of her cargo, bar iron for conversion into slave hatches. Her slave shackles had been concealed under the cargo, and the Havana Customs House had given false clearance for her water casks as intended for palm oil. Large quantities of rice and farina were entered in her manifest as cargo, and her slave boiler was so attached to the usual ship's copper that it could easily be removed and ditched in the event of interception by a cruiser. The Spanish slavers were already contriving means of evading the Equipment Clause.[16]

At dawn on 6 July, *Buzzard* was off the mouth of the San Bartolomeo River, to the west of the Bonny, when she sighted a schooner leaving the river. She intercepted and took her at 0800 without a chase or resistance. The schooner was the Spaniard *Famosa Primeira* of Santiago de Cuba, empty but fitted for slaving, and she was eventually condemned in October. It was discovered that her master and alleged owner, Mateo Moya, had been second mate of the Portuguese schooner *Nympha* which had fired on *Conflict*'s boats in 1830. He was handed over to the authorities in Lisbon and imprisoned for four years, but the punishment was certainly not salutary; he went straight back to slaving on release.

A trio of captures for the month was completed by an arrest by *Buzzard* on 22 July. Lieutenant Campbell was again served well by intelligence, this time on a Portuguese slaver in the Old Calabar River, and he manned and armed the gig to reconnoitre the river under the command of Mr Samuel Wooldridge, a mate lent from *Thalia*. The boat was given a tow upstream for 20 miles by the Fernando Po steamer *Quorra*, commanded by Mr Beecroft, which happened to be passing. On reaching the Panot Islands, where the river takes a sharp turn, the gig was cast off. She immediately encountered a brigantine, correctly assumed to be the Portuguese slaver, and boarded her. Wooldridge had only six men with him, and found himself faced by a crew of 23 officers and men, reinforced by eight Spanish passengers. Suspecting that the Portuguese were making up their minds to resist, Wooldridge rapidly hoisted out the slaver's boat, put 16 of the crew into her without oars and took her in tow. Leaving one man to prevent the tow rope from being cut, he and the remaining five hands took the prize over the bar and out of the river. *Buzzard* herself was unable to come up with the prize and relieve Wooldridge in his precarious situation until 11 hours after the capture. The brigantine was then found to be the *Joven Carolina*, with 421 slaves on board. After a relatively swift passage of 21 days to Freetown, the brigantine was condemned with 383 slaves. It was subsequently discovered that she had been bought from an American at Princes, where her papers were acquired, and that her master was Da Souza's son-in-law. On receiving the report of the case, Lord Palmerston sent his thanks to Beecroft for his assistance in the capture.

The capture of this Portuguese vessel gave substance to Captain Wauchope's belief that preparations were being made to renew the slave trade under the Portuguese flag in the Bights. Lord Palmerston, too, was concerned. In response to information from the Admiralty that Brazilian vessels were being transferred to the ownership of Portuguese subjects at Rio, he instructed the Ambassador there to urge the Brazilian Government to forbid the equipping of vessels for the slave trade, the transferring of ownership of vessels so equipped, and the entry and departure of such vessels. He also deplored the continuation of the "deliberate and open manner in which Slaving adventures are continued by all the Colonial Authorities of Portugal from the highest to the lowest".[17] There was, however, a ray of hope on this last score. As a result of representation from London, a new, and, it was to be hoped, less corrupt governor was being sent to Princes Island, and it was reported that a Portuguese man-of-war was being deployed to cruise off that island against slavers.

Lieutenant Miall, still kicking his heels in *Forester* at Freetown, wrote to the British commissioners in early August on behalf of the 12 Prize Masters in charge of the slowly reducing number of prizes lying in the Sierra Leone River. The inept introduction of action under the Spanish Equipment Clause was causing a breakdown in the financial arrangements for captured slavers prior to condemnation, arrangements designed to avoid expense to the Treasury, whatever difficulty might be imposed thereby on the captors. Expectation of bounties to be paid on slaves seized usually enabled commanding officers to make allowances of between 2s.4d and 4s.6d per day to Prize Masters to cover their expenses while waiting at Sierra Leone. There were, of course, no slaves (and therefore no bounties) involved in Equipment Clause arrests, so no allowances had been paid. The Prize Masters, junior officers whose pay was "wholly inadequate to their maintenance as gentlemen whilst remaining in this harbour", were in serious difficulty during the extended wait for adjudication. The response, which offered little hope to these gentlemen, was that Miall's request for help would be brought to Lord Palmerston's notice.

At the end of August Lord Palmerston gave his opinion that the Mixed Commission might have safely proceeded to adjudication of the Spanish vessels at Sierra Leone in May, when the Senior Naval Officer had officially handed to the British commissioners a copy of the new Act of Parliament. This contradicted an earlier opinion that legislation was necessary only if British vessels were arrested under the treaty, not to enforce the treaty against Spanish slavers. The commissioners meekly apologised for having awaited instructions from Lord Palmerston.[18]

Nearly two months elapsed after the capture of *Joven Carolina* before the next arrest, and then there was a clutch of three in three days. The first two demonstrated the next, and entirely predictable, step taken by Spanish owners to evade the Equipment Clause: nominal sale to a Portuguese citizen, usually a resident of the Cape Verde Islands, while the former Spanish master, who was often also the owner, remained on board in a supposedly subordinate capacity. The first such slaver to be taken was the Portuguese brigantine *Esperança*, which had notionally shed her Spanish nationality at the Cape Verdes in June. *Pylades* seized her, with 477 slaves, off the Bonny on 17 September. There was a chase of some hours, but no resistance to capture. She was condemned in October, with 417 slaves, and it later transpired that she was then sold to a Mr Robert Hornell. In December she cleared out of Sierra Leone in ballast for Cadiz with an English master and crew, and it was believed that Hornell had sold her to Pedro Blanco, the notorious slave trader at the Gallinas.

There was a virtually identical encounter the following day when, at about 1500, *Thalia* sighted the Portuguese brig *Felix* leaving the Bonny River with 591 slaves.* The brig endeavoured to escape, steering for Havana, but the frigate, which must have been very skilfully handled, managed to catch her four hours later and board without resistance. The *Felix* was similarly condemned, and the 481 surviving slaves were emancipated. Each of these two slavers had been sold by their Spanish owners at the Cape Verdes for "Three Contos of Reis".

Another vessel, a schooner, had crossed the Bonny bar at about the same time as the *Felix*, and *Thalia*, having seized the Portuguese brig, continued in chase of the schooner, being joined in the pursuit by *Buzzard*. The schooner made no alteration of course during the night, and they caught her the next morning at 0700. She was a Spaniard, the *Atalaya* of Havana, with 118 slaves. There was a heavy toll of 25 Africans during the passage to Freetown, and only 88 lived to be condemned with the slaver.

Lynx paid a ten-day visit to the Cape in early September before returning via Ascension to the West Coast, and at the end of that month two welcome reinforcements arrived. On 1 October the famous Symondite brig-sloop *Columbine* (18) reached Simon's Bay, having made her way from England via the West Coast and Ascension. By now she was ten years old, and had already spent nearly three years on the West Indies Station rigged as a barque, but in 1834 the Admiralty had wisely decided to remove her third mast and refit her as a brig. In this new configuration she was regarded as a peerless sailer, particularly so in the hands of Commander Tom Henderson, who recommissioned her in 1834. Another Symondite of a later generation had preceded her at the Cape by one day. This

---

* Three sources, two of them reports from the Sierra Leone Commissioners, give markedly different figures for slaves seized in this prize: 603 (C-in-C), 591 and 557. The discrepancy between the first two might be explained by a miscount at the time of capture, and 557 is probably a clerical error.

19. HM Barque *Columbine* off Lisbon

was the brig *Bonetta*, marginally smaller than *Columbine*, and commanded by Lieutenant H. P. Deschamps. Her armament of three 32-pounders, one of them an eight-foot gun on a pivot, was mounted in a novel arrangement which allowed them to be moved around the deck as circumstances demanded. She was a very fine command for a lieutenant.

The flow of prizes increased a little in October, beginning with a pair for *Curlew*. Her first, taken off the Bonny River on the third after a four-hour chase, was the *Esperança*, the second Portuguese vessel of that name taken in a little over a fortnight. Unlike the earlier prize, this was a brig, and she was captured with 438 slaves. She had been built in Baltimore, and, although she nominally belonged to a Portuguese lower-deck seaman, and was tried as such, she was actually owned by a Frenchman and a Belgian resident in Bahia. There were 30 deaths on passage to Freetown, and 396 slaves were emancipated when the brig was condemned. *Curlew*'s next prize was also Portuguese, and, unusually, a ship. She was the *Quatro de Abril* of 365 tons, taken about 50 miles south of Lagos on 19 October, the day after sailing, and loaded with 478 slaves from Whydah of whom 20 died before condemnation. Ten Kroomen were found among the slaves, kidnapped at Whydah. They had been loading palm oil into a vessel in the Benin Roads when they decided to abscond in their employer's boat and return to the Kroo Country, the best part of a thousand miles away. With provisions exhausted, they had landed near Whydah.

Forester had at last been released from her tedious watch over the prizes at Freetown, and on the afternoon of 20 October, after a chase which lasted much of the day, she took the Portuguese schooner *Victoria*, bound for Havana with 380 slaves, about 25 miles south-east of the New Calabar, where the cargo had been loaded. The slaver made every endeavour to escape, but offered no resistance when Miall's men boarded her. During her adjudication, which condemned her and emancipated 316 slaves, it was discovered that she had been the Spanish *Iberia* until January 1836, but had then been sold to a resident of the Cape Verde Islands, emerging thereafter for a successful slaving voyage under Portuguese colours, another beneficiary of a process which was becoming the standard means of evading the Spanish Equipment Clause.

Thalia had paid a visit to Ascension to load stores to replenish the cruisers in the Bights and, that task accomplished, she was making her way back to the Cape, leaving Commander Craigie in *Scout* as Senior Officer on the West Coast. *Columbine* and *Bonetta* were heading north for the Bights, and *Leveret* was steering in the opposite direction for Simon's Bay before being despatched to Mauritius.

Away from the Bights, *Charybdis* made a seizure on 21 October off Grand Bassa on the Grain Coast, catching, after a short chase, the empty Spanish schooner *Cantabra*, of Santiago de Cuba, on a course for Cape Palmas. The schooner's master deserted for the shore on the approach of Lieutenant Mercer's boat, and after some difficulty in establishing Spanish nationality in the master's absence, his vessel was condemned under the Equipment Clause. Then *Buzzard*, cruising the eastern part of the Bight of Biafra, added two more prizes to her already healthy score. Both were Portuguese schooners. The first, taken off the River Cameroons on the twenty-eighth, was the *Olympia* of Havana with 282 slaves, 252 of whom lived to be condemned and emancipated. The second was one of the small slavers owned by Princes Island residents and running between the mainland and the island. Initially there was no sign of slaves on board when she was intercepted in a squall off Fernando Po on 12 November, but the boarding officer, experienced in the tricks of the slavers, was suspicious of what might lie beneath the timber which apparently filled the hold to the hatch coaming. The answer was: 22 slaves, all but one of whom were emancipated when the schooner, the *Serea*, was condemned.

On 14 November *Columbine* announced her arrival in the Bights with a remarkable seizure which demonstrated the brig's fine qualities. On welcoming the newcomer on 10 November, Lieutenant Mercer of *Charybdis* bemoaned his failure, in company with three other cruisers, to catch a slaver in a 48-hour chase two days earlier. Commander Henderson immediately took up the challenge, "allowing *Columbine* to sail one-third faster than any slaver". Assuming that the slaver would aim to cross the Equator as soon as she could, he calculated a course to cut her off. He caught sight of her on the third evening after his meeting with Mercer. Next morning, 14 November, he brought her to, after a three-hour chase, about 40 miles west-north-west of Princes Island. She was the Portuguese brig *Veloz* with 508 slaves on board, taken from Lagos 12 days earlier and bound for Bahia, 48 of whom died before condemnation a month later.

It was as well that *Columbine* had caught her prize north of the Line. Three weeks earlier, the British commissioners in Freetown had received a despatch from London giving the Advocate-General's response to a query from Judge Macaulay. Portuguese vessels captured south of the Equator, although proved to have loaded slaves north of the Line, were, he said, not liable to condemnation.

Reports to the Foreign Office on recent cases at Freetown had stimulated thoughts of appointing a consul at the Cape Verdes. A provisional appointment of British Consul had, strangely, been held by an American, but (unsurprisingly) no useful information could be obtained from him, and Palmerston had decided in 1835 to supersede him with a Briton. No action had been taken on that decision, but for the time being there was an Acting Vice-Consul (Mr Merrill) in post, and he had reported to London that many Spanish vessels had recently been obtaining Portuguese papers and colours at the islands. The British commissioners strongly recommended that the Consul should pass information on slaving activity at the Cape Verdes directly to them at Sierra Leone, and reminded London that there was regular communication between the islands and the British settlement at the Gambia, and frequent vessels from there to Freetown.[19]

Shortly after *Columbine*'s success another first arrest was achieved, but this one had been much longer in coming. *Rolla* had been on the Coast since April 1835, without material success, and she had recently been keeping a fruitless watch on the rivers north of Sierra Leone. However, a visit to the Cape Verde Islands in November brought a change of fortune for Lieutenant Glasse. There he met the Spanish schooner *Luisita*, which was attempting, unsuccessfully, to change her nationality, and he followed her to the River Sherbro where, on 21 November, he arrested her under the Equipment Clause. The schooner's master later claimed, equally unsuccessfully, that his hatches, having been boarded over, qualified as closed hatches. On 2 December, in the same river, Glasse arrested the Spanish brig *San Nicolas*, also fitted for slaving. This vessel's fore and main hatches were positioned together, effectively forming one hatch, potentially causing a difficulty in adjudication because the treaty specified "hatches" in its equipment list. Fortunately this did not trouble the Court of Justice, and Palmerston later supported its view. Returning north during December, *Rolla* made up for lost time by taking two more Spanish prizes on the twenty-seventh, this time in the Rio Pongas. The first was the schooner *Esperimento*, empty but fitted for slaving, whose master had abandoned the *Gaceta* at Sierra Leone during the delay in adjudication; he also escaped his new command, with her papers, before *Rolla*'s boats arrived. The second prize, another schooner, was the *Lechuguino*, with 49 slaves. All, including the 49 Africans, were condemned.

The Mixed Court of Justice was steadily making its way through adjudication of the mass of Spanish prizes in the Sierra Leone River, finding little difficulty in condemning the majority of them on the strength of the Equipment Clause. However, the three prizes taken in November 1835 by *Curlew*, in controversial circumstances, were a different matter. In each case the initial seizure was an unlawful act and, the Court concluded, that act had thrown the prize into the power of the second seizor. *General Manso* and *Victorina* were therefore restored, but costs, which were heavy, fell to the masters and owners.* There was an added difficulty in the case of the *Josepha*. A considerable deficiency in her cargo had arisen while the vessel was in the charge of the Prize Master, Mr Reid, who had "proved himself a man of very imprudent habits, which brought upon him the full effects of the climate, and which led to his demise". It was the court's view that the seizor should not bear the cost of deficiencies caused by the neglect of the Prize Master over whom he had no control, but the court had no funds with which to make repayment to the vessel's owners for the £125 value of the missing cargo, and was therefore unable fully to restore the *Josepha*. The matter was referred to the Foreign Office.

Although the Foreign Office accepted the court's decision on liability for the *Josepha*'s losses, it would not regard the case as a precedent. In future, a blameless seizor would be held responsible for loss in a vessel owing to the misconduct of a Prize Master. Lord Stowell had stated that "it was a principle too clear to be doubted and too stubborn to be beat, that every Principal is civilly answerable for the conduct of his Agent."

Meanwhile, *Scout* too had taken her first slaver. On 5 December, her boats snapped up the 38-ton Spanish schooner *Gata*, which was attempting to cross the Bonny bar with 111 slaves. The boarding was made at night just inside the bar, and the boats were not detected by the slaver until they were alongside. There was no resistance. On board were a large number of casks, filled with water but,

---

* The Commissioners' report on these two cases ran to 191 pages and was covered by a 19-page letter to Lord Palmerston.

according to the clearance from the Havana Customs House, intended on the return voyage to contain "Palma Christie" or Castor Oil, a substance not available on the west coast of Africa. At Sierra Leone 101 slaves were emancipated, and the schooner was condemned.

On 14 December the brand-new Symondite brig *Dolphin*, sister of *Bonetta*, reached Ascension Island, joining the Station under the command of Lieutenant Thomas Lorey Roberts, who had been First Lieutenant of *Trinculo* and temporarily in command of *Buzzard*. She sailed the following day for the Cape, and before her arrival there she was to capture almost the largest cargo of slaves yet to fall into the hands of the West Coast cruisers. By happy chance, eight days out of Ascension, and a little short of a thousand miles south of the island, she encountered the ship *Incomprehensivel*, which, when encouraged by a round shot passing over her at the end of a seven-hour chase, showed Portuguese colours. When *Dolphin*'s men boarded her they found 696 slaves, the survivors of 785 taken from various ports in Mozambique. She was homeward-bound for Rio, where she had been sold in April to a Rio resident. The original owner was on board as a passenger, and he, as well as the new owner, had signed the instructions for the voyage. A third signatory was the supercargo appointed at Rio, who had become the master for the return voyage. The flags of most nations were discovered on board. The Anglo-Brazilian Mixed Commission, guided by Lord Palmerston's rebuke after the *Maria da Gloria* case, condemned her, to the subsequent approval of the Queen's Advocate. However, by the time she arrived at Freetown, via Ascension, on 27 January 1837 there remained of her cargo only 506 slaves to be emancipated.

There had been exchanges of despatches between the Foreign Office and the Sierra Leone commissioners on certain points of the new treaty, and in December the commissioners found it necessary to clarify for Lord Palmerston the financial arrangements for slaver prizes. They explained that in the case of a condemnation the expenses of bringing a vessel to adjudication, including reception, maintenance and care of the detained vessel, slaves and cargo, and the execution of sentences, were defrayed from funds arising from the vessel's sale, although the Seizor's Proctor was employed by the seizor. If the prize was restored, the expenses fell to the captor. They pointed out that responsibility for provisioning the crews of detained vessels before adjudication lay with the captor, and with the Commissariat Officer thereafter. These arrangements were all well and good when there were proceeds of a sale and when slave bounties were to be paid, but Spanish slaver prizes were now to be destroyed, and for vessels taken under the Equipment Clause there were no slave bounties accruing.

By the beginning of 1837 there had been six cases in which the proceeds of the sale of vessel and cargo had failed to cover the cost of condemnation, and drafts had been drawn on the military chest of Sierra Leone to pay the shortfall. It was becoming apparent that not all of the implications of the treaty had been considered.[20]

Two of the Sierra Leone commissioners' despatches had displeased Lord Palmerston. The first reported in early September that on 5 May the Governor had permitted a vessel to clear out of Freetown with some of the crews of vessels awaiting adjudication; this, wrote the Foreign Secretary, should not have been allowed. However, bearing in mind that it was mismanagement by the Foreign Office which had caused the adjudication backlog, this chastisement seems a little unreasonable. The second source of dissatisfaction concerned the destruction of condemned slavers. "I think it would be proper to adopt at once a more complete & effectual mode of breaking these vessels up," wrote Palmerston, "although it may be somewhat more expensive." The commissioners responded by proposing that vessels should be cut through the deck and deck-beams along the centre line from stem to stern, cut down through the stem and stern to the waterline, and cut athwartships across the middle of the vessel and through the keel to divide the hull into two parts. This method would avoid the need for a bond from purchasers. Palmerston was satisfied with this solution, and the Foreign Office conveyed clearance from the Board of Trade for the Collector of Customs to register vessels built from recovered materials, as long as they were not reconstructions of the original vessels.

The Foreign Office had also taken exception to a paper circulated by Admiral Campbell to his cruisers giving answers and opinions, apparently derived from the British commissioners, on queries he had raised with them. The Advocate-General declared that he did not concur with what had been written concerning captured Spanish vessels claiming to be Portuguese. These vessels, he said, should be brought before the Anglo-Spanish court if they had infringed the 1835 Treaty, notwithstanding any false claim to Portuguese nationality. The commissioners were firmly rebuked by Lord Palmerston for expressing their opinion on points concerning the slave trade other than when pronouncing judgment in court on specific cases. Deviation from that rule, he believed, "may be productive of much mischief". He was concerned that the commanding officers of the cruisers had been misled. Judge Macaulay received the rebuke while on sick leave in London, and, protesting that he was unaware of the paper, assumed that it had resulted from a private conversation that had taken place in his home at Freetown with Captain Wauchope. He claimed to have offered no opinions, and he feared that Wauchope had misconstrued the thoughts that he had voiced. In May 1837 Campbell expressed his regret over the misunderstanding. It had not been his intention that the paper should have been generally distributed, and it was not issued as a formal instruction. Wauchope had apologised. That was the end of the matter, but the fact was that clarification and instruction on a number of points arising from the treaty certainly should have been circulated to commanding officers, and the blame for misunderstandings lay squarely with the Foreign Office and the Admiralty for failing to communicate the necessary information and advice to the Commander-in-Chief.

On 10 December 1836 there was issued at Lisbon a long-overdue decree by the Queen of Portugal prohibiting the exportation and importation of slaves, by land or sea, in Portuguese dominions north and south of the Equator, although there were exemptions for movement of restricted numbers of slaves between Portuguese possessions in Africa and to the Atlantic islands. It authorised severe punishments for masters and mates of slave vessels and those dealing in slaves or conniving in such dealing; and it specified penalties for authorities, particularly colonial authorities, failing to take action under the decree or conniving in slaving; and for Customs House chiefs found negligent in prevention of slaving. Enforced with determination, the decree would have dealt a massive blow against the Trade, but it remained to be seen whether Portugal had either the will or the means to enforce it at all. When a translation of the decree, sent from London on 27 January 1837, reached the British commissioners at Freetown on 23 March, they urged that the Anglo-Portuguese Treaty be "extended in the true spirit of the decree". Jurisdiction on national legislation was no business of the Mixed Commissions, but, had the decree reflected the true intentions of Portugal, the geographic limitations in the treaty should have been lifted and an Equipment Clause introduced. No such action was taken.[21]

On 1 January 1837 *Fair Rosamond* sailed from Ascension for England, and on the twelfth *Pylades*, also leaving the West Coast but in the opposite direction, arrived at the island on passage to the Cape and then to Mauritius to relieve *Pelican*. *Forester* too was making her way home via Sierra Leone, and left Ascension on 17 February. It would not be long before both *Fair Rosamond* and *Forester* returned. Meanwhile *Thalia*, which had returned to Simon's Bay on 16 November, had rehoisted Admiral Campbell's flag, and on 22 December headed north again to visit the slave coast. She called at Ascension on 11 January and then made first for Sierra Leone. There the Commander-in-Chief learned that two American men-of-war, the frigate USS *Potomac* and the brigantine USS *Dolphin*, had been on the Coast to visit the settlement of Liberia, and the *Dolphin* had called at Freetown in January.

There was a smattering of prizes during January and February of 1837, the first three of them falling to Commander Craigie in *Scout*. For ten weeks Craigie had been anxiously watching two large Portuguese brigs lying in the Bonny River, and on 11 January, with a strong northerly breeze and a thick Harmattan, they made an attempt to escape. The sloop's pinnace and gig had been sent into the river under the command of Lieutenant John Price, and they were lying five miles below Jew Jew Point at 0900 that morning when they saw one of the brigs under all sail, studdingsails aloft and alow, standing directly for them. Price ordered the gig, which was armed with a Congreve rocket, to a position lower down the river in case the brig passed or ran down the pinnace, and he, in the pinnace, coolly awaited the brig's approach. His boat was loaded very deep with water and stores and with a carronade in her bows, but, when the brig was within 100 yards and running at seven knots, she dashed under the vessel's main chainwale and succeeded in hooking on with grapnels. Expecting to run the boat down, the brig's crew had neglected to man her guns until the pinnace was alongside, and a discharge of round shot and grape from the boat's carronade, aided by a well-aimed shot from the pistol of Mr Arthur Barrow, Mate, at the man holding the trigger-line of a gun directly above his head, dispersed them. Next moment the pinnace's crew was on board and had taken possession of the brig, the *Paquete de Cabo Verde*, with 576 slaves.

Price had seen the second brig two miles upriver, and leaving Barrow in charge of the prize with four men, he immediately set off against a strong ebb tide to take her. She was removing slaves and surrounded by large canoes, each pulling 60–80 paddles, but the pinnace was soon alongside and had taken possession of what was found to be the *Esperança*, with 108 slaves. The canoes had succeeded in landing 500 Africans, and it would have been more had Mr John Smith, Gunner, not arrived in the gig and turned back five of the canoes, in one of which was the slaver's crew. The two prizes were armed with four 18-pounder long guns and two 9-pounders, and their combined crews numbered 63 men. Craigie justifiably commented that "no effort was lost on the part of the noble fellows I had the honor to command". Lieutenant Price received further commendation from the commissioners at Freetown for the "extremely clean and creditable condition" of the vessels and Africans when he arrived with the two brigs as Prize Master of *Paquete de Cabo Verde*. Nevertheless, only 452 slaves lived to be emancipated from his prize when she was condemned, and 89 from *Esperança*. It was not long before the *Paquete de Cabo Verde* was again slaving, under the ownership of an agent of Pedro Blanco.

January's third victim fell also to *Scout*, but less dramatically, as she attempted to enter the Bonny on the fourteenth. She was the Spanish schooner *Descubierta* of Puerto Rico, condemned under the Equipment Clause after an exceptionally slow passage to Sierra Leone, having arrived there on 29 March. On 20 January Lieutenant Deschamps in *Bonetta* claimed the next prize, the brigantine *Temerario*, recently reflagged as Portuguese, with 349 slaves bound for Havana, another unremarkable capture and also immediately off the Bonny River. The unusual aspect of this incident, however, was the great deal of violence among the slaves on the passage to Freetown, and the Prize Master had to take strong measures to restrain them to protect the vessel and crew. *Bonetta* followed her prize to Sierra Leone to see her condemned and the badly depleted cargo of 236 slaves emancipated.

Commander Tom Henderson and *Columbine* returned to the limelight in February with the only two arrests of the month, two very similar incidents on the fourth and tenth. Both prizes were Portuguese schooners bound for Havana, and each was taken after a five-hour chase, finished in a calm by the sloop's boats. The first was the *Latona*, carrying 325 slaves, and she was taken 40 miles south of Whydah, where she had loaded her cargo, having procured Portuguese identity at Princes to replace Brazilian papers from Bahia. The second prize was the *Josephina*, with 350 slaves embarked a few days earlier at Lagos, and she was brought-to by *Columbine* 45 miles south-south-east of Whydah. In contrast to January's prizes, these two suffered very light losses before condemnation: five in *Latona*, and four in *Josephina*.

British merchants and the masters of British merchantmen trading legitimately in the rivers of the slave coast continued to be sources of valuable intelligence for the cruisers and the British commissioners in Freetown. Even though their information was frequently received too late for effective action to be taken against particular vessels, it contributed to the general picture on the intensity of the slave trade in the various locations. Mr Benjamin Campbell, a merchant on the River Nunez, for example, wrote to Judge Macaulay in February 1837 apprising him of Spanish slaving on the Nunez and Pongas rivers, and he mentioned that he believed 15 vessels from Cuba had received Portuguese papers at Porto Praya within the past two months, having offered "douceurs" of 15 per cent on the value of vessel and cargo. "From what you know of Portuguese Governors," he wrote, "you may judge what is the probability of any of them having the virtue to resist the temptation of 1500 or 2000 dollars, freely given by any Spanish slaver for a set of Portuguese papers."[22]

Renewed friction with the French began in March when *Curlew* arrived at Sierra Leone and Lieutenant Norcott heard that an act of piracy had been committed on a British merchant vessel in the Rio Nunez by a schooner called *Niger* under French colours. Norcott hastened north, caught the *Niger*, investigated the matter and took the alleged pirate to Freetown for trial by the Vice-Admiralty court. *Curlew* then sailed for the Gambia in accordance with previous orders. There followed a correspondence between the Commanding Officer of the French corvette *La Triomphante* and the Sierra Leone Government in which it was pointed out to the Frenchman that the case was in the hands of the court. When word of the trouble reached London, the Foreign Office complained to the Admiralty of "vexatious and arbitrary proceedings" by the naval commander, and Norcott was required by the Admiralty to explain not only his actions against the *Niger* but also his conduct towards the French vessels *La Dorade* and *Henri*, the latter at Goree. Having dealt with the *Niger* incident, Norcott denied any knowledge of the *Henri*, *Curlew* not having been to Goree. He then reported that he had boarded a schooner under French colours on 9 April after a chase during which the vessel had shown no colours despite the firing of two muskets. She hove to after the second shot, but *Curlew*'s boat had closed to 100 yards before French colours were hoisted. The schooner's papers were offered to the boarding officer, but he did not examine them, and, after identifying his own vessel, he returned to *Curlew*. This Frenchman was probably *La Dorade*. There seems to have been no further action, but it is clear that an atmosphere of distrust persisted between French and British on the Coast.

The Commander-in-Chief continued his progression on the Coast in *Thalia*, arriving at Cape Coast on 22 February, and while Captain Wauchope inspected Cape Coast Castle the Admiral himself visited the British establishment at Accra. Both places were found to be in good order except that nothing had been done to replace the guns found defective at earlier inspections. There was no sign of slave trading on the Gold Coast. Passing through the Bights, the flagship transferred supplies to *Scout* and *Columbine*, and by 14 March she was off Princes Island, where she was joined by *Dolphin* on her return from the Cape, bringing despatches sent out in the troopship *Atholl*, which had arrived at Ascension on 5 February. The Commander-in-Chief found a considerable slave trade still operating in the Bights, mainly under the Portuguese flag in vessels believed to be Spanish. Spanish colours were rarely to be seen on the Coast. Campbell considered that his cruisers had done much to check the Trade, and he was pleased with the quality of the vessels recently deployed to the Station, believing that much might be expected of them. After calling at

Ascension, *Thalia* arrived at Simon's Bay on 22 April, finding there the *Pelican* on her return from Mauritius.

On the passage south the Commander-in-Chief had taken the opportunity to respond to renewed pestering by the Colonial Office about the northern rivers which suggested that "a man-of-war's boat may occasionally visit the Rio Nunez and neighbouring rivers". Admiral Campbell pointed out that *Curlew* was currently cruising that part of the coast and visiting the rivers, having relieved *Rolla*, and that the Gallinas was often visited, not only by the vessel stationed on the northern rivers, but also by cruisers passing to and fro. He mentioned, in particular, that one cruiser in every three requiring provisions in the Bights was sent to Freetown to replenish, and was ordered to check the Gallinas on passage both up and down the coast. He also explained that communication between Sierra Leone and the neighbouring rivers was so easy that the slavers in the rivers were aware of man-of-war movements and could generally take action to avoid arrest, the Portuguese by landing their slaves.[23]

Late in March the British commissioners received from the Foreign Office a response to their December representation concerning the financial implications of condemning and destroying Spanish vessels fitted for slaving, and, in particular, for some monetary allowance for the Prize Officers kept waiting at Freetown in these slavers. They were told that consideration was being given to granting a bounty based on tonnage of vessels condemned without slaves. Consideration was all very well, but more than a year would elapse before relevant legislation was enacted, too late for many of those entitled to such bounties.

The arrest of the 70-ton Spanish schooner *Cinco Amigos* by *Bonetta* on 30 March revealed a new trick to evade the Equipment Clause. She was a Puerto Rican vessel, taken off New Sesters, undoubtedly a slaver but found to have hatches of the size fitted in legitimate merchantmen, too small for adequate light and ventilation for slaves. The discovery raised fears that this attempted evasion might become common, to the serious detriment of slave cargoes. The schooner had landed her slaver-sized cooking copper to avoid its being used in evidence, but Lord Palmerston had directed that a caboose capable of housing a slave copper or boiler was adequate to demonstrate infringement of the Article. *Cinco Amigos* was condemned. She was the only prize taken during March, and the following month was no more successful. On 19 April *Dolphin* intercepted the schooner *Dolores*, one of few Spanish slavers under her proper colours, about 30 miles south-west of the Old Calabar River entrance. She was carrying 314 slaves loaded in the river five days earlier, and was steering for Cuba. By the time the schooner was condemned, 28 slaves had died, 21 of them from dysentery.

An achievement of a quite different and novel kind was concluded on 9 March when a convention for the "protection of trade" was signed at the King's House at Bonny between Great Britain and King Pepple of Bonny. In exchange for an undertaking to restrict the slave trade from his territory, the King was promised protection for legitimate trade. The signatories on behalf of Great Britain were Commander Craigie and Lieutenant Dyke Acland of *Scout*, Lieutenant Huntley of *Lynx* and Lieutenant Roberts of *Dolphin*. It is unclear whether or not these officers had been granted any authority to exercise such powers, and it is likely that they acted on their own initiative. Nevertheless, in June 1839 Lord Palmerston recommended to the Treasury that it should sanction the annual payment, for five years, of 2,000 dollars' worth of goods in exchange for total and final abolition of the export trade in slaves throughout Bonny. The following month the Treasury agreed to make the payment. Clearly there would be a need for a firm hand to keep the Bonny chiefs to their agreement, but Craigie's diplomatic coup was the precursor of many such local treaties negotiated by the Navy which, in due course, were to have a most important bearing on the anti-slaving campaign.[24]

His part in this treaty and the arrest of the *Dolores* were the last notable acts of the life of the Commanding Officer of *Dolphin*. Ten days after his final capture, Lieutenant Roberts died of fever at Fernando Po. His death was the first blow to the Squadron of a season of serious sickness in which the cruisers were to lose 77 officers and men from a total mean strength of about 720.[*] *Dolphin* was not the worst hit, but the final toll on board her was 13 from a ship's company of 80. A Mr Pike had been appointed temporarily from *Raven*, which was surveying on the Coast, to be Second Master of *Dolphin*, and at the time of Roberts' death Pike was about to rejoin his ship. Such was the extent of the fever epidemic in *Dolphin*, however, that Pike shortly found himself in command of the brigantine, and he was ordered to take her to Ascension.

There were many instances of sickness and death at the Gambia and Sierra Leone, and *Curlew*, cruising off the northern rivers, suffered more loss than any other cruiser. Lieutenant Norcott hired a vessel in the Gambia for the accommodation of his sick, and 20 of his 75 people died. *Columbine* was healthy until late May when the rains set in, and in June there were eight cases of fever, mostly from the crews of boats sent into the Cameroons and Bimbia rivers. Of these the Master and two seamen died, and she later lost another man. She was sent

---

[*] Alexander Bryson in *Report on the Climate and Principal Diseases of the African Station* (W. Clowes & Son: London, 1847) gives a Total Mean Force of 815 for 1837, but that is assumed to include the survey vessels *Aetna* and *Raven*, whose nominal ship's companies were 67 and 30 respectively.

to Ascension to recuperate. In June the men of *Buzzard* experienced great discomfort when she was hauled up on the beach at Fernando Po for work on her hull, exposing them to a combination of heavy rain and intense sunshine which produced a temperature of 86 °F in the shade. Fever claimed the Master, the Assistant Surgeon and two men. The brigantine's boats, meanwhile, were working for several weeks amongst the mangrove swamps of the River Cameroons, entirely exposed to the weather. When they returned, the incidence of fever rapidly increased until there were 23 on the sick list, of whom three died. *Buzzard*'s loss finally amounted to eight.

Pelican arrived at Ascension on 3 June, after six months in the Indian Ocean, en route to assume the duties of Senior Officer on the West Coast, and she found *Scout* lying there. After leaving the Bonny, *Scout* had been badly struck with fever. She had lost Lieutenant Charles Acland, Mr N. E. Locker, Mate, and Mr John Smith, Gunner, some of her remaining officers were ill, and the crew was debilitated. Commander Craigie had already written to the Commander-in-Chief regarding the death of his Gunner, who had left a widow and two young children wholly unprovided for, asking that they should receive the "favourable consideration of Their Lordships".[25] In addition to her sickness, *Scout* was in need of new lower and topmast rigging, and Commander Popham of *Pelican* ordered her to the Cape for a refit. She arrived there on 10 July with her health much improved but with a final death toll of six. *Scout* sailed again for the Coast on 31 August, refitted and restored.

Pelican, meanwhile, returned to Fernando Po, and after two days at Clarence Cove she sailed for the Bonny on 7 July. Within the next 18 days there were 13 cases of fever on board. Worst hit were those who had been ashore at Fernando Po, almost all of whom contracted fever or suffered "deranged state of the biliary system". During the damp weather of July and August, with its frequent thunderstorms and heavy rain, the ship's company was almost constantly clothed in the "blanket dresses" introduced by Commodore Collier in *Sybille* as protection against fever. Surprisingly, *Pelican* suffered only two fatalities.

Fair Rosamond returned to the Coast from England in September, under the command of Lieutenant William Oliver. On arrival in the Bights she went into the River Benin, and remained there from 23 September until 3 October. During that period the ship's company were allowed a quarter-gill of rum mixed equally with water, night and morning, in addition to the daily ration, as a prophylactic. Six men fell sick and one died while the schooner was in the river, and there were subsequently three more deaths. *Waterwitch* remained healthy until October when she took station off the mouths of the Bonny and Calabar rivers, remaining almost constantly at anchor close to the shore, and sending her boats on detached service up the rivers. She too paid the price of lengthy proximity to the shore, losing five men.

Lynx escaped sickness almost entirely. After leaving the Gold Coast and the Bights in May, she called at Ascension and arrived at Simon's Bay on 2 June with returns for the Commander-in-Chief from the cruisers on the West Coast. Pausing for only four days, she sailed for Spithead, touching again at Ascension and at Freetown to land her Kroomen. Lieutenant Huntley had preceded her home, having been sent from the Bights by Commander Craigie to take the documents on the Bonny trade agreement to England. Lieutenant Birch had been transferred from *Scout* to take temporary command, thereby escaping the depredations inflicted by the fever on his own ship. The remaining cruisers on the Coast, *Bonetta* and *Charybdis*, lost six and four men respectively. *Forester*, which had left the Coast in February and did not return until December, lost four during the year. *Pelorus*, which arrived at Simon's Bay from England on 1 June before sailing for Mauritius, passed through the Station untouched by disease.

There had been a number of reappointments of officers during these months, occasioned not only by losses to fever but also by requests to exchange. The first of these voluntary exchanges, in April, was between Lieutenant Patrick Campbell, in command of *Buzzard*, and Lieutenant L. R. Stoll of *Thalia*. The reason is unknown, but it may have been a gesture of friendship by Campbell, who had been fortunate to be given *Buzzard* on the death of Lieutenant McNamara in 1835 and had achieved appreciable success in her. The death of Lieutenant Roberts in *Dolphin* led in July to the promotion to lieutenant of Mr John Macdougall, Mate, of *Thalia*, and his appointment to command the brigantine. Only a short time later, however, Macdougall and Patrick Campbell requested an exchange, which was approved. Campbell won the almost-new Symondite *Dolphin*, a clear improvement on *Buzzard*, but it seems strange that he, an already proven commanding officer, had not been appointed directly to her. It was perhaps a scheme to secure promotion for the deserving Macdougall. Another death-replacement was that of Frederick Archibald Campbell, College Mate, of *Thalia*, and no doubt another relative of the Admiral, to be a lieutenant in *Scout* in place of Charles Acland.

A handful of prizes were taken during the fever season. On 11 May Lieutenant Mercer in *Charybdis* detained the *Lafayette* under Portuguese colours about 50 miles south-east of Lagos. She was carrying 448 slaves loaded the previous day at that port. The vessel was unusual on two counts: her national identity was probably legitimate, and she was a three-masted schooner. She showed a passport from Lisbon but had sailed from Bahia, and she initially declared that she was returning there. Later, however, she admitted that she was bound for Havana. The *Lafayette* was condemned, and 441 of her slaves lived to be emancipated.

Dolphin, despite her heavy losses to the fever, managed two arrests during the summer, both under the command of Lieutenant Joseph Bates.* The first, the Portuguese-flagged schooner *Cobra de Africa*, which she took on 27 May about 50 miles south of Bimbia Island, had undoubtedly shed her Spanish identity at the Cape Verde Islands. However, as she was carrying slaves, 162 of them, embarked at the River Bimbia on the day before capture, she was adjudged and condemned by the Anglo-Portuguese court. There was a sad death toll of 61 of her slaves before emancipation. This success was closely followed by another. *Dolphin*, cruising between Fernando Po and the Old Calabar River on 1 June, caught another schooner, the *Providencia*, also under Portuguese colours but only recently bought by a Portuguese at Princes. She was bound from Lagos to Bahia with 198 slaves and, by stark and happy contrast with *Cobra de Africa*, lost only five of them before emancipation and condemnation.

In the course of the next two months there was only one arrest, that of the Havana brig *General Ricafort*, boarded a few miles off British Accra on 26 June by *Charybdis*. She was under Spanish colours and, although she was a large and roomy merchantman of 238 tons, unlike a slaver in appearance, she was carrying slave irons and the components of a slave deck in addition to her cargo of tobacco, rum and muskets. She was condemned without difficulty, and the sale of her contents and materials produced the unusually high proceeds of £3,137.18s.9d. The *General Ricafort* could claim the distinction of being the first slaver prize of the Victorian era, King William IV having died six days before her arrest.

In June Admiral Campbell gave the Admiralty an updated list of the commanding officers to whom he wished Instructions, now occasionally being referred to as Warrants, to be issued under the 1835 Treaty with Spain. The cruisers shown were *Thalia, Pelican, Scout, Pylades, Columbine, Bonetta, Buzzard, Charybdis, Curlew, Dolphin, Leveret, Lynx, Rolla* and *Waterwitch*. For some reason the brigantine *Viper* (6), which had joined the Station earlier in the year under the command of Lieutenant William Winniett, was not listed.† With the demands of the Cape

---

* The circumstances under which Bates found himself in command of *Dolphin* at this point are not known, and the only apparent evidence of the change lies in the reports on adjudications from the British Commissioners to Lord Palmerston. Only in the six-month tabular summary of cases is the name of "Joseph Batt Esq" mentioned for the captor of *Cobra de Africa*. However, it is clear from the report on the case of *Providencia* that Lieutenant Bates arrested her, and "Batt" would seem to be a clerical error.

† Curiously, the *Viper* also fails to make an appearance in the magnificent *Sailing Navy List* by David Lyon (Conway Maritime Press: London, 1993), but there is no doubting her existence or her presence on the Coast.

and Indian Ocean to be met, the flagship usually lying at Simon's Bay together with one vessel undergoing refit there, and three or four on passage between the slave coast and Sierra Leone, Ascension or the Cape, it was unlikely that as many as eight cruisers would be actively employed against the slavers at any one time. That number would include the Gambia guardship and any vessel visiting Fernando Po or Princes Island for stores or maintenance.

Resistance to arrest had been a rarity for some years, but evidence that this was not always the case, and also that boarding operations were not invariably successful, even when a suspected slaver had been brought-to, was shown in a "memorial" addressed to the Commander-in-Chief by Commander Popham of *Pelican*. Popham requests confirmation in appointment for his Acting Master, Mr James Russell, who had been severely wounded in his left arm by a volley of musketry fired into one of *Leveret*'s boats during a failed endeavour to board a Spanish brig in September 1836.[26] As it happened, before the Admiral could take action on the request, *Columbine* lost her Master to fever, and Russell was sent to replace him.

Popham was less complimentary when called upon in August to give judgment on the unsatisfactory Second Master of *Rolla*: "It appears to me", he wrote to Admiral Campbell, "he had the misfortune to be actuated by feelings and principles of a class not coming under the denomination of Gentleman; consequently he had not the power of reason to conduct himself like an officer."

Serious offences of a more specific nature were occasionally brought to the attention of Popham as Senior Officer on the Coast, as when Lieutenant Deschamps of *Bonetta* reported the conviction of one of his able seamen for "attempting to commit an unnatural crime on some of the Boys". Misconduct of that ilk was intolerable in the Service, and delay in dealing with it was unacceptable. Without reference to the Commander-in-Chief, Popham immediately ordered the man's discharge from the Service with disgrace.

After an entirely fallow month in July there was only a single arrest in August, and that was made by *Waterwitch*. On the promotion of Lieutenant John Adams, Lieutenant William Dickey was appointed in command of her, but at the time that she took the Portuguese brig *Amelia* on 6 August she was commanded by Lieutenant W. B. Marsh. The slaver was 60 miles south-east of Lagos when Marsh caught her, and was making for Bahia, three days out of Lagos, with 359 slaves, 345 of whom lived to be emancipated.

The sparsity of prizes did not reflect any reduction in the quantity of slavers. They continued to arrive on the Coast in considerable numbers and almost all under Portuguese colours. Admiral Campbell reported to London that they were all "well provided to evade the supposed new Portuguese treaty", revealing not only his misunderstanding of the purely national nature of the Queen of Portugal's decree but also, once again, a disgraceful absence of authoritative information and guidance to the Commander-in-Chief from the Admiralty and the Foreign Office. As far as the evasive measures were concerned, Campbell reported a new and worrying development which further undermined the 1835 Spanish Treaty. The empty Portuguese-flagged slavers were arriving with assorted general cargoes and carrying nothing to incriminate them. Their water casks, slave-deck timbers, slave irons and cooking coppers were being brought to the Coast in American vessels which, of course, were inviolate.

Lord Palmerston's patience with the perfidy of the Spaniards in their use of the Portuguese flag, and the apparent inability, or unwillingness, of the Portuguese authorities to prevent it, was wearing wafer-thin. In May 1837 he wrote a note to be sent to Lord Howard de Walden, the British Minister in Lisbon, who was to state to the Portuguese Government "that if this fraudulent use of the Port (*sic*) Flag shall Continue to be permitted Port Govt must not be surprized (*sic*) if a Flag thus prostituted to such base Purposes should no longer be respected by British Cruizers (*sic*)". This was pugnacious language, and it foretold strong action.[27]

Sierra Leone did not escape the ravages of disease. Indeed, many saw the place as the source of the cruisers' contagion. In June it was reported that there had been fever in the colony for two months, and in September the commissioners' despatches to the Foreign Office described "lamentable mortality" in Freetown. The fortunate Judge Macaulay gained Palmerston's permission in June for an extension to his leave in England until the end of September, and finally returned to his duties in mid-November. Perhaps contrary to his expectations, Mr Lewis survived the epidemic and, by the time of Macaulay's return, he had acted as Judge for 17 months. Lieutenant-Governor Campbell had acted as Arbitrator until June when he was replaced as Governor and in the Mixed Commission by Colonel Richard Docherty. The faithful Lewis was, in turn, given permission in September for home leave, but it was some months before he departed for England. By then he had the satisfaction of declaring to Palmerston, who had professed himself satisfied with the modified procedure for cutting-up condemned Spanish slavers, that a reduced cost for the work had been negotiated. The charge would now be three shillings per ton for the first 60 tons, and one shilling and sixpence for each further ton.

Late in September there was a brief upturn in the rate of success, beginning on the twenty-third with a resisted boarding by *Fair Rosamond*. The Portuguese brig *Velos* had been leaving the River Benin on 18 September with a full cargo of slaves when she was pounced upon by the schooner just to seaward of the bar and chased back into the river. There she rapidly landed her slaves, and when a boat attempted to board her she fired into it, killing one seaman. On 23 September Lieutenant Oliver took *Fair Rosamond* into the river, fired on the brig to deter any further resistance, and seized her. Some of the *Velos* crew had swum for refuge to another Portuguese vessel, the brig *Camoes*, which was awaiting a cargo of slaves, so Oliver boarded and arrested the *Camoes*, seizing the *Velos* men as pirates and putting them in irons on board *Fair Rosamond*. When the master and supercargo tried to leave *Camoes*, they were picked up by a boat from *Fair Rosamond* and also confined in irons. After being held for 36 hours on the grounds that she had given shelter to pirates, the *Camoes* was released, but then Oliver foolishly arranged with her agent ashore that, with assistance from *Fair Rosamond*, the slaver should be loaded with 138 slaves from the cargo landed by *Velos*. That done, she was again arrested on the twenty-eighth. *Fair Rosamond* meanwhile remained at anchor in the river, prevented by neap tides from crossing the bar and still with the master and supercargo of *Camoes* incarcerated on board.

When *Fair Rosamond* finally sailed on 2 October it was Oliver's intention to bring a charge of piracy against the master of *Velos* and his crew, but, after a number of delays, the *Velos* was finally brought to adjudication by the Anglo-Portuguese Mixed Commission in late April 1838. Her master and mate admitted shipment of 228 slaves on 18 September, in breach of the Additional Articles of the 1817 Convention with Portugal, and the brig was condemned. As far as the *Camoes* was concerned, Oliver had allowed anger and frustration to cloud his judgment, as had others before him, and he paid the price. The slaver arrived at Freetown in a particularly squalid condition and 22 slaves had died of dysentery, partially the result of the drunkenness of the Prize Master, *Fair Rosamond*'s Master's Assistant, who was subsequently discharged and left ashore when the schooner arrived at Ascension on 4 December. Oliver had no one more senior to put into the *Camoes*. When the court dealt with her in January 1838, it inevitably concluded that the Africans had been put aboard the *Camoes* not for the slave trade, but purely as a pretext for seizing her. She was therefore restored, the slaves were freed but not emancipated, and a decree was announced in favour of the slaver of £1,734.14s against Oliver. An expression of Their Lordships' disapprobation of his conduct added to Oliver's pain.

Lieutenant William Dickey in *Waterwitch* took another loaded Portuguese slaver on 25 September after she had cleared from Bimbia for Rio. She was nominally the schooner *Vibora de Cabo Verde* when *Waterwitch* caught her off Fernando Po, but she had probably been the American *Viper* until her transfer to Portuguese colours at the Cape Verdes. Her master claimed to have served in HMS *Maidstone* during the American War. Of the 269 slaves captured, only 221 lived to be emancipated when the brig was condemned.

On the same day, about 90 miles west of Princes Island, *Dolphin*, again under the command of Lieutenant Campbell, took the Portuguese schooner *Primoroza*, with 182 slaves shipped at the Bonny for Havana. Losses on passage to Sierra Leone were heavy in this slaver too, but, although the surviving 136 slaves were emaciated on arrival, the commissioners remarked on the clean and orderly state of the schooner, which "reflected great credit on the zeal and humanity of the Prize-Officer, Mr G. E. Burslem, whose conduct on similar occasions previously has been very praiseworthy." Indeed, Burslem had been similarly commended when he brought in the *Conde de los Andes* in October 1835.[28]

Late in September one of the more colourful and enthusiastic officers in the Service, Commander The Hon. Henry Keppel, arrived in his brig-sloop *Childers* (18) at Sierra Leone. After coasting south and east, making a number of boardings but no arrests, he arrived in the Bight of Benin to find *Waterwitch*, *Viper* and, also newly arrived, the "coffin brig" *Saracen* (10), Lieutenant Henry Hill. Keppel recorded his opinions not only of *Childers*' fellow cruisers but also of the men against whom they were pitted. "Most of the captains of these slavers are superior men," he wrote, "some belong to good Spanish and Portuguese families: generally young, I believe many of them take command of these vessels for the excitement of the service." The commanding officers of these little cruisers were largely starved of the fellowship of men of equal rank, and it was not surprising that they enjoyed the company of amiable and intelligent young slavers when they met on neutral ground. Their determination to make prisoners of these "superior men" when circumstances allowed was, however, undiminished.

The Foreign Office continued to worry at the matter of slaves landed at Fernando Po and subsequently emancipated in their absence. In a further attempt to allay concern in London, Commander Popham made a further investigation, with a result similar to that produced by Commander Castle nearly two years earlier. Only 50 of these emancipated Africans were living in Clarence village; 54 children had died as well as some of the adults, and others had gone willingly into the interior. All were "Practically free, contented and happy", and they strongly wished not to be moved. This seemed to be the final word on the matter, but it would not necessarily satisfy bureaucratic minds in London who could not bear a loose end.[29]

Bonetta fell in with the Portuguese schooner *Ligeira* on 10 November, about 70 miles south of Cape Formoso, and arrested her with 313 slaves. The prize had been given clearance by the Havana Customs House for a large boiler and materials for a slave deck, and she had acquired a passport at the Cape Verdes. The Mixed Commission condemned her, and emancipated 280 slaves. Ten days later *Scout*, of which Keppel wrote that she "loomed large in the Mosquito fleet: she was clean and very nice inside", made the first of two similarly routine captures. About 50 miles south-east of Lagos she boarded the Portuguese brigantine *Deixa Falar* with 205 slaves bound for Bahia, although the slaves were due to be landed 60 miles south of the port. The prize was satisfactorily condemned and the surviving 186 slaves were emancipated. Commander Craigie's second capture of the month, only three days later, was another Portuguese, the brig *Gratidão*, with 452 slaves embarked at Lagos. She too was bound for Bahia and was about 55 miles south-east of Lagos when caught. Her master swore that he lived in Havana, although his passport declared him a resident of Bahia. At the condemnation of the brig 380 slaves were emancipated.

Occasionally boarding attempts resulted in gallant failure, and a particular case in point occurred on 13 December. Lieutenant Dickey had stationed two of *Waterwitch*'s boats in the mouth of the Bonny River, and early that night they sighted a vessel subsequently identified as the Portuguese slave schooner *Donna Maria*, working her way out of the weather channel in a fresh breeze. The cutter, commanded by Mr J. A. Pritchard with 12 men, just missed the schooner, but kept up a smart fire which was steadily returned by the slaver. However, Midshipman W. E. Voules managed to make fast to the schooner in the four-oared gig. The discharge of a musketoon into the boat killed one man and dangerously wounded another, but Voules bravely held on until he saw that the cutter was rapidly dropping astern and there was no hope of support for him and his two remaining hands. He then wisely cast off and rejoined Pritchard. It was later learned that the schooner, which was carrying 350 slaves, had several men wounded. Admiral Campbell recorded his approbation of the two young officers.[*][30]

Leveret had been maintaining a remarkably low profile since her return to the Coast from the Indian Ocean, with no mention of her activities by either the Senior Officer in the Bights or the Commander-in-Chief, let alone an arrest to add to her successes of January 1836. On 6 December she arrived at Simon's Bay from Ascension, and ten days later sailed once again for the Mozambique Channel. The number of cruisers on the slave coast was maintained, however, by

---

[*] Lubbock writes in *Cruisers, Corsairs and Slavers* that Voules (he calls him Bowles) calmly picked off the slaver's crew with a fowling-piece, killing eight of them. This seems a far-fetched piece of embroidery, and the sober account by Voules's Commanding Officer is preferred for probable accuracy, if not for colour.

the return of *Forester* from England under the command of Lieutenant George Rosenberg. She arrived at Sierra Leone on 7 December.

There was one more arrest before the close of the year. On 26 December Edward Norcott in *Curlew* snapped up the Portuguese schooner *Princesa Africana* off the sea bar of the River Sherbro as she emerged with 222 newly embarked slaves on passage to Puerto Rico. As was to be expected, she had sailed from the Caribbean under the command of a Spaniard from Puerto Rico for the Cape Verde Islands, where a nominal sale had taken place, and it came as no surprise to find that her log was missing 26 pages recording the period prior to her arrival at the Cape Verdes. The Portuguese master had died since the sale. All 222 Africans survived to be emancipated when the Anglo-Portuguese court condemned the schooner.

Prior to this arrest, Norcott had found himself in conflict with the Governor of Sierra Leone. When *Curlew* anchored at Freetown on 6 December the Collector of Customs reported that she had yellow fever on board, and she was told to shift berth. Norcott responded that his only visitor since arrival had been the Surgeon of the Royal African Corps who, *Curlew* not having had a Surgeon for six months, had called to examine the brig's two sick men, neither of whom had fever, and he refused to move. Lieutenant-Governor Docherty remained unconvinced, largely because *Curlew* had recently taken to sea the colony's Queen's Advocate, sick with yellow fever, in the hope of saving his life, but within 48 hours of his being put ashore again on 6 December he was dead. Supported by a report from the Assistant Surgeon of *Forester* that there was no serious illness in *Curlew*, Norcott asserted that the claim by the Collector of Customs was based solely on idle rumour, and he dug in his heels.

The friction between Norcott and the Governor took a new turn in early January after *Curlew* had returned with her prize. Five British citizens had been seized from the Gambia by a Foulah chief, and on 3 January Docherty asked Norcott to take his brig to the River Nunez to release the five who were being held there as slaves. Norcott declined the request on the grounds that not only had he duty elsewhere but also he believed that there was currently fever on the Nunez and he had responsibility for the health of his 68 men. He would, he wrote, be happy to comply when circumstances permitted. Notwithstanding this refusal, when Norcott subsequently asked the Governor for the loan of a medical officer from the garrison, Docherty, although short of medical staff, sent on board the Staff Assistant Surgeon, who needed a passage to the Gambia. After his vessel's traumatic experience off the Gambia during 1837, Norcott's stance deserves some sympathy, but he was certainly tactless (if not unnecessarily uncooperative) in this series of exchanges with the Governor, and Docherty later complained to the Commander-in-Chief of Norcott's conduct.[31]

The year 1838 began on a sombre note as yellow fever re-emerged with renewed vigour. *Forester* joined Commander Popham, the Senior Officer on the Coast, in *Pelican* on 5 January and relieved *Columbine*, which was ordered to Plymouth, and Lieutenant Rosenberg brought a sad tale of loss. The brigantine was carrying the handful of survivors from prize crews at Sierra Leone; of seven officers and men sent by *Waterwitch* three returned, *Bonetta* had also lost four out of seven, and all of *Fair Rosamond*'s prize crews sent in *Velos* and *Camoes* had died. The survey cutter *Raven* was at Cape Coast having lost four officers and several men from her complement of 30, and her Commanding Officer was in a parlous condition. *Forester* herself had lost five men on passage. She left Popham on 6 January to take up her station off the Bonny, but when Lieutenant Dickey of *Waterwitch* met her and saw the sickness of her ship's company he very properly ordered George Rosenberg to proceed to Ascension without delay. It was too late for Rosenberg, who died at sea on the twenty-third, and by the time she reached Ascension *Forester* had buried 16 seamen, marines and boys.

Bonetta was in a similar state, and she too made for Ascension. She arrived on 30 January having lost eight men, including her Assistant Surgeon, on the passage. Another Assistant Surgeon joined her on 3 February and found the Commanding Officer, Master, Assistant Surgeon, Purser and 28 seamen and marines lying about the deck in a "most helpless and melancholy state, three with black vomit, and to all appearance beyond the aid of medicine. The vessel was in a very filthy condition, the stench from the holds being almost insupportable, and totally incompatible with health." The Indian corn and yams she was carrying from Accra for the Ascension garrison had rotted. Tents were erected ashore on the island and the entire ship's company was landed. The few remaining Europeans and three Africans fell sick, bringing the total to 39, of whom a further eight died, among them the Commanding Officer, Lieutenant H. P. Deschamps. Three more were invalided.

These two were not the only fever victims to seek refuge at Ascension. Captain Vidal, in the survey ship *Aetna*, was obliged to retreat there from the slave coast when almost all of his ship's company of 67 were stricken and 25 had died. At Ascension Vidal ordered *Bonetta* to sail for the Cape to recover, which she did on 13 March. Vidal recommended to the Commander-in-Chief's consideration Mr Roberts, Master's Assistant of *Bonetta*, who was her only officer fit for duty for three weeks,[*] and Mr Elliot, Assistant Surgeon of *Buzzard* (late of *Columbine*),

---

* "Officer" in this context would have included warrant officers.

who had removed into two sickly vessels on the deaths of their Assistant Surgeons, and who had cared for *Bonetta*'s sick ashore at Ascension.[32]

Sickness had also struck the slave-trade suppression cause at the Cape Verde Islands. Following a decision by Lord Palmerston as long ago as 1835, a British Consul, Mr Egan, had finally been installed early in 1838. At the end of February a new Portuguese Commissioner of Arbitration, Mr Gomes, arrived at Freetown bearing the news that Egan had died within a fortnight of arrival.

By late February Commander Popham was contemplating a further depletion of his force on the Coast. *Viper*, *Fair Rosamond* and *Saracen* were low on provisions, and *Viper* would have to leave for Ascension in March to replenish. She would be followed in early April by *Waterwitch*, which was in need of provisions, refit and convalescence, and by *Dolphin* later in the month. *Pelican* herself was also in need of maintenance, and Popham intended to take her to Ascension in May. After 11 months of cruising and 14 without refit, her copper, put on in 1835, was in a very bad state, and *Waterwitch*'s was no better. Furthermore, after a very short deployment, *Childers* was soon to leave for home. All of this would leave few cruisers on the Coast, but Popham was not greatly concerned; he expected little activity in the slave trade from April to June, and his plan would give him a rested and refitted force for the beginning of the new slaving season. Also, he was about to be rejoined by Commander Castle in *Pylades*, returning from Mauritius, who had sailed from Simon's Bay on 9 February, and *Forester* would be returning from Ascension when she had recovered.

For all its shortcomings on the introduction of the new Spanish Treaty and its failure to make headway with Portugal, the Foreign Office was continuing to inch forward in other areas of diplomacy. Although of no great significance to the suppression campaign, two new signatories, the Hans Towns and the Grand Duke of Tuscany, had been induced to accede to the Anglo-French Treaty, and their agreements were ratified on 12 September 1837 and 2 March 1838 respectively.

Meanwhile, anti-slaving operations had been continuing at a low ebb, with only two noteworthy incidents. On 2 March *Saracen* fell in with a schooner under Portuguese colours about 160 miles south-west of Sherbro Island and made her first arrest. She was unusually far into the Atlantic for a cruiser, but perhaps Lieutenant Hill was trying to catch the Equatorial Countercurrent to help him into the Gulf of Guinea rather than using the common practice of coasting in order to examine the rivers of the Windward Coast. The boarding party found ample evidence that the schooner, the *Montana*, had made a passage from

Havana to the Cape Verdes under Spanish colours, and the master admitted that Portuguese papers had then been procured for 400 dollars from a merchant of the islands. She was fitted for slaving, and there seems to have been no reason why she should not have been condemned under the Equipment Clause. However, *Saracen* encountered *Buzzard* on 7 March,* and the inexperienced Hill appears then to have decided not to take his prize for adjudication. He explained that he had expected a low probability of condemnation, and had doubts about the Mixed Commission acting on the Queen of Portugal's decree of January 1837. This sad confusion in Hill's mind, and probably that of the Commanding Officer of *Buzzard* too, led to a very poor decision.

Saracen continued to Freetown nevertheless, and there, on 11 March, she met *Waterwitch*. Lieutenant Dickey told Hill that he had boarded an empty slaver under Russian colours between Cape Palmas and Sherbro. The master had been Russian, but many of the crew were Spaniards. Commander Popham had already reported to the Commander-in-Chief that he had heard of a vessel slaving under Russian colours and another with either Tuscan or Austrian papers. He also lamented that many slavers were eluding interception thanks to the excellence of their information on the movements of cruisers. There were few places between Sherbro Island and Cape Lopez at which slaves were not embarked, he wrote, and, instead of sailing from fixed points, the slavers were marching their captives along the coast and embarking them "whenever safe opportunities for escape are almost certain".[33]

The second significant operational incident of these early months of 1838 was the arrest of the Portuguese brig *Felicidades* by Commander Craigie in *Scout* off the Old Calabar River on 8 March, the sole capture before early April. This "splendid new brig", armed with two 12-pounders, was carrying 559 slaves with whom she had sailed three days earlier, bound for Havana. She was undoubtedly owned by a Spaniard, but she had acquired a Portuguese passport at the Cape Verdes shortly after news of the Equipment Clause had reached the Coast. Losses on passage to Sierra Leone were appalling, and only 425 slaves were landed on 7 April. A further 14 died before emancipation, and three absconded. Five had thrown themselves overboard at sea.

To replace the recently deceased commanding officers, Admiral Campbell selected two of *Thalia*'s mates and appointed them to be lieutenants and in

---

* It is probable that *Buzzard* was on passage to England, and it is not known who was in command of her at the time of this encounter. Lieutenant Stoll replaced Deschamps in *Bonetta*, and, although the date of his move is uncertain, he probably took her from Ascension to the Cape; he was certainly in *Bonetta* at Simon's Bay in mid-April.

command: Colin Yorke Campbell to *Forester* and Henry Barnett Davis to *Bonetta*. Campbell took up his command, but it appears that Davis did not. A subsequent report on the issue of slave-trade Warrants to new commanding officers lists Lieutenant John Stoll, late of *Buzzard*, in *Bonetta*, and it was Stoll who wrote from *Bonetta* at Simon's Bay on 17 April to ask the Commander-in-Chief for an issue of warm clothing for the liberated Africans in his ship's company. They had been given only light clothing by the Liberated African Department at Freetown and were suffering from the cold of the southern latitudes.[*]

These appointments were virtually the last official acts of Rear-Admiral Sir Patrick Campbell as Commander-in-Chief. On 7 March the Third Rate battleship *Melville* (74) arrived at Simon's Bay from Ascension wearing the flag of Rear-Admiral the Hon. George Elliot. Five days later, Elliot relieved Campbell in command of the Cape Station, and *Thalia* sailed for Ascension and home.

* * *

The long-awaited Equipment Clause in the 1835 Anglo-Spanish Treaty initially caused consternation and mayhem among the slavers on the Cuba trade, but ultimately it was a disappointment to the Navy and the British commissioners in the Mixed Courts. The duplicity and venality of the Portuguese authorities, particularly at the Cape Verde Islands and Princes Island, provided too easy an avenue for the Spanish slavers to evade the provisions of the treaty, although, to its credit, the new Mixed Court of Justice at Freetown was receptive to evidence of fake sales and false papers.

The introduction of the new treaty measures had been chaotic, thanks to scandalous negligence by the Foreign Office, and the Admiralty was almost equally to blame. It is astonishing that, so much effort having been expended over such a lengthy period in persuading Spain to acquiesce to an Equipment Clause and slaver destruction, officials in London should have failed to attend to the imperative and perfectly obvious requirement immediately to report the treaty signing to the principal protagonists in the campaign at sea: the commanders-in-chief on the Cape Station and the North America and West Indies Station, and the British commissioners at Freetown and Havana. The Sierra Leone commissioners exacerbated matters by delaying adjudication of cases brought under the Equipment Clause until formal notification and instructions had been received from Lord

---

[*] It is to be hoped that the Africans did not have to wait for completion of the bureaucratic process of C-in-C to Admiralty to Colonial Office to Admiralty to C-in-C before their clothing was issued!

Palmerston, despite having been shown authoritative evidence of the treaty signing and ratification and of the subsequent British legislation. They deserve some sympathy, however. Messages received from London on the requirement for an Act of Parliament to bring the new treaty into force had been confusing if not contradictory, and, in a more general context, the Foreign Secretary, however uncritical he may have been of his own department's performance, was quick to chastise the commissioners for any misjudgment or failure on their part. They were, understandably, unwilling to risk Palmerston's displeasure.

The primary area of operations had remained the Bights, although the Windward Coast and northern rivers had not been neglected. Of the 85 vessels arrested on the Station during the past three years, 16 had been taken to the west of Cape Palmas. There had been an occasional foray to the south of the Line too, but the inexplicable directions from London not to interfere with the Portuguese traffic from southern African ports to Brazil would have discouraged further effort in that region even if more cruisers had been available. Nevertheless, four Spaniards had been captured south of the Equator, as had one Brazilian. Eighteen seizures had followed significant chases, some of them of considerable duration, not all of them by the newer cruisers, and it would seem, from some of these successes at least, that the deployment of *Columbine*, *Waterwitch* and the Symondites had achieved an improvement. Nevertheless, the great majority of arrests took place, as in the past, in the rivers and their approaches rather than in open water.

Resistance to arrest had become rare, thanks no doubt to a widespread appreciation of the fighting qualities and determination of the British Bluejackets, and the great killer with which the Navy had to contend was fever. The cruisers were still no better armed against its ravages, and nor were they any more capable, however hard they tried, of preserving the lives of captured slaves against the array of diseases which caused such carnage on voyages to Freetown.

Strategy and tactics on the Coast had not advanced one jot, and the only real progress had been diplomatic, in the Equipment Clause and slaver destruction measures included in the 1835 Anglo-Spanish Treaty. Portuguese colonial authorities and citizens were doing their best to enable the Spanish slavers to evade those new obstacles, and there were signs that Lord Palmerston's patience with Portugal was wearing thin.

CHAPTER 14

Obduracy and Obfuscation: Western Seas, 1835–8

Nobody speaks the truth when there's something they must have.
ELIZABETH BOWEN

THERE WERE A NUMBER of prerequisites for the commanders-in-chief on the North America and West Indies Station and the South America Station to fulfil their remit to suppress the slave trade. In essence, these were: watertight treaties coupled with the political will to enforce them; supportive administrations in Cuba and Brazil; integrity and disinterest on the part of the Mixed Commissions in Havana and Rio de Janeiro; and cruisers of the quality to catch and take slavers, in the numbers necessary to cover the approaches to Cuba and the slaving regions of the Brazilian east coast. None of these requirements were fully met on either Station, and in most aspects the reality fell woefully short of the need.

The addition in 1835 of an Equipment Clause greatly improved the Anglo-Spanish Treaty, but the inadequacy of Britain's treaty with Portugal offered the Spanish slavers a means of evading that clause, although that was availing them little in the courts at Sierra Leone. The shortcomings of the Anglo-Portuguese Treaty, particularly the absence of an Equipment Clause, were exacerbated south of the Equator by Britain's inexplicable reluctance to enforce the terms of the 1817 agreement by which slave trading was permitted in the southern hemisphere only between Portuguese territories. Despite the subsequent achievement of independence by Brazil, slaving to that erstwhile Portuguese colony from Portugal's African empire south of the Line continued unchallenged. An obvious means of evasion of the Anglo-Brazilian Convention was therefore open to Brazilian slavers: adoption of Portuguese nationality. Enforcement of the Portuguese Treaty was made even more difficult by the absence of an Anglo-Portuguese Mixed Commission court on the West Atlantic seaboard.

In Freetown the Anglo-Portuguese and Anglo-Brazilian Mixed Commission courts and the Anglo-Spanish Mixed Court of Justice were operating tolerably well, as was the Mixed Court of Justice in Havana. The same could not be said of the Anglo-Brazilian court in Rio de Janeiro, which was compromised not only by biased Brazilian commissioners but also by interference from the Brazilian authorities, interference which contravened the Anglo-Brazilian Treaty but which the British commissioners seemed not to have the determination to resist.

The third Mixed Commission court in the west, the Anglo-Netherlands court at Surinam, was still in existence, but with the virtual demise of the Dutch slave trade no case had been brought before it for many years.

In both Havana and Rio the Mixed Commission courts were working amid some degree of hostility to their existence. The Cuban colonial government and the island's merchants and planters unanimously wished the court and the British cruisers gone and to be allowed to pursue the slave trade without let or hindrance. They were also particularly anxious to avoid the dangerously disruptive and troublesome influence on the island of freed slaves emancipated by the court. The attitude in Brazil was more ambivalent. There were abolitionist elements in the central government, and Brazilian cruisers were conducting operations against inbound slavers, but regional authorities on the coast were compliant in the Trade, and there was widespread resentment against Britain's efforts to interfere in the affairs of this young nation by seeking to suppress its slave trade. This resentment was reflected in resistance to the authority of the Mixed Commission court at Rio.

These factors were well beyond the ability of the British Admiralty to influence, but shortage of cruisers was, on the face of it, a different matter. The reality, however, was that the Royal Navy, limited by typical peacetime constraints on funding, had heavy, worldwide commitments in protection of trade and the nation's other vital interests. Priorities had to be set, and in these burdensome circumstances suppression of the slave trade did not take precedence. Even the two commanders-in-chief charged with action against the slavers in the western Atlantic were likely to have higher priorities, and they certainly lacked the resources to fulfil all of their responsibilities effectively.

On the North America and West Indies Station in early 1835, Vice-Admiral The Rt Hon. Sir George Cockburn had under his command one Fourth Rate, two Fifth Rates, two Sixth Rates, ten sloops, three schooners and two steam vessels. This was his force for the western half of the North Atlantic and the Caribbean, an area bordered by the east coast of North and Central America and by the north coast of South America, and including the islands of Bermuda, the Bahamas and the West Indies. Even more stretched was Rear-Admiral Sir Graham Hamond on the South America Station, whose area of responsibility included both the Atlantic and the Pacific shores of the continent, and who had at his disposal one Third Rate, one Fifth Rate, five sloops, two brigantines and a gun-brig. A few of the Sixth Rates and sloops on both Stations were Symondites, but the eight

West Indies schooners of the early 1830s had been whittled away and Hamond lacked these invaluable anti-slaving cruisers entirely.

This, then, was the unpromising background against which the campaign continued in western waters in mid-1835.

* * *

Admiral Cockburn's sloops were responsible for the disappointingly few remaining arrests of the year in the West Indies. The Symondite brig-sloop *Serpent* (16), Commander Evan Nepean, achieved the first of these successes when she took the Spanish schooner *Tita* on 29 June a little over 40 miles south-west of Hogsty Reef in the approaches to the Old Bahama Channel. This Matanzas schooner had started her return from Whydah with 402 slaves, and she still had 394 of them at her capture. One died before condemnation, and on 6 August a Spanish brigantine sailed from Havana for Trinidad carrying 268 of the *Tita*'s emancipated female slaves.

It had been the intention to send the Africans from the three previous prizes to Trinidad, but the plan had been thwarted by the appearance of cholera at Havana. The Governor of Trinidad had insisted on a condition that there should have been no cholera in Cuba for at least three months before a consignment of emancipated slaves might be despatched to his island. The British commissioners were concerned that, if this condition was rigorously applied, none would ever be sent. They asserted that cholera was no longer contagious and was no longer appearing in Cuba in epidemic form, and the *Tita* women were sent on their way.[1]

Plans for removing emancipated slaves from Cuba were also threatened by a message in July from HM Superintendent at Honduras, Colonel Cockburn, reporting that, on account of the conditions laid down for the removal of emancipated Africans from Havana, the mahogany cutters did not wish to pursue their request for such Africans. In October, however, Colonel Cockburn conveyed a request from the settlers in Belize for 216 emancipated slaves, on the same conditions as those applying to Trinidad, as long as only two-fifths of them were female.

The total of six slavers taken in the first six months of the year failed to reflect the level of the Cuba traffic. The commissioners told Lord Palmerston at the end of July that the slave trade at Havana had reached an unprecedented height. During that month no fewer than 17 vessels had sailed for Africa and two had arrived safely in Havana. As usual, representations to the local authority had been of "no use whatsoever". The price of colonial produce had risen remarkably since the emancipation of slaves in the British colonies in 1833, and this had been one of the principal reasons for the rise in demand for slaves in Cuba. Also there had been a marked decrease of cholera on the island, and the planters were anxious to replace the Africans who had perished in the epidemic of 1833 and 1834. Perhaps most significant was the universal belief among the planters that the Trade was at last drawing near to its close, and they were buying before (as they supposed) it was too late. The British commissioners believed that there was yet one further factor: the "extraordinary increase" of the past 12 months was, they wrote, "principally owing to the impetus given by the establishment of [...] unprincipled insurance Companies".[2]

When the first rumours of a new treaty circulated on the island, the premiums on departing slavers rose to upwards of 40 per cent, but soon afterwards, following the receipt of private intelligence from Spain, they fell back to the standard premium of 22 per cent. When newspaper reports on the signing of the treaty were received, the insurance offices in Havana refused to insure any Africa-bound slaver against capture on her outward passage. Three Spanish vessels sailed from Havana in August and three at the beginning of September, but after 4 September the outward-bound Havana traffic was paralysed by the effect of the news of the new treaty on traders and insurers. Subsequent information from England persuaded the slave merchants that the newspapers had been incorrect, and five Spaniards and one Portuguese sailed in the second half of October. Four returning slavers arrived in Havana during August, having landed their cargoes elsewhere on the coast, and there were another four in September and two in October. By November panic had fully subsided and the traffic was back to a normal level, with six departures during the month and six arrivals.

In fact the risk of capture in the Caribbean region, new treaty or not, was low, and it is perhaps surprising that the slavers were not more aware of how few Admiral Cockburn's cruisers were. Apart from the schooners, they had in reality only the more modern of the larger vessels to fear, and these in the latter half of 1835 were the Symondite Sixth Rate *Vestal*; and the sloops *Racer* and *Serpent*, also by Symonds; *Champion*, designed by Captain Hayes; and *Larne* from Professor Inman's School of Naval Architecture. Cockburn was so concerned about his shortage of schooners for work off Cuba after the loss of *Firefly* that he sacrificed the services of *Pincher* as the flagship's tender to return her to anti-slaver duties, and his force of these small vessels, so useful in the hazardous waters around that island, was increased to four by the arrival of the elderly ex-American *Pike*.*
Considering the wide scope of duties to be fulfilled on the Station by small and

---

* *Pike* had been the American *Dart*, taken in 1813. She was unusually heavily armed, with two 6-pounders and twelve 12-pounder carronades.

handy men-of-war, including *Skipjack*'s employment in conveying the Governor of the Bahamas around his islands in April, the chances of a slaver encountering one of these nine in the approaches to Cuba were fairly slim.

As it happened, it was Captain William Jones in *Vestal* who made the next arrest, but not in the approaches to Cuba. On 7 October the Spanish schooner *Amalia* of Trinidad de Cuba was unfortunate to encounter *Vestal* off Grenada, and, after a chase during which she jettisoned her pivot-gun, she was taken 60 miles to the west of the island with 203 of the slaves she had loaded in the River Congo.* The prize reached Havana on 26 October, having lost three slaves, but it was not until 21 December that *Vestal* arrived and adjudication could begin. *Amalia* was at last condemned and her remaining 200 slaves were emancipated.

This case raised again the vexed question of the necessity of the presence of the captor at the proceedings of the Havana court, a presence still insisted upon by the Spanish commissioners, but, as long demonstrated by the Sierra Leone court, not required by the treaty. In this instance, Captain Jones was in command of the Windward Division of the Station, and the disturbed state of Venezuela required his presence on that coast, but he had been dragged twelve hundred miles or more off station thanks to the obduracy of the Spaniards. The British commissioners told Lord Palmerston that it would be timely to prevail upon the Spanish Government to recall the offending instructions to its Havana commissioners. Operational inconvenience to the Royal Navy was clearly not a persuasive factor as far as Madrid was concerned, but the commissioners pointed out that delays in adjudication were also "prejudicial to the Spanish Treasury".³

It was unhelpful to the British argument that there was currently no means by which the court could know in the absence of a cruiser whether or not she held the requisite anti-slaving Instructions. The Foreign Office belatedly asked the Admiralty to instruct the Commander-in-Chief to introduce the obviously necessary measure of formally announcing to the British commissioners the names of vessels and commanding officers issued with the Instructions, but in mid-January 1836 the list had still not reached the Commission.

As the year 1835 drew to a close, the commissioners in Havana had still received no official notification of the new Anglo-Spanish Treaty, signed six months earlier. The first they heard from a reliable source came from Commander Robert Fair of *Champion* when he arrived with the Spanish schooner *Diligencia*, which he had detained on 7 December. The slaver had originally sailed from Nuevitas for Mayumba, where she loaded 210 Africans. By then she had lost her master, and

---

* One source mistakenly gives the date of *Amalia*'s capture as 12 October. The true date of 7 October is confirmed by *Vestal*'s log.

20. HMS *Champion*, 18-gun sloop

there was appalling mortality on her return voyage. When she was intercepted by *Champion* off the north-eastern coast of Cuba she ran herself ashore near Punta de Mula, about 50 miles west of Cape Maisí, and four of the crew escaped in a boat with 16 slaves. Fair managed to salvage the schooner, and brought her into Havana on the twelfth, by which time a further 11 slaves had died since capture, and only 120 survived to be landed to a temporary lazaretto for quarantine. It was to the credit of the Spanish commissioners that, when quarantine expired on 23 December, they were prepared to sacrifice their Christmas holiday at the request of their British colleagues to deal with the case, condemning schooner and slaves. A further 26 slaves died before receiving their Certificates of Emancipation.

Commander Fair told the British commissioners that there were cruisers on the coast holding Instructions to seize Spanish vessels under the terms of the 1835 Treaty, and he particularly mentioned *Pincher*, Lieutenant George Byng. However, in the absence of orders to the contrary from the Foreign Office, the commissioners remained empowered only to adjudicate cases as agreed by the Treaty of 1817 and the Additional Articles of 1822. If, at this stage, a cruiser had, perfectly legitimately, brought in a prize taken under the 1835 Equipment Clause, the Havana Mixed Commission court would have restored her and condemned the captor in demurrage, damages and costs. No appeal was permitted against the decisions of the Mixed Commission courts, and, had this situation arisen, an extraordinary legal impasse would have been caused, thanks to the astonishing failure of the Foreign Office to pass timely revised instructions to its commissioners, and of a similar, but less surprising, failure by the Spanish Government. It transpired that copies of the new treaty were despatched to Havana from London on 29 October, four months after its signing and two months since ratification. Further instructions were sent on 16 November.

At the end of October the Commander-in-Chief reported that he had sent Captain Thomas Bennett in the Sixth Rate *Rainbow* to replace Commodore Pell as Senior Officer at Jamaica, but, in the event, Pell continued in the post, with responsibility for operations in Cuban waters for at least another year. His force was briefly enhanced by the arrival of the Symondite brig-sloop *Snake* to join her sister *Serpent* in the Caribbean, but the Station lost the services of *Racer*, which sailed for England from Bermuda at the end of November, and of *Vestal*, which, on sailing from Havana after the *Amalia* case, made for Bermuda and then left for home in mid-December. The Jamaica Division also gained the rather less useful ship-sloop *Cruiser*, Commander John McCausland, at the turn of the year.

Champion brought 1835 to an end with an unusual arrest, indeed the first case of its kind at Havana. While the *Diligencia* was in quarantine, Commander Fair gained permission from the British Commissary Judge to take some of his men out of the prize, and he sailed on a cruise. On 23 December he returned with the Spanish schooner *General Laborde*, which he had detained that morning in a river mouth 30 miles to the east of Havana. The schooner was empty of slaves, but, in Fair's opinion, "from every appearance of the vessel there was the strongest presumptive proof of her being a regular Slave Trader and of her Cargo having been very lately landed from her". It was initially assumed in Havana that she had been arrested under the terms of the new treaty, but this was not so, although she was certainly fitted for slaving, and it seems probable that Fair had not received the revised Instructions. The Captain-General demanded that she should be freed, there being no case to answer under the 1817 Treaty. Judge Macleay responded that no prize could be restored without a decree from the Mixed Commission following adjudication, and that the arrest had been made by right of the Additional Articles of 1822. The argument continued, but Macleay finally prevailed, although the court could then find no "clear and undeniable proof" that the vessel had carried slaves on her current voyage. She was restored and the captor was ordered to pay demurrage and other costs. *Champion* sailed for Jamaica on 6 January 1836, leaving for the Mixed Commission a letter enclosing a certificate from the obviously fair-minded master and owner of the *General Laborde* refusing to claim any compensation. Macleay felt some satisfaction that he had maintained the jurisdiction of the Mixed Commission, but he was not pleased with Fair for what he considered, perhaps rather harshly, an ill-judged seizure.[4]

During 1835, so the British commissioners reported, 80 slavers had sailed from Havana, two of them Portuguese, and a further seven had sailed from the island's outports of Trinidad, Santiago and Matanzas.[*] The commissioners had also counted 50 returning to Havana, and were aware of nine arrivals at the outports. They estimated that 15,000 Africans must have been landed. A more recent assessment is that 25,700 slaves were imported in that year, a total which had been exceeded only in 1817 and 1818 and had not been remotely approached since 1820.

On 10 January 1836 the British commissioners were at last able to acknowledge receipt of copies of the 1835 Treaty, but an enclosed note from the King's Advocate left them in a state of confusion. It informed them that it would be unlawful to proceed against British slavers without sanction of legislation, but that it might perhaps be lawful, by virtue of the treaty alone, to adjudicate Spanish vessels fitted for the slave trade in advance of an Act of Parliament to bring the treaty into force. For the time being, however, this inexcusable indecision was of no practical consequence because there had been no word from Madrid to the Captain-General and the Spanish commissioners in Cuba. It was still not possible, therefore, for the Mixed Court of Justice to be formed in Havana. It should hardly have been necessary for them to do so, but Macleay and Schenley impressed on Lord Palmerston the need for the British Minister in Madrid to insist on the treaty being officially communicated to the Captain-General forthwith.

After a lengthy fallow period, Admiral Cockburn's schooners began to make their mark once again in the New Year. On 7 January, Lieutenant George Byng in *Pincher*, probably relishing his freedom from the duties of tender to the flagship, took the Spanish brigantine *Ninfa* as she was making for Matanzas with 450 slaves. The slaver, although armed with three 12-pounder carronades, offered no resistance when *Pincher* caught her about ten miles east of Salt Cay, on the edge of the Great Bahama Bank, as she was heading for the Old Bahama Channel. Of her cargo of 578 loaded at Bimbia, the survivors consisted of 132 men, 92 women and 226 children. Another 17 were to die before the *Ninfa* and the remaining slaves had completed their six days of quarantine and been condemned under the 1817 Treaty.

The newcomer *Pike*, Lieutenant Arthur Brooking, was the next to bring a slaver into Havana, but this was a more problematic prize, the empty Portuguese brig *Esperança*. She claimed to have been on passage from Trinidad de Cuba to Lisbon when *Pike* caught her on 12 January, 20 miles south of the Isle of Pines,

---

[*] St Jago, the alternative version of Santiago, was still in occasional use in official correspondence.

that fruitful hunting ground of previous years. The *Esperança* was fitted for slaving, and Brooking had heard that a brig of her description had landed a cargo of slaves near Trinidad. He arrested her, it apparently having escaped his mind that there was no Equipment Clause in the agreement with Portugal. He was advised by Macleay that there was no point in taking her to Sierra Leone, and so, on the twenty-third, he sailed for Jamaica with the intention of taking his prize before the Vice-Admiralty court on a charge of carrying eight guns without authority from any established government. This course of action was in accordance with orders relating to piracy as well as slaving given to Brooking on 3 December by Commodore Pell, still in command at Jamaica. The fate of the brig is not known, but she may never have been brought to judgment because, on 5 February, *Pike* was wrecked on Bare Bust Cay, Jamaica.[*]

To exacerbate this loss of one of the four schooners, the Jamaica Division was deprived of its two Symondite sloops, *Snake* and *Serpent*, which sailed for Bermuda in early January. Hardly in compensation, there arrived in late February the *Nimrod*, an *Atholl*-Class Sixth Rate which had been "razeed" to become a 20-gun sloop.

In February the colonists in Honduras reconsidered their decision of the previous year not to take up the offer of emancipated Africans. Colonel Cockburn asked of the British commissioners that some should be sent from Havana under revised conditions, and the Captain-General was therefore told that he might despatch 300 of those recently emancipated. Shortly afterwards Lord Palmerston responded to a request from the Colonial Office that freed Africans from Cuba should be sent to the British West Indies colonies and to British Guiana by instructing the commissioners to afford every facility to meeting the request. These developments clearly came as some relief to Macleay and Schenley, who were concerned that most of the slaves lately released by the Mixed Commission had been sent to the interior of Cuba as "country labour" at a third of the price of slaves. These emancipated Africans were given for a period of five or seven years, but in reality they could be considered to have been sold. The British commissioners regarded the periodical inspections of these workers, required by Cuban regulations, to be a mockery.[5]

Also in February there was a development which was to prove distinctly unhelpful to the suppression campaign: the arrival in Havana of a new American Consul, Mr Nicholas Trist.[†] He immediately posted outside the consulate a notice

---

[*] The circumstances of this loss are not known, but the wreck was subsequently sold, which might indicate that there may have been little or no loss of life in the accident.
[†] Trist, who owned property in Cuba, had been secretary to Thomas Jefferson and had married Jefferson's granddaughter.

reminding all concerned of the restriction placed by United States legislation on the importation of black or coloured people into the USA. It pointed out that such importation was illegal unless the people involved were "to all intents and purposes" free. He then directed his energies to facilitating the illegal importation of enslaved black people into Cuba.

Champion was on passage from Honduras to Bermuda on 2 March when she fell in with a Spanish brig off the north coast of Cuba and ran her ashore 36 miles east of Havana. The master and almost all of the crew escaped with the vessel's papers and most of her cargo of slaves, although 32 of the Africans were drowned in the attempt to get them ashore when the brig grounded. Commander Fair managed to refloat the vessel and seize six whites and 188 Africans. The Spanish prisoners swore that they were all passengers, and that the brig's name was *Ricomar*, although she was known in Havana to be the *Zafiro*, and she was condemned under the 1817 Treaty as the *Ricomar*. A private letter found on board indicated that she had embarked a cargo of more than three hundred slaves at Whydah. Of these, 186 were emancipated.

Fair was intensely irritated that both *Champion* and the prize, both of which were healthy, were placed in quarantine for seven days on arrival at Havana. On protesting, he was told that that new regulations had been imposed by the Superior Board of Health regarding vessels from Africa, and that, as *Champion* had sent men on board the *Ricomar*, she had become subject to those regulations. Fair then complained to the Captain-General of a want of the usual courtesy shown to British men-of-war on arrival, and of the absurdity of captured slaves being placed in quarantine when slave cargoes were regularly being landed illegally in the immediate vicinity of Havana and sometimes marched through the city. Officers of British cruisers naturally supposed that the measures were really intended to inconvenience them. The Captain-General, in turn, complained of Fair's conduct, and there was clearly a strong antipathy between the two, but the commissioners told Palmerston that Fair's objection was well founded, if less than tactfully expressed, and that the complaint against him was overstated.[6] As a result of this exchange, in September 1836 Madrid ordered the Captain-General to give free pratique at the ports of Cuba to all HM Ships entering in charge of slave vessels.

The episode of *Champion* and the *General Laborde*, although it resulted in restoration of the prize, and the knowledge in Havana that the Mixed Commission had received copies of the new treaty, "certainly paralysed the traffic" in January, but that check lasted only for a short period, and in March eight slavers were despatched to Africa from Havana.

The Symondite sloop *Racer* had returned to the Station from England by the beginning of March, still commanded by James Hope, and she was about 70 miles to the east of the island of Dominica when, on the eighth, she intercepted the Portuguese brigantine *Vigilante* with a cargo of slaves. Hope sent her off to Nassau in the charge of Lieutenant Chambers, and there she landed the 231 slaves surviving from the cargo of 321 she had loaded at the Bonny River.* Despite the brigantine having been seized under the terms of the Anglo-Portuguese Treaty, the civil authorities at Nassau decided on her arrival on 17 March that, instead of proceeding to Sierra Leone for adjudication, she should be charged with piracy in the Bahamas Vice-Admiralty court. While awaiting the trial, Chambers cleared the slaver's hold and formed the opinion that the vessel was unseaworthy, a view supported by a local survey. He proceeded to "dismantle" her, reporting his actions to the Commander-in-Chief and Commodore Pell. It seems that the Vice-Admiralty court reconsidered its decision on jurisdiction at this point, and returned the Portuguese prisoners to Chambers, but the condition of the vessel was not reassessed until *Racer* arrived at Nassau. Finding the survey vessel *Thunder* there, Commander Hope arranged with her Commanding Officer that the officers of the two vessels should re-examine the prize, and he concluded that she could originally have been repaired in a few days at little expense. It was now too late, and the *Vigilante* was still lying at anchor when the flagship arrived on 5 July. The Commander-in-Chief removed *Racer*'s people and the Portuguese crew, and the brigantine was eventually sold at Nassau.

Chambers and witnesses were sent, via England, to Sierra Leone, and the case was at last brought before the Mixed Commission in November 1836. It was recalled that *Vigilante* had appeared before the Commission in 1835, and been condemned as *El Manuel*, prize of *Forester*. She had then been bought by a British merchant who sold her to the present master, who was an agent of the Gallinas slaver Pedro Blanco. Over nine months had elapsed since the arrest of *Vigilante* before she and the 231 slaves landed in the Bahamas were condemned, an inexcusable delay caused by the not only inappropriate but also illegal intervention by the Nassau Vice-Admiralty court.

Piracy seems to have been something of an obsession among the authorities in Nassau, and the Lieutenant-Governor asked the Commander-in-Chief to attach to him an iron steam vessel to deal with the perceived threat. Admiral Cockburn, however, believed that there had been no piracy in the vicinity of the Bahamas for a considerable time, and supposed that armed slavers may have been giving some cause for alarm. In any case, he had no vessel of the description requested,

---

* The number of slaves on board at the time of capture is not known.

and he emphasised that he always stationed to cruise in the Old Bahama Channel and in the approaches to Cuba as much of his force as he could spare.

On 17 March the Mixed Commission received a note from the Captain-General authorising the Spanish commissioners to act under the 1835 Treaty, at least until the new Mixed Court of Justice was installed in accordance with the new treaty. This welcome news introduced a new difficulty in the minds of the British commissioners. The 1835 Treaty required that the Mixed Courts of Justice should hand over emancipated slaves to the appropriate national authority without discrimination of gender, and that would interfere with the agreement made between the British commissioners and the Cuban Government for sending freed Africans to Trinidad. The commissioners asked Palmerston whether the agreement should be considered cancelled. Lord Palmerston did not want that, and nor did the Captain-General.[7]

Another Portuguese, the schooner *Criolo*, was the next slaver taken, and the episode was extraordinary only because the captor was the old *Cruiser*-Class sloop *Gannet* (16), launched in 1814 and converted from brig to ship rig in 1831. Commander John Maxwell caught his prize off the island of Haiti on 1 April with 315 slaves. There were eight deaths on the seven-day passage to Nassau, where the surviving 307 slaves were landed, and the schooner made her way to Sierra Leone. She had been Spanish at the beginning of her slaving voyage, but on her arrival at the Gallinas, under the name of *Carissimo*, her master heard the news of the new Equipment Clause and took her to the Cape Verdes in December 1835 to change her name and nationality. The Anglo-Portuguese court condemned vessel and slaves.

On 26 April Lord Palmerston despatched to Havana the appointments of William Sharp Macleay and Edward Wyndham Schenley as Judge and Arbitrator of the Anglo-Spanish Mixed Court of Justice, together with copies of the Act of Parliament which carried the 1835 Treaty into effect. Four days earlier, however, Macleay had sailed for New York and England to take leave for the first time since his arrival in post in 1825. It was the beginning of an unhappy period for the Havana Commission.

There was change too in command of the Station. On 1 May 1836 there arrived at Bermuda the elderly battleship *Melville* (74), flying the flag of Vice-Admiral Sir Peter Halkett. Three days later Vice-Admiral Cockburn handed over command of the North America and West Indies Station to Halkett and sailed in *President* for Spithead.

The new Commander-in-Chief had brought with him the *Harpy*, a ten-gun "coffin brig" commanded by Lieutenant The Hon. G. R. A. Clements, to replace her sister vessel *Savage* which had not been long on the Station. *Comus*, one of Professor Inman's ship-sloops, returned home after a deployment of only a few months, and the elderly and defective sloop *Scylla* had also been sent to England. There had been two valuable recent arrivals on the Station: the Symondite brig-sloop *Wanderer*, Commander Thomas Dilke, and the schooner *Lark* (6), Lieutenant Edward Barnett.

It was appreciated in London that the British commissioners were experiencing difficulty in ensuring that slaves emancipated at Havana were properly and humanely relocated by their Spanish custodians. By the terms of the treaties, freed slaves were not the concern of the Mixed Court once they were emancipated; at that point they became the responsibility, in the case of Cuba, of the Captain-General. However, as the Spanish authorities had agreed that the freed Africans should be sent to British colonies, it was accepted that Britain should deal with the matter, and the commissioners had, perforce, shouldered the task. This could not continue, and it was decided in May to appoint a Superintendent of Liberated Africans to undertake the work. The man chosen was Dr Richard Robert Madden, and Lord Palmerston told the commissioners to make no new arrangements for the transfer of Africans before Madden's arrival.

Palmerston had promised that someone from England would be appointed to act with Schenley in Macleay's absence, and in July he adopted the easy (and cheap) solution of promoting Schenley temporarily to the post of Judge and naming Madden as temporary Arbitrator. This was a poor decision. Not only did it undermine the purpose of appointing a Superintendent of Liberated Africans, but also it introduced dual accountability for Madden. As Arbitrator he was answerable to the Foreign Secretary, and as Superintendent he reported to the Colonial Secretary. To make matters appreciably worse, Schenley and Madden were incompatible in character. Madden was an arrogant, self-satisfied and resentful man, much given to standing upon his rights and dignity and who, as Arbitrator, was not prepared to subordinate himself to Judge Schenley or to cooperate with him. In addition, he was aggressively determined to maintain his independence as Superintendent. Schenley, well-meaning but uncomfortable in his more elevated role, lacked the strength of character to handle Madden. To the great annoyance of the Foreign Secretary, Madden, with little or no justification, repeatedly refused to sign despatches from the commissioners, and he bombarded Palmerston with closely written, seemingly interminable and almost unintelligible letters of complaint against Schenley, which were followed by letters of self-justification from his harassed colleague.* Palmerston, unfairly, blamed them equally for plaguing him. From late August 1836 onward the British Commission in Havana was not a happy organisation.

In July the Mixed Commission received from Admiral Halkett a list, dated mid-May, of HM Ships on the Station issued with Instructions under the new treaty with Spain. The Spanish commissioners, however, were still not prepared, in the continuing absence of direction from their Government to that effect, to admit production of the Admiral's letter as evidence of legality of capture or to dispense with the presence of the captor at adjudication. Added to this infuriating difficulty was the continuing absence of instructions to the Captain-General regarding the new treaty, although the Spanish Government claimed to have dispatched them in July 1835.[8]

Lieutenant George Byng in *Pincher* brought in another prize at the end of July, and the capture soon proved to be a particularly fortunate one on account of the passengers found in the slaver. Byng had made his capture on the morning of 12 July about 40 miles north-east of Matanzas, the port to which the prize, the Spanish schooner *Preciosa*, was returning from the Rio Pongas with 287 slaves, four African "servants" for the four passengers, and five Kroomen apparently hired to man the schooner. The vessel was not only condemned but also ultimately destroyed, the first to meet that fate at Havana in accordance with the new treaty, and the slaves were emancipated; but the verdict was not a straightforward matter.

Byng was holding copies of the 1835 Treaty and Act of Parliament bringing it into force, but not the appropriate Instructions. There was therefore no proof that *Pincher* was authorised to make the arrest, and the Commander-in-Chief's list of vessels so authorised did not show the names of commanding officers. Surprisingly, the British Judge managed to overcome this obstacle by promising to represent the circumstances to his Government, and he asked that the Commander-in-Chief's lists should in future include commanding officers' names and that the anti-slaving Instructions should have annexed orders from the Admiralty on carrying them into effect. The Spanish commissioners then accepted that the presence of a captor would no longer be necessary in court as long as his Instructions were presented. This did not immediately solve the problem, because once a commanding officer had surrendered his Instructions

---

* Palmerston was impatient with Madden's letters, and a secretary wrote in December 1836 to "advise privately" that Madden should write in a larger hand when communicating with Palmerston, "condensing, if possible, in a less number of Words the substance of what you wish to communicate". Palmerston's notes to his officials on the matter were more abrupt.

to a Prize Master he had lost his authority to make a further arrest, it being necessary for him to show his Instructions to the master of a prize on making the capture. The British commissioners therefore requested that cruisers should each be given two copies of the Instructions.[9]

One of the slaves died during quarantine, six more were lost after sentence but before landing, and it was decided to send the *Preciosa* to Honduras with the remainder of the Africans, apart from several too sick to go. It was intended that the Kroomen would be returned to Sierra Leone. The 261 due to leave in the condemned schooner, manned from *Pincher*, were issued with clothing before departure: a cap and blanket to each of them, a frock to each female, and a shirt and trousers to each male. Continuing in command of the vessel was Mr Richard Pridham, Admiralty Mate of *Pincher*, who had earned a commendation from Judge Schenley: "[to the] humane and judicious arrangements of this officer I in a great degree attribute the present healthy state of the Negroes." Palmerston later reluctantly approved of this use of the *Preciosa*, but he was concerned about overcrowding and, in particular, that any delay in the breaking-up of condemned Spanish slavers might provide an excuse for the Spanish authorities to defer (or defect from) other provisions of the treaty. Clearly there should be no repetition.

The *Preciosa*'s passenger list had attracted the keen interest of the British commissioners. Three of the four names, Josef, Ormond and Curtis, seemed familiar. It was concluded that Ormond, a 14-year-old boy, was the son of the late John Ormond, notorious slave trader on the Rio Pongas, and the Foreign Office later noted that the description of Curtis fitted that of the slave dealer who, in 1822, had directed the attack by natives on the boats of *Thistle* in the Rio Pongas, killing an officer and a seaman and wounding several more men. It seemed that the identity of Curtis would be impossible to prove, but this was not the case with Josef, apparently the owner of the *Preciosa*. Schenley issued a warrant for Josef's detention, and Byng held him in *Pincher* where he wrote a letter remonstrating at Judge Schenley's conduct, foolishly signing it "Edward Josiffe", confirming other strong evidence that he was the long-sought British slave trader.

Pincher sailed for Jamaica on 29 July to hand the prisoner over to Commodore Pell, and over the next ten months the case further diverted Byng from cruising against slavers. It had been found that there were among the *Preciosa*'s emancipated slaves several who claimed to have been kidnapped by Josiffe, and, on Admiralty orders, *Pincher* was sent in January 1837 to recover them from Belize and to take them, as well as the five Kroomen, to Sierra Leone as prosecution witnesses at Josiffe's trial. She also collected Josiffe himself for the passage, and reached Freetown on 23 April. After delays caused by difficulty in assembling witnesses, the trial finally took place before the Sierra Leone Vice-Admiralty court in August, and Josiffe was sentenced to 14 years' transportation.

The latter half of September 1836 brought a remarkable run of success for Captain William Jones in *Vestal*. The Sixth Rate had returned to the Station in the spring, and on the Commander-in-Chief's orders had sailed from Halifax in mid-August to patrol the channel between Trinidad and Grenada in the hope of intercepting the slave traffic entering the Caribbean by the southerly route on the Guiana Current. This initiative brought rapid benefit. On 20 September Jones took his first prize about 25 miles south-east of Grenada, and on the twenty-eighth he seized his second and third, 60 miles south-south-west of Grenada and 20 miles south-west of the island respectively. All three slavers were homeward-bound for Cuba.

His victim of 20 September, which he took after a short chase, was the Portuguese schooner *Negrinha* with 336 slaves loaded at the Gallinas, all but one of whom survived to be landed at Grenada before the schooner was despatched to Sierra Leone. There the Mixed Commission condemned her for a second time (she had appeared in 1835 before the Anglo-Spanish court as the *Norma*). The 335 slaves were emancipated.

The first of the two prizes seized on 28 September was the Spanish brig *Empresa*, arrested with 434 slaves from the River Congo. She was sighted about 15 miles to the south of Grenada, and *Vestal* chased her for four hours before boarding her. Lieutenant Tindal took her to Havana, arriving on 25 October with 418 slaves, after calling at Montego Bay in Jamaica to buy provisions and water for the Africans. The contentious matter of the captor's presence was again raised, although Tindal presented Captain Jones's papers. Judge Schenley pointed out that *Vestal* would not make Havana in much under a month, and that further delay would cause great suffering to the slaves. That obstacle overcome, another difficulty arose. On release from the seven days' quarantine three key witnesses – the master, chief pilot and second pilot – managed to escape during the night, and there was no hope of recovering them. However, as adequate evidence was still available, the Spanish Judge, Conde de Fernandina, agreed to proceed with the case, and the *Empresa* and 407 surviving slaves were condemned.

The last of Captain Jones's trio of prizes, and the second of 28 September, was another Portuguese, the brigantine *Fenix*, which had sighted *Vestal* at 1700 and mistaken her for a merchantman. After dark the cruiser managed to cross the slaver's tack and soon brought her to, without resistance, about 20 miles southwest of Grenada. The brigantine was carrying 484 slaves from Little Popo, and Jones landed them at Grenada on 6 October. The vessel herself was despatched for adjudication at Sierra Leone, but Jones ordered the Prize Master to call at Halifax to allow the Commander-in-Chief to inspect her as a possible replacement for the army's worn-out trooper brigantine *Duke of York*. On her arrival at Freetown on 9 December the *Fenix* was condemned and her slaves were emancipated.*

The capture of these three slavers, particularly the Portuguese vessels, presented Captain Jones with a manning problem. In a faintly sycophantic letter of congratulation to Admiral Halkett following the recent success, he explained that his full complement was 190 and that he had already been ten short before sending 40 officers and men away in prizes.† Having the added difficulty of 80 prisoners to guard, he had engaged 20 "stout Negroes [...] accustomed to the sea" and would probably take a further ten. If the requirement should arise to send any more Portuguese slavers to Sierra Leone, losing the prize crews for many months, he would certainly need to hire even more.[10] He had promised to land these people at Grenada when *Vestal* left the West Indies, and that is almost certainly where he had engaged them. At least he was able to shed the load of prisoners on arrival at Havana on 21 November, diverted a distance of 1,500 miles to leeward of his cruising station, unnecessarily, as it happened, and to the annoyance of the Commander-in-Chief.

During the late autumn Jones found himself engaged as a diplomat on behalf of none other than Captain-General Tacon. At the end of September the Commandant of Santiago de Cuba proclaimed independence for his province, and the Captain-General responded by blockading the port and cutting off all communication. *Vestal* had called into Port Royal on her passage to Havana, and found there a letter from the Consul at Santiago expressing concern for British interests in the town. Jones hastened there and, finding the situation volatile and threatening, he extracted from the Commandant an assurance of the safety of British persons and property, offering himself, unsuccessfully, as a mediator. In Havana the Commandant was preparing a military expedition to crush the revolt, and several influential landowners begged Judge Schenley to persuade Tacon to make a last bid for a peaceful resolution of the crisis. They also asked him to offer Captain Jones as an intermediary. Tacon agreed, and Jones was empowered to offer Commandant Lorenzo and his key supporters safe conduct out of the island together with a promise that their property would not be interfered with. He was also furnished with orders from the Captain-General to all heads of department in Cuba to assist and obey him if called upon to do so. *Vestal*, her Commanding Officer invested with these extraordinary powers,

---

* *Fenix* is often referred to in reports as *Phoenix*.
† Lyon's *The Sailing Navy List* gives the complement of these Sixth Rates as 240.

was towed to sea in late November, and proceeded on a successful mission. In recognition of this service, Jones was subsequently presented with a gold sword by the Captain-General, although the Admiralty declined to permit him to wear it with his uniform.

The adjudication of the *Empresa* generated another confrontation between the British commissioners and the Captain-General. Schenley requested, in almost obsequious terms, that emancipated slaves currently unfit for a further voyage might be temporarily landed in the care of the Superintendent of Liberated Africans, and asked for the Captain-General's support to enable Madden to perform his duty. He promised that the Africans would be removed to British colonies with the least possible delay. The Captain-General replied that he could not allow landing of "foreign people of that class" under the charge of another foreigner without understanding why they could not be removed. Schenley protested that this refusal was in contravention of the treaty in that it impeded the function of the Superintendent, an official appointed by agreement between Britain and Spain. The Captain-General, unmoved, pointed out that the regulations stemming from the 1835 Treaty required, in a change from the 1817 agreement, that slaves emancipated should be "delivered over to the government to whom belongs the cruizer (*sic*) which made the capture". He wrote, not unreasonably, that he had not failed in that and had no intention of doing so.

Conflict also erupted between the Mixed Court of Justice and Dr Madden. Lieutenant Tindal, the *Empresa* Prize Master, had asked that the slaves, who were without any covering, should be given blankets or clothing to protect them from rain and a strong north wind. With the agreement of the Conde de Fernandina, the request was passed to Madden, but the Superintendent refused to supply anything until he could oversee distribution. That was not immediately possible because the slaves could not be handed over to the British commissioners to be passed to the Superintendent until the Spanish authorities had completed registration, a process which took several days. On 11 November the Certificates of Emancipation were issued by the Captain-General, and Tindal was allowed to deliver the Africans to the Superintendent. At last they were clothed and immediately transferred to the Hamburg barque *Cuba* to be taken to Providence in the Bahamas. Schenley gave the barque a passport requiring HM Ships to give her free passage and any assistance necessary.

Madden was incensed by the denial of what he saw as his right to have slaves landed into his control on arrival, and he wrote a very long and barely coherent protest to Lord Palmerston, as well as another letter, similar in style, complaining of the inefficiency and rudeness of the apparently blameless Schenley. In response,

Palmerston made it clear that captured slaves remained under the control of the Mixed Court until it pronounced sentence of liberation, at which moment the Africans found in prizes of British cruisers came under the exclusive care of the Superintendent. Palmerston had not appreciated the further delay caused by the Cuban authorities' insistence on registration.

The capture of the *Empresa* achieved an immediate effect on the easily panicked businessmen of Havana. Insurance companies declined all further risks, giving as their reason the superior sailing qualities of the cruisers on the Station; and their fear of *Racer*, currently cruising off the north coast of Cuba, was so great that they hired (at considerable expense) some fast vessels in ballast to intercept and warn two notorious slavers expected to be closing the island.

During August and September several new schooners of between 50 and 150 tons, built in the USA, had arrived at Havana for sale. They were rigged as New York pilot boats, very lightly constructed, unarmed, fitted with 30 sweeps and with shallow draught. There was apparently a plan for them to sail for Africa in groups of three or four so that, if they were pressed in chase, one might be sacrificed to enable the others to escape. Two were bought and sailed under Spanish colours, but two more left in early October wearing the Stars and Stripes, heading for the Cape Verde Islands and fitted and laden for slaving. While these vessels were preparing for departure there was more than one visit to Havana by an American man-of-war.

This news worried Lord Palmerston who, in December, scrawled a note to his Foreign Office staff on the remains of the envelope from Schenley's despatch:

> Extracts to Mr Stevenson for communication to his Govt (Let me see extracts before they go) & state that H. M. Govt cannot entertain a doubt that the Govt of the US will take prompt & effectual steps for preventing the Flag of the Union from being used for the Protection of a Traffic which has been solemnly denounced as a Capital? Crime by the Laws of the United States.[11]

In October Judge Schenley had written to Consul Trist protesting at the continuing slave trade by Havana merchants in American vessels, under the Stars and Stripes, in connection with United States citizens and with crews under American protection. Trist, on return from a visit to Washington, sent the letter back to Schenley on the grounds that his Government refused to become a party to any discussion of the suppression of the slave trade. It was reasonably assumed that Trist was acting on the instructions of Washington, but a more encouraging

---

* Andrew Stevenson was the Virginia-born American Minister in London.

reaction came from Commodore Dallas, commanding the US Navy squadron in the area. During a visit by Dallas to Havana, Schenley "communicated very freely with him upon the scandalous abuse of the flag of his nation at this port as regards slave trading enterprises". Dallas promised to use every exertion to capture vessels slaving under American colours, and to detain any carrying Africans to Texas in contravention of United States law. Schenley regarded Dallas as a man of sincerity.

Early in November *Racer* demonstrated that the concern of the Havana insurers was well founded, although her two arrests resulted in disappointment. On the evening of the sixth, Lieutenant Hunt brought into harbour the schooner *Constitucaō*, under Portuguese colours, taken by Commander Hope on her way to Havana, clearly having landed her slaves on the coast of Cuba a day or two earlier. The crew had felt so secure under the Portuguese ensign that they had made no effort even to throw the slave shackles overboard. Hope and the British commissioners believed the schooner to be Spanish property and fitted-out in Cuba, and in all probability she had been given two sets of papers. No proof of Spanish nationality could be found, however, and Hunt handed the vessel back to her master, in accordance with his Commanding Officer's orders should the British Judge decide that the prize was outside the jurisdiction of the Havana court. Palmerston criticised the commissioners for failing to judge the case, but Schenley insisted that the necessary evidence was not available, although Madden felt that a more thorough search might have been made. Both agreed that it would have been impossible to find witnesses to Spanish nationality; it would have been suicidal to offer such evidence.

Two days later, Hope sent in the Spanish schooner *Manuelita*, arrested directly off Havana harbour under the terms of the Equipment Clause. She had the appearance of a slaver, but none of the listed fittings or implements could be found on board. The Prize Master, Lieutenant Seymour, continued his search all night in the hope of finding something incriminating under the ballast of sand and shingle, but he had no luck, and he handed the vessel back to her master. After landing her slaves the schooner had apparently been to one of the small harbours between Matanzas and Havana to be meticulously cleared of evidence. It had probably been her intention to claim damages from *Racer* in the event of prosecution, and possibly to divert the cruiser from her patrol at a time when a number of returning slavers were expected.

It seems that Commodore Pell in *Forte* was not permanently stationed at Jamaica as Senior Officer of the Division. It was Commander John Frazer of the sloop *Nimrod* (20) who, at Port Royal on 18 October, ordered *Pincher* to take up station in the eastern approaches to the Old Bahama Channel, between Cape Maisí and Punta Maternillos, and directed her to return to Port Royal after two months. *Champion* met her there on taking up her own station off the north-east end of Cuba in mid-November, apparently as ordered by the Commander-in-Chief, and it was *Champion* which made the first arrest, having shifted her patrol to the south of the island.

On 1 December, about 55 miles west-south-west of Cape Cruz, she seized the 60-ton Portuguese schooner *Carlota*, 50 days out of Whydah with 203 slaves and with papers obtained from the Provisional Government of Princes and St Thomas. Commander Fair initially intended to send her to Nassau, but her poor condition and the onset of a strong "Norther" persuaded him to bear up and run with her to Belize, arriving on 8 December. The prize was in a thoroughly unseaworthy state: the partners of her foremast had gone, some of her chainplates had carried away, the pintles of her rudder were insecure, one pump was disabled, and her rigging, tackling and general equipment were in a bad way. Added to that, she had only four days' water and provisions remaining. This may have been a particularly serious case, but the indications are that poor seaworthiness was not uncommon in Portuguese slavers. Fair was impressed by the prompt and humane treatment given to the slaves by Major Anderson, the Acting Governor of Belize. By noon on the day after their arrival they were all clothed, landed and comfortably housed and provided for. None had been lost since capture. A Prize Officer with the slaver's papers and her master and mate were sent to Sierra Leone in *Pincher*, already loaded with the prisoner Josiffe and witnesses, and on their arrival on 23 April 1837 the *Carlota* and all 203 slaves were condemned. The slaver was subsequently sold in Honduras.

Before sailing on her passenger-carrying voyage to Freetown, *Pincher* became engaged in an infuriating episode in her proper employment. On 20 December she brought in the Spanish brigantine *General Laborde*, which she had detained on the sixteenth off Gibara on the north coast of Cuba. This was the vessel taken by *Champion* in December 1835 and released, but now, with a modified rig, she was charged under the Equipment Clause. She was outward bound, supposedly for Gibraltar, with 828 bales of tobacco; but hidden amidst the tobacco were slave shackles, mess tins and tubs, large water casks, a stock of biscuit, cooking coppers, beams and numbered planks for a slave deck, and grating hatches. Any one of these items should have provided adequate evidence to condemn her as fitted for slaving, and it would have done so at Freetown. The Spanish Judge, however, disagreed with Schenley and gave his utterly perverse opinion that the vessel was employed at the time in a legal pursuit, apparently on the grounds that she had a woman and children on board.

The case went to arbitration, and the lot fell to the Spanish Arbitrator who, inevitably, supported his Judge, although, probably shamefacedly, he agreed that the arrest had been justified. Again the *General Laborde* was restored to her master, Schenley recording his dissent, and was freed to continue her slaving voyage. Completion of adjudication on 18 January 1837 had been followed by a three-day delay in releasing the prisoners, and the enraged Admiral Halkett protested to the Admiralty not only about the apparent pointlessness of the 1835 Treaty, but also at the wasteful detention of *Pincher* in Havana for five weeks. In 1838 the Spanish Government eventually agreed to adopt measures proposed by Lord Palmerston to prevent a similar incident and to avoid the outcome of the case becoming a precedent.[12]

The saga was not yet over for *Pincher*. The supercargo of the *General Laborde*, who had his wife and children on board, levelled a charge of gross misconduct against George Byng and his crew. The Captain-General forwarded the accusation to Schenley who, although believing it to be nonsense, promised to investigate. Byng produced a robust rebuttal, describing the charge as a "tissue of falsehoods and misrepresentations", together with affidavits, and the Captain-General quashed the allegation.

By then, *Pincher* had been despatched to Belize to collect witnesses for the Josiffe and *Carlota* cases, before embarking Josiffe himself for the voyage to Sierra Leone.* This valuable little vessel, one of only three schooners remaining on the Station, therefore continued to have her services frittered away. Indeed she made no further contribution to the suppression campaign, and she was wrecked in 1838. The choice of *Pincher* to make the purely administrative chore of a voyage to Freetown, instead of one of the sextet of old sloops or the "coffin brig" on the Station, is probably an indication of the priority given to slave-trade suppression among the Commander-in-Chief's multitude of tasks.

Duplicates of the Anglo-Spanish anti-slaving Instructions were issued to Admiral Halkett's cruisers in mid-December, and he listed 17 vessels as having received them: *Melville, Madagascar, Belvidera, Rainbow, Vestal, Nimrod, Racehorse, Racer, Wasp, Champion, Wanderer, Gannet, Cruiser, Harpy, Skipjack, Pickle* and *Pincher*. The schooners apart, only four of these could reasonably be expected to chase a slaver successfully: *Vestal, Racer, Champion* and *Wanderer*. Halkett felt

---

* The timing of *Pincher*'s departure for Belize is unclear. Halkett clearly believed that she had remained in Havana until at least 21 January, but the British Commissioners wrote to Palmerston on the sixth reporting that she had been sent to Belize.

that triplicate copies of the Instructions might be necessary in some cases, but they would certainly have been wasted on more than these seven. The difficulty was finally overcome in March 1837, when the Spanish commissioners in Havana received instructions from the Foreign Ministry in Madrid that they should consider a captor duly authorised if his name and that of his vessel appeared on the list sent from time to time by the Commander-in-Chief to the British commissioners.

The Spanish Government also agreed at the end of 1836 that the Royal Navy's cruisers should be exempt from quarantine in Havana, and the Captain-General was ordered to grant them immediate pratique. It was disappointing that the exemption did not extend to slavers which were free of contagion. The seven days of quarantine required, during which the Africans were cut off from aid, frequently caused much suffering.

Another small step forward was achieved by the arrival at the Mixed Court of Justice, in early January 1837, of a copy of the instructions developed by the commissioners at Sierra Leone for the breaking-up of condemned Spanish slavers, together with directions from Lord Palmerston to apply them. The *Preciosa* was accordingly destroyed.

In their January 1837 report to Lord Palmerston, the British commissioners explained that since the signing of the 1835 Treaty the publicity previously given to the sailing of vessels from Havana for Africa had been forbidden. The red flags customarily flown by slavers departing or arriving were no longer seen, and sailing after dark, normally an offence attracting a heavy penalty, was now permitted for slavers. Consequently it was difficult to ascertain the numbers of vessels which had cleared for the coast of Africa, but it was believed that the figures for 1836 were 28 Spaniards, five Portuguese and five Americans. This compared with 80 during 1835, two of which had been Portuguese. There was concern that new impetus was being given to the Trade by Americans openly sailing for Africa under US colours to ship slaves to Havana in order to supply the vast territory of the new independent Republic of Texas. This had become a "most lucrative and extensive" traffic, probably connived at by the local Government (although the Captain-General denied knowledge of it), and certainly known about by US Consul Trist.

At the beginning of February the Commander-in-Chief sent *Wanderer* to Port Royal to complete her provisions and then to cruise off Cuba against the slave trade, and Commodore Pell in *Forte*, which had been away from Jamaica, was ordered back to Port Royal for provisions and then to proceed to England. Pell was replaced as Senior Officer of the Jamaica Division by Captain Sir John Peyton in *Madagascar* (46).

It was hardly surprising that Mr Macleay, on his arrival in England, applied for retirement after long and exemplary service on the Havana Mixed Commission, and Lord Palmerston selected Mr James Kennedy to replace him as Judge at the Mixed Court of Justice. Kennedy would not arrive in post until late September, however, which meant that personal hostilities between Madden and Schenley continued, to the great annoyance of Palmerston.

Thomas Dilke in *Wanderer* had reached his cruising ground off the south coast of Cuba in mid-February but had sighted nothing suspicious, and on the twenty-second he went into Santiago de Cuba, where he learned that a Portuguese brig had landed 416 slaves early in the month while *Wanderer* had been in company with the Commander-in-Chief off Port Royal. Two slavers were fitting-out in the harbour, and it was expected that they would sail under Portuguese colours. Dilke also heard that the French admiral on the Station, arriving at Santiago on 12 February in the *Didon* for a two-day visit, had grounded in the entrance and remained embarrassingly stranded for some hours. Dilke then headed west, intending to cruise off the Isle of Pines until it was time for him to return to Jamaica, via another call at Santiago.

The British Consul complained to the Foreign Office that two large slave cargoes had recently been landed in the vicinity of Santiago in the absence of a cruiser, which, he believed, had been called away to join a blockading squadron off Cartagena; this was presumably before the arrival of *Wanderer*. Dilke's cruise too was brief and fruitless, and by the beginning of April *Wanderer* was in the North-West Providence Channel on passage to Nassau. There she had more luck.

At 1430 on 3 April she sighted a suspicious schooner to windward and made sail in chase. Nearly five hours later, and after firing a number of shots, she took possession of the *Flor de Tego* ten miles to the east of the southerly tip of Grand Bahama. The schooner, Baltimore-built, was bound for Havana, with 417 slaves remaining of the 486 she had embarked at the Gallinas seven weeks earlier. She was under Portuguese colours, but had transferred nationality at the Cape Verdes, coming under the ostensible ownership of a Portuguese master with her previous Spanish master as second-in-command. The slaves were sickly and weak, and the vessel had only five days' provisions remaining, but Dilke was delayed in reaching Nassau to land them because the slaver's crew had cut every rope they could lay their hands on when they had shortened sail for the boarding, and her foresail had been badly damaged by one of *Wanderer*'s shots and could not be set. Consequently she was obliged to lie to at night. It was initially thought that the schooner would not be fit to be sent to Sierra Leone, but, after survey at Port Royal, she made the voyage and was condemned at Freetown with her slaves.

Intelligence sources had led Commodore Peyton to expect a high probability of finding a slaver in the North-West Providence Channel, and it was presumably he who had directed *Wanderer* to take that route to Nassau from Jamaica. Dilke believed that his success was partially due to the short notice given of his destination. He was sure that the slaver was waiting for information from the slave dealers' spies at Kingston and other places in the islands to keep an eye on the whereabouts of cruisers.

On 3 April the Commander-in-Chief returned to Bermuda from his foray to Jamaica and was joined by the ship-sloop *Racehorse* (18), Commander Sir James Home, fresh from a thorough refit at Ireland Island.* The following day *Racehorse*, a lengthened version of a design by Sir William Rule and not an ideal choice, was despatched to join the Barbados Division to cruise against slavers. No doubt encouraged by *Vestal*'s success in those waters, Halkett seems to have been keen to maintain a patrol in the south-eastern approaches to the Caribbean, and any available sloop would have to suffice. At least *Racehorse* was probably newly coppered. On her new station she came under the command of Captain Charles Strong in *Belvidera* (42), the Senior Officer at Barbados. The Commander-in-Chief, meanwhile, made for Halifax in *Melville*.

An unusual capture was made a little to the east of Dominica on 25 April, unusual in that it was achieved by a "coffin brig". The captor was *Griffon* (3), Lieutenant John D'Urban, one of those improved as a brigantine and new to the Station. At dawn, about 0530, 30 miles to the east of Martinique and with the high ground of that island in sight, she detected a suspicious brig and made sail in chase. At noon she caught and boarded the Portuguese slaver *Don Francisco*, 52 days out of Whydah with 433 Africans. The prize was in a shocking condition, leaking so badly that pumping was necessary for 30 minutes each hour, much worm-eaten and with her sails and rigging in a wretched state. Her one boat was unseaworthy, she had no spare sails or rope, and there was only a 50-fathom light chain on her sole bower anchor. What little water remained was brackish, only one week's worth of provisions was left for the slaves and one cask of salt beef for the crew of 32 and five passengers. D'Urban was concerned that if he hauled to the wind to make Barbados the brig would open up and founder in the frequent squalls, and with Dominica under his lee he decided to head there and land the slaves. When the Prize Officer and witnesses arrived at Freetown in November the brig, which belonged to Da Souza at Whydah and had papers from Princes Island, was condemned with her slaves, but the Mixed Commission was displeased

---

* The Royal Navy's dockyard at the north-western tip of Bermuda until late in the twentieth century.

to discover that the Africans had been apprenticed out at Dominica for a term of seven years, a fate they considered quite improper. The Vice-Admiralty court at Dominica ordered the brig to be sold. Having given all due credit to D'Urban, it is perhaps reasonable to conclude that it was the appalling state of the slaver which had allowed *Griffon* to catch her.[13]

A rumour took hold in Cuba that Britain and Portugal had signed a new treaty which included an Equipment Clause, and the Havana slave traders declared that if such was the case they would adopt the United States flag for their vessels. As it was, almost all slavers sailing from Havana at this stage did so under the colours of either Portugal or the USA, and those with either real or adopted Portuguese identity took particular care on departure not to expose themselves to arrest. When news arrived on 4 April that the supposed treaty was actually the Lisbon Decree of December 1836, these Portuguese slavers set to work to equip themselves fully for their purpose.

Poor Judge Schenley was smarting yet again under the wrath of Lord Palmerston for failing to overcome the persistent refusal of the obstructive Arbitrator Madden to add his signature to commissioners' despatches when, to add massively to his distress, his wife died after giving birth prematurely at the end of April, and he was denied the opportunity to take her remains home to her family. However, an unusual incident would soon need to be handled by the court.

Vestal was on passage from Port-au-Prince to Santiago de Cuba on 4 May when, just before sunset, she sighted a suspicious schooner at anchor in the entrance to Guantanamo harbour. Captain Jones had reason to suspect piracy in the area, and he hauled inshore to examine her. As *Vestal* approached with colours flying, the schooner "showed evident symptoms of alarm" and tried to run up the harbour. Jones fired several rounds wide of her to induce her to stop, and sent two boats to board. When the boats had covered a third of the distance, a small schooner appeared at the east end of the harbour entrance and fired a shot at them. Undeterred, they boarded the suspicious vessel and discovered that she was the *Matilda*, prize to the Spanish pilot boat *Teresita*, the schooner which had fired at them. When the *Teresita* arrived, her Commanding Officer, Lieutenant Cruz, an officer of the Spanish Navy, told Lieutenant Watson, the Boarding Officer, that he had arrested the schooner at 1100 that day for being equipped for slaving and that he was taking her to Santiago. *Vestal* proceeded on her business, and Jones, having no reason to doubt Lieutenant Cruz, made no further investigation, but he later told Judge Schenley of the incident. It was as well that he did so, because the *Matilda* failed to appear at Havana for adjudication.

Jones, seething over this "serious and glaring infraction of the treaty", was unable to wait for developments, and deposited an affidavit with the Mixed Court. The *Matilda*, meanwhile, had been taken to Santiago, where she was examined by the Admiralty Court. Thereupon the Judges of the Mixed Court protested to the Captain-General that they had sole jurisdiction in the case, but a month passed before the Captain-General, in the face of opposition from local authorities, acceded to the demand that the vessel be handed over to the Court, a demand inexplicably opposed initially by Madden. It seemed, however, that the bird had already flown.

Following *Griffon*'s success in April, her half-sister brig *Harpy*, Lieutenant Clements, excelled herself in May. Off Martinique at dawn on the thirteenth she sighted a strange sail to leeward and immediately bore up in chase. The wind was light, but Clements carefully trimmed the brig and set his men to work at wetting her sails, and by about 1515 and approximately ten miles off the north-west coast of St Lucia, the chase was within range of the 6-pounder long guns. The first shot fell under her stern, and she hoisted Portuguese colours and hove to. She was found to be the New York-built Portuguese schooner *Florida*, owned by the notorious slaver Da Souza. The schooner was bound from Lagos for Havana with 275 slaves, and Clements judged that eight Africans listed as crew members were actually slaves too. The *Florida*'s master claimed to have been unsuccessfully chased by seven men-of-war since leaving the coast of Africa, and he had cut away the slaver's stern-boat, jettisoned water casks and knocked the wedges out of her masts to escape. She was now leaking and her spars were sprung, and on arrival at Grenada she was declared unseaworthy. Her poor condition, coupled with light winds, had led to her capture by a relatively sluggish adversary. In her absence she was condemned by the Anglo-Portuguese court at Freetown and was sold at Grenada. Three slaves had died before arrival, and the remaining 280 were landed after vaccination against smallpox, an epidemic being in its declining stage on the island. Most improperly, 112 of the men were "voluntarily entered" into the 1st West India Regiment at Trinidad even before they were emancipated.

Following these unlikely achievements of the two "coffin brigs", a much more capable cruiser re-entered the fray with the return of *Racer* to Cuban waters, and she again displayed her quality. During May she chased a suspected slaver under American colours, heading north from the Cuban coast, which escaped by crossing the shoals of the Great Bahama Bank. Soon afterwards, however, *Racer* encountered a man-of-war schooner of the United States Navy, and when the American Commanding Officer learned of the chase from Commander Hope he set off in pursuit of the slaver and caught her. She, the US merchant schooner *Emperor*, was charged with breaching US slaving legislation by shipping slaves from Havana to the coast of Florida.

More direct success came to *Racer* in early June. On the seventh, about 90 miles west-south-west of Cape Cruz, she seized the Portuguese-flagged schooner *Antonica* with 183 Congo slaves destined for Trinidad de Cuba. Four days later she arrested another schooner under Portuguese colours, the American-built *Traga Milhas*, about 60 miles east-south-east of Cape Cruz. This latter vessel, heading for Havana from the Sherbro with 283 slaves, was judged by Commander Hope to be unfit for a voyage of any length, and he took her to Port Antonio on the north-east coast of Jamaica to land the slaves. Then he took her to Port Royal, where she was found to be unseaworthy, and she was subsequently condemned by the Anglo-Portuguese court in Sierra Leone, although she had originally sailed from Havana as a Spaniard. Hope then set out on a time-wasting excursion to Nassau to land the slaves from the *Antonica* before dispatching her to Freetown. He had accepted her Portuguese colours at face value, and it was not until the slaves had been landed on 28 June that he discovered correspondence on board her showing clearly that she was Spanish. So he sent her off to Havana, where she finally arrived on 30 August. The Spanish Judge initially demanded that the slaves be returned to Cuba but was dissuaded, and the schooner and her slaves were condemned.

It had become clear that the officers arresting slavers in the West Indies were either accepting Portuguese colours and papers unquestioningly or were appreciably less adept than their counterparts on the coast of Africa at finding evidence of Spanish ownership in Portuguese-flagged vessels. This was compounded by the negative attitude of Schenley on the matter. He told the Foreign Office that there was no chance of pinning "Spanish character" on these supposed Portuguese slavers arrived at, refitted at, or sailed from Cuba, and that it was pointless to try to adjudicate them under the Equipment Clause. He had tried to persuade Madden to his view and to refrain from "a course of action which will inevitably end in acquittal at heavy expense". This defeatist view was strongly at variance with that of the Freetown commissioners, and must have been encouraging for the Havana slave merchants. It seems that Madden was right for once. The Foreign Office responded by asking the Admiralty to instruct officers to make more thorough searches of Portuguese prizes to find evidence of Spanish ownership.[14]

Schenley's morale was understandably low. He was desperate for the arrival of Kennedy to replace him as Judge and relieve him of the burden of the odious Madden as temporary Adjudicator, and he pleaded for leave as soon as Kennedy was in post. Nevertheless, he expressed a hope at the end of July that his replacement's arrival would be further delayed because yellow fever was raging in Havana worse than at any time in the past 20 years and Kennedy was coming with a young family.

Another key replacement had already arrived on the Station. Sir Peter Halkett's period in command had been unusually short, and on 10 July the elderly teak-built Third Rate *Cornwallis* (80), Captain Sir Richard Grant, arrived at Halifax flying the flag of a new Commander-in-Chief, Vice-Admiral The Hon. Sir Charles Paget GCH. Nine days later Halkett sailed for Spithead in *Melville*.

The new Commander-in-Chief was an officer of vast command experience, beginning with the sloop *Martin* at Camperdown when he was aged just 19. A week after that battle he was made Post. He then commanded four frigates in succession, and two 74-gun battleships, as well as two Royal Yachts. In *Superb* he had commanded the squadron blockading New London during the war with the United States, and in 1829 he had been Commander-in-Chief at Cork. He had been promoted to Vice-Admiral in January 1837.

In August there was a further notable new arrival. Following the refusal by the Captain-General to allow slaves from condemned slavers to be landed in Cuba, the British Government told Madrid that it proposed to send a hulk to Havana to accommodate liberated slaves until they could be despatched to the colony specified by the Superintendent.* In April the Spanish Government accepted the idea and also agreed to the establishment of a lazaretto ashore for the care of sick Africans. The Captain-General had no choice but to acquiesce, and he offered a berth in the Marimelena anchorage and a building on the adjacent wharf. The vessel chosen was the two-decker Fourth Rate *Romney*, launched in 1815 and employed as a troopship since 1822, and she arrived at Havana on 21 August in the company of the frigate *Seringapatam* (46), Captain John Leith. *Romney*'s Commanding Officer, Lieutenant Charles Jenkin, became subordinate to the Superintendent of Liberated Africans, Dr Madden, and therefore under the orders of the Secretary of State for the Colonies.

It was not to be expected that the hulk's arrival would be entirely to the liking of the Captain-General, and the first objection he made was to the presence on board of black soldiers of the 2nd West Indies Regiment, demanding that they be returned to Nassau and threatening to arrest them if they landed. Captain Leith responded that the soldiers were essential to the functioning of *Romney* and he could not consent to their removal, although he undertook that they would remain on board. The subject continued to be contentious.

---

* The suggestion appears to have originated with Lord Glenely, the Colonial Secretary.

Madden's contribution to these arrangements had been to make two ill-judged suggestions. He was concerned at the landing of slaves from Portuguese vessels immediately after capture at various islands, always necessitated (although he clearly failed to realise it) by the unseaworthiness of the slaver. His proposal to the Colonial Office was that a temporary landing place for these slaves should be established at either Salt Cay or the Anguila Islands on the eastern edge of the Cay Sal Bank, and that the place chosen should become the rendezvous for the cruisers. Even more removed from reality was his proposal that *Romney* should be given treaty Instructions so that she could act as a cruiser. On the subsequent briefing note to Their Lordships an unknown hand at the Admiralty wrote "*Quod est absurdum*".[15]

The next prize provided no grist to *Romney*'s mill. She was another Portuguese, the schooner *Ingemane*, taken with 82 Rio Nunez slaves on 12 September about 75 miles south of the Isle of Pines by the splendidly named Commander The Hon. Plantagenet Pierrepoint Cary in the recently returned *Comus*, one of Professor Inman's ship-sloops. Although the slaver was known to have sailed from Havana as the Spanish vessel *Lince*, no evidence could be found on board to disprove the Portuguese identity she had adopted at the Cape Verdes. After a two-day call at Havana, where he concluded that the prize was unfit for a voyage to Sierra Leone, Cary took her in tow for Nassau. There the surviving 79 slaves were landed and the *Ingemane* was surveyed and declared unseaworthy. The Prize Officer and witnesses arrived at Freetown in March 1838, but final condemnation and emancipation had to wait until the Mixed Commission received proof of the vessel's safe arrival at Nassau. That reached Sierra Leone in November.

On her return to Cuban waters *Comus* was sent to Cay Sal to evacuate a party of Spaniards whose presence threatened to become permanent occupation, a situation that Lord Palmerston was not prepared to permit. She arrived on 27 September, and the following day she transferred the Spaniards to *Pickle* for passage to Havana.

Mr Kennedy arrived at last on 27 September to relieve Schenley of his burden as temporary Judge, but Schenley's relief was short-lived. Kennedy did not speak Spanish, and it became rapidly clear that Schenley would be obliged to remain as Arbitrator for a further six months. In his disappointment, Schenley asked Lord Palmerston that at the end of that period he might be exchanged on the grounds of "much domestic affliction", and he wrote sadly that "my utmost endeavours to give satisfaction have only called forth reproof".[16]

Kennedy's first case was a difficult and deeply unsatisfactory one. On 14 October Commander Horatio Stopford Nixon in the Symondite brig-sloop

Ringdove intercepted the Spanish schooner *Vencedora* about 35 miles north-east of Matanzas and arrested her for having 26 slaves, fresh from Africa, concealed on board. The Judges could not agree on a verdict, and the Spanish Arbitrator, Don Juan Montalvo, selected by lottery to make a decision, sided with his Judge and the vessel was restored. Disagreement arose because the *Vencedora* had landed the slaves for eight days at Puerto Rico before re-embarking them for shipment to Cuba. It was the British contention that the passport for the reshipment, acquired at Puerto Rico, was not only inadequate but also fraudulent. The Spanish commissioners declared that, despite the illegality of the slaver's voyage to Puerto Rico, the passport was sufficient to protect her for the subsequent voyage between two Spanish territories. A sum of about £430 for demurrage and deterioration of cargo was awarded against Nixon, despite the belief of the British commissioners that he had been entirely right to seize the vessel, but Palmerston directed that the claimant would have to take steps to recover that amount from Nixon or, if he should default, from HM Government. It seems likely that the money was never paid.

It was Kennedy's fear that the outcome would provide a damaging precedent allowing slave trading from Puerto Rico (clandestine up to that point) to be carried on openly, but, strangely, the Spanish Judge later wrote a letter repudiating his reasons for demanding *Vencedora*'s restoration. Perhaps he was wise to do so, however. In April 1838 Lord Palmerston told the commissioners that, in the opinion of the Law Officers of the Crown, Kennedy's view had been correct, and in October he pointed out that although the 1835 Treaty was an extension of that of 1817, rather than a replacement for it, the limited permission for the Spaniards to carry slaves included in the former convention had been set aside by the statement in the 1835 Treaty that, in all parts of the world, the slave trade by Spain was "henceforward totally and finally abolished". Slave traffic between Spanish territories was therefore illegal, and the Madrid government had been asked to send modified guidance to its commissioners.

There was another matter to be resolved on the *Vencedora* case. The master of the slaver claimed compensation for items allegedly stolen while she was under *Ringdove*'s control, and Kennedy jumped prematurely, if not entirely incorrectly, to the conclusion that the Navy appeared to claim the right to take from prizes anything it needed, such as charts, chronometers and sextants. In this instance, Nixon had sensibly removed several casks of wine for safekeeping during the passage to Havana, but the master had refused to accept them when they were returned after the vessel's release. Nixon also acknowledged that two of his marines had stolen wine or spirits and had been flogged, but insisted that nothing else had been taken. An enquiry instigated by Kennedy later concluded that the theft charges were "utterly discredited".[17]

As late as January 1840, Palmerston delivered a scathing criticism of the court's performance in the case, accusing it in particular of failing to examine all papers available or to question Africans from the vessel's cargo, despite two written requests from Commander Nixon that it should do so. *Vencedora* provided a baptism of fire for Judge Kennedy, and tragedy struck the poor man in the midst of this lengthy case. His "young and devoted" wife died only a month after arriving in Cuba.

Command of *Romney* was never going to be an easy appointment, and a difficulty, perhaps predictable, presented itself to Lieutenant Jenkin in November. An escaped slave contrived to hide himself on board, and when the stowaway was discovered Jenkin decided, probably against his personal inclination, to give up the man to the local authorities. A more humane course of action would have endangered *Romney*'s continued presence at Havana, and Palmerston considered that the decision was correct. When the occurrence was reported to Madrid the Spanish Government, unsurprisingly, expressed itself much pleased by Jenkin's conduct.

It became apparent to all at the beginning of December that not even the Senior Officer of the Jamaica Division, Commodore Peyton, was above incurring the anger of the Commander-in-Chief. In the summer he had, in Admiral Paget's view, deserted his post at Port Royal for 120 days, having taken *Madagascar* on a mission to the Gulf of Mexico for which *Satellite* was available and suitable. On 5 August, to compound his offence, he had put his ship aground on the Bank of Compeche, off the north coast of the Yucatan Peninsula, remaining there for 37 hours. In order to refloat her he had jettisoned 12 guns and a quantity of shot, had lost a bower anchor and cable, and had dragged most of the ship's length over a coral reef, claiming thereafter that there had been no damage to her bottom. He and *Madagascar* were sent home in disgrace.

By contrast with *Ringdove*'s experience in October, the Symondite brig-sloop *Snake*, which had returned to the Station in the summer under the command of Commander Alexander Milne, had a very satisfactory few days in November and December. She was engaged in the somewhat undignified chore of towing the tank vessel *Fountain*, apparently back to Port Royal, when, having made Cape San Antonio at dawn on 23 November, she sighted a strange sail. It could be seen from the masthead that the stranger was using sweeps, which confirmed her as a slaver, and Milne cast off his tow and made sail for the more congenial business of a chase. *Snake* closed rapidly in a fresh north-westerly breeze, but at noon it fell calm and the sweeps were manned. As soon as the chase was within range Milne fired several rounds from his two long-guns, and the stranger hoisted Portuguese colours and shortened sail. At 1340, about 25 miles south of Cape San Antonio, *Snake* took possession of the brigantine *Arrogante* which, it transpired, had cleared from Havana as the Spaniard *Urraca*. Lieutenant Miller was ordered to take her and her 407 River Gallinas slaves initially to Port Royal.

Snake, meanwhile, headed for the Cuban harbour of Xagua on the bight to the east of the Isle of Pines to land the 25 crew and passengers she had taken out of the brigantine, reporting the circumstances to the governor of the local fort. Heading south the following day she fell in with another Portuguese slaver, a schooner, which was bound for Havana from the Cape Verdes but had apparently landed a cargo of Africans near Cape Cruz. Having found *Fountain* and again taken her in tow, Milne stood south-east for Jamaica, but he lost ground on 3 December owing to a strong westerly current and, having decided to stand over towards Cuba, he fell in with *Arrogante*. Finding that there was serious sickness on board, he told Miller to land the slaves at Montego Bay on the north coast of Jamaica. By the time that was done the next day 75 of them had died since capture, exceeding the toll of 68 on the passage from Africa. Miller then took the prize to Port Royal for repairs before making for Sierra Leone. There she and her slaves were condemned, but a request that the vessel might be used to take Miller and his men back to the West Indies and sold there was refused. Sensible though this use of the prize might have appeared, it would have contravened the Anglo-Portuguese Treaty.

Commander Milne was not yet finished with the slavers. At 1800 on 4 December, as dusk was falling, he sighted a suspicious two-topsail schooner on the weather bow. She altered course to close the land, and *Snake* again cast loose the *Fountain* and tacked inshore in chase. Milne managed to keep the schooner in sight with his night-glass until 2300 but then lost her. At 0140, however, she was seen again from the maintop, close on the weather bow. *Snake* bore up, made sail, got inshore of her and again tacked. On sighting the cruiser the chase also tacked, but at 0300 *Snake* came up with her, about 30 miles east-south-east of Cape Cruz, and took possession. The prize was none other than the Spanish slaver *Matilda*, which had evaded justice at the hands of *Vestal* thanks to Lieutenant Cruz of the Spanish Navy. She was carrying 259 slaves for Santiago de Cuba from Ambriz but had overrun her destination. It was a fatal error; there would be no escape this time.

Lieutenant Jauncey took her to Havana, losing 15 slaves on the six-day passage, and the Mixed Court of Justice quickly condemned her and the slaves while the

Cuban authorities were still arguing about jurisdiction following the episode in May. She was then broken-up. The court found that the real owner of the schooner was probably an American, and it noted that her armament of three guns, one of them a long 18-pounder, and an extraordinary outfit of firearms and ammunition led to a suspicion of piracy. Lord Palmerston had already instructed the British Ambassador in Madrid to request a Spanish Government enquiry into the original incident; Cruz claimed that he had taken *Matilda* to Santiago because she was unseaworthy, and blamed the Commander of Marine there for failing to send her on to Havana. Palmerston thereupon demanded that Cruz be dismissed from the Spanish service, but the outcome was merely a reprimand.[18]

Before resuming her towing duties, *Snake* landed her *Matilda* prisoners 15 miles to the west of Santiago. They told Milne that, among others, a brig carrying 600 slaves had sailed for Cuba two days ahead of them. The cruiser then chased another suspicious schooner, and boarded her to find that she was the Portuguese slaver *Constitution* which, to Milne's disappointment, had already landed her cargo.

The Symondite brig-sloop *Sappho*, sister of *Ringdove*, had been launched as recently as February 1837 and deployed on completion to the West Indies under the command of Commander Thomas Fraser. She was passing Cape Tiburon, at the south-western extremity of Hispaniola, at 1900 on 5 December when she fell in with the Portuguese schooner *Isabelita*, with 160 slaves from the River Gallinas bound for Cuba. Food and water for the slaves were almost exhausted, and Fraser took the schooner into Port Royal where, like so many of the Portuguese-flagged prizes, she was found to be unseaworthy. The 159 surviving slaves were landed at Kingston and the vessel and stores were left in the charge of the Commanding Officer of the coal depot ship *Galatea* to await a court decision. The Prize Master and witnesses took passage to Freetown in *Snake*'s prize *Arrogante*, and *Isabelita* and her slaves were condemned.

The final arrest of 1837, similar to that of the *Vencedora*, was made by *Ringdove* on 15 December as she was on passage from Havana to Jamaica at the end of her cruise off Cuba. That morning she sighted two suspicious vessels, a brig and, ahead of the brig, a brigantine. *Ringdove* went in chase of them and closed the brig at about noon. She showed the colours of the Spanish Navy and replied to Commander Nixon's hail that she was the *Marte*, brig-of-war, and replied to a query about the brigantine that she was a merchantman. At 1340, about ten miles south-east of Santiago de Cuba, the cruiser came up with and boarded the brigantine, finding that she was the Spanish vessel *Vigilante*, bound from Puerto Rico for Santiago. She had three Africans, not newly imported, on deck and

21. HM Brig *Ringdove*, 16-gun Symondite sloop

another 18, apparently fresh from Africa, stowed in her main hold.* She presented passports from Puerto Rico for all of the Africans similar to those held by the *Vencedora*. The *Marte* then came up and claimed to be escorting the *Vigilante*, to which Nixon replied that he was obliged to detain the brigantine but that if he had been told of the situation earlier he would have invited *Marte*'s Commanding Officer to accompany him in the boarding.

The vessel was quite unfit for a voyage to Havana and Nixon took her into Santiago, where he found that adequate repair would be uneconomical. *Ringdove* was short of provisions and had to continue her passage to Port Royal, so Nixon's options were severely limited. He decided to restore the vessel to her owner, but gained the assurance of the Governor that the disposal of the slaves would be subject to a decision by the Mixed Court of Justice, and sent a declaration and the *Vigilante*'s papers to Havana in *Snake*. *Ringdove* was then delayed by the investigation into an attempted assassination of one of her boat's crew and the subsequent legal proceedings.

In the absence of witnesses and proof of *Ringdove*'s right of detention the court at Havana felt itself unable to take action on the case, but Kennedy considered that

---

* The adjective being used in the West Indies to describe slaves newly imported from Africa was "Bozal".

Nixon had acted wisely. Lord Palmerston disagreed. He was of the opinion that the case should have been thoroughly investigated, and, although he acknowledged the court's difficulty in the absence of the captor and witnesses, he felt that this obstacle could have been overcome if Nixon had not returned the brigantine to her owner. He clearly failed to comprehend *Ringdove*'s practical difficulties; in any case, Nixon could hardly be blamed for not risking a repetition of his recent experience with *Vencedora*, with its potential financial penalties.

Slavers under Portuguese colours seem frequently to have been in poor condition, but there is no reason to suppose that they were generally worse than Spanish-flagged vessels. Seaworthiness after arrest was usually a matter of concern to the British authorities only in the case of those prizes which would be required to recross the Atlantic to face adjudication, namely the Portuguese, and it can be assumed that captors were anxious to avoid detaching prize crews to take vessels to Sierra Leone. They would not see their people again for many a month. It seems reasonable to suppose that the captor would persuade the surveyors to use the slightest pretext to declare such a prize unseaworthy. That would give valuable saving in manpower, and the vessel when condemned *in absentia* would raise as much money or more when sold in the West Indies as in Sierra Leone.

As 1838 dawned, the British commissioners reported that they believed that 70 slavers had sailed from Havana in the past 12 months: 40 Portuguese, 19 Spaniards and 11 Americans. This large increase over the 38 sailing in 1836 did not mean, however, that the slave trade was flourishing. Many traders had experienced considerable losses resulting from captures on the coast of Africa, and insurance premiums had risen, it was said, to more than 40 per cent. The price of slaves had, however, risen to $480 each owing to the anxiety of the planters quickly to stock their properties in the belief that Britain would eventually be successful in inducing other nations to agree to treaties similar to that of 1835 with Spain.[19]

By mid-1835 at Rio de Janeiro, Rear-Admiral Hamond had shifted his flag to the Fourth Rate frigate *Dublin* (50) and was attempting to nurture and defend British interests on both the Atlantic and Pacific shores of South America, as well as to police the Brazil slave trade. His force for this array of tasks consisted of his flagship, one Fifth Rate frigate, four Sixth Rates, five sloops, a gun-brig and two brigantines. Another Sixth Rate, *Challenger*, was wrecked on the coast of Chile in May; one of the sloops, *Beagle*, was engaged in surveying; and another, *Sparrowhawk*, was antiquated, having been launched only two years after Trafalgar. None of the Sixth Rates and only two of the sloops, *Rover* and *Snake*, were Symondites, and *Snake* was about to sail for home. Vessels regularly exchanged between the east and west coasts, but the centre of gravity of the South America Station was beginning to shift to the Pacific. Action against the slavers was at a low ebb.

In June there were several arrivals at Rio de Janeiro. The Sixth Rate *North Star* returned from Pernambuco on the fifteenth, having left *Rover* there, and touched at Bahia on her voyage south. Captain Octavius Harcourt reported that all was perfectly quiet at both places. Two days later the ketch *Basilisk* joined the Station, and on the twenty-first the Sixth Rate *Talbot* returned from a wasteful errand to Trincomalee, during which she had spent only 30 days at anchor since her departure on 20 October 1834.

In order to shake down a new ship's company in the flagship, and with the intention of "showing that the force is active on the Coast", the Admiral took *Dublin*, *North Star*, *Basilisk* and the gun-brig *Rapid* to sea on 6 July, leaving Captain Follet Pennell of *Talbot* as Senior Officer at Rio. At sunset he detached *North Star* and *Basilisk* for Valparaiso to join Commodore Mason of *Blonde*, Senior Officer on the Pacific coast, and, having despatched *Rapid* to visit Bahia, he headed north to visit Pernambuco. He had no intention of remaining long in "this wild anchorage", but there were signs of civil unrest in the town, where "parties run high and are violent". However, Hamond assured himself that British interests at Pernambuco were in no immediate danger, and he sailed for Bahia on 20 July. *Rover* and *Rapid* were awaiting him when *Dublin* arrived on the thirtieth, and, when these two had completed their provisions to three months, he sent *Rapid* to keep an eye on Pernambuco and left *Rover* to cruise against slavers in the vicinity of Bahia. *Dublin* sailed on 2 August and came to her anchor at Rio on the tenth. The short expedition had been without incident, and was unlikely to have achieved the slightest discouragement of the slave trade.

Brazilian cruisers made a handful of captures between June and September, but in only one instance was a detained slaver claiming to be Portuguese found by the Mixed Commission to be Brazilian and condemned. In the case of the Brazilian smack *Novo Destino*, arrested in July by a Brazilian brig-of-war and restored, possibly incriminating correspondence was found sealed on board, but the Government insisted that it should be forwarded unopened, and Lord Palmerston agreed that the British commissioners had no authority to oppose that insistence.

The British commissioners lamented the impossibility of obtaining convictions in the Brazilian courts against slave traffickers, and reported to Lord Palmerston that those involved with the two prizes taken by the Brazilian Navy in February and March had been acquitted.[20] They also reported that the Brazilian

Government had heard from its Chargé d'Affaires at Lisbon that the Portuguese Foreign Secretary had directed colonial governors not to issue passports to vessels with slaves. This was pointless, they said, because the practice of the slavers was to sail in ballast and then embark their Africans off the bar, a policy of "literal compliance and practical violation". They were probably unaware that the general practice of Portuguese colonial governors was simply to ignore inconvenient instructions from Lisbon.

The Sixth Rate *Acteon*, Captain The Right Hon. Lord Edward Russell, returned to Rio on 17 September from the River Plate, where she had left *Talbot*, and a few days later she sailed for the northern ports, where she was to be stationed for the next three months.* On 5 October she arrived at Bahia, where Russell intended to remain for a fortnight before heading for Pernambuco to relieve *Rapid*. On the night prior to the ship's arrival several slaves had been found armed in the town. They had been seized and were to be transported back to Africa.

Rapid's return to Rio on 27 October brought news of an insurrection by "people of colour" in the northern Brazilian province of Para, within the bounds of the West Indies and North America Station. White people who had not fled in time had been massacred, but British residents had taken refuge in *Racehorse*, one of Admiral Cockburn's sloops, and two English merchant brigs. Admiral Hamond feared that this "war of colour" would spread, but he felt unable to provide further support with his severely limited resources, and he expected similar disturbances in his own area of responsibility. His inclination was to seek additional force rather than deplete what he had.

This trouble in the north of Brazil was soon followed by disorder in the southern province of Rio Grande, caused by a separatist movement rather than African insurrection. Hamond thought the disturbances there would lead to a blockade, as had the Para revolt. It was another distraction for him, and slave-trade suppression moved yet lower in the list of priorities.

Nevertheless, an arrest, the first and last of the year, was achieved on 17 December by Commander Robert Smart in the ship-sloop *Satellite*, which had just returned from the Pacific and was making for Bahia on her way home. About 145 miles north-east of Cape St Tome, after a nine-hour chase, he caught the brig *Orion* under Portuguese colours. *Satellite* had sighted the brig on her weather bow at 0800, and the chase had tried to evade capture by hauling to the wind and heading west instead of holding her course for Rio and sailing free. When just outside gun-range at 1400 the sloop fired two rounds to induce the brig to hoist

---

* The spelling used in the *Navy List* is *Actaeon*, but Lyon shows *Acteon* in *The Sailing Navy List*, and this is probably the spelling in use by the Admiralty.

22. HM Brig *Rapid*, 10-gun "Coffin Brig"

colours, but to no effect, and, with the two vessels on opposite tacks, the distance was increasing. At 1700, however, *Satellite* managed to come up with her quarry and board her. She was found to have on board 245 slaves from Mozambique, mostly boys and girls, and had originally fitted-out at Rio. Her owner, a "violent and desperate character" according to Hamond, had sailed from Rio as master but was now still on board as a passenger.

On 20 December the prize was taken into Rio by Lieutenant Anson with Mr Biddlecome, Master, who was a passenger in *Satellite* and had volunteered to navigate the brig, Mr Gorton, Mate, and 14 men. It was a strong prize crew, and, as events transpired, that was just as well. In coming to the critical decision that it had jurisdiction over this Portuguese-flagged slaver, the Mixed Commission took into consideration two crucial points made, at last, "upon the authority of the Portuguese Government itself". The first was that "all Trade whatever in Slaves became illicit, on the part of Portugal, from the moment of her separation from Brazil." The second, no less important, was "that no vessel is entitled to be considered as Portuguese which does not conform to the requisitions of the Commercial Code of Portugal." Furthermore the Brazilian Government had conceded, at the insistence of Britain, that Portuguese subjects resident in Brazil and found to be engaged in the slave trade were answerable to the Anglo-Brazilian Mixed

Commission. This agreement had not been made public in Brazil when, a month after her arrival, the *Orion* was condemned with the surviving 243 slaves, and the outcome of the trial caused a considerable sensation. The youthfulness and good health of the slaves made the cargo of more than ordinary value, and one of the principal advocates in the city made strenuous efforts to have the vessel restored.

When sentence was passed the vessel's owner asked for a "demurrer" or embargo, a period of ten days allowed by Brazilian law for an appeal to be made. As had always been the case previously, this was permitted by the court, despite a strong and entirely justifiable protest by the Commander-in-Chief, who reminded the court, in terms which the British commissioners found "regrettable", that the Anglo-Brazilian treaty specifically prohibited appeals. The Admiral might have added that Brazilian law had nothing whatsoever to do with the affairs of the Mixed Commission. However, the British Judge, George Jackson, had weakly allowed this transgression in the past, although he had complained to the Foreign Office about it, and felt unable to refuse it in this instance. The sentence was finally confirmed on 29 January 1836.[21]

Jackson insisted that not a moment of unnecessary delay had been incurred, and that all the commissioners were "anxious to afford Sir George every facility and explanation". That was not how Admiral Hamond saw it. When he heard the sentence on 19 January he applied to the court to turn the Africans over to the Brazilian Government to be released from the confines of the 139-ton slaver and to allow removal of the prize crew. He was told that he would have to wait another fortnight for the appeal to be heard and then for the necessary forms of judicial execution of the sentence, which, in contravention of the treaty, was not operative until backed by the authority of Brazilian law. The vessel, her crew and the slaves were finally taken into the charge of the Government on 6 February, 48 days after her arrival.

Throughout this time the captor, *Dublin* in *Satellite*'s absence, had to guard and provide for the slaves, to hunt for evidence against the vessel when all in Rio were in favour of the slave trade, to carry crew witnesses as prisoners back and forth to the court under constant risk of their being rescued, and to row guard boats all night, every night, frequently soaked by tropical rain, to prevent the Africans being kidnapped. As the Commander-in-Chief pointed out, it was fortunate that the slaves were healthy; if there had been disease on board half of them would have died before emancipation. It was extraordinarily hot; *Dublin* recorded 96 °F under the shade of an awning on deck. The prize crew in the *Orion*, with no proper accommodation, were in misery, and Hamond wrote to the Admiralty, "the Capture is a downright misfortune to them [...] I confess that I dread the sight of one of these vessels coming in for trial". He saw no good being achieved by the work of the Mixed Commission, and the commissioners had admitted to him that the emancipated Africans never attained liberty because of the corruption in the country.[22]

Lieutenant Anson and his prize crew sailed for England in the packet *Nightingale* on 7 February, with the exception of Mr Gorton, who was kept in *Dublin* for medical treatment. The *Orion* was sold and put to sea again on 16 March under Sardinian colours and with the name of *Defendente*. She had cleared out for Montevideo, but the fort of Santa Cruz fired a gun at her to detain her for sailing without a register, and she anchored outside the bar late in the evening. In the morning, as was to be expected, she was nowhere to be seen, and no doubt she was soon engaged again in the slave trade.

Lord Palmerston later expressed approval of the verdict on *Orion*, but he was entirely in accord with the Commander-in-Chief on the matters of appeal and delay. He directed the commissioners to abide most strictly with the terms of the Treaty of 23 November 1826 and its Additional Articles which stipulated that the Mixed Commission was to judge without appeal, and the British commissioners were "not in any instance, or under any circumstances, to deviate in the least degree" from the treaty. Furthermore, he pointed out that the regulations urged the Mixed Commissions to pass sentence as summarily as possible and practicable, generally within 20 days, and always within two months of the arrival of the vessel. He also wished to know from the Admiralty what measures it had taken in consequence of the ratification in May 1835 of Sardinia's accession to the Anglo-French Treaty. The Admiralty confessed that nothing had been done as yet.

Admiral Hamond was correct in the case of the *Orion*, but he was definitely wrong in his next confrontation with the Mixed Commission. On 7 January 1836 he heard that a suspicious vessel was hovering off the coast to the north of Rio, and Hamond, realising that he would have to act quickly if he was to catch her, sent out the only available vessel, the brigantine *Hornet*, Lieutenant Francis Coghlan, to investigate. *Hornet*, which had returned from the River Plate on 2 January, was being employed as a branch packet, and, as surely the Commander-in-Chief should have known, she, unlike the rest of the South America squadron, had not been issued with slave treaty Instructions.* On the evening of 8 January Coghlan found the Sumaca *Vencedora*, under fraudulent Portuguese colours, at anchor off the Marica Islands, a little to northward of Rio. She was fully fitted for slaving and it was evident that she had just landed a cargo of slaves, so Coghlan

---

* A packet was a vessel employed to carry mails. A branch packet provided the service within the confines of the Station.

took her into Rio. Hamond claimed that as he, as Commander-in-Chief, was furnished with copies of the Instructions he had the authority to order any of his commanding officers to make a seizure, but the Mixed Commission was unmoved by his protest and quite correctly declared the arrest illegal. In fact the Admiral must have been wrong in asserting that he held Instructions; only commanding officers of men-of-war, the flagship among them, were so entitled. The court acknowledged that (the crucial matter of Instructions aside) the seizure was justified, but as an aside, the British commissioners raised the question of whether the slaver, which Coghlan reckoned was anchored two miles offshore, was "within the roadstead of either of the High Contracting Powers or within Cannon Shot of the Batteries on shore". If that had been the case the arrest would have contravened the treaty, and the commissioners thought it fortunate that the question had not arisen in court.

The Sixth Rate *Cleopatra*, Captain The Hon. George Grey, joined at Rio on 20 January, *Acteon* sailed for the River Plate the same day, and *Talbot* returned from the Plate on the following day before proceeding to the Pacific on 1 February. On 19 February the survey vessel *Sulphur*, Captain F. W. Beechey, and her tender, the cutter *Starling*, Lieutenant Henry Kellet, arrived at Rio, but the Commander-in-Chief was not there to greet them; he had put to sea on 2 February to exercise *Dublin*'s ship's company and for the good of their health.

The flagship cruised off Cape Frio from 2 February until the eleventh of the month, and then headed south-west to anchor off St Catherine Island on the eighteenth. There she completed her water and weighed anchor on the twenty-fourth to return northward, coming to anchor off St Sebastian, where she spent two days before returning to Rio on 9 March. She had sighted three probable slavers lying in various bays of the wide Bay of St Catherine, but none had slaves on board and they were "within control of the shore", so *Dublin* did not meddle. The flagship sighted no suspicious vessels at sea during her 35-day cruise.

The cruise had probably done little to salve the Commander-in-Chief's frustration which he had expressed to the Admiralty before sailing. Apart from the well-recognised problem of fraudulent adoption of Portuguese colours by Brazilian slavers, Hamond was aware that several vessels had sailed for Africa with trade goods, and that another had departed with empty water casks and illegal slaving gear to equip the others on their arrival on the African coast. They purported to be Portuguese, but none carried the papers required by Portuguese law, and the Portuguese Consul in Rio was determined to aid evasion by affording every assistance in changes of flag in defiance of the Queen of Portugal's decree of October 1835. There were four more such vessels, the Commander-in-Chief wrote, currently lying in the mouth of the harbour, cleared for Loando; "it is painful to witness their impudent boldness". He wished to know from Their Lordships what measures could be taken against them. They were sailing with slave cargoes and returning to harbour in Brazil in ballast, and on the coast of Brazil, hundreds of leagues in extent, there was little or no chance of preventing them from landing their slaves. They might, however, be stopped on the way *out* by a cruiser off the port if so authorised. Their Lordships did not appear to have useful advice to offer.

In an attempt to establish some degree of control over the flow of slaves from the African coast, the Brazilian Government appointed a Consul-General to reside in Loando, nominating to the post a naval lieutenant who had been zealous against the Trade while commanding a cruiser. The idea was not to the liking of the authorities in Loando, who refused to receive the officer on the grounds that the territory was a colony, and Lisbon supported their action.

In July 1835 Captain Hope of *Dublin* had been granted permission to give up command and return home, and he had been temporarily replaced by his First Lieutenant, William Puget, in the rank of commander. Captain George Willes was then appointed in command, but was ill from almost the moment that he joined, and Puget continued effectively to command the ship. The return of *Rover* from the Pacific in May allowed *Rapid* to be sent home, taking the invalided Willes with her. It also offered the opportunity for the Commander-in-Chief to appoint Commander Charles Eden of *Rover* to command *Dublin* and to send his Flag Lieutenant, one Andrew Hamond, to command *Rover* in the acting rank of commander "until Their Lordships' pleasure be known". In what appears to have been a cynical piece of nepotistic manoeuvring by the Admiral, poor Puget, after six months acting as Flag Captain, was bypassed. Faced with reversion to his previous post, he decided that "the Service and ship would not benefit from his returning to First Lieutenant's duties", and he was allowed to return home.

Cleopatra had arrived at Bahia from the River Plate on 5 April to relieve the ship-sloop *Harrier*, Commander W. H. H. Carew, of the duty of watching the northern ports. During her three months on that station *Harrier* had visited all the minor ports between Bahia and Pernambuco, as well as those two major harbours, and she had cruised off all the points where the slave trade was supposed to be most active. Carew reported that he had chased and examined every vessel that he had fallen in with, but none had been liable to detention.

The return of *Harrier* to Rio on 16 April brought one of a rash of disciplinary problems demanding the Admiral's attention in April and May. The continued drunkenness of the sloop's Carpenter required trial by court martial, but there was no prospect of convening a court at Rio. The required number of captains and commanders were simply not available. The Carpenter became another passenger for England in *Rapid*. A similar case was presented by Captain Beechey of *Sulphur*. He reported the "scandalous drunkenness" of his Purser who, at Montevideo, had to be hoisted out of a boat "to the disgrace of the ship" and, with the ship on the point of sailing, the vouchers could not be closed. This was not the first such occurrence, and the Purser had not benefited from a public reprimand following a previous similar offence. *Sulphur* was on her way to the Pacific, and the Commander-in-Chief instructed Commodore Mason, in command on the Chilean coast, to convene a court martial.

A more serious case came with the return of *Rover* from the Pacific. On the evening of 29 December the sloop's Master had been found in the Captain's cabin "taking indecent liberties with the person" of a Boy of the First Class. He had then attempted, by bribes and threats, to silence the Captain's cook, who had witnessed the offence, and he had tried to buy the boy's silence. He had then left the ship without permission and had been arrested in a merchant vessel bound for England. By the time of *Rover*'s arrival at Rio the Master had been in confinement for four months and, with no possibility of a court martial on Station, he too became an unwilling passenger in *Rapid* to face the displeasure of Their Lordships. As it happened, the boy he had allegedly abused shared the passage with him, having fallen from *Rover*'s mizzen-top and been ruptured.

Rapid had been on the Station for the best part of three years, and the Commander-in-Chief was probably sorry to lose her. He wrote to the Admiralty that the brig did credit to her Commanding Officer, Lieutenant Frederick Patten, who had been in her throughout her deployment and whom he found to be "a very active, zealous and efficient officer".

The arrival in England of the miscreants drew from the Admiralty a very stupid reproach, directing the Commander-in-Chief to convene courts martial himself whenever practical. Hamond replied in forthright terms that he could indeed have done so after a delay of at least six weeks, but it would have meant recalling from her station every vessel on the coast over an extent of "many hundreds of Leagues North and South". During their absence the ports of Bahia and Pernambuco and the River Plate would have been abandoned by British cruisers. He considered that he had exercised sound discretion.[23]

In mid-May *Acteon* arrived at Rio from Buenos Aires carrying the new British Minister, Mr Hamilton, and his suite. She remained for ten days before sailing to join Commodore Mason on the west coast, a few days astern of *Rover*. Only three cruisers, including the flagship, remained on the east coast. *Rover* arrived at Valparaiso on 16 July after a particularly stormy passage of 57 days during which her mainmast was sprung below the hoop to which the futtock shrouds were set up. The mast was secured by lowering the topmast and lashing it to the mast above and below the injured part, and also fishing the mast with capstan and hatch bars.

Cleopatra was due to leave Bahia for Rio on 1 July, and a day later the Admiral sailed in *Dublin* for a short cruise to the northward. The flagship anchored at Bahia after cruising for 18 days along the coast. No slaver or suspicious vessel had been sighted. All was quiet at Bahia, and *Dublin* was back at Rio on 2 August after an uneventful run south.

All remained quiet at Bahia, but the slave trade was very active in its vicinity, and the Admiralty drew the Commander-in-Chief's attention to a letter on the subject that it had received from the Consul at the port. At the time of the letter *Harrier* had been at Pernambuco, and Hamond insisted that every effort was being made against the slave trade, but he had very few vessels, and their duties in protection of British mercantile interests prevented as active cruising against the slavers as he would wish. The extent of the coast involved in the Trade was enormous and would require at least a dozen of the swiftest small cruisers for the suppression task alone.

To give emphasis to the Admiral's point, news of disturbances at Montevideo, caused by the declaration of the President of the province against the Government, had just been received in Rio, and Hamond felt obliged to send *Cleopatra* south to protect British citizens and property. Captain Grey was instructed to keep *Harrier* in the River Plate if absolutely necessary, but that meant that *Dublin* at Rio was the only cruiser on the Brazilian coast north of Montevideo. If *Harrier*'s services could be dispensed with Grey was to send her to the Commander-in-Chief, who was "greatly in need of cruisers in this direction". *Cleopatra* arrived at Montevideo on 12 August after a six-day passage from Rio, and Grey, judging that there was no immediate danger to British interests, therefore despatched *Harrier* to rejoin the Commander-in-Chief. She arrived at Rio on 24 August, and there she found *North Star*, which had returned from Valparaiso five days earlier, carrying the invalided Captain Beechey of *Sulphur*.

Earlier in August Hamond had returned to the charge on the matter of delays in adjudication. In a letter to the Admiralty he accepted that the Mixed Commission did not cause unnecessary delay, but holidays and "the Law and Custom of the Country" added to "impediments by forms and ceremonies after the decision of the court" were largely to blame. His real complaint was that the slaves had to remain on board in Rio for a period longer than the voyage itself.

There was no hospital for them or place of security ashore, and no one other than the prize crew to provide care or take charge of them. The Prize Master and his men were exhausted by the time they arrived, and should be relieved:

> they live on the open deck of the miserable vessel under such an awning as they can rig to keep off the burning sun by day and heavy dews by night, armed, and at all times on watch against surprise by the desperate outlaws who comprise her crew.

In harbour any relaxation of watchfulness would result in the slaves being carried off. In the two most recently condemned vessels "one officer was reduced to the point of death and another actually became for a time raving mad, from the effect on mind and body, of the distress to which they were exposed as Prize-Masters." The Africans, he felt, suffered more from the incarceration than if they were taken immediately into slavery. He argued passionately that the Brazilian Government or the Mixed Commission should become responsible for "these poor people" from the moment of arrival.[24]

Hamond had also written to Mr Hamilton to plead for action to prevent "the shameful facility given by inferior authorities at Rio to the slave trade", which was "absolutely undisguised". He was particularly concerned that homeward-bound slavers, having landed their cargoes along the coast, were cleared to return to port in ballast "through connivance of all inferior authorities". He believed that the British Minister was continuing his efforts on this score, but was being thwarted by the corruption of the administration, and the only response from the Brazilian Government was that it had a sincere desire to stop the contraband slave trade, but could not do so because men-of-war were not available from other duties.[25]

On 30 September the Sixth Rate *Imogene* (28), Captain Henry Bruce, arrived at Rio and was taken under the Commander-in-Chief's orders. She then sailed on 6 October for the River Plate to relieve *Cleopatra* and replenish her to four months' provisions. *Cleopatra* was then to proceed to the Falkland Islands and return to Rio at the beginning of February 1837.

Before leaving Rio, however, *Imogene* was involved in a brief drama. At 1920 on 1 October the flagship received a message from a colonel of the National Guard that the Treasury was on fire and requesting urgent help from HM Ships. In half an hour, according to the Admiral, 150 men from *Dublin* and *Imogene* were on the spot, about a mile from the landing place: a remarkable response. They left the scene at 2230 with the fire extinguished, having confined the damage to half of one wing of the building. A party had also landed from the French frigate

Heroine. The Commander-in-Chief received an expression of thanks from the Brazilian Secretary of State on behalf of the Regent.

The Branch Packet *Hornet* arrived at Rio from Montevideo on 13 September with the news that all was peaceful at the port when she left it at the end of August, but that the Brazilian province of Rio Grande was in a disturbed state. By mid-November, however, Hamond was able to report that the leaders of the insurrection in Rio Grande had been captured and brought to Rio and that the province had probably been brought to heel. He was also able to forward news, sent by Captain Grey before *Cleopatra* sailed from Buenos Aires for the Falklands, that government troops of the Banda Oriental State had subdued the forces of the dissident General Riviera, and there was no longer concern over the state of affairs in Montevideo. With peace restored in the southern part of Brazil, there was some hope that suppression of the slave trade might assume a higher priority.

In June the British commissioners had expressed gratitude to Lord Palmerston for his firm instructions on the matter of delays, or embargoes, experienced after sentences were passed by the Mixed Commission. The precedent had been established before the arrival of the current commissioners and they had understood that the procedure was acceptable. Armed with his comments they would attempt to prevent repetition, but held out little hope of success. They explained that the Mixed Commission was not the executor of its own sentences. The Municipal Judge with that duty refused to accept a warrant for the purpose from the Department of Justice for a period of 15 days, the general embargo period allowed by Brazilian law. Furthermore, the Brazilian commissioners argued that the treaty did not require that a sentence should be carried into execution within a specified time, merely that the court should reach a decision.

In reply, Palmerston forwarded Admiral Hamond's letter describing the consequences of delay and required the commissioners to bring the matter formally before the court to formulate a means of alleviating the difficulty, possibly to use a merchantman, berthed close to the flagship for security, to accommodate captured slaves prior to emancipation. They were to acquaint the British Ambassador with the arrangement agreed with the Brazilian commissioners so that he could lend support with the Brazilian Government.[26]

The matter of national identity of slavers, which seemed to have been satisfactorily clarified on the authority of the Portuguese Government during the *Orion* trial, was raised again in July in a letter from the Brazilian Minister for Foreign Affairs to the British commissioners acquiescing to the views of the British Government. That view was again stated by the Foreign Office in a letter to the Admiralty in September, giving the opinion of the King's Advocate:

Vessels though under the Portuguese Flag and Papers, and belonging to a native of Portugal, but who is resident in Brazil, and carries on his Trade from and to that Country, are to be considered as Brazilian Vessels, and liable to be prosecuted against as such before the Mixed British and Brazilian Commission.[27]

The Brazilian Government avoided issuing an official notice of this understanding, and when the commissioners expressed a hope that there would be a similar outcome in the case of the embargoes they received the reply: "That was the Act of one Minister, and already His Successor has declared himself of a different opinion."

In mid-August, and before the arrival of Palmerston's despatch with Hamond's letter, the Minister for Foreign Affairs firmly rejected the British argument against embargoes, a reaction viewed by the British commissioners as further evidence of the jealousy felt by Brazil at the existence of the Mixed Commission and of the particular sensitivity of the country on the question of the slave trade. In January 1837 the Brazilian Judge, João Carneiro de Campos, rejected the suggestion that Brazil should provide a hulk to accommodate captured slaves. He "declined to recommend to his Government objects which he conceives to be solely British and involving considerable trouble, responsibility and expense to his Government".

There also appeared to be little chance that Brazil would accept a British suggestion that emancipated slaves should be sent to Trinidad, or an American offer to take them in Liberia. Apparently the Brazilians were considering re-exporting free Africans to Africa in the hope of establishing a colony there.

No doubt exhausted by these hostile exchanges, Judge Jackson reported in November that he was about to avail himself of the leave granted some time previously by Lord Palmerston, and that HM Consul in Rio, Robert Hesketh, would act as Commissary Judge in his absence. Before leaving his post, Jackson took a moment to inform Lord Palmerston that the slave trade was currently being carried out "with unusual activity" on the Brazilian coast, unhampered by the presence of cruisers, either British or Brazilian. Palmerston passed this assessment to the Admiralty, without, it would appear, eliciting any response.

The arrival of an American vessel, the *Commodore*, at Rio, where she was sold and fitted-out for the slave trade before sailing under Portuguese colours, initiated a suggestion from the frustrated Commander-in-Chief that such vessels might be arrested and taken to Lisbon to be tried for navigating under that flag without authority. The idea did not find favour with Lord Palmerston, as he made clear to the Admiralty in December.

An addition to the squadron arrived at Rio on 6 December and was taken under Hamond's command. She was the *Fly* (18), Commander Russell Eliot, one of Professor Inman's ship-sloops. The following day *Sparrowhawk* arrived, 42 days from Valparaiso, but she delayed for only four days before sailing for Spithead, carrying treasure of 220,000 dollars.

Dublin and the Admiral began another of their forays to the north on 3 January 1837, primarily for the health and training of the ship's company, and with the vain hope of catching a laden homeward-bound slaver. Two days later *Harrier* sailed for the Pacific, and *Imogene* was expected at Rio in a few days' time from the River Plate. *Imogene* was to refit during the Commander-in-Chief's absence and would be ready for service on his return.

The flagship's cruise followed its usual uneventful pattern, with arrival at Bahia on 24 January, sailing for Pernambuco on the thirtieth, four days off Pernambuco from 15–19 February, and returning to Rio on the twenty-eighth. There was, however, an hour or two of excitement on the evening of 7 February, unusual for *Dublin*. A brig under Portuguese colours was sighted to windward, standing to the south. It seemed likely that she was a slaver, and the relatively stately frigate tried to close her, but, as was to be expected, without success. On arrival at Pernambuco it was learned that the brig had sailed from there for Africa on the sixth. *Dublin* did examine several other vessels during her cruise, but none was suspicious.

By this time Captain Robert Tait, late of *Spartiate*, had taken command of *Dublin*, releasing Commander Eden to return to *Rover* and thereby ousting Acting Commander Andrew Hamond to revert to the rank of lieutenant in the flagship.* This loss of rank in the Hamond family received compensation in early April when the Commander-in-Chief received news that he had been promoted to Vice-Admiral of the Blue Squadron.

On his return to Rio, the Admiral found that the Gunner and Carpenter of *Imogene* were being held on charges which warranted trial by court martial, but there was still no chance of convening a court on the east coast. Clearly he was anxious not to lose two valuable warrant officers by sending them home for trial, and probably also wished to avoid further confrontation with the Admiralty on the subject of courts martial. In any event, he decided that a verbal lambasting by the Commander-in-Chief would suffice on this occasion.

The story on the east coast of South America for the next eight months amounted to little more than a catalogue of ship movements, and no headway

---

* No record has been found of when the battleship *Spartiate* sailed for home. It is apparent that the C-in-C shifted his flag to *Dublin* during the early months of 1835, but *Spartiate* is shown in the quarterly *Navy List*s as remaining on the Station until the time of the June edition.

against the slave trade. The branch packet *Hornet* was relieved by her sister brigantine *Cockatrice* in March. On 13 March the teak-built Sixth Rate *Samarang* (28), Captain William Broughton, anchored at Rio and was taken under command.* The end of the voyage from England must have come as a relief to her ship's company because she had no cook on board, and it is to be hoped that improved feeding arrangements had been provided before she sailed for the River Plate. Her arrival did not represent a reinforcement; she replaced her sister *Talbot*, which passed through Rio in April on passage home from Valparaiso. Another exchange began on 14 April with the joining of the Fifth Rate *Stag* (46), commanded by Captain Thomas Sulivan, who had hoisted his Commodore's broad pendant in preparation for relieving Commodore Mason on the west coast. She sailed for the Pacific on 23 April.

Fly took up the station off the northern ports on 27 May to protect trade and cruise against slave vessels. In her 14 weeks in the vicinity of Bahia and Pernambuco she fell in with several vessels which were apparently outward-bound slavers, and she boarded one which had undoubtedly landed slaves on the coast, but in no case were there grounds for arrest. In mid-August she was relieved on the northern patrol by the ten-gun "coffin brig" *Wizard*, Lieutenant Edward Harvey, which had joined the Station on 5 August. *Fly* underwent a refit at Rio in September, particularly to replace her sprung foremast, before sailing for the Pacific to replace *Acteon*.

The old *Sparrowhawk* appeared again at Rio on 6 August after her brief respite in England, now commanded by Commander John Shepherd. Ten days later she sailed for the Falklands and then the River Plate to relieve *Samarang*, which returned to Rio. Heading in the opposite direction, *Blonde* and Commodore Mason arrived at Rio on the twenty-second, and a week later they departed for Spithead bearing 1.2 million dollars of specie for England. The final new arrival of the year was the ketch-rigged cutter *Sparrow* (10), Lieutenant Robert Lowcay, which joined at Rio on 17 October and, provisioned to four months, sailed for the Falklands at the beginning of November.

It was found that a Portuguese merchantman which arrived at Rio from Liverpool in late July had on board a cargo of British manufactured goods destined, it was believed, for Africa. The revelation must have raised doubts in the minds not only of Brazilian officials, merchants and slavers but also of the officers of the Royal Navy's cruisers about the extent and strength of Britain's commitment to suppression of the slave trade.

---

* This, no doubt, was the William Broughton who took the *Veloz Passagera* when he was in command of the sloop *Primrose* in September 1830.

An extraordinary source of information presented itself to the British commissioners in July. A French slaver, in remorse for his involvement in the Trade, resolved to give the British all the information at his disposal in order to help the suppression campaign, and he approached Consul Hesketh, who passed the intelligence to the commissioners. He revealed that large numbers of slaves were being landed at Isla Grande, 60 miles or so to the west of Rio. They were destined for three estates, two on the island and another on the mainland just to the westward. The slavers apparently approached during daylight but hauled off at night, remaining under sail. The Frenchman recommended that cruisers should keep out of sight during the day to the north of the island and at night approach the entrances of the channel between island and mainland, where they would find slave vessels. This intelligence was sent to Lord Palmerston, and it may be assumed that it was passed directly to the Admiral, but there is no evidence that Hamond took action on it, almost certainly because he lacked the necessary cruisers. One suspects that, in the Admiral's position, one of the more thrusting commanding officers in the Bights might have manned and armed the flagship's boats and sent them away for the task.

Early in September Mr Hamilton conveyed to the Commander-in-Chief a request from the Brazilian Foreign Secretary that he should deploy his cruisers against slavers off Rio de Janeiro. Hamond predictably responded that he was employing his few vessels against the slave trade as far as was consistent with their other duties, and he had no capacity to concentrate on the area of Rio. He had no intention of withdrawing cruisers from the northern ports or the River Plate, and pointed out that the Brazilian Government had available the simple expedient of refusing clearance for departure or entry of any vessel which lacked the correct papers for the flag she flew.

A serious blow to the suppression campaign fell at the beginning of November with a change of Brazilian Government. The previous administration had appeared to want to put down the slave traffic, both as a matter of principle and on account of the agreement with Britain. The new Government, according to its Minister of Justice, proclaimed that the slave trade was indispensable to the country, released those who were under prosecution for involvement in slaving, and "set at naught" the anti-slaving treaty with Britain.[28] At the end of December the British commissioners reported that there were continuing efforts by members of the Government, apparently with the sanction of the Regent *ad interim*, to abrogate the law passed against the Trade in November 1831 and revert to the open and unlimited importation of African slaves. The commissioners were aware, however, that there was a strong body of public opinion resisting the move and supporting the maintenance of the Anglo-Brazilian agreement.

Whatever attention the Commander-in-Chief was giving to the slave trade was diverted at the end of November by the arrival at Rio of news from Captain Broughton of *Samarang*, who had joined *Wizard* off Bahia, that the port was in the hands of insurgents. Hamond immediately sent instructions to *Sparrowhawk* to return from the River Plate to reinforce Broughton, if the British Minister in Buenos Aires felt that her presence could be temporarily dispensed with despite continuing unrest at Montevideo. Hamond had no spare vessel on the east coast, and he dismissed any idea of withdrawing a vessel from the west coast. Not only was *Acteon* on her way home from the Pacific, but also Commodore Sulivan had his hands full with hostilities between Chile and Peru.

Sparrowhawk sailed for Rio on 17 January, but before she did so she was able to provide valuable assistance to the French brig-of-war *Alerte*, which had run onshore in the River Plate and beaten off her rudder. With great difficulty Commander Shepherd got the *Alerte* into Montevideo and then recovered her rudder and cables. It appeared to the Admiral that without *Sparrowhawk*'s help the brig would have been lost.

When *Acteon* arrived at Rio at the end of December she was joined by Mr Hamilton, now described as HM Envoy Extraordinary, for passage to England, and on 3 January 1838 the Sixth Rate sailed for Spithead.

Two days later the schooner *Olive Branch*, flying the Stars and Stripes and with an American master and crew, arrived in ballast at Rio. She was 20 days from Angola, carrying a very small quantity of palm oil and wax, clearly for the sake of appearance, and undoubtedly having landed a cargo of slaves along the coast. Although she underwent a form of detention by the police, that was merely to obtain a fee for release. To the Commander-in-Chief's knowledge, she was the second US vessel to import slaves to the vicinity of Rio, but US-built slavers under Brazilian or Portuguese colours were numerous. As it happened the United States Navy Commodore on the Station had just called on Hamond in *Dublin*, and he promised to make enquiries.

Yet another disciplinary matter demanded the Admiral's attention in early January, and again it involved a warrant officer. The brigantine *Spider* (6), Lieutenant John O'Reilly, had been on the Station since mid-1836 and, a sister of *Hornet* and *Cockatrice*, had probably been employed as a packet. She was now at Rio, and O'Reilly applied to the Commander-in-Chief for a court martial for Mr Patrick Maitland, Mate, who had become drunk while in charge of the brigantine on 5 January, and consequently, as some officers of *Dublin* discovered at 2300, the entire ship's company had become "drunk and riotous". There was no chance of a court martial, and Hamond ordered that Maitland should be disrated to midshipman until his conduct should warrant application for rerating to mate. Maitland failed to learn his lesson, and a couple of weeks later he left the brigantine without permission, taking a boat and detaining it ashore. There he got drunk and behaved in a disgraceful manner to the boat's crew until he was forcibly taken back on board and placed under arrest. He then broke arrest, became drunk and violent, drawing a sword-stick and threatening injury. O'Reilly had to send to the flagship for help. This time the Admiral had Maitland removed to *Dublin*, and ordered O'Reilly to disrate him to able seaman and discharge him to *Dublin* until Their Lordships' pleasure was known.

Sparrowhawk reached Rio on 30 January and sailed for Bahia a couple of days later. On joining *Samarang* she found that the Brazilian Government had imposed a blockade on the port. Royalist forces had asked Captain Broughton and the commanding officers of other foreign men-of-war off Bahia for muskets and military accoutrements, but all had refused, insisting on maintaining neutrality. She also learned that, on 12 January, Lieutenant Edward Harvey, Commanding Officer of *Wizard*, had died at Pernambuco while being cared for ashore in the British Vice-Consul's house. When this news reached the Commander-in-Chief he gave Edward Tatham, a mate in *Dublin*, a commission as Lieutenant and Commander of *Wizard*. It appears, however, that in the meantime Broughton had put Lieutenant Bower, one of his own officers, into the brig.

On 17 February there appeared at Rio a stark indication of the declining emphasis given by the Admiralty to affairs on the east coast of South America and to action against the Brazilian slave trade. The Fourth Rate frigate *President* (52) arrived flying the flag of Rear-Admiral Charles Ross, appointed as Commander-in-Chief Pacific Ocean and relieving Admiral Hamond of his responsibilities on the west coast of South America.* The vessels on the west coast – *Imogene*, *Cleopatra*, *Fly*, *Harrier*, *Basilisk*, *Sulphur* and *Starling* – were placed under Ross's command by Hamond, and *President*, with not only the new Commander-in-Chief but also his wife and family embarked, sailed for the Pacific on 1 March.†

Hamond did not leave for England immediately. It was questionable whether or not there was still a Station for him to command, but he did retain command of the vessels on the east coast and of *Rover*, which was expected at Rio

---

* This ship was built to the lines of the USS *President*, taken in 1814, and after being hulked in 1861 she became the Royal Naval Reserves training ship in London.
† Presumably the reason for omitting *Stag*, Commodore Sulivan, from this list was that Sulivan was being replaced by Ross and was expected soon to be leaving the Station.

in mid-March from the Pacific. *Samarang* had arrived from the north, and on 9 March the Admiral, perhaps not best pleased, ordered Captain Broughton to return to Bahia with all possible despatch to protect British mercantile interests and to keep his small squadron there until the port was more settled. He followed in *Dublin* on 15 March, and on the morning of the twenty-first he fell in with *Wizard*, which had been sent by Commander Shepherd of *Sparrowhawk* to carry the news to Rio that Bahia had been recaptured from the rebels. Lieutenant Bower continued on his way with orders to Commander Eden not to proceed to Bahia in *Rover* as originally intended but to sail for Spithead.

Dublin anchored at Bahia on 22 March, a day astern of *Samarang*. Hamond found that *Sparrowhawk* and *Wizard* had offered all assistance in their power to British citizens in Bahia during the troubles, but only two or three ladies had taken refuge in *Wizard*, the others preferring not to leave the town. It had been diplomatically unacceptable to land men.

On 24 March the Admiral gave recommendations to Captain Broughton, as Senior Officer on the east coast, on the dispositions of his three remaining cruisers if all remained quiet at Bahia. The following day he sailed in *Dublin* for England, and at sunset on 17 May, shortly after anchoring at Spithead, he struck his flag.

* * *

The past three years had seen no concrete progress against the slave trade at its western terminals. Admittedly there had been a trickle of arrests of slavers bound for Cuba, but they had been few in number by comparison with the volume of the traffic, and in 1835, when the Trade to the island was at its peak, only three vessels had been condemned. There had been consternation among the Cuban slave merchants at the news of the Equipment Clause in the 1835 Treaty between Britain and Spain, but the adoption of false colours and the resistance of Portugal to any amendment to the very limited measures of its treaty with Britain had rendered that new agreement virtually ineffective against outward-bound Spanish slavers.

Although the argument on defining the nationality of slavers had at last been resolved to Britain's satisfaction, thereby denying claims of Portuguese citizenship to slave traders operating in Cuba or Brazil, the Havana Mixed Court of Justice had been insufficiently diligent in seeking evidence of Spanish ownership in Portuguese-flagged slavers. Spanish vessels equipped for slaving but masquerading under Portuguese colours had, in consequence, occasionally escaped adjudication. Others similarly under false colours but caught with slave cargoes had to be taken back across the Atlantic to the Anglo-Portuguese Mixed Commission at Freetown instead of being tried much more conveniently, and without incurring a long absence of prize crews, in the Anglo-Spanish court at Havana. The lack of a court on the western Atlantic seaboard to deal with Portuguese-flagged prizes, of both true and false identity, was a serious difficulty, but there was no territory upon which one might be established even in the unlikely event that the Portuguese Government had conceded the requirement.

With never more than two cruisers patrolling the immediate approaches to Cuba, returning slavers were finding it all too easy to land their slave cargoes at remote points on the coast before making for Havana or Santiago de Cuba, and, as the cases of the *General Laborde* and the *Manuelita* demonstrated, it was no easy matter to achieve condemnation of vessels after they had landed their slaves. The difficulty was even more acute in Brazilian waters, where cruisers were hopelessly thin on the ground and the scope for landing slaves undetected was vast.

At Havana the Mixed Court of Justice was, even in the period of conflict between its British members, generally supportive of the cruisers' work, and the Spanish commissioners mostly conducted themselves with integrity. The same could not be claimed for the Rio de Janeiro Mixed Commission. The attitude of the Brazilian commissioners was obstructive, and the British commissioners allowed themselves to be browbeaten into transgressions against the Anglo-Brazilian Treaty, in particular in allowing the intrusion of Brazilian judicial procedures into the business of the Commission, to the detriment of the condemnation and emancipation process and the severe inconvenience of the captors. Only the rarity of arrests prevented this from becoming a serious obstacle.

The population of Cuba, like its merchants and officials, remained unremittingly hostile to the suppression campaign, but in Brazil there was an element of ambivalence. There was general resentment in the country at what was seen as arrogant British action against its trade, and local authorities were mostly active in facilitating slave landing as well as profiting by it. However, there was increasing disquiet about the threat of slave uprisings, and for this, and more altruistic reasons, there was a degree of opposition to the slave trade among the public and politicians. Indeed a bill against the Trade was introduced, unsuccessfully, into the Brazilian Assembly in 1837.

The sole initiative shown by naval commanders had been the stationing, for a period, of a Sixth Rate in the south-eastern approaches to the Caribbean, but otherwise strategy and tactics had remained essentially unchanged and ineffective. Commanders-in-chief had, among the multifarious tasks of their Stations, given no high precedence to slave-trade suppression, and for this they cannot be strongly criticised. They had not been given the resources with which to prosecute the campaign vigorously, nor were they armed with treaties of adequate stringency. The Commander-in-Chief on the South America Station had been particularly poorly equipped, and, although Admiral Hamond might perhaps have shown more imagination in the use of his handful of cruisers, he was understandably distracted by the constant unrest threatening British commercial interests in that turbulent continent. His replacement by a Commander-in-Chief, Pacific made clear the priorities as far as London was concerned.

The slave traders in Cuba had occasionally been temporarily inconvenienced or disconcerted, but disruption of the slave trade in the western Atlantic remained a remote hope.

CHAPTER 15

High-Handed Action: Eastern Seas, 1838–9

God so commanded, and left that command
Sole daughter of his voice; the rest, we live
Law to ourselves, our reason is our law.

JOHN MILTON

MANY OF THE OFFICERS on the west coast of Africa may well have agreed with the opinion expressed by Commander Henry Keppel of the brig-sloop *Childers* some time after his return to Portsmouth at the end of April 1838:

> It appears to me that while cruisers were not allowed by treaties with Spain and Portugal to capture vessels fitted for the slave trade without slaves onboard, we did more harm than good.
>
> Along the coast negroes were brought from the interior and confined in pens and, when closely watched by our cruisers, are frequently starved to death.
>
> If a slaver is captured with slaves onboard, the price rises on the other side of the Atlantic, which is immediately followed by the increase in the number of vessels that come out.[1]

His conversations with young slaver owners, sons of wealthy Spaniards, had revealed that some of them, having purchased American clippers, went to the slave coast as much for pleasure as for business, and they had amused themselves by using their yacht-like vessels to distract the cruisers from loaded slavers leaving the Coast.

This pastime had become laden with risk with the introduction of the Spanish Equipment Clause and since the arrival of the new generation of cruisers which, particularly when commanded by the likes of Commander Tom Henderson, Lieutenant John Adams and Lieutenant Thomas Roberts, could outsail the best of the Baltimore clippers. The commanding officers of *Columbine*, *Waterwitch* and the Symondites relished the sailing races in which they engaged whenever opportunity offered, and their men were no less enthusiastic. Captains and Masters honed their own skills in these competitions while working up their crews to a level of efficiency which no slaver could emulate. The officers and ships' companies of the older and slower cruisers were no less determined to extract the best possible performances from their vessels.

The regrettable fact remained, however, that the cruisers were still far too few for the task with which they were faced. The volume of the slave trade to Cuba, supplied principally from the Bights and the rivers of the Windward Coast, had reached a peak in 1835. Although the immediate effect of the new 1835 Treaty had been to bring the Spanish traffic virtually to a halt, it had proved merely to be a temporary achievement. Not only were means of sidestepping the treaty constraints rapidly evolved, but also, with a degree of warning, it remained all too easy to evade the cruisers, and the slavers were generally well served with intelligence on the movements of the men-of-war. Blockade was not an option with the size of force available, and, there being so many points of slave embarkation, the holes in the net were much too big.

For all the improvements in ship design and treaty measures, identical tactics had now been employed against the slavers for 30 years, and the inevitability of their lack of success should have been clear to all. A change of direction was desperately needed, and the tentative agreement made by the Senior Officer on the Coast with the King of Bonny in April 1837 perhaps indicated a promising avenue.

However, the attention of the cruisers was currently being distracted and their effectiveness compromised by their greatest enemy, the curse of yellow fever. It was far from finished with them.

* * *

The new Commander-in-Chief on the Cape Station, Rear-Admiral The Hon. George Elliot, was the second son of the first Earl of Minto. He had served with Nelson, Foley, Hyde Parker and Thomas Masterman Hardy, and he had been at St Vincent, The Nile and Copenhagen. His record of command at sea was long and distinguished, and he was also a respected ship designer.

He would have received his first intimation of the sickness afflicting his cruisers on the slave coast with the arrival at the Cape in April 1838 of the stricken *Bonetta*, and worse was to come as the year progressed. The next victim was *Waterwitch*, but she had been healthy until April when she arrived at Ascension. There she found fever ashore, but it did not appear on board until she sailed again for the Coast on 3 May. By 4 June 60 of her ship's company of 52 Europeans and 18 Africans had been attacked, and 15 had died. Only three white men escaped the infection. Lieutenant Dickey died off Princes Island on 29 May, and the Assistant Surgeon also succumbed. In the period 4–6 June there were only three men fit enough to work the brig. William Austen, Mate, took command of her and wisely made for Ascension, reaching the island on 9 June. As soon as 25 of his white ship's company were fit for work, Austen sailed for England, taking his Kroomen with him. He was made lieutenant soon after arriving home, and the Africans were returned to the Coast in *Columbine*.

Ascension, which had long been the cruisers' invaluable refuge from the diseases of the African coast, had itself been badly hit by this yellow fever epidemic, probably introduced by *Bonetta*. Nearly 30 people had died, among them that fine officer Captain Bate of the Royal Marines, who had consistently provided such admirable support to the West Africa cruisers during his years in command of the garrison.

Fair Rosamond had been at the island with the homeward-bound *Childers* and the survey cutter *Raven* when *Thalia* passed through on her way to England at the beginning of April, and she then made her way to Freetown. There she embarked Lieutenant-Governor Docherty and sailed with him for the Gambia on 5 May. She soon had 16 fever cases, of whom five died. Returning to Sierra Leone in early June, she collected the prize crews languishing there. Of *Scout*'s people she embarked an officer and three men, leaving six more sick ashore; 12 others had died. For *Forester* she recovered an officer and three men, and for *Bonetta* one man. By 11 June six of these passengers had been seized with fever, and four subsequently died.

Dolphin cruised in the Bight of Benin for the first six months of the year, with a break for refit at Ascension in May when the island was in the grip of its fever epidemic. She suffered eight cases, and two died. Thereafter, however, she remained in good health. *Curlew* was watching the Rio Pongas and cruising off Sierra Leone throughout May and June, and into July. She contracted 25 cases of fever, although none proved fatal. The majority, and the most severe, occurred among boats' crews working in the rivers, whose dangerous practice was to lie concealed on the riverbanks, night and day, with no shelter other than a boat's sail. *Lynx* also lost no one, but during February, March and April, in two phases, 18 men fell sick. The fever contracted while she was at Sierra Leone was fairly mild, but as she later lay at anchor off Ascension the attack was more severe.

A difficulty of a quite different sort was worrying Commander Popham, still the Senior Officer on the West Coast. When *Thalia* called at Ascension in April on his way home, Admiral Campbell had received a letter from Madame Ferreira at Princes Island telling him of the death of her husband, Don Joze Ferreira Gomez, Governor-General of Princes and St Thomas. A self-constituted junta had then withheld news of the death to retain usurped powers in order, she wrote, to strip her of her property. She owned a very large estate at West Bay where, for many years, the West Coast cruisers had procured wood and excellent water, and where Gomez and his wife had frequently provided live cattle and "refreshments". As Campbell told the Admiralty, "their general kindness and friendly attention to the Officers" was well known. Madame Ferreira threw herself on the protection of the Commander-in-Chief.[2]

Popham had become aware in January of these developments, and no one was more conscious than he of the serious implications for the cruisers if the facilities of West Bay were denied them. He headed for Port Antonio in *Pelican*, with *Scout* and *Fair Rosamond* in company, and arrived on 28 January. *Dolphin* joined them the following morning. Popham's initial remonstration had received a very uncivil response, but he then gained an audience with the junta at which he was accompanied by all the commanding officers. At this meeting the attitude was apologetic, but Popham believed that the junta, which was composed entirely of Africans, was not to be trusted. "The present people", he wrote, "are, if possible, more deficient in moral feeling than even Portuguese."

The threat to Madame Ferreira, and to the hospitality at West Bay, seemed to have retreated for the time being, but Popham's primary concern now was for the safety of British merchantmen at Princes. That this concern was well founded became clear to Commander Craigie in *Scout* on his arrival at West Bay on 22 March. The masters of two British merchant vessels complained to him of fraudulent behaviour and threats of extortion by the island's authorities, still the unelected junta. The protective presence of *Scout* and *Pylades* induced a resolution of the affair, but Popham was convinced that the junta was beginning an attempt to expel the cruisers from West Bay. He noted too that in the past six months seven Spanish slavers had undergone false sales and obtained Portuguese papers at Port Antonio.

The relatively low level of slaving in the early months of the year, combined with the reduction in the number of cruisers on the Coast, thanks to the fever epidemic and requirements for replenishment and refit, had resulted in a low incidence of boardings, and arrests had been rare. On 29 March off Cape Palmas *Saracen* had boarded a schooner called *Veloz* which Lieutenant Hill believed to be Portuguese, but the master boasted that the vessel was under Russian colours and that he had sailed for Havana with a cargo of 400–500 slaves. When Hill sought advice on the action appropriate to such a situation, Admiral Elliot forwarded his letter to the Admiralty, complaining that if the Mixed Commission "considers acting only on the slave treaties and cannot pay any attention to the

Decree of the Queen of Portugal" the slave trade was "open to any who care to purchase Portuguese papers" for 200–500 dollars. It should have worried both the Admiralty and the Foreign Office that the Commander-in-Chief had such a flawed notion of the function of the Anglo-Portuguese Mixed Commission and of the relevance of the Portuguese decree, but neither seems to have been moved to clarify the position for him.³

At the same time Elliot did, however, re-emphasise an important point which, it seemed, the Admiralty might be intending to ignore. Whenever an empty slaver was condemned it was always doubtful whether the sum raised by her sale or, in the case of a Spaniard, the sale of her components would clear the cost of condemning her. If it did not, the additional expense fell on the commanding officer bringing her to adjudication. This, felt Elliot, would discourage officers from detaining empty vessels.

Saracen had not so far distinguished herself, and reports on the incidents in which she had been involved suggest that Lieutenant Hill, unlike the great majority of commanding officers of these small cruisers, may have lacked self-confidence. That the brig was not a happy ship was becoming apparent to Commander Popham, who wrote to the Commander-in-Chief from West Bay in April that he had for some time been concerned that the officers of *Saracen* were conducting themselves towards Hill "in a manner bordering on a combined determination to oppose and to break him with contumely". Following an enquiry on board by Commanders Craigie and Castle, ordered by Popham, Midshipman Holmes was removed as a "prisoner at large" to *Scout* to be taken to Simon's Bay. In addition, Lieutenant Hill was told to place Mr John Angelly, clerk, under arrest, although he had to be released to have charge of the stores. Popham then inspected *Saracen* in company with Craigie and Castle on 13 April, and read the Printed Instructions on the quarterdeck.*

His inspection completed, Popham sent *Saracen* to join the Flag, thereby depriving himself of a sorely needed cruiser, a measure which reflected his view of the seriousness of the episode. He told the Commander-in-Chief that insubordination by some of the junior officers in the cruisers was a frequent occurrence, and there had been instances of drunkenness among them too. He believed that a severe example was necessary. At the end of June Admiral Elliot ordered that Angelly, judged to be the leader of the offending clique, and Holmes should be dismissed from *Saracen* for misconduct and sent to England, and he removed Midshipman Hill from *Saracen* to *Bonetta*. He then expressed

---

* The "Printed Instructions" may have been either the relevant Articles of War or parts of the *Regulations and Instructions Relating to Her Majesty's Service at Sea*.

his disapprobation of Lieutenant Hill's conduct on the matter of two punishments. The first he regarded as totally unjustified and in direct disobedience of Admiralty orders regarding corporal punishment, and the second as unreasonably harsh. This insubordination and excessive harshness in *Saracen* were classic symptoms of weak leadership, and it is perhaps surprising that Hill retained his command, although it was probably judged important that the "combined determination" described by Popham should not be seen to have succeeded in putting paid to Hill.[4]

After her initial period of recuperation at Ascension, *Forester* returned to her station off the River Bonny under the temporary command of Lieutenant Francis Seymour Nott, and it was he who seized the schooner *Dous Irmaos* under Portuguese colours on 2 April. The prize had shipped her 305 slaves the previous day in the Bonny, and was taken close inshore a few miles to the east of the river mouth. *Viper* was in sight at the time at a distance of about seven miles. The schooner was carrying Portuguese papers from the Minister for Foreign and Maritime Affairs at Lisbon, but there was every reason to suppose she was Spanish and owned at Havana. The Customs House there had cleared casks and 3,000 feet of planking as her cargo for this, her fourth slaving voyage, without requiring a bond. Dysentery claimed 55 of the slaves before arrival at Freetown, six threw themselves overboard, and only 241 were finally emancipated when the schooner was condemned by the Anglo-Portuguese court.

Experienced cruisers could usually recognise slaving vessels at some distance, but they were occasionally mistaken. Commander Castle of *Pylades*, who had an exceptionally practised eye, boarded a "suspicious-looking" brig on 17 April, but discovered that she was an innocent French palm-oiler on passage to Gabon. He gave the master a certificate of the boarding and entered the incident in the brig's log, but there was no subsequent complaint.

It was to be a further six weeks before Castle achieved a more satisfactory boarding. He had stationed Lieutenant John Price in a five-oared gig in the mouth of the Old Calabar River before anchoring *Pylades* out of sight, and on 3 June Price took the Portuguese-flagged schooner *Prova*, which was about to get underway for Havana. The schooner, which mounted two guns and had a crew of 19, had just loaded a cargo of 225 slaves. Although she had Portuguese papers from the Cape Verdes she was undoubtedly a Spaniard, and she had been cleared from Havana with casks and a slave deck as part of her lawful cargo. Castle landed the slaver's crew at the mouth of the river with two days' provisions, and sent their vessel to Freetown, where she was condemned by the Anglo-Portuguese court with the 194 surviving slaves.

During June and July *Pylades* boarded a number of French merchantmen of suspicious appearance, but all were found to be legal. One of them, on 26 June, was bound for Accra, and Castle had to chase her, but the others were at anchor; the first two, a little earlier in June, were at Accra and the last one he found at Port Antonio on 28 July. There seem to have been no diplomatic repercussions from these incidents, but the Commander-in-Chief had to report to the Admiralty at the end of June that he had received a complaint from the Commanding Officer of the French corvette *Triomphante* that an officer of *Childers* had boarded the French schooner *Africaine* at anchor off British Accra and had behaved offensively, and that the master of the schooner had been put in irons.

Commander Keppel had apparently been required to give his version of events and explained that that he had sent his Master, Mr Coaker, to examine the suspicious-looking schooner and had discovered the main hatch covered. When Coaker presented his papers they were refused in an insolent manner, and he then attempted to look down the companionway to the cabin. At that he was attacked and struck by two men, and the gig's crew leapt on board to defend him; thereupon the schooner's master brought on deck a tin box which Coaker seized and took to *Childers*. Having collected three marines from the sloop, Coaker returned to the French vessel and removed the two men who had hit him. They were put in irons overnight in *Childers*.

After her brief return to her station in April and May, *Forester* had visited Sierra Leone and had again been attacked with fever. In the Bights on 30 June *Fair Rosamond* fell in with her and found her again in need of medical assistance: her Purser and two seamen were sick, and her Assistant Surgeon was dying. Having transferred a returning prize crew to *Fair Rosamond*, and with *Pylades* in company, she made once more for Ascension, quarantine and convalescence. From there she was sent to the Cape, where she remained until late October when, having recruited replacement hands from merchantmen, she sailed again for the Bights. Her total loss during the year was 19 officers and men, a third of her ship's company. *Pylades* had suffered too. She had returned to the Bight of Biafra from the Cape in good health, but when the boat's crews had returned on board from work in the rivers they brought fever with them. Six men sickened and one died, and she headed south as far as St Helena to recuperate before returning to cruise off the Bonny.

Another temporary loss to the West Coast force was *Scout*, which departed for the Cape in late spring on her way to relieve *Leveret* in the Indian Ocean. However, the strength of the West Coast force was improved in the early summer by the return from England of the brigantine *Lynx* (3), now in the hands of Lieutenant Henry Broadhead, and, in June, of her sister, the brand-new *Termagant*, commanded by Lieutenant Woodford Williams. They were followed by *Nautilus*, Lieutenant George Beaufoy, another of the same class but one of those retaining the original ten guns and brig rig. A fourth new arrival was the Symondite brig-sloop *Lily* (16), only a year old and under the command of Commander John Reeve.

A distasteful episode attended the initial stage of *Lynx*'s re-engagement in the campaign. While she lay at anchor off Accra during June, a boat's crew under the charge of Mr Hector McNeil, Mate, boarded the Portuguese merchantman *Maria*, and during the search a corporal of marines stole a spyglass and a pair of pistols. It appeared that, during the return to *Lynx*, the corporal presented his loot to McNeil, who took it and promised recompense. The theft did not come to light until it was reported to Lieutenant Winniett of *Dolphin* when she visited Accra in September,[*] and in October at West Bay, on receipt of a letter from Lieutenant Broadhead, the senior officer present, Commander Elliot, investigated the incident.[†] The result was that McNeil was ordered to be discharged into the first man-of-war sailing for England and to be dealt with as Their Lordships saw fit, and the corporal of marines was disrated. However, Broadhead examined the matter further when the owner of the articles was later arrested in a slaver, and it seemed that McNeil was less culpable than had initially been believed. He was regarded as "a well-disposed young man and attentive to his duty", and his punishment was reduced to a severe reprimand, not least because he was the only officer, other than Broadhead, able to do duty in *Lynx*.[5]

In mid-summer the Admiralty despatched orders for a number of promotions and reappointments: Elliot was advanced to Rear-Admiral of the White Squadron, Williams of *Termagant* was made commander, and, most deservedly, Popham was promoted to captain.[‡] Perhaps as a result of the evaporation of Admiral Campbell's patronage, a Lieutenant Napier was appointed to command *Forester* in place of Colin Campbell, who was to join *Melville*, but it is not clear when Campbell had actually assumed command of *Forester*. He may have joined the brigantine either when she called at Freetown in late May or at Ascension when she returned in July with her ship's company sick.

---

[*] Winniett was Commanding Officer of *Viper*, but, for unexplained reasons, he was apparently temporarily in command of *Dolphin* at the time of her visit to Accra in September.
[†] The Senior Officer on the Coast was still Popham in *Pelican*, but Commander Elliot, newly arrived in *Columbine* and unaware that Popham was still on the Station, investigated the incident.
[‡] The *Navy List* despaired of knowing who was in command of *Forester* for the nine months after Rosenberg's death.

At all events, he was certainly in command when she arrived at the Cape on 18 August.*

After an interval of nearly six weeks, there was at last, on 13 July, another slaver arrest. Three days after she had embarked 195 slaves at the Bonny River, the Portuguese brigantine *Felis* was seized by Lieutenant Oliver in *Fair Rosamond*, apparently without resistance, ten miles or so south of the river mouth. The slaver had cleared from Havana as the Spanish vessel *Ceres*, and had acquired her Portuguese papers from the Governor of the Cape Verde Islands. The Governor had also given her master, a Spaniard, a passport stating that he was a shipwrecked Spanish sailor, but the man continued to direct the voyage. The prize was condemned by the Anglo-Portuguese court, and 187 slaves were emancipated.

Admiral Elliot was clearly annoyed by a copy he had received of Lord Palmerston's directive of 30 April to the British commissioners at Rio de Janeiro. It referred to "vessels owned by Brazilian subjects or Portuguese subjects resident in Brazil" and to those "which shall be found carrying Slaves from Africa to Brazil". As Elliot pointed out in frustration to the Admiralty, this left glaring gaps. There were no instructions covering the West Africa problem of nominal Portuguese subjects, nor on the matter of the Portuguese traffic from Africa to the West Indies, and he questioned how it could be proved that a slaver was bound for Brazil unless she was caught in the act of landing Africans on its coast. He was convinced that unless the Mixed Commission courts were empowered to act on the 1837 Decree of the Queen of Portugal (which, of course, they could not) and to throw the responsibility of proving nationality onto the detained vessel instead of the captor, there would be no interruption of the Trade under the Portuguese flag, nor detention of empty Portuguese slavers. Elliot still failed to understand the limitations of Mixed Commission jurisdiction, and he was overstating the difficulty of dealing with false Portuguese nationality. Nevertheless, in complaining that no equivalent directive had been given to the Freetown commissioners, he was right to indicate the need for a comprehensive set of Instructions, to commissioners and commanders-in-chief on both sides of the Atlantic, on dealing with the Portuguese slave trade in all its convolutions. Where he made a serious omission of his own was in not demanding removal of the unjustified veto on seizure of laden Portuguese slavers south of the Equator.[6]

Amid the enthusiasm for the 1835 Spanish Equipment Clause, two very obvious anomalies, indeed injustices, introduced by the new Anglo-Spanish Treaty

---

* It is not clear when Popham's promotion took effect, but it was probably not until he relinquished command of *Pelican*.

seem to have been ignored by all except those upon whom they inflicted direct disadvantage. Under the 1817 Treaty those cruisers fortunate enough to achieve condemnation of Spanish prizes laden with slaves were granted bounties on the slaves captured and, after the deduction of costs, the British share of the proceeds from the sale of the condemned vessels. Now that condemned Spanish slavers were being broken-up, the proceeds of sales, particularly after the additional costs of destruction were subtracted, were reduced to little or nothing. Of greater concern in the case of a slaver taken under the Equipment Clause was the absence of slave bounties. In these latter circumstances the dangers and discomforts remained, as did the financial risk for commanding officers, but the potential financial compensations had gone. The absence of these customary rewards was not only an injustice but also a disincentive; there is, however, no evidence that any cruiser forewent capture of an empty Spanish vessel in the hope of seizing her later when she had loaded her cargo of slaves. At last, on 27 July 1838, this fault was rectified by what became known as the "Tonnage Act".

By this Act (1 and 2 Vict., c. 47), for vessels broken-up after condemnation, in addition to payment of proceeds and slave bounties, a further bounty of £1.10s per ton of the vessel's tonnage would be paid to the captor. For vessels seized and condemned but without slaves on board there would be an additional bounty of £4 per ton. The tonnage of such vessels was to be estimated or ascertained according to the British system of measurement, and was to be certified by the commissioners making the condemnation. For vessels having slaves on board, if the bounty calculated on the number of slaves was to be less than that calculated on the tonnage, the captor could choose to take the latter instead of the former. Bounties were to be paid out of the Consolidated Fund of the United Kingdom of Great Britain and Ireland. The tonnage bounty on an empty prize would be unlikely to match £5 per slave if she had been full, but it was certainly better than nothing.

It gave emphasis to Commander Popham's concern about the conduct of the junior officers in the cruisers that, on arrival at Simon's Bay, Lieutenant Campbell of *Forester* reported to the Admiral how, on joining the brigantine, he had found Midshipman Chambers under arrest for repeated drunkenness. He had hoped to "reclaim" the young gentleman, but had been told, even by Chambers' messmates, that such an attempt would be hopeless. Campbell requested the offender's removal from *Forester*, but Elliot, no doubt to make an example, ordered discharge from the Service. Considering the tribulations of *Forester* during the previous few months, it is perhaps possible to feel a little sympathy for Chambers.

Nearly five weeks had elapsed since the capture of the *Felis* when there was a brief flurry of arrests in August, led by Lieutenant Kellett in *Brisk*, who was cruising the Windward Coast. On the fifteenth the brig *Diligente*, under Portuguese colours, became his first victim while lying empty at anchor in the River Gallinas. In a different guise she had been condemned at Sierra Leone in March 1837, purchased from the court and taken initially to Havana and then to Cadiz. There she had undergone a nominal sale to the agent of a Cape Verde Islands resident, and she then left Spain for the coast of Africa fitted for the slave trade and with her erstwhile owner still on board. The falsity of her papers was emphasised by their claim that her Spanish owner had sold her at Cadiz before the date on which he had bought her at Freetown. She was condemned by the Anglo-Spanish court as fitted for slaving, and, the Queen's Advocate having subsequently approved of the court's decision, Palmerston told the British commissioners that they should follow the same course of action in further such cases.

On the following day, and also in the Gallinas, Kellett seized another Spaniard under Portuguese colours. This was the schooner *Ligeira*, and she too was empty but fitted for slaving with all the forbidden equipment. Although she was holding a Portuguese passport from the Cape Verdes, the Anglo-Spanish court traced her ownership to a merchant resident in Havana, and she was condemned.

A few days later in the Bights it was again the turn of *Fair Rosamond*. On 21 August she found another schooner masquerading under Portuguese colours, lying at anchor in Accra Roads, and Lieutenant Oliver arrested her on the grounds that she was Spanish property and equipped for the slave trade. When the schooner, the *Constitucão*, was boarded she was found to be armed with an 18-pounder on a pivot amidships, two other long 18-pounders, six blunderbusses, 20 muskets and 20 cutlasses. She was prepared for action with grape, chain and bar-shot, and other ammunition placed around the deck. Her crew numbered 37 officers and men. Oliver believed that if he had encountered her at sea she would probably have offered determined resistance, and he assessed that she was set on piracy. On board she had not only a Cape Verdes passport but also inaccurate documents issued in Cuba by the American Consul, Mr Trist. She was condemned by the Anglo-Spanish court.[7]

This was by no means an isolated case of Mr Trist aiding and abetting a slaver. In several recent adjudications at Freetown it was found that, in the absence of a Portuguese Consul at Havana to issue false papers to departing Spanish vessels, the American Consul had taken it upon himself to do so. There would have been many more instances which remained undetected.

Late in August, Commander Popham, still the Senior Officer on the slave coast, reported to the Commander-in-Chief that a new Lieutenant-Governor at Port Antonio, rather strangely a Pole, appeared sincere in an intention to prevent the illegal issuing of Portuguese papers to Spanish slavers. However, Popham expected that, in the impoverished state of Princes and St Thomas, lesser officials were unlikely to refuse bribes and the practice would continue. The Portuguese schooner-of-war conveying the new Lieutenant-Governor to Princes Island had been unable to beat there from Fernando Po and, having been found unseaworthy, was beached at Clarence Cove. The voyage had to be completed, embarrassingly, in a slave brig. Popham also reported his belief that many slaves were being forced to make "the most harassing and cruel marches" to embarkation points south of the Equator to enable the Portuguese slavers to take advantage of their continuing immunity in the southern hemisphere. There were no fewer than eight vessels constantly in the Trade from Cape Lopez, and, although the newly arrived Governor of St Paul de Loando had ordered slavers out of his port, they continued to load at places both east and west of it.

The Commander-in-Chief had apparently decided on *Saracen*'s arrival at Simon's Bay that she needed more corrective treatment than he was able to administer, and he ordered her back to England. This unwelcome depletion of his force was then exacerbated by a serious disciplinary incident in *Lily*. A lieutenant and "his boy" were under arrest, and on 13 August Elliot initially ordered them, together with witnesses, to be sent to England in *Pelican* when *Lily* relieved her to return home. However, he soon decided to keep Popham on the Station and instead sent *Lily* to England, hoping, as he told the Admiralty, that she would return to the slave coast in December. He now wanted Commander Craigie in *Scout* to replace Popham as Senior Officer in the Bights, but *Scout* had only just left the Cape to head eastward. So *Nautilus* was despatched to Madagascar to relieve *Scout* and instruct Craigie to return forthwith to Simon's Bay. Elliot complained that the only cruisers now on the west coast were *Pelican*, *Bonetta*, *Dolphin*, *Viper*, *Pylades*, *Lynx* and *Fair Rosamond*, and that the last three were all sickly. He had, however, forgotten *Brisk*, which was serving him particularly well on the Windward Coast.

Bonetta, returned to the Coast after her recuperation at the Cape, was blockading a Portuguese schooner in Bimbia Creek in late August, the *Secundo Orion*, but Commander Popham, very short of cruisers, sent her to take a look at the Bonny River. Having found all quiet there, Lieutenant Stoll returned to anchor at Bimbia on 14 September, after a fortnight's absence. Discovering the schooner still in the creek, he continued his blockade until the twenty-ninth, when once again he was required to leave. This time it was to deliver returns to Popham at Princes, but the schooner had by now bent on her sails and embarked most of her slaves, and Stoll stationed three boats with 26 officers and men to maintain the blockade. The *Secundo Orion* never appeared before the Mixed Commission, and she either managed to evade the blockade or, more likely, simply waited until *Bonetta* was obliged to depart. With so few cruisers available and so many embarkation points to watch, blockade was neither economical nor could it be adequately watertight.

The next success for *Brisk* was on 21 September, when she boarded the schooner *Eliza* in the River Sesters. There were no slaves on board and the vessel had Portuguese colours and papers, but Lieutenant Kellett was sure she was Spanish. A Spanish ensign was then discovered, and the master admitted that he was the sole owner of the schooner, that he was a resident of Havana, and that his vessel was employed in the slave trade. She was condemned as fitted for slaving.

Less satisfactory was Kellett's boarding of a schooner under United States colours, the *Mary Hooper* of Philadelphia, carrying a cargo shipped at Havana and consigned to a notorious slave dealer at the Gallinas. She had a crew of seven and was carrying nine passengers who would undoubtedly supplement the crew once slaves were on board. All of these men were either Spanish or Portuguese, the supercargo was a Spaniard, and the master had twice been arrested for slaving. Kellett was convinced that the schooner's recent call at Porto Praya had been to collect false papers, but he did not feel justified in further searching a vessel under the Stars and Stripes.

There was no difficulty with *Brisk*'s next victim. On 30 September she surprised an empty Spanish schooner off New Sesters just in time to prevent her loading her cargo of slaves, for whom all preparations had been made. It would probably have been to the benefit of the slaves and *Brisk* if the arrest had been delayed until after loading, but Kellett was not to know the situation until boarding; and in any case he could not risk being tied down to a blockade akin to *Bonetta*'s recent failed effort at Bimbia Creek. The prize, the *Constitucão*, was under Portuguese colours, but following the discovery that the owner was a Havana company, and that the crew had promised to acknowledge the authority of a Spaniard embarked as a passenger rather than that of the Portuguese shown in the papers as master, the schooner was condemned by the Anglo-Spanish court.

These two arrests by *Brisk* signalled the beginning of a richer period of captures which indicated a seasonal upsurge in the Trade. The first was by *Termagant*. Woodford Williams was still in command, and his promotion had yet to take effect, when the brigantine captured the schooner *Prova* on 9 October about 120 miles south of the Bonny River. The prize was showing Portuguese colours, but had sailed from Havana and was bound there with the 326 slaves she had embarked some days earlier at the River Nun. The master admitted to the Anglo-Portuguese Mixed Commission that he was the sole owner of the brigantine and a resident of Havana, although his papers, certified by Mr Trist in Havana, declared that the owner was a well-known resident of the Cape Verde Islands. As had been the recent practice with captured Spanish slavers with false Portuguese identity and carrying slaves, *Prova*'s nationality was not disputed by the Anglo-Portuguese court, and she was condemned with her surviving 295 slaves. The obvious disadvantage of not proving Spanish nationality in such cases and shifting them to the Anglo-Spanish court for condemnation was that the vessels were not subsequently broken-up.

Kellett continued to build upon the success of his cruise on the Windward Coast with two seizures in quick succession on 17 October. Both victims were Spanish schooners, empty but equipped for slaving, and both were under Portuguese colours. The first was the *Veloz*, found just off the Gallinas River. She had undergone a nominal sale at Porto Praya and was supposedly under the command of a Portuguese, but her real master, a Spaniard, was still on board as a "passenger". The *Veloz* was actually owned by Pedro Martinez's notorious Havana company. The second, *Josephina*, was lying in the Gallinas when Kellett arrested her. She had cleared out of Havana as the Spanish vessel *Ramoncita*, and her owner there had instructed the master to use her Spanish papers only when entering a Spanish port. There was no difficulty in establishing the true nationality of either prize, and both were condemned by the Anglo-Spanish court.

On 27 October *Brisk* boarded another schooner under American colours about 25 miles off Freetown. She was the *Mary Anne Cassard*, and this time Lieutenant Kellett was confident that, despite her American papers, the vessel would be condemned under the Spanish Equipment Clause, and he arrested her. In the space of two months she had passed through the hands of five owners at Matanzas in Cuba, the only American on board was the nominal master, who was sailing under a false name, and five Spaniards embarked as "passengers" were clearly the true master and four ship's officers. St Thomas was falsely specified in her papers as her destination, and she was taken on her way to the Gallinas. For some unexplained reason the schooner appeared before the Anglo-Portuguese Mixed Commission, which happened to be suffering a lengthy absence of the Portuguese commissioners, and the court disclaimed competence to adjudge a vessel furnished with United States colours and passport. When this decision was reported to Palmerston he protested that the Commission should have tried and condemned the schooner as Spanish.[8]

For all her success, *Brisk* was having no luck with loaded slavers, but there was better fortune on that score in the Bights. On 31 October Commander Popham of *Pelican* seized the Portuguese-flagged schooner *Dolcinea* about 130 miles south of Lagos, where she had embarked her 253 slaves. The obliging Mr Trist had certified the slaver's false Portuguese papers in Havana, and establishing Spanish nationality would have been a simple matter, but the Anglo-Portuguese court judged it unnecessary and condemned her and her 249 surviving slaves. She was then bought for £310 at public auction by an English merchant, and he sold her for £600 to a man named Sassette who had commanded two slavers condemned at Freetown. *Dolcinea* was subsequently seized at the Sherbro, but insufficient evidence was found to charge her.

On 1 November Lieutenant Broadhead in *Lynx* was even more fortunate than Popham, taking a cargo of 591 slaves in the brig *Liberal*, also under Portuguese colours. She was well armed and strongly manned, but offered no resistance when boarded a short distance offshore about 30 miles west of Whydah. The brig had sailed from Havana but had been given a new register by the Provisional Government of Princes declaring that she was owned by a Cape Verde Islands resident, justifying Commander Popham's scepticism on the matter of acquiring false papers at Princes. Only eight slaves died among this very large cargo on the passage to Freetown, and the commissioners, on condemning prize and slaves, commended the Prize Master, remarking that "This minor loss reflects the greatest credit on Mr Slade's humanity and attention."

A native war in the region of the rivers Pongas and Nunez required *Brisk*'s presence to safeguard British merchants on the rivers, and Lieutenant Kellett had landed on 8 November to demand of the local chief the release of some British subjects seized during the conflict. In his absence in the Nunez, his Second Master, in temporary command of the brig, caught the Portuguese-flagged Spanish schooner *Maria* in the Pongas, just before she loaded her slaves, and held her to await his Commanding Officer's return. Kellett had not succeeded in his mission, and he set about lightening *Brisk* by six inches to enable her to edge upriver to within musket range of the chief's town, but he strongly suspected that the captives had already been sold into slavery. The prize schooner had shipped her cargo at Havana, and she was bound there with her slaves. Her owner and master was a Spaniard, as were all her officers, and the Mixed Commission had no hesitation in condemning her as Spanish and equipped for slaving.

Kellett found that there were several British subjects involved in the slave trade on the Pongas, and he discovered that the *Maria*'s slaves were being supplied by a Mrs Faber (or Taber), one of the Sierra Leone immigrants from Nova Scotia, who had been the mistress of an American slaver on the river. When her lover returned to the USA she continued the business, assisted by a young Maroon man, formerly a clerk in Freetown. Both were British citizens. A local merchant, Mr Campbell, had reported to Governor Docherty in October that Mary Faber and her native allies were at war with the rival slaver, John Ormond of Bangalang, and his mixed-race friends, and not only had she sworn to exterminate all the people of mixed race but also she had attacked Campbell's factory.[9]

Columbine had been recommissioned with Commander George Elliot in command and, on despatching her to the slave coast, the Admiralty had ordered her to proceed directly to the Bights rather than, as was the general practice, to call first at Freetown. As a result, Elliot arrived at West Bay, Princes, in mid-October ignorant of who was Senior Officer on the Coast, of which cruisers were deployed there or of where they were stationed. In fact Popham would be in the Bights for another two months or so before sailing for England, although the Commander-in-Chief was under the impression that he had already departed as ordered by the Admiralty, and he was concerned that no commander was available to replace him. *Pylades* had not returned from convalescence, *Scout* was still in the Indian Ocean, and Elliot was shortly to take *Columbine* onward to the Cape.

Bypassing Sierra Leone presented Elliot with another problem. Among the supernumeraries brought out by *Columbine* for the Squadron were the Kroomen taken to England by *Waterwitch*. The Admiralty had still not appreciated, even after 30 years of anti-slaving operations on the west coast of Africa, that home for the Kroomen employed in the cruisers was Krootown, a district of Freetown, and that they would not take their discharge at any other port. The best that Elliot could do for these *Waterwitch* people was to land them at Princes with a month's subsistence money and orders to cut wood for the Squadron until they could be sent to Sierra Leone. Their wait was seven or eight weeks.

The long-standing difficulty of recovering prize crews from Sierra Leone caused not only protracted shortages of men in the cruisers but also had contributed to the fever outbreaks in parent vessels, and it is surprising that no attempt had been made to establish a system for returning these crews to their ships in a timely fashion. It was the disease risk which particularly concerned Admiral Elliot because the very long waiting periods, combined with the tendency to disobedience by Prize Masters in allowing men to go ashore from the accommodation hulk, held a very high risk of contracting fever.

Early in November he therefore asked the Admiralty to send him a small vessel for carrying these crews from Freetown to Ascension. The initial reaction of Their Lordships to this eminently sensible request was favourable, and they instructed the Surveyor of the Navy to recommend a suitable vessel. However, apparently unmoved by the recent fever losses in the West Africa force, they subsequently replied that no suitable vessel was currently available, but that "one will be selected when opportunity offers". Nevertheless, the matter was not forgotten, and in May 1839 a letter was despatched to Admiral Elliot directing him to purchase a vessel for the purpose, a sensible solution which might have been ordered at the outset.[10]

An improved arrangement for prize vessels at Freetown was, however, introduced by the Mixed Commissions in October, to the advantage of prize crews. Concerned at the number of thefts from detained vessels and the clandestine landing of goods and stores by their Spanish and Portuguese crews, the commissioners directed that the Marshal of the Courts should take charge of prizes immediately upon the "witnesses in preparatory" being landed from them, thereby releasing prize crews to shift to *Conflict*, the accommodation hulk. Apart from Prize Masters required for court proceedings, these men would immediately become available for return to their ships. The slaver crews were to be sent ashore, and the few needing to be detained as witnesses, rarely more than three, would receive a daily allowance from the Marshal for board and lodgings; 3s.6d. for a master or mate, and 2s. for a seaman.

Since the death of Lieutenant Roberts, *Dolphin* had suffered a number of changes of commanding officer, and it was over a year since she had made an arrest. That famine ended in spectacular fashion with the seizure of four prizes in one day. Her new Commanding Officer was Lieutenant Edward Holland, an officer of some anti-slaver experience having commanded the schooner *Nimble* and, briefly, the schooner *Firefly* in the West Indies. On 16 November, in obedience to orders transferred from his predecessor, he looked into Lagos Roads. There he found four vessels fitted, ready and waiting for slaves. Three of them, the brigs *Victoria* and *Dous Amigos* and the brigantine *Ligeiro* were under Portuguese colours, but Holland was sure they were Spanish. The fourth, the *Astran*, seemed to have neither colours nor papers. Holland sent Mr Rowlatt, Mate, to take possession of all four and to remove their crews, more than 100 men, to other vessels in the anchorage. There was a tornado coming on, and, although one of the prizes had already departed for Freetown, Holland delayed the sailing of the remainder until the following morning, bringing two to anchor within musket-shot and ordering the other to anchor under *Dolphin*'s guns. All of this was carried out by Rowlatt, described by Holland as a "very zealous and indefatigable officer [...] in whose praise I cannot speak too highly".[11]

After this fine haul, Lieutenant Holland continued to cruise off Lagos in the hope of intercepting the ship *Venus*, which was lying in the roadstead under American colours. He had heard that she had a Portuguese master and papers waiting ashore, and that she would shortly be sailing with 600 slaves. At 0700 on 28 November a ship was sighted 13 miles to windward and was quickly identified as the *Venus*.* *Dolphin* made all sail in chase and gained ground during the day, but, exasperatingly, a thick Harmattan came on, and she lost her quarry early in the night. Holland then made for Sierra Leone to deal with his prizes.

Buzzard had been away in England since March, and on 17 November, about 100 miles south-south-west of Freetown, she announced her return, under the command of Lieutenant Charles Fitzgerald, by seizing the Spanish schooner *Sirse*, fitted for slaving and showing Portuguese colours. The Anglo-Spanish Mixed Commission failed to establish that the vessel was owned in Havana, but decided that she had "a Spanish course of trade", which had been adopted by the court as one of the tests of national character. The commissioners believed that this lay within the spirit of the treaty, and explained that they had deduced a doctrine "by strong principles of equity and propriety, that there is a traffic which stamps a national character on the individual, independent of that character which mere personal residence may give him". They condemned the schooner, but with some trepidation, and later learned that the Queen's Advocate had not concurred with their verdict.

The British commissioners were well justified in describing *Brisk* as that "indefatigable cruizer" after her next arrest. In 1837 the Spanish brig *Veterano* had left Sierra Leone for London, where she had been reregistered by her master and professed owner, John White. She had taken cargo from West India Dock and Falmouth before sailing for Cadiz, where she had been handed over to a Spanish master and mate before making for Cuba. On 27 September she sailed from Havana for the Gallinas where, on 18 November, just off the river, she was seized by *Brisk* for being fitted for slaving. These actual movements of the vessel were wildly at variance with those shown in her papers, and the commissioners regarded both White and the Portuguese Consul-General at Cadiz as parties to the fraud. The real owner was discovered to be Pedro Martinez & Co. of Havana, and a Spanish "passenger" was her real master. This condemnation drew no criticism from London.

---

* This would not have been as difficult as it might appear. Ship rig was a great rarity in these waters, and there was not even a ship-rigged cruiser in the British force in the Bights.

Columbine's stay on the Coast had been brief, and by mid-November she was at the Cape. There concern was raised about the state of her main rigging and the security of the mast. Overstrain had caused the rigging to become "long-jawed", allowing the mainmast too much play in rough weather, and, in view of the imminence of a passage to Mauritius in the hurricane season followed by a potentially rough voyage to Sydney, the brig was supplied with an additional shroud on each side. There had been a similar problem with *Lily*, and the Commander-in-Chief concluded that it was a general weakness in these modern brigs that their rigging gave less support to mainmasts than to foremasts. The latter stood perfectly well.

The top hamper of the brigantines in general, and *Forester* in particular, was also of concern. On arrival at Simon's Bay, Lieutenant Campbell complained that the working of his mainmast prevented the setting up of stays and backstays in a swell, and the Commander-in-Chief permitted him to land the square sails of the mainmast and their gear, as well as the gaff-foresail, and supplied him with an additional jib to use as a main-topmast staysail and a 42-gun frigate's jib to try as a main staysail. These, together with the gaff-topsail and topmast staysail already on board, gave an increase in canvas, a reduction in weight and the ability to carry canvas longer. The brigantines, thanks to their cutting-down and reduction in armament, sailed one foot lighter than the ten-gun brigs, and it was Admiral Elliot's opinion that they should be given the shorter foretopmast and topgallant masts as fitted in the brigs to increase their handiness and allow them to carry sail longer.[12]

On 27 November *Buzzard* took a prize on *Brisk*'s cruising ground off the Gallinas River. Lieutenant Fitzgerald sighted the slaver during the night about 35 miles to the west of the river mouth, and, having immediately boarded her, he arrested her on discovery of 467 slaves. She was the brig *Emprendedor*, under Portuguese colours but carrying false documents from the Portuguese Consul-General at Cadiz. The *Emprendedor* had been condemned ten years earlier with the same name and the same master, but under Spanish colours. She was condemned again as Portuguese, along with 458 slaves.

The Africans found in the *Emprendedor* were in a dreadful state of emaciation and disease despite their having been loaded only six days earlier. This was evidence that the capture of empty slavers under the Spanish Equipment Clause was causing the slave barracoons to become grossly crowded, and the expense of feeding the exceptionally large numbers of captives had induced traders to reduce allowances of food to starvation level.

While making her way from the Bights to the Windward Coast on 3 December, *Bonetta* recorded her first capture since her frustration off the Bimbia Creek in September, and, indeed, since her return from the Cape, when she took the empty schooner *Isabel*, under Portuguese colours. Lieutenant Stoll made his arrest unusually far from the mainland, just south of the Equator and 330 miles to the west of St Thomas where the slaver, which had sailed from Havana as the *Hyperion*, had changed her name. She had been American, but had been sold to a Spaniard at Havana and had acquired Portuguese registration at Princes. It seems not unlikely that, after these transactions, she was returning to the Windward Coast to load slaves. The Anglo-Spanish court accepted witness statements that she was destined for Havana and was Cuban-owned, and it condemned her.

Verdicts of restoration had become rare at Freetown thanks to the experience and general good sense of commanding officers, and to the wisdom of the Mixed Commissions in dismissing the falsehoods often presented to them by slavers. However, Lieutenant Kellett in *Brisk* overstepped the mark when he arrested the Portuguese schooner *Aurelia Felix* off Bolama, one of the Bisagos Islands, on 9 December. He claimed that a slave boy found on board rendered the vessel liable to arrest, but the boy had not been "shipped for the purposes of the traffic" as was specified by the treaties. African sailors and servants did not give sufficient cause for detention, and the boy had been entered in the muster roll as a cabin boy. The schooner was restored and costs, damages and expenses of £109.3s.10d were awarded against Kellett.[13]

The *Aurelia Felix* had, however, been of secondary concern to Kellett at Bolama. He had been asked by the Governor of Sierra Leone to visit the island, a British possession, in response to information that Portuguese were ashore there. Therefore, on the evening of 9 December, he transferred his senior mate, William White, his clerk-in-charge, Kenneth Sutherland, and 25 seamen and marines to the prize, took two boats in tow, and, leaving *Brisk* well out of sight, he set off for the island. He arrived at 0400 the following morning, and landed at daybreak. The island was being used as a slave station by the Portuguese, the Portuguese flag was flying, and there was a garrison of 16 Portuguese troops. The slaves being held in the barracoons had been driven into the bush, but Kellett's men rounded up 211 to be shipped to Freetown, the Portuguese were ordered off the island, the barracoons were burned and the Union Flag was hoisted in place of the Portuguese colours.

This small victory was, however, a costly one for *Brisk*. The Commander-in-Chief visited Sierra Leone in mid-January, and he recorded that by the time he left Freetown, of the 16 men of the brig who had become infected with fever during the Bolama operation seven had died and others were in hospital, some of them unfit for further service, perhaps permanently so. They had been away from *Brisk* for only 28 hours. Insult was then piled onto injury by the Judge of the Vice-Admiralty court in adjudicating the case of the 211 liberated Africans. These people were not slaves, he declared; they had, in law, become free the moment they were landed on the British territory of Bolama Island. Kellett's protest that they were being held in chains as slaves, and were about to be carried across the Atlantic as such, was waved aside and his claim for a slave bounty was dismissed. Moreover, Kellett was called upon to pay not only expenses to the court but also for the provisioning of the released captives. He intended to appeal against this outrageous (if perhaps legally correct) judgment, but the result is not known.

During his January visit to Freetown the Commander-in-Chief was asked by Governor Docherty to send cruisers into the northern rivers to resolve the disputes which were blighting trade in the area, but Elliot declined. He cited several reasons, not least the risk to the health of his men, and *Brisk*'s recent losses gave emphasis to that concern. He suggested that a preferable option would be to procure a colonial vessel with a native crew, and his recommendation was agreed by the Admiralty and forwarded to the Colonial Office. Simultaneously, Docherty wrote to the Marquess of Normanby at the Colonial Office to request permission to buy a vessel to patrol the neighbouring rivers against "the inveterate traffic in slaves", and he took the opportunity to praise the efforts of *Brisk*, writing that he "cannot sufficiently commend the exertions of Commander Kellett on this part of the Coast". The Governor, as temporary Commissioner of Arbitration in the Mixed Courts, would have fully understood how limited the role of a colonial vessel would be against the slavers.

The departure of *Pelican* for home was much delayed, but she still had service to perform. On 17 and 18 December she took two final slaver prizes. The Portuguese schooner *Magdalena*, with 320 slaves shipped three days earlier at the River Brass, was the first of them, taken about 50 miles north-west of St Thomas. She was an American-built vessel and had been under Spanish colours prior to acquiring Portuguese papers at the Cape Verdes. Together with 302 surviving slaves, she was condemned. On the following day, about 100 miles further west, Commander Popham seized the Spanish schooner *Ontario* with 219 slaves loaded at the River Nun on 14 December. She had been protected by American colours and papers on her outward voyage, but on 1 December she had been nominally sold to her Spanish master. On condemning her and emancipating her 200 slaves, the commissioners commented, rather optimistically, that if she had been taken by a United States cruiser while carrying slaves under American colours her crew would have been liable to execution.

Anticipating *Pelican*'s departure from the Station, Admiral Elliot had written to Popham in November expressing warm appreciation:

> The whole of your conduct and arrangements as Senior Officer of the Squadron employed on the West Coast for the suppression of the Traffic in Slaves and the protection of British Interests have continued to afford me (as did the late Commander-in-Chief) the most entire satisfaction.[14]

It was an accolade well deserved after 18 months as Senior Officer, and at the end of the year Popham sailed for England.

The first three prizes taken by *Dolphin* off Lagos in November reached Freetown at about the same time as their captor on 22 December, and Holland was gratified to learn that in two of them a total of 9,427 dollars and 460 doubloons had been found. The excellent Mr Rowlatt arrived in the schooner *Astran* on 26 December bringing with him the schooner *Amalia*, which he had taken off the Gallinas River under Portuguese colours and equipped for slaving. He had also found in her a further 300 dollars and 335 doubloons. Rowlatt was, of course, not authorised to make such a seizure, and it was fortunate that his Commanding Officer was on hand to make the arrest legal. The Mixed Commission recorded the capture as having been made off Cape Sierra Leone by *Dolphin* on 27 December, while acknowledging where the arrest had actually taken place, and, having judged her to be Spanish, condemned the *Amalia* under the Equipment Clause.

The Anglo-Spanish court found no difficulty in condemning under the Equipment Clause *Dolphin*'s three prizes taken off Lagos while wearing false Portuguese colours. All had sailed from Havana. The *Ligeiro* belonged to a Havana firm and was widely known as the Spanish vessel *Galgo*. She had acquired a Cape Verdes passport and a Portuguese to act as nominal master. The *Dous Amigos*, formerly an American, had obtained her Portuguese passport at Bahia but was owned by a Spanish merchant at Havana. Shortly before arriving at Lagos she had purchased guns and powder at Cape Coast. The *Victoria*, also Havana-owned, had Portuguese papers from Princes, and she had been allowed to clear out of Havana with casks, slave boilers and slave decks shown as articles employed in lawful traffic. It was no surprise that her clearance certificate had been signed by Mr Trist, who asserted that "full faith and credit are due" to it. In their report on this case the British commissioners told Palmerston that the United States Government should condemn the conduct of Trist and his Vice-Consul, Mr Smith, who openly assisted in despatching vessels which, as they were well aware, were intending to ship slaves or were pirates.* 15

Lieutenant Holland had hoped that his fourth prize, the *Astran*, which was found to have Portuguese colours but no papers, would be dealt with by the Vice-Admiralty court. However, unless she could be proved to be British, that court would have had no jurisdiction in her case, and the Mixed Commissions could not adjudge her unless she could be shown to be either Dutch or Spanish and therefore subject to an Equipment Clause. No evidence remains of her fate, and it is reasonable to assume that she was released without penalty to Holland.

No similar difficulty was experienced in condemning another empty schooner named *Victoria*, taken off Princes Island under Portuguese colours by Lieutenant Broadhead in *Lynx* on 24 December. She was from Havana, and the Mixed Commission was satisfied with the evidence that she was Spanish in character. *Lynx* was perhaps fortunate to have survived to make this arrest. Her Gunner, Mr Thorne, probably a victim of a combination of tedium, the climate and drink, had been charged not only with drunkenness ashore on duty and with stealing wine from the gunroom, but also with the shockingly dangerous crime of attempting to take a lighted lantern into the magazine. With no possibility of convening a court martial on the Station, he was removed from *Lynx* and sent home.

The last two arrests of 1838 were made on the same day, 28 December. The first of these was by *Brisk*, her eleventh of her autumn cruise on the Windward Coast, and it was one which would have given Arthur Kellett particular satisfaction. She was the schooner *Violante*, under false Portuguese colours, and not only was she the first of *Brisk*'s excellent haul of prizes to have slaves on board, but also she was the vessel Kellett had taken and subsequently released in October while under American colours, then going by the name of *Mary Anne Cassard*. This sweet revenge came 20 miles south of the Sherbro, the river in which the schooner's 191 slaves had been loaded the previous day. The slaver had made use of the criminal services of the US Consuls in Cuba and had acquired Portuguese papers at Bissau, but, as the court commented on condemning her and all 191 slaves, she was no more Portuguese than American. The commissioners wrote angrily to Lord Palmerston that, although the Spanish and Portuguese Governments might not approve of "connivance and treachery" by their functionaries, they nevertheless tolerated them, despite binding themselves to punish "such violation of law and treaty".

---

* Trist's conduct was eventually investigated by a United States minister in Madrid, Alexander Everett, and he was condemned and dismissed.

The second capture on 28 December was by *Bonetta*, on passage to Sierra Leone. Her prize, the schooner *Gertrudes*, wearing Portuguese colours but with papers of dubious authenticity, had also shipped her 168 slaves at the River Sherbro, and Lieutenant Stoll arrested her to seaward of that river. There were no deaths before the schooner was condemned and the slaves were emancipated.

Saracen had made precious little contribution to the suppression effort since her deployment to the Coast in September 1837. She had not brought in a single prize before, with disciplinary problems, she had been sent first to the Cape and then to England. On 31 December she arrived again at Sierra Leone, still, perhaps surprisingly, commanded by Lieutenant Henry Worsley Hill.

The final incident of 1838 was a distasteful one. The senior mate of *Buzzard*, a Mr Aldrich, boarded the brigantine *Eagle*, under US colours, at Lagos on 31 December and forcibly opened her hatches, seized her papers and insulted her master. When restrained by his Commanding Officer, Lieutenant Fitzgerald, Aldrich's behaviour was insubordinate in the presence of the boat's crew, and Fitzgerald arrested him. That night Aldrich, ignoring the constraints of his arrest, joined the other officers in their New Year festivities, and his singing could be heard throughout the brigantine. On 5 January Commander Craigie of *Scout*, the senior officer present at Princes, interviewed Aldrich and pointed out to him that his conduct had been not only damaging to discipline in *Buzzard* but also liable to cause a diplomatic incident. Aldrich replied that his only regret was that he had not knocked the American master down. He was sent home "for Their Lordships' disposal".[16]

Frustration and annoyance were inevitable in the circumstances of the cruisers' work, but the bullying attitude revealed by Aldrich had become worryingly common. Craigie expressed to the Commander-in-Chief his concern that the prevailing opinion among junior officers of the Squadron was "that civility and forbearance do not form essential points in their duty in boarding Foreign Merchant Vessels". This oafish disposition was not only guaranteed to generate unnecessary trouble but also it flew in the face of the traditional, and respected, courtesy and correctness of Great Britain, and of the Royal Navy in particular. It was an ugly tendency which needed to be stamped on.

The Squadron had lost 93 men to disease during 1838, 69 of them in four cruisers: *Forester*, *Waterwitch*, *Bonetta* and *Scout*. *Scout*'s casualties had been suffered through the infection of her prize crews in Freetown, the cause of so many losses, and it was of interest that her boats' crews suffered no ill effects from being detached on a number of occasions during January in the River Bonny or lurking near the swamps at the river entrance. The sloop was almost constantly at sea in the Bights for the first half of the year, but the boats were again detached in March, this time up the Old Calabar for 13 days. Again the men returned in good health. The season was a dry one, but a further explanation was that, while they were away, the crews were dosed every night and morning with bark mixed (in the absence of wine) with rum. In June *Scout* left behind her the danger of fever as she made for the Cape, touching at Ascension and St Helena, and she returned to the slave coast in December. *Buzzard*, *Columbine*, *Curlew*, *Pylades* and *Saracen* suffered a few losses, but escaped relatively lightly.

The new year of 1839 was only four days old when Lieutenant Colin Campbell in *Forester* made the first arrest of the year, and the circumstances were symptomatic of the beginning of an immense difficulty for the British cruisers and the Anglo-Spanish courts. *Forester* was at anchor off Cape St Paul's when, at 1510, she sighted a suspicious schooner, weighed anchor and made all sail in chase. Three hours later the chase anchored between Cape St Paul's and the mouth of the River Volta, and Campbell boarded her. She was the *Hazard*, empty and under American colours, but Campbell was convinced that she was Spanish. It transpired that she had been registered at Baltimore, but had then immediately sailed for Havana, where a Spaniard supposedly acquired power of attorney but undoubtedly became her owner. She had shipped an American as so-called "Captain of the Flag" and a Spaniard who was nominally first mate but actually the master, and she had been cleared for St Thomas by US Vice-Consul Smith at Havana before making for Lagos. The American "captain" had died, and there was no other American on board at the time of her capture. Campbell detained her under the Spanish Equipment Clause, but the Mixed Commission declined to adjudicate.

Lord Palmerston later pointed out to the Admiralty, with obvious irritation, that the *Hazard* had been detained by a cruiser while bearing the flag of a state which had not conceded Right of Search to Britain. Six days later, however, he criticised the Anglo-Spanish court for not having condemned as Spanish the *Mary Anne Cassard*, taken in October the previous year under very similar circumstances. In an attempt to clarify matters, the Foreign Office told the Admiralty that HM cruisers were not authorised "to visit vessels which are, both in appearance and in fact, the property of American citizens". However, if there was proof that the flag and papers of a vessel were false and fraudulent, and she really belonged to a nation with which Britain had a slave-trade treaty conceding Right of Search and condemnation, such a vessel, although carrying an American or other foreign flag, would be liable to be condemned by the relevant Mixed Court. If a vessel of a nation with no treaty with Great Britain were to be seized on suspicion which turned out to be unfounded, "the captor would in such case incur a serious responsibility."[17] The question for the cruisers was what would suffice as proof of false colours and papers in the short period available to them to decide whether or not to detain a suspicious vessel. As the Foreign Office made clear, the penalty for error was likely to be heavy.

On 10 January *Dolphin* was involved in an equally unsatisfactory incident. After the adjudication of his November prizes, Lieutenant Holland sailed from Freetown to resume his station off Lagos, and he intended to call at Dixcove, Cape Coast, Anamabo and Accra, having been told that any vessel found equipped for the slave trade within three miles of any of those places could, subject to an Equipment Clause or not, be condemned in a Vice-Admiralty court.* He had reached only as far as Cape Mesurado when he found the Spanish schooner *Merced* and detained her under the Equipment Clause, but his prize demonstrated that the practice now being adopted by Spanish slavers of landing incriminating slaving gear immediately on arrival on the Coast and re-embarking it on the eve of embarking slaves was well worth the inconvenience. She was restored to her owner in the absence of adequate proof, and damages and costs of £85.15s were awarded against Holland.

Clearly the commissioners were as confident as was Holland that the *Merced* was slaving, and they suggested to Lord Palmerston that Spain should be urged to accept a new treaty clause which would allow condemnation under the Equipment Clause if there was clear and undeniable proof that slaving equipment had been on board during the voyage on which she was detained. In theory this was a good idea, but the difficulty of acquiring the necessary proof would have been considerable. Nevertheless, Palmerston felt it worth passing the proposal to the British Chargé-d'Affaires at Madrid.[18]

There then followed three more cases which, with subsequent similar incidents, demonstrated that, despite the risks to their pockets and careers, the cruiser commanding officers were determined to bring to book the Spanish slavers masquerading under the Stars and Stripes. The first of these involved the schooner *Florida*, taken in the Gallinas River on 13 January by the boats of *Saracen*, which had anchored outside the bar. The schooner had been registered at Baltimore and then immediately sent to Havana for sale. There she was supposedly sold to an American, but a paper found on board after her arrest showed that the American was receiving a stipend from the real owner in Havana for, by his name and his

---

* Holland did not reveal who gave him this advice, but they were probably Vice-Admiralty court officials, and clearly it was based on the assumption of British jurisdiction over the territorial seas off these four possessions.

presence on board, concealing the vessel's Spanish identity. At the time of the capture, no colours were flying and the American was not on board.

Two days later, however, the American returned and begged Lieutenant Hill in *Saracen* for protection against the schooner's Spanish crew who were threatening his life, and he admitted that the bill of sale to him was false. The Mixed Commission felt that even this evidence was inadequate proof of Spanish ownership, and declared that it was not competent to try the case. The American had no intention of returning the vessel to her real owner, and, making the most of his fraudulent authority, he had the schooner cut-up, and sold the materials and the vessel's gear at auction to his own benefit.

The British commissioners told the Foreign Office that they were anxious that "measures be taken to rectify the scandalous and increasing abuse of the flag of the United States," and suggested, unrealistically, the introduction of joint cruising by British and US men-of-war.

On the day following the *Florida*'s arrest, the brig *Eagle* was boarded in Lagos Roads by Commander John Reeve in *Lily*. She had already been visited by *Pelican*, *Pylades* and *Buzzard*, but Reeve found that, although the brig was under US colours, the only American on board was her master, and he decided to detain her under the Spanish Equipment Clause. Her story was very similar to that of the *Florida*, and, although it could not be established who her real owner was, she ostensibly belonged to her master. The American Consul and Vice-Consul at Havana had given every assistance in obtaining her clearance for the coast of Africa, and she had made a maiden slaving voyage from Lagos to Bahia. Then the American Vice-Consul at Bahia, Mr Foster, ignoring the slaving equipment on board, her arrival "in ballast" and her lack of papers for the voyage to Bahia, had added 12 men to her crew and despatched her for Africa with a ballast of large barrels for stowing water for slaves. Also embarked was $20,000 worth of tobacco, consigned to a Havana merchant at Lagos by the longest-established slave dealer in Brazil. The Anglo-Spanish court felt that it could not deal with her, but her Prize Master clearly felt unauthorised to release her, and she set off on several months of wandering.

Lieutenant Henry Frowd Seagram, who had recently replaced the newly promoted Woodford Williams in command of the brigantine *Termagant*, was the next to tackle an American-flagged slaver. He boarded and searched the schooner *Jago* close to Cape St Paul's on 21 January. He found that she was commanded by a US citizen, but detained her on the grounds that she was really the property of a Spaniard and equipped for slaving. He could not unearth sufficient evidence to persuade the Mixed Commission to bring her to trial, but the British commissioners commented that, as in the cases of *Mary Anne Cassard*, *Hazard* and *Eagle*, there was good reason to believe "that fraudulent adoption of United States colours would have been proved by examination in court".

Shortly afterwards, a similar vessel was detained, the schooner *Mary Cushing*, but the date, the place and the name of the captor were not recorded, and, as with all except *Eagle*, she was released.

In reporting the *Eagle* case in February, Commissioners Macaulay and Docherty took the opportunity to point out to Palmerston that, three years earlier, they had foretold this fraudulent use of American identity by Spanish slavers, and they wondered:

> how the people of the United States will bear to hear that under their flag at this moment the slave trade of the Whole World finds protection; that their home Government tacitly acquiesces in the monstrous wrong; while the representatives of their commercial interests in foreign countries openly and avowedly lend the whole influence of their official situations to encourage and extend the evil.[19]

Unfortunately the situation was not one of which those American people who were likely to object would readily become aware.

The Commander-in-Chief would have found out about this disquieting development when he visited Freetown in *Melville* in mid-January. He had headed north in December and called at Ascension at the end of the month. While at the island he resolved a difficulty over command of the garrison following the demise of Captain Bate in April 1838. In June Captain Roger Tinklar had been appointed in command, but a Captain Evans RM, who happened to be on the island at the time of Bate's death while awaiting passage home with his family, and who had no connection with the garrison, had taken command. Evans may have seen it as his duty to do so as the senior officer on the island, but Admiral Elliot clearly regarded it as intolerable opportunism, and, declaring this self-appointment to be irregular, he reaffirmed Tinklar's appointment. The Commander-in-Chief also detected an unexpected increase in the size of the garrison, and his recommendation for a reduction from 75 men to 65 was later approved by the Admiralty.

The Commander-in-Chief probably found *Bonetta* at Freetown, but *Melville*'s stay was not long, and by the end of January she was at Accra for a brief call before making for West Bay at Princes Island. She was at sea again on 13 February, heading south, when Elliot took the opportunity to write to the Admiralty on the subject of the fraudulent use by slavers of not only American colours but also the

Russian flag. He had heard that a vessel named *Goloubtchick* was currently on the Coast and masquerading under the latter ensign. After a short visit to Ascension, where Captain Tinklar was installed as Commandant, the flagship arrived at Simon's Bay in March. Apparently the Commander-in-Chief had encountered Lieutenant Holland at some point on his rapid tour, because shortly after his return he wrote to Commander Craigie in *Scout*, now the Senior Officer on the West Coast, to tell him that *Dolphin*'s copper was in a bad state and that she had consequently lost her sailing qualities. Craigie was therefore to take a convenient opportunity during May and June to send the brigantine to Plymouth for refit.

After a number of disappointments during January, there were three successful arrests before the end of the month. On the twenty-second, off the River Gabon, *Fair Rosamond* detained the brig *Matilde*, wearing Portuguese colours. She was without slaves, but Lieutenant Oliver believed her to be Spanish. The brig had been cleared from Havana with barrels which were supposedly for palm oil, a boiler declared to be for clarifying palm oil, and 3,600 boards claimed to be for building a hut to store palm oil, but the Anglo-Spanish court agreed with Oliver that she was indeed Spanish and that her barrels, boiler and boards were slaving equipment.

Matilde was condemned, as was the next prize, the brigantine *Maria Theresa*, also empty and under false Portuguese colours. She was seized close to the coast 20 miles to the east of Accra, also on 22 January, by Commander John Reeve in *Lily*, for being Spanish and equipped for slaving. The brigantine was proved to be trading for the benefit of Cuban residents, and her voyage was being directed by a Spanish supercargo. Accounts found on board from the notorious Cuban slave merchants Martinez & Co. confirmed the suspicion that the Havana Government was receiving, as "duty on entrance", half a doubloon for every slave landed in the vicinity of the city.

Then on 31 January, under similar circumstances, *Fair Rosamond*, still off the River Gabon, took the Portuguese-flagged brig *Tego*, and Oliver had her condemned as Spanish and equipped for slaving.

While at Ascension at the beginning of March, the Commander-in-Chief despatched to the Admiralty a summary of recent boarding lists he had received from cruisers during his visit to the Coast. It does not cover a single defined period, but it shows that in late 1838 and early 1839 the 12 cruisers boarded suspected slavers on 141 occasions. *Lynx*, for example, made 12 boardings between 1 October and 31 December. Over the same period *Saracen* made four, *Fair Rosamond* eight, *Scout* 13, and *Pylades* as many as 16. Between 2 January and 2 February *Lily* boarded seven, *Scout* another five in January, and a further five were recorded by *Fair Rosamond* between 6 January and 8 February. Although the total number of vessels stopped and searched seems fairly high, the maximum number of boardings by a cruiser in a single month appears to have been seven, and that does not represent high intensity. In the four months to 31 January 1839 the number of arrests was 28. What these figures indicate is that even in a relatively busy period, as this four months was, the frequency of incidents was low, and they were interspersed with many days of tedious patrolling and searching.

This tedium would have been one of the major causes of indiscipline in the Squadron, and one such instance was reported to the Commander-in-Chief during his tour of the Bights. Commander Reeve of *Lily* had placed his First Lieutenant and Surgeon under arrest for disobedience, and the Surgeon had made countercharges against Reeve. Reeve requested a court martial to clear his name, and, as it was impracticable to assemble a court on the Coast, Elliot decided to send *Lily* home. It was a loss which could ill be afforded.

Admiral Campbell had asked the Admiralty that vessels joining his Squadron should be sent directly to the Cape so that he could direct their deployment, but this arrangement had never been implemented. It was the turn of Commander Popham, while Senior Officer on the West Coast, to complain that much time was being wasted by newly joining cruisers searching for him in the Bights. He requested that vessels sent to the coast of Africa should call initially at Sierra Leone to collect Kroomen, and then proceed to West Bay, Princes, where they would find information on the whereabouts of the Senior Officer. Admiral Elliot was happy to pass this eminently more sensible proposal to the Admiralty.

Lieutenant Henry Broadhead of *Lynx* had reported that his pivot-gun mounting was defective and virtually useless.[*] It had worked well on her leaving Sheerness, but orders had been given that changes were to be made at Plymouth. The authorities at Plymouth had ignored Broadhead's objections, and the weapon was now so inefficient as to be dangerous. Furthermore, *Lynx*'s two carronades had stood four feet above the deck, and could be cast loose only in smooth water and with no slope on the deck. With the Commander-in-Chief's approval Broadhead had cut down the carriages, and the carronades were now usable, if unhandy.

These were not the only problems found with weapons in the small cruisers. It had been discovered that *Fair Rosamond*'s carronades had been mounted so that the centres of the bores were lower than the sides of the vessel, restricting their minimum elevation to about 5°. There was also some undefined concern about *Termagant*'s guns, and the carronade carriages in *Brisk* and *Forester* were

---

[*] A diagram showing the pivot-gun arrangement in *Lynx* appears in Appendix J. The mountings in other cruisers are likely to have been similar.

lowered while they were refitting at Simon's Bay. Elliot anxiously called Their Lordships' attention to the necessity of ensuring the efficiency of guns in these small vessels, which had so few of them to deal with slavers that were often well armed. He explained that these brigs and brigantines heeled very much under sail, often 12–15°, and under those circumstances not one of the carronades fitted in England could be cast loose without upsetting. It was Elliot's belief that the heavy pivot-gun in these cruisers was inferior to the same weight of metal in two bored-up guns on common carriages which could be run to any part of the deck, and which would permit keeping a boat inboard on the booms, but commanding officers seemed generally to have approved of having the one heavy gun.[20]

The fitting of pivot-guns also conflicted with a recent Admiralty order that sloops and brigs should not carry quarter boats.* Elliot pointed out to Their Lordships the seemingly obvious fact that boats on booms were incompatible with centre-mounted pivot-guns kept at short notice for action, as was required on the slave coast, and, even more obvious, that cruisers in chase could not stop to hoist out boats from the booms. Slavers were boarded, almost invariably, by cruisers' boats, and boats were often used to cut slavers off to windward, a tactic usually practicable in the moderate weather on the Coast. It was frequently necessary to send boats away twice in an hour, and the Admiralty's order would present an entirely avoidable obstacle in the way of capturing slavers. The Commander-in-Chief therefore suspended its execution. No damage had ever been done to quarter boats in "these smooth-water seas" on the Coast, and it seemed that the order had resulted from an accident in *Bonetta* on passage to the Cape. Not only had she left a boat on quarter-davits in the stormy southern latitudes, but also she had fitted her davits without the expertise of a Carpenter's crew. Elliot requested that the order be rescinded.[21]

Concern about emancipated Africans at Fernando Po surfaced once again in the Foreign Office. According to its records, Captain Owen had landed 88 men, 52 women and 64 children there during 1828 and 1829, and *Trinculo* had added another 30 from the Congo in 1830. Lord Palmerston wanted them to be removed to Sierra Leone, and in mid-February Commander Craigie despatched Lieutenant Fitzgerald in *Buzzard* to the island to investigate. By the time that Fitzgerald was ready to make his report at the beginning of April, Craigie had been relieved as Senior Officer on the Coast. Although Fitzgerald did not mention

---

* These were boats carried on davits, at one or both quarters of the vessel, rather than inboard on the booms amidships. The former arrangement allowed the boats to be lowered rapidly and while the parent vessel was still making way. The latter required a relatively slow evolution involving the use of the yards as derricks.

numbers to be moved, he explained that a transport would be required for the job; there was much furniture to be shifted, including four-poster beds. In June, however, the whole matter was again put in abeyance by Lord Palmerston. As HM Government was in negotiation with Spain for the purchase of Fernando Po, he felt that it would be best if the Africans should stay there for the time being.

The one prize taken in February 1839 was the brig *Braganza*, detained by *Termagant* on the ninth, just off Cape St Paul's. The brig was flying Portuguese colours, but Lieutenant Seagram was sure she was Spanish, and he arrested her for being fitted for slaving. He was quite right; she had been the Spaniard *Vigilante*, but had undergone a nominal sale at Lisbon. She had then been fitted-out at Corunna for a slaving voyage to Lagos, but Seagram took her before she reached her destination, and she was condemned.

In March the pace of arrests quickened a little, but until the very end of the month the prizes were monopolised by *Forester*, which was patrolling the Grain Coast. Her first capture was not a success. She took the empty schooner *Ligeira* under Portuguese colours just off Trade Town on 6 March, and Lieutenant Campbell, deciding that she was Spanish, sent her to Freetown. However, his Proctor felt that there was insufficient evidence to prosecute her under the Equipment Clause, and decided to restore her immediately.

An interception close inshore off Sanguin on 11 March was better judged. The schooner *Serea*, owned in Cuba and with Portuguese colours and a passport obtained at the Cape Verde Islands, was trading along the coast for provisions on her way to the Gallinas when *Forester* caught her, and there was no difficulty in condemning her as Spanish and fitted for the slave trade.

Heading north along the Grain Coast on 21 March, *Forester* sighted a large schooner at anchor in the Gallinas Roads. Despite her United States colours Campbell hove to and boarded her, finding that she was the *Rebecca*. Having anchored close to the schooner, Campbell sent a party on board for a full search which revealed that she was fitted for slaving, and, although her papers appeared to be correct, found correspondence which indicated that she was Spanish. The crew was shifted to *Forester*, and Campbell told the American who claimed to be her master that he would be arrested because, by the laws of his country, he was a pirate. On the afternoon of the twenty-second the cruiser sailed with the schooner in company until, at 1700, she hove to and Campbell removed his prize crew, giving up the schooner to her master, planning to rearrest her on more satisfactory terms.

The American had treated the threat of arrest lightly at first, but had then become frightened, and when *Forester* hoisted her ensign and pendant at 1710, he declined to show colours in response. Campbell's men again boarded and the American told Campbell that he disowned the schooner, admitting that, for her protection until slaves were embarked, he had been paid by Spaniards to take the vessel to the Gallinas, and on arrival there he was to give her up to the mate. The mate explained that, although she had been American, the schooner had been sold to a resident of Havana and had sailed from there. Campbell detained her and sent her to Freetown under the command of his Gunner, rightly confident that she would be condemned as Spanish and fitted for slaving. That was the end of Colin Yorke Campbell's command of *Forester*; on arrival at Freetown he found that Lieutenant Francis Bond had come to relieve him.

Four new cruisers joined the Cape Station at about this time, bringing the number of vessels available to Rear-Admiral Elliot to 22. Three of the newcomers were brig-sloops by Symonds, whose cruisers already on station had been proving their value. These Symondites, none of them more than three years old, were *Acorn*, Commander John Adams; *Harlequin*, Commander The Rt Hon. Lord Francis Russell; and *Wolverene*, commanded by the new Senior Officer on the slave coast, Commander William Tucker. The fourth new arrival, *Nautilus*, Lieutenant George Beaufoy, was a "coffin brig", but at least she had been launched as recently as 1830.

The month of March closed with an arrest by *Saracen* in the Rio Pongas. On the thirty-first she surprised the Portuguese schooner *Labradora* in the act of embarking a slave cargo for Havana, and there were 253 Africans already on board when Lieutenant Hill's boat made its appearance. Before *Saracen*'s boarding party could intervene, however, two passengers and a seaman made off with the schooner's boat and two slaves. The *Labradora* had been chartered to load 300 slaves who had originally been intended for the *Maria*, seized by *Brisk* in the previous November. Clearly this was an unlucky cargo for the slavers; the *Labradora* too was condemned, as were her surviving 248 slaves.

The brigantine *Eagle*, taken under United States colours by *Lily* in January, and refused for adjudication by the Anglo-Spanish court in Freetown, was initially taken by her prize crew to Lagos. From there she made her way to Fernando Po, presumably in the hope of finding *Lily*. Instead she discovered *Buzzard* at Clarence Cove. *Eagle*, with *Lily*'s prize crew still on board, was still flying her American ensign, and, in a strange but perhaps necessary move, Lieutenant Charles Fitzgerald of *Buzzard* rearrested her on the grounds that her American master had freely admitted to Spanish ownership.

On 13 March *Buzzard* sailed for the River Bonny, with *Eagle* in company, and lay at anchor off the river mouth on the sixteenth. She then rounded Cape

Formoso, and two days later Fitzgerald sent his boats, armed, up the River Nun. At 0930 the following morning a schooner was sighted coming out of the river with *Buzzard*'s boats in tow. The prize was the *Clara*, another empty American-flagged slaver. Fitzgerald's initial intention was to take both prizes to Sierra Leone, but first he called at Princes. There he met *Wolverene* and Commander Tucker. Tucker, as Senior Officer on the Coast, decided that the pair should be sent to New York. He hoped that the American authorities, faced with these cases, would take measures to prevent the false use of their country's flag by foreign slavers, and his decision was endorsed by the Commander-in-Chief. Although it meant losing a cruiser for some months, Tucker ordered *Buzzard* to escort her two prizes to New York and to bring them before the American courts.

The New York District Attorney was initially of the opinion that there was a reasonable prospect of conviction in both cases under the United States slaving and piracy legislation, considering that their arrest wearing American colours provided sufficient evidence. However, having subsequently become convinced that they were both Spanish property, he washed his hands of them. Fitzgerald was required to remain in New York as a witness in the prosecution of one of the American masters, but the court appearance, planned for 18 September, was postponed on account of the death of the prisoner's mother. In the meantime, *Eagle* dragged her anchor in a gale and was blown ashore on Staten Island, and to compound Fitzgerald's woes the Spanish mate and steward, his two remaining witnesses, absconded ashore. *Eagle* was refloated, and after she had been expensively but shoddily repaired, Fitzgerald finally sailed on 8 November to take *Eagle* and *Clara* to Bermuda in the hope of bringing them before the Vice-Admiralty court.

On 12 November *Buzzard* and the two slavers arrived at Bermuda, and found there the Commander-in-Chief, Vice-Admiral Sir Thomas Harvey. Having apparently discovered further incriminating papers in the prizes, Fitzgerald rightly concluded that the Mixed Courts of Justice again offered the best chance of a conviction, but, as he explained to the Commander-in-Chief, he expected difficulties if he were to go to Havana, and preferred the option of Sierra Leone which, furthermore, was on his proper Station. Harvey, who had clearly been helpful, therefore ordered him back to the coast of Africa.[22] On Christmas Day 1839 *Buzzard* arrived at Freetown, alone. Shortly after sailing from Bermuda, *Clara* had parted company, and soon after that *Buzzard* and *Eagle* had been hit by a severe gale in which *Eagle* foundered, although Fitzgerald had, with difficulty, saved his prize crew. *Clara* disappeared, and it was assumed that she had been lost in the gale which claimed *Eagle*. *Eagle* was subsequently condemned as a Spaniard fitted for slaving, a verdict approved by Lord Palmerston, and the court postponed *Clara*'s case for a year in the hope that some news of her might appear. The postponement was wise.

On 18 December 1840 documents arrived in Sierra Leone from Jamaica, via Lieutenant Fitzgerald's agent in London and the Captor's Proctor at Freetown, revealing that, after losing sight of *Eagle* and *Buzzard*, the Prize Master in *Clara* had decided to bear up for Antigua and had reached the island safely but with the vessel "much shattered and disabled". Finding her there on 2 February 1840, Admiral Harvey had sent her to Port Royal, where she was partly dismantled and laid up as unseaworthy. On the evidence of certificates signed by Commodore Douglas, the Senior Officer at Jamaica, the Mixed Commission condemned the schooner for being a Spanish vessel fitted for slaving.

Little, if anything, seems to have been gained by this costly and ultimately sad episode. The scandal of false United States colours had perhaps been drawn more forcefully to American attention, and to the British commissioners at Freetown there appeared to be a "feeling of indignation loudly expressed by the American public at the dishonour done to their flag by its employment in this commerce". This was probably wishful thinking, but they hoped that this supposed reaction, coupled with the reported removal of the perfidious US Consul Trist from Havana, might end the abuse. However, a complaint from Admiral Elliot in June 1839 that there had been an almost general adoption of American colours for the protection of Spanish slavers was, regrettably, more indicative of the future.[23]

The concept of negotiating anti-slaving treaties with the native chiefs on the slave coast was gradually taking hold, and Commander Craigie, after a visit to the River Bonny in March to discuss the idea, wrote a report to the Admiralty. He concluded, unsurprisingly, that compensation, in the first instance, was probably the only means of inducing the chiefs to end the Trade. Admiral Elliot, in commenting on the report, expressed his inclination to think that the chiefs, in agreeing to accept 2,000 dollars' worth of goods, and Craigie, in his estimate of the probable increase in the legitimate trade which might result from an end to slaving, had also overlooked the fact that a considerable proportion of the British goods currently being received by the coastal chiefs were going into the interior as payment for slaves, and that the inland tribes had no other produce to offer in barter.

Slaves were still being carried to Princes Island and St Thomas with the connivance of the local Portuguese authorities, usually in small vessels. These craft were often in a poor state, and when Commander Tucker intercepted the Portuguese schooner *Passos* on 8 April, about 20 miles east-north-east of Princes, he considered that "during many years acquainted with the slave trade [he] had never boarded a vessel so unseaworthy or so badly fitted or found". He viewed the shipping of her 87 slaves as a "gross act of inhumanity by the owner, an inhabitant of Princes". The prize could not be sent to Sierra Leone, and *Wolverene* towed her towards West Bay, but was delayed in getting her in by calms and by being obliged to cast her off for a chase. When she was brought to the island on the tenth her slaves and all removable stores and gear were transferred to *Dolphin* for passage to Freetown where, after the loss of six slaves, they were condemned. Appearing in the case, in the guise of a Customs House Inspector at Port Antonio, was a notorious slave trader from the Cape Verdes. Meanwhile, on 11 April, Tucker took the *Passos* to sea, and burned and sank her.

Wolverene was making her way south, and on 24 April she anchored at Ascension to embark provisions. She found *Scout* and *Bonetta* lying there, and two days later *Pylades* arrived and *Scout* and *Bonetta* sailed. The following day *Wolverene* departed for the mainland, initially to cruise off St Paul de Loando and Ambriz Bay.

Corruption among Portuguese colonial officials was by no means uncommon, but the next arrest revealed particularly disgraceful behaviour. On 14 April, Arthur Kellett in *Brisk* detained the Portuguese schooner *Liberal* in the Bissau Channel, between the Bisagos Islands and the mainland, and found 41 Africans on board. It was explained to the boarding officer that 38 of them were slaves being taken to the Cape Verdes in accordance with the regulations in the Lisbon Decree of 1836, for whom he was presented with bonds and certificates, and that the other three were crew members. Kellett was unconvinced and sent the schooner to Freetown. Examination of the papers revealed that all ten of the slaves on one bond and four of the ten on another were being shipped by the Government Secretary and Director of the Customs House at Bissau, and the three supposedly part of the crew, for whom there was no bond, were owned by the Governor of Bissau.[24]

Shipping of Portuguese slaves north of the Equator was permissible under the Lisbon Decree only if five criteria were fulfilled: it must take place between one Portuguese territory and another, the slaves had to be bona fide domestics, they had to have been purchased before January 1815, the party shipping them had to be the bona fide owner, and he had to be shipping them on his final retirement from the territory where the slaves were embarked in order to establish himself at the territory to which they were being taken. The circumstances fell well short of those requirements in this instance, but the Mixed Commission's business was to judge on the terms of the Anglo-Portuguese Treaty, not the Lisbon Decree. Having no doubt that these Africans were being shipped for the slave trade, it condemned the *Liberal* and emancipated the surviving 40 slaves. It hoped that the future was not bright for the two Portuguese officials involved.

With Bolama Island close at hand, Kellett took the opportunity for a second visit to discover whether his action of 10 December had achieved a lasting effect. It had not. He landed with 25 men and found that the barracoons had been rebuilt and that the resurrected slaving operation was being protected by 18 Portuguese soldiers. Again the barracoons were burned, and the soldiers were disarmed. Thirteen Africans claimed Kellett's protection, and he took four others out of a canoe on the Nunez.

When news of the initial Bolama operation reached the Foreign Office, officials demanded more information, and in particular wished to know on what grounds and by whose orders the removal of the Africans had taken place. This letter to the Colonial Office was followed a week later by one from Lord Palmerston to the Admiralty approving Kellett's action, but acknowledging that the sovereignty of Bolama was a matter of dispute between Britain and Portugal.

An agent of a Portuguese slaver had been removed from Bolama by Kellett on his second landing, and this man, a Senhor Escaelottia, was committed for trial at Freetown and released on bail. Kellett and his clerk-in-charge were bound over as prosecution witnesses, but *Brisk*'s fever epidemic was worsening, and the Governor placed her in quarantine. Kellett immediately put to sea, having written to the Governor asking that the Queen's Advocate should be directed to delay the trial. The Governor gave that direction, but Kellett found on his return that Escaelottia had been allowed to leave the colony without being called into court. Kellett protested, and with justification, at the lack of support he had been given by the law officers of the Crown. He was probably glad that, after a considerable period watching the northern rivers and islands in collaboration with *Saracen*, *Brisk* was leaving for the southern slave coast.

There had been some concern that the Cuba slave merchants, as the effectiveness of false Portuguese identity diminished, might resort to Russian colours to disguise Spanish nationality. So far there had been little evidence to support that concern, but on 19 April, just off the Gallinas River, Henry Hill in *Saracen* detained the Russian-flagged brig *Goloubtchick* on suspicion that she was Spanish. She was equipped for slaving, but the Mixed Court declined to hear her case in the absence of sufficient evidence that she was Spanish. A further search by Hill uncovered papers which would probably have provided the necessary proof, but the court regarded this search as illegal and referred to a judgment, made years earlier, by Lord Stowell that "if these facts are made known to the seizor by his own unwarranted acts, he cannot avail himself of discoveries thus unlawfully produced, nor take advantage of the consequences of his own wrong." It was therefore decided to send the brig to England, and she was handed over to the Russian Chargé d'Affaires as Russian property, at his request. On hearing of the discovery of the incriminating papers, Lord Palmerston was at a loss to understand how a search by the commanding officer of a British cruiser holding the requisite Instructions could be construed as illegal. No doubt Hill was similarly at a loss.

The inexperienced Russell in *Harlequin* suffered similar frustration when he sought to prosecute the schooner *Traveller*, which he had detained under American colours on 30 April in the vicinity of New Sesters. Even if she had been admitted by the court, Russell probably would not have proved Spanish identity. Between late December and mid-March the schooner had been boarded in turn by *Brisk*, *Buzzard*, *Bonetta*, *Forester* and *Saracen*, one of whose seasoned commanding officers would have sent her to Freetown if adequate evidence had been available.

Sailing as a passenger in the *Traveller* was the former owner of the *Ontario*, Eleazer Huntingdon, and he was still carrying the papers of that vessel. When Russell handed the papers to the British Commissary Judge, Macaulay, with obvious relish, wrote across each document "in a large hand":

> The within named schooner *Ontario* of Baltimore was sold and transferred by Eleazer Huntingdon to Jose Maria Mendez of the River Nun on the Coast of Africa on the 1st day of December 1838, and was captured a few days after by Her Majesty's sloop *Pelican* with two hundred and twenty slaves on board, who had been embarked in the aforesaid river. The *Ontario* was subsequently condemned at Sierra Leone where she was cut up and entirely destroyed.

Macaulay signed each endorsement and recorded with satisfaction that Huntingdon's rage on receiving his papers so endorsed proved his intention to make further improper use of them.[25]

Boat expeditions into the rivers had become fairly rare, but *Lynx* anchored under Corisco Island on the morning of 5 May with the intention of examining the adjacent rivers with the boats. Lieutenant Broadhead was told that a slaver would shortly be sailing from the River Gabon, and, on the informant promising to return with word of the sailing date, despatch of the boats was delayed. Early on the seventh Broadhead weighed anchor for a chase, joining *Forester* in the process, and, finding himself close to the mouth of the Gabon, dropped the gig and the cutter, commanded by his senior mate, to detain any suspicious vessel in the river and await the return of *Lynx* next day. The chase was apparently unsuccessful, and Broadhead spent the night and following morning working upstream to Prince William Town, the capital of King Denny, four miles upriver. There he found the gig in possession of an empty Spanish slave schooner. The two boats had pulled 15 miles upstream the previous day, but had become separated, and there had been no sign of the cutter since then.

King Denny immediately provided boats and men to help in a search, sending one boat 50 miles to Cape Lopez for information, and three more upriver. The boats and crew of the French merchant barque *Jeune Frederic* also joined the search, and the detained schooner made her way 20 miles upstream. King Denny then heard that the cutter had been attacked and seized at a village 60 miles up the river. The King and the very experienced master of the French merchantman advised that the only safe way of extracting the captives would be in a native boat, and both volunteered to accompany Broadhead to the rescue. That evening they started on their way in the prize, and, at midnight, transferred to a canoe. The following noon they reached the village, about 70 miles from the river mouth.

It transpired that the cutter, under the command of the Gunner, had carried on up the river after losing touch with the gig, and the crew were told by natives that there were two vessels lying further upstream. Having come to anchor close to the village, the boat was fired on by about 30 muskets from the mangroves, and the gunner's mate was immediately shot dead. As the cutter tried to escape, about 50 hostile canoes emerged from the banks. Two more men were killed, the remaining four were badly wounded, and their 80 rounds of ammunition were expended. The Gunner surrendered, but the survivors were cruelly treated. There had been no dispute or even communication before the attack.

Through the King, the four men were rescued and the boat and its equipment recovered, but their weapons could not be found. Late on 13 May the party returned to the anchored prize schooner, and she was despatched to Sierra Leone the following morning. At noon she was followed by *Lynx*. King Denny had paid for the prisoners with cloth, muskets, spirits and tobacco, but would accept no recompense. Both Denny and Monsieur Armouroux of the barque received warm thanks by letter from Broadhead and Commander Tucker, and the King was awarded a gold medal by a grateful British Government. The prize schooner never came to adjudication in the Mixed Court, and her fate remains a mystery.

Tucker was not inclined to take punitive action, and the Commander-in-Chief agreed with him. The natives who had made the attack claimed that they thought the cutter was Spanish, and the incident was attributed to misunderstanding. It was Tucker's intention to accompany Broadhead back to Prince William Town, on *Lynx*'s return from Sierra Leone, in the hope of concluding a treaty with King

Denny for the suppression of the slave trade on his territory, and of holding a palaver with the chief of the area in which the attack had taken place to prevent a recurrence.

There was further frustration for Russell of *Harlequin* when his second prize arrived at Freetown on 6 May. She was the *Merced*, which had been taken in January by *Dolphin* and restored with damages by the Anglo-Spanish court. On being rearrested at the beginning of May not far from Sierra Leone, her condition was identical to that when she was released, and Russell's Proctor wisely advised him not to take the case to the Mixed Court. Henry Seagram was no more successful when, three days later, *Termagant* arrived towing the stripped and empty hull of the schooner *Catalana*, which had been found by the boats a considerable distance up the River Sinou and towed out on 27 April. The hulk had been painted green as camouflage, and every effort had been made to conceal her in the mangroves. Two men had remained on board while the remainder of the crew lived ashore and traded with the natives. There was evidence that she had cleared out of Puerto Rico in December, but insufficient evidence to incriminate her, and the legal advice to Seagram was to return her whence she had come.

On 10 May *Forester* initiated a spate of arrests when her boats found the schooner *Raynha dos Anjos* in the River Nazareth, concealed while her crew made up and filled water casks and laid a slave deck. Her Spanish master, together with her trade goods, had been landed at Cape Lopez, 15 miles south of the Nazareth, to barter for slaves. There had apparently been a Portuguese Captain of the Flag, but he had disappeared, and, as she had been taken without slaves, it was necessary to prosecute the vessel as Spanish if she was to be condemned. Despite the discovery of a Lisbon passport, probably belonging to a craft which no longer existed, all the officers of the prize had confessed to Lieutenant Bond that they were Spanish. The chief mate and cook, sent to Freetown as witnesses, both died on passage, but Spanish "course of trade" was established to the satisfaction of the court and the schooner was condemned as fitted for slaving.

At last, on 16 May, Russell and *Harlequin* made a legitimate arrest, the first of three in quick succession. The schooner *Constanza*, flying Portuguese colours and with a Portuguese Captain of the Flag, was caught while attempting to run into the Gallinas River, where a slave cargo awaited her. Her water casks had been filled, her slave deck had been laid, and she had cooking coppers, large quantities of rice and other slave foodstuffs as well as general slaving equipment. Her supposed owner was a Cape Verde citizen, regularly employed in the role, but here was clear evidence that the vessel was actually owned by Pedro Martinez and Co. of Havana. Russell achieved his first condemnation.

That case was straightforward, but the aftermath of *Harlequin*'s next capture was far from it. The prize was the brigantine *Wyoming*, under American colours, which Russell initially found and boarded while she was at anchor off the Gallinas River on 30 April. He had heard that there were 24 slaves on board, but was unable to conduct a search because the vessel's hatches had been sealed and caulked. On 17 May she was still there, and a more thorough examination revealed no slaves but a full outfit of slaving equipment. The brigantine was clearly almost ready to embark slaves, and Russell was sure that she was Spanish.

Whether the Mixed Court would similarly be convinced was another matter, and Russell did not want to risk losing another case at Freetown. He decided to call the slaver's bluff, and, probably aware of the course of action that Commander Tucker had ordered for the *Eagle* and the *Clara*, he instructed his Prize Master, Lieutenant Beddoes, to take the brigantine to New York. There he was to communicate with the American authorities, through the British Consul, so that the slaver might be dealt with according to United States law. *Wyoming* arrived safely, and the New York District Attorney dealt with her case in parallel with those of *Eagle* and *Clara*. His initial opinion was that all three could be proceeded against with a reasonable prospect of conviction under US law, but then he concluded that they were Spanish property and declined to adjudicate. From that point the proceedings of *Wyoming* remain a mystery. It seems likely that Beddoes sailed for either Havana or Sierra Leone, but there is no record of the brigantine's arrival or of her case being presented in any court.

Russell commented to the Commander-in-Chief that there had been an increase in slaving under the Stars and Stripes "from all rivers and factories to the northward about Sierra Leone" by vessels undoubtedly Spanish, but *Harlequin*'s next prize was under Tuscan colours, a novel disguise. She was the schooner *Bella Florentina*, fitted for slaving and taken off the River Sesters on 20 May. Seizure of Tuscan slavers was legitimate, but the Prize Master was, on arrival at Freetown, unable to assure the commissioners that his Commanding Officer held the Instructions required by the Anglo-Tuscan Treaty ratified in March 1838. *Harlequin* had headed southward for a long cruise, and the Prize Master was advised that in the absence of the necessary information the case would not be allowed into court. He therefore decided to hand the schooner back to her master, but she was in such a poor condition that she could not sail until repairs had been made. Work on her was still underway when *Harlequin* returned and Russell presented his Tuscan warrant which, in the event, was not necessary. When the case was then belatedly taken before the Anglo-Spanish court the vessel was found to be Spanish property, given false papers by Mr Pluma, the

Tuscan Consul-General at Havana. She was condemned in September, and the money raised by sale of her cargo went towards the cost of her repairs, a total waste because she was, of course, destroyed.*

Forester had continued to work her way south since her arrest of *Raynha dos Anjos*, and she was at about 3° S when she detained the Portuguese-flagged schooner *Carolina* in Mayumba Bay on 22 May. When the prize reached Freetown she was prosecuted as Spanish, but shortly after the case began the Captor's Proctor asked for the case to be withdrawn. He had decided that there was insufficient evidence of Spanish nationality, the master having died and the first mate having been left ashore at Mayumba. The second mate and cook, the remaining witnesses, had professed complete ignorance on all salient points, and the prosecution was abandoned.

Resistance to arrest had become a rarity, but the Spanish felucca *Si* was determined not to be taken when she sailed from the Gallinas River with 360 slaves on 27 May. As she emerged she was sighted by the equally determined Lieutenant Matson in *Waterwitch*, and a five-hour chase began. The Spaniard was armed with two heavy guns, was carrying a large number of muskets and cutlasses, and her upper works were protected by a thick layer of cork between inner and outer planking, providing a breastwork impervious to musket balls. *Waterwitch* had arrived in the area a few days earlier, and it was fortunate that she was one of the cruisers on the northern coast swift enough to catch the felucca. Of about 30 rounds of gunfire aimed at the slaver during the chase seven hit her, but she refused to heave to until her bowsprit had been badly damaged, a round shot had passed clean through her exceptionally thick mainmast, her mainyard iron jeer block had been shot away, and one man had been killed and two wounded.† She and her cargo were condemned after the loss of one slave.

On the night of the same day, *Dolphin* boarded another vessel claiming American nationality, the schooner *Jack Wilding*, as she lay at anchor in Accra Roads. There was an American Captain of the Flag, but one of her "passengers" was a well-known Spaniard who had been master of the brigantine *General Manso*, condemned in 1836, and a log in Spanish, kept by the mate, was found concealed. The American "captain" admitted that he was aware of United States legislation

---

* It is not clear which of the treaty courts had jurisdiction over prizes under Tuscan colours, and it was probably Russell's intention that the *Bella Florentina* should immediately be adjudicated as a Spaniard with a false identity. It appears that he did not make that plain, and the Prize Master was given poor legal advice, presumably by the Captor's Proctor. The British Commissioners might have been more helpful, but both were overworked at the time and sick.

† Jeer blocks were the heavy blocks used for hoisting the lower yards. In some cases, and this appears to be one of them, the jeer block also supported the yard.

requiring a proportion of the crew of an American vessel to be American, but all of the schooner's men were Spanish. Apparently the US Vice-Consul at Havana had told him to ship whoever he liked. In any case, the vessel was equipped for slaving and her American colours gave her no protection in the British waters of Accra. Edward Holland arrested her, and she was condemned by the Anglo-Spanish court.

Somewhat ahead of *Forester* as she headed south in mid-May was *Wolverene*. Commander Tucker's cruise south of the Line took him initially to St Paul de Loando, which he reached on 27 May, and there he called on the Governor-General of Angola and Dependencies. He told His Excellency that four days earlier *Wolverene* had found at Ambriz a Portuguese brigantine ready to embark slaves, and had learned that during the past seven weeks six large brigantines had sailed from there with full slave cargoes. His Excellency ordered a corvette to sea to investigate, and claimed that he was opposed to the slave trade but that he had insufficient force for effective action against it. Tucker suspected, however, that the Governor-General's officers lacked the necessary energy; he had two corvettes under his orders, but they had been in harbour at St Paul since 3 April.

Tucker saw here an opportunity to secure an agreement for cooperation south of the Line between the Portuguese authorities and British cruisers against slaving under Portuguese colours, and his offer was readily accepted by the Governor-General. A provisional convention was arranged by which, south of the Equator, British cruisers were permitted to arrest Portuguese slavers with slaves on board and send them into St Paul de Loando to be dealt with under Portuguese law, the Lisbon Decree in particular. Furthermore, Portuguese and British cruisers would, when in sight of one another, cooperate in the capture of Portuguese slave vessels. The prize money arising from condemnations would be deposited in the National Bank of Angola until the two governments decided on its allocation. This was an admirable initiative on Tucker's part, but activation of the convention depended on agreement by the respective governments.[26]

By 1 June 1839, *Wolverene*'s cruise south of the Line had taken her to the River Congo, in the mouth of which she found the brig *Vigilante* at anchor.[*] The brig was fitted for slaving and was masquerading under Portuguese colours with a passport from the Cape Verdes. Tucker detained her as Spanish, but the cruiser's Carpenter had to construct a new rudder for the prize before she could be despatched to Sierra Leone the following day. At her adjudication the master contradicted the passport which described him as the owner, and revealed that

---

[*] The report by the British Commissioners on this case gives the date of arrest as 23 May, but *Wolverene*'s log shows that it took place on 1 June.

the true owner was a French merchant in Havana. The brig was consequently condemned by the Anglo-Spanish court.

Tucker had sent his pinnace, armed and victualled, and commanded by Lieutenant Newlands, to reconnoitre as far as Puerto da Lenhas, 25 miles or so upriver, and she returned on 2 June to report that she had found a schooner alongside the bank, to which she was secured from the masthead as well as the hull. As Newlands had pulled closer, with ensign and pendant flying, he had been fired on by a gun and muskets on the shore. He and Mr Thorburn, Mate, returned musket-fire before wisely withdrawing. On 3 June *Wolverene* made sail upriver, and the following day she found the schooner, the *Tres Emanuel*, scuttled. On speaking to the local chiefs, Tucker discovered that it had been the crews of both the *Tres Emanuel* and men landed from the *Vigilante* who had fired on his boat, and that the slavers, having stripped and sunk the schooner, had escaped upriver, intending to return and raise the vessel when the coast was again clear. On 5 June, while *Wolverene* headed further upstream with the sweeps, her Carpenter boarded the schooner and destroyed her by cutting away her masts and cutting her sides and decks down to the copper and across to the hatchways. She was too saturated to burn, and, concerned for the safety of the natives, Tucker declined to use gunfire.

This was the same vessel, with the same master, which had beaten off *Waterwitch*'s boats in the Bonny about two years earlier, and had made 24 successful slaving voyages. Not satisfied with destroying her, Tucker wrote to Judge Macaulay, giving his view that the schooner had been not only a Spanish slaver but also a pirate, and asking to prosecute her under the Equipment Clause. In the absence of either papers or witnesses, however, the Mixed Commission could not admit the prosecution. It also complained that it was improper of Tucker to address a letter to one of the Judges.

It was time for *Wolverene* to return to the Bights. On 7 June she was at Cabinda Bay, where she met *Forester*, and on the thirteenth she was off the River Nazareth, before recrossing the Equator and making for Whydah. On arrival there on the twenty-third she found a French barque, *La Felicie*, employed in the palm-oil trade, and a brig under Portuguese colours. As *Wolverene* entered the anchorage, Tucker saw a boat from the barque go alongside the brig, which he took to be a slaver, and pass some boxes up to her. He placed *Wolverene* between the two vessels and sent a boat with a lieutenant to each to prevent anything leaving them. On boarding the brig he found that she was supposedly the *Emprehendedor*, fitted for slaving and with a cargo of tobacco and "aquadente".[*] Correspondence found

---

[*] In various communications the British Commissioners refer to this vessel as *Emprehendedor* and *Emprendedor*. Commander Tucker uses the former spelling, and he sways the balance.

on board showed that the man claiming to be the mate was actually the master, and that he had been in communication with a merchant in Havana. Among the correspondence were instructions to the master that he would find at Whydah a French vessel from which he would receive some arms. Tucker also discovered a logbook in Spanish, and there were grounds for his believing that the vessel was really the *Rapido* of Cadiz.

Tucker detained the brig, and just before he despatched her to Sierra Leone he was approached by the master and mate of the French barque with a demand for the return of the 162 pistols and muskets they had loaded, no doubt legitimately as trade goods, into the brig. Tucker refused on the grounds that it was illegal to remove anything from a prize, and he left it to the Mixed Court to decide on the disposal of the weapons. It is not recorded what was concluded on that score, but the Anglo-Spanish court did decide that Tucker had failed to establish that the prize was Spanish. It seemed to the commissioners that she was Brazilian, and to the Anglo-Brazilian Mixed Commission she was passed for adjudication. The absence of slaves was now the difficulty, but the court carefully considered the clause of the 1826 Convention making it "unlawful for the subjects of the Emperor of Brazil to be concerned in carrying on the slave trade under any pretext or in any manner whatsoever". It concluded that under this all-embracing stipulation Brazilian vessels fitted for the slave trade were liable to condemnation, the absence of a specific Equipment Clause notwithstanding, and the *Emprehendedor* was condemned. Palmerston naturally approved, and an important precedent was established.

Lynx, meanwhile, was at the River Gabon where, on 8 June, she detained the schooner *Perry Spencer* under the flag and passport of the USA. Papers found on board by Lieutenant Broadhead clearly showed that the owner and master was a Spaniard and that American nationality was nothing more than cover, but the Anglo-Spanish court considered that recognition of US nationality by the Cuban authorities and the US Consul in Havana made it improper for the court "to take notice of discoveries made by exercise of rights of search denied by the United States". Why such scruples should have surfaced in this instance when the Mixed Court of Justice had accepted evidence found in apparently similar circumstances in previous cases is unclear. The American masquerading as master seized his opportunity for personal gain, as had his equivalent in the *Florida*, and had the schooner cut into four parts before selling her and her masts, spars, gear and stores at public auction.

Brisk was cruising south of the Equator and unusually far into the Atlantic when, on 14 June, about 280 miles to the west of Loango Bay, she encountered what

Lieutenant Arthur Kellett described as the Portuguese brigantine *Jacuhy*. Having detained her, however, with 203 slaves on board, he found that she was owned in Rio de Janeiro and was therefore Brazilian. She was bound for Rio d'Ostras, just north of Cape Frio, to land her slaves before entering Rio de Janeiro in ballast. *Jacuhy* had sailed from Cabinda only four days earlier, but had already lost 37 of her slaves. As he despatched her to Sierra Leone, Kellett feared that her further losses on passage would be severe, but 196 survived to be emancipated when the Anglo-Brazilian court condemned the prize. After the arrest, *Brisk* removed the slaver's surplus crew and proceeded to the River Congo to land them.

Kellett then headed up the river with his boats. On 22 June he boarded five vessels under Portuguese colours, one of which, fitted for slaving, was clearly Brazilian. He declined to arrest her, perhaps because he was uncertain of the current interpretation of the Anglo-Brazilian Treaty. There was also a Spaniard in the river, but she had landed her slaving gear.

Having returned to sea and headed north to Mayumba Bay, *Brisk* found a more troublesome opponent. At 1040 on 28 June she sighted a stranger and made all sail in chase. The breeze was light and the hands laboured at the sweeps until 1600, when Kellett hauled to the wind in the expectation of intercepting the chase at daybreak. However, it was not until 1230 on the following afternoon that the stranger was again sighted, and *Brisk* hauled up to renew the chase. His elderly "coffin brig" not living up to her name, Kellett decided that his cutter would have a better chance of success in the prevailing light airs, and he sent her away in the charge of his Acting Second Master, Mr William Dix. *Brisk* lost sight of the boat at sunset, but Dix forged on, and at 2115, about 60 miles west-north-west of the mouth of the Yumba River, he caught the chase, and, against musket fire, he succeeded in boarding and taking her.

The prize, a brig named *Matilde*, claimed to be Portuguese, but the only colours at hand were a piratical red flag and a Spanish ensign, and she was bound from Havana for the Congo. The supposedly Portuguese master admitted that the brig was owned by a Spanish merchant in Havana. The vessel was not only fitted and prepared for slaving, but also she was armed with two long 18-pounders and she carried an unusually large crew of 27 men and a boy, one of whom had been killed in the boarding and another wounded. She was undoubtedly a piratical slaver, and the Anglo-Spanish court had no hesitation in condemning her. The slaver's crew had been held prisoner by Kellett for 54 days, but they expressed contrition and begged not to be sent to Portugal for trial, so he released them. The Commander-in-Chief drew Their Lordships' attention to the gallant conduct of Dix, who had persevered with his single boat despite having lost sight of *Brisk* and consequently been beyond hope of support. Nevertheless, he had carried the slaver at the cost only of a slight wound to one seaman.[27]

Kellett did not escape the fever which had afflicted *Brisk*. In August he was invalided home, accompanied by a letter of commendation from Admiral Elliot, who suggested to the Admiralty that if it wished to know more about the difficulties being encountered on the slave coast it should question Kellett, who was "an officer with particular experience in the obstacles" encountered in suppression of the slave trade.

Curlew returned to the Coast during June, now under the command of Lieutenant George Rose, who had been the first Commanding Officer of *Fair Rosamond* after she had been commissioned as a man-of-war. Rose may have doubted that *Curlew*, a ten-gun "coffin brig" (albeit one of the more modern of her class), represented advancement for him, but he certainly brought with him valuable experience.

While her prize *Emprehendedor* was being dealt with in Freetown, *Wolverene* made for Accra for a refit. There Commander Tucker allowed his men shore leave by watches, and on 30 July, five days after departure, the price began to be paid. Of 18 men attacked by fever, nine died. *Wolverene* was by no means the only victim during the year. *Forester* was healthy until July, when eight fever cases appeared among a prize crew at Sierra Leone and three died. Then, at the beginning of November, 16 officers and men were sent up the Sierra Leone River for four days and all of them were attacked by fever. On their return *Forester* retreated south, but during her passage to Ascension she lost eight men.

Other cruisers were not as badly hit. *Buzzard* used the bark and rum prophylactic for her boats' crews in the Biafra rivers during the period January to April with a degree of success, but over that three months 27 of her ship's company of 72 suffered fever, and four died. The newcomer *Curlew* lost two men from ten fever cases between June and August, but escaped disease during the latter months of the year when she spent a good deal of time at anchor at the River Bonny and other places in the Bight of Biafra. *Saracen* frequently had boats in the northern rivers during April and May, and the brig herself went up the River Nunez for ten days, but she remained healthy until the last quarter of the year, when two of her nine cases were fatal.

Viper was similarly healthy until September, when her boats blockaded Port Antonio, Princes Island. Although rain awnings were spread the crews were regularly wet, and eight fever cases resulted in one death. *Columbine* escaped very lightly. She was at the Cape early in the year, and was then mostly on the slave coast south of the Equator. Between September and November her boats' crews saw much service off Angola and in the River Congo, but there were only a few mild infections among one crew which spent nine days in the river. Prize crews at Freetown frequently suffered heavily, and were liable to convey fever to their parent vessels or to the cruiser returning them to the Bights. *Lily* took on board three such crews when she sailed from Sierra Leone on 7 November, and, although she escaped herself, of about 25 passengers 11 were struck by fever and two died.

Conditions for prize crews waiting at Freetown were downright dangerous. The state of the accommodation hulk *Conflict* had deteriorated, and a building in the town, known as "The Barn", had been appropriated as quarters for the seamen and marines. Officers were accommodated elsewhere and generally lost control over their men who, as Dr Bryson complained in his report, being left to their own devices, "at once plunge into every kind of excess with all the characteristic carelessness of British seamen". Bryson did acknowledge that not all were entirely dissolute, but concluded that "Few entirely escape the danger of this ordeal, even if they be of the most orderly and temperate habits."

This problem of the welfare of prize crews was, however, being addressed, and not before time. On receipt in August of an Admiralty instruction to purchase a vessel to convey prize crews from Freetown to Ascension, the Commander-in-Chief ordered Commander Russell, currently the senior officer in the vicinity of Sierra Leone, to buy a condemned slaver of 50–80 tons. He was to rename her *Ascension*, and man her with a mate, one "steady English Seaman" who should be competent to navigate and was to be rated first-class petty officer, a gunner's or boatswain's mate, and ten or twelve Kroomen, one of whom was to be a second-class petty officer. Mr W. Hallett, a mate in *Melville*, was appointed in command. The vessel was to be a tender to the flagship, and she was to be given an anchor berth in the healthiest position in the lower part of the Sierra Leone River. The difficulty now was to find a suitable condemned prize which was not required by treaty to be destroyed.[28]

In the second half of June *Dolphin* seized a pair and *Harlequin* a quartet of prizes on the Windward Coast, but the first of *Dolphin*'s arrests was not a success. On the seventeenth Lieutenant Holland detained the schooner *Euphrates* just off the Grand Bassa River. The interception took place at night and therefore, as was common practice, the schooner was wearing no ensign at the time. Holland found that she had American papers provided by the US Vice-Consul at Havana, and that her alleged master and owner was a naturalised American, but, discovering that she was fitted for slaving, he arrested her as a suspected Spaniard. The Mixed Court refused to admit the case, however, and pointed out to Holland that although he had a right to board the vessel and examine her papers, he then had no right to detain her.

Holland's next attempt, which began on the evening of the same day, was a great deal more satisfactory. He stopped the Spanish schooner *Merced* as she was running into New Sesters, but was disappointed to find that her equipment was as it had been when he failed to have her condemned in February. However, just after the boarding party had allowed her to proceed, *Dolphin*, in the darkness, heard the cries of a man in the water. Boats were lowered, and a male African slave was hauled from the sea. He explained that *Merced*'s crew had set him and ten other slaves adrift in a boat with four black boatmen, but the overloaded boat had capsized, and the other slaves, all boys, had been unable to swim. He and the boatmen managed to right the boat and bale her out, but the boatmen had prevented him from reboarding by hitting him about the head with an oar, and had rowed away. He had swum towards the land, but had then heard a gunshot and seen a light and turned towards it.

Holland, doubtlessly delighted, set off again in chase. He caught and detained the slaver early on the morning of 18 June. For his earlier, improper, arrest of the *Merced* he had been charged costs, damages and expenses, but this time the schooner was condemned for having carried slaves on her current voyage, and was destroyed.

This result would have pleased Lord Francis Russell of *Harlequin*, who had recently brought the *Merced* into Freetown but had decided not to prosecute her. Then, on 20 June, he had his own success to celebrate. He took the brigantine *Emprendedor* off the Gallinas River, and, although she was wearing a Portuguese ensign, the crew admitted that the owner was a resident of Havana. The Mixed Court condemned her as fitted for the Trade, and alarm was subsequently expressed by the Spanish slave dealers at the lack of protection being provided by Portuguese colours. The British commissioners also heard that vessels equipped for slaving were now using the Canary Islands as a refuge from "the harassing and dangerous visitation and search of British cruisers".

A week later, Russell claimed three more prizes in the space of three days. The first of these was the schooner *Victoria de Libertade*, which he took on 26 June as she lay at anchor off the River Sesters. She too was under Portuguese colours, but was condemned as Spanish and fitted for slaving. The commissioners remarked that it was evident that the Consul-General of Tuscany, currently holding the consulship of Portugal at Havana, was "zealously, and not unsuccessfully, labouring to surpass his notorious predecessors in rendering active aid to the Havana slave-dealers".[29]

Harlequin's next victim, a brigantine found off New Sesters on 27 June, was unusual only in claiming Danish rather than Portuguese nationality. This offered no obstacle to Russell, who held Instructions under the Anglo-Danish Treaty, and on boarding her he discovered that she was fitted for slaving and was supposedly the *Cristiano*. He was right in believing, however, that she was Spanish. Her true nature was later proved, largely thanks to Midshipman Jackson, the Prize Master, who found Spanish papers concealed in the sleeve of a jacket, and it was shown that she had cleared from Puerto Rico under the name of *Carranzano* before adopting her false identity at the Danish island of St Thomas.

This run of success for *Harlequin* came to an end on 28 June with the seizure of the schooner *Sin-ygual* off the Gallinas River. She was masquerading under Portuguese colours, but Russell had no difficulty in persuading the Mixed Court of Justice that she was Spanish, and, as she was equipped for slaving, it condemned her. Her owner was the notorious Pedro Martinez of Havana, and she had long been known on the Coast as *Tres Manuelas*.

Meanwhile, in the Bight of Biafra, *Fair Rosamond* was cruising between Fernando Po and the mainland, and on 25 June she seized two small Portuguese slavers 20 miles south-east of the island. As had generally been the case for the traffic between the mainland and Princes Island, the vessels were not only small but also dreadfully crowded. The larger of the two, the 35-ton schooner *Pomba d'Africa*, was crammed with 155 slaves from Old Calabar, a crew of 13 and two passengers. Lieutenant Oliver landed the passengers and all but two of the crew at Princes, and took a third of the slaves into *Fair Rosamond* to ease the crowding for the voyage to Freetown, but 29 died before arrival.

The schooner's consort was the sloop-rigged boat *Sedo ou Tarde*, 29 feet in length and 9 feet in beam, with 23 slaves, also from Old Calabar, a crew of five and a passenger. The vessel was unfit for a voyage to Sierra Leone, and so Oliver loaded her slaves into *Fair Rosamond* and sold her by auction for £65 at Princes. When both of these prizes and their slaves were condemned in the Anglo-Portuguese court the irregularity of the sale was excused on the grounds of the necessity of buying provisions for the slaves. Two of the *Sedo ou Tarde*'s Africans died before emancipation.

At the end of June, Lord Palmerston enquired of the Admiralty whether it was clearly understood by the commanding officers of the cruisers that vessels under the Portuguese flag suspected of slaving could, under the existing treaty, be stopped and searched wherever they were met with, both north and south of the Equator. This was not clear to the Admiralty, let alone the cruisers' officers. Confusion still reigned over rights of search and detention of Portuguese slavers south of the Line, and the blame lay squarely at the door of the Foreign Office; although the Admiralty should, long since, have demanded clarification. The entire situation regarding action against Portuguese slavers was, however, about to be greatly simplified.

The exceptional pace of arrests during June slowed during July, and Edward Holland in *Dolphin*, patrolling the Windward Coast, was the next to strike, on the sixth. His victim, the schooner *Casualidade*, was unusual in that her Portuguese colours were disguising French interest in her voyage. French subjects had been involved in the loading of her 88 slaves at the Sherbro River, and the Freetown commissioners would willingly have laid the papers before the commanding officer of any French man-of-war visiting Sierra Leone. There was no opportunity to do so, however, and the schooner and all 88 slaves were condemned by the Anglo-Portuguese court.

Spanish slavers masquerading under Portuguese colours usually shed the disguise after loading slaves, at which point it became valueless, but a prize taken by *Waterwitch* about 70 miles west of Cape Formoso on 8 July maintained the pretence. She made every effort to escape, but Henry Matson was not to be thwarted. The chase was long, and the slaver, the schooner *Constitucaõ*, hove to only when she came within range of the cruiser's guns. She had on board 344 slavers taken from Lagos for Cuba two days earlier, and the loss of only two on the passage to Freetown was, in the view of the Anglo-Spanish court which condemned the vessel, "highly creditable to the humanity and care of Mr Clarence Taylor, the officer to whose charge the prize had been committed". Four more slaves died after arrival.

The abnormally large number of arrests made during the year had placed a heavy load upon the Freetown courts, which they were in poor condition to handle. Commissioners Macaulay and Lewis were conducting Spanish and Portuguese adjudications alone, and it was now only in the Anglo-Brazilian Mixed Commission that they had the support of a foreign judge. Since his return to the colony from sick leave in February, Mr Lewis had been constantly ill, and he was obliged again to leave rapidly for England on 22 July in order to save his life. Macaulay too was unwell, but stuck to his post. Governor Docherty was once more sworn in as temporary Commissioner of Arbitration.

It was the Anglo-Brazilian court which dealt with the next prize. Commander Tucker in *Wolverene* was still uncertain about that court's attitude to Brazilian-owned slavers under Portuguese colours, and he had decided against sending in five such vessels which he had intercepted during his cruise south of the Line. He was, however, concerned at the extent of the Trade by Portuguese-flagged slavers which, so he understood from the Foreign Office letter of 1 September 1836, properly came under the jurisdiction of the Anglo-Brazilian Mixed Commission. So when, on 25 July, he encountered the brig *Firmeza* at anchor off Whydah, wearing a Portuguese ensign and equipped for slaving, he sent her to Freetown as a test case. Finding that the brig was owned by a resident of Bahia, the court had no hesitation in condemning her.

The Netherlands slave trade had been virtually extinct for many years, but the Dutch at Elmina were still implicated in the traffic. In 1832 Lord Palmerston had called the attention of the Netherlands Government to the illegal supply by Elmina of canoes and canoemen to foreign slavers, but the Dutch Foreign Minister had attempted to discredit the allegation. It would have given Palmerston some satisfaction that Lieutenant Broadhead of *Lynx* furnished proof of the practice when he detained the Brazilian brigantine *Simpathia* at anchor off Popo on 27 July. There was found on board, in addition to her slaving equipment and canoes, a muster roll, prepared and signed by the Governor of Elmina, showing that 19 canoemen, a Headman and a Second Headman had been supplied to the *Simpathia*. She was subsequently condemned.[30]

Bounty payments on vessels condemned as equipped for slaving depended upon assessment of tonnage, and commanding officers were often disappointed by the results. At the end of July Commander Tucker wrote, on their behalf, to the commissioners to complain that cruisers were being "deprived of due rewards" by incorrect returns of tonnages being made "by ignorant officials". He represented also that sales of condemned vessels were being made to a few individuals instead of being sold openly at well-advertised auctions to obtain better prices. Furthermore, officers were having difficulty in extracting information from the courts on measurements, proceeds of sales, expenses and charges.

The response to this courteous letter to Judge Macaulay was a robust, indeed acerbic, letter from the Registrar, Mr Bidwell. He indignantly denied a suggestion by Tucker that Brazilian prizes equipped for slaving had escaped condemnation, and forcefully pointed out that tonnages were assessed according to the method used for British vessels, in accordance with the treaties, and the responsibility for the assessment lay with the Principal Officer of Customs, not the courts. Sales, he asserted, were conducted at a public auction room, were always well advertised, and the crowd, largely of liberated Africans, was always so large that "it requires the most strict and energetic measures to preserve order and regularity among them". As far as availability of information was concerned, he, as Registrar, was accountable solely to the Judges, but, he wrote, his office was most willing to give all information to Agents and Proctors of naval officers. He referred Tucker to Parliamentary papers to which none of the cruisers had access.[31]

Tucker, replying from the Bight of Benin, was conciliatory, but explained that commanding officers, and the Captors' Proctor, Mr Dougan, were under the impression that, until very recently, Brazilian vessels detained for being fitted for slaving would not be condemned. As the cost of sending a prize to Sierra Leone fell to the captor, doubts about liability to condemnation had deterred commanding officers from sending such vessels for adjudication. He lamented that he had no knowledge of instructions to the Mixed Courts beyond the contents of the treaties, and felt, as Senior Officer, that if he possessed fuller information he would be "a more able assistant in the cause of humanity, in my endeavour to put a final stop to the horrid traffic in human blood, a cause I have had for many years at my heart".

The commissioners acknowledged Tucker's "handsome and courteous letter", and regretted that he did not have the "Series of Printed Correspondence […] presented annually to Parliament" on all cases adjudicated by the Mixed Courts. If commanding officers had access to these papers, many illegal seizures would, they felt, have been avoided. They, rather less than helpfully, pointed out that copies were available, free of charge, from the Foreign Office in Downing Street.

A more sensible suggestion from the commissioners was that all cruisers on first visiting Freetown should submit to the courts all the slave papers (variously referred to as Instructions or Warrants) that they held in accordance with the various treaties. They had found that hardly any two cruisers had been issued with the same set of papers, and there had been occasions on which prizes had been delayed at anchor in the Sierra Leone River awaiting the arrival of a captor to prove his possession of the Instructions legalising the arrest. A record in the court registry of the Instructions held by each man-of-war would eradicate such delays.

The first arrest in August came to nothing. *Forester* detained the brig *Mary* off Cape Mount on the tenth, believing her to be Spanish and equipped for slaving, despite her American colours. The court declined to admit the case, but the opinion of HM Advocate-General, communicated to the commissioners 17 months after the arrest, was that good evidence was available that the *Mary* was Spanish and that her case should have been investigated. Lord Palmerston emphasised that, although there was no right to search American vessels on the High Seas, the commissioners might properly conduct an investigation if there was good reason to suppose that an American-flagged vessel was really Spanish. This incident apart, the month of August belonged to *Dolphin*.

The first of Lieutenant Holland's August captures, on the twelfth, was very similar to Francis Bond's two days earlier. The schooner *Catherine*, equipped for slaving, was under American colours when she was arrested off Lagos and, although she called at Freetown for water (very briefly, because of the yellow fever raging in Sierra Leone), Holland had no intention of bringing her before the Mixed Court of Justice. He had decided to send her to New York, under the command of Mr Dundas, his Senior Mate. In the opinion of the British commissioners, Holland had shown good judgment in choosing this case for representation to the United States authorities; in no other case of abuse of the American flag had exposure of fraud been more complete. Vice-Consul Smith in Havana had provided the false identity.

During the following week *Dolphin* cruised westward, and she was close off Cape St Paul's when she encountered the brig *Intrepido* under Portuguese colours on 19 August. There was a Portuguese Captain of the Flag who was alleged to be the owner, but papers found on board showed that the actual owner was a resident of Bahia, and the Anglo-Brazilian court condemned her for being fitted for slaving.

On reporting this case the British commissioners asked Lord Palmerston to request the Admiralty to instruct commanding officers on the Coast to search for, and present to the courts, any papers they could find. Although such documents might not be relevant to the current case, they might be found of value in other adjudications. This was clearly a sensible request, but it is difficult to understand why the commissioners took this long, slow route for it instead of writing directly to the Senior Officer on the Coast, and to all commanding officers calling at Sierra Leone.

Saracen had recently been in the River Nunez, not in pursuit of slavers but to resolve a quarrel between the native chiefs which had brought legitimate trade to a halt. Merchantmen of all nationalities had been fired on, preventing them from trading and even from extracting their possessions from the river. Lieutenant Hill had negotiated an agreement on behalf of them all and restored order.

The dissatisfaction of commanding officers on the question of tonnage assessment, as represented by Commander Tucker, would appear to have concentrated the attention of the British commissioners on the question, and when they examined new legislation on British tonnage they raised concerns with Lord Palmerston. They felt that the new system would be disadvantageous to captors because of the very shallow draught of the slaver schooners and brigantines. On the other hand, the old regulations gave undue advantage because of "the extreme length and breadth of these Baltimore Clippers, if taken without reference to depth of hold". They recommended that the foreign tonnage stated in vessels' papers should be used.[32]

On 26 August, *Dolphin*, still operating off Cape St Paul's, took another equipped slaver under the Stars and Stripes and sent her to New York. She was the schooner *Butterfly*, and the Prize Master gave the British commissioners the opportunity to examine her papers while she was taking water at Freetown before crossing the Atlantic.

Holland made a more satisfying arrest on the following day when *Dolphin* overhauled the Portuguese-flagged schooner *Dous Amigos*, about 20 miles off the mouth of the River Volta, after a 12-hour chase. She was equipped and prepared for immediate embarkation of slaves, and she presented forged papers apparently provided by the Governor of the Cape Verde Islands. According to her master, her voyage, rather unusually, had begun in Havana and was to end in Brazil. It was proved that she was Spanish-owned, and she was consequently condemned and destroyed.

There was, however, an aftermath to this case. Papers had been found in the prize showing that she had purchased arms and ammunition at Cape Coast Castle. The Foreign Office demanded an explanation of the Colonial Office, and in August 1840 that department confessed that merchants at Cape Coast had indeed supplied goods to slave vessels, but that it was very difficult to prove that they "knowingly and wilfully violated the provisions of the Slave Trade Abolition Act".

Lord Palmerston had shown uncharacteristic patience with Portugal in spite of its failure to prevent false use of its colours by Spanish slavers and of its refusal to extend its treaty with Britain to include an Equipment Clause and agreement to arrests south of the Equator. By the spring of 1839, however, that patience was exhausted, and the Foreign Secretary introduced in Parliament a most extraordinary Bill to eliminate Portuguese obstruction. At the first attempt the Bill passed the House of Commons but was defeated in the Lords, largely thanks to the opposition of the Duke of Wellington. Slightly amended, it was reintroduced and became law on 28 August.

The Act allowed anyone in Her Majesty's Service, acting under the authority of the Lords Commissioners of the Admiralty, to seize slave vessels and slaves and to bring them before the High Court of Admiralty, or any Vice-Admiralty court, provided they were British or Portuguese, or were unable to show that they were entitled to claim the protection of a flag other than British or Portuguese. Not only such vessels carrying slaves but also those fitted for slaving were liable to arrest, and the definition of "fitted for slaving" was as laid down in earlier Equipment Clauses. Seizure was permitted both north and south of the Equator. Slaves liberated by such arrests were to be landed in the nearest British settlement, and slaver crews were to be handed over to the courts of their own countries. All persons issuing orders in accordance with the Act or carrying them out were indemnified against legal proceedings.

Not surprisingly, this high-handed action did not find universal support in Parliament. The Duke of Wellington regarded it, with reason, as a violation of international law and as having "a criminal character". Perhaps influenced by his affinity with the Portuguese, he viewed it as an affront to an old ally and felt that it would have been better to declare war against Portugal. Even less surprising was the Portuguese reaction. The Minister for Foreign Affairs in Lisbon addressed a circular to the signatories of the Vienna Convention of 1815 complaining that Britain's conduct towards Portugal was "unprovoked, oppressive and unjust, and as being in a flagrant violation of the Law of Nations, and a direct attack upon the Rights of an independent State". Palmerston was unmoved, and responded only by sending to the addressees of the Portuguese circular copies of papers showing the substance of negotiations between Britain and Portugal on the matter.[33] He was equally unreceptive to a Portuguese offer to sign a treaty along the lines of the Act if Britain would stop pressing for repayment of her debts. Palmerston was no longer in the mood to make concessions.

The Foreign Office pre-empted the legislation by writing to the Admiralty on 15 August requesting that orders be given to cruiser commanding officers along the lines of the Bill currently before Parliament, but it was not until 1 October that it sent a copy of the new Act to the Admiralty. Clearly there was no confidence at the Foreign Office that instructions via the Admiralty had reached their ultimate destinations, and on 2 November it wrote to all the British commissioners in the Mixed Courts repeating the directive of 15 August, despite the commissioners having no authority to give orders to commanding officers. As it transpired, Admiral Elliot received his instructions regarding the Portuguese Treaty on 10 October from the Admiralty, and on the thirteenth he sent out orders to the Squadron to the effect that boarding and searching of Portuguese vessels north and south of the Equator were "recognised in the fullest measure".

Another arrest off Cape St Paul's took place on 5 September, that of the Brazilian barque *Augusto*, taken with slaving equipment and a large quantity of farina on board. Also found were letters useful to the commissioners at Freetown, who were beginning to accumulate knowledge of the Brazilian slave-trading houses, particularly those at Bahia, and of the vessels, masters and agents on the African coast engaged in the traffic to Brazil, as well as their methods of transacting business. She was condemned without hesitation by the Anglo-Brazilian court. Apart from the acquisition of useful intelligence, the arrest was, in a small way, notable. It was the final capture off West Africa by that stalwart *Fair Rosamond*.

Since falling victim to *Black Joke* in November 1830 and purchase by Commodore Hayes early in 1831, *Fair Rosamond* had, first as a tender and then as a commissioned man-of-war, captured 18 slavers and been instrumental in the release of 1,452 slaves. Although she was never to return to the coast of Africa, her career subsequently took her to the Caribbean, and she was ultimately paid off and broken-up in 1845.

Bonetta was cruising south of the Line in early September, and on the seventh she enjoyed a day of remarkable success in the River Congo, taking three Spanish vessels equipped for slaving. The slavers seem to have been extraordinarily unalert, allowing Lieutenant Stoll's boat crews to board them while *Bonetta* lay at anchor off Wood Point, Punta de Lenâ. Perhaps the schooner *Josephina* and the brigantine *Liberal* still had confidence in their Portuguese colours, but both were proved to be owned by a French merchant in Havana. The third vessel, the schooner *Ligeira*, was arrested for having on board a cooking boiler of excessive size, and she allowed *Bonetta* to see her landing another. Her Portuguese Captain of the Flag claimed that the vessel's owner resided in Bahia, and she might have been tried as Brazilian, but the clear evidence that she was Spanish enabled her, as well as the other two, to be destroyed on condemnation.

The next prize was truly Brazilian, the Baltimore-built brigantine *Pampeiro* from Bahia, taken at anchor off Lagos by *Wolverene* on 12 September. In addition to her slaving equipment, there was found on board a list of 41 canoemen supplied to her by the authorities at Elmina. She and her cargo were condemned to Commander Tucker by the Anglo-Brazilian court, and, at the same sitting, the court condemned the brigantine *Golphino* to Lieutenant Seagram of *Termagant*. *Golphino*, also fitted for slaving, was caught on the nineteenth as she ran eastwards along the coast towards Lagos and 100 miles short of her destination. Thanks to letters found in the *Firmeza* and *Augusto*, the commissioners were already aware of the shippers and consignees of her cargo.

After this succession of empty prizes there came the capture of a slave cargo on 27 September, and there was the additional satisfaction that the slaver, by then named *Sete de Avril*, had been the *Mary Cushing*, taken under American colours early in the year and released. On this second encounter, Lieutenant Henry Matson in *Waterwitch* caught her about 40 miles south-west of Lagos, with 424 slaves whom she had loaded there the previous day. At 144 US tons she was very crowded, and the slaves were already in a wretched state. She had been under US colours until the slaves were embarked, and the American Captain of the Flag was on board as a passenger. His post had been assumed by a Portuguese, but he admitted that the owners were Spaniards at Havana. The American was removed by Matson, pending instructions from the Commander-in-Chief, and the schooner was condemned in the Anglo-Spanish court after seven slaves had died on passage and a further two before emancipation.

Another Brazilian was seized on 29 September. She was the brig *Destemida*, arrested by Lieutenant Broadhead in *Lynx* for being equipped for slaving. The brig still had some of her trade goods on board when he intercepted her between Winnebah and Accra under Portuguese colours. It was admitted that her owner lived in Brazil, and the Anglo-Brazilian court condemned her when she was brought in by Mr Frederick Slade with a number of slaves from a prize taken by *Nautilus*.

There was a slower trickle of prizes during October, the first pair falling to Lieutenant George Beaufoy and *Nautilus*. The first of these two, the Portuguese schooner *Andorinha*, was unusual in that she was carrying eight African children aged between three and four years. When the schooner was boarded on the second, about 20 miles east of St Thomas in the Bight of Biafra, the master made the preposterous claim that the children were servants, and Beaufoy took the extraordinary decision to allow the man to land at Princes Island with two of the five children he professed to own, a decision which the Anglo-Portuguese court subsequently felt "was to be regretted". The prize then called at Cape Coast and Cape Palmas for provisions; at the latter she encountered *Harlequin*, which not only gave her supplies but also replaced her sick Prize Master. On reaching Freetown she was condemned for slaving and three of the children were emancipated, but there was no explanation of the fate of the remaining three.

On 4 October the 16-ton launch *Vencedora*, carrying 61 slaves from Cape Lopez, was probably fortunate to be found by *Nautilus* just off St Thomas, her intended destination, half-full of water and entirely unseaworthy. Lieutenant Beaufoy wisely decided to leave her at Princes, and he divided the slaves between *Destimida* and *Andorinha* for passage to Freetown. The commissioners on the Anglo-Portuguese court complained that both of Beaufoy's cases were "very imperfect and confused" in their presentation, but perhaps had sympathy for the two young officers bringing the cases to court: Midshipman John Milbourne of *Harlequin*, who had been the replacement Prize Master of *Andorinha*, and Mr Hunt, the young officer who had fallen sick and been removed from that prize by *Harlequin*. They condemned the *Vencedora*, after 11 of her slaves had died, and issued a commission for the sale of the launch at Princes.

There followed two unremarkable captures which both resulted in condemnation for being fitted for slaving. On 16 October Henry Hill in *Saracen* took the Portuguese-flagged schooner *Brilhante* off the River Gallinas, and she was found to be owned in Havana. On the twenty-seventh Henry Matson in *Waterwitch*, cruising the Bight of Benin, followed suit with the Brazilian schooner *Calliope* about 25 miles south of Little Popo.

Matson's next encounter, five days later, provided a greater challenge to *Waterwitch*'s prowess. At 0540 on 1 November she sighted a suspicious brig to the east-north-east, and at 0600 she made all sail in chase. She gained ground slowly, and at 0900 Matson tried a long-range shot, whereupon the brig began to jettison guns, anchors, boats, hammock nettings, sleeping berths and spare gear. At 1230 the range had closed sufficiently for effective fire by *Waterwitch*, and the chase hoisted Portuguese colours. This failed to delude Matson, and by 1330, her masts badly damaged and her rigging shredded, the brig could no longer carry sail, a state achieved by *Waterwitch* with only 18 full charges and 16 rounds.[*]

On boarding his prize, the *Fortuna*, just offshore about 40 miles east of Lagos, Matson found that she was fully prepared to receive slaves, and a log and papers discovered in the chest of a passenger, apparently the "principal director of the voyage", showed that she was owned in Havana. Her crew of 41 men and a boy were made prisoner and victualled in *Waterwitch*, at two-thirds of the daily rate. Twelve pairs of slave irons were borrowed from the prize to secure the captives as necessary, and as the cruiser had bread remaining for only four days, Matson, with retrospective approval from the Mixed Court, removed 1,150 pounds of the slaver's rice to feed his ship's company. These arrangements completed, *Waterwitch* made sail at 1600, and the *Fortuna*, her third slaving voyage prematurely ended despite being regarded as a superior sailer, headed for Freetown to be condemned and destroyed.

The brigantine *Viper*, now commanded by Lieutenant Godolphin Burslem, had recorded no arrests during more than two years on the Station, and it is probable that she was employed primarily as a packet. Nevertheless, she held a copy of the Instructions, and on 11 November, about 40 miles east of Cape Palmas, she took the schooner *Magdalena*. The master and supercargo, both Spaniards, readily admitted that, although the vessel was masquerading under Portuguese colours, she was owned by a merchant in Havana. She was fitted to

---

[*] Matson may have fired two long-range shots with double charges, or two charges may have been spoiled by spray or some other cause.

receive a cargo of slaves at the Gallinas, and it seems that she had missed her landfall, with fatal consequences. Having been condemned as Spanish, she was destroyed at Sierra Leone.

Also taken off Cape Palmas was the Brazilian brigantine *Sociedade Feliz*, found equipped for the slave trade by Lord Francis Russell in *Harlequin* on 21 November. The 130-ton, American-built vessel was so full of cargo that a thorough search after capture was impracticable, and she had large quantities of farina, jerk beef, beans and Indian corn stowed under her cabin floor, as well as four cooking boilers and water casks with a capacity of 1,808 gallons. Also found on board was a letter addressed from Bahia to a correspondent in Lagos which reported, erroneously, that "the English Government has entered into a Treaty with that idiot the Queen of Portugal, whereby all vessels belonging to her nation will be looked upon as pirates, if met with on any part of the Coast of Africa". The British commissioners commented that "Condemnation of this rich prize will be a severe blow to the already suffering slave-dealers of Bahia."

There were three further captures in the Atlantic on three consecutive days during November. Henry Broadhead in *Lynx* seized the schooner *Lavandeira* about 70 miles west of the Gallinas River on the twenty-seventh, finding a Portuguese Captain of the Flag but ample evidence that the owner was a resident of a Spanish port and that she was bound from and to Havana. She was condemned as fitted for slaving, as was the next prize, the Brazilian brig *Conceiçao*, taken 20 miles west of Whydah the following day by *Termagant*, Lieutenant Seagram. It seemed that the brig was returning to Whydah having lost an anchor and been driven to sea. The third of these prizes, and the final one of 1839, was another for *Termagant*. She was the Brazilian brigantine *Julia*, boarded off Whydah on the twenty-ninth, flying Portuguese colours and holding a false Portuguese passport issued by the authorities at Princes Island. *Julia* too was condemned as equipped for slaving.

There had been a number of arrivals and departures at the Cape. Commander Craigie and *Scout* sailed for England on 29 August, and *Columbine* arrived on 23 October. *Brisk*, now commanded by Lieutenant Whaley Armitage, had been sent to Angola on 26 August with the Commander-in-Chief's instructions to amend the agreement made there by Commander Tucker, and she returned on 13 December. *Bonetta* and *Curlew* both arrived from the west coast on 5 December. In mid-December Armitage was given permission to return to England, being replaced by Lieutenant George Sprigg of *Harlequin*, and a few days later, *Brisk* suffered the further loss of her main yard when *Bonetta* got foul of her. On 27 December *Melville* arrived.

Dolphin was under orders for England in early December, but her departure was delayed when a complaint was made about Lieutenant Holland's conduct, and then serious charges were brought against Lieutenant Rose of *Curlew*. Rose was suspended from duty and was to be sent home. The circumstances of these complaints and charges remain a mystery, but it was sad that two officers with excellent records of action against the slavers should become mired in some sort of misconduct, actual or alleged.

The new freedom of action against the Portuguese was slow to bear fruit, and a private letter at the end of January 1840 from the Registrar of the Sierra Leone Mixed Courts to a friend in the Foreign Office revealed that, by then, just one Portuguese slaver had been adjudicated by the Freetown Vice-Admiralty court under the new legislation. No further information is available on that prize, except that she had been equipped for slaving and was taken south of the Line, but there were certainly three other such arrests during December. Although these three prizes were engaged in the Atlantic slave trade, they were taken in the Indian Ocean, an area in which the slavers had hitherto been undisturbed.

On 5 November, the ship-sloop *Modeste* (18), Commander Harry Eyres, which had recently joined the Station from the West Indies, had sailed from Simon's Bay steering eastward, briefly in company with *Columbine*. She headed for the coast of Mozambique, and on the twenty-sixth, about 25 miles off the slaving port of Quelimane, she arrested the Portuguese brig *Anna Feliz*, with 56 slaves on board. A more spectacular capture came on 9 December. After a long chase, the brig *Escorpião*, armed with 18 "long carronades", was brought-to about 90 miles east-south-east of the mouth of the River Zambezi and found to be carrying 756 slaves, the greatest number yet seized in one vessel. Eyres was not yet finished. Three days later, and a further 120 miles to seaward of his second capture, he chased and caught the brig *Arab*, with a more modest cargo of 26 slaves. The three prizes were sent off to the Cape, but on the twenty-sixth Eyres found the *Arab* wrecked in Mayumba Bay, 250 miles north-north-east of the position of her capture. Her people were all saved, but there is no explanation for her having headed north, and it may be that she was retaken.

The *Anna Feliz* reached Simon's Bay on 26 December with 51 slaves, and the *Escorpião* also arrived safely, although it is not known how many of her extraordinary cargo survived the voyage. Adjudication by the Vice-Admiralty court was delayed because, even by mid-January 1840, no order had been received by the court on the subject of the Portuguese Act. All three prizes were eventually condemned, and Admiral Elliot sought to have the *Escorpião* bought into the Navy, but no decision had been made by the end of August 1841 when she broke from her moorings in a gale and was driven onto Seal Island in False Bay, becoming a total wreck. Her demise did not close her story, however. The legality of her condemnation was questioned because Eyres had not presented authority from the Admiralty to seize Portuguese vessels, as required by the Act. Legal quibbling continued until late in 1848, when an award to the captor was finally decided.

The weight of work caused by the extraordinary number of prizes, combined with the usual debilitating effects of the Sierra Leone climate, was taking a heavy toll on the health of the British commissioners. In October Arbitrator Lewis had been granted an extension of sick leave in England until he had recovered, and Judge Macaulay had requested return home on account of his health. Palmerston's response was that Macaulay was free to leave when Lewis returned to Sierra Leone, but that he was to make provision for the business of the Mixed Commissions and Mixed Court of Justice in his absence.

On 31 December a despatch from the Foreign Office to the British commissioners enclosed a copy of a Brief issued by His Holiness the Pope enjoining all Roman Catholics to abstain from the slave trade. Had this instruction been heeded by the Pope's flock the Atlantic slave trade would have been stopped in its tracks, but the words of His Holiness fell, of course, on deaf ears.[34]

* * *

As 1839 came to an end, the harassed British commissioners at Freetown might have been forgiven for concluding from their casework load that the slave trade to Cuba was all but destroyed and that the Brazil Trade had been severely depleted. By comparison with previous years, the number of arrests by the Cape Station cruisers in the two-year period 1838–9 had been extraordinary: 98 slavers had been taken and one scuttled, and 52 of these, being Spanish, had been destroyed on condemnation. The rate of success in 1839 alone had been prodigious: 68 slavers captured and one scuttled. Of those, 55 were Spanish and Brazilian vessels taken as fitted for slaving, and only 12 were captured with slave cargoes.

This last pair of figures is particularly significant. It shows clearly the importance of the Spanish Equipment Clause and of the more controversial acceptance by the Anglo-Brazilian Mixed Commissions of the illegality of equipping Brazilian vessels for the slave trade. More than that, however, it gives the lie to past allegations that the British cruisers were in the habit of allowing vessels inward-bound to the slave coast to pass unhindered so that they might be more profitably arrested when they subsequently sailed with slaves embarked.

These superficially encouraging statistics can be explained partly by the increased number of high-quality cruisers on the Coast and by the consequent ability of the Senior Officer to deploy them simultaneously, albeit very thinly, to all the main slaving areas north of the Equator and, not infrequently, south of the Line too. By the end of 1839 there were 20 men-of-war on the Station, 18 of them cruisers. Six were Symondites, four were the much-improved "coffin brigs" rigged as brigantines, four were younger members of the "coffin brig" class, one (*Modeste*) was a new ship-sloop designed by Admiral Elliot, and finally there were the elderly but still excellent *Fair Rosamond* and the splendid ex-yacht *Waterwitch*. However, a number of them would, at any one moment, be on passage between the slave coast and the Cape or Ascension, in refit at Simon's Bay, or deployed to the Indian Ocean.

The second major factor in the increased rate of arrests was the intensity of the slaver traffic. It was not possible for the British representatives in Cuba and Brazil to detect all slaver sailings, but, as will be seen in the next chapter, the British commissioners reported that in the years 1837–9 a total of 200 slave vessels departed from Havana alone, and that 53 sailed from Rio de Janeiro in 1839. This increase over earlier years would have been replicated at the other Cuban harbours and at Bahia.

The result, shown by recent research, was that for all the apparent success of the West Africa cruisers, slave imports to Cuba in 1838 showed no significant change from the previous year, and fell by only a thousand in 1839 to a total of 19,900. In Brazil there was a small drop in imports in 1838, but that reduction was reversed the following year to produce a figure of 54,400.

What was clear at the time was that the new "Palmerston Act", to empower the Navy to seize any Portuguese vessel suspected of slaving, was having little immediate effect. In his letter to an acquaintance in the Foreign Office in January 1840, Mr Bidwell, the Freetown Commission Registrar, expressed disappointment that only one vessel had so far been brought before the Sierra Leone Vice-Admiralty court. More had been expected in the first two months of this change in operations, considering "the novelty of the measure, the suddenness of enforcement, and the surprise with which slavers would be overtaken".[35] In due course, however, this high-handed Act, for all its questionable legality, would bear its intended fruit.

Of particular concern as 1839 came to a close was that the flag of the United States was to be found in every slaver haunt on the Coast, a consequence of the Spanish Equipment Clause and the expressed determination of the USA not to be a party to any convention on the slave trade. This was a heavy cloud on the horizon.

CHAPTER 16

Forbearance Exhausted: Western Seas, 1838–9

There is, however, a limit at which forbearance ceases to be a virtue.
EDMUND BURKE

RECENT EFFORTS of the Royal Navy's squadrons against the slave trade in the western Atlantic had produced no evidence of progress. The hopes attached to the Equipment Clause in the 1835 Anglo-Spanish Treaty had been dashed by the evasions practised by the Cuban slavers, and Portugal's resistance to any improvement of its treaty with Britain not only facilitated those evasions but also was partly responsible for the lack of effective action against the traffic into Brazil. Acceptance of an Equipment Clause by Portugal was certainly sorely needed, but it was the fault of the British Foreign Office that naval action against laden Portuguese slavers south of the Equator had not been authorised. By its treaty with Britain Portugal was permitted to trade slaves only between its own territories, and, as even the Government in Lisbon acknowledged at the time of the *Orion* trial in Rio, its traffic in the southern hemisphere had become illegal at the moment of Brazil's independence. The point seems clear, but Britain dithered and withheld authority from its cruisers to interfere with that traffic. This indecision, or perhaps forbearance for some obscure diplomatic reason, is now difficult to comprehend.

Nevertheless, the point was made almost academic by the dearth of cruisers on the Brazilian coast. Most of the vessels on the South America Station had been employed on the Pacific seaboard of the continent, and only enough cruisers had been retained on the eastern coast to support British mercantile interests in the Atlantic ports. Slave-trade suppression had become a sideline attracting little effort and no initiative. Relative priorities, as viewed in London, had now been emphasised by the replacement of the South Atlantic Station by a Pacific Station, and the matter of command on the Brazilian coast was yet to be clarified.

The situation was barely more encouraging in the seas around Cuba. The Mixed Court of Justice in Havana had been marginally busier than the Mixed Commission in Rio de Janeiro, but the numbers of arrests bore no relation to the enormous volume of slaving into the island. The handful of cruisers patrolling the approaches to Cuba was entirely inadequate for the task.

There had been a massive rise in the number of slaves landed in Brazil during the past three years. Recent research shows that imports in 1834 amounted to 21,500 slaves, and the figure almost doubled in 1835 to 40,900. The increase continued in 1836 to a total of 51,800, although the numbers landed at Bahia and to the northward actually dropped markedly during the year. In 1837 the figure for landings south of Bahia, the great majority, held steady, and there was a recovery in numbers for Bahia and the north to produce a total of 54,000. At the end of that year the British commissioners noted that there had been a dip in the traffic, possibly thanks to the Portuguese Decree, but that had been followed by an increase, which probably reflected the change in government.

Cuba had seen a reduction in imports from the peak of 25,700 slaves in 1835 to between 20,000 and 21,000 in the following three years, but this drop had resulted from market forces rather than the efforts of the British cruisers.

To add to these dispiriting circumstances, there had been an ominous increase in American involvement in the Cuban slave trade. Its consuls on the island had been giving aid and encouragement to the slavers seeking to evade the Spanish Equipment Clause, its shipbuilders were providing high-quality vessels to the Havana merchants, and the Stars and Stripes had made a significant appearance as an alternative to the Portuguese flag for false colours.

* * *

In January 1838 *Wanderer* and *Serpent*, two almost-sister Symondite brig-sloops, arrived off the southern coast of Cuba: the former, Commander Thomas Bushby, from Bermuda, and the latter, Commander Richard Warren, from Port-au-Prince. They were in company about ten miles south-east of Santiago de Cuba at 0800 on the twenty-first when Bushby ordered *Serpent* to examine a schooner which was seen to be heading for the coast. At 0900, just as *Wanderer* finished speaking to a French brig, Bushby saw the schooner bear up for Santiago, and *Wanderer* steered to cut her off. *Serpent* opened fire at the schooner's masts, and, in the hope of obliging the sloop to heave to, the chase put several Africans into a sinking boat and cut it adrift. However, *Serpent* was equipped with quarter boats, and she was able to lower and slip them without shortening sail, thereby rescuing five slaves. Seeing no prospect of escape, the slaver ran herself ashore about eight miles east of Santiago and almost immediately went to pieces. By the "very great exertions" of the officers and boats' crews of the two sloops, 121 more slaves were saved and two were found drowned. Neither papers nor colours were found in the wreck, and the schooner's officers and crew had escaped. With no idea of the slaver's nationality or identity, and in the absence of witnesses from her crew, Bushby was unable to bring the case to the Mixed Court of Justice, and before continuing his cruise he landed the slaves at Port Antonio in Jamaica on 25 January. There, presumably, they were emancipated by the Vice-Admiralty court.

At the end of the month a most unusual arrest took place, and its aftermath became something of a maritime saga. At 1630 on 30 January HM Steam Packet *Flamer* was about 150 miles to the south of Puerto Rico and making her way to Jamaica when she came across a schooner which, as her Commanding Officer, Lieutenant John Potbury, put it, "appeared to be desirous of avoiding me or even exchanging colours as usual with vessels meeting at sea". Having commanded the schooner *Nimble*, Potbury had a keen nose for a slaver, and he put a shot across the stranger's bow. She hove to and hoisted Portuguese colours. As "no diversion or delay would be incurred", Potbury sent across his Master, Mr Strutt, and was not surprised to hear that the schooner, the *Feliz*, had on board 326 slaves from the Gulf of Guinea for Havana as well as two masters, one Portuguese and another Spanish.[*] Undeterred by his lack of slave treaty Instructions, Potbury boarded the slaver himself and sent all of her crew to *Flamer*, apart from the master, mate and cook, whom he divested of weapons; took charge of the vessel's papers, for which he gave the master a receipt, together with a certificate showing the number of slaves; and gave Mr Cotter, the Second Master, a prize crew of five hands. Bearing in mind his duty to avoid delay to the mail on board *Flamer*, he later assured the Commander-in-Chief that all had been completed within an hour.

Potbury had no illusions about the illegality of his arrest, and when he arrived at Port Royal he reported the circumstances to Commodore Sir John Peyton of the frigate *Madagascar*, Senior Officer of the Jamaica Division. Peyton did hold treaty Instructions, and on 5 February he sent out his Master to meet the prize, take possession of her 20 miles to the east of Port Royal and bring her in. There the surviving 324 slaves were landed and, on 18 February, the *Feliz* sailed for Sierra Leone in the charge of Mr Baugh, a mate of *Madagascar*, to be taken before the Anglo-Portuguese court.

Some time later Potbury despatched a somewhat tongue-in-cheek letter to the Commander-in-Chief to explain his actions. He did actually have a set of Instructions on board, he wrote, but they had been supplied to his predecessor, who had handed them over to him on his taking command. The experienced

---

[*] Potbury's initial report to the C-in-C states that *Feliz* was boarded by the Second Master, but his statement to the Mixed Commission describes Mr Strutt as the Master.

Potbury well understood that this document gave him no authority to make an arrest, but he explained to Paget that, being in possession of these papers, and having no orders not to take slavers, he could not feel himself justified in passing by the slaver when she was actually in *Flamer*'s wake. He then solicited "your kind interference in behalf of the Officers and Crew of the *Flamer*" in requesting valid slave papers. It seems unlikely that Potbury was given the Instructions, but Paget must surely have been disarmed by the letter.[1]

Nothing more was heard of Mr Baugh and the *Feliz* until 29 April, when a merchant vessel arrived in the River Gambia from Goree and transferred some passengers to the brig *Curlew*: Baugh, his prize crew and two prisoners. They had a harrowing tale to tell. During her passage the *Feliz* had spoken to several merchant vessels to compare longitudes and to check her chronometer, and had always found that she was to the west of her reckoning. Baugh deduced that the prize had been experiencing a strong westerly current, and he was confused when land was sighted at noon on 4 April. He calculated that he was on the latitude of, and somewhat to the west of, the island of Sal in the Cape Verdes, but the land did not answer the description of that island. He anchored and went ashore with Midshipman Fairholm, but the surf prevented their return. During the night they walked for three miles until they found some natives who returned to attempt to launch their boat, but it was swamped. By now Baugh and Fairholm had realised that they had found mainland Africa.

The breeze freshened and *Feliz* let go a larger anchor, but the stock broke and she began to drag, so she got underway and stood further out to anchor alongside the French brig-of-war *Lancier*, which had just arrived. On the morning of 9 April Fairholm managed to pull out through the surf in the damaged boat, but the natives kept Baugh ashore. Fairholm then went aboard the *Lancier* to beg her Commanding Officer for a four-oared boat to recover Baugh. He was offered provisions and other assistance, and told that he was 40 miles north of the River Senegal, but the boat was refused. *Feliz* was too far offshore for a small boat to pull to the beach and back, so Fairholm weighed anchor, stood inshore to a depth of seven fathoms, launched the boat and put his helm down; but thanks, he said, to a strong current from the north, he missed stays twice. He then anchored in three fathoms, but the prize was not brought-to immediately and drifted close to the breakers. Fairholm put a spring on the cable, made sail, cast to port and cut the cable, but before the schooner could gain headway she struck. Her rudder was carried away and she became unmanageable.

As *Feliz* was swept into the breakers, Fairholm began to build a raft, but the schooner was carried so far up the beach that the crew were able to walk ashore.

They managed to save some valuables, but the Africans then pillaged the vessel and left the men with nothing but the clothes on their backs. They were marched to the native encampment, held there until 21 April, and then marched south to the River Senegal, where a passing government vessel rescued them and carried them 40 miles upriver to the capital, arriving on the twenty-fifth. The French authorities provided passage first to Goree and then to the Gambia, and *Curlew* carried the fugitives to Freetown.[2]

Baugh had contrived to save the prize's papers and Potbury's statement to the Mixed Commission, but the legitimacy of the secondary arrest by *Madagascar* was dubious. The cases of *General Manso* and *Donna Barbara*, not dissimilar from that of *Feliz*, were examined by the commissioners, and they noted that their condemnation of *Donna Barbara* had been approved of by the Foreign Office as being in the spirit of the treaty. The legal advice in London in those earlier cases was that there was nothing to prevent a second arrest, following an initial illegal capture, as long as the second seizor was not complicit in the first arrest. Consequently the *Feliz* was condemned and her slaves were emancipated, and the Queen's Advocate later concurred with the verdict.

Had a case of this kind come before the Havana or Rio court the outcome might well have been less satisfactory. Not only did the British commissioners in those cities seem less confident in giving captors the benefit of any doubt, but also they lacked the advantage enjoyed by their opposite numbers in Freetown in that their foreign colleagues were frequently absent from Sierra Leone on account of the unhealthy climate.

There was a postscript to this episode. In May 1839 a demand reached the Foreign Office from France for 1,833 francs and 23 centimes, the money laid out by the French authorities for the ransom and maintenance of the prize crew. Palmerston enquired of the Admiralty whether any lack of hospitality by the French had been a contributory factor in the affair, but the Admiralty could offer only the refusal of the *Lancier* to send a rescue boat.

The Commander-in-Chief had spent the winter in Bermuda, flying his flag in *Cornwallis* and, when *Cornwallis* was absent at Halifax and then Barbados, in *Wanderer* and *Pearl*. In late February he headed south, once again in *Cornwallis*, and arrived at Port Royal on 6 March. There he found *Seringapatam*, *Crocodile*, *Rainbow*, *Wanderer*, *Sappho*, *Comus*, *Ringdove*, *Magnificent*, *Hornet* and the steam vessel *Alban*, as well as three French men-of-war. Two days later *Sappho* and *Ringdove* sailed for a cruise off Cuba, and *Wanderer* followed them on the sixteenth to form an unusually strong presence in those waters. They were replacing *Nimrod* and *Champion*, which returned to Port Royal on 9 and 10 March. *Rainbow* sailed for England on the sixteenth, and, after a visit to Jamaica of only six days, Admiral Paget departed for Bermuda, where he intended to remain until the end of May before making for Halifax.

In Havana permission had still not been granted for the black soldiers in *Romney* to go ashore, and the Foreign Office had taken up the matter with Madrid. Captain-General Tacon had been vehemently opposed to allowing the men to set foot on the island, but he had resigned his post and it was yet to be seen what position his expected successor, Don Joaquin de Espeleta, would take. Clearly the soldiers were not the only men anxious to be out of the *Romney*. Her Commanding Officer, Lieutenant Jenkin, reported that one of his petty officers had deserted and was believed to have engaged on board the American schooner *Dido*, which had sailed on about 3 March for the coast of Guinea. Wage offers by the slavers of $40 or more per month gave great temptation for desertion.

Lord Palmerston was still trying to persuade the Portuguese Government to take effective action against the use of its colours to shield the Spanish slave trade, but all that he could extract from Lisbon was a promise that enquiries were being made and that orders were being passed for redoubling precautions against such "frauds and abuses".

Another abuse was causing concern in London. In March the Admiralty issued a circular to commanders-in-chief on foreign stations directing them to instruct their commanding officers to desist from the practice (to which it appeared that they claimed a right under the Articles of War) of taking from prizes navigational instruments for the use of the capturing vessel. This directive was clearly aimed at the anti-slaving cruisers, although the malpractice was not as apparent in the Caribbean as it had been on the African coast.[3]

The sloop *Racehorse* had been sent home for refit, and she returned to Bermuda in mid-April. No doubt Admiral Paget was expecting to welcome a reinvigorated cruiser, but in this he was disappointed. He complained to the Admiralty that she was "more in need of being sent again to a dockyard in England than to a foreign station for three years". The Admiralty called on the officers of the dockyard at Devonport to explain the poor state of the sloop, but what excuses they offered in response are not known.

After two fallow months April produced some reward for the Cuba cruisers, although the first arrest resulted from a chance interception. On the twenty-fourth Commander The Rt Hon. Lord Clarence Paget in *Pearl* took the Portuguese-flagged brig *Diligente* with 480 slaves about 155 miles north-north-west of Cape San Antonio, unusually far into the Gulf of Mexico for a Cuba-bound

23. Capture of the Spanish schooner *Opposiçao* by HM Sloop *Pearl* on 28 April 1838

vessel.* The brig was a notorious slaver, renowned in Havana for her fast sailing, and, relying on her speed for safety, her owners had not insured her. Paget initially took her into Havana, and arrived there on 28 April with two prizes. The second was the schooner *Opposiçao*, which had been making for Havana under Portuguese colours having apparently landed slaves on the Cuban coast, and, still equipped for slaving, she was careless enough to fall in with *Pearl* not far from the harbour.

Hoping to prove that the *Opposiçao* was Spanish, and apparently suspecting that the Freetown commissioners might be more open to persuasion than those at Havana, Paget took both prizes to Nassau, where the surviving 475 slaves were landed from the *Diligente*. He then made for Bermuda with the intention of sending both vessels to Sierra Leone.† While at Bermuda the *Diligente* was found to be unseaworthy, and her papers and witnesses were sent to Freetown in the *Opposiçao* under the command of Mr Donald McKenzie, Extra College Mate. There the Anglo-Portuguese court condemned the *Diligente* without difficulty and emancipated her slaves. The court experienced barely greater difficulty in also condemning the *Opposiçao*. She had clearly been fully equipped for the Trade, she had sailed from Havana and was returning there, her owner admitted that he was a merchant resident in Havana, and all her officers except the nominal Portuguese master were Spanish. This provided adequate proof that she was entirely employed in the Spanish slave trade and should be adjudged as Spanish.

The Symondite brig-sloop *Sappho*, Commander Thomas Fraser, had visited Havana in mid-April after an unsuccessful cruise, and Judge Kennedy recommended to Fraser that he should patrol to the east of Cuba. This was on the principle that the best slavers with the most able officers might opt for the more dangerous approach to the island through the Old Bahama Channel, in order to avoid cruisers, rather than the easier southerly route past the Isle of Pines. *Sappho* sailed on 27 April, and three days later she made an arrest. Her victim, the Portuguese brig *Camoens*, had probably approached from the north through the Bahamas rather than risking the Old Bahama Channel, and the sloop fell in with her a few miles south-east of Cay Sal. Despite *Sappho*'s youthfulness it required a five-hour chase to achieve the capture. The brig had 572 slaves on board, all that remained of a massive cargo of, apparently, 802 loaded at Lagos. Fraser took his prize to Nassau, where the surviving 569 slaves were landed and the vessel

---

* *Pearl* was designed and built by a Mr Sainty at Colchester, and clearly she sailed well.
† A glance at the wind and current maps of the North Atlantic will show why Bermuda was not a diversion from the optimum route from Nassau to Sierra Leone.

was caulked in preparation for her voyage to Sierra Leone. At Freetown she was condemned by the Anglo-Portuguese court, and the slaves were emancipated.

It seems that the disgraced Commodore Peyton was not immediately replaced as Senior Officer of the Jamaica Division, but in April the vacancy was filled by Captain John Leith of the frigate *Seringapatam* (46). A few weeks later there was change also in Havana. On 10 May, the exhausted Schenley, whose departure on leave had been delayed by the lengthy *Vencedora* case, embarked in *Pearl* for England; but not before he had been in acrimonious disagreement with Judge Kennedy. It must have come as no surprise to Schenley that Lord Palmerston directed that he should not return to Havana. Instead he was exchanged with the Arbitrator at Surinam, Mr Campbell James Dalrymple.

Word reached Judge Kennedy in early May that a Russian brig had landed slaves between Havana and Matanzas, and a steam vessel had then carried her 300 slaves into Havana, landing them in broad daylight. On the same day a further 700 slaves were similarly brought into the harbour by permission, as Kennedy indignantly complained, of the "inferior authorities of the island in disregard of the treaties between Spain and Great Britain". Following this successful landing from the Russian brig, later learned to be the *Gollupk*, Kennedy heard that several Spanish vessels were on their way to Odessa, and Havana slave dealers were heard to brag that "even if they should be deprived of the Portuguese Flag, they had another reserved with which the English would not dare to interfere," apparently referring to Russian colours.[4]

The British commissioners in Havana were understandably anxious to have comprehensive information on shipping movements in Havana, and as much as possible on that in the other major ports of the island. Better placed than they were to acquire this information was the British Consul, Mr Tolmé, but he had been consistently difficult about providing it, claiming with some truculence that Kennedy's request was an "instruction" which the Judge had no right to issue. Finally, Kennedy appealed to the Foreign Office to intercede with its difficult subordinate, and pointed out that Tolmé was essentially a clerk in "the house of a Second rate Spanish Merchant", a merchant inevitably connected in some way to the slave trade. Clearly Kennedy had concluded that this employment explained the Consul's lack of cooperation.

Lord Palmerston had demanded that Madrid should take action to prevent the frauds committed by the Customs House authorities in Cuba to assist slavers, and the Spanish Government directed that certificates should not be issued to vessels carrying casks for the palm-oil trade without exacting a bond against the use of the casks for illegal purposes, such as stowing water for slaves. It was a measure of negligible value. Palmerston also deplored the failure of the Captain-General to investigate the recent alleged landings of slaves on the Cuban coast, and commented that General Espeleta seemed inclined to follow the example of his predecessor in not enforcing the Anglo-Spanish Treaty. The Spanish Government promised to call on the Captain-General for an immediate and detailed report, effectively shelving the matter.

News arrived in Havana in June of the loss of a loaded slaver on the coast of Jamaica. When the Portuguese-flagged schooner *Estela* was wrecked the crew escaped ashore, not only leaving their 300 slaves on the reef but also failing to report the disaster for several days. Consequently, by the time that a search was eventually made all the Africans had perished. Some of the crew alleged that others had stabbed slaves attempting to climb into the boats, and those accused were held by the Jamaican authorities on a charge of murder. A second tragedy unfolded in July when the Spanish brig *Esplorador* arrived in Havana. She had gone to Madagascar and Mozambique for slaves, turned to piracy, and collected a cargo of 560. On her return voyage she had encountered a violent storm which lasted for two days. The hatches had been battened down, denying food and air to the Africans, and when the wind abated it was found that 300 had died. Only 200 remained alive when the brig reached Cuba.

On 13 July Commander Fraser of *Sappho*, described by the British commissioners in Havana as "that active and meritorious officer", took the Portuguese-flagged schooner *Rosália Habanera* about 35 miles south-west of Grand Cayman. The prize, carrying 247 slaves, was in a poor state, and Fraser tried for five days to work her the 300 miles to windward to make Port Royal, but water for the slaves was almost exhausted and he was eventually obliged to run for Belize. On arrival there was smallpox among the slaves, 80 of whom had died since being shipped at New Sesters, and Fraser landed them. He then gave the slaver a partial caulking, supplied her with half-worn rope and repaired her sails, which were in rags. That enabled her to reach Port Royal, where she was declared unseaworthy. When the papers and witnesses arrived in Freetown the Anglo-Portuguese court condemned the absent prize and emancipated the 223 slaves disembarked at Belize.[5]

The sugar trade in Cuba was booming. With the decrease in production in the British West Indies the Cuban planters were expecting to benefit from an extended market, and in the season just closed in July 1838 about 100,000 boxes of sugar, of 400 lb each, had been exported, more than in any previous year. Kennedy was reliably informed that 40 new estates had recently opened on the island, and there was inevitably a great demand for labour to sustain this expansion. In response to a request from the Judge, Lieutenant Jenkin of *Romney* carried out an investigation in June to ascertain how many vessels were fitting-out in Havana for Africa. He counted 15, and was sure that there were others. Although Consul Tolmé had still provided no information, Kennedy believed that 13 slavers had sailed in June.

One of the measures increasingly being taken by the slave merchants to avoid interception of their transatlantic vessels in Cuban waters was to use Puerto Rico as a staging depot. Having been offloaded there, the slaves were then taken to Cuba, primarily Havana, in small vessels which not only were unlikely to be sighted but also could claim, if caught, that they were engaged in an internal traffic. However, as Palmerston pointed out in October, the limited power granted by the 1817 Treaty to carry slaves between Spanish possessions had been set aside by the statement in the 1835 Treaty that in all parts of the world the slave trade on the part of Spain was "henceforward totally and finally abolished". Failure to apply this clause had led to mishandling of the cases of *Vencedora* and *Vigilante*.

There was also a coastal traffic of steamboats carrying slaves from their points of arrival on the island to the places where they were required, and it seemed that Consul Tolmé was a member of a company owning one of these steamers. When challenged by Kennedy, Tolmé denied any interest or share in the company and claimed no knowledge that the vessel was being so employed. Further investigation by Kennedy revealed that Tolmé was actually a planter who had recently spent 8,000 dollars on buying slaves, and the Judge represented to Lord Palmerston that the man's interests in Cuba were not consistent with the duties of British Consul.[6]

The sloop *Comus*, Commander The Hon. P. P. Cary, arrived at Havana on 19 July from Jamaica, bringing Mr Dalrymple to take up his duties as Arbitrator. Cary had been given the further task, on the orders of the Admiralty, to reach an amicable arrangement with the Cuban authorities on the question of shore leave for the black soldiers in *Romney*. Kennedy's advice was that further appeal to the Captain-General would be fruitless. His view was that *Romney* should be removed, to allay local ill feeling, and that alternative accommodation should be sought ashore from the new Captain-General. Cary wisely agreed to leave the matter in the Judge's hands. In September Lord Palmerston indicated a softening of the Spanish Government's position on the matter and gave the British commissioners authority to make the necessary arrangements with the Captain-General for occasional landing of detachments.

An appreciable stir was caused in Havana by the arrival on 4 August from Baltimore of the new American ship *Venus* of 460 tons. She had been described in the Baltimore press as

> a noble corvette ship [...] pierced for 18 guns, built in this city on foreign account [...] She is, we learn, the sharpest clipper built vessel ever constructed here, and, according to the opinion of nautical men, must outsail anything that floats.

The *Venus* was destined for Mozambique, and had the capacity to carry 1,000 slaves. In September she cleared for Bahia to acquire Portuguese ownership. Kennedy's discussions on the affair with the American community in Havana were not encouraging. They were not prepared to accept any interference by Britain in their trade, and said that "England may as well close down the workshops in Birmingham which, they say, make the bolts and shackles, as call on America to forbid sailing of vessels equipped with them". When informed of this remark, Palmerston directed Kennedy to tell Mr Trist, the US Consul, that HM Government "will feel most sincerely obliged" for any information he could provide which would enable the law to be enforced against British citizens involved in the slave trade. Kennedy was also asked to remind Trist that, by the Treaty of Ghent, the United States and Great Britain mutually engaged to "use their utmost endeavours to promote the entire Abolition of the Slave Trade".*

The sailing of the *Venus* was only the latest development in the increasing involvement of the United States in the Cuban slave trade. During July and August seven slavers left Havana for Africa under American colours, and two of the five which sailed in October and three of the six in November wore the Stars and Stripes. When the Portuguese Consul was suspended in early August the American Consul temporarily assumed his duties, and, as Kennedy complained, efforts to prevent fraudulent use of Portuguese colours would be of little avail when the slavers could so easily procure the protection of the American flag.

Early in October tragedy befell the sloop *Ringdove*, which had been the captor of the *Vencedora* and the *Vigilante* in 1837. Commander Nixon, who had behaved well in those two unfortunate cases, was accused by one of the sloop's boys of an attempt "to commit breaches of the 29th Article of War on the person of the said Boy". Admiral Paget told the Admiralty that there was every reason to believe the accusation to be false, but Nixon clearly felt unable to face the consequences of the accusation and, on the night of 6/7 October, he shot himself in the head.

---

* In response the Commissioners eventually received from Trist a letter of 260 closely written pages.

Paget ordered that the boy should be tried by court martial on a charge of false accusation, but, in the absence of court martial records for the period, the outcome is not known. The implication from the Commander-in-Chief is that the boy was convicted, but, in his report to the Admiralty, Paget regretted that the court martial terminated very unexpectedly.[7] The circumstances can only be guessed at. Lieutenant The Hon. Keith Stewart of *Cornwallis* was promoted into *Ringdove*. Nixon left a wife and child.

There was a happier note to an episode involving boys in the Sixth Rate *Vestal*. Two of them had been injured in accidents on board, one losing a leg and the other a hand. The ship's company asked Captain Carter that they should be allowed to donate two days' pay to the boys. Admiral Paget, who was flying his flag in *Vestal* at Bermuda in October, supported the request and passed it to the Admiralty.

Over four months had passed without an arrest on the Station when, on 26 November, the Symondite brig-sloop *Wanderer*, sister of *Ringdove* and *Sappho*, took the Portuguese slaver *Escorpion* after an exciting chase to the south-west of the Isle of Pines. Commander Thomas Bushby had first sighted his quarry at 0830 the previous day. It was blowing hard from the west-south-west and *Wanderer*, under close-reefed topsails, saw a brig to windward standing to the west under similar canvas. *Wanderer* was gradually closing, and Bushby did not want to alarm the stranger by making sail, but at noon, when the two were about 80 miles south-west of the Isle of Pines, the chase hoisted American colours to a passing English barque and then tacked to the east, shook out her reefs and set topgallants. *Wanderer* tacked in pursuit and made all sail but, as the wind began to moderate, appeared to be losing ground. At 1530 Bushby trimmed the sloop using the guns and piped hammocks down, and at 1600 the chase was seen to cut away her stern-boat, a desperate measure which proved that she was not American. *Wanderer* gained on the brig only when the wind was strong, and Bushby was concerned that she would escape when the moon set.

Wanderer's topmast rigging and stays were in a poor condition, and Bushby set up runners and tackles to the mastheads as preventer backstays, which permitted him to carry whole topsails, but his jib and flying jib both split. However, he managed with difficulty to keep the chase in sight with his night-glass, and he fired several rounds during the night, three of which fell close. At dawn it was seen that *Wanderer* had gained considerably, and another shot passed through the brig's boom mainsail. That was enough, and at 0530, about 45 miles east-south-east of the Isle of Pines, the brig lowered her sails. The prize, a beautiful vessel designed to carry 1,100 slaves, had only 212 on board, having loaded 500 at Mozambique and a number of other African ports. She had been chased on the west coast of Africa by *Pylades*, *Curlew* and *Fair Rosamond*, and claimed to have escaped with ease. Her track into the Caribbean had been south of Grenada.

The prize was initially taken to Nassau, where the surviving 190 Africans were landed, and then to Port Royal. There she was judged to be unseaworthy and left at anchor while Lieutenant Lawless made his way to Freetown with one witness and the brig's papers, her master having absconded in the West Indies. Although the crew of the *Escorpion* had declared that she had sailed from Cadiz, it was believed that she was owned by the notorious slave dealer Forcade of Havana, and the Anglo-Spanish court felt able to condemn her and the slaves landed at Nassau.

Also in November, the name of Forcade came to the attention of Judge Kennedy in another context, thanks to the stupidity of Lieutenant Jenkin of *Romney*. Having heard that Jenkin had hosted a party in his ship at which Forcade and his clerk had been present, Kennedy asked Madden to tell Jenkin that such an association was "exceedingly objectionable". That conversation was overheard and reported to Jenkin, who stormed into the Judge's office and told Kennedy that he had no right to interfere in his private arrangements. He asserted that he had no idea that Forcade was a slaver and claimed that "if he were to shun all persons suspected of being engaged in the slave trade he must give up going into any society in the place". Faced with this aggressive and puerile stance, Kennedy explained to Jenkin in forthright terms that his conduct had been discreditable and damaging and that it would give the people of Havana "a poor idea of our principle of action". He said that he did not intend to take the matter further, but he then heard from Commander Milne of *Snake*, which departed Havana on 19 November, that Jenkin had foolishly written to the Commander-in-Chief to request an enquiry, and also that Jenkin had been regularly socialising with Tolmé. Kennedy's report on the affair was passed to the Admiralty by the Foreign Office, but, fortunately for Jenkin, his request seems to have been ignored.

Evidence appeared in December that Washington was concerned about the abuses of its flag. An American had taken a Spanish brig to Key West, swore an affidavit that he had purchased her, and then, with this affidavit instead of a Ship's Register, he presented himself to the US Consul at Havana. The Consul, not satisfied with the document, and learning that the vessel's crew of 18 were all Spaniards and that her destination was the coast of Africa, arranged with the Commanding Officer of the visiting US Sloop-of-War *Ontario* to hand the brig over to the Captain-General. Espeleta agreed to detain her until he received instructions from the US Government, and *Ontario* was ordered to remain at Havana in order to prevent further abuses of American colours. It was heard also that two more United States Navy sloops were being sent to the coast of Africa with the same objective. These were most encouraging signs, and Lord Palmerston sent a personal note to Mr Fox, the American Minister in London, to thank the US Government for its action, although Washington would certainly not have wished it to be supposed that it had taken these measures to please Great Britain.

The Commander-in-Chief, having shifted to *Cornwallis* on her return to Bermuda, sailed for Port Royal at the end of November. He arrived on 9 December with *Vestal*, *Racehorse* and *Ringdove* in company. There he joined Commodore Peter Douglas of *Magnificent* (72), who had been appointed Senior Officer of the Jamaica Division in October, and found the newly joined *Edinburgh* (74), *Pique*, *Modeste* and *Rover*, which had arrived the previous day. This force was gathering for an expedition into the Gulf of Mexico in support of a diplomatic mission to prevent conflict between France and Mexico, and *Madagascar*, *Seringapatam*, *Andromache* and *Snake* were yet to arrive. Paget, however, was ill and, on the advice of the Surgeon of *Cornwallis*, he delegated command of this powerful squadron to Douglas, who hoisted a First Class broad pendant in *Cornwallis*. Captain Grant of *Cornwallis* and the Commander-in-Chief shifted to *Magnificent* and remained at Jamaica. On 12 December *Pique* sailed for Vera Cruz with HM Minister Mr Pakenham, and she was followed four days later by the main body of the Squadron with the exception of *Seringapatam* and *Andromache*, which were still to join. Little was left for other tasks on the Station.

However, Lieutenant Philip Hast in the schooner *Pickle* was still on the lookout for slavers, and at 1050 on 18 December he sighted a suspicious-looking maintopsail schooner hove to about ten miles east of Santiago de Cuba. As *Pickle* closed her, the stranger stood off the land with a boat in tow, and, on boarding her at 1230, Hast found that she was the American-built *Victoria*, 42 days from Lagos and under Portuguese colours which she had apparently acquired at the Cape Verde Islands. It was clear that she had very recently landed a cargo of slaves in contravention of the Lisbon Decree. Hast, unsure of what case he might have, took her to Port Royal to await instructions from the Commander-in-Chief. It appears that he was then advised to take the matter to Freetown, but the prize, declared unseaworthy, remained in the charge of the Gunner of *Magnificent*. The Prize Master, Mr Archibald Jolly, finally arrived in Sierra Leone with papers and three witnesses in early October 1839, via England and the Cape, having been passed from cruiser to cruiser up the coast of Africa. The *Victoria* was judged to be Spanish and equipped for slaving, and she was condemned.

The British commissioners at Freetown expressed their regret that such waste and delay should be unnecessarily incurred to bring cases of this kind to Freetown when they could be prosecuted at Havana. It seemed to them either that West

Indies commanding officers failed to understand that Portuguese-flagged slavers with Spanish characteristics could be tried as Spanish in Havana, or that prizes of this sort were not being properly pursued by the Havana judges. This latter circumstance they could not believe. The record of the Havana court had in fact not been good on that score, but it had rarely been tested because the immediate reaction of captors on taking a slaver under Portuguese colours had generally been to head for Nassau to land the slaves and then to resort to the Anglo-Portuguese court at Freetown.[8]

In December the problem of recreation for the black soldiers in *Romney* was resolved, for the time being at least, by the arrival of a despatch from Lord Palmerston laying out the conditions set by the Spanish Government for the soldiers to go ashore. Only six were to be allowed out of the ship at any one time, and they were to be accompanied by a British or Spanish officer; a place for their relaxation was to be designated; and the periods for their shore-going were to be fixed. The British were to ensure that there was no contact between the soldiers and slaves ashore. However, it was not until June 1839 that the British commissioners in Havana were able to report to Palmerston that the area for the soldiers' recreation had been established. It was to be within closely defined limits on the eastern side of the harbour to half a mile inland. Dr Madden then complained to the Colonial Office not only that the area was swampy and uninhabitable but also that the soldiers were forbidden to communicate with any of the inhabitants of Cuba. However, Judge Kennedy told Palmerston in October that the complaint was without foundation.

At the end of 1838 the British commissioners in Havana reported that 71 slavers had sailed from the port during the year, and that 50 were known to have returned and landed slaves. Despite this level of traffic, not a single case had come before them in the past six months. Indeed the number of arrests had been pathetically small, but the false claims of nationality by the slave vessels operating from Havana partially explain why the Anglo-Spanish court was so little employed. Of the 71 sailing, 42 were supposedly Portuguese, 19 American, one French and one Brazilian. Only eight admitted to being Spanish. Of those returning, 44 wore Portuguese colours, one Russian, one Brazilian and only four Spanish. The commissioners also noted that an astonishing number of new estates had been established in the past two years, a happy development for the Cuban slave traders.

The slaver *Venus* caused another stir in Havana by her return in January, less than four months after her departure. She had carried 860 slaves and had cleared a profit of $200,000 on her first voyage. It came as no surprise that when she reappeared she was claiming Portuguese nationality, although she was principally owned by a Havana slave dealer and a Frenchman, and her name had been changed to *Duquesa de Braganza*.[9] In preparation for her second voyage she was rerigged as a brig, and Lieutenant Jenkin thought it likely that she would again wear US colours on her outward passage to avoid molestation by British cruisers.

Admiral Paget's health appeared to be improving, but he was then struck by yellow fever. When, in mid-January, he seemed to be regaining strength, he asked to be taken to Bermuda and was put on board HM Steam Packet *Tartarus* for the passage. The packet reached the vicinity of Bermuda on the twenty-sixth, but could not find the island. She continued on the latitude of Bermuda until the twenty-eighth, but then, with it blowing hard from the north-west and the weather threatening, she ran for St Thomas. However, at 1400 the following day the Commander-in-Chief died. His body was subsequently carried by the steam packet *Flamer* to Bermuda for burial, and his family was taken home by *Cornwallis*.

Judge Kennedy and Arbitrator Dalrymple in Havana, frustrated by the continuing lack of any significant interruption to the slave traffic, told Lord Palmerston that, if progress was to be made, a squadron of ten men-of-war should cruise off Cuba. That number was probably not wide of the mark if a worthwhile cordon was to be formed, but the commissioners may not have appreciated that if such a force level was to be permanently maintained in Cuban waters at least five or six more cruisers would be necessary to allow for absence for maintenance, replenishment and passage time. However, they suggested that, if ten was considered to be an impracticable number, there should at least be cruisers watching six key areas. These were: Cape San Antonio; off the Isle of Pines towards Trinidad de Cuba and the new and flourishing port of Xagua; off Cape Cruz and towards Santiago de Cuba; off Cape Maisí and towards Tortuga and Great Inagua; and both the eastern and western approaches to Havana. These were sensible suggestions, but there was no chance that sufficient cruisers would be deployed to fulfil them. As it was there were rarely more than two vessels patrolling off Cuba, and often even fewer.

The dispute between France and Mexico had been resolved by mid-March, and Commodore Douglas, now temporarily in command of the Station, brought the majority of his force out of the Gulf of Mexico, leaving *Satellite* and *Comus* at Vera Cruz to protect British interests. Captain William Henderson in *Edinburgh* became the Senior Officer of the Jamaica Division when Douglas sailed for Bermuda, and he, Henderson, was off Cape San Antonio, with *Andromache* and

Pique in the vicinity, when, in the early hours of 25 March, his ship grounded briefly, thanks, he claimed, to a four-and-a-half-knot current. Whether it was for this mishap or not, he was ordered to proceed to Bermuda as soon as he was relieved by Commander Barren, sent by Douglas to command *Magnificent* pending Douglas's return. On 5 May the new Commander-in-Chief, Vice-Admiral Sir Thomas Harvey KCB, arrived at Bermuda in the frigate *Inconstant* (36) and relieved Douglas in command. Douglas hauled down his First Class Commodore's pendant, rehoisted that of the Second Class, and took passage in the sloop *Modeste* to Port Royal to resume command of *Magnificent* and the Jamaica Division. *Modeste* herself was shortly to leave the Station to apply her abilities to more valuable effect on the coast of West Africa.

At last, on 4 June, the first slaver arrest of the year was made. *Pickle*, with Lieutenant Frederick Holland newly in command, was patrolling off the Isle of Pines when, at 0900, she sighted a strange schooner, with her foretopmast missing, standing in for the Bight of Pines. Holland made sail in chase, and by 1300 he was close enough to fire several shots to bring the chase to, but without success. At 1400 the schooner hoisted Portuguese colours but still refused to heave to or shorten sail, and ran directly for the Isle of Pines. In a final attempt to stop her, *Pickle* fired several rounds of grape, but at 1455 the chase ran onshore, and the crew and 20 or 30 Africans were then seen making their escape. A boarding party discovered that the prize was the *Sierra del Pilar*, and found two Spaniards and one Portuguese still on board with 176 slaves.* Holland managed to refloat the schooner and hoped to take her to Havana, but she was making too much water, and on the eighth, off Cape San Antonio, he was obliged by deteriorating weather to transfer the slaves to *Pickle*, no mean task, and to set fire to the prize, abandoning everything on board, including two of *Pickle*'s spars. The slaves, who were from Onim, were very emaciated on arrival at Havana, and three had died since capture, but by the end of their stay in *Romney* they were much recovered. Correspondence found on board the prize proved that she was Spanish, and she was condemned. A vessel was chartered to take the emancipated Africans to Grenada on 30 June, and, by chance, she was escorted on her way by *Nimrod*, on passage to Bermuda and home.

Wanderer and *Madagascar* were also homeward-bound in July, as was Lieutenant John Robinson of *Skipjack*. Robinson had commanded the schooner for three years, but she had been denied the success against the slavers that had come the way of her one confederate, *Pickle*, and had generally been achieved in

---

* The original count was 180, but only 176 were later found. It was not unlikely that an error was made in the first count, but four may subsequently have been lost overboard.

earlier years by the West Indies schooners.* He had requested to be relieved, and the Commander-in-Chief provisionally appointed in his place Lieutenant Henry Wright of *Vestal*, an appointment which was later confirmed by the Admiralty.

The second and final capture on the Station in 1839 was made in an unusual area, about 50 miles north-north-east of the western end of Puerto Rico. The victim was the Spanish schooner *Caridad Cabana*, and, being without a chronometer, it was probable that she was approaching the islands of the West Indies with no more than a vague notion of her longitude. Perhaps she was fortunate to be sighted by chance on 3 July by the Symondite brig-sloop *Snake*, Commander John Hays, which was on passage from Bermuda to Port Royal. *Snake* took her after a short chase and found 174 slaves on board, loaded at Bissau and intended for Santiago de Cuba. Hays decided to escort his prize to Havana, but the Prize Master, Lieutenant Jauncey, discovered a shortage of water and provisions, and so a diversion to Jamaica was made to land the slaves, among whom a number of smallpox cases were appearing.† The Havana court condemned the schooner and the surviving 171 slaves. Only one had died prior to capture and three thereafter.

Judge Kennedy was becoming aware that a significant number of Africans kidnapped in Sierra Leone were being taken to Cuba as slaves, and he mentioned his concern to Commander Bushby when *Wanderer* was in harbour. Bushby told him that there had been several British Africans among the *Escorpion*'s cargo whom he had landed in the Bahamas, and he was chastised by Kennedy for not reporting the fact at Nassau. The Judge believed that the crews of slavers would be subject to the jurisdiction of British courts if suspected of kidnapping Sierra Leone people, and he suggested to the Foreign Office that commanding officers should make appropriate investigations in prizes.

Dr Madden, still the Superintendent of Liberated Africans, also had a concern to bring to the attention of the Foreign Secretary. He had heard in Havana that there were those involved in the slave trade who intended to bring vexatious proceedings against British officers who detained slavers under American colours. In passing this warning to the Admiralty, Lord Palmerston gave his opinion that naval officers must "be borne harmless against Expenses of any Actions at Law brought against them for Acts done in performance of their Duty to Her

---

* There was a third schooner on the Station, *Lark*, but she was employed as a survey vessel.

† The court records claim that the slaves were landed at Port Antonio, but Hays reported that he had taken the prize to Port Royal. It is unlikely that both statements were correct, unless the prize remained at Port Royal instead of being taken to Havana, which is most improbable. Mention of Port Antonio is probably erroneous.

Majesty". This was a sentiment to be welcomed by the Navy, but whether it would carry any weight in the event of a prosecution was open to question.[10]

The British Commissary Judge at Rio de Janeiro, Sir George Jackson, had asked to have his already lengthy leave extended, but a peremptory note from Lord Palmerston told him that "no further liberality" would be allowed. In January 1838 he returned to Rio and to the duties which Mr Hesketh, the British Consul, had been performing since November 1836. Jackson's first three months were occupied solely in recording the number of slaving vessels arriving at Rio in ballast having landed their slaves elsewhere on the coast. There were 16 of them in January, February and March, and they conducted themselves with total impunity. The Government found itself powerless to act against their lawless activity in the face of the support of the slave trade by subordinate authorities.

Three months after Jackson's return, Lord Palmerston addressed a forthright letter to the British commissioners of the Rio Mixed Commission complaining of the delays which had been endemic in their activities. He emphasised that the Commission's proceedings were not dependent on the laws of Brazil but on the Treaty of 23 November 1826 under which its powers were given. The agreed regulations required that sentences should be passed as summarily as possible and normally within 20 days from the date of arrival of the prize. The excuses of holidays and of embargoes sanctioned by Brazilian law were unacceptable, and Palmerston suggested that the Rio court should adopt the Havana court's practice of waiving the observance of holidays. Delay, he wrote, was inconsistent with the intention of the treaty, highly prejudicial to the interests of the captor and extremely serious for the Africans, whose release was one of the treaty's objects. Returning to the matter of embargoes, he stressed that the procedure provided opportunity for revision of sentences and thereby admitted a principle in direct contravention of the treaty.[11]

Hard on the heels of this directive was an equally emphatic letter on the subject of slaver nationality. Palmerston wished the commissioners to understand that vessels owned by Portuguese subjects resident in Brazil found carrying slaves for sale from Africa to Brazil might properly be brought before the court of Mixed Commission and condemned under the 1826 Treaty. He pointed out that the Portuguese Government had declared that no vessel could be considered Portuguese unless she had been built in Portugal or its dominions and had borne only Portuguese colours prior to the Decree of 16 January 1837, or, if a steamer, had been purchased within three years of the decree, belonged to Portuguese citizens and was navigated in accordance with the laws of Portugal.

These criteria were to be adopted by the commissioners as the definition of Portuguese nationality.¹²

These enjoinders were still to be received when the Mixed Commission's quiet existence was interrupted by Commander Charles Eden of the ship-sloop *Rover* (18). On 11 April he detained the schooner *Flor de Loanda* which he caught, wearing Portuguese colours, about seven miles off the Marica Islands, 40 miles or so east of Rio. She was carrying 289 slaves taken from Cabinda for whom, as events transpired, the arrest was a great misfortune. Two days later, five miles off the Marica Islands, Eden detained a second slaver, a brigantine, after a short pursuit during which *Rover* had fired at the chase but failed to induce her to hoist colours or heave to. A little later, however, a boat full of men was seen to be leaving the brigantine and pulling for shore, and it appeared to the boarding party that the entire crew had deserted, leaving 207 slaves on board. No evidence was found to identify the vessel or her nationality, but a Brazilian ensign was discovered bent on.

After the arrival of the two prizes at Rio it was learned by the Mixed Commission that the brigantine was the *César*, and the court was able to accept her nationality as Brazilian and condemn her and her surviving 202 slaves as a good and lawful prize. Having concurred on that decision, the judges then disagreed on a charge of piracy, under Brazilian legislation, against two Brazilian citizens found on board after arrival at Rio. They were eventually convicted, but it is likely that they suffered little punishment. By the standards of the Rio court the *César* had been a straightforward case, but the *Flor de Loanda* was a different matter.

It had seemed in 1836 that the Brazilian Government had accepted the British definition of nationality as applied to slave vessels, the principal criterion being that the nationality of the vessel was that of her owner, which was to be taken from the country in which he resided and conducted his business. It then became clear that a change of Brazilian administration had led to a rebuttal of that principle, and false Portuguese identity was re-established as a cover for Brazilian slavers. With this disgraceful obstacle again in place, and the master of *Flor de Loanda* having presented the passport issued to him by the Governor of Loanda, the Anglo-Brazilian court decided that it had no jurisdiction in the case.

The master of the prize had made a nuisance of himself from the moment the vessel had entered the harbour. He had hailed a boat and, when he tried to communicate with it, the Prize Master told him to desist. He reacted violently, saying that he was captain of the vessel until she was condemned and would do as he pleased, so he was taken on board *Rover* and put in irons overnight. He then complained to the court not only that he had been ill-treated himself but also that the British had been maltreating the slaves and had been secretly selling six or eight of them each night. An enquiry established that all slaves were accounted for, 53 having died by that time, and the master admitted that he had never seen either a boat coming alongside or a slave being taken out of the schooner. The allegation was dismissed as an "error".

Commander Eden was determined not to allow his prize to go free, but *Rover* was under orders to sail for home, and he left Lieutenant Whaley Armitage in charge of the schooner. After the Mixed Commission had washed its hands of her on 19 June, she remained in harbour while both the Brazilian Government and the Portuguese Consul refused to take charge of her, and by 20 July the slave cargo had been depleted to 210.* On 24 August she sailed for Sierra Leone, in company with the cutter *Sparrow*, for Armitage to try his luck with the Anglo-Portuguese Mixed Commission, but four or five days later she limped back in a sinking condition. Fortunately the weather had been moderate or else she would have foundered. Armitage then set off to England, taking the slaver's owner, her master and one member of the crew as prisoners in irons.

In May 1839 Armitage arrived at Freetown in the brig *Waterwitch* and presented himself to the Anglo-Portuguese court, but he had with him neither witnesses nor papers. It had been the Foreign Office's advice to the Admiralty that the three prisoners should go to Sierra Leone for the trial of their vessel and then be sent to Lisbon to face Portuguese justice, but for one reason or another they did not reach Freetown. The Captor's Proctor strongly advised Armitage that, even if the difficulty of papers and witnesses could be overcome, he should not pursue his prosecution. The circumstances were exactly similar to those in the tragic case of *Maria da Gloria* in 1833; the vessel was clearly not Portuguese, and, the British commissioners said, had the *Flor de Loanda* been tried initially in the Anglo-Brazilian court at Freetown she would have been condemned, but the Freetown court could not overturn the decision of the Rio Mixed Commission.

When the Admiralty asked Lord Palmerston in December what should be done with *Flor de Loanda*'s unfortunate slaves he gave his opinion that they should not be left in Rio but should be taken to the nearest British territory and set free. An instruction to this effect was passed by the Admiralty to Commander Eden, who (as the Admiralty obviously should have realised) was no longer in a position to do anything about it, and on 24 April 1839 the British commissioners

---

* The refusal of Brazilians and Portuguese to take charge of the slaver does not fully explain this two-month delay. The probable answer was that he needed an escort for the decrepit slaver on her ocean passage and had to wait for *Sparrow*'s departure for England.

in Rio reported that the slaves had been landed and the schooner's hull had been sold at public auction.

Palmerston castigated Jackson and Grigg, protesting that not only by the British definition of nationality but also by the criteria specified by Portugal the *Flor de Loanda* could not be considered Portuguese. She was owned by a Rio de Janeiro resident and she was trading to that port. The *Flor de Loanda* should have been condemned as Brazilian and the slaves emancipated. This chastisement crossed with a dispatch from the commissioners, who wrote that events surrounding the schooner since the court had released her had done more to harm the British cause and "to indispose even those most favourable to the Suppression of the Traffic than any Event within our Recollection". Palmerston was furious, and demanded to know "how and why?" they had made this interpretation of the outcome.

The commissioners, clearly unabashed, replied that the damage had been caused by observation of the horrors involved in the delay between the vessel's release and the refusal of other authorities to take charge of her, and of her eventual return to Rio. There had been taunting that this was "Proof of the Benefits to Humanity resulting from [Britain's] Interference". The real causes of this sad and disgraceful episode had been the unjustifiable obstructionism of the Brazilian Government and the lack of judicial integrity on the part of the Brazilian commissioners, but the feebleness of Sir George Jackson's performance had made its contribution, and for this he expressed no contrition.[13]

Just a month after the arrest of the *César*, Lieutenant Bower in the "coffin brig" *Wizard* brought more work for the Mixed Commission and more difficulty for the Navy. On 13 May he arrested the brig *Brilhante* close inshore at the western end of Isla Grande with 250 slaves from Loanda and, of course, flying Portuguese colours. The case came to court four days later and, with Jackson perhaps showing a little more determination, the Judges disagreed on the vessel's nationality. The brig was American-built and had been sold at Rio in 1836 to a Portuguese citizen resident in Rio. By both British and Portuguese criteria she was therefore Brazilian. Lots were drawn by the Arbitrators and the lot fell to Mr Grigg, who concurred with Jackson. *Brilhante* was therefore condemned as Brazilian and her remaining 245 slaves were emancipated, but, inevitably, the owner applied for an embargo on the sentence, delaying execution until the local Brazilian court had reviewed the case.

Captain Herbert of *Calliope* was the senior officer in Rio during these proceedings and was doing his best to keep abreast of what was happening. In this he was greatly hampered because letters from the Secretary of the Mixed Commission were in Portuguese. Herbert, who had no Portuguese speaker on board, made a request to the Commission for either translation of the letters or the services of a translator, but his plea was met with a flat refusal.[14]

While the matter of *Brilhante* dragged on, the question of naval command on the coast of Brazil was resolved. Commodore Thomas Sulivan was in command on the Pacific coast in the frigate *Stag* (46) when the new Commander-in-Chief arrived at Valparaiso, and Admiral Ross ordered him to sail for Rio to take charge on the Atlantic coast. *Stag* rounded Cape Horn on the evening of 5 June, and next morning her bowsprit was found to be badly sprung where old damage had been exacerbated, probably by the resistance of the spritsail yard when striking the water while the ship was pitching heavily in the swell down the west coast. The bowsprit was fished, but, with an easterly wind and heavy swell, Sulivan abandoned his intention to visit the Falklands and made directly for Rio, which he reached on 29 June. There he found the Sixth Rate *Calliope*, the brig-sloop *Lily*, *Wizard* and *Sparrow*, and on 1 July he wrote to acknowledge Instructions from the Admiralty, dated 16 December 1837, requiring him to take command of Her Majesty's Ships and Vessels on the eastern coast of South America. His bailiwick extended from Cape Horn to Cape St Roque, and he described himself as "Commander-in-Chief", but no new independent Station was designated, and it seems unlikely that Their Lordships acknowledged that title.

Two days before the Commodore's arrival the British commissioners had despatched 19 pages in response to Lord Palmerston's April letter on embargoes. They claimed to have thought that a despatch from Palmerston in October 1836 had tacitly, if reluctantly, acquiesced in the practice of embargoes which, they explained, the Brazilian authorities were unwilling to forego. They feared that unless the procedure was adhered to the officer charged with executing a sentence would simply return it to the court. However, they understood that the Brazilian Judge had finally yielded to the persuasion of his colleague, the Arbitrator, to reject embargoes altogether and to decree final execution of sentences. If this indeed was what the Brazilian Judge had agreed he was shortly persuaded to change his mind again, and in July the commissioners reported dejectedly that the Brazilian Government was determined to persist with embargoes.

It was not long before Sulivan became embroiled in a confrontation with Judge Jackson. On his arrival he found the three prizes languishing at their anchors, *Brilhante* having been there for nearly seven weeks, and *César* and *Flor de Loanda* for over two and a half months; thanks, in the first two cases, to the routine delays of the Mixed Commission and to arguments about embargoes on the sentences. He immediately became aware of the extreme difficulty of caring for the slaves in the prizes during these gross delays. Not only disease but also kidnapping by Brazilian slave traders was a constant threat, and efforts by Mr Ouseley, the British Minister in Rio, to procure a hulk from the Brazilian Government as slave accommodation to alleviate the problem, had made no headway.

When the master, the carpenter and a passenger of the *Brilhante*, convicted of piracy and being held in *Stag*, appeared to have contracted pneumonia Sulivan sent them ashore to hospital, but they escaped. An ill-tempered correspondence between the court and the Commodore ensued, letters to Sulivan being in Portuguese. The commissioners protested that the Commodore had failed to inform them of his proposed action and that he had allowed the prisoners to escape. Sulivan retorted that he had allowed nothing of the sort, that he was not having anyone in his ship with a disease that might be communicated to his ship's company, and that the prisoners had been the responsibility of the hospital authorities. The self-righteous Jackson then complained to Lord Palmerston about a "lack of cordiality" by naval officers.

Unwisely climbing onto his high horse, Sulivan then wrote that "I have yet to learn that, as Senior Officer of Her Britannic Majesty's Ships on this Station, I am at all under the controul (*sic*) of the Mixed Commission court." He was sharply told that as soon as a capture was validated and referred to the Commission the prize and her slaves became subject to it pending adjudication. The court also pointed out that it was concerned only with Prize Masters and captors, not with him as Commander-in-Chief. Sulivan responded that, if the commissioners could prove that detained vessels were under their authority from the time of arrival, "I will most certainly be obliged to them to send proper Persons to take charge of them immediately that they are reported to them."[15]

Of course, no such "proper Persons" were forthcoming, and responsibility for the care, custody and security of prizes and slaves remained with the captors. As in most cases of separation of authority and responsibility, this situation with prizes and Mixed Commissions was deeply unsatisfactory, but it was only at Rio that it had given rise to serious difficulty for the Navy. It had not become a significant problem at Sierra Leone, where the courts were efficient and the colonial authorities supportive, or at Havana, where not only were the commissioners generally helpful but also the arrival of *Romney* had lifted the burden on captors.

Mr Ouseley had eased matters a little by acquiring a vessel to be used as a hospital ship, and, by the end of August, 32 of the *Brilhante*'s slaves had been sent to her. Six had died, and those remaining in the slaver were growing weaker. There seemed to be no end in sight for their ordeal because an impasse had been reached in the adjudication. Lord Palmerston was demanding that embargoes imposed by the Brazilians on the sentences of the Mixed Commission should, in accordance with the treaty, no longer be allowed, and the Brazilian Government was insisting that the system should continue.

The deadlock was released by a proposal by Ouseley that an embargo should be accepted in the *Brilhante* case as long as the Brazilian Government did not regard the admission as a precedent. Jackson awaited instructions and Ouseley interceded with the Brazilians. In mid-September the Brazilian Foreign Minister assured Ouseley that neither past embargoes nor admission of an embargo in this instance would form precedents, and Ouseley told Jackson to get on with the adjudication. On 1 October, five months after *Brilhante* had been arrested, the British commissioners reported to Palmerston that the slaver had been condemned, that 229 slaves had been emancipated, and that "only" 22 had died. When the prize was sold, to her original owner and for a sum undoubtedly below her true value, the proceeds failed to cover the cost of disbursements, and the court demanded that the captor should pay for the blankets issued to the slaves before emancipation.

Awaiting the Commodore on his arrival at Rio was a letter from Captain Bruce of *Imogene* forwarding a request from his petty officers, seamen and marines that they might be permitted to contribute two days' pay each for the relief of Ordinary Seaman John Milsome, who had been blown from a gun during an exercise and suffered amputation of his right arm. This request, a corporate act of generosity similar to that by the ship's company of *Vestal* at Bermuda the previous year, was forwarded to the Admiralty by Sulivan, and it seems likely that it was approved.[16]

In August the Commodore heard that there were as many as 160 vessels constantly employed in the Brazilian slave trade, 80 per cent of them belonging to Rio de Janeiro. During the past month 30 had sailed for Africa from Rio and there were 20 more fitting-out there. All were under Portuguese colours, as were the 19 vessels which had entered Rio in ballast in the four months to August, having undoubtedly landed slave cargoes. All of these slavers had undergone the usual process of examination by a Justice of the Peace, then declared to have incurred no criminality, and almost immediately released.

To contend with this enormous traffic Sulivan had five men-of-war, including his flagship which he generally needed to keep at Rio. The state of affairs at the River Plate obliged him to station two vessels there, and all he could manage against the slavers was one cruiser between Bahia and Pernambuco while another was refitting. To emphasise the inadequacy of this force level he reminded the Admiralty that the coast on which slaves were landed extended over 4° of latitude, and by that he was referring only to the shore on either side of Rio. He pleaded for "a few 10-gun brigs".

The situation had been further exacerbated by an accident to the one vessel on the northern coast, the brig *Wizard*. She had sailed for a cruise from Bahia on 17 July, but on the following afternoon, in a light breeze and a big easterly swell, she took a heavy lurch and the foretopmast went in the cap, taking with it the main topmast. As it struck the bends the main topgallant mast carried away, as did the fore and main crosstrees. Despite this crippling damage, Lieutenant Birch was able to save all the gear and to return to Bahia that night. The Brazilian Government eventually agreed to provide the few spare parts he needed, and the brig was ready to sail on 4 August for a cruise which included visits to Pernambuco and Maceió.

Captain Herbert in *Calliope*, in company with the old ship-sloop *Sparrowhawk*, Commander John Shepherd, was keeping watch on the River Plate; the Sixth Rate *Cleopatra* passed through Rio on her way to England in September; and on 17 October, with the *Brilhante* case resolved, Sulivan sailed in *Stag* for Bahia. He found *Wizard* when he arrived on the twenty-fourth, and two days later the packet *Alert* passed through on her way to Rio. On 29 October *Stag* sailed for Pernambuco and then returned to Rio. Waiting there was the ship-sloop *Electra*, which Sulivan had been ordered to send to the Pacific on the promised arrival of the ageing ship-sloop *Orestes*. However, for reasons he does not explain, the Commodore was obliged instead to send *Electra* to England. Also expected to join the east coast squadron was the Sixth Rate *Acteon*, and on her arrival *Sparrowhawk* was to be sent to join Admiral Ross in the Pacific. Sulivan therefore sent for *Sparrowhawk* in November to return from the River Plate to Rio to prepare for the passage round Cape Horn.

Also at Rio in November was the schooner packet *Spider*, Lieutenant John O'Reilly, one of whose officers unwittingly incurred embarrassment for the Commodore and Mr Ouseley. Her boat, with the officer and three of *Stag*'s "Young Gentlemen", was seized by an armed Brazilian guard boat as the occupants attempted to board the British merchantman *Morayshire*, presumably on a social visit, and the four were taken on board the guardship. Sulivan sent Lieutenant Robinson to recover them, but the Guard Officer refused to release them unless he was paid 100 Milreis.[*] Robinson declined to pay and ordered the officers into his boat. When Mr Ouseley complained to the Brazilian Foreign Minister he was told that it was against port regulations to board a merchantman in the harbour and that a fine of 100 Milreis was payable for each offender. It was, said the Minister, an act of forbearance by the Guard Officer to release them. Contrition

---

[*] Four Milreis equalled roughly £1.

was Ouseley's only avenue of retreat, and the incident gave further evidence of ill feeling between the Brazilian authorities and the Royal Navy.

In October there were two promising developments. The Brazilian Government at last accepted the British ruling on the nationality of Portuguese slaving vessels, which accorded with the Portugal's Lisbon Decree. This left the matter of embargoes as the last major stumbling block for the Rio Mixed Commission. Then there was a helpful reaction by Lord Palmerston to Sulivan's representations on the difficulties experienced in caring for slaves awaiting adjudication. He proposed to the Admiralty that a hulk should be sent to Rio as a Receiving Ship for slaves and slaver crews; in November Sir John Barrow at the Admiralty replied that a vessel would be despatched as soon as possible, and that measures would be taken for the security of the slaves as well as accommodation for the sick.[17]

The ketch *Arrow* was visiting Berkeley Sound in the Falkland Islands in November and ran aground on a sandbank in Port Pleasant. Her Commanding Officer, Lieutenant Bartholemew Sulivan, did not succeed in refloating her for three days, and finally managed to heave her off only after landing guns, stores and ballast. Thanks to perfectly smooth water the ketch was undamaged, as, so it transpired, was the younger Sulivan's career. When Commander John Monday, second-in-command of *Stag*, requested his discharge on account of very urgent family affairs, the Commodore appointed Sulivan as a commander in his place. However, before he could take up this appointment, Sulivan had to be invalided home.

Electra was well out into the Atlantic on her way home when, on 1 December, the brigantine *Diligente* was unfortunate enough to cross her path while on passage from Benguela to Rio with 302 slaves. The chase had been wearing Montevideo colours when she was first sighted, but, by the time Commander Preston arrested her about 540 miles east-south-east of Bahia, she had hoisted a Portuguese ensign. Lieutenant Heseltine brought her into Rio on the eighth, and, only two days later, the Mixed Commission, having concluded that her owner resided in Rio, condemned her and emancipated the surviving 246 slaves. The Captor's Proctor, Mr Stevenson, had been threatened with murder and withdrew from the case, and so Heseltine brought the papers into court himself without legal assistance. Thereafter HM Consul assumed the duties of Proctor.

That was not the end of the *Diligente* case, of course, because the master's Proctor immediately petitioned for an embargo on the sentence. The Brazilian Judge pressed for admission on the grounds that, until some fresh agreement was reached between the two governments, embargoes, by the laws of Brazil, could not be denied. Jackson refused to agree and the case entered a state of limbo. Fortunately the slaves were healthy.[18]

Well before it had finished with the *Diligente* case, the Mixed Commission was presented with more business. *Wizard*'s deployment to the north had been fruitless, but her return to the waters off Rio brought her some success. On 27 December, 50 miles south-south-west of Guanabara Bay, she took the schooner *Feliz* with 229 slaves. The slaver had no colours hoisted when she was boarded, but her master told Lieutenant Birch's men that she was Portuguese. The court decided, however, on the evidence of her owner's residence in Rio, that she was Brazilian, and it condemned the schooner and emancipated the 229 slaves. Then this case too grounded on the matter of an embargo.

The crew of *Diligente* were all Portuguese and were therefore not convicted of piracy under Brazilian law. However, Sulivan chose to send most of them to England so that they might face legal action of some kind. On their arrival Lord Palmerston gave his opinion that they should be tried under Portuguese law, and to Portugal they went. Lord Howard de Walden, the British Minister at Lisbon, expected that the Portuguese would refuse to accept them, although, in the event, they did so. He later reported that the Foreign Minister, Viscount de Sa da Bandeira, was very annoyed and complained (with total irrelevance) that the slaver had been taken south of the Line. "The affair", the Foreign Minister said, "would make a great noise at Lisbon and in the Cortes and would be extremely inconvenient and embarrassing to the Government." Walden replied that Britain would hardly care about that, but could

> easily understand how inconvenient it was to him and to the Government to have their sincerity in enforcing their famous law for abolition of the slave trade and punishment of those guilty of the Traffic thus brought practically and publicly before the world and put to the test.[19]

The 14 men were absolved of wrongdoing on the grounds that the Queen of Portugal's Decree of December 1836 had not been made law in Angola, the nearest Portuguese possession to the *Diligente*'s point of departure on her slaving voyage. The Viscount de Sa da Bandeira then demanded "satisfaction for the insult offered by British Cruisers in the illegal capture of the brigantine *Diligente*, and the surrender of that vessel to her owner with full Indemnity for the loss and Damage incurred". It appears that Britain simply ignored this blustering nonsense.

Difficulties with prisoners continued. The master and pilot of *Diligente* were sick, and instead of despatching them to England, Sulivan had sent them ashore to the hospital. He had so informed the Mixed Commission, as the commissioners had previously demanded, but to their irritation, his letter arrived on the day following the transfer, probably as intended by Sulivan. To make matters worse, the pilot escaped, although the master's condition improved and he was returned safely to *Stag*. When Palmerston learned of the escape of the *Brilhante* prisoners he commented that Sulivan seemed not to understand that subordinate agents of the Brazilian Government sympathised with the slave traders, that the responsibility of that Government was almost nominal, and that no remedy could be expected from it. Sulivan probably understood well enough, but that did not ease the difficulty presented by sick prisoners. He was, however, overreacting when he complained to the Admiralty that the commissioners and the Brazilian Government were "throwing every obstacle" in the way of his sending diseased men to hospital from the slavers, and that it appeared to be with the intention of preventing HM Ships from bringing in more detained vessels.[20]

It was not until October 1839 that Lord Palmerston approved the hiring of a hospital vessel by Mr Ouseley, and he directed that the hulk's expenses should be paid from the proceeds of the sale of the prizes of which the slaves and crews were being accommodated in the Receiving Ship which, even then, had not arrived. It would appear, however, that by this time the vessel hired by Ouseley was no longer available. Perhaps the hire had to be terminated in late 1838 for lack of money, and, even after Palmerston's directive, the proposed source of funds would probably have been inadequate because of the fraudulent practice in Rio of selling prizes at well below their proper value.

The British commissioners and the Foreign Office were well aware of corruption in arrangements for the sale of condemned slave vessels at Rio, a process nominally the responsibility of the Mixed Commission but conducted by a subsidiary Brazilian authority. What gave the commissioners greater concern, however, was the fate of emancipated slaves. Once supposedly freed, they were hired out as "apprentices", and it was the commissioners' opinion that in this condition, in which they had no financial value to their employers, the Africans were worse off than as slaves. Jackson and Grigg suggested that it might be possible to take them to a British possession instead, as was happening at Havana. Palmerston was probably sympathetic to the idea, but it seems to have been shelved as impracticable.

In light of this concern, an accusation that the British commissioners themselves were corruptly involved in hiring out the apprentices came as a shock. A Dr Cullen wrote to the Foreign Office in February 1839 alleging that when the

Brazilian Government advertised freed slaves for 14-year apprenticeships the only people to receive allocations were those involved in the hiring out, and that the only Englishmen to obtain Africans were those belonging to the Mixed Commission.[21] In response to Lord Palmerston's demand for an explanation, Judge Jackson admitted that he had indeed hired three emancipated slaves as servants, largely because of his sympathy for them. Palmerston replied that he did not question Jackson's kindness, but pointed out, as Jackson should surely have realised himself, that the arrangement was open to misinterpretation and might lead to abuse.

It had been necessary for the Commodore to remain in Rio in his capacity as captor of the prizes awaiting adjudication, but there was nothing more for him to do in the two outstanding cases.* So, on 15 January 1839, he sailed in *Stag* to take a look at affairs at the River Plate and the Falkland Islands. Having delayed for four days to complete with wood at St Catherine's, on the southerly coast of Brazil, he reached Montevideo on the twenty-ninth. There he found *Orestes*, Commander Peter Hambly, who complained that the authorities ashore had failed to return his gun salute on the sloop's arrival on the thirteenth. An apology was proffered and accepted.

The packet *Cockatrice* was also in the anchorage, and the Sixth Rate *Acteon*, Captain Robert Russell, had been at Buenos Aires earlier in the month.† While there, Russell had declined a request from Acting Commander Mackenzie of the USS *Fairfield* to surrender one of his quartermasters who had been recognised as a deserter, three years earlier, from the United States Navy. Mackenzie was not best pleased, but the matter was amicably resolved when Commodore Nicholson, the USN Senior Officer on the coast, acknowledged that the man was a British citizen and gave him his discharge from American service.

In Sulivan's absence there was most welcome news at Rio. On 15 January the British commissioners were able to report to London that the Brazilian Government had instructed the Brazilian commissioners not to request admission of embargoes on the sentences of *Diligente* and *Feliz*. The way became clear to conclude those two cases, and it was to be hoped that the subject of embargoes would not be raised again.

There was an interesting development a week later when Mr Ouseley, perhaps acting as unofficial locum for the Commodore, informed the commissioners that

---

* The Commodore was acting on behalf of the actual captors in order to release them from Rio during the long court delays.

† *Acteon* was completed as a survey vessel, and she was probably on the Station in that capacity. She does not appear to have been employed as a cruiser.

Captain Nias of the Sixth Rate *Herald*, while on passage to Rio, had boarded a vessel fitted for the slave trade and wearing Brazilian colours. Nias had released her in the belief that, in the absence of slaves, she was not liable to arrest. The commissioners, however, told Lord Palmerston that they considered this belief erroneous. Their understanding was that the Anglo-Brazilian Convention of 23 November 1826 gave the Mixed Commission authority to adjudge all cases of Brazilian slave trading, and they quoted the cases of *Paquete do Sul*, taken by *Satellite* in 1834, *Dois de Marco* and *Aventura*, taken by Brazilian cruisers in 1834 and 1835, and *Vencedora*, taken by *Hornet* in 1836. In all of these instances the vessels had been restored, but in every case on grounds other than the absence of slaves.

This view took the Foreign Office by surprise, and the Queen's Advocate could not agree with the commissioners' interpretation of the Convention. However, Palmerston was not going to look a gift horse in the mouth, and, acknowledging that the wording of the convention was to some degree ambiguous, he replied that if the Brazilian Government accepted that interpretation then the Queen's Advocate could see no reason why it should not be adopted. He also wrote that he felt that it was certainly in accordance with the intentions of the contracting parties, and in this he was perhaps stretching a point. By the Convention of 1826 the Brazilians had undertaken to abide by the Anglo-Portuguese Treaty of 1815 and its Additional Article of 1817, in which there was certainly no Equipment Clause, and to abolish its slave trade in 1829. Perhaps there was room for manoeuvre in the wording of this last aspect. In any event, the commissioners were instructed by Palmerston to consult with the Brazilian commissioners and the Brazilian Government to secure agreement, but if that was not forthcoming they were not to act purely on the basis of their own opinion. That instruction, however, was not despatched until the end of August, too late to have a bearing on the first relevant case.[22]

In late March *Electra* returned to the Station, and as she approached Rio at the end of her passage from England she snapped up two prizes in quick succession. The first was the brigantine *Especulador*, taken on the twenty-fifth about 110 miles east of Cape Frio with 278 slaves shipped at Anha in Benguela, and bound, undoubtedly having disposed of her cargo elsewhere on the coast, for Rio. She was wearing no colours when she was seized by Commander Preston, and the only ensign found on board was Portuguese. Commodore Sulivan consequently expected to have to send her to Sierra Leone for trial, but the Rio Mixed Commission found that the brigantine was both Brazilian-built and Brazilian property. It convicted the master, a Brazilian, of piracy, and it condemned the vessel and emancipated the surviving 268 slaves.

Two days after taking the *Especulador*, *Electra* intercepted the Portuguese-flagged brigantine *Carolina* about 100 miles south-east of Cape Frio, and she too was condemned as Brazilian, together with the 211 slaves surviving from the 214 captured. In neither of these cases was the subject of embargoes raised, but there was still controversy. The Commodore returned to Rio on 2 April after his cruise to Montevideo and to Berkeley Sound in the Falklands, and he declined to give up the *Carolina* for sale because (so he said) he wanted to use her as a hospital ship. In fact he was acting at the instigation of Mr Ouseley, who had represented to the Foreign Office the gross corruption in the sale of condemned slavers at Rio, and, wanting to prevent the *Carolina* from following the almost invariable path back to the former owner, he advised Sulivan to hold on to her until a response had arrived from London.

The Mixed Commission was entirely justified in protesting that Sulivan had no authority to prevent execution of the sentence passed on the *Carolina*; it was the court's duty to sell her after condemnation, and the Commodore caused further irritation to the commissioners by sending to England as a prisoner a Portuguese "passenger" found in the prize. Sulivan remained unmoved, and Lord Palmerston, supporting his action, wrote to the British commissioners that it "does not in the least follow that the Naval Officer should not use a Slave Ship as a Receiving Ship, or that Portuguese subjects taken on board a slaver should not be sent to Europe".[23]

The Foreign Secretary was undoubtedly wrong in law on the former point, but the representation from Ouseley may have encouraged him to take the line he did. It seems that he later accepted that he was in error; a letter from the Foreign Office to the Admiralty in August reminded that department that officers in command of Stations had no right to retain condemned vessels. It is also probable that Palmerston's inclination to side with the Commodore was reinforced by his exasperation with Judge Jackson's string of complaints and the British commissioners' failure to take a robust line on the various difficulties presented by the Brazilians. His opinion would not have been improved by a letter from Jackson in May complaining of the conduct toward him of Mr Ouseley, the man who had given him an escape route from the embargo impasse. Palmerston was also irritated by the failure of the British commissioners to send him full reports on every case adjudicated, as did the Freetown and Havana commissioners. He required of them copies of all papers presented, witness statements and arguments given by all members of the court, to enable the Foreign Office to review the proceedings.

Responsibility for the care of prisoners remained a bone of contention between the Commodore and the commissioners, and when Sulivan wrote to Jackson and Grigg to ask for arrangements to transfer to hospital several of the prisoners from *Especulador* and *Carolina* they again washed their hands of the problem. They refused even to pass the letter to their Brazilian colleagues because, they wrote, the only possible response was that the Commission "had no means at its command of complying with your request". Sulivan sent the prisoners to the hospital ship hired by Ouseley for sick slaves.

On his return to Rio on 2 April 1839, Sulivan had found awaiting him, in addition to *Electra* and her two prizes, the brand-new Symondite brig-sloop *Grecian* (16), Commander William Smyth, and the 1816 vintage, teak-built, "coffin brig" *Cameleon* (10), Lieutenant George Hunter. *Grecian* was a most valuable addition to the squadron, and she was sent out on the fourth to cruise for six weeks against slavers between Rio and 2° to the northward of Cape Frio. *Cameleon* sailed on the tenth for the Rio Grande to collect despatches, and, on the same day, *Electra* departed to join Admiral Ross in the Pacific. Clearly *Electra* had to take her prize crews with her, although she did leave a mate and a seaman in *Especulador* in the expectation of the slaver having to cross the Atlantic for adjudication, and so marines from *Stag* and some *Wizard* hands were sent to guard *Electra*'s prizes.

Grecian did not take long to demonstrate her worth. At 0600 on 7 April, about 80 miles to the east of Cape Frio, she sighted a suspicious brig about five miles to leeward.* The brig was standing in for the land on the larboard tack, and *Grecian* gradually edged down towards her. The stranger then took fright, bore up and made all sail with the wind on her larboard quarter, and Smyth crowded on sail in chase. At 0845 the sloop fired a few rounds at the brig, and she shortened sail and showed Portuguese colours. *Grecian* ran alongside her, boarded and took possession, discovering that the prize was the *Ganges*, a new vessel with 419 slaves. Smyth supplied her with water, took out her crew and sent her into Rio with a prize crew commanded by Mr Mowle, Mate. The Mixed Commission found that the slaver was owned by a Rio resident and condemned her as such, but they did not do so until early June, and during the inordinate delay 33 slaves died before emancipation.

Grecian's next encounter was less than completely satisfactory, but it began well. At 1630 on 11 April she sighted a brig and a schooner about 15 miles off Cape Frio. They were apparently communicating with each other and, both being "very rakish-looking", Commander Smyth took them to be slavers. He first tackled the brig, which was under Portuguese colours, and on boarding her he found that she was the *Seal*, with 364 slaves loaded in Loanda. Having taken possession of

---

* Sulivan reported to the Admiralty that the position was 20 miles off Cape Frio, but the longitude given by the Commissioners' report, presumably an accurate reproduction of the report to the court by Smyth, produces a distance of 80 miles.

her and sent her on her way to Rio under the command of Lieutenant Andrews, Smyth turned his attention to the schooner.

It was 1900 and very dark by the time *Grecian* closed on her intended victim, and Smyth hailed her master, ordering him to bring his vessel under the sloop's lee. There was no sign of compliance, and so a musket loaded with ball was fired ahead of the schooner. This had no apparent effect, and so Smyth ordered more muskets to be fired clear of her to make her hold her position until a boat could go alongside her. She was boarded by *Grecian*'s Second Lieutenant, who discovered that she was the innocent Danish schooner *Charlotte*. The master was very civil and offered to take the sloop's mail into Rio.

Although he made no complaint at the time, the master's perception of the incident was rather different. He had not detected *Grecian* until she was close at hand, and the next he knew was that a shot was fired at him. In a subsequent complaint to the Commodore he wrote that he ran under the sloop's stern and hove to, and that when he was ordered to follow the cruiser's movements he obeyed, but only slowly because he was short-handed. The firing continued nevertheless, although he had hoisted a lantern, and in the morning he had discovered that his main shrouds were damaged. A stupid complaint was later received by the Foreign Office from Denmark that the *Charlotte* had been visited and detained by a cruiser not holding the necessary warrant, in contravention of the Convention between Britain, France and Denmark of 26 July 1834.

It is easy to see how misunderstanding of this sort could occur at night, but there is no doubt that Smyth should have shown more caution and patience. On the African slave coast it was a reasonable presumption that a strange sail was probably a slaver, but that was not the case in the approaches to a seaport such as Rio, and *Grecian*'s approach, particularly in darkness, should have been less aggressive. It seems also (and this may have been a common naval failing) that Smyth, used to the smart sail-handling of a ship's company in a man-of-war, did not appreciate that a merchantman, likely to be short-handed, would be much slower in reacting to his orders, particularly when taken unawares at night. The incident did, however, produce a beneficial side effect; it reminded the Foreign Office that it had never made an application to Denmark for warrants for British cruisers under the 1834 Convention, and it rapidly asked the Admiralty for a list of qualifying vessels so that the omission could be rectified. In November 33 warrants were received from Denmark.

Lieutenant Andrews took the *Seal* into Rio on the 19 April, but the Mixed Commission did not pass sentence on her until 17 June. She was found to be Brazilian, and not only was the vessel condemned but also the owner, master and pilot were declared to be pirates. The slaves were emancipated, but 45 had died since capture. One African found on board declared himself to be a kidnapped Krooman, and Sulivan entered him on *Stag*'s books. This drew forth a complaint from Arbitrator Grigg that Sulivan had failed to deliver up four Africans from the *Seal*, but the other three discrepancies probably resulted from confusion between the slaves from *Seal* and *Ganges*. Four Portuguese prisoners arrested in *Ganges* and *Seal* were sent by the Commodore to England in the packet *Penguin*.

There was great resentment in Rio against these arrests, there had been assaults by the mob against the Mixed Commission court, the police and sailors, and a *Ganges* prisoner being returned on board from the court had been rescued from his escort by a large and violent crowd on the jetty. It was not safe for the men-of-war to send boats ashore, and when an officer took *Stag*'s pinnace to collect *Especulador* prisoners from the Mixed Commission, the mob, determined to release the slavers, collected on the wharf and stoned the boat. The Brazilian authorities initially sent the prisoners to the Arsenal for safety, under a strong guard, and then returned them on board in two armed launches. The Commodore was sure that the crowd had been stirred up by those involved in the slave trade and was largely composed of slavers.

A Brazilian paddle steamer had made a dangerous practice of charging into the Rio anchorage in a fashion apparently designed to alarm and provoke the cruisers and prizes lying at their anchors, and she had recently run on board the slaver *Especulador*, stoving in a boat and tearing another from her stern.[*] On the night of 21 April her reckless behaviour caused a most unpleasant incident which further poisoned the already unfriendly relationship between the Royal Navy in Rio and the Brazilian authorities. The Commodore was watching from a quarter-gallery in *Stag* as the steamer narrowly cleared the launch secured to the frigate's stern and headed directly for the starboard bow of the slaver *Ganges*, anchored a short distance away. Following the recent prisoner escape ashore, Sulivan was convinced that this was an attempt to ram the prize and seize slaves and prisoners, and the conviction was reinforced when the steamer stopped her paddles but did not go astern. He hailed Mr Mowle, who was in charge of the *Ganges*, telling him to be ready to open fire when he, Sulivan, gave the order, as much to warn off the Brazilian as to alert Mowle to the approaching danger. A few moments thereafter he was astonished to hear a shot.

Mowle, conscious that previous attempts had been made against slaver prizes, had already been aware of the immediate threat. At the front of his mind was

---

[*] A Foreign Office letter gives the name of the steamer as *Especuladora*, but there must be a suspicion that there is confusion here with the name of the slaver prize.

also an instruction from the Commodore that musket fire was to be used only as a last resort. He decided that this point had arrived, and he ordered his Royal Marine sentry to fire over the steamer. At the first attempt the weapon flashed in the pan, at which the steamer's master laughed and clapped his hands. The sentry quickly reprimed his musket and fired. Unfortunately he appeared, in his haste, to have forgotten Mowle's order, and he fired into the steamer and hit and killed a passenger.[24]

The Brazilian authorities immediately complained to Ouseley who, describing the incident as "a wanton act of barbarity", demanded that the authorities be allowed to conduct enquiries on board *Stag*. Sulivan replied that he would give assistance in the *Ganges* to the Brazilians appointed by their Government to enquire into the incident, and would allow the British Judge-Conservator to conduct an enquiry, also in the prize, but he could not allow any such proceedings in *Stag* without authorisation from the Admiralty. He also forwarded to Ouseley statements from those involved.

Enquiries ashore revealed that the master of the steamer had bet with a passenger that he would go so close to the prize that a person could step on board her, and he had been heard to boast that he "would frighten some of these slavers and break them adrift". Furthermore, two merchants of the city were able to swear that the master had been bribed to run down the *Ganges*. The furore gradually died down, and Ouseley, at great length and with admirable diplomacy (as an approving Palmerston described it), explained to Brazilian Minister Oliveira that warships were entirely within their rights to use force to defend themselves in a foreign harbour, and that the right extended to their prizes. Palmerston also approved of the Commodore's decision not to hand over the sentry to the Brazilians, and he agreed with Ouseley that the greatest blame for the incident attached to the steamer's master, who had steered his vessel "in so unwarrantable a manner against the *Ganges*".[25]

The difficulties caused to the Navy by the repeated refusal of the Brazilian Government and the Portuguese authorities in Rio to take charge of slaver prisoners, and of the Mixed Commission to take responsibility for the slaves, continued to plague the Commodore and divert his resources during the disgracefully lengthy deliberations of the court. At the beginning of June he made another plea to the Admiralty for a vessel to be sent out as a depot for slaves awaiting adjudication, together with a complement of guards. Meanwhile, every opportunity was taken by Sulivan, with Ouseley's support, to send to England the Portuguese crew taken out of prizes so that they could be submitted to whatever justice the Portuguese Government was prepared to mete out.

Between late March and mid-September Sulivan had to provide the following from *Stag*: the Prize Masters, guards for prisoners, carers for slaves, and Royal Marine sentries for never less than two (and at times four) prizes and the hospital hulk, as well as ensuring the security of the flagship herself. The people so employed were not always reliable in their tedious tasks, and when seven prisoners escaped from the hospital ship on 2 May it was found that Mr Johnson, Master's Mate, who was in charge of the hulk, had not been on board at the time. He was sent home and subsequently discharged, with "an expression of Their Lordships' disapprobation", for neglect of duty.*

One of the prisoners, a member of *Carolina*'s crew, was identified by three men from *Wizard* as having been on board the condemned prize *Brilhante*, and as one of those who had escaped after having been convicted of piracy. The allegation was examined by the Mixed Commission, but, despite the sworn evidence of identity from *Wizard*'s marine and two seamen, no action was taken. The prisoner later confirmed his guilt by escaping from the hospital hulk.

An encounter on the evening of 7 May gave further evidence of the antipathy of Brazilians towards the Royal Navy. Extraordinarily, it involved Brazilian soldiers. Captain Eliott and Commander Preston, in uniform, were crossing Palace Square on horseback in the company of an English civilian when they were confronted and attacked by an officer and three men armed with muskets and fixed bayonets. The sailors' robust response drove off the attackers, and the Brazilian authorities expressed regret and promised an enquiry, but clearly the streets of Rio were hostile territory for the Navy.

Rio Bay was not much better. The Commodore complained in mid-May that rarely did a night pass without an attempt being made on a slaver prize. Boats frequently prowled around the detained vessels in the anchorage, and were hailed to keep away on pain of arrest. At about 0300 on 14 May, however, a boat went alongside the *Ganges*, despite warning hails, and a man was in the act of climbing the side when a sentry fired at him. The boat immediately pulled away, leaving the wounded man clinging to the side of the prize. He was hauled on board and taken to *Stag*, where he admitted that the boat had come to steal slaves. He was a slave himself, owned by a passenger in the slaver *Especulador* who was then a prisoner. The other occupants of the boat were white Portuguese.[26]

In a conversation with the Commanding Officer of *Grecian*, Judge Jackson found that he was as puzzled as was Commander Smyth by the legal implications of an incident during the sloop's previous cruise. Smyth had stopped and boarded a small vessel under Brazilian colours but commanded by an Englishman. There

---

* Johnson rejoined the Navy and served, finally, as Acting Master of *Pantaloon* until 1848.

were 70 or 80 Africans on board, but they were not new slaves being imported. They already belonged to an Englishmen named Platt, a resident of Rio, and they were being taken to his estate along the coast. The vessel was not therefore engaged in the slave trade, and there seemed to be no grounds under either the Anglo-Brazilian Treaty, or under British slave trade abolition legislation, on which the vessel could be arrested. Equally, it seemed that the Englishmen involved had evaded the British slavery abolition law of 1833 by conducting their business on Brazilian territory and in a Brazilian registered vessel. Smyth clearly felt that there ought to have been some legal justification for bringing these Englishmen to justice, but Jackson could not help him to find one, and his reluctant decision to allow the vessel to proceed seems to have been correct.

Grecian sailed for another cruise on 25 May, and only five days later she reappeared with two prizes at heel. They were the schooner *Recuperador*, taken on the evening of 28 May about 30 miles south-west of Cape Frio, bound from Rio to Benguela and Angola; and the barque *Maria Carlota*, seized the following day, 12 miles or so south-east of the entrance to Rio Bay, on passage from Rio to Quelimane. Both were under Portuguese colours, and both were fitted for slaving. They were also carrying, between them, 250 barrels of gunpowder and 17 cases of muskets not included in their cargo manifests. These were the first arrests of slavers outward bound from Rio, and the British commissioners' belief that the Anglo-Brazilian Treaty conveyed authority for condemnation of vessels fitted for slaving was about to be tested.

The Brazilian Judge would not agree with Jackson that the Mixed Commission was empowered to adjudicate the two prizes. At the end of July, however, his Government directed him to proceed to adjudicate these two cases, but it did not accept the principle of liability to detention and condemnation of Brazilian vessels fitted for the slave trade but without slaves on board. It was Lord Palmerston's wish that if this dubious principle remained in dispute the British commissioners should not act on the basis solely of their own opinion, but his instructions on the matter had yet to reach Rio. The cases accordingly went ahead.

As far as the *Maria Carlota* was concerned, the one point of agreement was that the vessel was Brazilian, but the Brazilian Judge continued to protest that the seizure was illegal. His opposition was justifiably founded on Articles 5 and 6 of the Anglo-Portuguese Convention of 28 July 1817, which was embraced in the Anglo-Brazilian Agreement of 23 November 1826 and stated "detention of vessels suspected of carrying on the Illicit Traffic in Slaves can only take place in the Sole Case of there being found Slaves onboard." The counterargument employed by Jackson is not recorded, but it must have been grounded on the Brazilian undertaking, in the 1826 Treaty, to abolish its slave trade in 1829. Eventually lots were drawn for arbitration; the final decision fell to Grigg, who supported Jackson. The vessel was consequently condemned on 13 September, and seven Brazilian members of her crew were declared to be pirates. Palmerston, notwithstanding his reservations about the liability to condemnation of such vessels, approved of the outcome, but he was less than pleased by the result of the *Recuperador* trial.

The *Recuperador* had been seized by *Wizard* in December 1838 under the name of *Feliz*, and, having been condemned, had been sold back to slavers, as was usually the case at Rio, and retained the same master. All of the slave equipment carried by *Feliz* had remained in the vessel and was found on board by *Grecian*. This was the evidence on which Smyth arrested her, but the new owner claimed that the equipment was merely being used as ballast and would have been put up for sale on the coast of Africa. It was hardly surprising that the Brazilian Judge accepted this specious nonsense, but it was astonishing that Jackson should have agreed that the presence of the gear was not proof of an intention to engage in the slave trade. *Recuperador* was consequently restored to her owner. Palmerston was predictably angry. As he saw it, the verdict was entirely at variance with that passed on the *Maria Carlota*, and the slaver's defence was "a weak and shallow pretence" inconsistent with the facts. He saw the grave danger that if the case was used as a precedent it would give immunity to every vessel charged with being equipped for the Trade.[27]

During a search of the *Maria Carlota* a "most diabolical correspondence" between the Rio slave merchants and their agents on the coast of Africa was discovered. Letters to the agents instructed them to leave a cask of poisoned wine on the deck of every slave vessel and to poison some of her water, in order to kill the prize crew if she were to be arrested. Sulivan asked the Admiralty to warn the commanding officers on the coast of Africa.

At the end of May *Cameleon* departed again for the northern ports after a break in Rio, and on 3 June *Grecian* sailed for another cruise between the entrance of Rio harbour and Campos. *Wizard*, which had been keeping an eye on Bahia in the absence of *Cameleon*, arrived at Rio on the fourteenth, as did *Orestes* the following day from a cruise. Towards the end of June *Wizard* was sent to patrol against slavers between Cape Frio and St Sebastian Island, and *Orestes* to cruise off Cape Frio. For once there was a useful force level in the approaches to Rio.

There was then a setback in early July. *Wizard* anchored in Port Cabo Frio to cut fuel, and in attempting to leave the confined harbour she missed stays and struck a submerged rock, damaging her rudder. In the process of refloating the brig, Lieutenant Birch was obliged to slip both bower anchors and a kedge anchor.

His own boat managed to recover the kedge and its cable, and the keeper of the lighthouse kindly despatched two boats to help, although one needed repair by *Wizard*'s men to keep it afloat. While the bower anchors and cables were being recovered, the Carpenter made temporary repairs on the rudder. The brig was not making any water, and she returned to Rio without difficulty, but the Commodore decided to send her to Plymouth for rectification of her rudder damage and to have other defects made good.

Grecian too was in the wars in mid-July when she sprung her foreyard and main gaff, and, further afield, *Acteon* experienced a strong gale and heavy seas off Montevideo and lost one of her cutters. *Acteon* and *Calliope* were still keeping an eye on the troubled region of the River Plate where, earlier in the year, war had been declared in Montevideo against the Argentine Republic.

On 6 July *Calliope* rendered assistance to the American merchant barque *Richard*, which was wrecked that night. Commodore John Nicholson of USS *Independence* wrote to Sulivan that the services of *Calliope*'s men would "ever be remembered with gratitude, and should an opportunity occur, the officers and crews of the American National Vessels will feel proud to emulate their gallant conduct and reciprocate their generous kindness". Nicholson was grateful to Captain Herbert personally for "the flattering manner in which he mentions the Navy of the United States in his correspondence", and he told Sulivan that he intended to forward this exchange of letters to his Government.[28]

Sulivan intended that, when repaired, *Grecian* should pay a visit to the Falklands, despite a plea from Ouseley that she should continue to operate against the slave trade. The Commodore pointed out that he had many tasks on the Station, and that he had to attend to all of them as best he could. There was reinforcement at hand, however, thanks to Ouseley. Sulivan had changed his mind about the *Carolina*'s future, and had made her available for sale, although, to the annoyance of Judge Jackson, he did not hand her to the Mixed Commission to be subjected to its corrupt sale procedure. In order to prevent her falling into the hands of her previous owner, she was bought by Mr Ouseley for about £800, a fraction of her true value.* Consequently the *Ganges* and the *Seal* were priced considerably higher than would have been usual to discourage Ouseley from buying them too. The new owner of this particularly fine brigantine immediately suggested to the Commodore that he should take her as a tender.

---

* It is not clear whether Ouseley bought the *Carolina* with his own or public money. The proceeds of the sale went to the Mixed Commission, in accordance with the treaty rules, to be divided between the two governments.

Sulivan accepted the proposal with alacrity, and on 27 July the ex-*Carolina* sailed under British colours and the name of *Fawn* on a cruise against her erstwhile confederates. She had been given a ship's company of about 40 from *Stag* and was probably armed with two 24-pounders and one 32-pounder.* When this news eventually reached the Admiralty, the Lords Commissioners were not at all sure that it was right and proper for Sulivan to have a tender, and a secretary's note informed the Board that no regulation on the matter could be found in the Admiralty.[29] The precedents from the successful use of tenders on the coast of Africa had obviously been forgotten, and no policy had been developed as a result of that experience. It is clear, however, that Their Lordships had lost none of their earlier doubts about the idea.

Lieutenant John Tyssen, the Commanding Officer of *Fawn*, was given treaty Instructions by the Commodore, and Mr Ouseley put his name to the papers to convey the added authority of the British Minister. The question then remained of whether the Mixed Commission would accept the tender's status as a cruiser. The matter was put to the test when, on 28 August, she caught the brig *Pompeo* under Portuguese colours about 35 miles south of the mouth of Guanabara Bay after a three-and-a-half-hour chase. The prize had sailed only that day from Rio for Mozambique, and Tyssen sent her in to be charged with being fitted for slaving. A barque had sailed at the same time as the *Pompeo*, and *Fawn* chased her too, but she managed to gain the protection of the guns of the fort outside Rio before she was caught. Nevertheless, the capture of *Pompeo* discouraged her to the extent that she ran back into the harbour and was unloaded and dismantled. Sulivan later heard that this success against outward-bound vessels had achieved more towards suppression of the Trade than any other measure since his arrival.

When the Commodore took the *Pompeo*'s papers into court the commissioners immediately deduced that there were irregularities in the case. To begin with, there had been a delay of ten days while the detained brig was unloaded and reloaded in order to search her for slaving gear, without the owner being present, a procedure which contravened prize law. Slaving equipment had been found, and the vessel was clearly Brazilian, so the case rested on two points. The first was whether the authority of the Commodore, backed by that of HM Representative in Rio, was sufficient to make *Fawn* duly authorised as a tender. The second was whether the arrest took place so close to *Stag* that the tender might be considered, in making the capture, to be acting under the control of *Stag*.

It appears that the prosecution failed on the latter point, which rendered

---

* It seems that *Fawn* was added to the *Navy List* in 1840, and the figures on men and guns, about which some doubt remains, are taken from Lyon's *The Sailing Navy List*.

Fawn virtually useless, and the former point seems not to have been tested. The brig was declared illegally detained, and both vessel and cargo were restored to the owner. Then arose the question of indemnity. Jackson opposed it because the vessel had clearly been engaged in slaving and had escaped condemnation only through irregularities in capture. The Brazilian Judge considered that indemnity should be allowed, and he was supported by his Government. The decision went to arbitration, and the lot fell to the Brazilian Arbitrator, who agreed with his Judge. So indemnity was allowed against Sulivan.*[30]

The Commodore was understandably furious. He wrote to the Secretary of the Admiralty that "A Manifest Injustice has been committed towards myself, my officers and Ship's Company on the part of the British and Brazilian Commissary Judges." He protested against the verdict, and demanded that the circumstances be brought to the attention of HM Government. In this instance, however, Jackson and Grigg could hardly be blamed. The fault lay with the unreasonable guidance to the Mixed Commissions on the use of tenders, apparently originating in the Foreign Office, and, on the matter of indemnity, with the hostility of the Brazilian commissioners and Government. Lord Palmerston made no comment on the outcome of the case, remarking in November, apparently with satisfaction, that the *Pompeo* was "a remarkably fast-sailing vessel" and supposed to be able to escape any cruiser.

In a letter to the Admiralty, Palmerston did, however, mention two points which gave encouragement for *Fawn*'s future employment. He noted, first, that the slaver *Carolina* had been bought into the Service, and, second, that the tender had been provided "with the necessary papers". From this it might be supposed that *Fawn* had been accorded similar status to that enjoyed by *Black Joke*, although whether the Admiralty concurred in this position is unclear. Meanwhile, the Commodore decided to make less controversial use of his tender after she returned to Rio on 10 September from a cruise, and, with Lieutenant Robertson in command, she sailed ten days later to seek news of the state of affairs on the Rio Grande.

Late in September, Lieutenant Hunter in *Cameleon*, arriving at Pernambuco from Rio, was told that some slavers from Africa were expected shortly in the offing. He sailed to watch for them, but as he again approached the harbour on 8 October, empty-handed, the British Consulate tried to communicate with him to tell him the whereabouts of two slavers in the vicinity, one of which was landing a cargo at an inlet. It was unable to pass on the message, and the frustrated Consul asked the Foreign Office that he should be given a "set of Marriott's signals" to

---

* The amount of this indemnity is not known.

enable him to convey such intelligence to cruisers, pointing out that the consulate had an extensive and uninterrupted view of the coast south from Olinda.³¹

On 20 September *Imogene* returned to Rio after a spell in the Pacific, and on the following day the brig-sloop *Clio* (16), Commander Stephen Fremantle, arrived from England. *Grecian*, meanwhile, was making her way back from the Falklands via the River Plate, and she reached Rio on 6 October. Three days after that *Clio* sailed for a cruise against slavers, and *Grecian* was then obliged to retrace her steps to the River Plate. She had brought for the Commodore a letter from Captain Herbert of *Calliope*, still the guardship at Montevideo, representing the need to keep "a respectable naval force" in the River Plate where the French were maintaining a blockade of the ports, and a confrontation between the French brig-of-war *Sylphe* and the packet *Cockatrice* gave emphasis to his advice. So, on 17 October *Grecian* headed south to join Herbert. Admiral Ross had instructed the Commodore to send *Calliope* to the Pacific unless affairs in the River Plate made it essential that a ship of her strength should remain there. Sulivan decided to retain her until the expected arrival of the Sixth Rate *Curaçao*, and to send *Orestes* to the Pacific instead once her refit, which had been delayed by the inability of the caulkers to work during the wet weather, had been completed.*

The Commodore directed Commander Smyth to make use of *Grecian*'s passage to the Plate to cruise against slavers, and shortly after leaving Rio on 17 October Smyth decided to run between Isla Grande and the mainland to examine the inlets and creeks off the enclosed Marambaya Bay. No sooner had the sloop entered the eastern channel than she was rewarded with the sight of a brig standing out with a boat in tow. Smyth hauled up towards her, hoisted his colours and fired a gun. The brig immediately showed Portuguese colours, but the two were too close to land to lie to with safety, and *Grecian* shortened sail to allow the brig to clear Punta Castillanos. Smyth then hailed her and sent across an officer, who signalled that he thought her a slaver. Smyth boarded her himself and learned that she was the *Dom João de Castro* from Mozambique to (her master claimed) Montevideo. She had been in the bay for two days, and the master could offer no answer to the question of why she was 700 miles off her supposed track.

The brig was empty of slaves, but she was carrying slaving equipment, there was a bulkhead between male and female slave decks, she had a farina room and many large water casks, and the smell of slaves was as strong as if they had still been on board. Smyth sent three officers to make a detailed inspection and report, and they had no doubt that she had landed a cargo of slaves during the past two days. Clearly she had no intention of making for Montevideo. Smyth sent her

---
* *Curaçao* had been cut down in 1831 from a 36-gun Fifth Rate.

to Rio under the command of Mr Mowle, and, as she waited, inexplicably, into December for adjudication, Mowle again found himself in trouble. The slaver's crew had been transferred to the hospital hulk from which five of them had contrived to escape, and the Mixed Commission summoned the unfortunate Mowle, who had responsibility only for the prize, to explain the circumstances.

The case of *Dom João de Castro* dragged on into 1840, to the increasing irritation of Lord Palmerston, and even the British commissioners were exasperated by the repeated delays. They complained in November to Palmerston about "impediments thrown in our way by the Brazilian Authorities", the "System of Procrastination which pervades every Branch of the Public Services", and the "want of Authority, on the part of the Commission, to enforce its own orders, or to carry its Sentences into Execution". They were, however, about to receive encouragement. The Brazilian Judge continued to protest about the adjudication of vessels without slaves, but in mid-December he admitted that the principle was by then so far settled that he would not henceforth hesitate to proceed to adjudication of such a case. Finally, on 28 January 1840, *Dom João de Castro* was condemned as Brazilian property.[32]

The Foreign Office had pre-empted this admission by the Brazilian Judge. In a letter of 23 November it told the Admiralty that the Mixed Commission courts at Freetown and Rio de Janeiro had come to the conclusion that, in accordance with the spirit and content of the Convention of 1826 between Great Britain and Brazil, which had declared that from three years after ratification slave trading by Brazilian subjects would be unlawful, and that the offence would thenceforth be regarded as piracy, Brazilian vessels fitted for slaving and proceeding from Brazilian ports on slave-trading voyages on Brazilian account were liable to condemnation. It requested that British cruisers should be so instructed.

The year 1839 ended with another piece of good news for the cruisers on the Brazilian coast, although it would be some while before they would hear of it. As long ago as October 1838 Lord Palmerston had proposed to the Admiralty that a hulk should be sent to Rio as a Receiving Ship for captured slaves and crews pending adjudication, but there had been no apparent action on the suggestion despite pleas from Sulivan and Ouseley. The vessel hired by Ouseley as a hospital hulk had provided only a partial and temporary solution, and it had been at the expense of the captors. On 18 December, however, the Foreign Office was informed by the Admiralty that the frigate *Crescent* was being fitted for the reception of liberated Africans, and would shortly be ready to sail for Rio.*[33]

---

* This was the frigate launched in 1810, not the command of Captain James Saumarez when he took the French *Réunion* off Cherbourg in 1793.

* * *

The lamentable rate of success off the coasts of Cuba and Brazil was bearing no relation to the vigorous slave traffic north and south of the Equator. The reasons were broadly the same in both theatres, and the primary blame lay with the crippling shortage of cruisers. Of the invaluable flotilla of schooners on the North America and West Indies Station only two remained, and the Commander-in-Chief had a myriad of tasks for his sloops and Sixth Rates other than cruising against the slavers. On the eastern shore of South America Commodore Sulivan was never allocated more than eight vessels at any one time (and generally fewer) to support British interests over more than 3,000 miles of virtually lawless shore, as well as the Falkland Islands, and including the frequently troubled estuary of the River Plate. It was rare that he could task more than two of his cruisers against the Brazilian slave traffic.

Inadequate command arrangements exacerbated the difficulties. North of the Line, successive commanders-in-chief occasionally visited Jamaica, but it was further to the north that their presence was more pressingly required, and the flagship of the North America and West Indies Station was usually to be found at Bermuda or Halifax. Senior Officers were appointed to the southern Division of the Station, based at Port Royal, but there was little continuity in the appointment and it carried scant authority for deploying the Station's resources. Arrangements on the coast of Brazil were scarcely more satisfactory. Commodore Sulivan was more independent than the Senior Officer at Port Royal, but he could not exert the weight of a commander-in-chief of flag rank, and, once the Pacific Station had been established, the Brazilian coast was remote from the support of a higher authority.

It is probable that the ineffectiveness of the treaties in the face of the fraudulent use of Portuguese colours by Spanish and Brazilian vessels fitted for slaving discouraged the deployment of greater numbers of cruisers to the Cuban and Brazilian coasts. Neither the Admiralty nor the Commander-in-Chief North America and West Indies would have wanted to waste the time of precious men-of-war on what they might, not unreasonably, have perceived by now to be a hopeless task.

Time was wasted in both theatres. Cruisers detaining suspected slavers under Portuguese colours in the approaches to Cuba invariably took or sent them to the Bahamas to land slave cargoes and then despatched them across the Atlantic for adjudication in the Anglo-Portuguese court at Freetown. In doing so they removed themselves from their proper station for weeks and lost their Prize Masters and prize crews for many months. This problem was not helped by the unwillingness of commanding officers to trust the Havana Mixed Court of Justice to detect and condemn Spanish slavers falsely claiming Portuguese identity.

At least the Havana court was generally supportive in its attitude to the Navy's task. That could not be said of the Mixed Commission at Rio. Judge Jackson seems to have been permanently at loggerheads with the Commodore, and the greater blame for that lay with Jackson, who leaves the impression of a weak but self-important figure. For his part, Sulivan appears to have made little allowance for the difficulties the British commissioners experienced with their Brazilian colleagues, and tended to be aggressive in his relationship with the court. That conceded, the appalling delays in the proceedings of the Rio Mixed Commission, and the acceptance by Jackson and Grigg of the embargoes imposed by the Brazilians in contravention of the Anglo-Brazilian Treaty reflect badly on the British commissioners. It is hardly surprising that the Navy regarded the Mixed Commission as obstructive and pro-slaver.

The Cuban administration remained unmistakably hostile to the suppression campaign, although the Spanish commissioners seem to have been admirably impartial. Attitudes in Brazil were more ambiguous. Abolitionists in the Government held sway at times, but, for reasons of financial self-interest, if nothing else, local authorities along the coast ensured that there was no obstruction to slaving operations. Almost all, however, were united in their resentment against the British for their determination to impose abolition on Brazil.

The only steps forward in the campaign in the western Atlantic had been two important decisions by the Rio Mixed Commission. The first was the belated acceptance of the residency criterion for slaver national identity, thereby dismissing the defence of false Portuguese colours. The other, taken with remarkably little political resistance despite its questionable legality, and largely on the initiative of the British commissioners, was to prosecute Brazilian vessels fitted for slaving.

Otherwise, the scene was bleak. Persistence in the measures employed for the past two decades and more were clearly achieving little or nothing against the slave trade as a whole, and the passing of the extraordinary "Palmerston Act" against the Portuguese showed that Britain's patience with the slaving nations was virtually exhausted. It remained to be seen whether that Act would bring any benefit, or whether other action might be attempted against the market end of the slave traffic.

PART THREE

Conclusion

SUMMARY

Taking Stock

> This is not the end. It is not even the beginning of the end.
> But it is, perhaps, the end of the beginning.
>
> WINSTON CHURCHILL

ALTHOUGH ITS WEST INDIES sugar islands were in decline by the beginning of the nineteenth century, it would have been in Britain's economic interests to maintain its leading position in the Atlantic slave trade. With its maritime dominance and mercantile strength, the country was well positioned to do so. As Spain, Portugal, the Netherlands and France were to demonstrate in the coming decades, the market for slaves in the Americas was likely to increase with the ending of the Napoleonic War, and large profits were to be made by those meeting the demand. Great Britain chose, however, to turn its back on that opportunity and, the heavy burden of a war of national survival notwithstanding, to direct its efforts towards destruction of the slave traffic.

It is scarcely surprising that its enemies and competitors at the time, and its detractors in a later, more cynical, age should attribute Britain's stance to self-interest. The country's success and power, and a not entirely unfair perception of its arrogance, were unlikely to win any friends or credit. However, it is difficult to detect any potential advantage to Britain in its decision to abolish its slave trade and to pursue a subsequent campaign of suppression. In the long political struggle leading to the Abolition Act of 1807, the arguments presented by the abolitionists were entirely humanitarian. It was those opposed to them, primarily the West Indies plantation owners and others with commercial interests in the slave trade, who presented arguments of economic advantage in continuation of the Trade. It was the gradual transfer of political power from the West Indies interest to the emerging industrialists which allowed the slave-trade abolitionists to win the day. Unpalatable though it may be to some, the truth is that Britain's abolition legislation and subsequent suppression campaign were motivated by altruism.

It seems most unlikely that the Government conducted any analysis to discern the optimum method for suppressing the newly illegal Trade. The necessary course of action would have been obvious to ministers and their advisors, and consideration of possible alternatives would have appeared to them to be superfluous.

It was, of course, the business of the Royal Navy to protect British trade at sea and to disrupt and destroy that of Britain's enemies. Slavers were outlaws who carried their proscribed merchandise by sea; prevention of this unlawful traffic, therefore, was simply just another task for the Navy. No doubt the Admiralty was given the opportunity to contribute to such debate as there might have been, but the attention that the Admiralty Board gave to the matter would necessarily have been brief, and there was no Naval Staff to assess the implications of the proposal. However, there were already cruisers on the coast of Africa to protect British trade and to prevent any interference by the French with British settlements, and their duties could readily encompass the new task. In all probability, Their Lordships would have concluded that the necessary deployment was already in place and that no further action was called for on their part.

Enforcement of its domestic abolition legislation was Britain's initial objective, and it was probably hoped, or even expected, that its example would lead to international abolition. This first stage proved not to be difficult. British merchants who had been engaged in the shipment of slaves, being generally law-abiding citizens, moved their business to alternative commodities when slave-trade abolition was enacted. It was against the residual seaborne traffic in the hands of renegade British subjects that policing action was necessary, and faced with a Navy which was currently squeezing the life out of the trade of Britain's continental enemies, the few slavers sailing under British colours would (it could safely be assumed) soon be eradicated. Indeed they were, but if there was a presumption that other nations engaged in the slave trade would follow Britain's lead it was doomed to disappointment.

The raging of the worldwide conflict against Napoleonic France and its allies was distorting all aspects of international trade, and would have prevented any accurate assessment by the British Government of the state of the slave trade, although Lord Castlereagh attempted at the outset to discover how matters stood on the northern slave coast of Africa. A lack of a clear appreciation of the suppression task, combined with the circumstances of maritime war, led to a lack of focus in early preventive action at sea. This was apparent in the absence of distinct delineation between the policing of the slave trade and the exercise of combatant rights against neutral vessels carrying contraband cargoes. Confusion would have been alleviated by precise instructions from the Admiralty, but in this, as in other operational and diplomatic matters, commanding officers were expected to act upon their own initiative within the broad constraint of their orders. On the subject of the slave trade in the early years of suppression, these orders were so broad as to be almost worthless.

At that juncture, Royal Navy officers at sea had spent much, if not all, of their professional lives at war, and although they had of necessity a sound grasp of the rules of war and prize law, the concept of policing action in accordance with the strict dictates of domestic law or international treaty was novel to them. They mostly had firm, if British and Christian, ideas of what was right and wrong, and they appear to have adhered to notions of "natural law" and "the Law of Nations". Neither of these concepts of legitimacy seems ever to have been defined, but, as Foreign Office communications show, the Navy was not alone in giving credence to them. Upon this basis, and knowing that not only Great Britain but also the United States of America had legislated against the slave trade, commanding officers perceived that they had a right and a duty to confront not only those contravening British anti-slaving law but also (mistakenly, of course) to enforce American legislation.

Naturally, slavers under the colours of Britain's wartime enemies could legitimately be seized, as could neutral vessels carrying war contraband, which, stretching a point, might include slaves. Then, after the signing of the unsatisfactory Treaty of Friendship and Alliance in Rio de Janeiro in 1810, the Navy's net fell around some of Portugal's slavers too, although many of them later escaped thanks to disagreement between the signatories on interpretation of the treaty. Others not in these categories fell victim to the Navy's enthusiasm, and the lack of legal expertise in the Vice-Admiralty courts gave encouragement to these errors. The concepts and regulation of prize of war and belligerent rights were well entrenched in naval minds, but the realisation that the arrest of foreign slavers should take place only within precise rules specified by international treaty was slow to gain hold.

There is no identifiable point at which Britain decided to progress enforcement of legislation against its own slave trade into a campaign to eradicate the entire international traffic in slaves, but the policy was formulated as the war against Napoleon came to an end, and its architects were the Prime Minister, Lord Liverpool, and his Foreign Secretary, Lord Castlereagh. Their ambitious aspiration became apparent at the Congress of Vienna, which Castlereagh saw as an opportunity to secure the agreement of the great powers to take coordinated action against the slave traffic in the Atlantic. Of course, none of the delegates at the Congress of Vienna was prepared to contradict the desirability in principle of abolition, but Britain found that it was the only country with the wish and stomach to confront the slavers, and was probably also the only one with the wherewithal to take action. The outcome of this sole attempt to reach agreement on multilateral action against the slave trade was a mealy-mouthed statement and a decision, effectively, to do nothing. In disappointment, Britain rightly concluded that bilateral treaties with the slaving nations offered the only avenue to its desired end of universal abolition of the Trade.

The five major nations with whom Britain had to deal in its abolition endeavours, the United States, France, Spain, the Netherlands and Portugal, who naturally resented what they saw as a British attempt to interfere in their affairs, represented the policy as a "grab for world power". Foreigners also suspected that, having destroyed its own slave trade, Britain felt it necessary for its economy that it should demolish that of its trading competitors. Neither view bears close examination, and foreign statesmen failed to understand what Hugh Thomas, in his history of the Atlantic slave trade, describes as "the quasi-religious enthusiasm which had come to possess Britain with respect to abolition". Of course there were abolitionists in all of these countries, but their ideals had not been espoused by their leaders as they had in Britain.

The fact was that slavery in their colonial possessions, and consequently the slave trade, were seen by the participating nations to be vital to their economies; they were not going to give them up at the whim of Great Britain, powerful though it was at the conclusion of the Napoleonic War. Nevertheless they saw in Britain's enthusiasm for the abolition crusade a lever with which they might extract money or concessions in exchange for limited agreements to curtail their slaving. The result in the postwar years was the signing of treaties with the Netherlands, Portugal and Spain which gave a basis for action by Britain but contained sufficient limitations and loopholes to ensure that the slave traffic in the hands of its co-signatories would continue.

At this point the British Government, finding that what had been a simple task of domestic law enforcement had transformed into a diplomatically and legally complex, and probably lonely, international campaign, might profitably have analysed the challenge with which it was faced, with a view to deciding what strategic policy to adopt. If it had troubled to do so it would probably have identified three broad options: to strike at the slave markets, to destroy the sources of supply, or to break the link between the two. The first and second, which might have involved a combination of diplomatic, military and economic methods, would have presented the difficulty that sovereign territory would be involved, and the likelihood of cooperation from the nations concerned would be slight. That difficulty would have been particularly acute in Cuba and Brazil at the market end of the chain, but might have been appreciably less so at the supply end in Africa. The third option presented no such obstacle, and would have appeared (in theory, at least) to be more straightforward and less risky.

The anti-slavery campaigners appreciated the need to generate legitimate commerce on the slave coast to provide an alternative to slaving, but it seems that no serious consideration was given to the destruction of the slave outlets on the African coast. Consequently the massive task of suppressing the slavers of all those nations with whom Britain could secure treaties fell on the shoulders of the Royal Navy alone, and was effectively limited to operations against slave vessels at sea and lying in the rivers and creeks of the African coast. There then followed a long period during which Spain, Portugal and the Netherlands exploited the loopholes in their abolition treaties in order to frustrate Britain's suppression efforts while evading their own obligations under the agreements. France, re-emerging as a slaving state after the war, declined at this stage to make any agreement at all.

Successive Foreign Secretaries consistently pursued abolitionist policies, and Britain's diplomats struggled to achieve improvements to the original treaties, very slowly managing to introduce indictable offences additional to the actual carriage of slaves at the moment of arrest. The most significant early innovation was the establishment of the bilateral Mixed Commission courts which, at Freetown and Havana at least, soon began to work effectively, although the judgments of commissioners tended too often to comply with the wishes of parent governments rather than to serve the cause of justice. It was a great weakness, however, that their jurisdiction extended only to slave vessels, leaving the trial and punishment of slaver crews in the hands of the nations to which they belonged, nations which almost invariably neglected this responsibility. Not only were these men therefore free to re-enter the slave trade, but also the vessels condemned for slaving were sold at public auction by the courts and generally returned to their criminal employment.

It was a particular frustration to the Navy that the early treaties prevented the arrest of vessels without slaves actually on board despite their showing clear evidence of the recent presence of slaves or of preparation to receive them. An expectation that closure of these loopholes would sign the death warrant of the Trade was greatly exaggerated, but the Dutch slave trade did indeed enter a terminal decline after the addition in 1823 of an Equipment Article or, as it became commonly known, Equipment Clause, to the Anglo-Netherlands Treaty of 1818. Spain accepted at an early stage that a vessel shown to have carried slaves during her current voyage might be condemned, but dragged her feet until 1835 on the matter of an Equipment Clause, and the value of that advance when it was eventually agreed was degraded by the persistent refusal of Portugal to accept any improvement to her 1817 Convention with Britain. At that point Spain also conceded that condemned slave vessels should be destroyed rather than sold at auction, an important, and obvious, step forward.

That inadequate agreement with Portugal was the weakest link in Britain's network of abolition treaties, and the inability of the Lisbon Government to ensure obedience from its venal colonial officials exacerbated the difficulty. By evading the more stringent constraints added to the Dutch and Spanish treaties, the Portuguese made their colours an attractive and readily available disguise for the slavers of other nations, the Spaniards in particular. It was only in the courts at Freetown, where the British commissioners were, during the 1830s, frequently unhindered by the presence of foreign commissioners, that false Portuguese identity rarely achieved its objective. Britain had only itself to blame, however, in allowing one of the original limitations in the Portuguese Treaty to persist. Slave traffic between Portuguese territorial possessions was permitted by the 1817 Convention, but that provision as it related to the massive trade to Brazil from Portugal's African colonies was automatically annulled by the recognition of Brazil's independence. Inexplicably, Britain remained silent on the matter, and that traffic continued unhindered.

Brazil inherited the Anglo-Portuguese Convention, but, although the Government in Rio de Janeiro was generally resistant to further constraints, it did gradually concede on the crucial matter of proof of national identity of detained vessels and on the illegality of slaving equipment. The Brazilian administration was by no means entirely opposed to abolition, largely because it was concerned at the growing proportion of Africans in its population, but it naturally resented being dictated to by Britain, and it was frequently unhelpful or obstructive towards Britain's suppression efforts. Its own attempts to restrict slave imports were routinely undermined by self-seeking regional administrators.

The colonial authorities in Cuba, the other great market for the transatlantic slave traffic, pursued a policy tailored solely to the economic and social welfare of the island, unhindered by proximity of any parent government. Successive Captains-General paid lip service to the Anglo-Spanish treaties, but, courteously, threw every available obstacle into the path of the suppression campaign. A difficulty which surfaced in Cuba considerably earlier than it did in Brazil was the unsettling presence of emancipated Africans on the island, and, although the arrival of the (otherwise unwelcome) accommodation vessel *Romney* in Havana harbour partially alleviated the problem, the determination of the authorities to expel freed slaves from the island presented an additional administrative task for Britain.

After re-entering the slave trade on the conclusion of the Napoleonic War, France, naturally resistant to any abolition treaty with Britain, gradually gained a prominent position in the northern hemisphere traffic. For a period at the beginning of the early 1830s, slavers under the White Flag of Bourbon were, to the intense frustration of the British cruisers, the most numerous on the northern slave coast. Unlike Spain or Portugal, however, France did eventually deploy men-of-war to bring to book slavers under its colours. There is evidence that the French slavers enjoyed provoking Royal Navy cruisers into unwise reaction, but relations between the two naval squadrons on the African coast appear to have been cordial, although, on the part of the French, sensitive to perceptions of unwarranted interference. A Convention was at last agreed in 1831 to allow mutual Right of Search by the cruisers of the two navies, but requiring adjudication in the courts of the alleged slaver's own nation. It was an agreement hedged around with so many restrictions that it was of little use in practice, but it did serve most usefully as a parent treaty to which other smaller nations could, and did, accede.

It was perhaps the greatest tragedy of the suppression campaign that no agreement was reached with the United States of America. It was hardly surprising that, the War of Independence not far in the past and with the war of 1812 poisoning relations between the two countries, Washington should be resistant to any compliance with the wishes of the British, but the 1807 Abolition Acts of the American Congress and the British Parliament had been achieved by related abolition movements, had been simultaneously enacted, and had been similar in objective, and it might have been hoped that the two nations would move forward in step to suppress the Trade. Of course, the individual states were far from unanimous in their attitude to slavery and the slave trade, and Washington, unlike London, lacked the political power to enforce its own abolition legislation. In 1824 the State Department and the Foreign Office negotiated a convention similar in aim to that between Britain and France, and apparently unobjectionable to either side, but Congress declined to ratify it, and a major amendment it proposed was unacceptable to London.

At times and for fairly brief periods, ships of the United States Navy were deployed to the west coast of Africa to police the American slave trade and support the new colony of Liberia, but, although the cruisers of the two navies remained on most amicable terms and as supportive of each other as practicable, cooperation between them was frowned on by the US Government. American capital, shipbuilders, trade goods and seamen continued to underpin the Cuban slave trade, and dishonest American consular officials in Havana aided the illegal traffic. Washington remained ultra-sensitive to any interference at sea with the

Stars and Stripes, particularly by the British, and American colours consequently became the preferred option for slavers seeking the protection of false identity after the usefulness of Portuguese ensigns and papers declined.

That decline was gradual as the Mixed Courts became more rigorous in investigating the true identity of slavers, but it became sudden and final with the extraordinary legislation enacted by Britain against the Portuguese slave trade in 1839 at the instigation of Lord Palmerston. British exasperation with the continuing refusal of Portugal to agree to eradication of the many means of evasion embodied in its treaty with Britain is understandable, and it had been endured for over 20 years, but the measure taken to rectify the matter was unacceptable by the norms of international relations. Palmerston and Parliament undoubtedly felt that it could be justified by the greater good of slave-trade suppression, but they would certainly have resisted any similar unilateral action by another nation. It was a striking illustration of the power that Britain could wield, and of its arrogance, particularly in what it believed was a good cause. It entirely changed the basis of action against the Portuguese trade, taking jurisdiction out of the hands of the Anglo-Portuguese Mixed Commission court at Freetown and giving it to British Vice-Admiralty courts at various British territories around the Atlantic Ocean. It remained to be seen how effective it would prove to be.

This then, with all its shortcomings and pitfalls, was the legal and diplomatic foundation upon which the Royal Navy's suppression operations were based, and the international background against which they were conducted, up to the end of 1839.

There is little doubt that the driving force of the suppression campaign lay in the Foreign Office under a succession of abolitionist Foreign Secretaries, the most prominent and determined of whom was Lord Palmerston. It is also clear that the Admiralty, the department controlling the force primarily engaged, did not share that enthusiasm. The half-heartedness of the Lords Commissioners of the Admiralty at the outset and early stages is easily explained: their department, tiny by today's standards, and with no Naval Staff, was contending with a worldwide maritime war of national survival, and it is unsurprising that it should not have welcomed any distraction from its primary concern or an additional task for its thinly stretched resources.* The coming of peace hardly alleviated the difficulties;

---

* In 1800, the clerical and administrative staff at the disposal of the Admiralty Board was only 28. This number included the First and Second Secretaries, considerable figures who, on relatively minor matters, could issue instructions on their own authority.

the nation's interests in every corner of the globe had to be protected by the Navy, but, even before cessation of hostilities, a massive reduction in the number of warships began, and severe financial constraints were imposed. It was all very well for the Foreign Office to enthuse about slave-trade suppression, but it was the Admiralty which had to carry the cost.

A letter in January 1831 from Sir James Graham, First Lord of the Admiralty, to Lord Palmerston laid out the concerns of his department. He explained that

> [my] earnest desire to reduce the Naval expenditure without diminishing the real efficiency of the Force has led me to pass in review our Squadrons on Foreign Stations and carefully to consider the necessity which may justify the maintenance of each on its present scale.

He had been "struck by the increasing charge of the Squadron on the coast of Africa, now amounting to £100,000 a year", and expressed his concern at the "deadly climate", with "men constantly employed in Boats under a vertical Sun, and where the mortality among them is as three to one when contrasted with the casualties on other Foreign Stations". He accepted that the expense, including that for the victualling establishments at Sierra Leone and Fernando Po, was necessary if the Squadron was to remain, but it had been found necessary to begin a large outlay at Ascension to provide refreshment for the men in a cooler climate. Furthermore,

> other considerations besides those of health make it expedient that ships should not remain at sea, as they now do on the coast of Africa, for three years without intermission, but that they should be annually relieved from England, and return at stated periods to harbour in this Country.

This would lead to yet greater expense.

Graham continued, bluntly, that the question remained:

> Wherein consists the necessity of a Squadron on this Station? The prevention of the slave trade its only employment; and does it effect its object? I fear it must be avowed it does not, and no force can ever really be effective while France remains a stranger to our efforts to repress this inhuman traffic.

After briefly surveying the Navy's current tactics and deploring the shortcomings and evasions of the extant treaties, he awarded the United States undeserved credit for declaring the slave trade to be piracy, and inaccurately laid the majority of blame for the continuation of the Trade on France. It was his opinion that "if France would cooperate we could withdraw the Squadron and a single cruiser would be more efficacious than the present Fleet". He was "confirmed in this opinion by the authority of the most experienced officers".[1]

This concern over expenditure was, of course, a recurring one in all periods of history and is deserving of sympathy, and Graham was not unreasonable in claiming that the West Africa Squadron was ineffectual. He had again, unintentionally, betrayed the Admiralty's lack of understanding of the magnitude of the task being faced on the coast of Africa by "the present Fleet" (actually, at the time, seven men-of-war and two tenders), or of the real reasons for the ineffectiveness of its efforts, but he had made clear the view of his department that the suppression campaign was far from cost-effective. Indeed it regarded it as a waste of money and resources, and considered that it would continue to be so until Britain could rely on the cooperation of the slaving nations. There is little to indicate that this attitude changed materially during these early decades of the campaign.

Lack of understanding and, by inference, of interest was apparent not only in the entirely inadequate number of cruisers deployed to the west coast of Africa, the principal theatre of suppression operations, but also in the wording of orders to successive commodores. Regular repetition in these orders of the phrase "the several Bays and Creeks" when referring to the 3,000 miles of coastline between Cape Verde and Benguela, much of it deeply indented, demonstrated (even allowing for the language of the period) a lamentable ignorance of the fundamental circumstances of the campaign. There was also a persistent assumption at the Admiralty that operations against the slavers were compatible with the protection and support of British settlements and trade on the slave coast, and for many years this latter function took precedence.

The Admiralty's neglect in its support of the cruisers is further illustrated in an exchange between the British Commission at Freetown and the Senior Officer on the West Coast late in 1839. The perceived method of measurement of prizes for "tonnage money" was causing annoyance among commanding officers, and the Registrar referred Commander Tucker to parliamentary papers which were necessary to an understanding of the system. The Admiralty had failed to issue these papers to the cruisers, and the commissioners, noting that the information they contained would have prevented many illegal seizures and avoided the expenses of restoration, recommended that commanding officers should acquire them from the Foreign Office, through their agents.

With Admiralty minds concentrating on economy and lacking an appreciation of the magnitude and difficulty of operations on the African coast, and with the depleted Navy stretched to fulfil its peacetime tasks worldwide, it was inevitable that the West Africa Squadron would be critically short of cruisers. The position was no better in the West Indies. That station was allocated no additional vessels to deal with the Cuba slavers, but, for a time, it did at least have its schooners, mostly built locally at the instigation of the Commander-in-Chief. Similarly there was no strengthening of the South America squadron when, at last, it became empowered to detain slave vessels. In these straitened circumstances the initiative of several commodores in purchasing condemned slave vessels as tenders should have been given every encouragement. When presented with this apparently admirable and low-cost means of boosting capability, however, the Admiralty's reaction varied between lack of enthusiasm and downright opposition, and only in the cases of *Black Joke* and *Fair Rosamond* was the idea sensibly embraced. There was nothing novel in the concept of tenders, and the perversity displayed by the Admiralty in this instance is difficult to fathom.

For the first quarter-century of the campaign the West Africa Squadron was hamstrung not only by its shortage of cruisers but also by the unsuitability of the vessels deployed, and both of these shortcomings would have been alleviated by a leavening of ex-slaver tenders. The Admiralty's persistence in deploying the notorious ten-gun "coffin brigs", of which it had a large number, despite pleas and criticism from officers on the Coast, was particularly mulish. Slow, unhandy, crowded and atrociously uncomfortable in the tropics, these brigs could hardly have been more badly suited to the suppression role.

Criticism of the use of Fifth Rate frigates has, however, been excessive. Their large ships' companies and big complement of boats gave them flexibility denied to lesser vessels, especially the ability to crew tenders and prizes and to support smaller cruisers; they had good endurance; they provided relatively comfortable living conditions; and, although slow by comparison with the Baltimore clippers in light and moderate conditions, they could carry canvas far longer than the slavers in heavy weather. It should not be forgotten that it was the elderly (albeit French-built) frigate *Sybille* which chased and caught the *Henriquetta*, later to become the splendid *Black Joke*. As far as a frigate's detectability was concerned, the factor attracting most criticism by historians, the height of eye offered by her tall masts would, if her upper sails were not set, give her the sighting advantage over a slaver under the habitual cloud of canvas.

The handful of tenders and the West Indies schooners apart, improvement began with the final batch of ten-gun brigs, a cut-down version under brigantine rig modified specifically for slaver-chasing, which performed fairly well. Following the purchase by the Admiralty of a couple of privately built yachts, real progress got underway with the arrival of William Symonds as Surveyor of the Navy. Whatever shortcomings his vessels may have had as general-purpose men-of-war, Symonds's brigs, gun-brigs and brigantines were ideally designed for work on the west coast of Africa, and his 16-gun brig-sloops brought a valuable advance in capability to the squadrons in the western Atlantic. So far there had been only one brief experiment with steam power on the African coast, and it would be well into the 1840s before the benefits of steam were applied to the campaign. Development of steam propulsion had as yet to reach a point at which it might give decisive advantage over sail, and the heavy consumption of fuel by the early steam plants introduced an unwelcome logistical problem.

Hampered though they were by their sailing limitations, the cruisers of the West Africa Squadron in the pre-Symondite period achieved a respectable capture rate, and that must be attributed to the skill and determination of commanding officers and crews. These men had learned their business in a Navy which was accustomed to keeping the sea in all weathers, an unforgiving school in which competence in seamanship was of supreme importance. The skill they acquired, allied to naval discipline and manning levels, produced a sailing prowess rarely achievable in a merchant vessel. This superior ability not infrequently overcame a theoretical speed disadvantage.

Handicaps associated with paucity and inferiority of materiel were not the only major troubles with which the cruisers had to contend. "Command and Control" was inevitably difficult in the age of sail, and it could be exercised with full effectiveness only within visual signalling range. Much had to be left to the initiative of individual commanding officers, and, although that to a degree was much to be welcomed, it led after a point to increasing inefficiency. Although his intended area of operations was grossly excessive, the Commodore commanding a squadron of a half-dozen cruisers on the coast of Africa could supervise operations tolerably effectively. The Commander-in-Chief at the Cape of Good Hope, with responsibilities extending from halfway across the Atlantic in one direction to halfway across the Indian Ocean in the other, could not. Yet the Admiralty, with blind indifference to the damaging effect on the suppression campaign, removed the Commodore and incorporated the entire African west coast into the Cape Station.

Try as he might, the Commander-in-Chief could not stay in touch with the work of the west coast cruisers, correlate and disseminate information, order the disposition of vessels, develop tactical doctrine, or direct any of the crucial operational matters which required personal presence at the centre of affairs. This remoteness of command, in combination with the lack of thought and analysis in London, led to stagnation in the mode of operations. Indeed, there was an astonishing absence of innovation in tactical methods during the first 30 years of the West Africa campaign, and it was only at the very end of that period that a small number of bright and energetic young commanding officers began to challenge the tactical and strategic status quo.

On the other side of the Atlantic command was organised differently, and on the South Atlantic Station the almost permanent presence of the Commander-in-Chief at Rio removed most of the difficulties until the subsequent formation of the Pacific Station left a poorly defined situation on the coast of Brazil. However, the appointment then of a commodore, still stationed at Rio, to command the ships on the Atlantic coast of the continent effectively retained the structure, although the Commodore could exert less influence with local authorities or with London than had the Commander-in-Chief. Further north, the amalgamation of the West Indies and North America Stations threatened a situation similar to that on the African coast, although operations in the approaches to Cuba were much more straightforward than those in the eastern Atlantic. However, the Commander-in-Chief, appreciating that he would be spending most of his time at Bermuda or Halifax, established the post of Senior Officer of the Jamaica Division, held by a commodore and based at Port Royal. That provided an adequate local command structure for the cruisers around Cuba, although officers holding the post changed rather too frequently and they had little scope for ordering deployments.

Most of the failures of the Admiralty in its direction and support of the suppression campaign can be excused, in part at least, on the grounds of being overstretched and economy being enforced, but there is one matter of wilful neglect with which it is difficult to sympathise: its refusal to aid commanding officers who fell foul of the slaver courts in the course of the performance of their duty. Such instances arose from captors' misjudgments, or though the perversity of commissioners in the Mixed Commission courts, which resulted in the restoration of prize slavers. In such a case the captor commanding officer was liable to be sued for costs and damages by the vessel's owner, and awards were occasionally severely, even shockingly, punitive. The circumstances of slaver arrests were often complex, but decisions had to be made rapidly, and although sometimes commanding officers were incautious, they invariably did what they believed to be right at the time. When adjudications went against them through the political inclinations of commissioners, or because cool analysis of evidence showed that they had had been mistaken in decisions taken under the stress of violent activity, they were owed, and deserved, backing from the Admiralty. However, the principle of personal responsibility of commanding officers was carried to an extreme, and such support was never given.

This was a campaign waged by minor warships, almost all of them Sixth Rates and unrated vessels, and the most numerous of them were small gun-brigs, brigantines and schooners. The commodores of the West Africa Squadron apart, the most senior of the commanding officers involved were junior captains, and the great majority were commanders and lieutenants. The wide extent of the theatres of operations meant that cruisers worked independently, there was negligible supervision, and mutual support was rarely available. It says much for the quality of these commanding officers that, as the tone of their reports shows, they relished this independence, and the self-confidence, initiative and decisiveness they generally displayed under the heavy responsibilities placed on their (often very young) shoulders were impressive, although these qualities would not have appeared remarkable at the time. The loneliness of command seems not to have weighed heavily on them, and in a gun-brig, brigantine or schooner, as the only commissioned officer on board, the Commanding Officer was indeed lonely.

It becomes clear from the narrative that the cruiser commanding officers gradually achieved a decisive psychological ascendancy over the slaver masters. In the early days many cornered slavers would defend themselves against capture, and some would be spoiling for a fight. Numerous though their crews might be, however, they almost never succeeded in overcoming the skill, resolve and discipline of the naval boarding parties. The will to win of the officers and men of the cruisers reflected the Royal Navy's tradition of victory founded on the best part of a century of almost unblemished success, and the possibility of failure probably never crossed the minds of the bluejackets and marines. The slavers had no answer to this confidence and determination, and, as the Navy's fighting ascendancy became clear to them, they generally decided against resistance.

The fighting spirit of the cruisers' men was matched by their seamanship. Not only were they accustomed to keeping the sea in all weathers, honing their skills in their daily work, but also they exercised to improve and maintain their sailing expertise and their judgment of their vessels' capabilities under all conditions. They clearly took immense pride in their sailing prowess, even if their vessels were not of the highest quality, and they took delight in sailing races whenever opportunity offered. The result was that an arrest was often achieved after a chase in which the slaver was theoretically the faster vessel.

Competitive sailing, chases, the occasional skirmish and arrest, and perhaps even back-breaking hours at the sweeps or at the oars of the boats in a calm, were welcome respites from the tedium and discomfort of cruising for weeks on end in the heat and humidity off the African slave coast. A sight of the shore, generally low-lying and dreary, provided no relief. Boredom occasionally led to indiscipline, for which punishment was generally (and customarily) severe, but it seems that the men tolerated the monotony fairly well. Boredom presents one of the greatest challenges to leadership, and the evident efficiency of the cruisers in these circumstances speaks well of the leadership of the officers. Neither slavers nor boredom were the greatest enemy on the slave coast, however. Fevers, of which the causes were still unknown, were a constant and terrifying threat, and epidemics were usually of appalling proportions. The fortitude of officers and men in the face of these diseases is astonishing, as is the rapidity with which stricken cruisers recovered efficiency after outbreaks.

Fortitude was a fundamental requirement in the Navy of the early nineteenth century, and, in that period, display of courage by officers and men was an expectation. Very rarely was that expectation disappointed during the suppression campaign, but one notable exception was during Mr Crawford's gallant defence of the prize slaver *Netuno* in 1826, when the greater part of the prize crew fled below as the *Netuno* came under attack by a pirate. If anything can be concluded from this blemish on an admirable episode, it might be that courage came more readily to men surrounded by their shipmates and under the eyes of trusted commanding officers and familiar superiors than it did in the unsettlingly strange surroundings of a prize.

It was unfortunate, but perhaps inevitable, that excessive self-confidence allied to self-righteousness produced a tendency to arrogance, and in some junior officers this degenerated further into high-handedness and bullying in their dealings with slavers. To the Navy's credit, when this unacceptable development appeared in the West African cruisers in the late 1830s it was detected and reported by more senior officers, and it can be assumed that corrective action was taken, although it was an unattractive characteristic probably not expunged entirely.

The evolution of taking a detained slave vessel from the position of her arrest to the place of adjudication has tended to be taken for granted, but it was usually a matter fraught with difficulty and danger, and it made heavy demands on a Prize Master and his crew. Only if the capturing cruiser was a frigate might there be a commissioned officer available as Prize Master, otherwise the task fell to a master's mate, a midshipman or a warrant officer, none of whom had the training or experience to exercise command unless he happened to have commanded a prize previously. He would be allocated a tiny number of hands as prize crew, and at only a few moments' notice would be transferred to an unfamiliar vessel to sail her perhaps as much as two thousand miles to the prize court, using only such navigational aids as were available in the prize.

On passage, critically short-handed, the Prize Master would have had to contend not only with sailing and navigating the vessel, and with very long hours of watch-keeping, but also with guarding the prisoners required as witnesses in the court proceedings and, of course, with the control and care of a cargo of several hundred slaves, many of whom were likely to be sick or dying. Apart from the dangers of squalls and foul weather, there was the constant threat that the slaver crew would attempt to retake the prize, and attack by pirates was an ever-present danger, in African waters especially. It was not a task for the faint-hearted or incompetent, and it should be remembered that the twenty-first-century equivalents of many of the young men who were thrust into it are schoolboys or undergraduates. It ought to be a matter of admiration that only about three per cent of these hazardous and exhausting voyages were not satisfactorily accomplished.

Amid the difficulties, hazards and boredom of cruising against the slavers, there was at least a prospect of bounty or tonnage money, and perhaps the proceeds of the sale of a prize vessel, in the event of a capture. The magnitude of these rewards should not be overestimated, however. The sums granted by the 1807 Abolition Act were generous, but subsequent legislation whittled away at them, and the Treasury ensured that as much as possible of the cost of dealing with prizes was set against the awards ultimately made to the captors. Court costs were extracted from the proceeds of prize sales, and Proctors' and agents' fees, payment for admeasurement of prizes and so on had to be found by captors. For the successful ships' companies these hard-earned supplements to pay were most welcome (when they eventually arrived), but there were no fortunes to be made.

The part in the suppression campaign played by the Kroomen, who formed a substantial proportion of each of the ships' companies of the West Africa cruisers, has been insufficiently appreciated. These natives of the Windward Coast were consummate seamen, particularly in handling boats on the hazardous surf-battered open beaches of the slave coast. They were also resilient to the energy-sapping and disease-ridden conditions of West Africa, and were able and willing to undertake laborious tasks which would quickly reduce white men to exhaustion. Not only did they shoulder a disproportionately heavy load of a cruiser's manual work, but also they fought bravely alongside their British shipmates in the encounters with slavers. It is certain that the cruisers on the west Coast of Africa could not have approached the pitch of efficiency they achieved without the loyal and wholehearted assistance of their Kroomen, and this immense contribution to the campaign should be acknowledged.

It is apparent from the records of the campaign and from the letters of those involved that, despite the tedium, difficulties and frustrations, the officers and men of the cruisers tackled their work with determination and dedication, and it is worth considering their motivation for doing so. To begin with, they were generally imbued with a strong sense of duty, and they believed that they owed it to the Service and their shipmates to perform any mission, whether they sympathised with it or not, to the best of their ability. Indeed that is what the Navy demanded, and pride would admit of nothing less. Of course they hoped to achieve advancement through outstanding deeds, the officers especially, and they were encouraged by the hope of financial reward in the shape of bounties on captured slaves, tonnage money and the proceeds of prize sales. They undoubtedly enjoyed pitting their wits and skills against those of the slavers, and, tough as they were, many of them also enjoyed a fight. However, there is strong evidence that there was more to their motivation than all of these.

Whatever prior knowledge the officers engaged in the suppression campaign may have had of the Trade, or whatever opinion they may have formed of it before coming face to face with its reality, it is obvious from their letters and reports that their first encounters with laden slave vessels left them appalled. These were men familiar with hard conditions, and, having fought in the French wars, many of them were accustomed to distressing sights which no civilian was likely to have seen. Nevertheless, they were shocked by what they found. Despite the requirements of the Articles of War that divine service should be regularly performed on board, there was probably little overt expression of religious belief in the average man-of-war, but most officers would have professed a Christian faith. The standards against which both officers and men, believers or not, judged good and bad behaviour were generally derived from Britain's Christian tradition and culture, and, for those who observed it in action, the slave trade offended against those standards to an extreme degree.

Hard though his own conditions were, the Royal Navy's sailor was a kindly creature, and the brutality, degradation and squalor he found in most slavers, as well as the notion that men, women and children were mere merchandise, horrified him. Few were not lastingly affected by the experience, and some officers became dedicated, with missionary zeal, to the eradication of the slave trade. The motivation of most of the men of the suppression cruisers consequently rose to a high moral plane, and it may not be unreasonable to assert, although it would probably not have occurred to the officers and men engaged, that the Royal Navy's part in the campaign was "muscular Christianity" in action. When Commodore Hayes, in a letter to the Secretary of the Admiralty in 1831, exclaimed: "Gracious God! is this unparalleled Cruelty to last for ever?", he was certainly not taking the name of the Almighty in vain.

There had been much endeavour during the first 32 years of the suppression campaign, and considerable expenditure in treasure and in lives, but it was unclear whether any real progress had been made towards Britain's objective. Incompleteness in the Admiralty Archives and a total absence of Vice-Admiralty court records for all but the early years of the period preclude a guarantee of entire accuracy in the statistics of the Royal Navy's achievements to date. However, there is reliable evidence, particularly in the comprehensive reports to the Foreign Office by the British commissioners on the Mixed Commission courts and Mixed Courts of Justice, that the Navy's cruisers, aided to a very small degree by colonial vessels and Letters of Marque in the early days, had arrested 767 slaving vessels, and this will be very close to the true number. The fate of some of these prizes is unknown, but at least 597 were satisfactorily condemned in Admiralty or Mixed courts. In most cases the number of slaves captured and the number subsequently emancipated or released are recorded, but in others only one or the other is known, and very occasionally neither figure is shown either in court records or in naval reports. Nevertheless, these incomplete figures reveal that at least 104,034 slaves were captured.

Respectable (even impressive) though these statistics may appear, they tell only a part of the story, and can be misleading. Against the number of slaves captured by the cruisers between 1807 and 1839 must be set the figure for the export of slaves from Africa between 1811 and 1839, as calculated by the American historian David Eltis. This he gives as 1,908,600, and the number exported in the last of those years he estimated to be 90,800, a little over twice the figure for 1811. Eltis also estimates that between 1811 and 1840 a total of 2,640 vessels embarked, or intended to embark, slaves for the transatlantic traffic. The comparison between the number of slave vessels engaged in the Trade and the number captured is less unfavourable than the equivalent figures for slaves largely because of the number of empty vessels arrested for contravention of Equipment Clauses, and in the early years a significant proportion of the vessels taken were carrying very few slaves.

It is apparent that the slave trade in the Atlantic remained buoyant, despite the increasing depredations inflicted by the Navy, and that its volume was probably regulated entirely by the size of demand in Cuba and Brazil rather than by suppression efforts. It was being demonstrated that a supply would be made available in Africa if the demand existed, and the means to convey the slaves across the Atlantic would be found, whatever treaties might be signed or obstacles, legal or naval, might be placed in the way. Further discouragement came from the realisation later on that interdiction by the cruisers might well be exacerbating the sufferings of the slaves. Arrests of slavers en route to Africa were leading to gross overcrowding in the slave coast barracoons as the waiting times for embarkation lengthened and new slaves continued to arrive from the interior, and the vessels which did sail with slaves were consequently even more tightly packed than had previously been the case.

It would be easy to conclude that nothing had been achieved by the struggle of the past 32 years, but that would be an injustice to the diplomatic exertions of the Foreign Office and to the endeavours, courage and sacrifice of the Royal Navy's cruisers. The initial objective of destruction of the British slave trade had been accomplished, slaving by the Netherlands had ended, the slave trade had been made considerably more difficult and expensive for those engaged in it, and encouragement had been given to the abolition movements in the slaving nations. Furthermore, but on a less positive note, it had at least been shown by hard experience that interdiction at sea was not the answer to slave-trade suppression, although it probably had a part to play. Nevertheless, there was no denying that the main aim was as far as ever from fulfilment, and it was high time that a new strategy was developed. What might be in doubt was whether Britain would be prepared to shoulder the increased cost likely to be incurred. It was, however, beyond question that the Royal Navy would continue to be central to its execution, whatever an improved strategy might involve.

24. A Royal Navy landing force destroying a slave factory (the event depicted was in Mozambique, but the picture is representative of similar raids on the west coast of Africa)

EPILOGUE

Until It Be Thoroughly Finished

> O Lord, when thou givest to thy servants to endeavour in any great matter, grant us also to know that it is not the beginning but the continuing of the same until it be thoroughly finished that yieldeth the true glory.
>
> SIR FRANCIS DRAKE

AT THE END OF 1839 THE CAMPAIGN against the Atlantic slave trade was very far from reaching the conclusion so fervently desired by the Royal Navy and the Foreign Office who, at this point, could reflect only on a third of a century of little but frustrated joint endeavour. However, change was afoot, and it had begun with Lord Palmerston's 1839 controversial legislation against the Portuguese traffic.

Armed with the new "Palmerston Act", the Royal Navy's cruisers proceeded, unfettered, to seize all suspected slavers under Portuguese colours, or wearing no colours at all. The details of these operations are unknown because all relevant reports from the Navy and from the courts adjudicating the cases have been lost, but by the end of 1841 a total of 65 condemnations under this legislation had been reported to the Foreign Office by the Vice-Admiralty courts at Sierra Leone, St Helena, The Cape and Barbados.[1] In 1842 the Portuguese Government capitulated and agreed to an improved treaty, which included an Equipment Clause. Consequently the "Palmerston Act", its purpose fulfilled, was repealed.

At last, in 1840, there emerged two significant new strategic developments on the West African coast, both on the initiative of Commander The Hon. Joseph Denman, who was senior commanding officer on the northern part of the slave coast. He was concerned that the long-employed tactics of cruising across the routes into and out of the embarkation points, combined with occasional examinations of the slave harbours, was allowing too many slavers to evade detection. Having rather more plentiful resources than had been the case until then, he established a close blockade of the principal harbours, subordinating the objective of capturing slavers to that of preventing embarkation. Although the work was exhausting for ships' companies and wearing on their vessels, a stranglehold on the selected places was achieved.[2]

For ten months a ceaseless blockade was maintained on the Gallinas River, and Denman, asked by the Governor of Sierra Leone to rescue two British citizens held by the King of Gallinas, seized the opportunity to mount the first serious attack on the slave trade close to its roots. Having prevented removal to the mainland of the slaves in the extensive Gallinas barracoons, he demanded that the King sign a document abolishing the slave trade in his dominions, permitting the destruction of the barracoons and expelling slave traders from his territories. That done, 841 slaves were liberated and the barracoons and the trade goods they contained were burned.[3]

Denman's action delighted his fellow officers on the Coast, was applauded by Ministers in London, and caused consternation in Havana. A similar operation was then conducted at the Shebar, and in the following year the slave factories on the Rio Pongas and south of the Equator at Cabinda and Ambriz were also destroyed. This initiative had the potential to inflict very serious damage on the Trade, but a delayed and ill-judged reaction by the Foreign Office halted it in its tracks, and incurred unfortunate repercussions for Denman.

Questioned by the new Foreign Secretary, Lord Aberdeen, on the legality of the Navy's actions, the Queen's Advocate-General gave his opinion that the activities could not be justified "with perfect legality". He reversed this opinion six years later, but by then the damage had long been done. Without proper consideration of the probable consequences, Aberdeen wrote to the Admiralty requiring that such operations were not to be repeated. The "Aberdeen Letter" was not intended for publication, but it reached the public domain in error. The slavers and chieftains on the African coast concluded that there had been a revolution in England resulting in a reversal of the anti-slaving policy, and traders who had lost their slaves and their trade goods brought massive claims for damages against Denman, and against Commander Matson, who had destroyed the Cabinda and Ambriz factories. The case against Denman, the test case, was not brought to court until 1848, when it was dismissed.[4]

The beneficial effects of Denman's innovative tactics were shown by the subsequent dramatic decline in the number of slaves imported to the Americas. In 1839 the figure for Cuba was 19,900, and for Brazil it was 54,000. In 1842 the numbers had declined to 4,100 and 20,000 respectively. Numbers recovered somewhat after 1842, but the pressure was not entirely relaxed.[5]

An additional detrimental effect of the Aberdeen Letter debacle was the stifling of further major initiatives by the cruisers, but one important development, initiated in the 1830s, progressed well. Commanding officers were encouraged to negotiate local treaties by which coastal chieftains undertook to abolish the slave traffic in their domains in return for modest payment. Many such treaties were achieved north of the Equator, and they were enforced by the threat of retribution by the cruisers.[6] Further agreements with these chieftains gave permission for barracoons to be destroyed in exchange for small subsidies, and these agreements seem to have been fairly easy to conclude.[7] By this combination of means the slave trade from the shores of Africa north of the Line was considerably reduced.[8]

At the beginning of 1840 the Brazil and Cape Stations were combined. Once again, a commodore was appointed to command the force on the West Coast of Africa, but was made accountable to the Commander-in-Chief at the Cape. Authority was further delegated to Senior Officers on the northern, Bights and southern divisions of the Coast. This made for more effective control of operations than had previously been possible since the Cape Station had assumed responsibility.

Effectively denied the use of Portuguese colours, the slavers required the cover of another false ensign, and they did not have to look far for an alternative. The adamant refusal of the United States to countenance any anti-slaving agreement with Britain, or to allow any Right of Search by foreigners of American-flagged vessels, had already led to the Stars and Stripes being used by Spanish slavers, but now the abuse rapidly became widespread.[9]

There had been a period in the early 1820s when cooperation between the United States Navy and the Royal Navy on the African coast had seemed to be a real possibility. Then treaty negotiations broke down, American cruisers were withdrawn from the slave coast, and by the late 1830s relations between the two nations became tense. US Navy cruisers returned to West African waters in 1840, but it seemed that their objective was more to prevent interference with the American flag than to inconvenience the slavers. A squadron of no more than four vessels was based at Porto Praya in the Cape Verde Islands, impossibly remote from a worthwhile cruising ground, and it made few arrests.[10] Nevertheless, there arose one example of close cooperation, agreed between a pair of commanding officers, which demonstrated what might have been achieved had the two navies worked together. When Washington became aware of the so-called Paine–Tucker Agreement, however, the US Government declared it unacceptable.[11]

The replacement of Lord Palmerston as Foreign Secretary by the more conciliatory Aberdeen, and the arrival of Daniel Webster at the US State Department, provided a more promising atmosphere for accord, and in 1842 the two countries concluded the Webster–Ashburton Treaty. This pact was principally concerned with national boundary matters, but it included an undertaking by each signatory to maintain a minimum force on the coast of West Africa. However, lacking any agreement on joint cruising, on a system of adjudicating prizes or on authorising cruisers to arrest slavers of any nation other than their own, it inevitably contributed little to slave-trade suppression.[12] Further progress was prevented, on the one hand, by British insistence on maintaining a right to impress its citizens found in foreign ships, and, on the other, by American refusal to permit any foreign right to search (or even to visit) vessels under American colours, however fraudulent those colours might be.

In 1841 a Quintuple Treaty was signed in London by France, Russia, Austria, Prussia and Britain by which the slave trade was declared to be piracy, and giving authority to any of those nations' warships, over most of the Atlantic Ocean, to stop and search any merchantman belonging to the signatory powers on suspicion of slaving. France then refused to ratify the agreement, largely owing to the pernicious influence of the United States Ambassador in Paris, the Anglophobe General Cass.[13]

The French had been keeping a small squadron on the Coast through the 1830s to police their own slavers, but Palmerston's efforts to increase cooperation and expand it to cover the whole Atlantic were thwarted, again probably by Cass. An incident at sea, in which the crew of a Nantes merchantman were alleged to have been roughly handled by a British cruiser, also inflamed a simmering anti-British feeling in France. Nevertheless, in 1845 the French agreed with Aberdeen that each nation would maintain a squadron of 26 cruisers off West Africa. Again, no Right of Search was conceded, and in 1848 Commodore Hotham complained that the French cruisers had been of no assistance because they spent most of their time in harbour.[14]

The Spanish Government had continued to encourage its representatives in Cuba to avoid complying with the Anglo-Spanish Treaty, and the authorities on the island had remained hostile and obstructive to Britain's efforts against the Trade. However, in 1841 a liberal, General Valdés, was appointed Captain-General. He took unprecedented action against the slave traffic, but was careful not to appear to be bending to British pressure, and improvement was sporadic.[15] His successor, General O'Donnell, who detested Britain, was much more to the taste of the planters and slavers. The Trade recovered under his supervision, but Cuba suffered a number of slave revolts in 1843 and 1844, which O'Donnell brutally suppressed, and there was even a widespread rumour that Britain planned armed intervention in the island. This was a period of political turmoil in Spain, and in 1844 a Foreign Minister friendly to Britain returned to power. He introduced in the Cortes a bill defining penalties for those convicted of involvement in the slave trade, and it became law despite much hostility. It caused panic among the Cuban planters, and for some years in the late 1840s the slave traffic into the island almost ceased.[16]

Lord Aberdeen made amends for his earlier misjudgment, in the eyes of the Navy, by seeking the advice of Denman and Matson on the future conduct of the campaign. Their recommendations, presented in 1844, were for inshore cruising, close blockade, destruction of barracoons, sudden concentrations of force, more steamers, faster sailing vessels and the use of transports to relieve cruisers of distracting tasks.[17] Destruction of barracoons had to wait, but almost all of the other ideas were adopted, although material improvements were slow in coming. Denman and Matson also believed that cruisers should be withdrawn from the Brazil and Cuba theatres and concentrated on the African coast, and although this recommendation was not adopted, the level of activity in the western Atlantic fell to a low ebb. Subsequently Denman was also instrumental in the issue to the cruisers of a pocketbook of instructions to guide officers through the growing maze of treaties.[18]

Difficulties then gathered in an unexpected quarter. From 1845 to 1850 the Government's suppression policy, and consequently the existence of the West Africa Squadron, came under almost continuous attack in Parliament. An unlikely coalition, usefully dubbed the "Anti-Coercionists" by Dr William Mathieson, forced a string of debates which in turn spawned a host of parliamentary committees and their reports. The five identifiable groups in this heterogeneous body were: the pacifists, those who believed that the current campaign was wasteful and that the only way forward was to civilise Africa and to encourage the growing of palm oil, those who considered slavery to be no bad thing, a few naval officers who were pessimistic about the likelihood of success, and, most important, the Free Trade movement. The outcome of this "Great Debate" was in doubt throughout, but in March 1850 opposition to the suppression policy crumbled, and the Government's cause was strengthened.[19]

For a period in the mid-1840s, thanks to an extraordinary error of judgment by Commodore Jones, patrolling south of the Equator was left entirely to the Portuguese.[20] This, combined with the expiry of the Anglo-Brazilian Treaty in the early 1840s, resulted in unrestricted slave importation to Brazil. The Brazilian Government was disinclined to negotiate a new treaty, and so Lord Aberdeen then took a leaf out of Palmerston's book, introducing a bill to enable the Royal Navy to seize Brazilian slavers and prosecute them as pirates in the Vice-Admiralty courts. The bill became law in 1845 as the Aberdeen Act. Under its authority the

Navy's cruisers again became active on the coast of Brazil, operating even inside territorial limits.[21]

The British suppression cause suffered another blow in 1846, and it was self-inflicted. There had long been a heavy duty on sugar grown other than in British colonies, but political pressure to reduce duties on imported foodstuffs, instigated by the Free Trade movement, led initially to a halving of duty on foreign-grown sugar, and shortly afterwards to a further reduction which allowed foreign, slave-grown, sugar to be imported on the same terms as that produced in the colonies. The result was a huge boost to the slave trade, particularly into Brazil.[22]

Inshore cruising and close blockade had become the standard practice on the slave coast north of the Equator, and, with the increasing number of local treaties with native chieftains, this achieved a reduction of slave imports to Cuba until, in 1846, the annual figure was only 1,000. The grip was then somewhat relaxed when a new Commodore, Hotham, no disciple of Denman's, discarded the policy of inshore cruising and blockade and encouraged the generally useless practice of cruising offshore as in the best interests of health.[23]

The number of cruisers stationed off West Africa remained steady at 12 or 13 until 1844, but then it climbed to about 20, and steamers were reintroduced in very small numbers in the early 1840s. By the beginning of the next decade the number of steamers equalled or outnumbered the cruisers under sail. A few of the better vessels were withdrawn from both shores of the Atlantic for a time during the Crimean War, but otherwise the numbers averaged about 25 until the end of the campaign.[24]

The focus of interest in the late 1840s was Brazil. The depredation of British cruisers on the coast, particularly close inshore, was imposing intense pressure on the Government, and political power was finely balanced between the slaving clique and the abolitionists. The recent large influx of slaves had heightened public concern about the danger of slave revolt, and it was touch-and-go whether the seizures by the British cruisers, resented by Brazilians on both sides, would strengthen the abolitionist cause or undermine it. The former was the result, and, with effective Brazilian legislation enacted and enforced, that nation's slave trade ended with remarkable speed.[25]

Thereafter Cuba was the only significant remaining market for new slaves, and by 1850 the only embarkation points on the slave coast north of the Equator available to the slavers were Whydah and Lagos. South of the Line, however, the Portuguese slaving ports were still open. The majority of the traffic during the 1850s consequently flowed from the southern coast of Africa into Cuba, and, as its volume increased to a level in 1859 approaching that at its peak in 1835, most of it was covered by the American flag.

The United States Navy squadrons off the African coast and in the approaches to Cuba provided the only force empowered to interdict this burgeoning traffic, but their cruisers in eastern waters were generally ineffective. They were still based at Porto Praya, it seems that they never included steamers, and their commitment to the task was sporadic, depending as it did on the political inclination of individual commanding officers. The hands of the British cruisers remained tied by the continuing refusal of Washington to allow them Right of Search or Visit, but, at the risk of diplomatic repercussions, boardings of vessels obviously slaving under false US colours did occasionally take place.[26]

In 1851 it was decided that the major slaving hotbed of Lagos had to be tackled by force, attempts at bringing its traffic to a halt by diplomatic means having failed. After an initial amphibious assault had been repulsed, a more thoroughly prepared attempt succeeded. That did not finally solve the problem, and in 1861 it was reluctantly decided in London to annex the city and surrounding territory in order to prevent a slaving resurrection. Porto Novo was burned, also in 1861, and the Trade north of the Equator was virtually destroyed.[27]

When President Lincoln arrived in the White House in 1861 the situation, domestic and international, changed rapidly. The US Abolition Acts of 1807 and 1820 were enforced, after half a century of neglect, and the American capital, ships and crews which had become essential to the Cuban Trade were withdrawn. Any hope that the Cuban planters had entertained of American annexation of Cuba to preserve slavery was dashed, and the Washington Treaty was signed between Britain and the United States. By this treaty, which was along the same lines as the 1835 Treaty with Spain, an Equipment Clause relating to American-flagged vessels was introduced, Mixed Commission courts were established at New York, Sierra Leone and the Cape, and the Royal Navy was permitted to board, search and arrest slavers under the Stars and Stripes.[28]

By that point Spain and the administration in Havana were inclined towards abolition, and, although the Trade struggled on in small numbers of ships under a variety of false ensigns, its volume decreased dramatically through the early 1860s, and by 1867 it had been effectively exterminated.[29]

In that year the West Africa Squadron was amalgamated with the Cape Squadron, and, 60 years after it had begun, Britain's suppression campaign came to an end.

* * *

The transatlantic trade in African slaves to the Americas became a deeply entrenched element of the commercial life of the western world, a matter of serious competition between the major maritime powers, and a crucial part of the economies of Spain, Portugal and, latterly, Brazil. In the second half of the eighteenth century, after 250 years of comfortable acceptance, the conscience of Christian society in Europe, the United States and Britain became increasingly troubled by the institution of slavery and by the traffic from West Africa which supported it. In the early 1800s abolitionist movements achieved anti-slaving legislation in several countries, most significant of which were France, the United States of America and Great Britain. Despite subsequent claims that Britain was motivated in its abolitionist stance by hope of economic gain, there is no evidence that the legislation in any of these nations stemmed from anything other than an altruistic desire to end a despicable trade.

For much of the next 50 years Britain stood alone among these nations in its determination to enforce its abolition law, which in its own case proved an easy task. Again alone, it then undertook suppression of the continuing slave traffic of other nations, and at the end of the Napoleonic Wars it was in a uniquely strong position to do so. Not only did Britain have unrivalled economic and naval power, but also political opinion in the country was essentially undivided on the question. However, as is generally so with supremely powerful and confident nations, the country was frequently (and occasionally justifiably) accused of arrogance in its international affairs, and it was inevitable that its stance on the slave trade, creditable though it was, would be unpopular in many quarters. The keen resentment of the slaving nations at Britain's efforts to impose its "holy philanthropy" upon them, and to disrupt their trade, was natural and unavoidable.

That was one of the factors which made the crusade such a difficult one, and it was one which it perhaps failed to appreciate fully. Britain did understand the determination of Spain, Portugal and Brazil to evade their responsibilities under the abolition treaties, and to place every available obstacle in the path of suppression, but it had neither sympathy nor patience with this viewpoint. It regarded the behaviour of these nations as morally and legally criminal, and, its contradictory (if understandable) reduction of duty on slave-grown sugar apart, successive governments, despite political dissent in the 1840s, were consistent in their belief that it was Britain's national duty to pursue to the end the destruction of the slave trade. In this they were surely right.

This belief was also strong in the Royal Navy's cruisers deployed against the Trade on both sides of the Atlantic. To their officers and men the struggle was not one of policy and theory, as it was to those in departments of state; they came face to face with the horrors perpetrated by the slavers, and to them it was personal. By contrast, British industry and commerce were concerned more with profit than philanthropy, and abolition legislation failed to prevent British trade goods from fuelling the Trade throughout the illegal period, undermining in foreign eyes the integrity of Britain's suppression policy.[30]

The first 30-year phase of the campaign, described in the main body of this book, was largely ineffective. This was partly because of the inadequacy of the resources allocated to the task, but primarily because the only part of the Trade under attack was the one least likely to deliver ultimate success, namely the Middle Passage. The markets in the Americas were beyond Britain's reach, but the African source was not, and the later use of local treaties showed that legal difficulties in striking at it could be overcome. The traffic on the Middle Passage initially seemed to be the easiest and most obvious target considering Britain's naval strength, but the Trade should have been attacked at its source as well once it was clearly demonstrated that a very high proportion of slavers were evading arrest. On that score it should certainly have been noted that, as objectives of offensive action, slave factories had the significant advantage over ships of being static.

It has repeatedly been demonstrated, not least in the illegal drug traffic of the modern age, that if a lucrative market exists for a commodity a means will be found to supply it. The slave trade was no exception. When the Navy began to attack the source end of the chain, by burning slave factories and concluding treaties with the coastal chieftains north of the Equator, the slave traders were, for the first time, seriously troubled; but the Trade did not cease until the destruction of the hardest target, the market, was achieved in Brazil at the beginning of the 1850s, and in Cuba over ten years later. It must be admitted, however, that, in the case of the Netherlands slave trade four decades earlier, it was primarily the interdiction of the Dutch seaborne slave traffic after the insertion of an Equipment Clause into the Anglo-Dutch Treaty which brought that trade to a halt.

The summary in the previous chapter of that first 30-year phase concludes with an assessment of the suppression campaign's admittedly modest achievements, but to those can be added the beginnings of a progressive increase in the price of slaves in the Americas. As the Navy's operations became more effective with the implementation of Equipment Clauses in the treaties, more refined tactics and improved and more numerous cruisers, difficulties and costs escalated for the traders and slave prices climbed. By 1861 the Cuban planters were paying $1,000 for a slave. This factor alone might, in time, have destroyed the Cuban market.

A damaging accusation levelled by the "Anti-Coercionists" was that the operations of British cruisers were not only useless and wasteful of public money and naval lives, but also were exacerbating the sufferings of the slaves. The abolitionists were naturally concerned that this last point might be valid, but it was surely only partially so. The policy of close blockade to prevent embarkation of cargoes certainly caused gross overcrowding in the barracoons, and consequently extreme privation for the slaves, but on the other hand the horrors of the Middle Passage were somewhat alleviated, at least in terms of duration. The slavers were obliged by the threat of arrest to invest in the swiftest vessels available, and, as the market value of the slaves increased, the traffickers would have done all they reasonably could to ensure that the lives and health of their cargoes were preserved.

Brazil's circumstances differed from those in Cuba. Its abolition movement had long exerted influence, occasionally in government, and its strength had gradually increased, probably as much from fear of slave revolution as from humanitarian considerations. The resentment against British pressure, already described, was strong in the country, but so was British influence in many aspects of Brazilian life. There is little doubt that intense action by the Royal Navy against slavers in Brazilian waters, permitted by the Aberdeen Act of 1845 and coupled with naval diplomacy, was decisive in bringing the abolitionists into power in the late 1840s, and enabled them to legislate against the slave trade and to enforce that legislation.

The slavers in Cuba were more tenacious, although they had been subjected to more determined suppression measures. The disparity between the British treaties with Spain and those with Portugal and Brazil, added to the consequent concentration of cruisers primarily against the traffic north of the Equator (at least until 1845), had achieved a check on the Cuban trade to some degree. However, by contrast with Brazil, there was no ambivalence in Cuba on support of the slave trade until the early 1860s. The availability of false colours, first Portuguese and then American, provided essential protection against greater depredation by British cruisers, and then American capital, ships and manpower fuelled a final bonanza. Lincoln's accession to the United States presidency brought the realisation in Cuba that its slave economy could not survive. The destruction of the island's slaving was ultimately ensured by the decision of the Lincoln administration in 1860 to enforce, at last, the American abolition legislation, and, by the Washington Treaty of 1862, to join its suppression efforts to those of Great Britain.

In the early 1860s the Cuban slave trade, the last vestige of the transatlantic traffic, gradually dwindled towards its end, an event for which no precise date can be identified. By the time of his death in 1865, Lord Palmerston knew that the Trade, which he had fought with such determination as Foreign Secretary, was a thing of the past, and in 1866 the British Consul in Havana reported that "The Cuban Slave Trade is virtually at an end." But there was no moment of victory to mark the satisfactory conclusion of Britain's enormous undertaking.

For the Royal Navy the campaign had been an extraordinary one, and not simply in its unique length. It was a campaign which had not formed part of any war, and no battle fleet had been engaged; indeed, no line-of-battle ship had been directly involved in it. It was a campaign fought by small vessels, the cruisers. Although it incurred very considerable cost in lives from innumerable actions and from disease, the Navy gained two significant benefits from the experience. First, several generations of junior officers developed the art of command in the hard school of complex suppression operations in gruelling conditions, and, second, the Navy for the first time developed an understanding of how to conduct operations in peacetime under a code of law other than that of belligerent rights.

Trafficking of human beings continues to this day in African coastal waters of the Atlantic Ocean under one guise or another, but by 1867 the transatlantic slave trade, begun in 1510, had finished. It cannot be claimed that Great Britain brought the Trade to an end, but it had led the way in suppressing it. Britain had struggled against it, almost entirely alone and in the face of obstruction by every slaving nation, and the contribution it had made to the destruction of this abomination, in endeavour, resources and lives, was incomparably greater than that of any other country.

Viscount Cardwell, during a debate in 1865, declared that "I own I do not know a nobler or a brighter page in the history of our country," and a more disinterested commentator, the nineteenth-century Irish historian William Lecky, remarked that "The unweary, unostentatious and inglorious crusade of England against slavery may probably be regarded as among the three or four perfectly virtuous pages comprised in the history of nations." As a more recent historian, Christopher Lloyd, has written:

> The heaviest burden throughout that long period fell upon the shoulders of the junior officers and men of the Royal Navy, and their work on all the coasts of Africa deserves recognition as one of the noblest efforts in our national history.

At the conclusion of this remarkable service to the civilised world, there was, however, little applause and scant praise for the British cruisers, and some would never forgive them for their persecution of the slavers. Nevertheless, posterity should conclude that there have been few episodes in the illustrious history of the Royal Navy which have been more deserving of true glory.

Guide to Abbreviations used in the Notes, Bibliography and Appendices

ADM   Admiralty
CO   Colonial Office
FO   Foreign Office
HCA   High Court of Admiralty
SP   State Papers
WO   War Office

APPENDIX A

Suspected Slave Vessels Detained 1807–39 by Royal Navy Cruisers, Colonial Vessels and Letters of Marque Vessels

Key

COLUMN A	*Date of arrest*
COLUMN B	*Name of vessel arrested*
COLUMN C	*Nationality of vessel arrested*

KEY:

Br	Brazilian
Fr	French
GB	British
Ne	Dutch
Po	Portuguese
Pr	Piratical
Ru	Russian
Sp	Spanish
Sw	Swedish
US	American

COLUMN D *Type of vessel arrested*

KEY:

B	Brig
Ba	Barque
Bn	Brigantine
C	Cutter
Co	Corvette
F	Felucca
G	Galliot
L	Launch
P	Polacca
S	Ship
Sc	Schooner
Sl	Sloop
St	Steamer
Su	Sumaca
X	Xebec

COLUMN E	*Arresting vessel* (all are commissioned warships of the Royal Navy, or tenders to such warships, unless otherwise indicated)
COLUMN F	*Geographical position of arrest*
COLUMN G	*Fate of arrested vessel*

KEY:

Condemned at:
- A   Antigua
- B   Barbados
- C   Cape of Good Hope
- E   Bermuda
- F   Freetown
- H   Havana
- HC   High Court of Admiralty
- J   Jamaica
- M   Bahamas
- O   Rio de Janeiro
- P   Plymouth
- T   Tortola
- V   St Vincent

- D   Released by captor
- G   Taken to Goree for trial
- K   Taken to foreign court for trial
- L   Condemnation reversed by London Commission
- N   Driven ashore
- R   Restored by British or Mixed Commission Court
- S   Seized by pirate
- U   Scuttled
- W   Case withdrawn
- X   Foundered
- Y   Adjudication refused
- Z   Retaken

COLUMN H	Number of slaves on board at capture
COLUMN I	Number of slaves emancipated or released
	(EC Vessel detained under an Equipment Clause)
COLUMN J	Page reference for capture of suspected slaver

\# indicates that the vessel is not mentioned by name in the narrative.

Suspected Slave Vessels Detained 1807–39   763

A	B	C	D	E	F	G	H	I	J
??.06.07	Busy	GB	B	Alexandria	West Indies	T	?	?	101
??.10.07	Nancy	US	Sc	Cerberus/Venus	Off St Thomas (West Indies)	T	77	77	105
??.12.07	Amedee	US	B	Swinger	West Indies	T	90	90	106
30.12.07	Minerva	?	?	Derwent	Freetown	?	?	?	112
02.02.08	Tartar	US	B	Ulysses	Off Martinique	B	?	?	106
05.03.08	La Jeune Laure	? Fr	?	Laurel, Grampus, Harrier	Off Cape of Good Hope	C	?	18	124
??.03.08	Baltimore	US	Sl	Derwent	? Off Sierra Leone	F	?	140	112
??.03.08	Eliza	US	Sc	Derwent	? Off Sierra Leone	F			112
15.06.08	La Fortuna	Sw	Sc	Subtle	West Indies	T	?	?	106
??.07.08	America	US	B	Latona	West Indies	R	?	?	106
??.07.08	Africa	US	S	Haughty	West Indies	T	?	?	106
23.08.08	Marie Paul	Fr	Sc	Derwent	? Off Sierra Leone	F	60	60	113
06.09.08	Two Cousins	US	?	Derwent	? Off Sierra Leone	F	?	4	113
??.10.08	São Joaquim	Po	Sl	Derwent	? Off Sierra Leone	F	0	0	113
??.10.08	São Domingo	Po	Sc	Derwent	? Off Sierra Leone	F	0	0	113
??.10.08	La Parsiphal	? Fr	?	Raisonable	Off Cape of Good Hope	C	?	? 19	124
05.10.08	La Prairie	? Fr	?	Leopard	Off Cape of Good Hope	C	?	27	124
19.10.08	La Souffleur	? Fr	?	Harrier	Off Cape of Good Hope	C	?	142	124
28.10.08	Marschal Dandels	? Fr	?	Otter	Off Cape of Good Hope	C	?	6	124
07.11.08	L'Esperance	? Fr	?	Leopard, Otter	Off Cape of Good Hope	C	?	59	124
22.11.08	L'Aventure	? Fr	?	Raisonable	Off Cape of Good Hope	C	?	5	124
??.12.08	Ceres	? Fr	?	Olympia	Off Cape of Good Hope	C	?	7	124
19.12.08	Gobe Houcha	? Fr	?	Nereide, Leopard, Harrier	Off Cape of Good Hope	C	?	35	124
??.1/2.09	La Paris	? Fr	?	Boudicea	Off Cape of Good Hope	C	?	3	124
??.02.09	Rapid	?	?	Derwent	? Off Freetown	R	0	0	113
??.02.09	Africaan	?	?	Derwent	? Off Freetown	F	?	3	113
20.04.09	Tilsit	? Fr	?	Charnell	Off Cape of Good Hope	C	?	8	124
??.5/6.09	Le Joseph	?	?	Argo	West Indies	J	?	?	124
??.6/7.09	Jane	?	?	Satellite	West Indies	J	2	2	124
30.07.09	Penel	Sw	Sc	Letter of Marque Minerva	? Off Sierra Leone	R	?	18	126
22.09.09	Le Trois Amis	? Fr	?	Raisonable	Off Cape of Good Hope	C	?	55	124
19.10.09	La Venus	? Fr	?	Boudicea	Off Cape of Good Hope	C	?	3	124
27.12.09	La Mouche	? Fr	?	Caledon	Off Cape of Good Hope	C	?	?	124

A	B	C	D	E	F	G	H	I	J
??.01.10	L'Urania	? Fr	?	Stork	Off Cape of Good Hope	C	?	?	124
09.01.10	La Charlotte	? Fr	?	Iphigenia	Off Cape of Good Hope	C	?	41	124
24.03.10	Rayo	Sp	B	Tigress	Off Rio Pongas	R	29	129	118
30.03.10	L'Amazone	? Fr	?	Otter	Off Cape of Good Hope	C	?	?	124
03.04.10	Lucia	?	?	Tigress	Off Rio Pongas	F	?	?	118
04.04.10	Polly	?	Sl	Crocodile	Matacong	R	o	o	120
04.04.10	Doris	US	Sc	Crocodile	Matacong	F	o	o	120
20.04.10	Mariana	? Sp	?	Crocodile	Off Sierra Leone	F	?	186	122
24.04.10	Esperanza	US	Sc	Crocodile	Shebar River	F	?	91	121
17.05.10	Ana	Sp	B	Crocodile	Off Cape Three Points	F	o	o	121
22.05.10	Donna Mariana	GB	B	Crocodile	Off Cape Coast Castle	F	o	o	121
02.06.10	Zaragozano	US	?	Crocodile	Off Sierra Leone	F	?	118	122
27.07.10	Santo Antonio Almos	Po	?	Letter of Marque Dart	? Off Sierra Leone	F	?	8	126
27.07.10	Flor Deoclerim	Po	?	Letter of Marque Dart	? Off Sierra Leone	F	?	8	126
??.08.10	St Jago	Sp	Sc	Crocodile	Off Sierra Leone	R	?	57	125
??.08.10	Pez Volador	Sp	Sc	Tigress	Off Isles de Los	R	?	82	128
??.08.10	Cirilla	?	?	Letter of Marque Dart	? Off Sierra Leone	F	o	o	126
04.08.10	Hermosa Rita	?	?	Letter of Marque Dart	? Off Sierra Leone	F	?	77	126
11.09.10	Diana	US	?	Crocodile	? Off Sierra Leone	R	?	84	127
22.09.10	Gallicia	GB	S	Amelia	Off Tenerife	P	?	?	130
22.09.10	Palafox	GB	Sc	Amelia	Off Tenerife	P	?	?	130
22.09.10	Marquis de Romana	GB	S	Tigress	Off Badagry	F	?	101	128
??.10.10	Emprenadadora	US	?	Crocodile	? Off Sierra Leone	R	o	o	127
??.10.10	Los Dos Amigos	US	?	Crocodile	? Off Sierra Leone	R	o	o	127
06.10.10	Fortuna	US	?	Melampus	Off Funchal	P	o	o	133
??.11.10	Merced	US	?	Col. Sc. George	River Gambia	F	o	o	125
??.11.10	Vincedor	US	?	Col. Sc. George	River Gambia	F	?	30	125
??.11.10	Maria Delores	GB	?	Col. Sc. George	River Gambia	F	o	o	125
??.11.10	Catalina	?	?	Col. Sc. George	River Gambia	R	o	o	125
??.11.10	Santa Barbara	?	?	Col. Sc. George	River Gambia	R	o	o	125
09.11.10	Maria	Fr	B	Protector	River Scarcies	F	?	?	129
??.11?.10	Vivilia	Sp	S	Crocodile	Off Sierra Leone	F	o	o	128
30.12.10	Mariana	Po	?	Letter of Marque Dart	? Off Sierra Leone	F	?	10	126
??.??.11	San Jose y Anemas	Sp	Sc	St Christopher	West Indies	A	?	211	144
??.??.11	Confianza	?	?	Col. Sch. George	Off Sierra Leone?	F	o	o	#
??.??.11	Jove	Sp	Sc	Persian	West Indies	R	?	?	145
??.01.11	Neptune	?	?	Wanderer	West Indies	R	?	140	144
10.01.11	Lucy	?	?	Crocodile	Off Sierra Leone	F	o	o	128

Suspected Slave Vessels Detained 1807–39   765

A	B	C	D	E	F	G	H	I	J
14.01.11	Aragansia Castellano	? Sp	B	Mutine	Mid-Atlantic	?	0	0	131
??.03.11	El Alrevido	Sp	?	Calibri	Bahamas	M	?	204	144
??.03.11	Empressa	? Sp	B	Calibri/Lille Belt	Off Bermuda	E	0	0	144
??.04.11	Edward	?	?	Arachne	West Indies	R	?	4	144
03.04.11	Elizabeth	US	?	Tigress	Off Cape Mount	F	?	87	128
??.05.11	Nos. Sen. de los Dolores	US	Sc	Myrtle	River Gambia	F	112	111	140
13.05.11	Roebuck	GB	Sc	Myrtle	Porta Praya	R	0	0	140
17.05.11	Gerona	GB	S	Myrtle	Off Cape Verde Islands	F	0	0	140
18.05.11	Santa Rosa	? US	?	Myrtle	Off Cape Verde Islands	R	0	0	140
??.06.11	Hawk	?	Sc	Arethusa	Off Iles de Los	F	0	0	141
??.06.11	Harriet	?	B	Arethusa	Off Iles de Los	R	0	0	141
25.06.11	Palomo	Sp	Sc	Protector	Off Cape Mount	F	0	0	147
??.6–9.11	Sancta Isabel	Po	?	Rattler	Bahamas	M	?	115	144
??.6–9.11	Joanna	Po	?	Decouverte	Bahamas	M	?	129	144
??.07.11	Capac	?	?	Tigress	Off Sierra Leone	F	?	?	139
27.07.11	Havanna	? GB	?	Thais	Off Trade Town	F	100	98	142
??.08.11	Falcon	?	?	Liberty	West Indies	R	?	?	144
26.08.11	Paquette Volante	Po	S	Tigress	Cabinda	R	?	38	143
26.08.11	Urbano	Po	Sc	Tigress	Cabinda	R	?	59	143
30.08.11	Venus	Po	B	Thais	Off Badagry	F	?	21	142
02.09.11	Calypso	Po	B	Thais	Off Lagos	R	?	13	142
30.09.11	Sally	?	?	Laura	West Indies	R	?	3	144
01.10.11	Scourge	?	?	Maria	West Indies	R	?	6	144
02.10.11	Industry	?	?	Maria	West Indies	R	?	11	144
02.10.11	Porpoise	?	?	Laura	West Indies	T	?	9	144
02.10.11	Espiegle	?	?	Laura	West Indies	T	?	3	144
04.10.11	Regent	?	?	Laura	West Indies	T	?	4	144
21.12.11	Gibraltar	?	?	Amaranthe	West Indies	T	?	2	144
30.12.11	Princessa de Beira	? US	Sc	Sabrina	Boa Vista	F	?	56	153
31.12.11	Adela	?	?	Laura	West Indies	T	?	3	144
31.12.11	San Jao	Po	B	Amelia	Cape Coast	R	?	10	146
07.01.12	Destino	Po	B	Amelia	Porto Novo	R	?	23	147
07.01.12	Dezanganos	Po	B	Amelia	Porto Novo	F	?	23	147
07.01.12	Felis Americano	Po	B	Amelia	Porto Novo	F	?	31	147
08.01.12	Flor de Porto	Po	Sc	Amelia	Off Lagos	F	?	116	147
09.01.12	Prezares	Po	B	Amelia	Off Lagos	R	?	204	147
09.01.12	Lindeza	Po	B	Amelia	Off Lagos	R	?	142	147
12.01.12	Il Pepe	? US	Sc	Sabrina	River Gambia	F	73	73	154
02.02.12	Maria Primero	Po	S	Protector	Off St Thomas	F	490	381	149
21.02.12	Vigilant	?	C	Kangaroo	Off St Thomas	F	63	53	150
21.03.12	Anthony	?	?	Laura	West Indies	T	?	1	145
22.03.12	St Joseph	?	?	Laura	West Indies	T	0	0	#
29.03.12	Jonah	?	L	Maria	West Indies	T	?	?	145
05.04.12	Urania	Po	B	Kangaroo	Whydah	F	0	0	151

A	B	C	D	E	F	G	H	I	J
05.04.12	S. Mig. Triumphante	Po	B	Kangaroo	Whydah	R	136	132	151
20.04.12	Penelope	?	?	Amaranthe	West Indies	T	?	2	145
28.04.12	Augustus	?	?	Laura	West Indies	T	?	2	145
??.06.12	Centinella	? Sp	B	Daring	Off Sierra Leone	F	o	o	155
??.06.12	San Carlos	? Sp	S	Daring	Off Trade Town	F	o	o	155
??.06.12	[Unidentified]	Sp	B	Daring	Off Trade Town	Z	?	?	155
22.06.12	Coque Maire	?	?	Maria	West Indies	T	?	3	145
24.06.12	Dolphin	US	Sc	Thais	S of Goree	F	?	79	157
19.07.12	Hope	US	Sc	Kangaroo	River Gambia	F	67	67	157
14.08.12	Carlotta	Sp	B	Thais	Loango Bay	F	o	o	157
29.08.12	Flor d'America	Po	B	Thais	Loango Bay	F	364	?	157
05.09.12	Orizonte	Po	Sc	Thais	Mayumba Bay	F	?	18	158
06.10.12	Andorinha	? GB	S	Amelia	Ambriz Bay	F	?	270	156
12.10.12	Prevoyant	?	?	Peruvian	West Indies	T	?	3	145
??.11.12	Nueva Constitución	US	?	Daring/Kangaroo	Off Sierra Leone	R	o	o	158
20.12.12	Triumpho de O'Nia	?	?	Amelia	? Off Sierra Leone	F	?	17	160
??.??.13	Amelia	?	?	Letter of Marque Kitty	? Off Sierra Leone	F	?	85	#
??.05.13	[Unidentified]	?	?	Colibri	Bahamas	B	?	3	#
28.05.13	Juan	US	Sl	Thais	Off Cape Sierra Leone	F	o	o	162
04.06.13	San Jose Triumfo	Sp	B	Letter of Marque Kitty (Thais)	Off Sierra Leone	F	?	96	162
04.06.13	Phoenix	Po	B	Letter of Marque Kitty (Thais)	Off Sierra Leone	F	?	1	162
31.08.13	Providencia	Po	B	Favorite	Porto Novo	F	o	o	165
01.09.13	San Josef Desforca	Po	B	Favorite	Badagry	F	?	48	165
07.10.13	St Joseph	Po	Sc	Favorite	Old Calabar River	D	59	48	165
16.10.13	Bon Jesu	Po	Sc	Favorite	River Gabon	F	55	47	165
??.11.13	El Dos de Mayo	Sp	Sc	Maria	West Indies	A	?	52	145
??.11.13	S. Francisco de Paula	?	?	Forester	West Indies	J	?	78	#
??.12.13	Union	?	?	?	West Indies	J	?	?	173
??.12.13	Carthagena	?	?	Sappho	West Indies	J	?	98	#
??.??.14	Nossa Sen. da Vitória	Po	S	Col. Sch. Princess Charlotte	? Off Sierra Leone	F	?	434	188
??.02.14	Laura Ana	?	?	Col. Sch. Princess Charlotte	? Off Sierra Leone	F	o	o	#
??.02.14	Yeavel	?	?	Col. Sch. Princess Charlotte	River Pongas	F	o	o	#
??.03.14	San Jose	?	?	Col. Sch. Princess Charlotte	? Off River Pongas	F	o	o	#
??.03.14	Teresa	?	?	Col. Sch. Princess Charlotte	? Off Sierra Leone	R	?	?	#
05.03.14	Carlos	? Sp	B	Pique	West Indies	?T	?	444	168
15.03.14	Concha	? Sp	Sc	Ister	West Indies	R	?	?	168

Suspected Slave Vessels Detained 1807–39   767

A	B	C	D	E	F	G	H	I	J
??.06.14	Maria Josepha	?	Sc	Col. Sch. *Princess Charlotte*	River Gallinas	F	?	56	#
01.06.14	Gertrudis la Preciosa	Sp	S	Creole/Astrea	? Windward Coast	F	?	477	167
03.06.14	El Josepha	? Sp	Sc	Maria	West Indies	A	?	278	168
30.06.14	Manuella	Sp	X	Mosquito	West Indies	T	502	314	168
??.07?.14	Bon Successo	Po	B	Creole/Astrea	? Windward Coast	L	?	251	167
??.07.14	Nostra Sen. de la Bella	?	?	Col. Sch. *Princess Charlotte*	Cape Mount	F	?	56	#
17.07.14	Venus Havannera	Sp	S	Barbadoes	West Indies	T	410	303	168
??.09.14	Dolores	?	B	Col. Sch. *Princess Charlotte*	Cape Mount	F	?	34	#
??.09.14	Vanganza	?	Sc	?	? Off Sierra Leone	F	?	43	180
28.09.14	Boa União	Po	B	Niger	400nm NE of Ascension Is.	Z	?	0	169
07.10.14	Candelaria	? Sp	?	Barossa	West Indies	T	?	172	168
??.11.14	Resurrexion	?	Sc	Col. Sch. *Princess Charlotte*	Cape Mount?	F	?	60	#
??.11.14	Golandrina	?	S	Col. Sch. *Princess Charlotte*	Cape Mount	F	?	144	188
08.12.14	Union	Sp	Sc	Brisk	Grand Bassa	F	0	0	173
07.01.15	Conceição	Po	B	Brisk	Off Porto Novo	F	0	1	174
13.01.15	General Silveira	Po	B	Brisk	50nm S of Cape Formoso	L	?	238	174
20.01.15	Alrevido	? Sp	?	Ister/Columbine	West Indies	B	?	?	168
15.02.15	São Joachim	Po	?	Cumberland	Off Cape of Good Hope	L	?	300	#
20.02.15	Dolores	Sp	Sc	Ulysses	Off Senegal	F	0	0	178
16.03.15	Dos Amigos	Po	Sc	Comus	Old Calabar River	F	?	1	175
24.03.15	Sophie	? Sp	Sc	Porcupine	River Gambia	R	0	0	175
25.03.15	Catalina	Sp	Sc	Comus	Duke Town	F	0	0	176
25.03.15	Carmen	Sp	Sc	Comus	Duke Town	F	?	120	176
25.03.15	Intrepide	Sp	B	Comus	Duke Town	F	?	245	176
25.03.15	Bon Sorte	Po	?	Comus	Duke Town	L	?	61	176
25.03.15	Estrella	Po	Sc	Comus	Duke Town	F	?	42	176
??.04.15	Le Cultivateur	Fr	?	Ulysses	Bight of Biafra	R	0	0	179
03.04.15	Santa Anna	Po	B	Comus	Old Calabar River	L	0	3	176
23.04.15	Maria Madelena	Po	Sc	Comus	Off Princes Island	F	0	0	177
??.11.13	El Dos de Mayo	Sp	Sc	Maria	West Indies	A	?	52	145
??.??.15	Correio	Po	?	Ulysses (Diligente)	Bights	?	345	0	180
??.??.15	Leal Portugueze	Po	?	Col. Sch. *Princess Charlotte*	? Off Sierra Leone	L	0	0	#
??.??.15	Reinha dos Anges	Po	?	Col. Sch. *Princess Charlotte*	? Off Sierra Leone	L	0	0	#
??.??.15	Gaveo	Po	?	Col. Sch. *Princess Charlotte*	? Off Sierra Leone	L	0	0	188
??.??.15	San Joaquim	US	B	Col. Sch. *Princess Charlotte*	? Off Sierra Leone	F	?	39	#
??.05.15	Bella Amazon	?	?	Col. Sch. *Princess Charlotte*	? Off Sierra Leone	F	0	0	#

A	B	C	D	E	F	G	H	I	J
??.05.15	*Triumfo Africano*	Po	?	Col. Sch. *Princess Charlotte*	? Off Sierra Leone	L	?	18	#
?3.05.15	*Diligente*	Po	Sc	*Ulysses*	River Gabon	F	?	29	179
12.05.15	*Dido*	Po	S	*Brisk*	Off St Thomas	L	?	327	174
??.06.15	*Nova Fragantina*	Po	B	*Comus*	Anamabo	L	o	o	188
15.07.15	*Abismo*	Po	Bn	*Comus*	Cape Palmas	L	o	o	188
15.07.15	*Palafox*	Sp	Sc	*Comus*	Cape Palmas	F	o	o	188
??.09.15	*La Belle*	Fr	S	*Barbadoes/Columbia*	West Indies	A	?	512	223
??.09.15	*Hermione*	Fr	B	*Barbadoes/Chanticleer*	West Indies	A	?	211	223
18.01.16	*Rosa*	US	Sc	*Bann*/Col. Sloop *Mary*	Off River Gallinas	F	276	276	190
13.02.16	*Guadeloupe*	Sp	Sc	Col. Sch. *Young Pr. Charlotte*	Off Sierra Leone	F	o	o	190
14.02.16	*Rayo*	Sp	B	Col. Sch. *Young Pr. Charlotte*	Off Sierra Leone	F	o	o	118
03.03.16	*Eugenia*	Sp	Sc	Col. Sch. *Young Pr. Charlotte*	Rio Pongas	F	49	49	191
03.03.16	*Juana*	Sp	Sc	Col. Sch. *Young Pr. Charlotte*	Rio Pongas	F	33	33	191
04.03.16	*Dolores*	Sp	Bn	*Ferret*	300nm N of Ascension Is.	F	250	249	193
05.03.16	*Temerario*	Sp	B	*Bann*	Whydah	F	17	17	192
11.03.16	*Selby*	US	Sc	*Zenobia*	West of C. Verde Islands	X	o	o	195
11.03.16	*Louis*	Fr	B	Col. Schs *Q. Ch'lte, Pr. Ch'lte*	Off Cape Mesurado	R	o	o	191
16.03.16	*S. Antonio Milagroso*	Po	B	*Bann*	Off Princes Island	L	548	505	193
20?.04.16	*La Neuve Aimable*	Sp	Sc	Col. Brig *Prince Regent*	Off Freetown	F	388	388	197
05.05.16	*La Paz*	Pr	?	Col. Brig *Prince Regent*	Off Cape Mesurado	F	?	108	197
21.05.16	*Carmen*	Sp	Sc	*Inconstant*	Off Quitta Fort	F	o	o	196
24.05.16	*Caveira*	Po	Sc	*Inconstant*	Off Lagos	L	10	10	196
27.06.16	*Dos Amigos*	Po	Sc	*Inconstant*	Corisco Bay	L	?	247	197
28.07.16	*Mt. de Carmotesta*		B	*Inconstant*	Off Freetown	L	o	o	197
??.08.16	*Scipiao Africano*	? Po	?	*Inconstant*	? Cape Coast	L	o	o	197
??.08.16	[Unidentified]	?	?	*Bermuda*	Bahamas	C	?	221	184
30.09.16	*Triumphante*	Sp	B	Col. Brig *Prince Regent*	River Cameroon	F	?	349	198
07.11.16	*Rodeur*	Po	Sc	Col. Brig *Prince Regent*	? Bight of Benin	L	?	32	198
07.11.16	*Caroline*	Po	Bn	Col. Brig *Prince Regent*	? Bight of Benin	F	?	18	198
07.11.16	*Santa Johanna*	Po	Sc	Col. Brig *Prince Regent*	? Bight of Benin	F	72	65	198
15.11.16	*Ceres*	Po	B	Col. Brig *Prince Regent*	? Bight of Benin	L	?	11	198
??.12.16	[Unidentified]	?	?	*Orpheus*	West Indies	V	?	152	223

Suspected Slave Vessels Detained 1807–39   769

A	B	C	D	E	F	G	H	I	J
17.01.17	*Esperanza*	Po	S	*Cherub*	Off Cape Lahou	F	413	408	200
31.01.17	*Louisa*	Fr	Sc	*Cherub*	Off Princes Is.	X	0	0	210
29.03.17	*Laberinto*	Sp	Sc	Col. Brig *Prince Regent*	Rio Pongas	F	?	28	207
12.04.17	*Gramachree*	Po	Sc	Col. Brig *Prince Regent*	Rio Nunez	F	0	0	207
18.04.17	*Esperanza*	Sp	B	Col. Brig *Prince Regent*	Rio Nunez	F	0	0	207
21.10.17	*Two Boats*	GB	B	Col. Brig *Prince Regent*	Off Rio Pongas	?	0	0	207
07.12.17	*S. Juan Nepomuceno*	Sp	Bn	Col. Brig *Prince Regent*	Off Rio Pongas/Rio Nunez	F	272	269	207
14.12.17	*Linde Africano*	Po	Sc	Col. Brig *Prince Regent*	Rio Pongas/Rio Nunez	F	7	5	207
27.12.17	*Concha*	Pr	Sc	*Cherub*	Princes Island	R	?	433	208
15.01.18	*Descubridor*	Sp	B	*Cherub*	Off Quitta Fort	R			209
21.01.18	*Hannah*	GB	Sc	Col. Brig *Prince Regent*	Off Rio Pongas	F	5	5	207
31.01.18	*Belle Machancho*	Sp	Sc	Col.Brig *Prince Regent*	Off R. Nunez ?	U	130	130	207
15.05.18	*Josepha*	Sp	S	*Cherub*	Off Cape Three Points	Y	35	35	209
??.06.18	*Alfred*	GB	S	*Tagus*	Off Madeira	R	0	0	212
09.02.19	*La Sylphe*	Fr	?	*Redwing*	Mid-Atlantic	R	388	364	214
25.03.19	*Armistad*	Po	Sc	*Tartar*	Princes Island	?	1	1	216
25.03.19	*Princessa*	? Sp	Sc	*Tartar*	Princes Island	?	6	6	216
30.07.19	*Novo Felicidade*	Po	Sc	*Pheasant*	Off River Campo	F	72	71	217
10.08.19	*Nostra Sen. de Regla*	Sp	Sc	*Morgiana*	Off Little Bassa	F	1	1	218
18.09.19	*Fabiana*	Sp	Sc	*Morgiana*	Off Cape Palmas	F	13	13	225
30.09.19	*Juanita*	Sp	Sc	*Snapper*	River Costa	F	9	9	220
06.10.19	*Vulcano do Sud*	Po	B	*Pheasant*	West of Cape Formoso	Z	270	0	221
09.10.19	*Eliza*	Ne	Sc	*Thistle*	Trade Town	F	1	1	219
10.10.19	*Virginie*	Ne	Sc	*Thistle*	River Sesters	F	32	31	220
26.10.19	*Cintra*	Po	Sc	*Morgiana*	Off River Gallinas	F	26	25	227
10.12.19	*N'a Sn. de los Nieves*	Sp	Sc	*Myrmidon*	N of River Gallinas	F	122	121	224
10.12.19	*Esperanza*	Sp	Sc	*Morgiana*	Off Little Bassa	F	41	40	227
25.01.20	*La Marie*	? Fr	Sc	*Morgiana*	River Gallinas	R	106	90	237
25.01.20	*San Salvador*	Po	Sc	*Myrmidon*	River Manna	R	1	1	237
30.01.20	*Francisco*	Sp	Sc	*Tartar/Thistle*	Rio Pongas	F	69	69	229
30.01.20	*Marie*	Ne	B	*Tartar/Thistle*	Rio Pongas	F	12	2	229
30.01.20	*L'Arrogante*	Pr	Bn	*Myrmidon/Morgiana*	Off River Manna	?	0	0	238
30.01.20	*Anna Marie*	Sp	Sc	*Myrmidon/Morgiana*	Off River Manna	R	0	0	238
30.01.20	*El Carmen*	Sp	Sc	*Myrmidon/Morgiana*	Off River Manna	R	0	0	238
03.02.20	*Prince of Orange*	Pr	Sc	*Morgiana*	Off Little Bassa	?	0	0	239
05.02.20	*L'Invincible II*	Pr	Sc	*Morgiana*	Off Grain Coast	R	28	27	239
15.02.20	*Prince of Brazil*	GB	S	*Morgiana*	Cape Coast	F	?	?	239
15.02.20	*Jane Nicholl*	GB	?	*Morgiana*	Cape Coast	R?	0	0	240

A	B	C	D	E	F	G	H	I	J
??.03.20	La Catherine	Fr	Sc	Tartar	S of River Gallinas	Z	50	0	240
02.03.20	Gazetta	Sp	Sc	Tartar	Off Trade Town	F	82	81	240
12.09.20	Two Sisters	Br	Sl	Thistle	Rio Pongas	F	15	15	246
16.10.20	Nra. Sn. de Montserrate	Sp	Sc	Thistle	Little Cape Mount River	F	85	84	246
14.02.21	Emilia	Po	Sc	Morgiana	Bight of Benin	O	398	352	247
23.03.21	Anna Maria	Sp	Sc	Tartar/Thistle	River Bonny	F	491	400	249
24.03.21	Donna Eugenia	Po	?	Tartar/Thistle	River Bonny	F	83	78	249
09.04.21	Constantia	Po	Bn	Tartar/Thistle	Old Calabar River	F	250	220	250
09.04.21	Gavião	Po	B	Tartar/Thistle	Old Calabar River	R	1	1	250
25.07.21	Adelaide	Po	Sc	Myrmidon/Pheasant	Off Cape Formoso	F	232	207	253
02.08.21	Conceição	Po	Sc	Snapper	Old Calabar River	F	56	54	255
31.08.21	La Caridad	Sp	B	Myrmidon	River Bonny	F	153	136	254
31.08.21	El Neuve Virgen	Sp	Sc	Myrmidon	River Bonny	F	140	106	254
11.01.22	Rosalía	Sp	Sc	Thistle	Rio Pongas	F	59	59	256
21.02.22	Conde de Ville Flor	Po	Bn	Iphigenia	Bissau	F	171	171	259
26.02.22	Joseph	Sw	Sc	Iphigenia (Augusta)	River Gallinas	F	0	0	260
17.03.22	Dichesa Estrella	Sp	Sc	Morgiana	Trade Town	X	34	29	261
01.04.22	Dies de Feverio	Po	B	Iphigenia	Appam	F	10	10	261
06.04.22	Nympha del Mar	Po	Sc	Iphigenia	Whydah	X	3	2	261
07.04.22	Esperanca Felix	Po	P	Iphigenia (Tender)	Lagos	U	187	180	263
15.04.22	Icanam	Sp	Sc	Iphigenia/Myrmidon	River Bonny	X	380	0	263
15.04.22	Vecua	Sp	Sc	Iphigenia/Myrmidon	River Bonny	F	300	325	263
15.04.22	Vigilante	Fr	B	Iphigenia/Myrmidon	River Bonny	?			264
15.04.22	Petite Betsy	Fr	B	Iphigenia/Myrmidon	River Bonny	?	808	?808	264
15.04.22	Ursula	Fr	Bn	Iphigenia/Myrmidon	River Bonny	?			264
15.04.22	Esperanca Placido	Po	B	Morgiana	Lagos	F	149	147	265
27.04.22	Defensora da Patrie	Po	Sc	Myrmidon	Old Calabar River	U	100	80	264
23.06.22	San Jose Xalanca	Po	Sc	Thistle	Old Calabar River	U	20	17	266
29.06.22	Estrella	Po	B	Thistle	Off Cape Formoso	F	298	292	266
19.08.22	Josepha	Sp	Sc	Driver	Off River Bonny	F	215	183	266
27.08.22	San Raphael	Sp	Sc	Bann	Whydah	R	0	0	267
07.09.22	Commerciante	Po	B	Driver	River Cameroons	F	179	167	267
29.09.22	Magdalena da Praca	Po	Sc	Bann	E of St Thomas	F	33	33	268
05.10.22	S. Antonio de Lisboa	Po	Bn	Bann	Whydah	F	335	317	268
14.10.22	Nova Sorte	Po	Bn	Snapper	West of Little Popo	R	122	122	269
23.10.22	Aurora	Ne	Sc	Cyrene	Off Cape Mount	F	180	178	269
23.10.22	L'Hypolite	Fr	Sc	Cyrene	Off Cape Mount	R	0	0	269
31.10.22	Juliana da Praca	Po	Sc	Bann	Off Porto Novo	F	114	99	268
10.11.22	La Caroline	Fr	Sc	Cyrene	Off Little Bassa	R	80	80	270
13.11.22	Conceição	Po	Sc	Bann	Off St Thomas	F	207	198	268
03.12.22	Sinceridade	Po	Bn	Bann	600nm west of St Thomas	F	123	123	268
10.06.23	Maria la Luz	Sp	Sc	Owen Glendower	New Calabar River	F	230	?230	275

Suspected Slave Vessels Detained 1807–39 771

A	B	C	D	E	F	G	H	I	J
16.06.23	Conchita	Sp	Sc	Owen Glendower	New Calabar River	?	58	? 58	275
14.09.23	Fabiana	Sp	Sc	Owen Glendower	River Bonny	F	170	118	276
30.01.24	Cerqueira	Br	B	Bann	Lagos	R	0	0	279
30.01.24	Minerva	Br	S	Bann	Lagos	R	0	0	279
30.01.24	Creola	Br	Sc	Bann	Lagos	R	0	0	279
10.03.24	Bom Caminho	Br	B	Bann	Bight of Benin	F	334	?	285
22.04.24	El Vencador	Br	B	Victor	Off Lagos	F	0	0	285
08.05.24	Maria Piquena	Po	Sc	Victor	Princes Island	F	17	16	285
11.08.24	Dianna	Br	Bn	Victor	100nm W of Princes Island	F	143	?	288
18.09.24	Dos Amigos Brazilieros	Br	Bn	Victor	70nm SW of Princes Island	F	260	?	288
26.09.24	Aviso	Br	B	Maidstone	S of Princes Island	F	465	431	289
23.10.24	Bella Eliza	Br	B	Bann	100nm NW of Princes Is.	F	371	359	289
14.12.24	Relámpago	Sp	Sc	Carnation (Lion)	SE of Cay Sal Bank	F	159	159	310
14.01.25	Bom Fim	Br	S	Swinger	100nm NW of Princes Is.	F	149	146	291
07.05.25	Espanola	Sp	S	Atholl	River Gallinas	F	270	270	292
19.05.25	Bey	Ne	S	Maidstone	River Gallinas	F	0	0	294
17.07.25	Bom Jesus dos Nav.	Br	Su	Esk	100nm W of Cape Formoso	F	280	267	294
31.07.25	Z	Ne	B	Maidstone	River Sombrero	F	0	0	295
01.09.25	La Venus	Ne	Sc	Atholl	Off Cape Formoso	F	0	0	295
09.09.25	União	Br	Sc	Esk/Atholl/Redwing	130nm SW of C. Formoso	F	361	249	296
29.09.25	Segunda Gallega	Sp	Sc	Maidstone	Off Lagos	F	285	285	296
05.10.25	Isabel	Sp	Bn	Lion	Off Nuevitas (Cuba)	H	10	10	388
06.10.25	Teresa	Sp	Sc	Redwing	Old Calabar River	X	248	0	296
06.10.25	Isabella	Sp	Bn	Redwing	Old Calabar River	S	273	45	296
11.10.25	Ana	Sp	Bn	Redwing	River Cameroons	F	106	85	297
04.11.25	Clara (or Clarita)	Sp	Sc	Brazen	50nm S of Cape Mesurado	F	36	36	298
12.11.25	Aimable Claudina	Ne	Sc	Atholl	Elmina	F	34	34	300
17.11.25	Ninfa Habanera	Sp	Bn	Brazen	Accra	F	236	228	298
22.11.25	Paqueta de Bahia	Br	B	Swinger	Whydah	F	386	385	299
28.11.25	S. João Seg'nda Rosália	Br	Bn	Atholl	130nm S of Cape St Pauls	F	258	186	300
19.12.25	Malta	GB	S	Brazen	Princes Island	F	0	0	299
19.12.25	Charles	Ne	B	Conflict	Old Calabar River	F	265	243	300
27.12.25	Iberia	Sp	S	Brazen	Bight of Benin	F	422	417	299
03.01.26	Hoop	Ne	Sc	Maidstone	Off River Gallinas	F	0	0	303
22.01.26	Vogel	Ne	Sc	Brazen (Black Nymph)	Off Cape Mount	F	0	0	299
22.01.26	Magico	Sp	Bn	Union	Near Manatí (Cuba)	H	179	179	388
28.01.26	Pilar	Br	?	Redwing	Whydah	O	0	0	304
01.02.26	Activo	Br	B	Atholl	S of Gulf of Guinea	R	165	0	303

A	B	C	D	E	F	G	H	I	J
03.02.26	Fingal	Sp	Sc	Ferret	Near Salt Cay	H	58	58	388
04.03.26	Esperanza	Br	Sl	Esk	River Benin	F	4	4	305
04.03.26	Netuno	Br	Bn	Esk	River Benin	F	92	84	305
05.03.26	Orestes	Sp	Bn	Speedwell	Grass Cut Cays	X	238	212	388
08.03.26	Cantabre	Fr	?	Redwing	Bights	D	0	0	305
02.04.26	Atalanta	Fr	B	Col. St. Vssl. African	Off R. Gallinas	?	0	0	310
18.04.26	Perpetuo Defensor	Br	B	Maidstone	Off Annobón	R	424	0	308
17.05.26	La Fortunee	Ne	Sc	Brazen	W of Princes Island	F	245	126	307
20.05.26	Nicanor	Sp	Sc	Maidstone (Hope)	Off Whydah	F	174	173	309
11.06.26	San Benedicto	Br	S	Brazen	Popo	R	25	25	308
05.08.26	Principe de Guiné	Br	B	Maidstone (Hope)	Off Whydah	F	608	579	309
21.08.26	Mexicano	Sp	St	Pylades	Havana–Matanzas	R	0	0	390
29.08.26	Nuevo Campeador	Sp	Bn	Aurora	Off Santiago de Cuba	H	263	217	390
28.09.26	De Snelheid	Ne	Bn	Brazen	Off St Thomas	F	23	23	312
10.10.26	Intrepida	Sp	c	Esk	50nm NW of Princes Island	F	290	235	313
17.10.26	Hiroina	Br	n	Maidstone	Badagry	F	0	0	311
06.12.26	Paulita	Sp	Sc	Maidstone (Hope)	Off Lagos	F	221	191	313
21.12.26	Invincival	Br	S	Esk	River Cameroons	F	440	254	313
06.01.27	Eclipse	Br	Sc	North Star	Whydah	F	0	0	314
09.01.27	Lynx	Ne	B	Esk	Off Princes Island	F	265	251	314
31.01.27	Emilia	Sp	Sc	North Star	River Bonny	F	282	177	314
06.02.27	Venus	Br	Sc	Esk	Whydah	F	191	191	315
07.02.27	Fama de Cuba	Sp	Sc	North Star	Old Calabar River	F	100	95	314
08.02.27	Dos Amigos	Br	Sc	Esk	SE of Princes Island	F	317	308	315
28.02.27	Independencia	Br	Sc	Conflict	Accra	F	0	0	316
04.03.27	Conceição de Marie	Br	Bn	North Star	Whydah	F	232	198	314
12.03.27	Silveirinha	Br	B	North Star	Old Calabar River	F	266	209	314
13.03.27	Trajano	Br	Bn	Maidstone	Whydah	F	0	0	315
14.03.27	Venturosa	Br	B	Maidstone	Badagry	F	0	0	315
14.03.27	Carlotta	Br	Sc	Maidstone	Badagry	F	0	0	315
14.03.27	Tenterdora	Br	Sc	Maidstone	Near Porto Novo	F	0	0	315
16.03.27	Providencia	Br	Bn	Maidstone	Lagos	F	0	0	315
22.03.27	Con. Paquete do Rio	Br	Sl	Maidstone	Off River Benin	F	0	0	315
03.04.27	Bahia	Br	B	Conflict	10nm E of Cape St Paul's	F	0	0	316
10.04.27	Creola	Br	Bn	Maidstone	N of Fernando Po	F	308	289	318
19.04.27	Tres Amigos	Br	Sc	North Star (Little Bear)	Off Cape Sierra Leone	F	3	3	317
15.05.27	Copioba	Br	B	Clinker	10nm E of Cape St Paul's	F	0	0	316
18.06.27	Toninha	Po	Sc	North Star (African)	Bisagos Islands	U	65	58	318
06.09.27	Henriquetta	Br	B	Sybille	Off Lagos	F	569	542	322

Suspected Slave Vessels Detained 1807–39

A	B	C	D	E	F	G	H	I	J
12.10.27	Diana	Br	Sc	Sybille	180nm W of Princes Island	F	87	82	322
23.10.27	San João Voador	Br	Su	Eden	Quitta	R	0	0	324
24.10.27	Vencedora	Br	Sc	Eden	Whydah	R	0	0	324
19.12.27	Guerrero	Sp	B	Nimble	Florida Reef	X	120	120	391
12.01.28	Gertrudis	Sp	Sc	Sybille (Black Joke)	Off River Gallinas	F	155	155	330
03.02.28	Felis Victoria	Sp	?	Eden	River St John	F	2	2	334
19.03.28	La Fanny	Ne	Sc	Sybille	115nm S of Cape Formoso	F	266	238	330
13.04.28	Esperanza	Po	Sc	Sybille	Off Lagos	F	0	0	331
14.04.28	Musquito	Sp	Sc	Eden	Old Calabar River	F	126	?	334
18.04.28	Voadora	Br	Sc	Eden	River Cameroons	F	234	45	335
20.04.28	Terceira Rosália	Br	B	North Star	Popo	F	0	0	331
16.05.28	Vengador	Br	B	Sybille (Black Joke)	70nm SE of Cape St Paul's	F	645	624	336
11.06.28	Emprendedor	Sp	Sc	Eden	60nm E of Princes Island	F	3	3	332
27.06.28	Xerxes	Sp	Sc	Grasshopper	Gulf of Mexico	H	405	405	393
04.07.28	Josephina	Br	Sc	Sybille	120nm SW of C. Formoso	F	79	77	336
28.07.28	Nova Virgen	Br	Sc	Primrose	Off Lagos	F	354	320	337
02.08.28	Intrepido	Sp	Bn	Skipjack	Off Haiti	H	153	135	393
05.08.28	Clementina	Br	B	Clinker	River Cameroons	F	271	156	336
08.08.28	Sociedade	Br	Sc	North Star	100nm NW of Princes Is.	F	0	0	341
11.08.28	Henrietta	Ne	B	Eden	N of Fernando Po	F	425	350	342
20.08.28	Voador	Br	Sc	Clinker	River Bimbia	F	0	0	336
27.08.28	Presidente	Pr	Sc	Sybille (Black Joke)	Off Whydah	X	0	0	339
27.08.28	Hosse	Br	B	Sybille (Black Joke)	Off Whydah	F	0	0	339
14.09.28	Zepherina	Br	Sc	Primrose/Black Joke	30nm S of Lagos	F	218	153	340
03.10.28	Penha da Franca	Br	Sc	Medina	Open ocean N of Line	F	184	169	341
17.10.28	Min'a da Conceição	Br	Sl	Plumper	250nm SW of C. Formoso	F	105	82	340
17.10.28	Santa Effigenia	Br	Sc	North Star	50nm S of Badagry	F	218	217	342
30.10.28	Estrella do Mar	Br	?	North Star (Little Bear)	River Cameroons	F	0	0	342
30.10.28	Arcenia	Br	Sc	North Star (Little Bear)	River Cameroons	F	448	269	342
01.11.28	Campeadora	Sp	Sc	North Star	Bonny Bar	F	381	364	342
12.11.28	El Juan	Sp	B	Medina	180nm SSW of Princes Is.	F	407	378	341
13.11.28	La Coquette	Ne	Sc	Eden?	Off Fernando Po	F	220	185	344
16.11.28	Venus	Pr	S	Eden (Cornelia)	Old Calabar River	?	10	10	343
21.11.28	Neirsee	Ne	B	Eden (Cornelia)	? Calabar River	Z	280	0	344
22.11.28	Firme	Sp	Bn	Grasshopper	Off Martinique	H	487	484	394
23.11.28	Triumpho	Br	Sc	Medina	90nm SSE of C. Formoso	F	127	?	341
29.11.28	Maria	Sp	Sc	Skipjack	9nm W of Havana	U	1	1	395
18.12.28	Bolivar	Sp	B	Eden (Portia)	Old Calabar River	F	426	426	345

A	B	C	D	E	F	G	H	I	J
06.01.29	Jules	Ne	B	Eden (Cornelia)	Old Calabar Bar	R	220	?	346
06.01.29	La Jeune Eugenie	Ne	Sc	Eden (Cornelia)	Old Calabar Bar	R	50	?	346
07.01.29	Bella Eliza	Br	Sc	Medina	80nm SW of C. Formoso	F	232	?	341
15.01.29	Vingador	Po	B	Primrose	Off River Cacheu	F	223	220	347
15.01.29	Aurelia	Po	G	Primrose	In River Cacheu	F	29	29	347
01.02.29	El Almirante	Sp	B	Sybille (Black Joke)	S of Lagos	F	466	416	350
06.02.29	União	Br	B	Sybille	90nm S of Cape Formoso	F	405	366	351
09.02.29	Adeline	Ne	Sc	Eden	Fernando Po	F	0	0	352
15.02.29	Mensageira	Br	Sc	Eden (Cornelia)	Bonny Bar	F	353	304	353
19.02.29	Andorinha	Br	B	Sybille	Lagos	F	0	0	351
19.02.29	Golondrina	Sp	Sc	Pickle	20nm E of Punto Manatí	H	1	0	396
26.02.29	Hirondelle	Ne	Sc	Eden	Mouth of River Calabar?	F	113	89	353
??.03.29	Diana	Sp	Co	Eden (Cornelia)	Off River Bonny	D	0	0	354
06.03.29	Carolina	Br	Bn	Sybille (Black Joke)	85nm SE of Lagos	F	420	?	354
15.03.29	Donna Barbara	Br	B	Sybille (Paul Pry)	Bight of Benin	F	376	351	352
23.03.29	Hosse	Po	B	Sybille	Off Whydah	F	182	166	355
07.04.29	Josepha	Sp	Sc	Monkey	Near Berry Islands	H	206	206	396
29.04.29	Panchita	Sp	Sc	Sybille	Bight of Biafra	F	292	259	355
06.06.29	Voladora	Sp	Sc	Pickle	Off Gibara	H	335	331	399
27.06.29	Midas	Sp	B	Monkey	Off Bimini Island	H	400	281	401
??.08.29	Ceres	Br	Sc	Plumper	90nm E of Princes Island	F	279	130	360
07.08.29	Santo Jago	Br	Sc	Medina	Bights	F?	209	?	359
16.08.29	Emilia	Br	Sc	Sybille (Dallas)	Off River Benin	F	486	?	360
17.08.29	Clarita	Sp	Sc	Medina	Off River Gabon	F	261	201	359
22.08.29	Restamador	Br	Sc	Plumper	River Cameroons	F	277	?	360
31.09.29	Emelia	Br	B	Clinker	80nm N of Princes Island	F	157	148	360
01.10.29	La Laure	Fr?	?	Atholl	130nm W of Isles De Los	F	249	?	362
01.11.29	Tentadora	Br	C	Sybille (Dallas)	50nm W of River Benin	F	432	320	360
10.11.29	Cristina	Sp	Bn	Sybille (Black Joke)	Scarcies Bank	X	348	216	363
16.11.29	Gallito	Sp	Sc	Nimble	Off Berry Islands	H	136	136	404
28.11.29	Ismenia	Br	Bn	Eden	River Cameroons	F	0	0	364
09.12.29	Emilia	Br	Bn	Atholl	50nm S of Lagos	F	187	128	363
10.12.29	Nao Lendia	Br	Sc	Medina	110nm SSW of R. Bonny	F	184	159	362
07.01.30	N. Senora da Guia	Br	Sc	Sybille (Dallas)	Off Lagos	F	310	238	365
15.01.30	Umbelina	Br	Sc	Sybille	150nm S of Porto Novo	F	377	163	365
23.01.30	Primeira Rosália	Br	Bn	Sybille	60nm SW of Lagos	F	282	242	365
02.02.30	Nova Resolução	Br	Bn	Medina	200nm S of Cape St Paul's	F	43	42	365

Suspected Slave Vessels Detained 1807–39 775

A	B	C	D	E	F	G	H	I	J
24.03.30	Maria de la Conc'ption	Sp	Sc	Primrose	Rio Pongas	F	79	79	367
24.03.30	Conchita	Sp	Sc	Primrose	Rio Pongas	F	79	79	367
27.03.30	Altimara	Sp	Sc	Clinker	Off Cape Formoso	F	249	198	367
01.04.30	Manzanares	Sp	Bn	Sybille (Black Joke)	200nm W of C. Mesurado	F	354	349	367
09.04.30	Santiago	Sp	Sc	Sparrowhawk	Off Santiago de Cuba	X	108	108	370
??.??.30	Madre de Dios	Sp	Sc	Sybille (Dallas)	Bight of Benin	F	360	?	369
12.05.30	Loreto	Sp	Sc	Plumper	25nm off Trade Town	F	186	183	370
11.06.30	Emilio	Sp	Bn	Victor	Off Santiago de Cuba	H	192	192	405
03.08.30	Santiago	Sp	Sc	Atholl	Off River Bonny	F	162	153	370
18.08.30	Atafa Primo	Sp	Sc	Medina	80nm SW of C. Formoso	R	0	0	370
07.09.30	Veloz Passagera	Sp	S	Primrose	Bight of Benin	F	556	530	371
??.10.30	Pajorito	Sp	Bn	Medina	Old Calabar River	F	239	?	374
17.10.30	Nueva Isabelita	Sp	Sc	Atholl	210nm W of C. Mesurado	F	141	139	374
07.11.30	Maria	Po	Sc	Plumper	Rio Pongas	F	35	35	375
09.11.30	Dos Amigos	Sp	Bn	Atholl (Black Joke)	River Cameroons	F	0	0	376
24.11.30	Nympha	Po	Sc	Conflict	100nm W of Rio Pongas	F	167	167	376
??.12.30	Caroline	Fr	Sc	Conflict	Sierra Leone coast	G	?	4	379
02.12.30	Destimida	Po	Sc	Druid	Brazilian coast	R	50	50	409
26.12.30	Maria	Sp	Sc	Plumper	60nm SW of C. Mesurado	F	505	497	379
22.02.31	Primero	Sp	Sc	Dryad (Black Joke)	SW of Cape Mount	F	311	?	381
26.04.31	Marinerito	Sp	B	Dryad (Black Joke)	15nm E of Fernando Po	F	496	439	416
18.06.31	Roza	Po	Sc	Pickle	Berry Islands	F	157	157	406
21.07.31	Potosi	Sp	Sc	Dryad (Fair Rosamond)	60nm SSE of Lagos	F	192	183	421
10.09.31	Regulo	Sp	B	Dryad (Black Joke)	River Bonny	F	207	164	424
10.09.31	Rapido	Sp	B	Dryad (Fair Rosamond)	River Bonny	F	2	2	424
15.02.32	Frasquita	Sp	Sc	Dryad (Black Joke)	50nm S of River Bonny	F	290	228	432
19.03.32	Segunda Teresa	Sp	B	Pelorus	140nm S of Badagry	F	459	445	433
06.04.32	Planeta	Sp	Sc	Speedwell	45nm SE of Isle of Pines	H	239	238	469
03.05.32	Prueba	Sp	Sc	Brisk	Off River Bonny	F	318	284	435
03.06.32	Aguila	Sp	B	Speedwell	Off Isle of Pines	H	616	604	470
25.06.32	Indagadora	Sp	Sc	Speedwell	Off Isle of Pines	H	134	134	470
13.07.32	Hebe	Po	B	Nimble	Off Isle of Pines	F	401	385	471
15.08.32	Carolina	Sp	B	Favourite	60nm SW of Fernando Po	F	426	369	438

A	B	C	D	E	F	G	H	I	J
13.09.32	Friendship	GB	Sc	Pylades	Bahia	K	0	0	490
21.11.32	Negrito	Sp	B	Victor	40nm N of Tobago	H	526	490	472
22.02.33	Desengano	Sp	Sc	Charybdis	Off River Bonny	F	220	209	440
15.03.33	Indio	Sp	Sl	Favourite	Four days from R. Bonny	F	117	108	441
29.03.33	Negreta	Sp	Sc	Nimble	65nm SSE Santiago, Cuba	H	196	195	473
23.04.33	Veloz Mariana	Sp	Sc	Curlew	Off Old Calabar River	F	290	265	441
05.05.33	Josepha	Sp	B	Pluto	50nm SE of River Bonny	F	278	193	441
04.06.33	Panda	Pr	Sc	Curlew	River Nazareth	X	0	0	443
07.07.33	Segundo Socorro	Sp	Sc	Trinculo	60nm WSW of C. Mount	F	307	307	442
18.09.33	Caridad	Sp	Sc	Trinculo	30nm SW of Cape St John	F	112	107	446
23.10.33	Virtude	Po	Bn	Brisk	Off Old Calabar River	F	350	314	446
28.10.33	El Primo	Sp	B	Isis	60nm NW of Princes Island	F	343	335	446
10.11.33	Joaquina	Sp	Sc	Nimble	Off Isle of Pines	H	327	321	476
15.11.33	Paqueta do Sul	? Po	B	Satellite	Off Rio de Janeiro	O	0	0	493
25.11.33	Maria da Gloria	? Po	Ba	Snake	70nm S of Rio de Janeiro	R	423	64	494
07.12.33	Manuelita	Sp	Sc	Nimble	Off Isle of Pines	H	485	477	477
25.12.33	Rosa	Sp	Sc	Despatch	450nm NE B. of St Marcos	H	292	290	478
27.12.33	Apta	Po	Sc	Trinculo	85nm E of St Thomas	F	54	54	447
28.12.33	St Rosario e Bon Jesus	Po	Sc	Trinculo	40nm SE of Princes Island	F	54	54	447
08.01.34	Vengador	Sp	B	Pluto	SW of River Bonny	F	405	376	447
16.02.34	Carolina	Sp	Bn	Isis	30nm SW of R. Forcados	F	350	323	448
26.04.34	La Pautica	Sp	Sc	Fair Rosamond	Off Old Calabar River	F	317	270	451
25.05.34	Desfrique	Po	Sc	Firefly	50nm S of Isle of Pines	F	215	205	480
14.06.34	Tamega	Po	B	Charybdis	40nm SSW of Lagos	F	444	434	451
15.06.34	Duquesa de Braganza	Br	Sc	Satellite	Off Isla Grande	O	270	270	497
30.06.34	Pepita	Sp	Sc	Pelorus	River Cameroons	R	179	153	452
05.08.34	Maria Isabel	Sp	Sc	Fair Rosamond	75nm SE of Princes Island	F	146	131	453
18.08.34	Felicidad	Po	Sc	Nimble	Off Cape Maisí	F	164	162	480
17.09.34	Arrogante Mayaques'na	Sp	Sc	Lynx	330nm NE of Ascension Is.	F	350	309	454
30.10.34	Carlota	Sp	Sc	Nimble	Punta de la Vaca	N	272	194	481

A	B	C	D	E	F	G	H	I	J
31.10.34	Indagadora	Sp	Sc	Griffon	75nm WSW of St Thomas	F	375	363	456
03.11.34	Clemente	Sp	Bn	Griffon	40nm SE of Princes Island	F	415	408	456
28.11.34	Rio de la Plata	Br	B	Raleigh	20° 58′ S″, 17° 49′ W″	O	521	221	498
17.12.34	Sutil	Sp	Sc	Pelorus	Off NE of Fernando Po	F	307	228	457
17.12.34	Formidable	Sp	B	Buzzard	Off Old Calabar River	F	712	418	459
25.12.34	Atrevido	Po	B	Lynx	100nm S of Whydah	F	494	482	461
03.01.35	Maria	Po	Sc	Fair Rosamond	Off River Gabon	F	48	48	462
14.01.35	Maria	Sp	Sc	Cruizer	Old Bahama Channel	H	346	342	483
15.01.35	Minerva	Sp	Po	Pelorus	Old Calabar River	F	675	469	463
22.01.35	Julita	Sp	Sc	Racer	Off Tortuga	H	342	340	483
02.02.35	Iberia	Sp	Sc	Buzzard	30nm off River Bonny	F	313	305	506
23.02.35	El Manuel	Sp	Bn	Forester	160nm SW of C. Formoso	F	387	375	507
12.03.35	Chubasco	Sp	Bn	Racer	Cay Sal Bank	H	253	253	483
20.03.35	Legitimo Africano	Po	Sc	Forester	Off Cape St Paul's	F	200	186	507
28.03.35	Bienvenida	Sp	Sc	Buzzard	420nm W of Princes Island	F	430	367	507
31.03.35	Joven Reyna	Sp	Sc	Arachne	90nm W of Havana	H	254	254	483
08.04.35	Marte	Sp	B	Skipjack	NW of Little Cayman	H	442	403	484
15.06.35	Numero Dos	Sp	Sc	Forester	Off River Bonny	F	154	141	510
29.06.35	Tita	Sp	Sc	Serpent	40nm SW of Hogsty Reef	H	394	393	561
29.07.35	Volador	Sp	Sc	Fair Rosamond	65nm NE of Princes Island	F	487	428	511
02.09.35	Semiramis	Sp	Sc	Buzzard	Mouth of River Bonny	F	477	426	511
07.10.35	Amalia	Sp	Sc	Vestal	60nm W of Grenada	H	203	200	563
11.10.35	Argos	Sp	B	Charybdis	30nm NW of Cape Lopez	F	429	366	512
18.10.35	Conde de los Andes	Sp	Sc	Britomart	35nm SE of River Bonny	F	282	269	513
16.11.35	Theresa	Po	C	Britomart	30nm W of Lagos	F	214	202	513
27.11.35	Norma	Sp	C	Buzzard	Off River Bonny	F	249	234	516
04.12.35	General Manso	Sp	N	Leveret	Off Freetown	R	-	EC	515
04.12.35	Victorina	Sp	C	Leveret	Off Freetown	R	-	EC	515
04.12.35	Josepha	Sp	C	Curlew	Sierra Leone River	R	-	EC	515
07.12.35	Diligencia	Sp	C	Champion	P. de Mula, Cuba	H	131	120	563
17.12.35	Orion	Br	B	Satellite	145nm N of Cape St Tomé	O	245	243	596

A	B	C	D	E	F	G	H	I	J
19.12.35	Tres Tomasas	Sp	C	Curlew	130nm SW of Freetown	F	-	EC	517
22.12.35	Isabella Segunda	Sp	N	Trinculo	Bonny Bar	F	347	334	517
23.12.35	General Laborde	Sp	C	Champion	30nm E of Havana	R	0	0	565
24.12.35	Ligera	Sp	Sc	Buzzard	40nm SE of Cape Formoso	F	198	192	518
25.12.35	Tersicore	Sp	B	Pylades	100nm W of Freetown	W	-	EC	519
28.12.35	Segunda Iberia	Sp	Sc	Fair Rosamond	160nm SSE of C. Formoso	F	260	238	518
02.01.36	Rosarito	Sp	B	Curlew	Accra Roads	F	-	EC	519
07.01.36	Ninfa	Sp	Bn	Pincher	10nm E of Salt Cay	H	450	433	566
08.01.36	Vencedora	Po	Su	Hornet	Off Marica Islands	R	0	0	599
12.01.36	Esperanca	Po	B	Pike	20nm S of Isle of Pines	?	0	0	566
14.01.36	Gaceta	Sp	Sc	Pylades	100nm SW of Cape Mount	F	225	169	519
21.01.36	Vandolero	Sp	B	Lynx	Off River Bonny	F	377	343	519
25.01.36	Atafa Primo	Sp	Sc	Leveret	20nm SE of Sanguin	F	-	EC	520
25.01.36	Zema	Sp	Sc	Leveret	20nm SE of Sanguin	F	-	EC	520
28.01.36	Felix Vascongada	Sp	Sc	Trinculo	River Bonny	F	-	EC	521
28.01.36	Eliza	Sp	Sc	Trinculo	River Bonny	F	-	EC	521
28.01.36	Diligensia	Sp	Bn	Trinculo	River Bonny	F	-	EC	521
28.01.36	Maria Manuela	Sp	B	Trinculo	River Bonny	F	-	EC	521
29.01.36	El Esplorador	Sp	B	Fair Rosamond	120nm SE of C. St Paul's	F	-	EC	521
05.02.36	Matilde	Sp	Sc	Charybdis	Off St Thomas	F	-	EC	521
06.02.36	Mosca	Sp	Sc	Britomart	Whydah Roads	F	-	EC	521
06.02.36	El Casador Santurzano	Sp	B	Waterwitch	Off Whydah	F	-	EC	522
08.02.36	Seis Hermanos	Sp	Bn	Thalia	140nm ESE of C. Palmas	F	189	171	523
09.02.36	Golondrina	Sp	Sc	Forester	Loango Bay	F	-	EC	523
09.02.36	Luisa	Sp	B	Forester	Loango Bay	F	-	EC	523
19.02.36	Tridente	Sp	B	Charybdis	Loango Bay	F	-	EC	523
02.03.36	Ricomar	Sp	B	Champion	36nm E of Havana	H	188	186	568
06.03.36	General Mina	Sp	Sc	Britomart	River Nun	F	-	EC	526
06.03.36	Dos Hermanos	Sp	Sc	Britomart	River Nun	F	-	EC	526
08.03.36	Vigilante	Po	Bn	Racer	70nm E of Dominica	F	?	231	569
10.03.36	La Mariposa	Sp	Sc	Fair Rosamond	Old Calabar River	F	-	EC	526
13.03.36	Galanta Josepha	Sp	Sc	Waterwitch	Off Little Popo	F	-	EC	526
14.03.36	Joven Maria	Sp	Sc	Waterwitch	25nm off Whydah	F	-	EC	526
14.03.36	El Mismo	Sp	B	Charybdis	Off Ambriz	F	-	EC	526
01.04.36	Criolo	Po	Sc	Gannet	Of Haiti	F	315	307	570
04.05.36	Mindello	Po	B	Buzzard	Off SE of Fernando Po	F	267	257	527
02.07.36	Felicia	Sp	Bn	Buzzard	River Bonny	F	401	355	529

Suspected Slave Vessels Detained 1807–39   779

A	B	C	D	E	F	G	H	I	J
06.07.36	Famosa Primeira	Sp	Sc	Buzzard	Off River Bartolomeo	F	-	EC	530
12.07.36	Preciosa	Sp	Sc	Pincher	40nm NE of Matanzas	H	287	286	572
22.07.36	Joven Carolina	Po	Bn	Buzzard	Old Calabar River	F	421	383	530
17.09.36	Esperanca	Po	Bn	Pylades	Off River Bonny	F	471	417	532
18.09.36	Felix	Po	B	Thalia	Off River Bonny	F	557	481	532
19.09.36	Atalaya	Sp	Sc	Thalia/Buzzard	Off River Bonny	F	118	88	532
20.09.36	Negrinha	Po	Sc	Vestal	25nm SE of Grenada	F	336	335	574
28.09.36	Empresa	Sp	B	Vestal	15nm S of Grenada	H	434	407	574
28.09.36	Fenix	Po	Bn	Vestal	20nm SW of Grenada	F	484	484	574
03.10.36	Esperanca	Po	B	Curlew	Off River Bonny	F	438	396	533
19.10.36	Quatro de Abril	Po	S	Curlew	50nm S of Lagos	F	478	458	533
20.10.36	Victoria	Po	Sc	Forester	25nm SE of New Calabar	F	380	316	534
21.10.36	Cantabra	Sp	Sc	Charybdis	Off Grand Bassa	F	-	EC	534
28.10.36	Olympia	Po	Sc	Buzzard	Off River Cameroons	F	282	252	534
06.11.36	Constitucao	Po	Sc	Racer	Off Havana	D	0	0	578
08.11.36	Manuelita	Sp	Sc	Racer	Off Havana	D	-	EC	578
12.11.36	Serea	Po	Sc	Buzzard	Off Fernando Po	F	22	21	534
14.11.36	Veloz	Po	B	Columbine	40nm WNW of Princes Is.	F	508	460	535
21.11.36	Luisita	Sp	Sc	Rolla	River Sherbro	F	-	EC	535
01.12.36	Carlota	Po	Sc	Champion	55nm WSW of Cape Cruz	F	203	203	579
02.12.36	San Nicolas	Sp	B	Rolla	River Sherbro	F	-	EC	535
05.12.36	Gata	Sp	Sc	Scout	Bonny Bar	F	111	101	536
14.12.36	Incomprehensivel	Po	S	Dolphin	1,000nm S of Ascension Is.	F	696	506	537
16.12.36	General Laborde	Sp	Bn	Pincher	Off Gibara, Cuba	R	-	EC	579
27.12.36	Esperimento	Sp	Sc	Rolla	Rio Pongas	F	-	EC	536
27.12.36	Lechuguino	Sp	Sc	Rolla	Rio Pongas	F	49	49	536
11.01.37	Paquete de C. Verde	Po	B	Scout	River Bonny	F	576	452	540
11.01.37	Esperanca	Po	B	Scout	River Bonny	F	108	89	540
14.01.37	Descubierta	Sp	Sc	Scout	Off River Bonny	F	-	EC	541
20.01.37	Temerario	Po	Bn	Bonetta	Off River Bonny	F	349	236	541
04.02.37	Latona	Po	Sc	Columbine	40nm S of Whydah	F	325	320	541
10.02.37	Josephina	Po	Sc	Columbine	45nm SSE of Whydah	F	350	346	541
30.03.37	Cinco Amigos	Sp	Sc	Bonetta	Off New Sesters	F	-	EC	543
03.04.37	Flor de Tego	Po	Sc	Wanderer	10nm off Grand Bahama	F	417	417	582
13.04.37	Florida	Po	Sc	Harpy	10nm NW of St Lucia	F	283	280	585
19.04.37	Dolores	Sp	Sc	Dolphin	30nm SW of Old Calabar	F	314	286	543

A	B	C	D	E	F	G	H	I	J
25.04.37	Don Francisco	Po	B	Griffon	E of Dominica	F	433	433	583
11.05.37	Lafayette	Po	Sc	Charybdis	50nm SE of Lagos	F	448	441	546
27.05.37	Cobra de Africa	Po	Sc	Dolphin	50nm S of Bimbia Island	F	162	101	547
01.06.37	Providencia	Po	Sc	Dolphin	Old Calabar/Fernando Po	F	198	193	547
07.06.37	Antonica	Po	Sc	Racer	90nm WSW of Cape Cruz	F	183	?	586
11.06.37	Traga Milhas	Po	Sc	Racer	60nm ESE of Cape Cruz	F	283	283	586
26.06.37	General Ricafort	Sp	B	Charybdis	Off Accra	F	-	EC	547
06.08.37	Amelia	Po	B	Waterwitch	60nm SE of Lagos	F	359	345	548
12.09.37	Ingemane	Po	Sc	Comus	75nm S of Isle of Pines	F	82	79	588
23.09.37	Velos	Po	B	Fair Rosamond	River Benin	F	0	0	549
25.09.37	Vibora de C. Verde	Po	Sc	Waterwitch	Off Fernando Po	F	269	221	550
25.09.37	Primoroza	Po	Sc	Dolphin	90nm W of Princes Island	F	182	136	551
28.09.37	Camoes	Po	B	Fair Rosamond	River Benin	R	138	116	550
14.10.37	Vencedora	Sp	Sc	Ringdove	35nm NE of Matanzas	R	26	0	589
10.11.37	Ligeira	Po	Sc	Bonetta	70nm S of Cape Formoso	F	313	280	551
20.11.37	Deixa Falar	Po	Bn	Scout	50nm SE of Lagos	F	205	186	552
23.11.37	Gratidão	Po	B	Scout	55nm SE of Lagos	F	452	380	552
23.11.37	Arrogante	Po	Bn	Snake	25nm S of Cape Antonio	F	407	332	591
05.12.37	Matilda	Sp	Sc	Snake	30nm ESE of Cape Cruz	H	259	244	591
05.12.37	Isabelita	Po	Sc	Sappho	Off Cape Tiburon	H	160	159	592
15.12.37	Vigilante	Sp	Bn	Ringdove	10nm SE, Sant'go de Cuba	D	21	0	592
26.12.37	Princesa Africana	Po	Sc	Curlew	Bar of River Sherbro	F	222	222	553
21.01.38	[Unidentified]	?	Sc	Serpent/Wanderer	8nm E of Sant'go de Cuba	N	126	?	679
05.02.38	Feliz	Po	Sc	Madagascar (Flamer)	20nm E of Port Royal	XF	326	324	680
02.03.38	Montana	Po	Sc	Saracen	160nm SW of Sherbro Is.	D	-	EC	555
08.03.38	Felicidades	Po	B	Scout	Off Old Calabar	F	559	408	556
02.04.38	Dous Irmaos	Po	Sc	Forester	Off River Bonny	F	305	241	620
11.04.38	Flor de Loanda	Po	Sc	Rover	40nm east of Rio	Y	289	?	698
13.04.38	Cesar	Br	Bn	Rover	Off Marica Is	O	207	202	698
24.04.38	Diligente	Po	B	Pearl	155nm NNW of St Antonio	F	480	475	688
28.04.38	Opposicao	Sp	Sc	Pearl	Off Havana	F	-	EC	685
30.04.38	Camoens	Po	B	Sappho	Off Cay Sal	F	572	569	685
13.05.38	Brilhante	Br	B	Wizard	Off Isla Grande	O	250	245	700

A	B	C	D	E	F	G	H	I	J
03.06.38	Prova	Po	Sc	Pylades	Old Calabar River	F	225	194	620
13.07.38	Felis	Po	Bn	Fair Rosamond	10nm S of River Bonny	F	195	187	623
13.07.38	Rosália Habaneira	Po	Sc	Sappho	35nm SW of G'nd Cayman	F	247	223	687
15.08.38	Diligente	Sp	B	Brisk	River Gallinas	F	-	EC	625
16.08.38	Ligeira	Sp	Sc	Brisk	River Gallinas	F	-	EC	625
21.08.38	Constitucao	Sp	Sc	Fair Rosamond	Accra Roads	F	-	EC	625
21.09.38	Eliza	Sp	Sc	Brisk	River Sesters	F	-	EC	627
30.09.38	Constitucao	Sp	Sc	Brisk	Off New Sesters	F	-	EC	627
09.10.38	Prova	Po	Sc	Termagant	120nm S of River Bonny	F	326	295	627
17.10.38	Veloz	Sp	Sc	Brisk	Off River Gallinas	F	-	EC	628
17.10.38	Josephina	Sp	Sc	Brisk	River Gallinas	F	-	EC	628
27.10.38	Mary Anne Cassard	US	Sc	Brisk	25nm off Freetown	R	0	0	628
31.10.38	Dolcinea	Po	Sc	Pelican	130nm S of Lagos	F	253	249	629
01.11.38	Liberal	Po	B	Lynx	30nm W of Whydah	F	591	583	629
08.11.38	Maria	Sp	Sc	Brisk	Rio Pongas	F	-	EC	629
16.11.38	Victoria	Sp	B	Dolphin	Lagos Roads	F	-	EC	631
16.11.38	Dous Amigos	Sp	B	Dolphin	Lagos Roads	F	-	EC	631
16.11.38	Ligeiro	Sp	Bn	Dolphin	Lagos Roads	F	-	EC	631
16.11.38	Astran	?	?	Dolphin	Lagos Roads	R	0	0	631
17.11.38	Sirse	Sp	Sc	Buzzard	100nm SSW of Freetown	F	-	EC	632
18.11.38	Veterano	Sp	B	Brisk	Off River Gallinas	F	-	EC	632
26.11.38	Escorpion	Sp	B	Wanderer	45nm SSE of Isle of Pines	F	212	190	690
27.11.38	Emprendedor	Po	B	Buzzard	35nm W of River Gallinas	F	467	458	633
01.12.38	Diligente	Br	Bn	Electra	540nm ESE of Bahia	O	302	246	705
03.12.38	Isabel	Sp	Sc	Bonetta	330nm W of St Thomas	F	-	EC	634
09.12.38	Aurelia Felix	Po	Sc	Brisk	Off Bolama	R	0	0	634
17.12.38	Magdalena	Po	Sc	Pelican	50nm NW of St Thomas	F	320	302	635
18.12.38	Ontario	Sp	Sc	Pelican	150nm NW of St Thomas	F	219	200	635
18.12.38	Victoria	Po	Sc	Pickle	10nm E of Sant'go de Cuba	F	0	0	637
24.12.38	Victoria	Sp	Sc	Lynx	Off Princes Island	F	-	EC	692
27.12.38	Amalia	Sp	Sc	Dolphin	Off River Gallinas	F	-	EC	636
27.12.38	Feliz	Br	Sc	Wizard	50nm SSW of Rio	O	229	229	706
28.12.38	Violante	Po	Sc	Brisk	20nm S of River Sherbro	F	191	191	637
28.12.38	Gertrudes	Po	Sc	Bonetta	Off River Sherbro	F	168	168	638
??.??.39	Mary Cushing	?US	Sc	?	?	Y	-	EC	642
04.01.39	Hazard	?US	Sc	Forester	Off Cape St Paul's	Y	-	EC	639

A	B	C	D	E	F	G	H	I	J
10.01.39	Merced	Sp	Sc	Dolphin	Off Cape Mesurado	R	-	EC	640
13.01.39	Florida	? US	Sc	Saracen	In River Gallinas	Y	o	o	640
14.01.39	Eagle	Sp	B	Lily/Buzzard	Lagos Roads/ Fernando Po	F	-	EC	641
21.01.39	Jago	? US	Sc	Termagant	Off Cape St Paul's	Y	-	EC	641
22.01.39	Matilde	Sp	B	Fair Rosamond	Off River Gabon	F	-	EC	643
22.01.39	Maria Theresa	Sp	Bn	Lily	20nm E of Accra	F	-	EC	643
31.01.39	Tego	Sp	B	Fair Rosamond	Off River Gabon	F	-	EC	643
09.02.39	Braganza	Sp	B	Termagant	Off Cape St Paul's	F	-	EC	646
06.03.39	Ligeira	Po	Sc	Forester	Off Trade Town	R	-	EC	646
11.03.39	Serea	Sp	Sc	Forester	Off Sierra Leone	F	-	EC	646
19.03.39	Clara	Sp	Sc	Buzzard	River Nun	F	-	EC	648
21.03.39	Rebecca	Sp	Sc	Forester	Gallinas Roads	F	-	EC	646
25.03.39	Especulador	Br	Bn	Electra	110nm east of C. Frio	O	278	268	709
27.03.39	Carolina	Br	Bn	Electra	100nm SE of C. Frio	O	214	211	710
31.03.39	Labradora	Po	Sc	Saracen	In Rio Pongas	F	253	248	647
08.04.39	Passos	Po	Sc	Wolverene	20nm NNE of Princes Is.	F	87	81	649
07.04.39	Ganges	Br	B	Grecian	80nm east of C. Frio	O	419	386	711
11.04.39	Seal	Br	B	Grecian	15nm off C. Frio	O	364	319	711
14.04.39	Liberal	Po	Sc	Brisk	In Bissau Channel	F	41	40	650
19.04.39	Goloubtchick	Ru	B	Saracen	Off River Gallinas	Y	-	EC	651
27.04.39	Catalana	Sp	Sc	Termagant	River Sinou	W	-	EC	654
30.04.39	Traveller	? US	Sc	Harlequin	Grain Coast	Y	-	EC	652
??.05.39	Merced	Sp	Sc	Harlequin	Grain Coast	W	-	EC	663
10.05.39	Raynha dos Anjos	Sp	Sc	Forester	River Nazareth	F	-	EC	654
16.05.39	Constanza	Sp	Sc	Harlequin	River Gallinas	F	-	EC	654
17.05.39	Wyoming	Sp	Bn	Harlequin	Off River Gallinas	?	-	EC	655
20.05.39	Bella Florentina	Sp	Sc	Harlequin	Off River Sesters	F	-	EC	655
22.05.39	Carolina	? Po	Sc	Forester	Mayumba Bay	W	-	EC	656
27.05.39	Si	Sp	Fe	Waterwitch	Off River Gallinas	F	360	359	656
27.05.39	Jack Wilding	Sp	Sc	Dolphin	Accra Roads	F	-	EC	656
28.05.39	Recuperador	Br	Sc	Grecian	30nm SW of C. Frio	R	o	o	716
29.05.39	Maria Carlota	Br	Ba	Grecian	12nm SE of Rio	O	o	o	716
01.06.39	Vigilante	Sp	B	Wolverene	River Congo	F	-	EC	657
04.06.39	Tres Emanuel	? Sp	Sc	Wolverene	River Congo	U	o	o	658
04.06.39	Sierra del Pilar	Sp	Sc	Pickle	Off Isle of Pines	H	176	173	695
08.06.39	Perry Spencer	? US	Sc	Lynx	River Gabon	Y	-	EC	659
14.06.39	Jacuhy	Br	Bn	Brisk	280nm W of Loango Bay	F	203	196	660
17.06.39	Euphrates	? US	Sc	Dolphin	Off River Grand Bassa	Y	-	EC	662
18.06.39	Merced	Sp	Sc	Dolphin	New Sesters	F	o	o	663
20.06.39	Emprendedor	Sp	Bn	Harlequin	Off River Gallinas	F	-	EC	658
23.06.39	Emprehendador	Br	B	Wolverene	Whydah	F	-	EC	663
25.06.39	Pomba d'Africa	Po	Sc	Fair Rosamond	20nm SE of Fernando Po	F	155	126	664
25.06.39	Sedo ou Tarde	Po	Sl	Fair Rosamond	20nm SE of Fernando Po	F	23	21	664

Suspected Slave Vessels Detained 1807–39   783

A	B	C	D	E	F	G	H	I	J
26.06.39	Victoria de Libertade	Sp	Sc	Harlequin	Off River Sesters	F	-	EC	663
27.06.39	Cristiano	Sp	Bn	Harlequin	Off New Sesters	F	-	EC	664
28.06.39	Sin-ygual	Sp	Sc	Harlequin	Off River Gallinas	F	-	EC	664
29.06.39	Matilde	Sp	B	Brisk	60nm WNW of R. Yumba	F	-	EC	660
03.07.39	Caridad Cabana	Sp	Sc	Snake	50nm N of Puerto Rico	H	174	171	696
06.07.39	Casualidade	Po	Sc	Dolphin	Off River Sherbro	F	88	88	665
08.07.39	Constitucao	Sp	Sc	Waterwitch	70nm W of Cape Formoso	F	344	338	665
25.07.39	Firmeza	Br	B	Wolverene	Off Whydah	F	-	EC	666
27.07.39	Simpathia	Br	Bn	Lynx	Off Popo	F	-	EC	666
10.08.39	Mary	? US	B	Forester	Off Cape Mount	Y	-	EC	667
12.08.39	Catherine	US	Sc	Dolphin	Off Lagos	?	-	EC	668
19.08.39	Intrepido	Br	B	Dolphin	Off Cape St Paul's	F	-	EC	668
26.08.39	Butterfly	? US	Sc	Dolphin	Off Cape St Paul's	Y	-	EC	669
27.08.39	Dous Amigos	Sp	Sc	Dolphin	20nm off River Volta	F	-	EC	669
28.08.39	Pompeo	Br	B	Stag (Fawn)	35nm south of Rio	R	0	0	719
05.09.39	Augusto	Br	Ba	Fair Rosamond	Off Cape St Paul's	F	-	EC	670
07.09.39	Josephina	Sp	Sc	Bonetta	River Congo	F	-	EC	671
07.09.39	Liberal	Sp	Bn	Bonetta	River Congo	F	-	EC	671
07.09.39	Ligeira	Sp	Sc	Bonetta	River Congo	F	-	EC	671
12.09.39	Pampeiro	Br	Bn	Wolverene	Off Lagos	F	-	EC	671
19.09.39	Golphino	Br	Bn	Termagant	100nm W of Lagos	F	-	EC	671
27.09.39	Sete de Avril	Sp	Sc	Waterwitch	40nm SW of Lagos	F	424	415	671
29.09.39	Destemida	Br	B	Lynx	Winnebah/Accra	F	-	EC	672
02.10.39	Andorinha	Po	Sc	Nautilus	20nm E of St Thomas	F	6	3	672
04.10.39	Vencedora	Po	L	Nautilus	Off St Thomas	F	61	50	672
16.10.39	Brilhante	Sp	Sc	Saracen	Off River Gallinas	F	-	EC	673
17.10.39	Joao de Castro	Br	B	Grecian	Off Isla Grande	O	0	0	721
27.10.39	Calliope	Br	Sc	Waterwitch	25nm S of Little Popo	F	-	EC	673
01.11.39	Fortuna	Sp	Bn	Waterwitch	40nm E of Lagos	F	-	EC	673
11.11.39	Magdalena	Sp	Sc	Viper	40nm E of Cape Palmas	F	-	EC	673
21.11.39	Sociedade Feliz	Br	Bn	Harlequin	Off Cape Palmas	F	-	EC	674
26.11.39	Anna Feliz	Po	B	Modeste	25nm off Quilimane	C	56	51	675
27.11.39	Lavandeira	Sp	Sc	Lynx	70nm W of River Gallinas	F	-	EC	674
28.11.39	Conceição	Br	B	Termagant	20nm W of Whydah	F	-	EC	674
29.11.39	Julia	Br	Bn	Termagant	Off Whydah	F	-	EC	674
09.12.39	Escorpião	Po	B	Modeste	90nm ESE of R. Zambezi	C	756	?	675
12.12.39	Arab	Po	B	Modeste	210nm ESE of R. Zambezi	C	26	?	675

APPENDIX B

Treaties, Conventions and Conferences Concerning Atlantic Slave-Trade Suppression, 1807–39

19 FEBRUARY 1810 — **Treaty** of Friendship and Alliance between **Great Britain** and **Portugal**. Portugal agrees to adopt the most efficacious means for bringing about a gradual abolition of the Portuguese slave trade. Slaving by Portugal on the coast of Africa to be permitted only within territories and ports belonging to Portugal, including Cabinda and Malembo. Signed in Rio de Janeiro.

3 MARCH 1813 — **Treaty** between **Great Britain** and **Sweden** for abolition of the slave trade. Signed in Stockholm.

14 JANUARY 1814 — **Treaty** between **Great Britain** and **Denmark** for abolition of the slave trade. Signed in Kiel.

30 MAY 1814 — **Additional Article** to the Treaty of Peace between **Great Britain** and **France**. Agreement to urge abolition of the slave trade at the forthcoming conference. Signed in Paris.

28 AUGUST 1814 — **Additional Article** to the treaty between **Great Britain** and **Spain** of 5 July 1814. Spanish promises regarding the slave trade. Signed in Madrid.

21 JANUARY 1815 — **Convention** between **Great Britain** and **Portugal**. Indemnification of Portuguese subject for detention of slaving vessels. Signed in Vienna.

22 JANUARY 1815 — **Treaty** between **Great Britain** and **Portugal**. Restriction of Portuguese slave trade and annulment of the Convention of Loan of 1809 (loan by Britain of £600,000). Signed in Vienna.

8 FEBRUARY 1815 — **Declaration** by the eight Allied Powers (**Great Britain, France, Prussia, Russia, Austria, Sweden, Spain** and **Portugal**) at the conclusion of the Congress of Vienna regarding abolition of the slave trade.

28 JULY 1817 — **Additional Convention** between **Great Britain** and **Portugal** for prevention of the slave trade. Geographical limits for Portuguese slaving. Signed in London.

11 SEPTEMBER 1817 — **Separate Article** to the Additional Convention between

	Great Britain and **Portugal** for prevention of the slave trade. Mixed Commissions established. Signed in London.
23 SEPTEMBER 1817	**Treaty** between **Great Britain** and **Spain** for abolition of the slave trade. Mixed Commissions established. Signed in Madrid.
4 DECEMBER 1817	**Conference** in **London** between **Great Britain, Austria, France, Prussia** and **Russia** concerning abolition of the slave trade.
4 MAY 1818	**Treaty** between **Great Britain** and **the Netherlands** for the prevention of the traffic in slaves. Mixed Commissions established. Signed in The Hague.
24 OCTOBER 1818	**Conference** in **Aix-la-Chapelle** between **Great Britain, Austria, France, Prussia** and **Russia** concerning abolition of the slave trade.
28 NOVEMBER 1822	**Resolutions** of the Conference of **Verona** between **Great Britain, Austria, France, Prussia** and **Russia** relating to abolition of the slave trade.
10 DECEMBER 1822	**Explanatory and Additional Articles** to the Treaty of 23 September 1817 between **Great Britain** and **Spain** concerning liability to condemnation of vessels engaged in the slave trade but detained without slaves on board. Signed in Madrid.
31 DECEMBER 1822	**Explanatory and Additional Articles** to the Treaty of 4 May 1818 between **Great Britain** and **the Netherlands** concerning liability to condemnation of vessels engaged in the slave trade but detained without slaves on board. Signed in Brussels.
25 JANUARY 1823	**Further Additional Article** to the Treaty of 4 May 1818 between **Great Britain** and **the Netherlands**. The Equipment Article (later known as the Equipment Clause). Signed in Brussels.
15 MARCH 1823	**Additional Article** to the Convention between **Great Britain** and **Portugal** of 28 July 1817. Liability to condemnation of vessels engaged in the slave trade but detained without slaves on board. Signed in Lisbon.
13 MARCH 1824	**Convention (Not Ratified)** between **Great Britain** and the **United States of America** agreeing mutual right to board and search the merchant vessels of the other signatory, and to send suspected slave vessels to the courts of their own country for adjudication. Signed in London.

6 NOVEMBER 1824 — **Treaty** between **His Britannic Majesty** and **The Kings of Sweden and Norway** for preventing their subjects from engaging in any traffic in slaves. Sweden and Norway undertake to enact further legislation. Mutual Right of Search, geographical limits and instructions to cruisers agreed. Mixed Courts of Justice to be established at Freetown and St Bartholomew. Agreement on indemnification against illegal detention and entitlement to reparation. An Equipment Clause included. Signed in Stockholm.

23 NOVEMBER 1826 — **Convention** between **Great Britain** and **Brazil** for abolition of the African slave trade. Agreed that the Brazilian slave trade would be illegal from 13 March 1830, and slave trading deemed to be piracy. Mixed Commissions established. Signed in Rio de Janeiro.

30 NOVEMBER 1831 — **Convention** between **Great Britain** and **France** for the more effectual suppression of the traffic in slaves. Agreement on geographical limits, mutual Right of Search, and adjudication of detained vessels by courts of their own nation. Other maritime Powers to be invited to accede to the Convention. Signed in Paris.

22 MARCH 1833 — **Supplementary Convention** between **Great Britain** and **France** for the more effectual suppression of the traffic in slaves. Establishing the procedure if a suspected vessel is under convoy. Liability to condemnation of vessels engaged in slaving but detained without slaves and vessels fitted for slaving. Signed in Paris.

26 JULY 1834 — **Treaty** between **Great Britain, France** and **Denmark**. The accession of Denmark to the Conventions of 1831 and 1833 between Great Britain and France. Signed in Copenhagen.

8 AUGUST 1834 — **Treaty** between **Great Britain, France** and **Sardinia**. The accession of Sardinia to the Conventions of 1831 and 1833 between Great Britain and France. Signed in Turin.

8 DECEMBER 1834 — **Additional Article** to the Treaty of 8 December 1834 to establish places of adjudication. Signed in Turin.

15 JUNE 1835 — **Additional Article** to the Treaty of 6 November 1824 between **Great Britain** and **Sweden**. Condemned vessels to be broken-up. Signed in Stockholm.

28 JUNE 1835 — **Treaty** between **Great Britain** and **Spain** for the abolition of the slave trade (ratified 27 August 1835). Agreement by Spain to promulgate, within two months of ratification, penal

law against the traffic in slaves. Vessels engaged in the slave trade but detained without slaves, and those fitted for the slave trade, to be liable to condemnation. An Equipment Clause added, and geographical limits established. Signed in Madrid.

7 FEBRUARY 1837 **Additional Article** to the Treaty of 4 May 1918 between **Great Britain** and the **Netherlands**. Condemned vessels to be broken up. Signed in The Hague.

9 APRIL 1837 **Convention** of Amity and Commerce between **Great Britain** and **Bonny**. Agreed between the King of Bonny and Commander Robert Craigie, Senior Officer on the West Coast of Africa, and others. Concerns the conduct of trade by British merchant vessels, but does not address the slave trade. Signed in Grand Bonny.

9 JUNE 1837 **Convention** between **Great Britain**, **France** and the **Hans Towns**. The accession of the Hans Towns to the Conventions of 1831 and 1833 between Great Britain and France. Signed in Hamburg.

24 NOVEMBER 1837 **Convention** between **Great Britain**, **France** and **Tuscany**. The accession of Tuscany to the Conventions of 1831 and 1833 between Great Britain and France. Signed in Florence.

14 FEBRUARY 1838 **Convention** between **Great Britain**, **France** and **The Kingdom of the Two Sicilies**. The accession of The Two Sicilies to the Conventions of 1831 and 1833 between Great Britain and France. Signed in Naples.

19 JANUARY 1839 **Treaty** between **Great Britain** and **Chile** (ratified 6 August 1842) for abolition of the traffic in slaves. Mutual Right of Search, geographical limits and instructions to cruisers agreed. Agreement on indemnification against illegal detention and entitlement to reparation. Mixed Courts of Justice established. An Equipment Clause included. Freedom of emancipated slaves guaranteed. Signed in Santiago.

15 MARCH 1839 **Treaty** between **Great Britain** and **Venezuela** for the abolition of the Venezuelan slave trade. Mutual Right of Search agreed, and geographical limits established. Slaving deemed to be piracy. Agreement on indemnification against illegal detention and entitlement to reparation. An Equipment Clause included. Adjudication to be in national courts, and condemned vessels to be broken up. Instructions to cruisers agreed. Signed in Caracas.

24 MAY 1839	**Treaty** between **Great Britain** and the **Argentine Confederation** for abolition of the slave trade. Mutual Right of Search, geographical limits and instructions to cruisers agreed. Agreement on indemnification against illegal detention and entitlement to reparation. Mixed Courts of Justice established. An Equipment Clause included. Freedom of emancipated slaves guaranteed. Signed in Buenos Aires.
13 JULY 1839	**Treaty** between **Great Britain** and **Uruguay** for abolition of the traffic in slaves. Mutual Right of Search, geographical limits and instructions to cruisers agreed. Agreement on indemnification against illegal detention and entitlement to reparation. Mixed Courts of Justice established. An Equipment Clause included. Freedom of emancipated slaves guaranteed. Signed in Montevideo.
23 DECEMBER 1839	**Convention** between **Great Britain** and **Haiti** for the more effectual suppression of the slave trade. The accession of Haiti to the Conventions of 1831 and 1833 between Great Britain and France, but with reservations. France not a signatory. Signed in Port-au-Prince.

APPENDIX C

Acts of Parliament and Orders in Council Relating to Suppression of the Atlantic Slave Trade, 1807–40

25 MARCH 1807	**47 Geo. III, c. 36.** An Act to abolish the African slave trade from 1 May 1807. Penalties for offenders established. Bounties authorised on slaves seized.
14 MAY 1811	**51 Geo. III, c. 23.** An Act for rendering more effective the Act of 1807 to abolish the African slave trade. Slave-trading declared a felony. Further penalties established. (This became known as The Felonies Act.)
28 MAY 1818	**58 Geo. III, c. 36.** An Act to carry into execution the Treaty of 23 September 1817 between Great Britain and Spain for preventing the traffic in slaves.
5 JUNE 1818	**58 Geo. III, c. 85.** An Act to carry into execution the Conventions of 28 July 1817 and 11 September 1817 between Great Britain and Portugal for preventing the traffic in slaves.
31 MARCH 1819	**59 Geo. III, c. 16.** An Act to carry into effect the treaty with the Netherlands relating to the slave trade.
	59 Geo. III, c. 17. An Act to amend the Act of 5 June 1818.
31 MARCH 1824	**5 Geo. IV, c. 17.** An Act for the more effectual suppression of the slave trade. Slave trading declared to be piracy.
24 JULY 1824	**5 Geo. IV, c. 113.** An Act to amend and consolidate the laws relating to the abolition of the slave trade. Repeated earlier legislation, widened the definition of slave-trading, established new penalties and revised bounties.
2 JULY 1827	**7 & 8 Geo. IV, c. 74.** An Act to carry into execution the Convention between Great Britain and Brazil for the regulation and final abolition of the African slave trade.
16 JULY 1830	**1 Will. IV, c. 55.** An Act to reduce the rate of bounties payable upon the seizure of slaves to £5 per man, woman and child slave.
28 AUGUST 1833	**3 & 4 Will. IV, c. 72.** An Act for carrying into effect two Conventions with France for suppressing the slave trade.
19 MARCH 1834	**Order in Council.** A new distribution of rewards for prizes, and for revenue, slave trade and piratical seizures.

9 SEPTEMBER 1835	**5 & 6 Will. IV, c. 60.** An Act for carrying into effect the Treaty of 8 August 1834 between Great Britain, France and Sardinia for suppressing the slave trade.
5 & 6 Will. IV, c. 61. An Act for carrying into effect the treaty of 26 July 1834 between Great Britain, France and Denmark for suppressing the slave trade.
3 FEBRUARY 1836	**Order in Council.** A new distribution of rewards for prizes, and for revenue, slave trade and piratical seizures.
30 MARCH 1836	**6 Will. IV, c. 6.** An Act for carrying into effect the treaty of 28 June 1835 between Great Britain and Spain for abolition of the slave trade.
27 JULY 1838	**1 & 2 Vict, c. 39.** An Act for carrying into effect the Convention of 9 June 1837 between Great Britain, France and the Hans Towns for suppressing the slave trade.
1 & 2 Vict, c. 40. An Act for carrying into effect the Additional Article of 15 June 1835 to the Treaty of 6 November 1824 between Great Britain and Sweden.
1 & 2 Vict, c. 41. An Act for carrying into effect the Additional Article of 7 February 1837 to the Treaty of 4 May 1818 with the Netherlands.
1 & 2 Vict, c. 47. An Act for the better and more effectual carrying into effect of the treaties and Conventions made with foreign powers for suppressing the slave trade. Tonnage bounties on slave vessels condemned and destroyed, as well as Her Majesty's moiety of proceeds of sale, to be paid to captors. (This became known as The Tonnage Act.)
10 AUGUST 1838	**1 & 2 Vict, c. 83.** An Act for carrying into effect the Convention of 24 November 1837 between Great Britain, France and Tuscany for suppressing the slave trade.
1 & 2 Vict, c. 84. An Act for carrying into effect the Convention of 14 February 1838 between Great Britain, France and The Kingdom of the Two Sicilies for suppressing the slave trade.
24 AUGUST 1839	**2 & 3 Vict, c. 73.** An Act for the suppression of the slave trade. Empowered the Royal Navy to seize Portuguese vessels and vessels of other nations under false colours on suspicion of slaving, and directed that they should be adjudicated in the High Court of Admiralty or Vice-Admiralty courts as though they were British. Authorised indemnification of those concerned in such seizures and adjudications.

7 AUGUST 1840   **3 & 4 Vict, c. 67.** An Act for carrying into effect the Treaty of 15 March 1839 between Great Britain and Venezuela for the suppression of the slave trade.

APPENDIX D

Commanders-in-Chief, Commodores and Senior Officers Appointed to Conduct Slave-Trade Suppression Operations, 1807–39

WEST COAST OF AFRICA

In command of the West Africa Squadron

1809	Commodore Edward Henry Columbine
1810–11	Captain Edward Henry Columbine
1811–13	Commodore The Hon. Frederick Paul Irby
1814–15	Commodore Thomas Browne
1816–18	Commodore Sir James Lucas Yeo KCB
1818–21	Commodore Sir George Collier, Bt
1822–23	Commodore Sir Robert Mends

(Until this point there was no continuity of command on the Station)

1824–27	Commodore Charles Bullen CB
1827–30	Commodore Francis Augustus Collier CB
1830–32	Commodore John Hayes CB

Commanders-in-Chief on the Cape of Good Hope Station

1832–34	Rear-Admiral Frederick Warren
1834–36	Rear-Admiral Patrick Campbell CB
1836–37	Rear-Admiral Sir Patrick Campbell KCB
1837–40	Rear-Admiral The Hon. George Elliot

WEST INDIES

Commanders-in-Chief on the Jamaica Station

1823–7	Vice-Admiral Lawrence Halsted
1827–30	Vice-Admiral The Hon. Charles Fleeming

Commanders-in-Chief on the West Indies, Halifax and Newfoundland Station

1830–1	Vice-Admiral E. G. Colpoys
1831–2	Vice-Admiral Sir E. G. Colpoys
1832–6	Vice-Admiral The Rt Hon. Sir George Cockburn GCB
1836–8	Vice-Admiral Sir Peter Halkett GCH
1838–9	Vice-Admiral The Hon. Sir Charles Paget GCH
1839	Vice-Admiral Sir Thomas Harvey KCB

BRAZIL

Commanders-in-Chief on the South America Station

1830–3	Rear-Admiral Thomas Baker
1833–4	Rear-Admiral Sir Michael Seymour, Bt, KCB
1834–7	Rear-Admiral Sir Graham Eden Hamond, Bt, KCB
1837–8	Vice-Admiral Sir Graham Eden Hamond, Bt, KCB

Senior Officer on the Coast of Brazil

1838	Commodore Thomas Sulivan

APPENDIX E

The Equipment Clause: Criteria for Arrest under the Equipment Clause in Various Bilateral Treaties

THE EQUIPMENT ARTICLE, or, as it became more commonly known, the Equipment Clause, incorporated in the various treaties between Great Britain and slaving nations, stated that a vessel should be condemned as fitted for the slave trade "if, in her equipment, there shall be found any of the things hereinafter mentioned", and listed incriminating criteria. It further stated that "Any one or more of these several circumstances, if proved, shall be considered as *prima facie* evidence of the actual employment of the vessel in the Slave Trade." The criteria were:

1. Hatches with open gratings, instead of the close hatches which are usual in merchant vessels.

2. Divisions or bulkheads, in the hold or on deck, in greater numbers than are necessary for vessels engaged in lawful trade.

3. Spare planks, fitted for laying down as a second or slave deck.

4. Shackles, bolts or handcuffs.

5. A larger quantity of water, in casks or in tanks, than is requisite for the consumption of the crew of the vessel, as a merchant vessel.

6. An extraordinary number of water casks, or of other vessels for holding liquid; unless the Master shall produce a certificate from the customs house at the place from which he cleared outwards, stating that a sufficient security had been given by the owner of such vessel that such extra quantity of casks or other vessels should only be used to hold palm oil or for other purposes of lawful commerce.

7. A greater quantity of mess tubs or kids than are requisite for the use of the crew of the vessel as a merchant vessel.

8. A boiler of an unusual size, and larger than requisite for the use of the crew of the vessel as a merchant vessel; or more than one boiler of the ordinary size.

9. An extraordinary quantity of either rice, of the flour of Brazil, of manire or cassada, commonly called farina, of maize or of Indian corn, beyond what might probably be requisite for the use of the crew, such rice, flour, maize or Indian corn not being entered on the manifest as part of the cargo for trade.

APPENDIX F

Declarations and Certificates: Forms Ordered to be Used by Captors on Detention of Suspected Slave Vessels

THE TREATIES made by Great Britain with Portugal on 28 July 1817, with Spain on 23 September 1817 and with the Netherlands on 4 May 1818, required that the commanding officer of a cruiser arresting a vessel on suspicion of slaving should make a written declaration on the state of the vessel at the time of capture, to be submitted to the adjudicating court, and should give to the master of the arrested vessel a certificate stating the circumstances of the capture. A further certificate was required if it became necessary to disembark slaves prior to arrival at the place of adjudication. The form of the declaration and certificates is given in the following extract from the Treaties, which, in this respect, were identical:

> When a Slave-ship shall be detained, the Master thereof and a part, at least, of the Crew, are to be left onboard, and the Captor is directed to draw up in writing an authentic Declaration, which shall exhibit the state in which he found the detained Ship, and the changes which may have taken place in it; and to deliver to the Master of the Slave-ship a signed Certificate of the Papers seized onboard such detained Vessel, as well as the number of Slaves found onboard. None of the Slaves are to be disembarked till after the Vessel shall have arrived at the Place, where the legality of the Capture is to be tried, unless urgent motives, deduced from the length of the Voyage, the state of health of the Negroes, or other causes, should make a disembarkation (entirely or in part) necessary before the Vessel's arrival: the Commander of the capturing Ship, however, takes upon himself the responsibility of such disembarkation, and the necessity thereof must be stated in a Certificate in proper form, and the following are considered as proper Declarations or Certificates, to be used as circumstances may arise.

FORM OF DECLARATION OF THE STATE
OF THE VESSEL AT THE TIME OF CAPTURE

I .... Commander of His Britannick Majesty's Ship ....hereby declare, that on this ...... day of .... being in or about latitude .... longitude .... I detained the Ship or Vessel named the .... sailing under .... Colours, armed with ...... guns, ... pounders, commanded by .... who declared her to be bound from .... to .... with a Crew consisting of ...... Men, ... Boys, ... Supercargo, ... Passengers, whose names, as declared by them respectively, are inserted in a List at foot hereof, and having onboard ...... Slaves, said to have been taken on board at .... on the ...... day of .... and are enumerated as follows, viz.

	HEALTHY	SICKLY
MEN	......	......
WOMEN	......	......
BOYS	......	......
GIRLS	......	......

I do further declare that the said Ship or Vessel appeared [or not] to be seaworthy, and was [or not] supplied with a sufficient stock of water [or not] and provisions for the support of the said Negroes and Crew on their destined Voyage to ....

I do further declare ....

[Here insert any observations of the state and condition of the Ship and Crew, and Slaves, which it may appear important to notice and record.]

To be witnessed by two Officers, of whom the Surgeon to be one, if on board.

(2.) – FORM OF CERTIFICATE TO BE GIVEN
TO THE MASTER OF A VESSEL CAPTURED

I .... Commander of His Britannick Majesty's Ship .... hereby certify, that on this ...... day of .... being in or about latitude .... longitude .... I detained the Ship or Vessel named the .... sailing under .... Colours, armed with ...... guns, ... pounders, commanded by .... who declared her to be bound from .... to .... with a Crew consisting of ...... Men, ... Boys, ... Supercargo, ... Passengers, and having on board ...... Slaves, viz.

	HEALTHY	SICKLY
MEN	……	……
WOMEN	……	……
BOYS	……	……
GIRLS	……	……

and that the Papers and Documents seized by me on board the said Ship or Vessel, being marked from No. 1.to No. …… are enumerated in the following List. [Here the List is to follow.]

(3.) – FORM OF CERTIFICATE OF THE NECESSITY OF DISEMBARKING SLAVES FROM A CAPTURED VESSEL

I ……………… Commander of His Britannick Majesty's Ship …………… hereby certify, that on this …… day of ……………… being in or about latitude ……… longitude ……… I detained the Ship or Vessel named the ………………… sailing under ……………… Colours, armed with …… guns, …… pounders, commanded by ………………… who declared her to be bound from …………… to …………… with a Crew consisting of …… Men, …… Boys, …… Supercargo, …… Passengers, and having on board …… Slaves, viz.

	HEALTHY	SICKLY
MEN	……	……
WOMEN	……	……
BOYS	……	……
GIRLS	……	……

I do further declare, that finding it necessary to disembark …… of the Slaves before the Vessel could arrive at …………… to which place it was my intention to send her for Adjudication on account of [Here insert the cause, such as there being insufficient quantity of provisions or any other circumstances to justify the disembarkation] I did on the …… day of …………… disembark …… of the Slaves at …………… where they remained.

To be witnessed by two Officers.

APPENDIX G

Mortality in the West Africa Squadron, 1825–39

THE FOLLOWING FIGURES for deaths from accident and disease in the West Africa Squadron are taken from Alexander Bryson's *Report on the Climate and Principal Diseases of the African Station* (1847), p. 177. They are repeated in Christopher Lloyd's *The Navy and the Slave Trade* (1968 edn), p. 288. It is assumed that the figures include the surveying vessels occasionally working on the coast.

YEAR	ANNUAL MEAN FORCE	DEATHS FROM DISEASE	DEATHS FROM ACCIDENT, ETC.	TOTAL DEATHS	RATIO OF DEATHS FROM DISEASE PER 1,000
1825	663	41	7	48	61.8
1826	1,043	57	6	63	54.7
1827	955	40	4	44	41.9
1828	958	81	3	84	84.6
1829	792	202	2	204	255.1
1830	667	72	4	76	107.9
1831	785	22	3	25	28
1832	512	18	3	21	35.1
1833	562	12	10	22	21.4
1834	620	18	8	26	29
1835	815	19	3	22	23.3
1836	965	16	4	20	16.6
1837	815	105	4	109	128.8
1838	885	115	3	118	129.9
1839	790	55	5	60	69.6

APPENDIX H

Annual Imports of Slaves to Cuba and Brazil, 1807–39

THE FIGURES BELOW are taken from *Economic Growth and the Ending of the Transatlantic Slave Trade* (1987) by David Eltis (Oxford University Press: New York, 1997), pp. 343–5. Eltis gives the following warning concerning accuracy: "For neither import nor export series is any claim made for completeness. It is unlikely, however, that bias in either direction was consistent over time. Given the determination of the British and the scale of the resources they devoted to the task, it seems unlikely that the real totals can be much greater than those developed here":

Numbers given are in thousands

YEAR	CUBA	BRAZIL SOUTH OF BAHIA	BRAZIL BAHIA	BRAZIL NORTH OF BAHIA
1807	3.4	9.7	7.9	6
1808	2.1	9.6	7.5	6
1809	1.5	13.2	7.5	5
1810	8.9	18.7	7.5	4.5
1811	8.5	17.4	5.8	5.1
1812	8.1	17.2	8.8	4
1813	6.4	17.1	6.8	4.5
1814	5.8	12.2	8.2	5.4
1815	12.1	14.8	6.8	5.3
1816	23.6	20.1	4.8	8.7
1817	34.5	18.2	6.1	11.2
1818	26.5	20.1	8.7	13
1819	20.2	17.1	7	15.5
1820	22.9	20.2	7.7	9.9
1821	4.5	20.4	6.7	10.1

Annual Imports of Slaves to Cuba and Brazil, 1807–39

YEAR	CUBA	BRAZIL SOUTH OF BAHIA	BRAZIL BAHIA	BRAZIL NORTH OF BAHIA
1822	4	27.8	7.1	9.4
1823	1.9	19.2	2.7	11
1824	7.7	26.2	3.1	4
1825	13.8	26.5	4.1	2.9
1826	4	30.4	7.9	4.5
1827	5	27.4	10.2	5.7
1828	12.9	43.4	7.8	6.9
1829	14.9	43.7	15	6.3
1830	14.4	31.2	7	2.8
1831	16.1	1	1	1.5
1832	13.6	4	3.3	3.8
1833	13.8	9	3.6	4.1
1834	16.7	13.8	3.6	4.1
1835	25.7	30	5.2	5.7
1836	20.2	46	2.9	2.9
1837	20.9	46	4	4
1838	21	42.8	4	4
1839	19.9	46	2.9	5.5

APPENDIX I

Comparative Values of Sterling, 1807–39

IT IS NOT POSSIBLE to be precise in the calculation of the historical value of sterling by comparison with its current value. The following table, however, gives approximations sufficiently accurate to allow an appreciation of the buying power of sums of money mentioned in the narrative. The figures are derived from the Bank of England Inflation Calculator, which uses the composite price index published by the Office for National Statistics. The composite price index is produced by linking together prices data from several different published sources – both official and unofficial.

The table shows the value early in 2014 of one pound sterling in each of the years shown.

YEAR	AMOUNT	YEAR	AMOUNT
1807	£82.10	1824	£98.00
1808	£78.90	1825	£83.50
1809	£72.10	1826	£88.60
1810	£70.10	1827	£94.40
1811	£72.10	1828	£97.10
1812	£63.50	1829	£98.00
1813	£62.00	1830	£102.00
1814	£71.10	1831	£92.70
1815	£79.50	1832	£100.00
1816	£87.10	1833	£106.30
1817	£76.50	1834	£116.10
1818	£76.50	1835	£113.50
1819	£78.30	1836	£102.00
1820	£86.30	1837	£100.00
1821	£98.00	1838	£99.00
1822	£113.50	1839	£92.70
1823	£106.30		

APPENDIX J

Pivot-Gun Arrangement in HMS Lynx

THE FOLLOWING DRAWING and explanatory notes were sent by Lieutenant Broadhead of *Lynx* to Commander Popham on 23 August 1838, and subsequently by Admiral Elliot to the Admiralty on 12 February 1839 (ADM 1/86). They are believed to be representative of the mountings of pivot-guns, the main armament, in all the smaller cruisers of the period. Clearly there were severe design faults in the arrangement which limited the safe and efficient operation of the weapon.

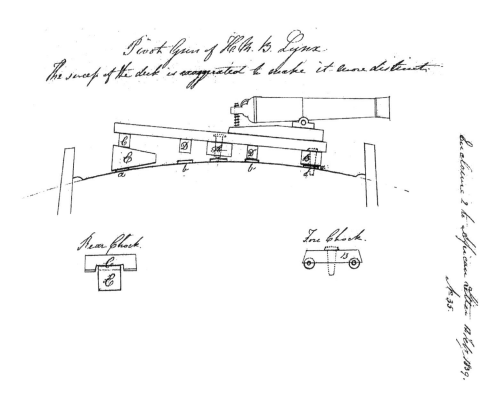

Pivot-gun arrangement, HMS *Lynx*

NOTES IN EXPLANATION OF THE SKETCH

A. A center Chock to pivot the Gun and lay it fore and aft.
B. chock runs on two Iron Trucks.
C. [Word unreadable] Bed, comes on the Deck when the Slide is fore and aft.
    D.D. Inner chocks to run on circle **b.b.**
    E. Quoin used to support **C**.
    a.a. Outer Iron Circle fixed to the Deck and standing about half an inch above it.
    b.b. Inner Iron Circle ——— Ditto ——— Ditto ———.
    d. Pivot Bolt.

When the Slide is pivoted on **B, a**, on the Gun running in, the bolt **d** would be thrown out of its short socket, and the slide, Gun and all would be entirely adrift, to prevent which, they place a Quoin in to wedge up **C**. This must be done after the Gun is pointed, by which they must very frequently lose their opportunity of firing, and the Quoin must be removed for further training. When the Gun runs in, the slide becomes very unsteady, the Quoin being necessarily in the center of Chock **C**.

On training the Slide pivoted at **a**, the chock **A** comes against the Iron circle **b**, and stops it, till lifted bodily up. In coming back, it does the same, and when there is a little more training, it is the same with the Iron Rollers, taking the edge of the circle **a**, here the whole weight of the Slide, Carriage and Gun, must be lifted, to get the rollers up, and they must train at the same time – all the chocks are shod with strong Iron.

If the Gun is worked, pivoted from the center, on running it out quickly on to leeward, it tilts the Slide enough to throw the Centre Bolt out of the socket, so that it is never secure.

Glossary

ADMIRALTY	The office of the Lords Commissioners of the Admiralty ("Their Lordships"), and the centre of naval administration. The Admiralty Board considered promotions and appointments of commissioned officers, the movements of fleets and individual ships, and the allocation of resources. Also included the Hydrographer.
BARGE	The second largest of a warship's boats, propelled by oars or sails.
BARQUE	A three-masted vessel, with fore and main masts square-rigged, and the mizzen fore-and aft rigged.
BENDS	An alternative name for the wales, the strong reinforcing planking extending the entire length of a vessel's side at various heights: the channel-wale, the upper-wale and the main-wale.
BILGED	The fracturing of that part of a vessel's hull which curves-in towards the keel and on which she would rest if aground.
BREECHING	The heavy rope passed around the breech of a gun and secured to the ship's side, which checked the gun's recoil.
BRIG	A two-masted sailing vessel, square-rigged on both masts.
BRIGANTINE	A two-masted vessel, square-rigged on the fore mast and fore-and-aft (schooner) rigged on the mizzen.
BY THE WIND	Sailing as close to the wind as possible without the luffs of the sails shaking.
CARRONADE	A short-barrelled, relatively light, short-range gun firing a heavy shot, which took its name from the Carron Iron Works at Falkirk, Scotland, where it was first made. It was introduced into the Royal Navy in 1779.
FELUCCA	A boat, to a design originating in the Mediterranean, with lateen sails.
GIG	A light, fast boat, with oars and sail, generally used by the Commanding Officer or Flag Officer.

JOLLY BOAT	A small ship's boat.
KEDGING	Moving a vessel by laying out an anchor and then hauling in the anchor rope.
LANGRIDGE	Shot consisting of jagged pieces of iron.
LAUNCH	The largest boat belonging to a warship, designed for pulling or, rigged as a sloop (one mast rigged fore-and-aft), for sailing.
LETTER OF MARQUE	The document issued by the Admiralty to a private man-of-war (privateer) or armed merchantman authorising hostile action against vessels of a specified enemy. An armed merchantman given this authority is referred to as a Letter of Marque to distinguish her from a privateer.
MASTER	[1] The Warrant Officer responsible for sailing and navigating a man-of-war (until 1843). [2] The commanding officer of a merchant ship.
MATE	[1] A warrant rank in the Royal Navy held by midshipmen passed for lieutenant and awaiting promotion. [2] A warrant officer in the Royal Navy responsible for assisting the Master with sailing and navigation (master's mate). [3] A rating in the Royal Navy assisting a specialist warrant officer (gunner's mate, boatswain's mate, etc.). [4] A deck officer in the Merchant Service subordinate to the master (first mate, second mate, etc.).
NAVY BOARD	Responsible to the Admiralty for the technical and financial aspects of naval administration; the dockyards and the maintenance of ships and buildings, ship design and building, and the appointment and examination of warrant officers.
NEAP TIDE	A tide occurring during the first and third quarter of the moon, when the gravitational pull of sun and moon are in opposition to each other, producing lower high waters and higher low waters than average.
PINNACE	[1] A warship's boat with six or eight oars. [2] A small schooner-rigged vessel with a transom-stern.
POLACCA	A large square-rigged merchant vessel with two or three single-piece masts.
PRIVATEER	A privately owned man-of-war, issued with a Letter of Marque to authorise her to operate against enemy vessels.
RAZEE	A vessel cut down to fewer decks than those with which she was originally built.

SCHOONER	Originally a vessel with two masts rigged fore-and-aft. If she carried square topsails on one or both masts she was a "topsail schooner". Later, similarly rigged vessels with up to six masts were built, and described as "three-masted schooners", and so on.
SENIOR OFFICER	"Senior Officer" denotes an officer specifically appointed in charge of a defined area. However, "senior officer" indicates an officer who merely happens to be the most senior, by rank and date of promotion, in a location.
SHIP	A seagoing vessel with three or more masts (usually three), square-rigged on all of them.
SPRING TIDE	Tides occurring in the second and fourth quarters of the moon when the gravitational pull of the sun and moon are in conjunction, so producing higher high water and lower low water than average.
SUMACA	(Spanish) A small schooner, generally used in the coasting trade.
SUPERCARGO	The agent of a vessel's owners who sailed with the vessel to attend to the commercial business connected with the cargo.
WARPING	Moving a vessel by hauling on a rope or ropes secured to a fixed position such as a jetty or buoy.

Endnotes

FOREWORD

1. The official commemoration included an important book produced by the Foreign Office: Keith Hamilton and Fairida Shaikh (eds), *Slavery in Diplomacy: The Foreign Office and the Suppression of the Atlantic Slave Trade* (FCO: London, 2007).
2. Christopher Lloyd, *The Navy and the Slave Trade: The Suppression of the African Slave Trade in the Nineteenth Century* (Frank Cass: London, 1968); William E.F. Ward, *The Royal Navy and the Slavers: The Suppression of the Atlantic Slave Trade* (Allen & Unwin: London, 1969).
3. Donald L. Canney, *Africa Squadron: The U.S. Navy and the Slave Trade, 1842-1861* (Potomac Books Inc.: Washington DC, 2006).
4. Hugh Graham Soulsby, *The Right of Search and the Slave Trade in Anglo-American Relations 1814-1862* (Johns Hopkins University Press: Baltimore, MD, 1933) is the standard account.
5. Leslie Bethell, *The Abolition of the Brazilian Slave Trade: Britain, Brazil and the Slave Trade Question 1807-1869* (Cambridge University Press: Cambridge, 1970), p. 190.
6. Canney, *Africa Squadron*.
7. Luise Martinez-Fernandez, *Fighting Slavery in the Caribbean: The Life and Times of a British Family in Nineteenth-Century Havana* (Routledge: New York, 1998).
8. Kathleen Mary Butler, *The Economics of Emancipation: Jamaica and Barbados 1834–1843* (University of North Carolina Press: Chapel Hill, NC, 1995).
9. David Murray, *Odious Commerce: Britain, Spain and the Abolition of the Cuban Slave Trade* (Cambridge University Press: Cambridge, 1980). John H. Schroeder, *Shaping a Maritime Empire: The Commercial and Diplomatic Role of the American Navy, 1829-1861* (Praeger: Westport, CT, 1985).
10. F. Egerton, *Admiral of the Fleet Sir Geoffrey Phipps Hornby* (William Blackwood: London, 1896) deals with the last years of the patrol.
11. Raymond Howell, *The Royal Navy and the Slave Trade* (Routledge: Beckenham, 1987) is the most recent study of the East African trade.
12. Andrew Lambert, 'Palmerston and Sea Power', in David Brown and Miles Taylor (eds), *Palmerston Studies II* (Hartley Institute: Southampton, 2007), p.61.

CHAPTER 2

1. Hugh Thomas, *The History of the Atlantic Slave Trade 1440–1870* (Picador: London, 1997), p. 289.

2   Ibid., p. 299.
3   Ibid., p. 316.
4   Ibid., p. 309.
5   Ibid., p. 376.
6   George Francis Dow, *Slave Ships and Slaving* (Cornell Maritime Press: Cambridge, MA, 1968), p. 16. Please note that this book was originally published in 1927 by the Marine Research Society, Salem, MA.
7   Ibid., p. 224.
8   Thomas, *The Slave Trade*, p. 394.
9   Ibid., p. 395.
10 Dow, *Slave Ships*, p. 224.
11 Ibid., p. 70.
12 Ibid., p. 161.
13 Thomas, *The Slave Trade*, p. 418.
14 Dow, *Slave Ships*, p. 191.
15 Ibid., p. 159.
16 Ibid., p. 191.
17 Thomas, *The Slave Trade*, p. 406.
18 Ibid., p. 417.
19 Dow, *Slave Ships*, p. 74.
20 Thomas, *The Slave Trade*, p. 425.
21 Dow, *Slave Ships*, p. 91.
22 Ibid., p. 202.
23 Ibid., p. 155.
24 Ibid., p. 160.
25 Ibid., p. 162.
26 Ibid., p. 181.
27 Ibid., p. 166.
28 Ibid., p. 168.
29 Thomas, *The Slave Trade*, p. 441.

CHAPTER 4

1   Basil Lubbock, *Cruisers, Corsairs and Slavers* (Brown, Son & Ferguson: Glasgow, 1993), p. 94.
2   George Francis Dow, *Slave Ships and Slaving* (Cornell Maritime Press: Cambridge, MA, 1968), p. 203.
3   Ibid., p. 198.
4   Hugh Thomas, *The History of the Atlantic Slave Trade 1440–1870* (Picador: London, 1997), p. 562.
5   Macarthy to Barrow, 23.4.1811 (ADM 1/4895).
6   Wilberforce to Colonial Office, 30.10.1807 (WO 1/352).
7   Wilberforce to Colonial Office, 30.10.1807 (WO 1/352).

8   Thompson to Castlereagh, 2.11.1808 (CO 267/24).
9   Order in Council, 16.3.1808 (WO 1/742).
10 Castlereagh to Admiralty, 15.8.1808 (ADM 1/4209).
11 Admiralty Boardroom Journal, 9.4.1809 (ADM 7/255).
12 Columbine to Admiralty, 20.7.1809 (ADM 1/1653).
13 Thompson to Castlereagh, 31.12.1808 (CO 267/24).
14 Liverpool to Admiralty, 28.12.1809, and Admiralty Note (ADM 1/1653).
15 Ludlam to Macaulay, 18.05.1808 (CO 267/24).
16 Macaulay letter, 18.9.1808 (CO 267/24).
17 State Papers 1812–14.
18 Columbine to Admiralty, 1.11.1810 (CO 267/28).
19 Wilberforce to Liverpool, 30.6.1810 (C O 267/28).
20 Columbine to Admiralty, 25.7.1810 (ADM 1/1659).
21 Sneyd to Admiralty – Enclosures, 9.10.1811 (ADM 1/2528).
22 Ibid.
23 Thomas, *The Slave Trade*, p. 579.

CHAPTER 5

1   Admiralty Boardroom Journal Jan–June 1811, p. 220 (ADM 7/259).
2   Sneyd to Berkeley, 2.10.1811 (ADM 1/2528).
3   *Tigress* log, 18–27.9.1811 (ADM 51/2904).
4   Macaulay to (probably) Liverpool, 25.5.1811 (WO 1/135).
5   Admiralty Boardroom Journal, 5.10.1811 (ADM 7/260).
6   Irby to White, 5.1.1812 (ADM 1/1996).
7   *Protector* log, 10–17.8.1811 (ADM 51/2683).
8   Irby to Croker, 8.6.1812 (ADM 1/1996).
9   Maxwell to Liverpool, 10.8.1811 (CO 267/30).
10 Irby to Croker, 23.11.1812 (ADM 1/1997).
11 Lloyd to Irby, 8.11.1812 (ADM 1/1996).
12 *Thais* log, 24.6.1811 (ADM 51/2913).
13 Irby to Croker, 20.11.1812 (ADM 1/1997).
14 Scobell to Irby, 18.11.1812 (ADM 1/1997).
15 Scobell to Croker, 14.3.1813 (ADM 1/2535).
16 *Thais* log, 31.3.1813 (ADM 51/2913).
17 Bickerton to Croker, 29.11.1813 (ADM 1/1227).
18 Maxwell [*Favorite*] to Admiralty, 25.2.1814 (ADM 1/2176).
19 Hugh Thomas, *The History of the Atlantic Slave Trade 1440–1870* (Picador: London, 1997), p. 581.
20 Tomkinson to Durham, 6.7.1814 (ADM 1/335).
21 Stopford to Croker, 29.1.1812 (ADM 1/64).
22 Irby to Croker, 27.2.1815 (ADM 1/1998).
23 Admiralty Boardroom Journal, 8.10.1814 (ADM 7/266).

24 Tailour to Browne, 25.3.1815 (ADM 1/1560).
25 *Ulysses* log, 15.6.1815 (ADM 51/2933).
26 David Eltis, *Economic Growth and the Ending of the Transatlantic Slave Trade* (Oxford University Press: New York, 1997), pp. 243–7.

CHAPTER 6

1 Theodore Canot (ed. Brantz Mayer), *Captain Canot; Or, Twenty Years of an African Slaver* (Project Gutenberg online edition, first published 1854), p. 61.
2 Hyde to Bathurst, 1.7.1815 (CO 267/40).
3 Maxwell to Bathurst, 29.3.1815 (CO 267/40).
4 MacCarthy to Bathurst, 22.7.1815 (CO 267/40).
5 Hugh Thomas, *The History of the Atlantic Slave Trade 1440–1870* (Picador: London, 1997), p. 588.
6 Fisher to Croker, 21.4.1816 (ADM 1/1813).
7 Dobree to Croker, 14.1.1817 (ADM 1/1742).
8 Admiralty to Yeo, 28.11.1816 (ADM 2/1327).
9 MacCarthy to Bathurst, 18.4.1816 (CO 267/42).
10 MacCarthy to Bathurst, 11.6.1817 (CO 267/45).
11 Foreign Office to Croker, 7.6.1817 (ADM 1/4237).
12 Castlereagh to Vaughan, 7.6.1816 (FO 84/1).
13 Home Popham to Admiralty, 12.4.1818 (ADM 1/269).
14 MacCarthy to Bathurst, 26.9.1818 (CO 267/47).
15 Willis to Croker, 1.2.1819 (ADM 1/2720).
16 Thomas, *The Slave Trade*, p. 592.
17 MacCarthy to Bathurst, 31.5.1816 (CO267/42).
18 Admiralty to Collier, 19.9.1818 (ADM 2/1327).
19 Scobell to Croker, 13.10.1818 (ADM 1/2546).
20 Strong to Admiralty, 24.8.1819 (ADM 1/2547).
21 MacCarthy to Bathurst, 25.12.1818. (CO 267/97).
22 Kelly to Admiralty, 8.3.1822 (ADM 1/2027).
23 Leeke to Collier, 13.1.1820 (ADM 1/1674).
24 Leeke to Collier, 13.1.1820 (ADM 1/1674).
25 Strong to Croker, 24.12.1819 (ADM 1/2548).
26 Collier to Admiralty, 13.1.1820 (ADM 1/1674).
27 Collier to Admiralty, 21.2.1820 (ADM 1/1674).

CHAPTER 7

1 Basil Lubbock, *Cruisers, Corsairs and Slavers* (Brown, Son & Ferguson: Glasgow, 1993), p. 115.
2 Collier to Admiralty, 19.3.1820, enclosing Sandilands letter 18.2.1820 (ADM 1/1674).
3 MacCarthy to Bathurst, 8.3.20, enclosing Collier letter 28.2.1820 (CO 267/51).

4   Collier to Admiralty, 7.3.1820 (ADM 1/1674).
5   Collier to Admiralty, 7.3.1820 (ADM 1/1674).
6   Collier to Admiralty, 26.5.1820 (ADM 1/1674).
7   Kelly to Collier, 3.9.1820 (ADM 1/2027).
8   Grant to Bathurst, 26.10.1820 (CO 267/51).
9   Haynes to Londonderry, 24.10.1821 (FO 131/1).
10 Collier to Croker, 19.3.1821 (ADM 1/1675).
11 Collier to Croker, 16.4.1821 (ADM 1/1675).
12 Collier to Croker, 17.6.1821 (ADM 1/1675).
13 John Marshall, *Marshall's Naval Biography* (Longman, Hurst, Rees, Orme and Brown: London, 1823–45), Vol. II, part ii, p. 540.
14 Knight to Croker, 5.8.1821 (ADM 1/2980).
15 Hugh Thomas, *The History of the Atlantic Slave Trade 1440–1870* (Picador: London, 1997), p. 595.
16 Collier to Admiralty, 27.12.1821 (ADM 1/1675).
17 Knight to Croker, 16.2.1822 (ADM 1/2027).
18 Clavering to Barrow, 21.12.1823 (ADM 1/1677).
19 Mends to Croker, 18.6.1822, enclosing Knight letter 30.5.1822 (ADM 1/2188).
20 Mends to Croker, 17.4.1822 (ADM 1/2188).
21 Hagan to Mends, 19.7.1822 (ADM 1/2947).
22 Mends to Croker, 27.3.1823 (ADM 1/2189).
23 British Commissioners, Freetown, to Foreign Office, 31.1.1823 (SP 22–3).
24 Foreign Office to Admiralty, 19.4.1823 (ADM 1/4239).
25 Mends to Croker, 29.3.1823 (ADM 1/2189).
26 British Commissioners, Freetown, to Canning, 23.2.1824 (SP 24–5).
27 Prickett to Barrow, 18.8.1824 (ADM 1/2358).
28 Filmore to Croker, 24.11.1823 (ADM 1/1815).
29 Courtenay to Croker, 22.2.24 (ADM 1/1678).
30 William Edward Burghardt Du Bois, *The Suppression of the African Slave Trade to the United States of America, 1638–1870* (Harvard University Press: Cambridge, MA, 1896; Dover Publications: Mineola, NY, 1999), p. 138.
31 Mends to Croker, 26.6.1822 (ADM 1/2188).

CHAPTER 8

1   Collier to MacCarthy, 1.6.1820 (CO 267/51).
2   MacCarthy to Bathurst, 27.8.1823 (CO 267/58).
3   Admiralty Boardroom Journal, 25.3.1824 (ADM 7/285).
4   Bullen to Croker, 22.7.1824 (ADM 1/1571).
5   Bullen to Croker, 2.9.1824 (ADM 1/1571).
6   Bullen to Croker, 23.9.1824 (ADM 1/1571).
7   Bullen to Croker, 19.3.25 (ADM 1/1572).
8   Bullen to Croker, 28.2.25 (ADM 1/1572).

9   Bullen to Croker, 18.6.1825 (ADM 1/1572).
10 Bullen to Croker, 12.9.1825 (ADM 1/1572).
11 Rendall to Canning, 21.3.1826 (SP 26–7).
12 Willes to Croker, 28.11.1825 (ADM 1/2724).
13 Bullen to Croker, 5.3.1826 (ADM 1/1573).
14 Turner to Canning, 20.7.1825 (CO 267/66).
15 Turner to Bathurst, 2.3.1826 (CO 267/71).
16 Bullen to Croker, 15.9.1826 (ADM 1/1573).
17 Bullen to Croker, 18.6.1826 (ADM 1/1573).
18 Purchas to Bullen, 18.9.1826 (ADM 1/1573).
19 Basil Lubbock, *Cruisers, Corsairs and Slavers* (Brown, Son & Ferguson: Glasgow, 1993), p. 154.
20 Elliot to Croker, 8.9.1826, enclosing Tucker letter (ADM 1/1773).
21 Macaulay to Bathurst, 21.4.1826 (CO 267/72).
22 Macaulay to Canning, 2.2.1827 (SP 27–8).
23 Bullen to Croker, 30.12.1826 (ADM 1/1574).
24 Campbell to Bathurst, 14.3.1827 (CO 267/81).
25 Collier to Croker, 30.6.1827 (ADM 1/1682).
26 Bullen to Croker, 2.7.1827 (ADM 1/1575).
27 Collier to Croker, 29.10.1827 (ADM 1/1682).
28 Croker to Backhouse, 3.11.1827 (SP 27–8).
29 Collier to Croker, 32.12.1827 (ADM 1/1682).
30 Bullen to Croker, 10.11.1826 (ADM 1/1574).
31 Admiralty Boardroom Journal, 27.6.1827 (ADM 7/291).
32 Owen to Collier, 25.1.1828 (ADM 1/2273).

CHAPTER 9

1   Turner to Collier, 4.4.1828 (ADM 1/1683).
2   Reffell to Lumley, 9.7.1828 (SP 28–9).
3   Owen to Collier, 1.2.1828 (ADM 1/2273).
4   British Commissioners, Freetown, to Palmerston, 10.2.1831 (SP 31–2).
5   Collier to Croker, 6.6.1828 (ADM 1/1683).
6   Collier to Croker, 8.9.1828, enclosing Turner letter 28.8.1828 (ADM 1/1683).
7   Collier to Croker, 13.9.1828 (ADM 1/1683).
8   Owen to Croker, 10.9.1828 (ADM 1/2273).
9   Owen to Croker, 18.12.1828 (ADM 1/2274).
10 Admiralty Boardroom Journal, 3.1.1829 (ADM 7/295).
11 Collier to Croker, 11.2.1829, enclosing Downes letter, 2.2.1829 (ADM 1/1684).
12 Collier to Croker, 28.1.1829 (ADM 1/1684).
13 Foreign Office to Croker, 2.7.1830 (CO 267/102).
14 Collier to Croker, 6.5.1829 (ADM 1/1685).
15 Collier to Croker, 10.5.1829 (ADM 1/1685).

16 Christopher Lloyd and Jack L.S. Coulter, *Medicine and the Navy 1200–1900*, Vol. IV 1815–1900 (E. & S. Livingstone: Edinburgh and London, 1963), p. 173.
17 Basil Lubbock, *Cruisers, Corsairs and Slavers* (Brown, Son & Ferguson: Glasgow, 1993), p.150.
18 Owen to Croker, 2.8.1829 (ADM 1/2274).
19 British Commissioners, Freetown, to Palmerston, 2.7.1831 (SP 31–2).
20 Foreign Office to Croker, 27.8.1830 (CO 267/102).
21 Collier to Croker, 25.5.1830 (ADM 1/1).
22 Gordon to Croker, 5.10.1830 (ADM 1/1867).
23 Foreign Office to Barrow, 21.12.1830 (CO267/102).
24 Hayes to Croker, 4.12.1830 (ADM 1/1).
25 Hayes to Squadron commanding officers, 10.12.1830 (ADM 1/1).
26 Hayes to Elliot, 31.1.1831 (ADM 1/1).
27 Hayes to Elliot, 20.1.1831 (ADM 1/1).

CHAPTER 10

1 Hugh Thomas, *The History of the Atlantic Slave Trade 1440–1870* (Picador: London, 1997), p. 637.
2 Basil Lubbock, *Cruisers, Corsairs and Slavers* (Brown, Son & Ferguson: Glasgow, 1993), p. 230.
3 Vives to Kilbee, 14.3.1825 (State Papers [SP] 25–6).
4 Macleay to Canning, 22.2.1826 (SP 26–7).
5 Macleay to Canning, 2.9.1826 (SP 26–7).
6 De Mayne to Hydrographical Office, 20.10.1827 (ADM 1/4537).
7 British Commissioners, Havana, to Dudley, 20.1.1828 (SP 28–9).
8 Macleay to Dudley, 26.7.1828 (SP 28–9).
9 Lubbock, *Cruisers, Corsairs and Slavers*, p. 241.
10 Macleay to Aberdeen, 1.1.1829 (SP 29–30).
11 Aberdeen to British Commissioners, Havana, 20.3.1829 (SP 29–30).
12 Macleay to Aberdeen, 3.7.1829 (SP 29–30).
13 Macleay to Aberdeen, 17.7.1829 (SP 29–30).
14 Jones to Croker, 28.10.1830 (ADM 1/3716).
15 Fleeming to Croker, 19.4.1830 (ADM 1/283).
16 Foreign Office to Croker, 10.7.1830 (CO 267/102).
17 Treasury Chambers to Admiralty, 5.8.1831 (ADM 1/4306).
18 Warren to Elliot, 21.1.1832 (ADM 1/74).
19 British Commissioners, Rio, to Aberdeen, 13.1.1831 (FO 84/120).

CHAPTER 11

1 Foreign Office to Croker, 2.7.1830 (ADM 1/4245).
2 Hayes to Elliot, 6.7.1831 (ADM 1/1).

3   Hayes to Elliot, 21.1.1831 (ADM 1/1).
4   Hayes to Elliot, 13.3.1831, enclosing Castle letter (ADM 1/1).
5   Hayes to Elliot, 6.5.1831, enclosing Ramsay letter (ADM 1/1).
6   Basil Lubbock, *Cruisers, Corsairs and Slavers* (Brown, Son & Ferguson: Glasgow, 1993), p. 215.
7   Hayes to Gordon and Gordon to Hayes, 22.4.1831 (ADM 1/1).
8   Lubbock, *Cruisers, Corsairs and Slavers*, p. 217.
9   British Commissioners, Freetown, to Palmerston, 20.6.1831, enclosing De Paiva letter (FO 315/21).
10 Foreign Office to Admiralty, 7.2.1832 (ADM 1/4248).
11 Harrison to Warren, 28.3.1833 (ADM 1/76).
12 Palmerston to Admiralty, 7.11.1831 (ADM 1/4247).
13 Warren to Elliot, 29.2.1832 (ADM 1/74).
14 Warren to Elliot, 1.2.1832 (ADM 1/74).
15 Colonial Office to Admiralty, 14.8.1832 (ADM 1/4249).
16 Lieutenant de Saumarez case, 16.6.1834 (ADM 1/5480).
17 Hayes to Elliot, 7.5.1832 (ADM 1/1).
18 Foreign Office to British Commissioners, Freetown, 27.12.1832 (FO 315/3).
19 Foreign Office to Admiralty, 17.8.1832 (ADM 1/4249).
20 Warren to Elliot, 23.3.1833 (ADM 1/75).
21 Bate to Elliot, 21.8.1833 (ADM 1/3333).
22 Solicitor's Office to Elliot, 8.1.1834 (ADM 1/3720).
23 Warren to Elliot, 6.5.1834, enclosing Sulivan letter (ADM 1/77).
24 Warren to Elliot, 13.2.1834 (ADM 1/77).
25 Order in Council, 19.3.1834, SP 33–4, p. 1208.
26 British Commissioners, Freetown, to Palmerston, 16.7.1834 (FO 84/149).
27 Warren to Elliot, 4.9.1834 (ADM 1/78).
28 Wauchope to Campbell, 30.10.1834 (ADM 1/78).
29 British Commissioners, Freetown, to Palmerston, 3.12.1834 (FO 84/148).
30 British Commissioners, Freetown, to Wellington, 10.2.1835 (FO 84/167).
31 Lubbock, *Cruisers, Corsairs and Slavers*, p. 303.
32 British Commissioners, Freetown, to Wellington, 4.3.1835 (FO 84/167).

CHAPTER 12

1   Macleay to Palmerston, 16.2.1832 (FO 313/12).
2   Colpoys to Elliot, 24.12.1831 (ADM 1/287).
3   Colpoys to Elliot, 4.6.1832 (ADM 1/288).
4   Colpoys to Elliot, 28.7.1832 (ADM 1/288).
5   Foreign Office to British Commissioners, Havana, 20.10.183. (FO 313/3).
6   Macleay to Palmerston, 1.1.1833 (FO 313/12).
7   Stanley to Elliot, 9.8.1833 (ADM 1/4252).
8   Cockburn to Elliot, 26.1.1834, enclosing Bolton letter 19.11.1833 (ADM 1/291).

9   British Commissioners, Havana, to Palmerston, 31.3.1834 (FO 84/150).
10 Court Martial report, 28.4.1834 (ADM 1/5480).
11 British Commissioners, Havana, to Palmerston, 27.11.1834 (FO 84/151).
12 Cockburn to Elliot, 16.5.1835, enclosing Ussher letter (ADM 1/294).
13 British Commissioners, Rio, to Palmerston, 27.7.1831 (SP 31–2).
14 Baker to Waldegrave, undated May/June 1831 (ADM 1/35).
15 British Commissioners, Rio, to Palmerston, 22.2.1832 (FO 84/129).
16 Foreign Office to Admiralty, 13.2.1832 (ADM 1/4248).
17 Ouseley to Palmerston, 25.3.33 (ADM 1/4252).
18 Palmerston to Hoad, 26.6.1833 (ADM 1/4253).
19 Seymour to Elliot, 13.1.1834 (ADM 1/42).
20 Palmerston to British Commissioners, Rio, 8.10.1834 (ADM 1/4256).
21 Seymour to Elliot, 10.4.1834 (ADM 1/42).
22 Grigg to Tait, 11.7.34 (FO 84/152).
23 British Commissioners, Rio, to Wellington, 23.3.1835 (FO 84/174).

CHAPTER 13

1   British Commissioners, Freetown, to Palmerston, 5.1.1835 (FO 84/166).
2   Campbell to Admiralty, 10.5.1835 (ADM 1/79).
3   Campbell to Admiralty, 25.5.1835 (ADM 1/79).
4   British Commissioners, Freetown, to Wellington, 30.5.1835 (ADM 1/4259).
5   British Commissioners, Freetown, to Palmerston, 26.11.1835 (FO 84/168).
6   Campbell to Wood, 14.12.35, enclosing Puget letter 20.11.1835 (ADM 1/79).
7   SP 34–5, p. 343.
8   Palmerston to British Commissioners, Freetown, 16.11.1835 (FO 84/166).
9   British Commissioners, Freetown, to Palmerston, 9.4.1836, enclosing Puget letter (FO 84/190).
10 Basil Lubbock, *Cruisers, Corsairs and Slavers* (Brown, Son & Ferguson: Glasgow, 1993), p. 308.
11 British Commissioners, Freetown, to Palmerston, 26.9.1836 (FO 84/192).
12 Campbell to Wood, 27.2.1836 (ADM 1/80).
13 Campbell to Wood, 2.3.1836 (ADM 1/80).
14 British Commissioners, Freetown, to Palmerston, 31.5.1836 (FO 84/190).
15 Handwritten note by Palmerston, 5.7.1836 (FO 84/191).
16 British Commissioners, Freetown, to Palmerston, 20.8.1836, enclosing Fox letter (FO 84/191).
17 British Commissioners, Freetown, to Palmerston, 5.7.1836 (FO 84/194).
18 British Commissioners, Freetown, to Palmerston, 31.10.1836 (FO 84/193).
19 Internal Foreign Office memorandum, 31.10.1836 (FO 84/193).
20 British Commissioners, Freetown, to Palmerston, 17.3.1836 (FO 84/190).
21 British Commissioners, Freetown, to Palmerston, 23.3.1837 (FO 84/214).
22 Macaulay to Palmerston, 26.4.1837 (FO 84/211).

23 Campbell to Wood, 21.3.1837 (ADM 1/81).
24 SP 36–7, 9.4.1837.
25 Campbell to Wood, 15.7.1837, enclosing Craigie letter (ADM 1/81).
26 Campbell to Wood, 17.7.1837 (ADM 1/81).
27 Palmerston note, 1.5.1837 (FO 84/211).
28 British Commissioners, Freetown, to Foreign Office, 30.10.1837 (FO 84/214).
29 Campbell to Wood, 3.11.1837 (ADM 1/82).
30 Campbell to Wood, 16.4.1838 (ADM 1/82).
31 Elliot to Wood, 26.6.1838 (ADM 1/83).
32 Campbell to Wood, 10.4.1838, enclosing Popham letter 13.11.1837 (ADM 1/82).
33 Campbell to Wood, 10.4.1838, enclosing Popham letters 5.1.1838 and 21.2.1838 (ADM 1/82).

CHAPTER 14

1 British Commissioners, Havana, to Governor of Trinidad, 4.8.1835 (FO 84/172).
2 British Commissioners, Havana, to Palmerston, 31.8.1835 (FO 84/172).
3 British Commissioners, Havana, to Palmerston, 25.11.1835 (FO 84/172).
4 British Commissioners, Havana, to Palmerston, 11.1.1836 (FO 84/196).
5 British Commissioners, Havana, to Palmerston, 17.2.1836 (FO 84/196).
6 British Commissioners, Havana, to Palmerston, 21.3.1836 (FO84/196).
7 British Commissioners, Havana, to Palmerston, 15.1.1836 (FO 84/196).
8 British Commissioners, Havana, to Palmerston, 2.7.1836 (FO 84/197).
9 British Commissioners, Havana, to Palmerston, 30.7.1836 (FO 84/197).
10 Jones to Halkett, 5.10.1836 (ADM 1/298).
11 Note by Palmerston, 9.12.1836 (FO 84/197).
12 Halkett to Admiralty, 3.3.1837 (ADM 1/299).
13 D'Urban to Strong, 3.5.1837 (ADM 1/300).
14 Foreign Office to Admiralty, 22.11.1837 (ADM 1/4269).
15 Madden to Glenely, 14.2.1837 (ADM 1/4267).
16 Schenley to Palmerston, 28.9.1837 (FO 84/217).
17 British Commissioners, Havana, to Palmerston, 21.4.1838 (FO 84/240).
18 Paget to Admiralty, 12.4.1838 (ADM 1/301).
19 British Commissioners, Havana, to Palmerston, 1.1.1838 (FO 84/240).
20 British Commissioners, Rio, to Palmerston, 3.8.1835 (FO 84/175).
21 British Commissioners, Rio, to Palmerston, 10.2.1836 (FO 84/196).
22 Hammond to Admiralty, 23.1.1836 (ADM 1/45).
23 Hammond to Admiralty, 27.8.1836 (ADM 1/45).
24 Hammond to Admiralty, 6.8.1836 (ADM 1/45).
25 Hammond to Admiralty, 25.8.1836 (ADM 1/45).
26 Palmerston to British Commissioners, Rio, 22.10.1836 (FO 84/199).
27 Foreign Office to Admiralty, 1.9.1836 (ADM 1/4264).
28 British Commissioners, Rio, to Palmerston, 17.11.1836 (FO 84/219).

CHAPTER 15

1. Basil Lubbock, *Cruisers, Corsairs and Slavers* (Brown, Son & Ferguson: Glasgow, 1993), p. 325.
2. Campbell to Wood, 15.4.1838 (ADM 1/82).
3. Hill to Elliot, 5.6.1838 (ADM 1/83).
4. Elliot to Wood, 25.6.1838 (ADM 1/83).
5. Elliot to Wood, 19.11.1838 (ADM 1/84).
6. Elliot to Wood, 27.7.1838 (ADM 1/83).
7. British Commissioners, Freetown, to Foreign Office, 20.10.1838 (FO 84/232).
8. British Commissioners, Freetown, to Foreign Office, 30.11.1838 (FO 84/237).
9. British Commissioners, Freetown, to Foreign Office, 21.11.1838 (FO 84/233).
10. Admiralty to Elliot, 10.5.1839 (ADM 1/86).
11. Holland to Elliot, 1.1.1839 (ADM 1/84).
12. Elliot to Wood, 9.10.1838 (ADM 1/84).
13. British Commissioners, Freetown, to Palmerston, 14.2.1839 (SP 39–40).
14. Elliot to Popham, 8.11.1838 (ADM 1/84).
15. British Commissioners, Freetown, to Foreign Office, 31.1.1839 (FO 84/268).
16. Elliot to Admiralty, 13.2.1839 (ADM 1/85).
17. Fox-Strangways to Admiralty, 28.9.1839 (SP 39–40).
18. British Commissioners, Freetown, to Palmerston, 14.2.1839 (SP 39–40).
19. British Commissioners, Freetown, to Palmerston, 12.2.1839 (FO 84/267).
20. Elliot to Admiralty, 12.2.1839 (ADM 1/86).
21. Elliot to Admiralty, 16.2.1839 (ADM 1/85).
22. Harvey to Admiralty, 14.11.1839 (ADM 1/304).
23. British Commissioners, Freetown, to Palmerston, 31.1.1840 (FO 84/308A).
24. British Commissioners, Freetown, to Palmerston, 13.5.1839 (FO 84/269).
25. British Commissioners, Freetown, to Palmerston, 13.5.1839 (FO 84/267).
26. Tucker to Elliot, 14.6.1839 (ADM 1/86).
27. Kellett to Tucker, 30.6.1839 (ADM 1/86).
28. Elliot to Admiralty, 27.8.1839 (ADM 1/86).
29. British Commissioners, Freetown, to Palmerston, 31.7.1839 (FO 84/270).
30. British Commissioners, Freetown, to Palmerston, 11.9.1839 (FO 84/271).
31. Bidwell to Tucker, 14.9.1839 (FO 84/273).
32. British Commissioners, Freetown, to Palmerston, 24.8.1839 (FO 84/270).
33. Foreign Office to British Commissioners, 12.10.1839 (FO 84/266).
34. British Commissioners, Freetown, to Palmerston, 17.3.1840 (FO 84/308A).
35. Bidwell to Foreign Office, 11.1.1840 (FO 84/307).

CHAPTER 16

1. Paget to Admiralty, 4.7.1838, enclosing Potbury letter 4.2.1838 (ADM 1/302).
2. Elliot to Wood, 4.11.1838 (ADM 1/84).

3   Palmerston to British Commissioners, 19.3.1838 (FO 84/239).
4   Kennedy to Palmerston, 19.6.1838 (FO 84/240).
5   Paget to Admiralty, 16.10.1838, enclosing Fraser letter 2.8.1838 (ADM 1/302).
6   Kennedy to Palmerston, 16.7.1838 (FO 84/239).
7   Paget to Admiralty, 20.11.1838 (ADM 1/302).
8   British Commissioners, Freetown, to Palmerston, 31.10.1839 (FO 84/271).
9   British Commissioners, Havana, to Palmerston, 19.1.1839 (FO 84/274).
10 Foreign Office to Admiralty, 31.12.1839 (ADM 1/4277).
11 Palmerston to British Commissioners, Rio, 18.4.1838 (FO 84/241).
12 Palmerston to British Commissioners, Rio, 30.4.1838 (FO 84/241).
13 British Commissioners, Rio, to Palmerston, 5.2.1839 (FO 84/275).
14 Herbert to Mixed Commission, Rio, 10.6.1838 (FO 84/241).
15 Foreign Office to Admiralty, 26.4.1839 (ADM 1/4275).
16 Sulivan to Admiralty, 27.6.1838 (ADM 1/2563).
17 Foreign Office to Admiralty, 13.10.1838 (ADM 1/4273).
18 British Commissioners, Rio, to Palmerston, 22.1.1839 (FO 84/275).
19 Foreign Office to Admiralty, 1.2.1839 (ADM 1/4274).
20 Sulivan to Admiralty, 14.1.1839 (ADM 1/2564).
21 Palmerston to British Commissioners, Rio, 5.12.1838 (FO 84/242).
22 Palmerston to British Commissioners, Rio, 31.8.1839 (FO 84/276).
23 Palmerston to British Commissioners, Rio, 8.5.1839 (FO 84/276).
24 Sulivan to Admiralty, 23.4.1839 (ADM 1/2565).
25 Foreign Office to Admiralty, 5.8.1839 (ADM 1/4276).
26 Sulivan to Admiralty, 14.5.1839 (ADM 1/2565).
27 Palmerston to British Commissioners, Rio, 23.11.1839 (FO 84/277).
28 Sulivan to Admiralty, 1.9.1839 (ADM 1/2566).
29 Sulivan to Admiralty, 27.7.1839 (ADM 1/2566).
30 British Commissioners, Rio, to Palmerston, 28.10.1839 (FO 84/277).
31 Foreign Office to Admiralty, 21.11.1839 (ADM 1/4277).
32 British Commissioners, Rio, to Palmerston, 29.11.1839 (FO 84/277).
33 Foreign Office to Admiralty, 31.12.1839 (ADM 1/4277).

SUMMARY

1   Graham to Palmerston, 11.1.1831 (FO 84/124).

EPILOGUE

1   James Bandinel, *Some Account of the Trade in Slaves from Africa as Connected with Europe and America* (Frank Cass & Company Ltd: London, 1968), p. 277.
2   Christopher Lloyd, *The Navy and the Slave Trade: The Suppression of the African Slave Trade in the Nineteenth Century* (Frank Cass: London, 1968), p. 94; William Law

Mathieson, *Great Britain and the Slave Trade 1839–65* (Longmans, Green & Company: London, 1929), p. 62.
3   Lloyd, *Navy and the Slave Trade*, p. 94; Mathieson, *Great Britain and the Slave Trade*, p. 62.
4   Lloyd, *Navy and the Slave Trade*, p. 97; Mathieson, *Great Britain and the Slave Trade*, p. 93.
5   David Eltis, *Economic Growth and the Ending of the Transatlantic Slave Trade* (Oxford University Press: New York, 1997), pp. 244–5.
6   Bandinel, *Some Account of the Trade in Slaves*, p. 297.
7   Hugh Thomas, *The History of the Atlantic Slave Trade 1440–1870* (Picador: London, 1997), p. 669.
8   Mathieson, *Great Britain and the Slave Trade*, p. 59.
9   Thomas, *The Slave Trade*, p. 659.
10 Lloyd, *Navy and the Slave Trade*, p 176.
11 Ibid., p. 53.
12 Thomas, *The Slave Trade*, p. 670.
13 Ibid., p. 663; Lloyd, *Navy and the Slave Trade*, p. 52.
14 Lloyd, *Navy and the Slave Trade*, p. 49.
15 Thomas, *The Slave Trade*, p. 665.
16 Ibid., p. 747.
17 Lloyd, *Navy and the Slave Trade*, p. 119.
18 Mathieson, *Great Britain and the Slave Trade*, p. 71.
19 Lloyd, *Navy and the Slave Trade*, p. 104.
20 Ibid., p. 119.
21 Ibid., p. 141.
22 Mathieson, *Great Britain and the Slave Trade*, p. 80.
23 Lloyd, *Navy and the Slave Trade*, p. 120.
24 Ibid., p. 281.
25 Ibid., p. 139.
26 Ibid., p. 176.
27 Ibid., p. 156.
28 Ibid., p. 174.
29 Ibid., p. 180.
30 Mathieson, *Great Britain and the Slave Trade*, p.64.

Bibliography

SOURCES USED

General

Material for the Prologue and the introductory section of the book, Part One: The Trade, has been drawn entirely from published work. That for the main narrative, Part Two: The Suppression Campaign, is mostly from primary sources held at the National Archives, which are identified below by "piece numbers".

Listed below, chapter by chapter, are the documents used.

Prologue and Part One

PROLOGUE AND CHAPTER 1

Only one source has been used:

Thomas, Hugh, *The History of the Atlantic Slave Trade 1440–1870* (Picador: London, 1997).

CHAPTER 2

Two sources have been used:

Dow, *George Francis, Slave Ships and Slaving* (Cornell Maritime Press: Cambridge, MA, 1968). Please note that this was originally published by the Marine Research Society, Salem, Massachusetts, in 1927.
Thomas, Hugh, *The History of the Atlantic Slave Trade 1440–1870* (Picador: London, 1997).

CHAPTER 3

The sources used are all published documents and charts, as follows:

The Admiralty Sailing Directions (current):
——*The West Indies Pilot*.
——*The South America Pilot*.
The Admiralty Manual of Seamanship (HMSO, London, 1981 edn).
The African Pilot (1856).

The African Pilot (1868).
Charts of the Atlantic Ocean & Coast of Brazil (Arrowsmith: London, 1805).
Charts of the West Coast of Africa (Laurie & Whittle: London, 1794; 1797).
The English Pilot (1753–9).
The Mariner's Handbook, 7th edition (United Kingdom Hydrographic Office: Taunton, 1999).
Ocean Passages for the World, 7th edition (United Kingdom Hydrographic Office: Taunton, 1987).
Sailing Directions for the West Coast of Africa (1849).
The South America Pilot (1860).
The Times Atlas of the World (Book Club Associates: London, 1979).
The West Indies Pilot (1859).
Lloyd, Christopher, and Jack L. S. Coulter, *Medicine and the Navy 1200–1900, Vol. IV 1815–1900* (E. & S. Livingstone: Edinburgh and London, 1963).

Part Two

Throughout Part Two regular use has been made of the following publications:

Eltis, David, *Economic Growth and the Ending of the Transatlantic Slave Trade* (Oxford University Press: New York, 1997).
Lubbock, Basil, *Cruisers, Corsairs and Slavers* (Brown, Son & Ferguson: Glasgow, 1993).
Lyon, David, *The Sailing Navy List* (Conway Maritime Press: London, 1993).
The Navy List (published quarterly; dates covered by narrative used).
Thomas, Hugh, *The History of the Atlantic Slave Trade 1440–1870* (Picador: London, 1997).

Additional publications of which occasional use has been made are listed below under chapter headings.

CHAPTER 4

Primary sources

ADM 1/4895, 1126, 1651, 1653, 1659, 1733, 2528, 4209, 4887, 4897.
ADM 2/1082, 1083, 1084.
ADM 7/254, 255, 256, 256, 258, 259.
ADM 12/143.
ADM 51/2291, 2041, 2198, 2470, 2683, 2890, 2904.
CO 142/25.
CO 267/24, 25, 27, 28, 30.
FO 72/103, 121, 138.
HCA 30/791, 793, 796.
HCA 49/97, 98, 101.
WO 1/352, 742.

Additional publications

Crawley, Charles William (ed.), *New Cambridge Modern History*, Vol. IX (Cambridge University Press: Cambridge, 1965).
Hill, Richard, *Prize Law and the Royal Navy in the Napoleonic Wars 1793–1815* (Sutton Publishing: Stroud, 1998).
Public General Acts
State Papers (SP) 1812–14, Vol. 1.

CHAPTER 5

Primary sources

ADM 1/64, 66, 335, 336, 501, 1227, 1560, 1562, 1567, 1674, 1771, 1996, 1997, 1998, 2176, 2528, 2535, 2546, 2612, 4227, 4231, 4897, 5003.
ADM 2/1084, 1327.
ADM 7/260, 261, 262, 265, 266, 267, 268.
ADM 51/2010, 2680, 2683, 2904, 2913, 2933.
CO 267/30.
FO 63/163.
FO 308/1.
HCA 30/791, 796.
HCA 49/97, 98, 101.
WO 1/135.
WO 41/74.

Additional publications

Du Bois, William Edward Burghardt, *The Suppression of the African Slave-Trade to the United States of America 1638–1870* (Harvard University Press: Cambridge, MA, 1896; Dover Publications: Mineola, NY, 1999).
Lloyd, Christopher, *The Navy and the Slave Trade: The Suppression of the Atlantic Slave Trade in the Nineteenth Century* (Longmans, Green & Company: London, 1949; Frank Cass: London, 1968).
Marshall, John, *Marshall's Naval Biography* (Longman, Hurst, Rees, Orme and Brown: London, 1823–45).
O'Byrne, William R., *O'Byrne's Naval Biography* (J. Murray: London, 1849).
SP 12–14, 14–15, 16–17.
Ward, W. E. F., *The Royal Navy and the Slavers: The Suppression of the Atlantic Slave Trade* (Allen & Unwin: London, 1969).
Woodman, Richard, *The Victory of Seapower: Winning the Napoleonic War 1806–14* (Chatham Publishing: London, 1998).

CHAPTER 6

Primary sources

ADM 1/68, 269, 1673, 1674, 1742, 1743, 1813, 1814, 1833, 2027, 2544, 2546, 2547, 2548, 2719, 2720, 2721, 2783, 2944, 2945, 4236, 4237, 4458, 4529.
ADM 2/1327.
ADM 7/268, 269, 272, 274, 276, 277.
ADM 12/210.
ADM 51/2195, 2206, 2487, 2680, 2853.
CO 267/40, 42, 45, 47, 49, 65.
FO 84/1.
FO 129/3.
FO 308/1.
FO 313/1.
FO 315/1.
HCA 30/787, 789.
HCA 49/97.
WO 41/74.

Additional publications

Brown, D.K., *Before the Ironclad: The Development of Ship Design, Propulsion and Armament in the Royal Navy 1815–60* (Conway Maritime Press: London, 1990).
Canot, Theodore (ed. Brantz Mayer), *Captain Canot; Or, Twenty Years of an African Slaver* (D. Appleton & Co.: New York, 1854; also available as an e-book on the Project Gutenberg website).
Chappelle, Howard Irving, *The Baltimore Clipper: Its Origins and Development* (Marine Research Society: Salem, MA, 1930; Dover: Mineola, NY, 2012).
Lloyd, Christopher, *The Navy and the Slave Trade: The Suppression of the Atlantic Slave Trade in the Nineteenth Century* (Longmans, Green & Company: London, 1949; Frank Cass: London, 1968).

CHAPTER 7

Primary sources

ADM 1/1571, 1674, 1675, 1677, 1678, 1679, 1815, 2027, 2188, 2189, 2190, 2721, 2722, 2947, 2980, 4238, 4239.
ADM 7/280, 281, 282, 283, 284, 285.
CO 267/47, 51, 53, 56, 58.
FO 83/2343.
FO 84/166, 168.
FO 131/1.
FO 315/22.

Additional publications

Crawley, Charles William (ed.), *New Cambridge Modern History*, Vol. IX (Cambridge University Press: Cambridge, 1965).
Du Bois, William Edward Burghardt, *The Suppression of the African Slave-Trade to the United States of America 1638–1870* (Harvard University Press: Cambridge, MA, 1896; Dover Publications: Mineola, NY, 1999).
Lloyd, Christopher, *The Navy and the Slave Trade: The Suppression of the Atlantic Slave Trade in the Nineteenth Century* (Longmans, Green & Company: London, 1949; Frank Cass: London, 1968).
Marshall, John, *Marshall's Naval Biography* (Longman, Hurst, Rees, Orme and Brown: London, 1823–45).
O'Byrne, William R., *O'Byrne's Naval Biography* (J. Murray: London, 1849).
SP 1820–1, 1822–3, 1823–4, 1824–5.

CHAPTER 8

Primary sources

ADM 1/1571, 1572, 1573, 1574, 1575, 1632, 1678, 1681, 1682, 1773, 2200, 2273, 2358, 2724, 3846, 3907.
ADM 7/285, 288, 290, 291, 292, 293.
ADM 12/248.
CO 267/51, 58, 60, 65, 66, 71, 72, 81, 82.
FO 84/42, 307.

Additional publications

Du Bois, William Edward Burghardt, *The Suppression of the African Slave-Trade to the United States of America 1638–1870* (Harvard University Press: Cambridge, MA, 1896; Dover Publications: Mineola, NY, 1999).
Hill, J. R. (ed.), *Oxford Illustrated History of the Royal Navy* (Oxford University Press: Oxford, 1995).
Lloyd, Christopher, *The Navy and the Slave Trade: The Suppression of the Atlantic Slave Trade in the Nineteenth Century* (Longmans, Green & Company: London, 1949; Frank Cass: London, 1968).
Lloyd, Christopher, and Jack L. S. Coulter, *Medicine and the Navy 1200–1900, Vol. IV 1815–1900* (E. & S. Livingstone: Edinburgh and London, 1963).
Marshall, John, *Marshall's Naval Biography* (Longman, Hurst, Rees, Orme and Brown: London, 1823–45).
Ward, W. E. F., *The Royal Navy and the Slavers: The Suppression of the Atlantic Slave Trade* (Allen & Unwin: London, 1969).
SP 24–5, 25–6, 26–7, 27–8, 28–9.

CHAPTER 9

Primary sources

ADM 1/1, 32, 76, 280, 1356, 1578, 1682, 1683, 1684, 1685, 1867, 2273, 2274, 2275, 3715, 4246.
ADM 7/293, 294, 295, 296.
CO 267/99, 98, 102.
FO 84/117, 118, 127.
FO 315/3, 21, 27.

Additional publications

Brown, D. K., *Before the Ironclad: The Development of Ship Design, Propulsion and Armament in the Royal Navy 1815–60* (Conway Maritime Press: London, 1990).
Bryson, Alexander, *Report on the Climate and Principal Diseases of the African Station* (W. Clowes & Son: London, 1847).
Lloyd, Christopher, and Jack L. S. Coulter, *Medicine and the Navy 1200–1900, Vol. IV 1815–1900* (E. & S. Livingstone: Edinburgh and London, 1963).
SP 28–9, 29–30, 30–1, 31–2.

CHAPTER 10

Primary sources

ADM 1/74, 280, 281, 282, 283, 284, 286, 2238, 3716, 3908, 4305, 4306, 4537.
ADM 2/1327.
CO 267/102.
FO 84/119, 120, 127.
FO 313/12.

Additional publications

SP 24–5, 25–6, 26–7, 28–9, 29–30, 30–1, 31–2.

CHAPTER 11

Primary sources

ADM 1/1, 36, 73, 74, 75, 76, 77, 78, 79, 80, 1582, 1868, 1966, 3332, 3333, 3475, 3720, 4245, 4247, 4248, 4249, 4251, 4252, 4253, 4254, 4255, 4256, 4258, 5478, 5479.
ADM 13/103.
FO 84/117, 127, 134, 148, 149, 166, 167, 194.
FO 315/3, 27, 21.

Additional publications

Du Bois, William Edward Burghardt, *The Suppression of the African Slave-Trade to the United States of America 1638–1870* (Harvard University Press: Cambridge, MA, 1896; Dover Publications: Mineola, NY, 1999).

Lloyd, Christopher, *The Navy and the Slave Trade: The Suppression of the Atlantic Slave Trade in the Nineteenth Century* (Longmans, Green & Company: London, 1949; Frank Cass: London, 1968).
Lyon, David and Winfield, Rif, *The Sail and Steam Navy List 1815–1889* (Chatham Publishing: London, 2004).
Marshall, John, *Marshall's Naval Biography* (Longman, Hurst, Rees, Orme and Brown: London, 1823–45).
O'Byrne, William R., *O'Byrne's Naval Biography* (J. Murray: London, 1849).
SP 30–1, 33–4.

CHAPTER 12

Primary sources

ADM 1/35, 36, 37, 38, 39, 40, 42, 43, 287, 288, 289, 291, 292, 294, 1819, 2861, 4248, 4249, 4252, 4253, 4256, 4257, 5480.
FO 84/120, 128, 129, 136, 137, 138, 149, 150, 151, 152, 153, 171, 174.
FO 313/3, 12, 23.

Additional publications

Marshall, John, *Marshall's Naval Biography* (Longman, Hurst, Rees, Orme and Brown: London, 1823–45).
O'Byrne, William R., *O'Byrne's Naval Biography* (J. Murray: London, 1849).
SP 31–2, 33–4, 35–6.
Walvin, James, *Black Ivory: Slavery and the British Empire* (HarperCollins: London, 1992).
Wareham, Tom, *The Star Captains: Frigate Command in the Napoleonic Wars* (Chatham Publishing: London, 2001).

CHAPTER 13

Primary sources

ADM 1/79, 80, 81, 82, 83, 4259, 4271.
FO 84/165, 166, 167, 168, 188, 190, 191, 192, 193, 194, 211, 212, 213, 214, 230, 235, 236.

Additional publications

Brown, D. K., *Before the Ironclad: The Development of Ship Design, Propulsion and Armament in the Royal Navy 1815–60* (Conway Maritime Press: London, 1990).
Bryson, Alexander, *Report on the Climate and Principal Diseases of the African Station* (W. Clowes & Son: London, 1847).
SP 34–5, 36–7.

CHAPTER 14

Primary sources

ADM 1/44, 45, 47, 48, 50, 294, 295, 296, 298, 299, 300, 301, 4261, 4263, 4264, 4265, 4267, 4268, 4269.
FO 84/172, 194, 195, 196, 197, 198, 199, 212, 214, 216, 217, 219, 230, 235, 237, 239, 240, 312.

Additional publications

Marshall, John, *Marshall's Naval Biography* (Longman, Hurst, Rees, Orme and Brown: London, 1823–45).

CHAPTER 15

Primary sources

ADM 1/82, 83, 84, 85, 86, 304, 4275, 4276.
ADM 7/617.
ADM 50/202.
ADM 51/3064, 3184, 3496, 3533, 3535, 3634.
CO 267/155.
FO 84/231, 232, 233, 234, 236, 237, 266, 267, 268, 269, 270, 271, 272, 273, 307, 308A, 344, 310, 312, 313.

Additional publications

Bryson, Alexander, *Report on the Climate and Principal Diseases of the African Station* (W. Clowes & Son: London, 1847).
Public General Acts
SP 39–40.

CHAPTER 16

Primary sources

ADM 1/84, 301, 302, 303, 304, 2563, 2564, 2565, 2566, 2567, 4277, 4273, 4274, 4275, 4276, 4277.
FO 84/216, 232, 237, 239, 240, 241, 242, 266, 268, 269, 271, 274, 275, 276, 277, 314.

Additional publications

SP 39–40.

Part Three

SUMMARY

Primary sources

FO 84/124.

EPILOGUE

All new material is drawn from the following publications:

Bandinel, James, *Some Account of the Trade in Slaves from Africa as Connected with Europe and America* (Frank Cass & Company Ltd: London, 1968).
Lloyd, Christopher, *The Navy and the Slave Trade: The Suppression of the Atlantic Slave Trade in the Nineteenth Century* (Longmans, Green & Company: London, 1949; Frank Cass: London, 1968).
Mathieson, William Law, *Great Britain and the Slave Trade 1839–65* (Longmans, Green & Company: London, 1929).
Thomas, Hugh, *The History of the Atlantic Slave Trade 1440–1870* (Picador: London, 1997).

FURTHER READING

Overviews

Davis, David Brion, *Inhuman Bondage: The Rise and Fall of Slavery in the New World* (Oxford University Press: Oxford, 2006).
Eltis, David and David Richardson, "A New Assessment of the Transatlantic Slave Trade", in David Eltis and David Richards (eds), *Extending the Frontiers: Essays on the New Transatlantic Slave Trade Database* (Yale University Press: New Haven, CT, 2008), pp. 1–60.
——— *Atlas of the Transatlantic Slave Trade* (Yale University Press: New Haven, CT, 2010).
Eltis, David and Stanley L. Engerman (eds), *The Cambridge World History of Slavery, Volume 3, AD 1420–AD 1804* (Cambridge University Press: Cambridge, 2011).
Klein, Herbert S., *The Atlantic Slave Trade* (Cambridge University Press: Cambridge, 1999).
Klein, Martin A. (ed.), *Breaking the Chains: Slavery, Bondage, and Emancipation in Modern Africa and Asia* (University of Wisconsin Press: Madison, WI, 1993).
Slavery & Abolition: A Journal of Slave and Post-Slave Studies, various issues.
The William and Mary Quarterly, Third Series, Vol. 58, No. 1, "New Perspectives on the Transatlantic Slave Trade" (January 2001).
The William and Mary Quarterly, Third Series, Vol. 66, No. 4, "Abolishing the Slave Trades: Ironies and Reverberations" (October 2009).

Origins and Effects of Abolitionism

Bender, Thomas (ed.), *The Antislavery Debate: Capitalism and Abolitionism as a Problem in Historical Interpretation* (University of California Press: Berkeley, CA, 1992).

Blackburn, Robin, *The Overthrow of Colonial Slavery, 1776–1848* (London: Verso, 1988).

Brown, Christopher Leslie, *Moral Capital: Foundations of British Abolitionism* (University of North Carolina Press: Chapel Hill, NC, 2006).

Davis, David Brion, *The Problem of Slavery in the Age of Revolution, 1770–1823* (Cornell University Press: Ithaca, NY, 1975).

Drescher, Seymour, *Capitalism and Antislavery: British Mobilization in Comparative Perspective* (Macmillan: Basingstoke, 1986).

——— *The Mighty Experiment: Free Labour versus Slavery in British Emancipation* (Oxford University Press: Oxford, 2002).

——— *Abolition: A History of Slavery and Antislavery* (Cambridge University Press: Cambridge, 2009).

Jennings, Lawrence C., *French Anti-Slavery: The Movement for the Abolition of Slavery in France, 1802–1848* (Cambridge University Press: Cambridge, 2000).

Kielstra, Paul Michael, *The Politics of Slave Trade Suppression in Britain and France, 1814–48: Diplomacy, Morality and Economics* (Macmillan: Basingstoke, 2000).

Miers, Suzanne, *Britain and the Ending of the Slave Trade* (Longman: London, 1975).

Oldfield, J. R., *Popular Politics and British Anti-Slavery: The Mobilisation of Public Opinion Against the Slave Trade, 1787–1807* (Manchester University Press: Manchester, 1995).

Slavery in Africa

Diouf, Sylviane A. (ed.), *Fighting the Slave Trade: West African Strategies* (Ohio University Press: Athens, OH, 2003).

Getz, Trevor R., *Slavery and Reform in West Africa: Toward Emancipation in Nineteenth-Century Senegal and the Gold Coast* (Ohio University Press: Athens, OH, 2004).

Klein, Martin A., *Slavery and Colonial Rule in French West Africa* (Cambridge University Press: Cambridge, 1998).

Lawrance, Benjamin N. and Richard L. Roberts (eds), *Trafficking in Slavery's Wake: Law and the Experience of Women and Children in Africa* (Ohio University Press: Athens, OH, 2012).

Lovejoy, Paul E., *Transformations in Slavery: A History of Slavery in Africa, Third Edition* (Cambridge University Press: Cambridge, 2011).

Miers, Suzanne and Igor Kopytoff (eds), *Slavery in Africa: Historical and Anthropological Perspectives* (University of Wisconsin Press: Madison, WI, 1977).

Miers, Suzanne and Richard Roberts, *The End of Slavery in Africa* (University of Wisconsin Press: Madison, WI, 1988).

Miller, Joseph C., *Way of Death: Merchant Capitalism and the Angolan Slave Trade, 1730–1830* (University of Wisconsin Press: Madison, WI, 1988).

Nwokeji, G. Ugo, *The Slave Trade and Culture in the Bight of Biafra: An African Society in the Atlantic World* (Cambridge University Press: Cambridge, 2010).

Shumway, Rebecca, *The Fante and the Transatlantic Slave Trade* (University of Rochester Press: Rochester, NY, 2011).
Sparks, Randy J., *The Two Princes of Calabar: An Eighteenth-century Atlantic Odyssey* (Harvard University Press: Cambridge, MA, 2009).

Slavery in the Americas

Bethel, Leslie, *The Abolition of the Brazilian Slave Trade: Britain, Brazil and the Slave Trade Question, 1807–1869* (Cambridge University Press: Cambridge, 1970).
Brown, Vincent, *The Reaper's Garden: Death and Power in the World of Atlantic Slavery* (Harvard University Press: Cambridge, MA, 2010).
Marques, João Pedro (trans. Richard Wall), *The Sounds of Silence: Nineteenth-century Portugal and the Abolition of the Slave Trade* (Berghan Books: New York, 2006).
Schmidt-Nowara, Christopher, *Slavery, Freedom, and Abolition in Latin America and the Atlantic World* (University of New Mexico Press: Albuquerque, NM, 2011).
Tomich, Dale W., *Slavery in the Circuit of Sugar: Martinique and the World Economy, 1830–1848* (The Johns Hopkins University Press: Baltimore, MD, 1990).

The Experience of Enslavement and the Middle Passage

Behrendt, Stephen D., David Eltis and David Richardson, "The Costs of Coercion: African Agency in the Pre-Modern Atlantic World", *Economic History Review*, New Series, Vol. 54. No. 3 (August 2001), pp. 454–76.
Christopher, Emma, *Slave Ship Sailors and Their Captive Cargoes, 1730–1807* (Cambridge University Press: Cambridge, 2006).
Falola, Toyin and Amanda Warnock, *Encyclopedia of the Middle Passage* (Greenwood Press: Westport, CT, 2007).
Rediker, Marcus, *The Slave Ship: A Human History* (Viking: New York, 2007).
Smallwood, Stephanie, *Saltwater Slavery: A Middle Passage from Africa to American Diaspora* (Harvard University Press: Cambridge, MA, 2007).

The Naval Campaign

Edwards, Bernard, *The Royal Navy versus the Slave Traders: Enforcing Abolition at Sea, 1808–1898* (Pen & Sword: Barnsley, 2008).
Rees, Siân, *Sweet Water and Bitter: The Ships that Stopped the Slave Trade* (Chatto & Windus: London, 2009).

Sierra Leone

Fyfe, Christopher, *A History of Sierra Leone* (Oxford University Press: Oxford, 1962).
Peterson, John, *Province of Freedom: A History of Sierra Leone, 1787–1870* (Northwestern University Press: Evanston, IL, 1969).

General Index

(Vessels arrested as suspected slavers are listed at Appendix A)

"Aberdeen Act", 751
Aberdeen, Earl of, Foreign Secretary, 332, 749, 751
"Aberdeen Letter", 748
Abolition Act
  British of 1807, 1
  French of 1830, 427
  United States of 1807, 16
  United States of 1820, 257
accommodation hulk/vessel, 434, 587, 705, 722
Accra, Chiefs of
Act of Union, 32
Adams, John Quincy, US Ambassador/Secretary of State, 160, 281
Admiral Owen, British merchant schooner, 375
Admiralty
  The, 108, 110, 136, 138, 165, 185, 231–2, 319, 328, 383, 428, 434, 464, 495, 511, 539, 738
  Board, 728, 734, 805
Adventurers Company, 28
Advocate-General, 201, 392, 413, 437, 491, 501, 748
Affriquain, Nantes slaver, 56
Affriquain, Surcouf slaver, 183
Africa, hired merchant schooner, 333
Africa Committee, 177

Africaine, French schooner, 261
African "servants", 572
African Institution, 103–4
African Survey Commission, 110–13
Albion-Frigate, slaver, 57
Alerto, Spanish piratical slaver, 291, 295
allowances, 521
Altavilla, Mr, Portuguese Commissary Judge, 237, 276, 302
Altrevida, Spanish schooner slaver, 294
Alvara of 1818, 205
American Revolutionary War, 36, 38
American War (1812–14), 159–60
Amiens, the Peace of, 15, 38
Amistad, Spanish schooner, 481–2
Amistad Habanera, Spanish slaver schooner, 477
anchors, shortage of, 251
Anderson, Major, Acting Governor, Belize, 579
Anglo-French relations, 202, 355, 413, 463
Anita, Spanish slave schooner, 240
"Anti-Coercionists", 751, 756

Apollo, English schooner, 198
Apollonia, natives of, 192
Appeal Court of the Privy Council, 106
Appleton, Major, temporary Governor, 188–9
apprentices, 212
"apprentices", 112, 169, 707–8
Arango, Francisco de, Cuban planter, 122
Arethuse, French frigate, 160–1
Armouroux, Monsieur, French merchant master, 653
Arnold, James, surgeon, 48, 54, 58
Articles of War, 509, 619, 683, 743
Ashanti
  King of, 70, 215
  nation, 34, 48
  slaving, 73
  War, 274–6, 280, 284–6, 298, 311, 326
Ashman, Mr, Cape Mesurado Agent, 267
asiento, 24, 26, 29, 31–2, 35
Assistant Surgeons, shortage, 244, 417
Atalanta, French slaver, 310
Atalanta, Spanish piratical slaver ship, 312

Badia, Don Pedro, slaver master, 448
Ballerney, Captain James, Liverpool merchant barque master, 424
Baltimore clippers, 184–5, 329, 414
Bandeira, Viscount de Sa, Portuguese Foreign Minister, 706
Bannister, John William, Chief Justice/Acting Governor, 332
bark, Peruvian or cinchona, 174, 358–9, 639, 661
"Barn, The", 662
Baron, Monsieur, slaver master, 270
Barra, King/Kingdom of, 407, 425–6, 430
barracoons, 47, 165, 633, 745
Barrow, Sir John, Secretary of the Admiralty, 705
base facilities, 412, 449
Bathurst, Lord, Colonial Secretary, 147, 219, 323
Bathurst settlement, 212, 234, 245, 341, 430, 432, 439, 455
Beecroft, Mr, in charge of Fernando Po establishment, 524, 530
Behn, Aphra, 1
Belfast, Earl of, 504
Benezet, Anthony, 4–8
Bidwell, Mr, Commissary Registrar, 666, 677
bilateral treaties, policy of negotiating, 192, 211, 231, 256, 730
black soldiers, Havana, 587, 683, 688, 693
Blanco, Don Pedro, slaver, 235–6
blanket dress, 358
blockade, of slave harbours, 310, 616, 626–7, 747–8, 751–2, 756
boarding list, 643
Board of Trade, 538
boilers/coppers, for cooking, 272, 530, 543, 795

Bonnouvrié, Mr, Dutch Commissary Arbitrator/Judge, 217, 276, 302, 311–12
Bostock, Robert, British slaver, 163, 215
bounties, 110, 203, 205, 371, 378, 449–50, 531, 537, 543, 624, 743
Boyle, Doctor, Surgeon to the Mixed Commissions, 356–7
brand, branding, 50
Brazil
 authorities, 269, 494, 497–501, 560, 604, 697, 701, 705, 713–14, 722–4
 declaration of independence, 408
 Government, 311, 408, 411, 436, 489–92, 597, 601–9, 698–709, 714
 Minister for Foreign Affairs, 247, 408, 492, 605–6
breaking bulk, 314
breaking-up of condemned prizes, 436, 492, 529, 538, 573, 581, 624
Brereton, Lieutenant-Colonel, Lieutenant-Governor, 189
brigantines, 428, 442, 464, 502, 633, 645, 738
brigs, 42, 136, 185–6, 307, 321, 360, 431, 633, 645, 704, 737–8
Britannia, slaver, 59
British Minister/Ambassador
 Buenos Aires, 610
 Lisbon, 103, 549, 706, 103
 Madrid, 134, 436, 566, 592
 Rio de Janeiro, 205, 408, 492–4, 497, 602, 604–5, 702, 719
broad pendant, 112
Brodie, British slaver, 189
Brodie stoves, 358
Bryson, Dr, medical historian, 545, 662

bullying attitude, 638
Burke, Edmund, Member of Parliament, 4, 11–12, 18
Burke, Lieutenant-Colonel, Governor, 246
businessmen, Cuban, 37, 577
Busy, English brig, 101

caboose, 543
Cacimbo, fog, 85
Campbell, Mr Benjamin, British merchant, 541, 630
Campbell, Major-General Sir Neil, Governor, 309, 315–16, 320, 323, 326
Campbell, Major Henry Dundas, Lieutenant-Governor, 508, 524, 528, 549
Campo, Don Juan, Spanish Commissary Arbitrator, 217, 276
Campos, João Carneiro de, Brazilian Commissary Judge, 606
Canning, George, Foreign Secretary, 13, 116, 256, 260, 270, 277, 311, 322
canoes, 46, 52, 65, 73–5, 147, 202, 248, 666
Canot, Theodore, slaver, 184
Canot, Joseph, slaver, 439
Cape Mesurado settlement, 259–60, 267
Captain-General
 Vives, 122, 134, 234, 385–7, 390, 392, 396, 468
 Ricafort, 468, 478, 480, 483
 Tacon, 480, 483, 566–76, 580–1, 585, 587, 683, 687–8
 Espeleta, 691
 Valdes, 750
"Captain of the Flag", 639
captor
 declaration by, 437, 796
 presence of, 392, 404, 483, 563, 572, 574

General Index 837

Captured Negro Department, *see* Liberated African Department
Cardwell, Viscount, parliamentarian, 757
Caridade, slaver smack, 297
Caroline, French slaver schooner, 379–80
Caroline, Spanish pirate brig, 306
carriage (gun) failure, 187, 191, 399, 644
Casamajor, Justinian, British Commissary Arbitrator, 222
casks, water, 134, 272, 521, 529, 549, 600, 620, 654, 674, 686, 721
Cass, General, US Ambassador, 750
Cassandra, American schooner, 312
Castor oil, or "Palma Christie", 510, 537
Castlereagh, Lord, Colonial Secretary, 108, 110–16, 171–2, 192, 205, 211, 231, 256, 728–9
castles, Gold Coast, 21, 26, 31, 48, 70, 234
Cedule
of 1804, 205–6
of 1817, 393
Certificates of Emancipation, 564, 576
Charles V, King of Spain, 23
Charlotte, Danish merchant schooner, 712
Chief Justice
of Sierra Leone, 139, 163, 189, 252, 308, 310, 332
Chisholm, Captain/Major, 343, 276, 278
Chisholm, Lieutenant-Colonel, Lieutenant-Governor, 189
cholera, 86, 276, 464, 473, 479, 482, 485, 561–2
Church Mission Society, 212
Churruca, Spanish slaver brig, 469

Clapperton, explorer, 299
Clarence, Duke of, 14, 17
Clarke, Henry, ship-owner, 102
Clarkson, Thomas, 7–9, 11, 13, 15–16, 18, 43
Clegg, Captain Thomas, Liverpool merchant-ship master, 424
Clerk of Ordnance Works, 529
coal, 310, 429
Cockburn, Colonel, 561, 567
Cochrane, Vice-Admiral Sir Alexander, Governor, 145
coffles, 46, 52
"coffin brig", *see* brigs
Cole, Thomas, Acting Governor, 364, 508
Collector of Customs, 363, 366, 523, 529, 538, 553
colours/papers
American, 578, 635, 648–9, 657, 689, 691, 696, 734, 750
Artigas, 238
Brazilian, 496
British, 132, 138, 490, 728
French, 191, 220, 240, 287, 361, 389
Portuguese, 107, 141, 408, 467, 496, 501, 534, 548, 586, 594, 610–12, 689, 697, 723–47, 749
Russian, 651, 686
Spanish, 106, 129, 134–5, 141, 159, 164, 168, 199–200, 206
Columbus, Christopher, explorer, 23
command structure, 119, 136, 139, 182, 412, 464, 739
Commercial Code of Portugal, 597
Commissariat Officer, 537
Committee for Abolition, the, 8, 10
Commodore, American slaver, 606

Company of Merchants Trading to Africa, 35, 104
compensation for losses, 204, 206, 222
Conference of Aix-la-Chapelle, 211, 285
Congress, United States, 159, 257, 733
Congreve
gun, 484
rocket, 540
Consolidated Fund, 624
contagion, 549, 581
contraband, 24, 32, 105, 138, 466, 604, 728–9
Convention(s)
with Brazil, 408, 492, 559, 659, 709, 722, 731–2
with Denmark, 712
with France, 428, 457, 733
with Portugal, 303, 206, 217, 222, 411, 413, 437, 467, 491, 501, 716
with USA, 487
of Vienna, 670
Cook, British slaver, 189
cooking coppers, *see* boilers
copper sheathing, 129, 331, 361, 434, 555, 643
Coralline, slaver, 49–50, 53
corruption, 116, 152, 599, 604, 707, 710
Cortes
Portuguese, 271, 706
Spanish, 134, 751
Costa, Antonio Julião da, Portuguese Commissary Judge, 222
cotton, 11, 39, 44, 105, 395, 467
Council of the Indies, 205
"country labour", 567
"course of trade", 632, 654
court expenses, 537, 635, 666, 736
courts
of Admiralty, Spanish, 393
of Appeal, 171
of Guildhall, 210
of King's Bench, 5, 215
of Oyer and Terminer, 163

Court of St James, 167
Crab Island, 413, 432
Crawford, English slave trader, 143
Crewe, English slave factory owner, 178–9
Crimean War, 752
Croker, The Rt Hon. J. W., Secretary of the Admiralty, 346, 351, 356, 361
Crow, Captain Hugh, slaver, 51, 57, 102
Crowther, Bishop Samuel, 253
Crundell, slave factory owner, 167
Cruz, Donna Maria da, Princes landowner and slaver, 255, 266, 285
Cuba, Hamburg barque, 576
Cullen, Dr, letter-writer, 707
Cunningham, Alexander, British Commissary Arbitrator, 223, 247, 409, 489
currents
  Benguela, 89
  Canary, 88, 97
  Equatorial Countercurrent, 88, 556
  Guinea, 88–9, 128
  North Equatorial, 88–90, 92, 96
  South Equatorial, 88–90, 92, 96, 149
  South Sub-Tropical, 89–90, 220
Curtis, British slaver, 143, 242–3, 573
Customs House authority, 528, 539, 650, 686
clearance, 521, 529, 537, 552, 620

Daendels, Dutch Governor, 200, 202
Daget, Serge, historian, 427
Dahomey
  Kingdom of, 31, 71, 73
  King of, 46, 123, 467

Dalrymple, Campbell James, British Commissary Arbitrator, 686, 688, 694
damages, 230–1, 564, 739, 748
Dawes, William, Commissioner, 110, 116, 120, 131
decrees
  of Lisbon of 1836, 584, 650, 657, 679, 705
  of the Queen of Portugal, 619, 623, 706
  Portuguese, of 1818, 486
  Portuguese, of 1837, 697
delays in adjudication, 497, 563, 573, 603, 605, 667, 697, 702, 722, 724
Denham, Lieutenant-Colonel Dixon, Governor, 332
Denny, King, 653–4
Department of Justice (Brazil), 497, 605
Desperado, pirate felucca, 369–70
deterioration of canvas, cordage, etc., 148
deserters, 402, 683
design of ships, 42, 328–9, 503
de Walden, Lord Howard, British Minister in Lisbon, 549, 706
Diana, Spanish piratical slaver, 354
Dido, American slaver schooner, 683
District Attorney, New York, 648, 655
Division
  Barbados, 583
  Jamaica, 565, 567, 578, 581, 586, 692, 694–5, 739
  Windward, 563
Docherty, Colonel Richard, Governor and Commissary Arbitrator, 549, 553–4, 617, 630, 635, 642, 665
Doctors' Commons, 192, 244
Dois de Marco, Brazilian slaver, 497, 709

Dolben, Sir William, 10
Dolben Act of 1788, 54
Dolphin, American slave schooner, 257–8
Doldrums, 76, 85, 92, 128
Dom Pedro, Prince Regent of Portugal/Emperor of Brazil, 271, 485
Don, Mr, Horticultural Society, 259
Don Pedro, piratical slaver brig, 296
Dougan, Mr, Captors' Proctor, 667
Drake, Captain Richard, slaver, 50
Droits of Admiralty, 139, 498
Dudley, Earl of, Foreign Secretary, 332
Duke Ephraim, Old Calabar headman, 176, 236, 250, 255, 266, 275, 343
Duke of York, trooper brigantine, 575
Dunbar, British slaver, 189
Dundas, Henry, Lord Melville, 12–15
Dupuis, Mr, British consul, 242
Duquesa de Braganza, Spanish/Portuguese slaver, 694
Dutch West India Company, 25–6, 30, 34, 43, 71
"duty on entrance", 643
dysentery/the "flux", 24, 58, 60, 282, 455, 509

East India Company, 180
East Indiamen, 180
Egan, Mr, British Consul at Cape Verde Islands, 555
Eliza, French slaver, 362
Elizabeth, British schooner, 131
Elizabeth, US merchantman, 245
Elizabeth I, Queen of England, 25

Elkins, British slaver, 143
emancipados, 403, 471, 478, 485
emancipated Africans, welfare, 211
embargoes, 598, 605–6, 697, 700–3, 705–6, 708, 710, 724
Emperor, US merchant schooner, 586
Equipment Article/Clause, 272, 514, 731, 785, 794
Erskine, Lord Chancellor, 16
Escaelottia, Senhor, Portuguese slaver, 651
Espeleta, Don Joaquin de, Captain-General, *see* Captain-General
Esperanza, Spanish slave schooner, 240
Esplorador, Spanish slaver brig, 687
Etoile, French frigate, 167
Estralia, Spanish slaver, 479–80
Evans, Captain, Acting Governor, 364

Faber, Mrs, slave merchant, 630
factories, 16, 35, 47, 179, 748, 755
Falconbridge, Alexander, surgeon, 9, 44, 47, 49–51, 53, 55, 58, 60
false colours/identity, 105, 132–3, 136, 234, 280, 290, 557, 612–13, 625, 640, 648–9, 651, 679, 693, 698, 724, 734, 749, 753, 756
false sale, 133–4, 618, 641
Fama de Cadiz, Spanish slaver corvette, 359
Fantees, 276, 308
farina, 296, 300, 512, 529, 670, 674, 795
Favorite, Spanish slaver schooner, 349
fee for release, 610

Félicité, French slaver brig, 377–8
Felonies Act, the, 132, 147, 789
Ferguson, Doctor, Surgeon to the Courts, 509–10
Fernandina, Conde de, Spanish Commissary Judge, 468, 574, 576
Fernando Po
  dismantlement, 431–2, 436, 441, 447, 449, 524, 551, 645–6
  purchase of, 423, 432
  settlement/base, 287, 323–4, 332–5, 343, 346, 356, 366, 382, 412–14, 428–9, 735
Ferreira, Madame, wife of Governor-General of Princes Island, 617–18
Ferreira, Silvestre Pinheiro, Portuguese Commissary Judge, 223, 247
Figaniere e Morão, J. César de la, Portuguese Commissary Arbitrator, 237
Findlay, Lieutenant-Colonel, Governor/Judge, 364–6, 374–5, 379–81, 426, 431–5, 439, 452
fitting out, 10, 15, 118, 132, 235, 280, 292, 402, 688, 703
Fitzgerald, Edward, Chief Justice, Judge and British Commissary Arbitrator, 217, 237, 252, 269, 274
Fleet, the, reduction of, 183
"flux", the, *see* dysentery
Forcade, slave dealer, 691
Forest, Captain, French slaver, 192
Fornaro, Guiseppe, slaver master, 394–5
Forrest, Captain Thomas, slaver, 102
forts, *see* castles
Fortuna, Portuguese slaver brig, 460–1

Foster, Mr, American Vice-Consul, 641
Fountain, tank vessel, 591
Fox, Charles James, political leader, 12–13, 16, 18
Fox, Mr, American Minister in London, 692
Fraser, Captain, Acting Governor, 364, 367
free pratique, 568, 581
Free Trade movement, 751–2
French squadron/cruisers, 289, 305, 358, 366, 463, 750
frigates, 111, 115, 141, 186, 737

Galatea, coal depot-ship, 592
Gallinas raid, 748
Gascoyne, Bamber, Member of Parliament, 11
Gavilina, Spanish piratical brigantine, 297, 453
General Turner, brigantine, 435
George, English slave vessel, 107
George and James, London "trade ship", 297–8
Gilbert, Captain Don Pedro, pirate, 444–5
Glenely, Lord, Colonial Secretary, 587
Gloucester, Duke of, President African Institution, 17, 104
Goderich, Viscount, Colonial Secretary, 426, 432
gold dust, 142, 150, 208, 380
Gomes, Mr, Portuguese Commissary Arbitrator, 555
Gomez, Joze Ferraira, Portuguese Governor-General, 216, 218242, 617
Governors-General/ Governors/ Lieutenant-Governors
  of Angola, 657
  of Bissau, 259, 318, 480, 650
  of Cape Coast Castle, 147, 242

Governors-General/
Governors/Lieutenant-
Governors (*cont.*)
of Elmina, 245, 300, 315, 666
of Princes Island, 216, 218,
244, 294, 355
of Senegal, 189, 430
of the Bahamas, 282, 485,
501, 563
of Trinidad, 478, 561
Graham, Sir James, First
Lord of the Admiralty,
504, 735–6
Grant, Major/Lieutenant-
Colonel, Lieutenant-
Governor/Acting
Commandant,
245–6, 293
Grant, Sir William, Master
of the Rolls, 107
gratings/grating hatches, 51,
57, 134, 272, 794
"Great Debate", the, 751
Grecian, British merchant
brig, 297
Gregory, Edward, British
Commissary Judge,
217, 252
Gregory, Thomas, British
Commissary Judge, 217,
237, 252, 274, 302
Grenville, Lord, Prime
Minister, 1, 16–17
Greville, Captain The Hon.
R. F., Royal Yacht
Squadron, 433
Grey, Viscount Charles, 12
Grigg, Frederic, British
Commissary
Arbitrator, 409, 498–9,
700, 707, 710, 713, 717,
720, 724
grog, 524–5
grounding, 115, 125, 141, 156,
161, 219, 254, 296, 388,
391, 424, 590, 705
Guadeloupienne, French slave
brig, 345–6
Guerrero, Spanish slaver (?)
brig, 391
Guinea Company, 28

Guinea-men, 42, 101
gum trade, 251, 273, 286, 293,
321, 326, 367

Hadden, Henry, ship-
wrecked mariner, 407
Hamilton, Daniel Molloy,
King's Advocate/
Acting Governor/
Arbitrator/Judge, 217,
274, 290, 293, 302, 310
Hamilton, Mr, British
Minister/HM Envoy
Extraordinary, 602, 604,
609–10
hammock cloths, 337
Hannibal, slaver, 50, 56–7
Harmattan, 77–8, 175,
540, 632
hatches, *see* gratings
Hawkins, John, explorer/
slaver, 25
Hayne, Henry, British
Commissary Judge,
223, 247
head money, *see* bounties
Henri, French merchant
vessel, 542
Henry, Prince, The
Navigator, Portuguese
explorer, 21–2
Heroine, French frigate, 605
Hesketh, Robert, HM
Consul at Rio, 498, 606,
609, 697
Hickson, Scottish slaver,
154, 189
High Court of Admiralty,
105, 126, 133, 155, 179, 192,
223, 352, 669
Hoad, Mr, Consul-General
at Montevideo, 493
Hogan, Dr Robert, Chief
Justice and Judge Sierra
Leone, 189, 192
Holiday, King, 102
Hospital at Kissy, Lower
and Upper, 523
hospital ships, 177, 702,
710–11
hot pursuit, 204, 269, 303

House of Commons, 5–6, 8,
11, 15, 104, 669
Huntingdon, Eleazer, slaver
owner, 652
Huskinson, Liverpool palm-
oiler, 424
Hyde, Mr, temporary
Governor, 188

Imprehendedor, Brazilian
Imperial brigantine, 304
indemnification, 170,
204, 786–8
Indian corn/maize, 51, 554,
674, 795
Industry, French schooner, 246
Infante Don Pedro, Brazilian
frigate, 167
inferior authorities, 604, 686
Inman, Professor, naval
architect, 503
inshore cruising, 751–2
insubordination, 433, 479,
509, 619–20
insurance, 16, 47, 108, 292,
562, 594
insurrection, 54, 403, 596, 605
intelligence, operational, 126,
136, 164–5, 179, 182, 225,
232–3, 281, 343, 385, 406,
541, 609, 616, 721
invaliding, 241, 232, 317
inventory, 356, 366
iron tanks, 422
Irwin, British slaver, 143
ivory, 70, 103, 131, 142, 149–50,
380, 420

Jackson, Sir George, British
Commissary Judge,
332, 337–40, 356, 364,
489, 497–9, 598, 606,
697, 700–3, 706–10,
715–20, 724
Jameson, Mr, British
Commissary
Arbitrator, 223
Jawegiu, Don Andres de,
Spanish Commissary
Judge, 388
Jeffcott, Chief Justice, 379

jettison, 342, 459, 477, 519, 673
Jeune Aimee, slaver, 61
Jeune Frederic, French merchant barque, 653
João III, King of Portugal, 23
John VI, King of United Kingdom of Portugal and Brazil, 271
Jorge, John, Portuguese Commissary Arbitrator, 222
Josiffe, British slaver, 379, 426, 479, 573–4, 579–80
Juan, sloop, tender, 162
junta, Princes Island, 617–8
jurisdiction, 135, 138, 158, 181, 216, 374, 406, 410, 412, 422, 466, 482, 492, 500, 514, 585, 597, 623, 640, 666, 696, 731, 734

Kearney, John, slaver, 224–5, 235, 237–8, 240
kedging, 143, 208
Kennedy, Mr James, British Commissary Judge, 582, 586–90, 593, 685–96
Kent, British palm-oiler, 334
kidnapping, 21, 25, 45, 128, 319, 696, 702
Kilbee, Henry, British Commissary Judge, 223, 313, 385–92, 396, 468
King, Robert, British suspected slaver, 405
Kingmore, British brig, 118
King
  of Bonny, 254, 616
  of Dahomey, 46
  of Gallinas, 748
  of Loango, 371
  of Norway, 290
  of Spain, 29, 31, 39, 205–6
  of the Netherlands, 200
  of Trade Town, 270
King's Advocate, 179, 217, 244, 275, 347, 566, 605
King's Proctor, 152
Kingston, Lieutenant-Colonel, Royal African Corps, 407

Kroomen, 68, 190, 209, 219, 259, 286, 303, 305, 337, 345, 357–8, 378, 435, 448, 462, 509

La Dorade, French vessel, 542
Lagos, annexation of, 753
La Jeune Estelle, French schooner, 239
La Meduse, French frigate, wreck of, 198
La Prothée, French slaver, 245
Law Officers of the Crown, 437, 445, 491, 589
"Law of Nations", 256, 281, 496, 670, 729
Lawrence, Mr, Sierra Leone citizen, 143
lazaretto, 564, 587
Le Charles, French privateer, 130
Lecky, William, Irish historian, 757
Le Fer, Don Francisco, Spanish Commissary Judge, 217, 236, 251–2, 276
legitimate trade, 63, 103, 199, 380, 544, 649
Les Deux Amis, slaver, 343
Letter of Marque, 102, 150, 162, 238, 354
Lewis, Thomas, slave, 5
Lewis, Walter, British Commissary Registrar/Arbitrator, 363–4, 439, 508, 528, 549, 665, 676
Liberated African Department, 320, 557
liberation, sentence of, 204, 577
Lightburne, Bermudan slaver, 154, 255
Lincoln, Abraham, US President, 753, 756
Lindsay, Lawrence, shipwrecked mariner, 407
Liverpool, British merchantman, 244
Liverpool, Lord, Colonial Secretary/Foreign Secretary, 116, 120, 152, 729
livestock, 65, 161, 334, 343, 345, 419
Lloyd, Christopher, historian, 368, 757
logistical support, 173, 244, 412, 407, 449, 487
London Slave Trade Commission 1819, 180, 204, 222
Lord Chief Justice, 4–5, 14
Lords Commissioners of the Admiralty, 108, 110, 346, 516, 669, 734, 805
Lorenzo, Commandant, 575
lots/lottery, 222–3, 308, 589
Louis incident, 191–2
Louis XVIII, King of France, 172, 183
Louis-Philippe, King of France, 426
Ludlam, Thomas, Governor/Commissioner, 110, 113, 116, 119–20, 122, 135
Lumley, Lieutenant-Colonel, Acting Governor, 324, 332–3, 347

Macaulay, H. W., British Commissary Arbitrator/Judge, 436, 508, 511, 527–8, 535, 538, 541, 542, 549, 652, 665–6, 676
Macaulay, Kenneth, Acting Governor, 302–3, 310
Macaulay, Zachary, Governor/Secretary, African Institution, 9, 107, 119–20, 134, 145, 212
MacCarthy, Lieutenant-Colonel Charles, Governor, 189–90, 199–20, 207–8, 211–12, 219–20, 231, 240, 243–6, 354–5, 258–9, 274, 278, 280, 283, 293, 311, 326

Mackenzie, Charles, British Commissary Arbitrator, 406, 468, 480
Macleay, Mr William Sharp, British Commissary Arbitrator/Judge, 389, 392–7, 402–6, 467–8, 480–3, 565–70, 582
McQueen, John, British slaver, 163, 215
Madden, Dr Richard Robert, Superintendent of Liberated Africans, 571–2, 576–8, 582, 584–7, 693, 696
Magico, Spanish slaver, 387
manacles/shackles, 52, 54, 249, 519, 529, 579, 689
malaria, 87, 166
Manuel, Spanish slaver schooner, 362
Marée, Major I. A. de, Dutch Commissary Judge, 276, 302
Margaret, Liverpool merchantman, 333
Maria, Portuguese merchantman, 622
Marsden, Alexander, British Commissary Judge, 222–3
Martinez, Colonel, Portuguese Government Agent, 318, 377
Martinez, Pedro & Co, Havana slave mechants, 632, 643, 654, 664
Mary Hooper, American schooner, 627
Marshal of the Courts, 631
Master's Proctor, 705
Mauri, José, slaver master, 453
Maxwell, Major/Lieutenant-Colonel Charles, Commandant/Governor, 114, 125, 131–3, 139, 150, 152–4, 189
measurement, British system, 624, 736

medicines/medical stores, 335, 356, 359
Melville, Lord, First Lord of the Admiralty, 15, 414, 503–4
Melville, M. L., British Commissary Registrar, 508, 528
memorandum, by Z. Macaulay, 120
Merrill, Acting Vice-Consul, 535
mess tins/tubs, 59, 272, 579, 795
Mexicano, Spanish slaver steamer, 390
Mexican, Salem merchantman, 443–4
Middle Passage, 14, 42, 44, 47, 54, 58, 88, 755–6
Milet, Mr, Dutch, Government Secretary, Elmina, 202
military chest, 538
Minerva, Spanish slaver schooner, 389–90
Minister for Foreign and Maritime Affairs, Lisbon, 620
miscarriages of justice, 222, 252
misconduct, 346, 521, 536, 580, 619, 675
Mississippi Company, the, 33
Mixed Commissions, courts
 Anglo-American, 753
 Anglo-Brazilian, 316, 337, 338, 353, 409–10, 422, 437, 466, 492, 559, 597, 665–6, 676, 699
 Anglo-Dutch/Netherlands, 229, 236, 560
 Anglo-Spanish, 230, 437, 454, 514
 Cape, the, 753
 Surinam, 211, 560
Mixed Court of Justice, Anglo-Spanish, 514, 528, 538, 559, 570, 613, 628, 659, 693

Monroe, US President, 257, 260, 283
Montalvo y O'Farrill, Brigadier-General Don Juan, Spanish Commissary Arbitrator, 406, 467
Montezuma, Spanish slaver, 419
Morão, Mr, Portuguese Commissary Arbitrator, 237, 276
Morayshire, British merchantman, 704
Morland, George, painter, 11, 61
mortality
 in men-of-war, 735
 in prizes, 200
 in slavers, 58–9, 403
Moya, Mateo, slaver, 530
Mungo Brama, village chief, 243
Municipal Judge, 497, 605
murder, 6, 221, 253, 274, 312, 334, 374–5, 687, 705
mutiny
 naval, 374, 525
 slave/slaver, 54, 254, 476

Napoleon
 escape and imprisonment, 175, 183
 slave trade and abolition, 15, 38, 183
Napoleonic War, 231, 235, 384, 727, 730, 733
national character, 496, 632
"natural law", 138, 729
navigational instruments/aids, 64, 589, 683, 742
Navy Board, 244, 287, 293, 337, 429
Neptune, Liverpool merchantman, 334
"neutral institution", 211
Neville, George, master, 151
Newton, Captain John, British slaver, 8, 11, 55, 75
Niger, French schooner, 542
night-glass, 381, 591, 690

Normanby, Marquess of, 635
"norther", 95–6
Novo Destino, Brazilian smack, 595
Nueva Diana, slaver, 364

O'Donnell, General, Captain-General, 750
Olive Branch, American slaver, 610
Oliveira, Mr, Brazilian Minister, 714
orders, 111–13, 139, 145–6, 173, 182, 195, 214, 228, 259, 272, 286, 319, 378, 384, 434, 488, 587, 670, 736
Order in Council, 15–16
Orestes, Spanish slaver brigantine?, 388
Ormond, John, slaver, 230, 235, 255, 439, 630
Orphée, French slave ship, 296
"ourselves alone" policy, 122–3
Ouseley, Mr, British Minister, 492, 702–4, 707–8, 710–11, 714, 718–19, 722
outports, 566
overloading, 199

paddles/paddle-blades/paddle-wheels, 429, 447–8
Paine-Tucker Agreement, 749
Paiva, Mr Joseph de, Brazilian Commissary Judge, 338, 364, 422, 452
Pakenham, Mr, British Minister, 692
"Palmerston Act", 677, 724, 747
Palmerston, Viscount, Foreign Secretary
  on
    Brazil, 422, 492, 605, 659, 697, 700, 702, 705–6, 709–10, 714, 722
    Commissioners, 454, 496, 517, 528, 531, 538, 558, 572, 577, 590, 652
    condemned slavers, 436, 528, 538, 549, 581
    Portugal, 437, 501, 606, 664, 669–70, 683, 734
    Spain, 436–7, 471, 474, 570, 588–9, 592, 640, 686–9
    The Netherlands, 666
    The USA, 577, 639, 667
  period in office, 406, 482, 514, 749, 757
palm oil, 48, 51, 103, 236, 529, 686, 751, 794
palm-oilers, 264, 266, 305, 423–4
Palowna, Spanish slaver brigantine, 389
Panda, piratical slaver schooner, 443–5
Park, Mungo, explorer, 46, 102
Pass-All, King, 443
passport, 224, 248, 264, 278, 285, 294, 311, 318, 354, 406, 453, 576, 589, 623, 625, 659
patronage, 446, 622
Pennell, Richard, British Consul at Bahia/Vice-Consul at Rio, 297–8, 489
petition, 2, 10, 202, 435, 516
petty officers, 132, 524–5, 683, 703
Pepys, Samuel, 28
Phillips, Captain Thomas, slaver, 49–50, 56
Pigott, Sir Arthur, 16
pilot boats, New York, 577
Pinillos, Spanish Commissary Judge, 388
piracy, 25, 184, 200, 208–10, 223–4, 231, 239, 256–7, 271, 281, 291, 374, 384–5, 390, 406, 410–12, 422, 648, 722, 750, 786
"piracy law", 292
Pitt, William, Prime Minister, 1, 8, 10–16, 18
pivot-gun, 350, 644–5
planters, 12, 14, 16, 26–9, 34–8, 60, 103, 134, 205–6, 235, 396, 466–7, 560, 562, 594, 687, 750–5
Platt, Mr, Englishman at Rio, 716
Pluma, Mr, Tuscan Consul-General, 656
plunder, 199, 260, 339, 528
Pope, His Holiness The, 3, 22, 676
Portland, Duke of, 504
Porto Novo, burning of, 753
Portuguese ambassador, 218
President of the United States, 445
Principal Officer of Customs, 666
Printed Instructions, 619
Prince Regent, 103, 123, 167, 170, 189, 271
privateers, 109, 114, 126, 133, 161, 167, 182
private slaving, 34, 48, 144, 291
Privy Council, 8, 106
prize of war/prize law, 109, 203, 353, 719, 729
prize money, 109–10, 203, 292, 336, 357, 449, 657
profit, 39, 61, 222, 255, 395, 693, 755
Providentia, Spanish pirate, 330, 359
Provisional Government
  of Bahia, 285
  of Pernambuco, 278
  of Princes and St Thomas, 579, 629
provisions, 161, 179, 192, 199–200, 227, 234, 247–8, 261, 265, 272, 287–8, 294, 311, 320, 323, 331, 356, 361–2, 543
punishment, naval, 157, 208, 305, 479, 509, 620, 622, 741
Purdie, Robert, Chief Justice, 163

Quakers, 2, 4–6
quarantine, 308, 357, 434, 483–4, 564–8, 581, 621, 651

quarter boats/davits, 645, 679
Queen's Advocate/
    Advocate-General, 537,
    553, 632, 651, 709, 748
Quesada, Colonel Don
    Rafael de, Spanish
    Commissary Arbitrator,
    406, 467
quinine, 322, 340, 358
Quorra, Fernando Po
    steamer, 446, 449,
    452, 530

rains, the, 81, 219, 244,
    276, 289
Rambler, American privateer
    brig, 162, 164
rebellion, slave, 1, 13, 34, 38,
    43, 56, 468, 472
Receiving Ship, 705, 707,
    710, 722
recuperation, 234, 321, 356
Reffell, Joseph, Colonial
    Secretary/Commissary
    Registrar, 302, 310,
    332–3, 346, 364
regiments, West Indies/
    black, 108–9, 153,
    158, 169
register, ship's, 113, 196,
    599, 629
registration, 529, 576–7, 634
Rendall, Lieutenant-
    Governor, 407, 432
rendezvous, 119, 129, 150, 366,
    370, 414, 487, 588
repairs, *Protector*, 148
repercussions, diplo-
    matic, 753
resale of condemned vessels,
    142, 203–4, 218, 336, 436,
    497, 619, 624, 707, 710,
    718, 742
restitution, 107
Ribello, Jose Silvestre,
    Portuguese
    Commissary Arbitrator,
    223, 247
rice, 51, 65, 103, 161
Richard, American merchant
    barque, 718

Richardson, William,
    slaver, 48
Ricketts, Major H. J.,
    Acting Governor/
    British Commissary
    Arbitrator, 347,
    353, 363–4
rigging, 144, 428, 633
Right of Search, 195, 211,
    260, 281, 283, 290, 427,
    457, 514, 639, 733, 749, 753
Riviera, General,
    Montevideo dissi-
    dent, 605
Robertson, Mr, *Prince of
    Brazil* expedition, 240
Robinson, Sir Charles,
    Judge High Court of
    Admiralty, 352
Rolla, Liverpool merchant
    barque, 424
Romano, pirate brig, 285
Royal Adventurers into
    African Company, *see*
    Adventurers' Company
Royal Africa Company,
    31, 58
Royal African Colonial
    Light Infantry, 274
Royal African Corps/
    African Corps, 125, 152,
    156, 286, 418, 553
Royal Fund
    (Portuguese), 169
Royal Yacht Club, 504
Ruby, slaver, 53–5
Rule, Sir William, Surveyor
    of the Navy, 583
rum, 44, 158, 173, 196, 227,
    270, 288, 337, 359, 524,
    545, 639, 661

Sabine, Captain, Royal
    Society, 259, 265
sailing
    races, 616, 740
    trials, 328, 377, 503–4
sails, wetting, 178, 421, 585
St Helena Squadron, 183, 214
Sainty, Mr, Colchester
    shipbuilder, 685

sale, nominal, 130, 531, 553,
    625, 628, 646
Samo, English slaver, 154
Samuel, Mr, slave-factory
    owner, 143
San Francisco, brigantine, 123
sanitary arrangements, 58
Santiago de Cuba,
    Commandant of, 575
Sardinian accession
    to Anglo-French
    Treaty, 457
Sassette, slaver owner, 629
Seppings, Sir Robert,
    Surveyor of the Navy,
    328, 428, 503, 527
Schenley, Mr Edward
    Wyndham,
    Commissary Arbitrator,
    482, 570–1, 575–88, 686
School of Naval
    Architecture, the,
    503–4, 562
Scott, Sir William, Judge
    High Court of
    Admiralty, 133, 155
scurvy, 58, 86
seamanship skill, 43, 143, 148,
    377, 738, 740
Segure, French slaver, 197
Senior Officer Northern
    Ports, 488, 490
Serra, French prize, 161
Seven Years' War, 35–6
Sharp, Granville, abolition-
    ist, 4–9, 15, 18
Sheridan, Richard Brinsley,
    abolitionist, 14, 18
shipbuilders, American, 103,
    123, 679, 733
ship design, 42, 328–9, 377,
    503–4, 527, 616
Sierra Leone
    Company, 9, 109–10
    transfer to the Crown, 108
Silveira, Matthew
    Equidio da, Brazilian
    Commissary
    Arbitrator/Judge, 452
Sirtema, Dow van, Dutch
    Commissary Judge, 452

General Index   845

Skull and Crossbones/black flag/piratical red flag, 151–2, 660
slaves
  imports, numbers of, 38, 122, 205, 500, 677, 679, 752
  price of, 61, 199, 206, 567, 594, 615, 755
  self-destruction, 47, 55, 224, 249, 334, 401, 448, 460, 516, 523, 556, 620
  suffering of, 58, 60, 193, 249, 360, 495, 574, 581
smallpox, 58–9, 209, 266, 288, 359, 365, 424, 434, 483–4, 512, 520, 523, 585, 687, 696
Smart, Samuel, King's Advocate/Acting Governor, 332, 347, 364
Smeathman, Dr Henry, 9
Smith, Adam, Professor of Philosophy, 3, 5
Smith, Alexander, Deputy-Governor, Chief Judge, 126
Smith, Governor Cape Coast Castle, 248
Smith, Mr, schooner master, 375
Smith, Mr, US Vice-Consul, 636, 639, 668
Smith, William, British Commissary Registrar/Arbitrator/Judge, 302, 332, 364, 408, 422, 436, 439, 452, 508
South Sea Company, 31–3, 50
Souza, Da ("Cha-Cha"), slaver, 236, 251, 507, 583, 585
Souza, M. Pereira de, Brazilian Commissary Arbitrator/Judge, 499
Special Instructions/Instructions/Warrants, 204, 206–7, 210, 223, 322, 324, 329, 352, 391, 414, 485, 495–6, 514, 516–7,

522, 547, 563–5, 572–3, 580–1, 599–600, 667, 680–1, 719
spies, 583
Stars and Stripes, 106–7, 327, 415, 577, 640, 655, 679, 734, 749, 753
stations
  Cape, 124, 169, 412, 428
  Jamaica, 105, 223, 384–5, 386, 405
  Leeward Islands, 168, 223, 405
  Pacific, 678, 723, 739
  South Atlantic, 488, 678, 739
  West Indies, Halifax and Newfoundland, 405
  West Indies and North America, 502, 596
steam vessel, 317, 429, 452, 682, 686
stern-boat, 459, 585, 690
Stevenson, Andrew, American Minister in London, 577
Stevenson, Mr, Captor's Proctor, 705
stores depots, 234
stove, cooking, 521
stowaway, 590
Stowell, Lord, Judge, 536, 651
sugar
  cane, 22–3
  duty, 752
  trade, 687
Superior Board of Health, 568
Surcouf, Captain, French slaver, 183
Surgeons/Assistant Surgeons, 87, 228, 241, 244, 277, 369, 417, 509
Surveyor of the Navy, 328, 502–4, 631, 738
Superintendent of Liberated Africans, 571, 576, 587, 696
supernumeraries, 228, 259, 358, 454

suppression campaign
  beginning of, 110
  end of, 757
survey, hydrographic, 64, 110–22, 131, 135, 302, 307, 380, 392
Susan, colonial tender brig, 301, 308, 310
Sutherland, Lieutenant-Colonel, Commandant on Gold Coast, 386–7
Sweden, King of, 290
sweeps, 143–4, 162, 191, 239, 299, 349, 369, 376, 399, 415–6, 421, 440, 453, 459, 469, 477, 480, 484, 577, 590–1, 658, 660, 741
sweep mounting, 448
Swift, American schooner, 239, 241
Symondites, 502, 504
Symonds, Commander/Captain Sir William, Surveyor of the Navy, 502–4, 562, 647, 738

Talleyrand, French Foreign Minister, 172
Telegraph, merchant brig, 139
Temple, Governor, 452, 495, 508
tenders, policy, 132, 322–3, 382, 414, 434–5, 464, 719–20, 737
theft, 208, 519, 590, 622
Thomas, London merchantman, 309
Thompson, Thomas Perronett, Governor, 116
Thorpe, Mr, Chief Justice and Judge Sierra Leone, 139, 189
Thurlow, Edward, Lord Chancellor, 13, 15
tobacco, 34, 39, 44, 403, 453, 547, 579, 641, 653, 658
Tolmé, Mr, British consul, 686, 688, 691
"Tonnage Act", 624
tonnage bounty, 624
top hamper, 633

topmasts, loss of, 117, 162
tornado, 264, 321, 343, 345, 354, 360, 461, 631
trade goods, 11, 17, 33–4, 42, 44–5, 107–8, 132, 200, 292, 297, 351, 377, 395, 436, 439, 452–3, 456, 474, 733, 755
"trade ship", 377
trade winds
  North-East, 76–7, 92, 95
  South-East, 85, 89–92, 149, 220
transportation (punishment), 132–3, 163, 189, 402, 574
Treasurer of the Navy, 12, 15, 357
Treasury, the, 203, 257, 320, 376, 531, 544, 742
Treasury Solicitor, 215, 405
treaties
  of Paris, 1763, 35
  of Paris, First, 167, 172
  of Tordesillas, 22
  of Utrecht, 31
  Quintuple, 750
  Webster–Ashburton, 749
  with Brazil, Nov 1826, 599
  with Denmark Jan 1814, 172
  with France, 1831, 427
  with King of Bonny, 649
  with Portugal, Jan 1815, 170
  with Portugal, of Friendship and Alliance, 123
  with Spain, Sept 1817, 206
  with the Netherlands, May 1818, 211
  with Tuscany, 555
  with USA, of Ghent, 160
  with USA, Washington, 753
Trenholm, George, American shipowner, 133
Tricolore, 191, 361, 463
Trist, Nicholas, American Consul in Havana, 567, 577, 581, 625, 628–36, 649, 689

Troubriant, slaver master, 345–6
Tsar Alexander, 172
Tsar Nicholas, 211
Tucker, slaver, 310
Tucketts, Judge in Vice-Admiralty court, 402
Turner, Major-General Charles, Governor, 290, 293–4, 301–2, 310, 320, 326–7

United Kingdom of Portugal, Brazil and the Algarves, 103
United States relations with Britain, 17, 37, 105, 144, 160, 246, 415, 457, 577, 641, 689, 708, 718, 733, 749–50, 753
unseaworthiness, 447, 518, 583, 594, 650, 687

vaccination, 585
van Sirtema, Dutch Commissary Judge, 217, 229–30, 236–7, 276
Van Tromp, Dutch slaver, 299
Vatican, the, 2, 29, 40
Veira, Spanish slaver schooner, 476
Veloz, slaver schooner, 618
Venus, American/Spanish slave ship, 689, 693
Verona, Congress of, 256
vexatious proceedings, 696
Vienna, Congress of, 172, 192, 211, 231, 729, 784
Vienna Declaration, 172, 192
Victualling Board, 356

warm clothing, 356–7
"war of colour", 596
War of Jenkins' Ear, 32
warping, 143
Warrants, *see* Special Instructions
watering, 65, 293, 339, 358, 367, 431

Webster, Daniel, US Secretary of State, 749
Wellesley, Marquis, British Minister in Spain, 134–5
Wellington, Duke of, Foreign Secretary, 171–2, 457, 482, 514, 669–70
Wesley, John, abolitionist, 4, 5, 11
West Indies/black regiments, 108–9, 153, 158, 169
White, Governor, Cape Coast Castle, 147, 200
White, John, shipowner, 632
White, Joseph, shipbuilder, 504
White Flag of France/Bourbon, 331, 361, 377, 463, 733
Wilberforce, William, Member of Parliament, abolitionist, 1, 8–15, 18, 108, 110–11, 120, 129, 134–5, 171–2, 175
Williams, Captain, slaver master, 53, 55
Williams, John Tasker, British Commissary Judge, 302, 311, 332
Wilson, Irish deserter/slaver, 154
Wilson, Dr John, 87
wine, 358–9, 429, 451–2, 454, 589, 637, 717
women/female slaves, 55, 104–5, 249, 561
Woodbridge, British merchantman, 287
wooding, 358
Woolman, John, abolitionist, 2, 4
worm damage, 325

Xavier, Major, Portuguese Military Commandant, 216

yams, 51, 76, 161, 250, 278
Yando Coney, King, 243

yellow fever, 87, 166, 272, 274, 307, 357, 360, 364, 368, 553–4, 587, 616–7, 668, 694
York, Duke of, President of Royal Adventurers into Africa, 28
Yorke, First Lord of the Admiralty, 175
Young, Captain Thomas, British slaver master, 299

Zong, slaver, 6, 13

Index of Naval Personnel

Including Royal Marines, Army Officers at Sea and Privateer Officers

(Royal Navy unless otherwise indicated)

Absolon, Lieutenant, 140
Acland, Lieutenant Charles B. Dyke, 544–5
Adams, Lieutenant/Commander John, 370–425, 522–48, 647
Addis, Lieutenant, 153
Aldrich, Master's Mate, 638
Allen, Master's Mate, 361
Andrews, Lieutenant, 712
Angelly, John, Clerk, 619
Anson, Lieutenant, 597–9
Arabin, Captain Septimus, 311–42
Armitage, Master's Mate, 476–7
Armitage, Lieutenant Whaley, 674, 699
Armsink, Lieutenant, 279
Arundel, Warrant Officer, 461
Ashley, Captain B., 167
Austen, Captain Charles, 390
Austen, William, Master's Mate, 617
Austin, Lieutenant, 288–93, 310

Badgley, Lieutenant/Acting Commander, 324–57
Bagot, Lieutenant Christopher, 479
Baker, Acting Commander James, 483
Baker, John, Master's Mate, 224
Baker, Rear-Admiral Sir Thomas, 409, 428, 485–91
Barham, Admiral Lord, 6, 18, 108
Barnett, Lieutenant Edward, 571
Barren, Commander, 695
Barrow, Lieutenant, 452
Barrow, Arthur, Master's Mate, 540
Bate, Captain, Royal Marines, 422–54, 617
Bates, Lieutenant Joseph, 547
Baugh, Master's Mate, 680–2
Beaufoy, Lieutenant George, 622, 647, 672
Beddoes, Lieutenant, 655
Beechey, Captain F. W., 600–3
Belcher, Commander, 63
Belcher, Lieutenant, 224
Bennett, Lieutenant James, 106, 388–9
Bennett, Captain Thomas, 565
Bentham, Midshipman, 374

Berkeley, Admiral, 140–1, 155
Blanckley, Commander Edward, 456, 488–90
Bligh, Lieutenant Francis, 106
Blythe, Lieutenant, 342
Bickford, Warrant Officer, 525
Biddlecome, Master, 597
Bingham, Commander Arthur, 144
Bingham, Acting Lieutenant, 254
Birch, Lieutenant, 546, 704, 717
Bolton, Lieutenant Charles, 473–82
Bond, Lieutenant Francis, 647, 654
Bones, Lieutenant Robert, 112, 118, 128–49
Booth, Commander James, 440–9, 507
Bosanquet, Master's Mate, 415–6
Bosanquet, Lieutenant Charles, 515
Boteler, Commander, 364
Boultbee, Lieutenant, 321
Bourchier, Commander Henry, 117–9
Bourne, Thomas, Master's Mate, 127

Bouvet, Captain (French Navy), 161
Bowen, Commander Charles, 276–7, 283
Bower, Lieutenant, 611–2
Broadhead, Lieutenant Henry, 644–74
Brooking, Lieutenant Arthur, 566–7
Broughton, Commander/ Captain William, 371–3, 608–12
Browne, Captain/ Commodore Thomas, 169, 173–80
Browne, Lieutenant, 360–1
Browne, Acting Master, 352
Bruce, Captain Henry, 604, 703
Buchanan, Lieutenant George, 429
Bullen, Commodore Charles, CB, 285–336, 386
Bullen, Midshipman, 302
Burney, Commander, 483–4
Burslem, G. E., Warrant Officer, 513, 551
Burslem, Lieutenant Godolphin J., 673
Burton, Lieutenant Richard, 117
Bushby, Commander Thomas, 679–80, 690, 696
Butterfield, Master's Mate/ Lieutenant E. H., 350, 367–73, 428–38
Byng, Lieutenant George, 564–80

Calger, Lieutenant, 300
Calvert, William, Seaman, 374
Campbell, Commander Colin, 166
Campbell, Lieutenant Colin Yorke, 557, 622–47
Campbell, Lieutenant Frederick Archibald, 546
Campbell, Rear-Admiral Patrick CB, 454–557, 617–8, 644
Campbell, Lieutenant Patrick, 510, 527–51
Card, Lieutenant, 296–7
Carew, Commander W. H. H., 601
Carnegie, Lieutenant William, 149
Carter, Captain, 690
Carter, Thomas, Seaman, 525
Cary, Commander The Hon. Plantagenet Pierrepoint, 588, 688
Castle, Lieutenant/ Commander William, 381, 415–20, 515–55, 619–21
Castles, Midshipman, 221
Chambers, Lieutenant, 569
Chambers, Midshipman, 624
Chrystie, Lieutenant John, 292, 312
Clarkson, Lieutenant (*Iphigenia*), 260–1, 275
Clarkson, Lieutenant John, 9
Clavering, Commander D.C., 255–60, 295–305, 313, 321
Clements, Lieutenant The Hon. G. R. A., 570, 585
Clerkson, Lieutenant Edward, 291, 294
Coaker, Master, 621
Cochrane, Vice-Admiral Sir Alexander, 145
Cockburn, Vice-Admiral Sir George GCB, 414, 474–5, 560–70
Coffin, Captain Holmes, 139, 141
Coghlan, Lieutenant Francis, 599–600
Colchester, Captain The Right Hon. Lord, 487
Collier, Commodore Sir George, Bart., 213–57, 283, 313
Collier, Commodore Francis Augustus, CB, 318–80, 419, 545
Colpoys, Vice-Admiral E. G., 405, 410, 471–4
Columbine, Captain/ Commodore Edward Henry, 111–32
Cooper, James, Seaman, 416
Cooper, Thomas, Master, 152
Cory, Lieutenant, 314
Cotter, Second Master, 680
Courcy, Commander Nevinson de, 130–1
Courteney, Lieutenant/ Commander George, 269–70, 277–86
Cowie, Acting Master, 255
Cox, Henry, Master's Mate, 521
Coyde, William, Master's Mate, 367
Craigie, Commander Robert, 527–56, 618–49, 674
Cramer, Captain John, 168
Crawford, Commander Abraham, 393–5
Crawford, master's mate/ Lieutenant Richard Burrough, 305–12, 428–42
Creedy, Mr, master of colonial schooner, 143
Creser, Lieutenant, 425, 434
Crofton, Lieutenant Thomas, 317
Crump, Richard, Seaman, 208
Cruz, Lieutenant (Spanish Navy), 584, 591–2

Dallas, Commodore (US Navy), 578
Daniell, Commander George, 478
Davies, Commander H. T., 163
Davis, Lieutenant Henry Barnett, 557
De Roos, Commander The Hon. John, 490

Denman, Lieutenant The Hon. Joseph, 494–6, 508, 515–16, 747–51
Deschamps, Lieutenant H. P., 533, 541, 548, 554
Deschamps, Midshipman/Admiralty Mate, 254, 300
Dewar, Lieutenant John, 173
Dickey, Lieutenant William, 548–56, 616
Dickson, Lieutenant W. H., 145
Dickinson, Commander T., 489
Digby, Captain, 124
Dilke, Commander Thomas, 571, 582–3
Dix, William, Acting Second Master, 660
Dobree, Commander Nicholas, 195
Douglas, Assistant Surgeon, 416
Douglas, Commodore Peter, 649, 692–5
Downes, Lieutenant Henry, 340, 349–54, 363
Dundas, Captain, 212
Dundas, Master's Mate, 668
Du Plessis, Commodore (French Navy), 258
D'Urban, Lieutenant John, 583
Durham, Admiral, 168

Eden, Commander Charles, 601, 607, 612, 698–9
Eliot, Commander Russell, 607
Elliot, Assistant Surgeon, 554
Elliot, Captain The Hon. George (Adm. Sec.), 378–80, 414–17, 445
Elliot, Captain William, 309, 311
Elliot, Commander George, 622, 630
Elliot, Lieutenant, 264
Elliot, Rear-Admiral The Hon. George, 557, 616–49, 661, 670–7
Ellis, Commander John, 166–7
Evans, Captain, Royal Marines, 642
Evans, Commander R., 124
Evans, John, Clerk, 224–5
Evans, Lieutenant Thomas, 246
Eyres, Commander Harry, 675–6

Fail, Boy, 416
Fair, Commander Robert, 563–8, 579
Fairholm, Midshipman, 681
Falconer, Samuel, Master's Mate, 304–5
Farquhar, Captain/Commodore Arthur, 471
Farrant, John, Marine, 351
Filmore, Lieutenant/Commander/Acting Captain John, 119–22, 277–9
Finlaison, Lieutenant/Commander William, 241–7, 261
Fisher, Captain William, 190–3, 200–2
Fitzgerald, Lieutenant Charles, 632–8, 645–8
Fitzgerald, Edward, Able Seaman, 462
Fleeming, Vice-Admiral The Hon. Charles, 391–3, 402–5
Fleming, Acting Commander John, 168
Foley, Acting Lieutenant, 373
Forrester, Surgeon, 228
Fowell, W. N., Master's Mate, 399, 407
Fox, Benjamin, Master's Mate, 529
Fraser, Acting Master, 373
Fraser, Commander Thomas, 592, 685–7
Frazer, Commander John, 578
Fremantle, Commander Stephen, 721
Frost, Seaman, 306

Gammon, Warrant Officer, 457
Giles, Acting Lieutenant J. C., 299
Gill, Commander Thomas, 404
Glasse, Lieutenant Frederick, 508, 535
Gordon, Captain Alexander, 362–81, 414–20
Gordon, Commander, 479
Gordon, Commander A., 144
Gordon, Midshipman Samuel, 267
Gore, Commander, 153
Gorton, Master's Mate, 597, 599
Grace, Commander Percy, 265, 269–74
Graham, Lieutenant P., 175–7, 249
Grainger, Marine, 479
Grant, Captain Sir Richard, 587, 692
Gray, Lieutenant William, 276–7, 303–4
Greer, Lieutenant John, 358, 370
Green, Lieutenant, 368
Grey, Captain The Hon. George, 600–5
Griffinhoofe, Commander Thomas, 323, 349, 367

Hagan, Lieutenant Robert, 175, 189–98, 207, 219–30, 242–66, 273
Halkett, Vice-Admiral Sir Peter, 570–5, 580–7
Halsted, Vice-Admiral, 386–91, 396
Hambly, Commander Peter, 708
Hamilton, Captain G. W., 409, 486–8

Index of Naval Personnel 851

Hamond, Rear-Admiral/Vice-Admiral Sir Graham KCB, 498, 560–1, 594–614
Hamond, Lieutenant/Acting Commander Andrew, 601, 607
Harcourt, Captain Octavius, 499, 595
Hardy, Mate, 350
Harris, William, Seaman, 227
Harrison, Acting Captain, 343
Harrison, Commander Joseph, 380, 419–26, 434–41
Harvey, Captain Booty, 173–4, 189
Harvey, Lieutenant Edward, 352, 360–9, 608–11
Harvey, Master's Mate, 330
Harvey, Vice-Admiral Sir Thomas KCB, 648–9, 695
Hast, Lieutenant Philip, 692
Hawker, Captain Edward, 133
Hayes, Commodore John, CB, 377–82, 414–19, 425–37, 503, 562, 671, 744
Hays, Commander John, 696
Hayman, Master, 197
Head, Lieutenant, 227, 239
Helby, Lieutenant, 260
Henderson, Lieutenant James, 201, 220, 241
Henderson, Commander Tom, 532, 535, 541, 615
Henderson, Captain William, 695
Herbert, Captain, 700–1, 704, 718, 721
Herbert, George, Mate, 374
Herd, Lieutenant, 284, 286
Heseltine, Lieutenant, 705
Higman, Commander Henry, 173–4
Hill, Lieutenant Henry Worsley, 551–6, 618–20, 638, 641, 651–2, 668, 673

Hill, Midshipman, 619
Hinde, First Class Volunteer, 416
Hobson, Lieutenant William, 388
Hoffman, Acting Commander Frederick, 101
Holland, Lieutenant Edward, 389, 391, 403, 631–43, 657–69
Holland, Lieutenant Frederick, 695
Holmes, Midshipman, 619
Home, Commander Sir James, 583
Hope, Captain Charles, 487–8, 601
Hope, Commander James, 483, 569, 578, 585–6
Hopkins, Lieutenant, 166–7
Hotham, Commodore, 750, 752
Hunn, Commander Frederick, 214–15, 240
Hunt, Lieutenant, 578
Hunt, Warrant Officer, 672
Hunter, Lieutenant Charles, 144
Hunter, Harry, Master (US Navy), 258
Hunter, Lieutenant George, 711, 720
Hunter, William, Second Master, 270
Huntley, Lieutenant Henry Vere, 381, 420–5, 432–3, 450, 454, 460–1, 544–1
Huntingdon, Captain the Earl of, 387
Hutchinson, Midshipman, 302
Hyne, Warrant Officer, 376

Inman, Midshipman Robert, 242
Irby, Captain/Commodore The Hon. Frederick Paul, 130, 146–61, 171–3

Jack Fryingpan, Krooman, 462
Jackson, Commander George, 389–90
Jackson, Master's Mate, 297
Jackson, Midshipman, 664
James, Lieutenant, 358
Jauncey, Lieutenant, 591, 696
Jeayes, Lieutenant, 310
Jellicoe, Lieutenant, 244
Jenkin, Lieutenant Charles, 587, 590, 683, 688–94
Jenkins, Master's Mate, 376
Johns, Stephen, Master's Assistant, 354
Jolly, Archibald, Warrant Officer, 692
Jones, Captain William, 563, 574–6, 584–5
Jones, Commodore, 751
Judd, Warrant Officer, 459
Julian, Master's Mate, 499

Keane, Commander Richard, 405
Kellett, Lieutenant Henry, 353
Kellett, Lieutenant Arthur, 625–37, 650–1, 660–1
Kelly, Commander Benedictus, 217–21, 242–58
Kennedy, Captain (US Navy), 415
Keppel, Commander The Hon. Henry, 390, 551–2, 615, 621
King, Captain The Hon. Edward, 101
King, Lieutenant John, 267, 276
Kingdom, Lieutenant John, 254
Kirby, Midshipman, 300
Knight, Lieutenant Christopher, 250, 255–7, 260–1, 265

Laborde, Rear-Admiral (Spanish Navy), 469
Lane, Acting Surgeon, 373

Lawless, Lieutenant, 691
Leeke, Commander Henry, 223–4, 237–8, 243, 248, 254–6, 264
Leith, Captain John, 587, 686
Leonard, Lieutenant, 435
Liardet, Lieutenant Francis, 386
Lloyd, Commander John, 146, 150–1, 156–8
Locker, N. E., Master's Mate, 545
Lockyer, Captain, 511
Lord, Marine, 227
Lowcay, Lieutenant Robert, 608
Lowe, Lieutenant A. B., 388–9
Lowe, James, Seaman, 157
Lyons, Midshipman Charles, 249

Macaulay, Alexander, Letter of Marque master, 126
McCausland, Commander John, 565
McCoy, Master, 254
McCulloch, Captain William, 168
McCulloch, Commander, 144
McDonell, Lieutenant John, 479–80, 482
McDonnell, Charles, Master's Mate, 438
Macdougall, Master's Mate/Lieutenant John, 546
McHardy, Lieutenant John Bunch Bonnemaison, 396–9
McKechnie, Alexander, Assistant Surgeon, 368–9
McKenzie, Donald, College Mate, 685
Mackenzie, Acting Commander (US Navy), 708
McKinnel, Robert, Surgeon, 322, 340, 356, 368–9

McKinzie, Captain G. C., 167
Maclean, Commander Rawdon, 386–7
Maclean, Midshipman, 261
McNamara, Lieutenant Jeremiah, 455, 510
McNeil, Hector, Master's Mate, 622
Madden, Lieutenant C., 391
Maitland, Captain The Hon. A., 168
Maitland, Patrick, Master's Mate, 610–11
Mansell, Midshipman, 227
Marsh, Lieutenant, 229, 249–50
Marsh, Lieutenant W. B., 548
Mason, Commodore, 595, 602, 608
Massieu, Commodore (French Navy), 305
Matson, Lieutenant George, 299, 307, 316, 336, 347, 444
Matson, Master's Mate/Lieutenant Henry James, 444–5, 656, 665, 671–2, 748–51
Maxwell, Captain John, 163–6
Maxwell, Commander John, 570
Maxwell, Midshipman, 166
Mayne, Commander, 402
Mayne, Anthony de, Master, 161, 392
Medley, Lieutenant Edward, 323, 339–40
Mends, Commodore Sir Robert, 258–65, 272–86, 313, 320
Mends, Midshipman, 278
Mercer, Midshipman/Lieutenant Samuel, 334, 343, 364, 451–2, 509, 512–13, 526
Meredith, Commander Richard, 431, 433, 445–6, 452–3, 457, 462, 505–6, 509–11

Miall, Lieutenant George Gover, 154, 439, 507–10, 523, 528, 531
Middleton, Captain/Rear-Admiral Sir Charles, *see* Barham, Admiral Lord
Milbourne, Midshipman John, 672
Mildmay, Lieutenant George St John, 259, 263–4
Miller, Lieutenant, 591
Miller, Midshipman, 279
Miller, Robert, Master's Mate, 404
Milne, Commander Alexander, 590–2, 691
Milsome, Ordinary Seaman John, 703
Milward, Lieutenant Clement, 455, 459–60, 507, 510, 525
Mitchell, Lieutenant John, 106
Mitchener, Lieutenant George, 127, 129, 149, 155
Monday, Commander John, 705
Moorhouse, Lieutenant, 166
Morton, Lieutenant, 318
Mowle, Master's Mate, 711, 713, 722
Muddle, Commander Richard, 168
Murray, Captain James, 292, 300, 303, 308
Murray, Mr, colonial vessel master, 310

Napier, Captain, 402
Napier, Lieutenant, 622
Nash, Lieutenant R. J., 224, 241, 243, 257
Naulty, Assistant Surgeon, 438
Nelson, Vice-Admiral Viscount, 14
Nepean, Commander Evan, 561
Nesham, Captain C. J. W., 106

Newlands, Lieutenant, 658
Nias, Captain, 709
Nicholls, Lieutenant-Colonel Edward, Royal Marines, 184, 288–9, 299, 317, 321, 413, 418, 443–9
Nicholson, Commodore John (US Navy), 708, 718
Nixon, Commander Horatio Stopford, 588–94, 689–90
Noddall, Warrant Officer, 457
Norcott, Lieutenant Edmund, 515–17, 542, 544, 553
Northrup, Job, privateer master/pirate, 239
Nott, Lieutenant, 389–90
Nott, Lieutenant Francis Seymour, 620

Ogle, Rear-Admiral Sir Charles, 405
Oliver, Lieutenant William, 545, 550, 623–5, 643, 664
Olivine, Master's Assistant, 306
O'Reilly, Lieutenant John, 610–11
Owen, Captain William Fitzwilliam, 63, 84, 87, 301–2, 307, 323–5, 332–5, 342–7, 352–66, 395

Paget, Captain Charles, 488, 490
Paget, Vice-Admiral The Hon. Sir Charles, GCH, 587, 681, 683, 689–92
Paget, Commander The Rt Hon. Lord Clarence, 683, 685
Pakenham, Lieutenant, 223
Parker, Commander Charles, 404
Parker, Commander Frederick, 112–15

Parlby, Lieutenant James, 439, 456–7
Parrey, Lieutenant/Acting Commander Edward Iggulden, 349, 363, 367, 370–1
Pascoe, Lieutenant W. R., 155, 159–60
Patten, Lieutenant Frederick, 602
Pearne, Warrant Officer, 365
Pell, Commodore, 484–5, 565, 567, 569, 573, 578, 581
Pengelly, Robert, Admiralty Mate, 309
Pennell, Captain Follet, 595
Peyton, Captain/Commodore Sir John, 581, 583, 590, 680, 686
Phillimore, Captain Sir John, 286
Phillips, Able Seaman, 479
Phillips, Lieutenant, 178
Phillips, Commander/Captain Charles, 265–9, 274, 473
Pierce, Lieutenant, 176–7
Pierce, Midshipman, 416
Piggott, Captain, 223
Pike, Second Master, 544
Pipon, Midshipman, 300
Pitt, James, Seaman, 374
Plumridge, Commander, 212
Poingdestre, Lieutenant, 294, 298
Polkinghorne, Captain James, 446–7
Popham, Commander Brunswick, 508, 545–56, 617–36, 644
Popham, Admiral Home, 206
Potbury, Lieutenant John, 471, 476, 680–1
Pratt, Midshipman/Lieutenant James, 243, 246
Preston, Commander, 705, 709, 715
Price, Lieutenant John, 540, 620

Prickett, Commander Thomas, 284
Pridham, Richard, Admiralty Mate, 573
Pringle, Commander George, 144
Pritchard, J. A., Warrant Officer, 552
Puget, Acting Commander Henry, 508, 510, 513, 517, 520–5
Puget, Lieutenant/Commander William, 601
Pulling, Lieutenant James, 393, 395
Purchas, Commander William, 292, 315, 325

Quashie Sam, Cape Coast Native Seaman, 221
Quin, Commander Michael, 498–9
Quin, Lieutenant William, 439, 441, 505–6, 513, 521–2, 526

Ranier, Captain Peter, 169
Ramsay, Dr James, Surgeon, 6–8
Ramsay, Lieutenant William, 366, 375–6, 381, 415–25
Rees, Assistant Surgeon John, 509
Reeve, Commander John, 622, 641, 643–4
Reid, Midshipman, 166
Reid, Warrant Officer, 536
Roach, Mr John, Privateer Master, 162, 166
Roberts, Carpenter, 430
Roberts, Master's Assistant, 554
Roberts, Thomas, Seaman, 208
Roberts, Acting Lieutenant, 364
Roberts, Lieutenant Thomas Lorey, 510–11, 518, 537, 544, 615

Robertson, Commander William, 493, 495
Robertson, Lieutenant, 720
Robinson, Lieutenant, 704
Robinson, Lieutenant John, 695
Robinson, Lieutenant William, 334
Robinson, Master's Mate/Acting Lieutenant, 424–5, 432, 434
Robinson, Midshipman Frederick, 455
Rose, Lieutenant George, 450–3, 462, 518–9, 521, 524, 661, 675
Rose, Master, 376–7
Rosenberg, Lieutenant George, 553–4
Ross, Commander Daniel, 172
Ross, Rear-Admiral Charles, 611, 701, 704, 711, 721
Rothery, Lieutenant T. H., 261, 267, 269
Rowlatt, Master's Mate, 528, 631, 636
Rowley, Rear-Admiral Sir Charles, 384
Russell, Captain The Rt Hon. Lord Edward, 596
Russell, Commander The Rt Hon. Lord Francis, 647, 652–5, 662–4, 674
Russell, James, Acting Master, 548
Russell, Commander/Captain Robert, 472, 708
Ruyter, Admiral de (Dutch Navy), 28
Ryves, Lieutenant, 225

St Vincent, Admiral Earl, 17
Saumarez, Lieutenant Philip de, 433
Saumarez, Lieutenant Thomas, 274, 277
Sanders, Lieutenant W. S., Royal African Corps, 152
Sandilands, Commander Alexander, 229, 237, 239–40, 243
Schomberg, Commodore, 428
Scobell, Captain Edward, 139, 142, 157–8, 161–3, 181, 215
Scott, Lieutenant James (*Myrtle*), 140
Scott, Lieutenant John, 277, 279, 284–6, 289, 307
Scriven, Commander Timothy, 217
Seagram, Lieutenant Henry Frowd, 641, 646, 654, 671, 674
Sealy, Midshipman, 114
Selby, Captain William, 106
Seymour, Lieutenant, 578
Seymour, Rear-Admiral Sir Michael Bt, KCB, 491–8
Shepherd, Commander John, 608, 610, 612, 704
Sherer, Lieutenant Joseph, 396, 401, 403
Simmons, Master, 345
Skyring, Commander, 64
Slade, Mate, 350
Slade, Frederick, Warrant Officer, 672
Smart, Commander Robert, 493, 497, 596
Smith, Lieutenant, 237
Smith, Lieutenant Edward, 388–90
Smith, John, Gunner, 540, 545
Smithers, Lieutenant George, 370, 376, 378, 380, 422, 434
Smyth, Commander William, 711–12, 715–17, 721
Sneyd, Captain Clement, 140–1, 153
Spearing, Lieutenant George, 106
Spence, Captain (US Navy), 280
Sprigg, Lieutenant George, 674
Stevenson, Assistant Surgeon, 509
Stewart, Lieutenant/Commander The Hon. Keith, 690
Stewart, Surgeon, 278
Stirling, Commander J., 193
Stokes, Lieutenant Pringle, 276–8
Stoll, Lieutenant John L. R., 546, 556–7, 626–7, 634, 638
Stopford, Lieutenant Edward, 468, 479
Stopford, Rear-Admiral, 169
Strong, Commander/Captain Charles, 217–9, 225, 228, 583
Strutt, Master, 680
Suckling, Commander W. B., 340–1
Sulivan, Captain/Commodore Thomas, 608, 610–11, 701–24
Sulivan, Lieutenant Bartholemew, 705
Sulivan, Lieutenant James, 425
Sulivan, Lieutenant Thomas Ross, 441, 447–8
Sutherland, Kenneth, Clerk, 634
Symonds, Commander T. E., 161

Tailour, Captain John, 173–9, 188
Tait, Captain Robert, 491, 498, 607
Taplen, Lieutenant, 406–7, 468, 479
Tatham, Master's Mate/Lieutenant Edward, 611
Taylor, Clarence, Warrant Officer, 664
Taylor, John, Gunner 508
Tetley, Lieutenant, 115
Thompson, Commander J., 144

Thompson, Commander T. Bouldon, 9
Thompson, Lieutenant/Acting Commander Josiah, 438, 440, 442, 446
Thompson, Thomas, Master's Mate, 287
Thompson, Mate, 197–8
Thorburn, Master's Mate, 658
Thorne, Gunner, 637
Tillard, Captain James, 153–5
Tindal, Lieutenant, 574, 576
Tinklar, Captain Roger, Royal Marines, 642–3
Tollevey, Lieutenant, 313
Tomkinson, Commander James, 168
Townshend, Captain The Right Hon. Lord James, 488, 490
Trenchard, Captain (US Navy), 246
Trotter, Commander Henry, 441, 443–6
Tryon, Lieutenant Robert, 520
Tucker, Lieutenant/Commander William, 303, 309–13, 647–61, 665–71, 674, 737
Turnbull, John, Seaman, 443
Turner, Lieutenant, 285
Turner, Lieutenant/Commander William, 323, 330, 336–40, 356, 377, 419
Tweed, Lieutenant, 190

Ussher, Lieutenant Sydney, 484–5

Vidal, Lieutenant/Captain Alexander, 63–4, 302, 554
Vilaret de Joyeuse, Captain/Commodore Alexis (French Navy), 346, 354, 426
Voules, Midshipman W. E., 552

Wadsworth, Captain (US Navy), 257
Wakefield, Lieutenant Arthur, 312, 315–6
Wake Walker, Lieutenant Baldwin, 299
Waldegrave, Captain The Hon. William, 487–8
Walpole, Captain, 476
Warren, Rear-Admiral Frederick, 407, 428–49, 454–5
Warren, Acting Commander/Commander Richard L., 446–7, 679
Warren, Lieutenant William, 468–71, 476
Watson, Lieutenant, 584
Wauchope, Captain Robert, 454, 524–5, 528, 538–9, 542
Webb, Commander/Acting Captain Edward, 355, 359, 365, 370–1, 415–26, 520
Wellesley, Captain The Hon. William, 469
Weston, John, Second Master, 509

Westopp, Commander A. J., 145
White, William, Master's Mate, 634
Wilkins, Lieutenant, 162
Willes, Captain George, 208–10, 232, 298–9, 305, 312, 601
Williams, George, Assistant Surgeon, 309
Williams, Lieutenant R., 144
Williams, Lieutenant Woodford, 622, 627, 641
Williamson, Acting Purser, 373
Willock, Commander Gore, 144
Wilson, Lieutenant, 279, 297, 304–5
Wilson, Corporal, Royal Marines, 508
Winniett, Lieutenant William, 547, 622
Wittman, Commander Josiah, 117
Wolrige, Commander Thomas, 265–7
Wood, Captain James, 106
Wooldridge, Samuel, Master's Mate, 530
Woollcombe, Lieutenant, 280, 284–9
Wright, Lieutenant Henry, 696

Yeo, Commodore Sir James Lucas, 195–202, 208–11

Index of Naval Vessels

Warships, Tenders, Colonial Vessels, Privateers and Auxiliaries

(British unless otherwise indicated)

Acorn (16), brig-sloop, 647
Acteon (26), Sixth Rate frigate, 596–602, 608–10, 704–8, 718
Adelaide, schooner, tender, 491
Adolphe, schooner (French), 261
Aetna (10), survey ship, 63, 544, 554.[1]
Africa, merchant schooner, tender, 333–4
African, colonial steam vessel, 310–11, 317–18, 324
Agincourt, transport, 114
Albacore (16), ship-sloop, 163, 166
Alban (8), steam vessel, 682
Albatross, survey schooner, 302–7
Alert, packet, 704
Alerte, brig (French), 610
Alexandria (32), Fifth Rate frigate, 101
Algerine (10), gun-brig/brig-sloop, 490
Alligator, schooner (American), 257–8, 280
Amaranthe (18), brig-sloop, 144–5
Amelia (38), Fifth Rate frigate, 130, 145–7, 153, 155–6, 160–1

Andromache (28), Sixth Rate frigate, 692–4
Arachne (18), ship-sloop, 391, 472, 483–4
Arethusa (38), Fifth Rate frigate, 139–41, 148
Arethuse, frigate (French), 160–1
Argo (44), Fifth Rate two-decker, 124
Argus (18), ship-sloop, 119
Ariadne (28), Sixth Rate frigate, 472–3
Ariel (18), brig-sloop, 172, 175, 186
Arrow, ketch, 705
Assiduous (3/4), schooner, 386–91
Astrea (36), Fifth Rate frigate, 167
Atholl (28), Sixth Rate frigate, 292–303, 362–81, 414–28
Augusta, schooner (American), 260–1
Aurora (46), Fifth Rate frigate, 390–1

Badger, mooring vessel, 527
Bann (20), Sixth Rate frigate/ship-sloop, 190–7, 265–92
Barbadoes (16), brig-sloop, 168, 223

Barham (50), Fourth Rate frigate, 391, 405
Barossa (36), Fifth Rate frigate, 168
Barracouta (10), brig-sloop, 63, 302–7
Basilisk (12), ketch-rigged cutter, 595, 611
Beagle (6), sloop, survey barque, 594
Beaver (10), gun-brig, 391
Bellona, brigantine-of-war (Spanish), 387
Belvidera, (42), Fifth Rate frigate, 580–3
Bermuda (10), gun-brig, 223
Black Joke (3), brig, tender, 323, 329–40, 349–81, 414–35, 737
Black Nymph, schooner, tender, 298, 322
Blanche (46), Fifth Rate frigate, 471
Blonde (46), Fifth Rate frigate, 595
Boadicea (38), Fifth Rate frigate, 124
Bonetta (3), gun-brig, 533–4, 541–54, 616–7, 626–7, 633–52, 671–4
Brazen (18), Sixth Rate frigate, 288, 298, 305–12

Index of Naval Vessels

Brisk (16), ship-sloop, 173–4, 179
Brisk (3), brigantine, 428–38, 446, 508–11, 625–37, 644–52, 659–61, 674
Britomart (10), gun-brig, 440–1, 505–13, 521–7
Bustard (10), gun-brig, 391
Buzzard (10), brigantine, 455–60, 506–34, 545–6, 556, 632–52, 661

Caledon, captured privateer brig, 24
Calliope (28), Sixth Rate frigate, 700–4, 717–18
Cameleon (10), gun-brig, 711, 717–20
Carmen, tender, 202
Carnation (18), brig-sloop, 314, 386–7
Cato, transport, 293
Caveira, tender, 197, 202
Cerberus (32), Fifth Rate frigate, 106
Challenger (28), Sixth Rate frigate, 594
Champion (18), ship-sloop, 358, 377, 503, 562–8, 579–80, 682
Chanticleer (10), gun-brig, 223
Charybdis (3), brigantine, 428–32, 438–42, 449–52, 511–13, 521–8, 534, 546–7
Cherub (26), Sixth Rate frigate, 170, 200–2, 208–10, 236
Childers (18), ship-sloop, 551–5, 615–21
Cleopatra (26), Sixth Rate frigate, 600–4, 611, 704
Clinker (12), gun-brig, 316–25, 331–6, 347, 354–60, 367
Clio (16), brig-sloop, 721
Cockatrice (6), brigantine, packet, 608, 708, 721
Colibri (16), brig-sloop, 144
Columbia (18), brig-sloop, 223

Columbine (18), brig-sloop, 168, 472, 503–4, 532–5, 541–7, 554, 617, 630–3, 661, 674–5
Comus (22), Sixth Rate frigate/ship-sloop, 173–7, 179, 186–8, 571, 588, 682, 688–94
Conflict (12), gun-brig, 292–321, 370, 376–9, 422–5, 434, 631, 662
Cornelia (3), schooner, tender, 343–6, 353–4, 365
Cornwallis (80), Third Rate battleship, 587, 682, 692–4
Creole (36), Fifth Rate frigate, 167
Crescent, slave accommodation ship, 732
Crocodile (22), Sixth Rate frigate, 115–22, 126–32, 682
Cruiser (18), ship-sloop, 483, 565, 580
Curacao (26/24), Sixth Rate frigate, 721
Curlew (10), gun-brig/brig-sloop, 441–5, 508, 515–22, 533–6, 542–53, 617, 661, 674, 681–2, 691
Cyane, Sixth Rate frigate (American), 265–78
Cyrene (20), ship-sloop, 265–78

Dallas, brig, tender, 351, 360–70
Daring (12), gun-brig, 155, 158–60
Dart, privateer, 126
Dauntless (18), ship-sloop, 117
Decouverte (12), schooner, 144
Derwent (16), brig-sloop, 112–15
Despatch (16), brig-sloop, 478
Diadem, hired merchant vessel, 324
Didon, French flagship, 582
Dolores, schooner, tender, 178, 188

Dolphin (3), gun-brig, 537–55, 617–43, 650, 656, 662–9, 675
Doris, schooner, tender, 120, 122, 125–7
Dragon, brig-of-war (French), 294
Driver (16/18), ship-sloop, 265–77, 286
Druid (46), Fifth Rate frigate, 391, 409, 486
Dryad (36), Fifth Rate frigate, 377–8, 417–35
Dublin (50), Fourth Rate frigate, 488, 594–612

Eden (26), Sixth Rate frigate, 323–4, 332–5, 342–5, 352–8, 364–6
Edinburgh (74), Third Rate battleship, 692, 694–5
Eleanor, brig, colonial vessel, 301, 310
Electra (18), ship-sloop, 704–11
Ellen, schooner, tender, 304
Erne (20), Sixth Rate frigate/ship-sloop, 217, 228, 254, 377
Esk (22), ship-sloop, 292–307, 315–30
Espiegle (18), ship-sloop, 391
Etoile (40), Fifth Rate frigate (French), 167

Fair Rosamond (2), brigantine, tender, 414, 420–5, 434–5, 444, 450–3, 462, 507, 511, 518–26, 539, 545–55, 617–26, 643–4, 664, 670–1, 677, 691–737
Fairy (10), gun-brig, 391
Falcon (10), gun-brig, 407
Favorite (16), ship-sloop/Sixth Rate frigate, 101, 163–6
Favourite (18), ship-sloop, 380, 419–25, 438–41, 449
Fawn (3?), brigantine, tender, 719–20
Ferret (14), brig-sloop, 193

Ferret (10), gun-brig, 388–91
Firefly (3), schooner, 403–7, 476–82
Flamer (3), steam packet, 680, 694
Fly (18), ship-sloop, 472, 607–11
Forester (3), brigantine, 439, 447–9, 507–14, 539, 546, 553–5, 617–24, 633, 639–61, 667
Forte (46), Fifth Rate frigate, 484, 578

Galatea (42), Fifth rate frigate, 402
Gannet (16), ship-sloop, 472, 570, 580
George (6), schooner, colonial vessel, 114–15, 125, 143, 147, 152
Grampus (50), Fourth Rate frigate, 124
Grasshopper (18), ship-sloop, 393–5
Grecian (16), brig-sloop, 711–21
Griffon (3), brigantine, 439–57, 505, 522, 583

Harlequin (18), ship-sloop (pre-1830), 390–1
Harlequin (16), brig-sloop (post-1835), 647–55, 663–4, 672–4
Harpy (10), gun-brig, 570, 580–5
Harrier (18), brig-sloop (pre-1830), 124
Harrier (18), ship-sloop (post-1830), 601–11
Haughty (12), gun-brig, 106
Hawke (18), brig-sloop, 117–19
Hay, tender's jolly boat, 324–5
Hermione, frigate (French), 430
Herald (28), Sixth Rate frigate, 709
Heroine, frigate (French), 605

Hope (5), schooner, tender, 303, 309–13, 322, 329
Horatio, schooner, tender, 324, 342–5, 353
Hornet (6), brigantine, packet, 599–608, 682
Hornet, sloop (American), 257
Huron, brig-of-war (French), 258

Icarus (10), brig-sloop, 402
Imogene (28), Sixth Rate frigate, 509, 604–11, 703, 721
Inconstant (36), Fifth Rate frigate, 195–7, 200–3, 208
Infante Don Pedro, frigate (Brazilian), 167
Iphigenia (36), Fifth Rate frigate, 124, 258–65, 272
Isis (58), Fourth Rate frigate, 381, 428–38, 446–55
Ister (36), Fifth Rate frigate, 168

Java, frigate, American, 415
John Adams, Sixth Rate frigate (American), 257
Juan, sloop-rigged tender, 162

Kangaroo (16), ship-sloop, 145–6, 150–2, 156–8
Kangaroo (3), schooner, 407
Kangaroo, survey brig, 392
Kitty, privateer, 162, 166
Kitty's Amelia, Letter of Marque, 102

La Bayonaise, corvette (French), 425
La Bordelaise, brig-of-war (French), 425
La Flore, frigate, (French), 305
La Meduse, frigate (French), 198
Lancier, brig-of-war (French), 681–2

Lark (6), schooner, 571, 696
Larne (18), ship-sloop, 562
Latona (38), Fifth Rate frigate, 106
La Triomphante, corvette (French), 542, 621
Laura (10), schooner, 144–5
Laurel (22), Sixth Rate frigate, 124
Laurel (36), Fifth Rate frigate, 169
Leopard (50), Fourth Rate frigate, 124
Leven (22), Sixth Rate frigate/ship-sloop, 63, 302–7
Leveret (10), gun-brig, 515–27, 534, 547–52, 621
Lille (or *Little*) *Belt* (14), ship-sloop, 144
Lily (16), brig-sloop, 622–6, 641–52, 661, 701
Lion (3/4), schooner, 314, 386–91
Little Bear, schooner, tender, 314–22, 342
Lively (46), Fifth Rate frigate, 309–11
Lynx (3), brigantine, 451–4, 460–1, 519–22, 532, 546–7, 621–9, 637, 643–4, 652–3, 659, 666, 672–4

Madagascar (46), Fifth Rate frigate, 580–1, 590, 680, 692–5
Magnificent (74), Third Rate battleship, 377, 682, 692
Magpie (4), schooner, 389–90
Maidstone (42), Fifth Rate frigate, 285–320, 331
Maria (10), cutter, 145, 168
Marte, brig-of-war (Spanish)
Mary, sloop, colonial vessel, 190
Medina (22), ship-sloop, 332, 341, 347, 354–74, 380, 370–4, 418–20

Melampus (36), Fifth Rate frigate, 133
Melville (74), Third Rate battleship, 557, 570, 580–7, 642–3, 674
Minerva, privateer, 126
Minx (3), schooner, 407, 472
Modeste (18), ship-sloop, 675–7, 692–5
Monkey (3), schooner, 391–407
Morgiana (18), ship-sloop, 217–29, 237–47, 260–1
Mosquito (18), brig-sloop, 168
Mutine (18), brig-sloop, 116, 130
Myrmidon (20), ship-sloop, 223–9, 237–56, 263–5
Myrtle (18), Sixth Rate frigate, 139–41

Nautilus (10), gun-brig, 622–6, 647, 672
Nereide (36), Fifth Rate frigate, 124
Niger (38), Fifth Rate frigate, 169
Nightingale (6), packet brig, 599
Nimble (5), schooner, 389–91, 403–7, 471–82
Nimrod (20), ship-sloop, 567, 578–80, 682, 695
North Star (28), Sixth Rate frigate, 311–25, 331–4, 341–2, 354, 472, 499, 595, 603

Olympia (10), cutter, 124
Ontario, sloop-of-war (American), 691
Orestes (18), ship-sloop, 503, 708, 721
Orpheus (36), Fifth Rate frigate, 223
Otter (16), ship-sloop, 124
Owen Glendower (42), Fifth Rate frigate, 272–86

Pallas (42), Fifth Rate frigate, 472–6

Pantaloon (10), gun-brig, 504
Parmela, transport, 430
Paul Pry, schooner, tender, 351–2, 360–1
Pearl (20), ship-sloop, 472, 479, 682–5
Pelican (18), brig-sloop, 509–11, 522–7, 539–47, 554–5, 618, 626–9, 635, 641, 652
Pelorus (18), brig-sloop, 431–8, 445, 452–63, 505–11, 546
Penguin (3), packet brig, 713
Persian (18), brig-sloop, 145
Peruvian (18), brig-sloop, 145
Pheasant (22), ship-sloop, 217–22, 242–65
Pickle (3), schooner, 391–407, 468–79, 580, 692–5
Pike (14), schooner, 562–7
Pincher (3), schooner, 391, 407, 475, 562–6, 572–80
Pique (36), Fifth Rate frigate, 168, 692–4
Plover (18), ship-sloop, 166
Plumper (12), gun-brig, 323–5, 331, 339–40, 358–60, 370–9, 425–34
Pluto (2), paddle steam-vessel, 429–31, 438–41, 447–54
Porcupine (22), Sixth Rate frigate, 173–5, 177
Portia, schooner, tender, 343–5
President (52), Fourth Rate frigate, 611
Primrose (18), ship-sloop, 323–5, 331, 337–40, 347–78
Prince Regent, colonial brig, 197–8, 207–8
Prince Regent, colonial schooner, 254, 267
Princess Charlotte (14), colonial schooner, 152–68, 186–203
Protector (12), gun-brig, 127–9, 141, 147–9

Pylades (18), ship-sloop, 389–91, 456, 488–90, 503, 515–29, 532, 539, 547, 555, 618–30, 641–3, 650, 691

Queen Adelaide, colonial schooner, 439
Queen Charlotte, colonial schooner, 190–2
Quiz, Letter of Marque schooner, 150–2

Racehorse (18), ship-sloop, 580–3, 596, 683, 692
Racer (16), brig-sloop, 483, 504, 562–9, 577–86
Rainbow (28), Sixth Rate frigate, 565–80, 682
Raisonable (64), Third Rate battleship, 124
Raleigh (18), ship-sloop, 498–9
Rapid (10), gun-brig, 499, 595–602
Rattler (16), ship-sloop, 144
Raven, survey ship, 544, 554, 617
Redwing (18), ship-sloop, 214, 295–7, 304–12, 321
Revenge, colonial vessel, 310
Renegade (3/4), schooner, 386, 391
Ringdove (16), brig-sloop, 589–93, 682, 689–92
Rolla (10), gun-brig, 508–12, 522, 535–6, 543–7
Romney, slave accommodation ship, 587–90, 683–93, 702, 732
Rover (18), ship-sloop, 594–5, 601–3, 611, 692, 698–9
Royal Admiral, pinnace, 333
Rubis (40), frigate (French), 160–1

Sabrina (20), Sixth Rate frigate, 153–5
St Christopher (18), sloop, 144
St Jago, tender, 143
Samarang (28), Sixth Rate frigate, 488–90, 608–12

Sapphire (28), Sixth Rate frigate, 469–72
Sappho (18) brig-sloop (pre-1830), 212
Sappho (16), brig-sloop (post-1836), 592, 682–7
Saracen (10), gun-brig, 551–6, 618–20, 626, 638–52, 661, 668–73
Satellite (16), brig-sloop, 124
Satellite (18), ship-sloop, 493–7, 590, 596–7, 694
Savage (10), gun-brig, 445, 571
Scorpion (18), brig-sloop, 153
Scout (18), ship-sloop, 527, 534–47, 556, 617–30, 639–43, 650, 674
Scylla (18), ship-sloop, 391, 571
Seaflower (4), cutter, tender, 377–81, 420–35
Semiramis (42), Fifth Rate frigate, 208–10
Seringapatam (46), Fifth Rate frigate, 487, 587, 682, 692
Serpent (16), brig-sloop, 561–7, 679
Shark, schooner (American), 257–60
Skipjack (3), schooner, 397–5, 407, 484–5, 563, 580–95
Slaney (20), ship-sloop, 404
Snake (16), brig-sloop, 493–5, 504, 565–7, 590–4, 692–6
Snapper (12), gun-brig, 219–21, 241–58, 265–76
Solebay (32), Fifth Rate frigate, 111–15
Sparrow (10), cutter, 608, 699
Sparrowhawk (18), ship-sloop, 404, 472, 594, 607–12, 704
Spartiate (76), Third Rate battleship, 491, 607
Speedwell (3/4), schooner, 381–91, 407, 422, 442, 468–70

Spider (6), brigantine, 610, 704
Spitfire (16), ship-sloop, 166–7
Stag (46), Fifth Rate frigate, 608–11, 701–15
Starling (4), schooner cutter, tender, 600, 611
Stork (16), ship-sloop, 124
Subtle, schooner, 106
Sulphur (10), survey ship, 600–2, 611
Sultane (40), Fifth Rate frigate (French), 167
Susan, colonial brig, 301, 308–10
Swift, Governor's cutter, 302
Swinger (12), gun-brig, 106, 277–307
Sybille (48), Fifth Rate frigate, 318–29, 336, 346–70, 737
Sylphe, Sixth Rate frigate (French), 721

Tagus (42), Fifth Rate frigate, 212
Talbot (28), Sixth Rate frigate, 595–600, 608
Tartar (42), Fifth Rate frigate, 213–16, 228–52
Tartarus (2), steam packet, 694
Teresita, pilot boat (Spanish), 584
Termagant (3), brigantine, 622–8, 641–6, 671–4
Thais (20), Sixth Rate frigate, 139, 157–9, 161–3
Thalia (46), Fifth Rate frigate, 454–5, 510–11, 522–47, 557, 617
Thetis (46), Fifth Rate frigate, 286–7, 489
Thistle (12), gun-brig, 219–20, 228–30, 237, 240–66, 273
Thunder, survey ship, 569
Ticklar (8), cutter, 117–8
Tigress (14), gun-brig, 112–14, 117–18, 127–8, 141–4, 149–50, 153

Trinculo (16), brig-sloop, 440–8, 478, 513–23
Tweed (16), ship-sloop, 161
Tweed (28/20), Sixth Rate frigate, 390, 472
Tyne (28), Sixth Rate frigate, 368, 487–8

Ulysses (44), Fifth Rate frigate, 106, 169, 173–4, 177–80
Undaunted (46), Fifth Rate frigate, 439, 446
Union (3/4), schooner, 386–91

Valorous (26), Sixth Rate frigate, 387–91
Venus, hired schooner, 106
Vernon (50), Fourth Rate frigate, 479, 504
Vesta (10), schooner, 154, 189
Vestal (26), Sixth Rate frigate, 562–5, 574–84, 690–2
Victor (18), ship-sloop, 284–94, 405, 472
Viper (6), brigantine, 547–55, 620, 626, 661, 673

Wanderer (16), brig-sloop, 571, 580–3, 679–82, 690–6
Wanderer (16), ship-sloop, 144
Warspite (76), Third Rate battleship, 485, 491
Wasp (16), brig-sloop, 580
Waterwitch (10), gun-brig, 504, 522–6, 545–57, 616–7, 656–8, 665, 671???, 699
Winchester (52), Fourth Rate frigate, 405, 468, 472
Wizard (10), gun-brig, 608–12, 700–6, 717–18
Wolverene (16), brig-sloop, 647–50, 657–71

Zenobia (18), brig-sloop, 195

Index of Places

Accra, 70, 150, 234, 245, 248, 253, 308, 315, 319, 506, 542
Ambriz Bay, 82, 657, 748
Anamabo, 48, 70, 234, 261, 640
Angola, 24–7, 30, 33–4, 38, 41–2, 45, 50, 89, 102, 141, 408, 412, 467, 503, 657, 706
Annobón, 37, 84, 119, 417, 512
Ascension Island, 84, 183, 214, 234, 242, 244, 251, 259, 265–6, 272, 274, 284–6, 288–9, 304, 321, 325, 356, 361–2, 431, 434, 442, 449, 453, 462, 534, 554, 617, 643, 662, 735
Axim, 25, 70, 200, 505–6

Badagry, 72, 234, 288, 308, 315, 342
Bahamas, 95, 144, 391, 397, 403, 485, 563, 569, 576, 685, 695, 723
Bahia, 25, 34, 90, 102–3, 156, 205, 221, 265, 274, 279, 297, 385, 408–10, 446, 467, 502, 603
Baltimore, 123, 408, 511, 518, 533, 582, 639–40, 652, 671, 689

Bance Island, 67, 101, 129, 289
Barbados, 26–8, 32, 35, 106, 145, 153, 186, 583, 682, 747
Bathurst, 212, 234, 245, 315–16, 341, 430, 432, 438–9, 441, 455
Benguela, 25–6, 63, 82–3, 146, 156, 195, 408, 412, 467, 503, 736
Benin, 22, 26, 31, 36, 63, 71–3, 127, 149, 171, 199, 214, 234, 354, 420, 467
Bissau, 12, 65, 159, 233, 235, 280, 318, 637, 650, 696
Boa Vista (or Bonavista), 153, 295, 407
Bonny, 24, 46, 74, 87, 102, 168, 199, 234, 253–4, 256, 287, 423–4, 467, 516, 520, 532, 544, 626, 638, 649
Brass River, 24, 73, 334, 367, 513, 635

Cabinda, 81, 123, 142, 158–9, 204, 248, 351
Cacheu River, 65–6, 233, 347
Cameroons, 33, 49, 75–6, 85, 193, 198–9, 234, 264, 312–13, 333, 342, 376, 440, 452, 544–5
Cape Coast Castle, 26, 28, 48, 117, 136, 142, 147, 165, 173, 215, 242, 274, 276, 278, 356, 455, 669
Cape Mesurado, 93, 68, 233, 261, 267, 283, 311
Cape Mount, 63, 68, 128, 199, 233, 238, 246–7, 367
Cape Verde Islands, 23, 28, 64, 153, 217, 219, 234, 251, 317, 382, 408, 534, 555, 557, 623, 669, 749
Cay Sal Bank, 95, 473, 483, 588
Christiansborg, 48, 70–1
Clarence Cove, 325, 332, 334–5, 347, 445, 626
Congo, 22, 24–5, 27, 30–1, 45, 50, 81, 86, 89, 102, 408, 412, 419, 447, 503
Congo River, 22, 81, 85, 89, 156, 512, 660, 662, 671
Costa Rica, 474
Cuba, 12–13, 17, 23, 35–8, 40, 62, 64, 92–7, 104–6, 123, 132, 135, 185–6, 206, 223, 235, 280, 284, 327, 351, 384–8, 393, 396–7, 405–10, 427, 446, 464–8, 474, 482, 500–3, 557, 559, 561–3, 567–8, 584, 587, 613–14, 616, 625, 651, 676–9, 686–8, 723, 730, 732, 739, 748, 750–3, 755–9

Dixcove, 244, 247, 261, 311, 640
Duke's Town, 74, 176, 250, 462, 526

Elmina, 22, 25, 48, 70, 102, 200, 202, 248, 308, 315, 338, 666, 671

Fernando Po, 37, 76, 84, 250, 253, 287, 323–4, 331–5, 341–7, 352–8, 366, 382, 412–20, 427–32, 436, 445, 447, 449, 524, 545, 645–6, 735
Formoso River, 72, 278–9
Freetown, 9, 67, 87, 109, 129, 139, 148, 161, 165, 188, 206, 211–12, 220, 233, 242, 245–6, 259, 269, 274, 276, 282, 289–90, 293, 285, 308, 323–4, 333, 353, 356, 358, 367, 412, 414, 434, 437, 452, 467, 508, 517, 528–9, 531, 630–1, 638, 662, 668

Gabon River, 75, 173, 179, 202, 512, 652, 659
Gallinas River, 67, 166, 173, 190, 223–5, 233, 235, 240, 248, 269, 291, 301, 310, 349, 426, 543, 640, 654
Gambia River, 64, 125, 140–1, 152, 158, 174, 212, 228, 234, 245, 251, 286, 316, 367, 375, 407, 425–6, 430, 432, 438, 441, 464, 505, 544, 553
Gold Coast, 22, 25–6, 28, 31, 33, 48, 63, 70–1, 78, 86–7, 136, 141, 200, 233–4, 274, 347, 542
Goree, 25, 64, 113–14, 125–6, 130, 133, 141, 174–5, 198–9, 212, 228, 233, 238, 246, 287, 380, 428
Grand Popo, 72, 234
Grand Sesters, 68, 209, 380
Great Bahama Bank, 94–5, 97, 388–9, 391–2, 401, 566, 585

Guadeloupe, 26, 92, 145, 214, 235, 345, 353, 402–3
Guinea Coast, 88, 92, 102, 184, 347
Gulf of Guinea, 22, 30, 38, 69, 77, 88–9, 123, 167, 428, 467

Havana, 35–6, 38, 94, 97, 105, 108, 129, 156, 168, 185, 198–9, 206–7, 223, 235, 257, 384–97, 402–3, 405, 410, 440, 468–73, 477–8, 485, 513–14, 529, 560–3, 566–8, 571, 575, 581, 584, 586–7, 590, 594, 625, 649, 663, 677, 679, 686, 689, 691, 693, 732, 748, 753, 757
Hispaniola, 23, 92, 97

Isle of Pines, 93, 469–71, 476–8, 480, 566, 582, 690, 694–5
Isles de Los, 66–7, 118, 120, 140, 148–9, 160–1, 212, 321, 439

Jamaica, 2, 28, 32, 35–6, 38, 206, 209–10, 384, 402, 407, 427, 453, 468, 472, 474, 484, 574, 586, 680, 739

Lagos, 21, 72, 220, 234, 279, 336, 467, 674, 753
Liberia, 260, 310, 540, 606, 733
Lisbon, 34, 40–1, 103, 205, 234, 271, 377, 409, 530, 538, 549, 596, 670, 678, 683, 706, 732
Little Popo, 72, 128, 234, 298, 574
Loango, 25, 28, 31, 36, 79–80, 102, 157, 371, 408, 412, 447, 503, 512, 523

Malembo, 81, 123, 203, 249, 265, 274, 278, 289, 316, 354

Martinique, 26, 92, 106, 145, 191, 235, 345, 353, 402–3, 427
Massachusetts, 2, 10, 26, 36
Mozambique, 26, 30–1, 62, 169, 205, 537, 597, 687, 690, 719

Nantes, 33, 37, 40–2, 56, 61, 280, 750
New Calabar River, 24, 50, 74, 234, 263, 275, 278, 333, 405, 456, 519
New Sesters, 519, 663–4, 687
New York, 10, 35–6, 133, 209, 518–19, 577, 585, 648, 655, 668–9, 753
Niger, 24, 36, 46, 72–3, 77, 86–7, 129, 149, 234, 236, 449
Nova Scotia, 8–9, 405, 630
Nunez, River (or Rio), 66–7, 128, 154–5, 158–9, 207, 220, 233, 235, 260, 290, 317, 349, 426, 439, 480, 541–3, 553, 629, 651, 661, 668

Old Bahama Channel, 94–7, 392, 481, 561, 566, 570, 579, 685
Old Calabar, 24, 74, 107, 149, 165, 168, 175–6, 222, 234, 236, 250, 255, 264, 266, 296, 330, 314, 316, 334, 343, 345, 376, 445, 459–60, 462, 530, 620, 639, 664

Pernambuco (or Recife), 25, 27, 90, 156, 170, 175, 360, 408, 469, 488–90, 502, 595–6, 607–8, 703–4, 720
Pongas, River (or Rio), 66–7, 118–19, 125, 140, 143, 154–5, 158–9, 166, 190, 199, 207, 220, 229, 233, 235, 242, 255, 257–8, 290, 317, 375–6, 379, 382, 436, 439, 629–30, 748

Index of Places

Portendic, 251, 273, 286, 293, 315, 317, 321, 511, 522
Porto Novo, 72, 129, 165, 174, 209, 234, 268, 753
Porto Praya, 153, 228, 309, 541, 627–8, 749, 753
Port Royal, 476, 579, 582–2, 586, 590–3, 649, 680, 687, 691–2, 695–6, 723, 739
Princes Island, 76, 117, 127, 146, 149, 192, 196, 208, 218, 250–1, 264, 354, 431, 438, 443, 448, 526, 531, 534, 557, 583, 617, 626, 649, 661, 664, 672, 674
Puerto Rico, 23, 209, 227, 397, 427, 500–1, 541, 553, 589, 592–3, 664, 688

Rhode Island, 2, 10, 15, 35–7, 42, 131
Rio de Janeiro (or Rio), 30, 34, 91, 103, 123, 131, 156, 204–5, 223, 247, 271, 311, 385, 408–11, 427, 436, 467
Rio Grande (West Africa), 65, 143, 179, 251

St Bartholomew, 145, 290, 339, 345, 362, 403

Saint-Domingue, 13, 33, 36–8, 60–1
St Helena, 156, 180, 183, 321, 346, 356–9, 361–2, 368, 370, 412, 511, 621, 747
St Paul de Loando, 82–3, 156, 317, 419, 471, 494, 512, 601, 626, 657
St Philip de Benguela, 83, 156
St Thomas (or São Tomé) (Gulf of Guinea), 6, 22–3, 26–7, 76, 119, 123, 146, 150, 161, 167, 169, 173, 188, 209, 234, 265, 268, 288, 338, 408, 447, 579, 617, 626, 649
St Thomas (West Indies), 12, 24–5, 30, 34, 38, 168, 353, 474, 664
Santiago (or St Jago) de Cuba, 93, 296–7, 370, 390, 404, 441, 453, 476, 521, 575, 582, 613
Senegal, 22, 31, 62, 64, 106, 114–15, 125, 152, 174–5, 180, 189, 198–9, 212, 228, 233, 258, 280, 283, 294, 331, 425, 430, 682
Sesters, 68, 209, 365, 380, 519, 543, 627, 652, 663–4, 687

Sette River, 79–80, 142
Shebar, River, 67, 121, 362, 748
Sherbro, River, 9, 67, 193, 224, 233, 235, 245–6, 257, 259, 291, 301–2, 309, 339, 553, 556, 637–8, 665
Sierra Leone, River, 67, 129, 157, 217, 301, 352, 378, 435, 439, 517, 527, 531, 536, 661–2, 667
Simon's Bay, 439, 446, 448–9, 453, 522, 527, 539, 546, 548, 557, 619, 624, 626, 633, 643, 645, 675, 677
Surinam, 2, 34, 211, 270, 560, 686

Trade Town, 142, 155, 225, 233, 241, 270, 312, 646
Trinidad, 471, 473, 478, 485, 561, 570, 585, 606
Trinidad de Cuba, 93, 472, 479, 563, 566–7, 586

West Bay, 431, 618–9, 622, 630, 642, 644, 650
Whydah, 44, 72, 123, 147, 151, 192, 214, 221, 234, 236, 240, 251, 299, 315, 338, 354–5, 359, 461, 507, 533, 659, 752

Lightning Source UK Ltd.
Milton Keynes UK
UKHW050818190521
383846UK00007B/23